# Personal Computers for Technology Students

### Charles Raymond
**Bessemer State Technical College**

Upper Saddle River, New Jersey
Columbus, Ohio

**Library of Congress Cataloging in Publication Data**

Raymond, Charles
   Personal computers for technology students / by Charles Raymond.
     p. cm.
   Includes index.
   ISBN 0-13-020791-8
   1. Microcomputers. 2. Microcomputers–Maintenance and repair. 3. Microcomputers–Equipment and supplies. I. Title.

TK7888.3 .R38 2001
004.16–dc21                                           00-061980

**Vice President and Publisher:** Dave Garza
**Editor in Chief:** Stephen Helba
**Assistant Vice President and Publisher:** Charles E. Stewart, Jr.
**Production Editor:** Alexandrina Benedicto Wolf
**Production Coordination:** Carlisle Publishers Services
**Design Coordinator:** Robin G. Chukes
**Cover Designer:** Marjory Dressler
**Production Manager:** Matthew Ottenweller
**Marketing Manager:** Barbara Rose

This book was set in Times Roman by Carlisle Communications, Ltd. It was printed and bound by The Banta Book Group. The cover was printed by Phoenix Color Corp.

Copyright © 2001 by Prentice-Hall, Inc., Upper Saddle River, New Jersey 07458. All rights reserved. Printed in the United States of America. This publication is protected by Copyright and permission should be obtained from the publisher prior to any prohibited reproduction, storage in a retrieval system, or transmission in any form or by any means, electronic, mechanical, photocopying, recording, or likewise. For information regarding permission(s), write to: Rights and Permissions Department.

10 9 8 7 6 5 4 3 2 1
ISBN 0-13-020791-8

# PREFACE

Compatible personal computers (PCs) were originally designed to be maintained and upgraded by individuals who possessed many technical skills. In fact, when I first started teaching computer upgrade and repair, the program was taught only to students who majored in electronics due to the extremely technical nature of the material.

Over the years, compatible computers have changed dramatically. The biggest change has been the downward spiral of the cost of the technology. This has made computer repair more costly than simply replacing the unit itself, and has drastically reduced the knowledge needed to maintain a PC. To repair a computer, the technician only needs to know how the components operate and how they interact with the other units of the computer. The skills required are now within the grasp of most technically-minded individuals. The primary goal of *Personal Computers for Technology Students* is to provide knowledge at a level that most technically-minded individuals can easily comprehend.

Along with the physical parts, another major aspect of a computer is software (programs). One major software component is the computer's operating system (OS), which controls the overall operation of the computer. Even though the OS controls the computer, the user controls the OS. Therefore, learning how to use the computer's OS proficiently is imperative for individuals who intend to use, build, upgrade, or repair PCs. This text contains chapters that introduce the most popular operating systems used in today's PCs, along with their configuration.

Along with hardware and software knowledge, users also need to learn how to communicate with other computer-literate people. With PCs now ingrained in almost every profession, it is important for those who want to excel in their chosen profession to also know how to converse with those who use and maintain PCs. Individuals who understand the skills required for their own craft, along with the skills and knowledge required to maintain PCs, are in high demand. Teaching the jargon required by a computer-literate individual is another goal of this text.

There are two major categories of PCs available in the market: the Apple™ computer and the IBM™-compatible computer. Due to the open technology architecture of the IBM-compatibles, these computers are by far the dominant PCs used in offices and homes throughout the world. Because of the dissimilarities between these two categories of PCs, textbooks that try to cover both usually are more confusing than informational to the reader. *Personal Computers for Technology Students* is geared toward the dominant IBM-compatibles.

## ORGANIZATION OF THE TEXT

Each chapter begins with learning objectives followed by a list of key terms and expressions relevant to the topics covered. At the end of each chapter is a set of review questions that should be used to determine if the chapter's objectives were met.

**Introduction: Basic Electricity and Electronics**  The introduction provides students with basic electricity/electronic knowledge that is required to repair PCs. Basic electronic devices are introduced with an explanation of their uses in the PC. Devices used to protect the computer against static discharge and faulty power sources are also discussed. Finally, this chapter discusses the power supplies used in the PC along with removal, installation, and troubleshooting procedures.

**Chapter 1: Computer Numbering Systems and Terminology**  This chapter covers the numbering systems used in PCs—which include decimal, binary, and hexadecimal—and provides the knowledge needed to easily convert among these numbering systems. The chapter also discusses the codes used to transfer information within the PC. The units used to represent numbers are also explained.

**Chapter 2: Microcomputer Basics** This chapter provides an introduction to microcomputers by providing explanations of a computer's various units. Microprocessor buses, memory, operating systems, applications, and programming are also introduced.

**Chapter 3: Microprocessors and the Compatible Computer** This chapter provides information on all of the microprocessors that have been and are currently used in compatible PCs. It discusses the evolution of the PC through the many Intel processors along with its major competitors. The technologies that were brought about by each processor's introduction are also explained.

**Chapter 4: Evolution of the PC and the Expansion Bus** Fully understanding a technology requires an individual to have knowledge of its history. This chapter discusses the evolution of the PC by providing explanations of the technologies introduced with each PC family. The evolution of the expansion bus is also covered in this chapter.

**Chapter 5: Operating System Basics, Including MS-DOS and Windows 3.X** Understanding the computer operating system is just as important as understanding the components that comprise the computer. This chapter introduces operating systems, including MS-DOS and Windows 3.X. It also provides insight into the configuration of MS-DOS and Windows 3.X. Information is also given on how to load the operating system into the computer's memory.

**Chapter 6: Windows 95/98/2000** This chapter provides basic working knowledge on the Windows 95, 98, and 2000 operating systems. The configuration of these operating systems is also explained. In addition, the chapter also provides the installation procedure for Windows 9X.

**Chapter 7: PC Ramdom-Access Memory** Detailed information on the different categories of RAM is provided in this chapter. DOS- and Windows-based memory management are also discussed. The chapter also gives the procedure for removing and installing memory in compatible PCs, and covers the steps necessary to troubleshoot the most common memory problems in PCs.

**Chapter 8: BIOS and the CMOS** This chapter covers the different types of ROMs used in compatible computers. It also explains the configuration of the BIOS—the programs contained within the system ROM.

**Chapter 9: Input/Output** Keyboards, serial ports, mice, parallel ports, joysticks, sound cards, and modems are discussed, as well as the computer resources available to I/O devices and the methods used to configure these resources. Interrupts and DMA are also introduced. In addition, the standard connectors and cables used to interface peripherals to the PC are also explained. Finally, the removal and installation procedure for a typical adapter card is explained.

**Chapter 10: Mass Storage Basics** Tape drives, floppy disk drives, CD-ROMs, CD-Rs, CD-RWs, and DVDs are the mass storage devices covered in this chapter. The format and methods used to save information on mass storage devices, including file allocation, are explained. Detailed information is also provided on formatting, removing, installing, and troubleshooting mass storage devices.

**Chapter 11: Hard Disk Drives** This chapter discusses hard drive evolution from the first hard drives used in compatible PCs to the present, including an in-depth discussion of the FAT16 and FAT32 methods of formatting a hard drive. The chapter discusses the procedure for setting partitions using the DOS utility FDISK. Detailed information on configuration, removal, installation, and troubleshooting hard drives is provided.

**Chapter 12: Video Adapters and Monitors** This chapter explains the many video standards, configurations, and specifications used in the compatible PC, along with a basic explanation of how the video system operates. It also provides the procedures for isolating and troubleshooting common video problems.

**Chapter 13: Printers** The printers used with the compatible PCs are covered in this chapter. Procedures that cover the installation of printer drivers into the Windows 9X operating system are included. The final section of the chapter provides troubleshooting procedures for the most common printer problems.

**Chapter 14: Computer Network Basics** Probably one of the fastest growing fields in PCs is networking. This chapter discusses different network media and topologies. Protocol suites and the procedures required to configure a network using Windows 9X and a discussion of the Internet are also provided.

**Chapter 15: Diagnostics** This chapter provides information on different diagnostic methods, including beep codes, POST codes, and high-level diagnostics. Viruses are also discussed in this chapter. Finally, a procedure is given to troubleshoot a computer that is totally dead.

**Chapter 16: Interrupt and DMA Basics** Two of the most important but least understood and discussed topics in other textbooks are the compatible PC interrupts and direct memory access (DMA). This chapter provides basic knowledge on how these technologies operate.

In addition to the chapters discussed, the textbook also includes appendixes, a list of acronyms, and a glossary of commonly used computer terms. The Appendixes contain tables that give the decimal equivalent for numbers from $2^0$ and $16^0$ through $2^{70}$ and $16^{27}$; include charts that compare the features provided by previous and current generation microprocessors; provide a DOS quick reference guide that contains commonly used DOS commands along with their syntax; contain a debug quick reference guide of commonly used debug commands along with their syntax; and provide the answers to the odd-numbered chapter review questions.

# ACKNOWLEDGMENT

I would like to thank the following reviewers for their useful comments and suggestions: Mohammed Abdulla, Wentworth Institute of Technology; John Bentley, Mesa Community College; Charlie Jones; Hong Lin, DeVry–North Brunswick; and Charlotte Turner, DeVry–Phoenix.

# WARRANTIES AND DISCLAIMERS

This publication is provided "as is" without warranty of any kind, either expressed or implied, including, but not limited to, the implied warranties of merchantability, fitness for a particular purpose, or non-infringement. The author and publisher assume no responsibility for errors or omissions in this publication or other documents that are referenced by this publication.

References to corporations, their services and products, are provided "as is" without warranty of any kind, either expressed or implied. In no event shall the author be liable for any special, incidental, indirect, or consequential damages of any kind, or any damages whatsoever, including, without limitation, those resulting from loss of use, data, or profits, whether or not advised of the possibility of damage, and on any theory of liability, arising out of or in connection with the use or performance of this information.

This publication may include technical or other inaccuracies or errors. The author may make improvements and/or changes in the product(s) and/or the program(s) described in this publication and in the publication itself at any time.

## Trademarks

AMD, the AMD logo, and combinations thereof as well as certain other marks listed at http://www.amd.com/legal/trademarks.html are trademarks of Advanced Micro Devices, Inc.

AMI is a registered trademark of American Megatrends, Incorporated.

Apple, Macintosh, and MAC are trademarks of Apple Computer Company.

Award is a trademark of Award Software Company.

Compaq is a registered trademark of Compaq Computer Corporation.

Centronics is a registered trademark of Genicom Corporation.

Sound Blaster is a registered trademark of Creative Technologies, Ltd.

Mylar is a registered trademark of E. I. Du Pont de Nemours and Company.

Hayes is a registered trademark of Hayes Microcomputer Products, Inc.

Hewlett-Packard is a registered trademark of Hewlett-Packard Company.

Intel, Intel logo, Pentium, Celeron, Itanium, MMX, Pentium®II Xeon, and Pentium®III Xeon, are either trademarks or registered trademarks of Intel Corporation.

IBM, IBM-PC, PC/XT, IBM-AT, Micro Channel, OS/2, OS/2 Warp, PS/2, and Personal System/2 are either trademarks or registered trademarks of International Business Machines Corporation.

Microsoft, MS-DOS, Windows, Windows NT, and Windows Me are either trademarks or registered trademarks of Microsoft Corporation.

Microcom is a registered trademark of Microcom Systems, Incorporated.

Netware is a registered trademark of Novell, Inc.

Disk Manager is a registered trademark of OnTrack Computer Systems, Inc.

QEMM is a registered trademark of Quarterdeck Office Systems.

RDRAM is a trademark of Rambus, Inc.

Cyrix is a registered trademark of Cyrix Corporation.

# BRIEF CONTENTS

| INTRODUCTION | BASIC ELECTRICITY AND ELECTRONICS | 1 |
|---|---|---|
| CHAPTER 1 | COMPUTER NUMBERING SYSTEMS AND TERMINOLOGY | 37 |
| CHAPTER 2 | MICROCOMPUTER BASICS | 55 |
| CHAPTER 3 | MICROPROCESSORS AND THE COMPATIBLE COMPUTER | 73 |
| CHAPTER 4 | EVOLUTION OF THE PC AND THE EXPANSION BUS | 129 |
| CHAPTER 5 | OPERATING SYSTEM BASICS, INCLUDING MS-DOS AND WINDOWS 3.X | 153 |
| CHAPTER 6 | WINDOWS 95/98/2000 | 197 |
| CHAPTER 7 | PC RANDOM-ACCESS MEMORY | 231 |
| CHAPTER 8 | BIOS AND THE CMOS | 269 |
| CHAPTER 9 | INPUT/OUTPUT | 283 |
| CHAPTER 10 | MASS STORAGE BASICS | 331 |
| CHAPTER 11 | HARD DISK DRIVES | 367 |
| CHAPTER 12 | VIDEO ADAPTERS AND MONITORS | 411 |
| CHAPTER 13 | PRINTERS | 439 |
| CHAPTER 14 | COMPUTER NETWORK BASICS | 451 |
| CHAPTER 15 | DIAGNOSTICS | 495 |
| CHAPTER 16 | INTERRUPT AND DMA BASICS | 513 |
| APPENDIX A | | 527 |
| APPENDIX B | | 529 |
| APPENDIX C | | 535 |
| APPENDIX D | | 547 |
| APPENDIX E | | 553 |
| LIST OF ACRONYMS | | 559 |
| GLOSSARY | | 565 |
| INDEX | | 579 |

# TABLE OF CONTENTS

| INTRODUCTION | BASIC ELECTRICITY AND ELECTRONICS | 1 |
|---|---|---|
| | Objectives | 1 |
| | Key Terms | 1 |
| I.1 | A Typical Water System | 2 |
| I.2 | Voltage and Current | 3 |
| I.3 | Protection | 8 |
| I.4 | Some Shocking Facts | 9 |
| I.5 | Direction of Current Flow | 10 |
| I.6 | Electronic Devices | 13 |
| I.7 | DC Power Supplies | 16 |
| I.8 | The Compatible Computer's DC Power Supply | 19 |
| I.9 | On or Off, That Is the Question | 24 |
| I.10 | Power Management | 25 |
| I.11 | Protecting Your Investment | 25 |
| I.12 | Power Supply Troubleshooting | 28 |
| I.13 | Power Supply Removal and Installation | 32 |
| I.14 | Conclusion | 33 |
| | Review Questions | 33 |

| CHAPTER 1 | COMPUTER NUMBERING SYSTEMS AND TERMINOLOGY | 37 |
|---|---|---|
| | Objectives | 37 |
| | Key Terms | 37 |
| 1.1 | Numbering System Basics | 38 |
| 1.2 | ASCII | 46 |
| 1.3 | Computer Numbering System Terminology | 50 |
| 1.4 | Conclusion | 52 |
| | Review Questions | 53 |

| CHAPTER 2 | MICROCOMPUTER BASICS | 55 |
|---|---|---|
| | Objectives | 55 |
| | Key Terms | 55 |
| 2.1 | Digital Computer Basics | 56 |

| | | |
|---|---|---|
| 2.2 | The Microprocessor Buses | 59 |
| 2.3 | Memory | 62 |
| 2.4 | RESET | 64 |
| 2.5 | Booting the Computer | 64 |
| 2.6 | Programming Terminology | 65 |
| 2.7 | Programming Basics | 65 |
| 2.10 | Conclusion | 69 |
| | Review Questions | 69 |

## CHAPTER 3  THE MICROPROCESSORS AND THE COMPATIBLE COMPUTER  73

| | | |
|---|---|---|
| | Objectives | 73 |
| | Key Terms | 73 |
| 3.1 | Intel | 74 |
| 3.2 | The Microprocessors | 75 |
| 3.3 | The Compatible Microprocessors | 103 |
| 3.4 | Motherboards | 108 |
| 3.5 | Motherboard Removal and Installation | 121 |
| 3.6 | Conclusion | 125 |
| | Review Questions | 125 |

## CHAPTER 4  EVOLUTION OF THE PC AND THE EXPANSION BUS  129

| | | |
|---|---|---|
| | Objectives | 129 |
| | Key Terms | 129 |
| 4.1 | The IBM-PC | 130 |
| 4.2 | IBM-Compatible Computers | 131 |
| 4.3 | IBM-XT | 132 |
| 4.4 | IBM-AT | 133 |
| 4.5 | IBM-PS/2 and Micro-Channel | 135 |
| 4.6 | The Split | 138 |
| 4.7 | EISA | 138 |
| 4.8 | VL Bus | 141 |
| 4.9 | Peripheral Component Interface (PCI) | 144 |
| 4.10 | PCMCIA (PC-Card) | 147 |
| 4.11 | ISA Lives On | 148 |
| 4.12 | What Is the Right Expansion Bus for Me? | 149 |
| 4.13 | Growing Pains | 149 |
| 4.14 | Conclusion | 149 |
| | Review Questions | 150 |

| CHAPTER 5 | **OPERATING SYSTEM BASICS, INCLUDING MS-DOS AND WINDOWS 3.X** | **153** |
|---|---|---|
| | Objectives | 153 |
| | Key Terms | 153 |
| 5.1 | Who Owns Software? | 154 |
| 5.2 | Similarities between DOS and Windows 3.X | 155 |
| 5.3 | Differences between MS-DOS and Windows | 165 |
| 5.4 | DOS | 165 |
| 5.5 | Windows 3.X | 170 |
| 5.6 | Summary of the Differences | 181 |
| 5.7 | Basic Boot Sequence | 181 |
| 5.8 | Customizing Your Computer | 183 |
| 5.9 | Conclusion | 194 |
| | Review Questions | 194 |

| CHAPTER 6 | **WINDOWS 95/98/2000** | **197** |
|---|---|---|
| | Objectives | 197 |
| | Key Terms | 197 |
| 6.1 | The Shell | 199 |
| 6.2 | My Computer | 203 |
| 6.3 | Windows 9X/2000 Filenames | 204 |
| 6.4 | Windows Explorer | 206 |
| 6.5 | Support for Windows 3.X | 207 |
| 6.6 | Support for DOS Applications | 207 |
| 6.7 | Windows 95/98/2000 Applications (Preemptive Multitasking) | 210 |
| 6.8 | The Major Differences among Windows 95, 98, and 2000 | 210 |
| 6.9 | Booting Windows 9X | 212 |
| 6.10 | Windows 9X/2000 Configuration | 218 |
| 6.11 | Installing Windows 95 | 225 |
| 6.12 | Conclusion | 227 |
| | Review Questions | 228 |

| CHAPTER 7 | **PC RANDOM-ACCESS MEMORY** | **231** |
|---|---|---|
| | Objectives | 231 |
| | Key Terms | 231 |
| 7.1 | Fundamental Storage Unit | 232 |
| 7.2 | Pages and Bytes | 232 |
| 7.3 | Memory in the PC | 233 |
| 7.4 | Types of RAM | 244 |
| 7.5 | DIPs, SIMMs, and DIMMs | 249 |

| | | | |
|---|---|---|---|
| | 7.6 | Memory Banks | 254 |
| | 7.7 | Memory Installation | 257 |
| | 7.8 | Cache Memory | 262 |
| | 7.9 | Windows 3.X Virtual Memory | 264 |
| | 7.10 | Troubleshooting Memory Problems | 265 |
| | 7.11 | Conclusion | 266 |
| | Review Questions | | 266 |

## CHAPTER 8  BIOS AND THE CMOS  269

| | | | |
|---|---|---|---|
| | Objectives | | 269 |
| | Key Terms | | 269 |
| | 8.1 | BIOS Configuration Using the CMOS | 270 |
| | 8.2 | Using the Setup Program | 272 |
| | 8.3 | Clearing the CMOS | 279 |
| | 8.4 | Upgrading the BIOS | 280 |
| | 8.5 | Conclusion | 281 |
| | Review Questions | | 281 |

## CHAPTER 9  INPUT/OUTPUT  283

| | | | |
|---|---|---|---|
| | Objectives | | 283 |
| | Key Terms | | 283 |
| | 9.1 | Computer Resources | 284 |
| | 9.2 | Configuring and I/O Port Resources | 289 |
| | 9.3 | Programmable Serial Ports | 295 |
| | 9.4 | Parallel Ports | 304 |
| | 9.5 | Keyboards | 308 |
| | 9.6 | Pointing Devices (Mice and Trackballs) | 310 |
| | 9.7 | Sound | 313 |
| | 9.8 | Joysticks and Game Pads | 315 |
| | 9.9 | Configuring a Hardset MI/O CARD | 316 |
| | 9.10 | Configuring the Softset Integrated Peripheral Ports | 318 |
| | 9.11 | Finding Out which Resources Are In Use | 319 |
| | 9.12 | Adapter Card Removal and Installation | 326 |
| | 9.13 | Conclusion | 328 |
| | Review Questions | | 328 |

## CHAPTER 10  MASS STORAGE BASICS  331

| | | | |
|---|---|---|---|
| | Objectives | | 331 |
| | Key Terms | | 331 |
| | 10.1 | Tape Drives | 332 |

| | | |
|---|---|---|
| 10.2 | Disk Drives | 335 |
| 10.3 | The Basics of Storing Data | 336 |
| 10.4 | Floppy Drives | 336 |
| 10.5 | The First Cylinder | 340 |
| 10.6 | Formatting a Disk | 348 |
| 10.7 | The Floppy Connection | 350 |
| 10.8 | CD-ROM | 353 |
| 10.9 | Digital Versatile Disc Read-only Memory (DVD-ROM) | 360 |
| 10.10 | The Chassis | 361 |
| 10.11 | Troubleshooting Floppy and CD-ROM Drives | 362 |
| 10.12 | Conclusion | 364 |
| | Review Questions | 365 |

## CHAPTER 11  HARD DISK DRIVES  367

| | | |
|---|---|---|
| | Objectives | 367 |
| | Key Terms | 367 |
| 11.1 | Read/Write Head Positioning | 369 |
| 11.2 | Controllers | 372 |
| 11.3 | EIDE and SCSI | 375 |
| 11.4 | CHS (Normal), ECHS (Large), and LBA | 378 |
| 11.5 | The Resources Required by EIDE | 379 |
| 11.6 | The IDE Connection | 380 |
| 11.7 | SCSI | 382 |
| 11.8 | Installing the Drive | 389 |
| 11.9 | The Software Side of a Hard Drive's Installation | 389 |
| 11.10 | Preventative Maintenance | 401 |
| 11.11 | Compression | 402 |
| 11.12 | Type Numbers | 403 |
| 11.13 | Troubleshooting Hard Drives | 405 |
| 11.14 | Conclusion | 408 |
| | Review Questions | 408 |

## CHAPTER 12  VIDEO ADAPTERS AND MONITORS  411

| | | |
|---|---|---|
| | Objectives | 411 |
| | Key Terms | 411 |
| 12.1 | Video Basics | 412 |
| 12.2 | Video in the Compatible Computer | 418 |
| 12.3 | Video Adapter Resources | 424 |
| 12.4 | Windows Accelerated Cards | 424 |
| 12.5 | Monitor Specifications | 425 |

Contents   xiii

| | | |
|---|---|---|
| 12.6 | Expansion Bus | 428 |
| 12.7 | The Video Driver | 431 |
| 12.8 | Video Troubleshooting | 433 |
| 12.9 | Conclusion | 435 |
| | Review Questions | 436 |

## CHAPTER 13   PRINTERS   439

| | | |
|---|---|---|
| | Objectives | 439 |
| | Key Terms | 439 |
| 13.1 | Terminology | 439 |
| 13.2 | Impact Printers | 442 |
| 13.3 | Nonimpact Printers | 444 |
| 13.4 | Emulation | 447 |
| 13.5 | Setting Up a Printer Using Windows 9X/2000 | 447 |
| 13.6 | Troubleshooting Printer Problems | 448 |
| 13.7 | Conclusion | 449 |
| | Review Questions | 449 |

## CHAPTER 14   COMPUTER NETWORK BASICS   451

| | | |
|---|---|---|
| | Objectives | 451 |
| | Key Terms | 451 |
| 14.1 | Medium (Media) | 452 |
| 14.2 | Network Protocols | 456 |
| 14.3 | Network Topology | 457 |
| 14.4 | Network Standards | 460 |
| 14.5 | Other Network Hardware Protocols | 467 |
| 14.6 | The Network Interface Card (NIC) | 468 |
| 14.7 | The Network Operating System | 470 |
| 14.8 | Protocol Stacks | 473 |
| 14.9 | Binding, Sharing, and Mapping with Windows 9X | 478 |
| 14.10 | The Internet | 480 |
| 14.11 | Modems | 481 |
| 14.12 | Conclusion | 491 |
| | Review Questions | 491 |

## CHAPTER 15   DIAGNOSTICS   495

| | | |
|---|---|---|
| | Objectives | 495 |
| | Key Terms | 495 |
| 15.1 | ROM-based Diagnostics | 496 |

| | | | |
|---|---|---|---|
| | 15.2 | Disk-based Diagnostics | 499 |
| | 15.3 | Windows 9X/2000 System Properties | 502 |
| | 15.4 | Windows 98/2000's System Information | 506 |
| | 15.5 | Dr. Watson | 507 |
| | 15.6 | Windows 2000 Troubleshooter | 508 |
| | 15.7 | Virus Scanners | 508 |
| | 15.8 | Troubleshooting a Dead Computer | 509 |
| | 15.9 | Conclusion | 510 |
| | | Review Questions | 511 |

## CHAPTER 16  INTERRUPT AND DMA BASICS  513

| | | | |
|---|---|---|---|
| | Objectives | | 513 |
| | Key Terms | | 513 |
| | 16.1 | Software or Hardware Interrupts | 514 |
| | 16.2 | Interrupts in Detail | 515 |
| | 16.3 | Interrupt Request | 518 |
| | 16.4 | Direct Memory Access | 522 |
| | 16.5 | Conclusion | 524 |
| | | Review Questions | 524 |

| | |
|---|---|
| APPENDIX A | 527 |
| APPENDIX B | 529 |
| APPENDIX C | 535 |
| APPENDIX D | 547 |
| APPENDIX E | 553 |
| LIST OF ACRONYMS | 559 |
| GLOSSARY | 565 |
| INDEX | 579 |

# INTRODUCTION

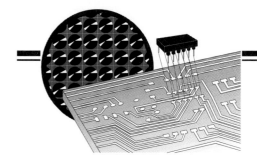

# Basic Electricity and Electronics

## OBJECTIVES

- Define voltage, current, resistance, and power.
- Describe the characteristics of a basic circuit.
- Describe the characteristics of direct current (DC).
- Describe the characteristics of alternating current (ac).
- Define frequency, cycle, and hertz.
- Make basic calculations used in electronics.
- State the differences among a conductor, a semiconductor, and an insulator.
- Describe the characteristics of round and flat cables.
- Define the over-current protection devices used in compatible computers.
- List safety procedures to follow when working on compatible computers.
- List the function of a DC power supply.
- Define basic terms and expressions used in electronics.
- Describe the characteristics of linear and switching power supplies.
- Describe the differences between mini-AT and ATX power supplies.
- Define the basic functions preformed by transistors, capacitors, inductors, switches, and transformers.
- Describe the function of a UPS.

## KEY TERMS

voltage source
load
circuit
volt
electrical current
ampere (amp)
conductor
insulator
resistance
ohms
omega (Ω)
watts
power

fuse
circuit breaker
alternating current
cycle
frequency
hertz
direct current
synchronous
digital clock
switch
resistor
inductor

capacitor
transformer
diodes
transistor
regulator
inverter
circuit common (common, or electrical ground)
Energy Star
uninterruptible power supply (UPS)
linear power supply
switching power supply

This book explains the basic operation and setup of hardware in a compatible computer to readers who do not have prior knowledge of either computers or electricity. However, to fully understand the operation and terminology used in the discipline of computers and in this book, it is necessary to introduce some of the basic concepts, formulas, and terms used in electricity and electronics.

**Figure I.1**  Typical Water System

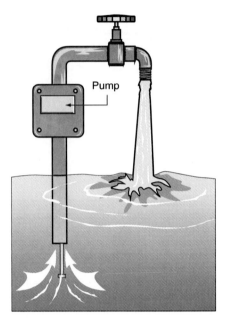

This chapter does not explain everything there is to know about electricity and electronics. This is only an introduction with just enough theory to aid in the explanation of terms used throughout this book. In fact, if you already understand DC and ac voltages, currents, and powers, you may want to jump ahead to section I.6.

Throughout the chapter, you will encounter mathematical formulas that are used for explanation only. It is not the intent of this chapter to have you try to memorize these formulas. In fact, the formulas introduced in this chapter that are used later in the book will be referenced back to the section in this chapter that explains their use.

Before we get started with our discussion on the fundamentals of electricity and electronics, we are first going to discuss something we all should understand—a basic water system. I have found that learning something new is made easier if it can be compared to something we already know.

## I.1 A TYPICAL WATER SYSTEM

A good analogy of the basic operation of electricity is to compare it to the operation of the local water system. The water company uses a pump to push water through pipes to residences and industries. The push supplied by the pump is measured in units called pounds-per-square-inch (PSI). The unit used to measure the amount of water flowing past a certain point is called gallons per minute (GPM). An illustration of a typical water system is shown in Figure I.1.

Getting water to flow through pipes into our homes is great and wonderful, but once it gets to its destination, it usually performs some kind of work. This work may be replenishing the moisture in the human body, cooking, cleansing, and many other things. In order for the water to perform this work, we must have some method to control the flow of the water, or we would have water all over the place. Devices that exhibit resistance to the flow are called *pressure regulators;* water valves control the flow of the water. Some of the water valves control the water flow over a wide range of flow rates, such as the valves on most faucets. Some of the valves simply turn the water off or on, like those found in most washing machines, ice makers, and water fountains. Once the water is used, it always returns to its source, the earth. Here the water waits until it is pumped out and is used over and over.

We will find out in the next few sections that electricity behaves in a similar manner to water flowing through pipes, except the terms used will be different. In electricity, the device that serves the same

# Basic Electricity and Electronics 3

**Figure I.2** A Basic Circuit

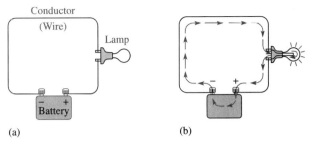

*Source:* Antonakos, Boylestad, Floyd, Mazidi, Mazidi, and Miller, *The CEA Study Guide* (Upper Saddle River, NJ: Prentice Hall, 1999) p. 49.

function as the pump is called a voltage source. Electrical current is similar to the water flowing through the pipes. And, electrical conductors are similar to the pipes carrying the water.

## I.1.1 The Comparison between Water Flow and Electricity

In electricity, a device called a **voltage source** is similar to the pump used in a water system. Batteries, solar cells, and electrical generators are examples of voltage sources. The function of the voltage source is to supply the force needed to push millions of electrons that belong to atoms through electrical conductors to some destination, similar to water flowing through pipes.

Once the electrons arrive at their destination, they flow through one or more electrical devices that are commonly called **loads.** The loads use the energy produced by the flowing electrons to perform some kind of work. Once the work is accomplished by the load, the electrons always return through another conductor back to the voltage source that originally pushed the electrons out. This is why all electrical wall outlets and batteries have a minimum of two connection points. The complete path the electrons take from the voltage source, through the source conductors, through the load, through the return conductor, and back to the voltage source is called a **circuit.** Figure I.2 illustrates a very basic circuit utilizing a battery (source), conductors, and a lamp (load).

## I.2 VOLTAGE AND CURRENT

The push that is provided by the voltage source is called an *electromotive force (EMF)*. EMF is measured in units called **volts,** and is similar to the PSI measurement used by a water system. One volt is defined as the amount of push required to move one ampere of current through one ohm of resistance. You are probably saying to yourself, "That is great but what is an ampere and an ohm?" The answers to this question will require further explanation.

The atom's electrons that are pushed by the voltage source through the electrical conductor are called **electrical current** or, simply, current. The measurement of current flow is given in units called **amperes,** or **amps.** Ampere is a unit that is similar to the GPM unit used in a water system. One amp of current flow equals 6,250,000,000,000,000,000 (pronounced 6 quintillion, 250 quadrillion, or simply $6.25 \times 10^{18}$) electrons flowing by a given point in one second.

### I.2.1 Conductors and Insulators

Electrical **conductors** carry the current that is pushed by the voltage source to and from the loads. Conductors are similar to the pipes that carry the water in a water system. Conductors are made from materials that offer very little resistance to the flow of current. Most metals exhibit this characteristic, but the best conductors are made from silver, copper, gold, and aluminum.

Contrary to popular belief, the best conductor of electrical current is not gold. In fact, the best known conductor of electrical current is silver. The second best electrical conductor is copper, followed by gold and then aluminum. Even though silver is the best conductor, it is very seldom used because of its high price. Of the four mentioned, the most popular conductors are copper and aluminum because they are both good conductors and are fairly cheap.

Even though gold is the third best conductor, it is still used quite often in electronics. Gold is not used as the primary conductor, but as plating (coating) for exposed electrical connections. The reason for this is that gold does not oxide (rust) like other metals. Metal oxides are usually not good conductors of electrical current. For this reason, a lot of electrical connection points and electronic component leads are plated with a very thin coating of gold to prevent oxidation upon exposure to the environment.

Some materials such as rubber, plastic, Teflon, and fiberglass have an almost infinite resistance to current flow. Materials that have this characteristic are called **insulators.** As you have probably noticed, most conductors are enclosed within a jacket made from an insulator. This is to prevent the conductor that supplies the current and the conductor that returns the current from coming in contact with each other and bypassing the load. When this happens, it is called a *short circuit,* which will usually cause a large amount of current to flow. This is because the load is bypassed and only the very small resistance of the conductors limits the current that is supplied by the generator. Devices that detect and protect against a short circuit condition will be discussed later in section I.3.

## I.2.2  More about Conductors

There are three different types of electrical conductors found inside computers. These types go by the names *wire, cable,* and *trace.* A *wire* is a term used for an individual conductor that is covered with its own insulator. In computers, the insulator is usually color-coded according to the electrical signal it carries. The color codes used in most of the compatible personal computers (PCs) are given in section I.8.2. The larger the diameter of the wire, the more electrical current it will be able to carry safely. Also, using more than one conductor to carry the same electrical signal will also increase the current carrying capability. Due to limited space inside the compatible PCs, multiple wires of the same size are usually used to carry the same electrical signal to increase the current carrying capability.

The size of the wire is given by a number called its *gauge.* The gauge number is determined by a standard called the *american wire gauge (AWG).* As strange as it seems, the larger the wire's gauge, the smaller the size of the wire. This means that an AWG 22-gauge wire is smaller than an AWG 20-gauge

**Table I.1**  AWG for Copper Wire at 70°F

| Gauge | Diameter | Ω/1,000 ft | Gauge | Diameter | Ω/1,000 ft |
|---|---|---|---|---|---|
| 10 | 0.102 | 1.02 | 20 | 0.032 | 10.30 |
| 11 | 0.091 | 1.29 | 21 | 0.028 | 13.00 |
| 12 | 0.081 | 1.62 | 22 | 0.025 | 16.50 |
| 13 | 0.072 | 2.04 | 23 | 0.024 | 20.70 |
| 14 | 0.064 | 2.57 | 24 | 0.020 | 26.20 |
| 15 | 0.057 | 3.24 | 25 | 0.018 | 33.00 |
| 16 | 0.051 | 4.10 | 26 | 0.016 | 41.80 |
| 17 | 0.045 | 5.15 | 27 | 0.014 | 52.40 |
| 18 | 0.040 | 6.51 | 28 | 0.013 | 66.60 |
| 19 | 0.036 | 8.21 | 29 | 0.011 | 82.80 |

# Basic Electricity and Electronics

wire. To help with this explanation, Table I.1 lists some of the most widely used gauges available with the AWG, along with the conductor's diameter and resistance per 1,000 feet.

Two more wires enclosed within a common insulating jacket are called a *cable*. Sometimes the insulator is the common jacket for the individual wires; sometimes the wires within the cable have their own insulators. The cables used in the compatible computer are categorized as either a round cable or a flat cable.

The round cables used by the compatible computer contain wires that have individual insulators. Round cables get their name because the common jacket that contains the wires is physically round. The only round cables inside the computer are used to connect the computer's power on/off switch to the power supply and the audio output of a CD-ROM to the PC's sound card. All of the other round cables are outside the compatible computer and are used to connect together devices such as the power outlet, printer, monitor, mouse, network, and telephone jack.

*Flat cables* use a jacket that is also the insulator for the individual wires. Flat cables get their name from the fact that the wires are placed side-by-side in a row within the jacket, which makes the cable flat. Due to their appearance, flat cables are also referred to as *ribbon cables*. The jacket that covers the wires is usually gray in color, with one edge painted with either a red or blue stripe. This colored edge of the ribbon cable identifies the pin-1 side of the cable. Flat cables are used inside the compatible computer because they can pass through narrow spaces. They are used to connect devices such as hard drives, CD-ROMs, floppy drives, zip drives, and tape backup units. A typical round cable and a flat cable used in the compatible computer are illustrated in Figure I.3.

All of the electronic devices that are discussed later in this chapter are connected together by conductors. These electronic devices need a lot more mechanical support than could be provided by simply attaching them to the end of a wire. To provide this extra support, electronic devices are usually mounted on a thin card made from fiberglass. The electronic devices are then connected together by thin, flat conductors that appear to be printed to the top, and sometimes the bottom and center, of the fiberglass board. Because the conductors appear to be printed on the board, these fiberglass cards are called *printed circuit boards (PCBs)*. The conductors on the PCB are called *traces,* or *foil*. Figure I.4 is an illustration of a PCB showing electronic devices that are connected together electrically using traces. Notice that the power supplied to all of the components uses the traces that are labeled *Vcc* and *ground*.

**Figure I.3** Typical Round and Flat Cables Used in Compatible Computers

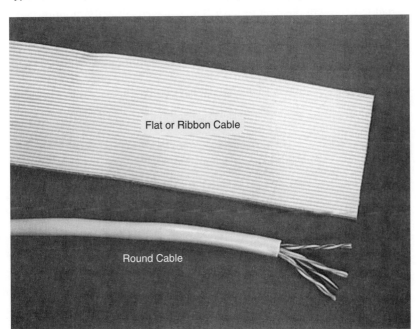

**Figure I.4** Motherboard PCB

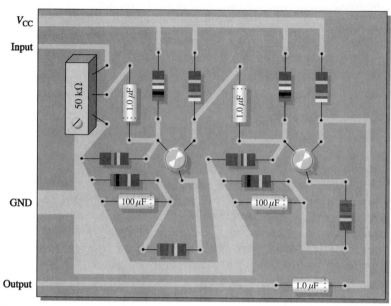

*Source:* Floyd, *Electronic Fundamentals, Devices, and Applications,* 4th ed. (Upper Saddle River, NJ: Prentice Hall, 1998) p. 809.

## I.2.3 Resistance, Voltage, Current, and Watts

All devices that use the flow of electrical current to perform work oppose the flow of the current. The opposition to current flow is called **resistance.** Good conductors of electrical current offer very little resistance to current flow. Poor conductors such as glass, wood, plastic, rubber, and paper offer a very high resistance to current.

Resistance to current flow is measured in units called **ohms.** The Greek symbol for **omega ($\Omega$)** is used to represent the amount of resistance (ohms). As an example, 250 ohms of resistance can also be represented as 250 $\Omega$. One ohm is defined as the resistance that will allow one amp of current to flow with one volt of push. The voltage is directly proportional to the amount of resistance and current. The term *directly proportional* means if resistance or current double, the voltage will double. When something is directly proportional, it can be expressed mathematically. The formula for calculating the amount of voltage required to produce a certain current flow *through* a certain resistance is $V = I \times R$, where $V$ represents volts, $I$ represents amps, and $R$ represents resistance. It should be noted that because voltage is also known as an emf, some texts use the formula $E = I \times R$, where $E$ represents volts. To avoid confusion, $V$ represents voltage throughout the remainder of this text.

**Problem I.1:** What voltage is required to cause 0.3 amps of current flow through 15 ohms of resistance?

**Solution:** Since $V = I \times R$, then $V = 0.3 \times 15 = 4.5$ volts

Using a little algebra, the formula $V = I \times R$ can be transposed to solve for either resistance ($R$) or current ($I$). The formula to solve for resistance ($R$) when voltage and current are known is $R = V / I$. The formula to solve for current flow ($I$) when the voltage and resistance are known is $I = V / R$.

**Problem I.2:** What will be the current flow through a resistance of 250 ohms that creates a voltage drop of 5 volts?

**Solution:** Since $I = V/R$, then $I = 5/250 = 0.02$ amps

# Basic Electricity and Electronics

**Problem:** What value of resistance will drop 2 volts with 0.01 amps of current?

**Solution:** Since $R = V/I$, then $R = 2/0.01 = 200$ ohms

Currently, every material has some resistance to the flow of current. Even good conductors such as copper wire can have a resistance of 0.01 Ω for a 1-foot length (the actual resistance of a conductor depends on the material, length, and circumference). Devices (loads) that are designed to perform work, such as light bulbs, amplifier circuits, and switching circuits, will have a higher resistance than the resistance of the conductors but a lower resistance than an insulator. As an example, a 100-watt light bulb has about 150 Ω resistance.

When current flows through resistance, work is produced. The work produced by current flowing through resistance is called *power*. We pay the power company for power, not for voltage and current.

## I.2.4 Power

**Power** is the term used for the rate of work performed by an electrical current. Power is important to understand because power companies bill for the amount of power used, not individually for voltage and current. However, the amount of voltage and current determine the amount of power delivered to a house or business. Power is measured in two units called the *volt-ampere (VA)*, or **watts.** We will discuss watts in this section and VA later in section I.11.5.

One of the methods used to calculate the number of watts (W) of power used is to multiply the number of volts providing the push by the amps of current flowing through the load, or $P = V \times I$, where $P$ is the power in watts, $V$ is the voltage in volts, and $I$ is the current in amps. This is illustrated in Problem I.3.

**Problem I.3:** What will be the power produced if 120 volts is pushing 8 amps of current through a vacuum cleaner?

**Solution:** Since $P = V \times I$, then $W = 120\,V \times 8\,A = 960$ watts

If the power and voltage are known, the current flow can be calculated by dividing the power, in watts, by the voltage. The mathematical formula for this calculation is $I = P/V$.[1] Problem I.4 is an example of the use of this formula.

**Problem I.4:** How much current flows through a 75-watt light bulb with 120 volts applied?

**Solution:** Since $I = P \div V$, then $I = 75\,W \div 120\,V = 0.625$ amps

Also, if watts and current are known, the amount of voltage required to produce the power can be calculated. The formula to perform this calculation is $V = P / I$. This formula is used in Problem I.5.

**Problem I.5:** What amount of voltage is required to produce 500 watts of power with 2.3 amps of current flow?

**Solution:** Since $V = P \div I$, then $V = 500 \div 2.3 = 217.4$ volts

## I.2.5 The Cost of Power

As previously stated, power companies charge their customers according to the amount of **power** used. The unit the power company uses to bill its customer is called the *kilowatt-hour (KWH)*. One KWH is equal to 1,000 watts of power used for one hour. The amount the power company charges per KWH is usually stated on your monthly power bill.

---

[1] Two additional formulas can be used to calculate power that are not discussed in the text:
$$P = V^2/R$$
$$P = I^2 \times R$$

The number of KWHs used is fairly easy to calculate if you know the number of watts consumed by a device and the length of time the device is turned on. Simply multiplying the number of watts times the number of hours used and then dividing the product by 1,000 is all that is required to make the calculation. Problem I.6 calculates the kilowatt-hours used by a 100-W light bulb left on for 24 hours.

**Problem I.6:** How many kilowatt-hours does a 100-watt light bulb left on for 24 hours consume?

**Solution:**  $100 \times 24 = 2,400$ watts $\div 1,000 = 2.4$ kilowatt-hours

When electrical devices perform work, heat is generated. The heat generated is proportional to the amount of power being consumed by the load. You have probably noticed the heat coming from the different electrical devices in your home. Some electronic circuits can start acting erratically, cease to operate, or be damaged by excessive heat. To dissipate the heat generated by electronic devices, devices called *heat sinks* and *cooling fans* are sometimes used. Heat sinks and cooling fans will be discussed again in chapter 3.

When more current is allowed to flow in a circuit than it is designed to handle, excessive heat is generated. It should be obvious that excessive heat can have devastating results. For this reason, it is imperative that some means be available to sense and prevent an over-current condition from occurring. This leads to the next section.

## I.3 PROTECTION

If something happens in a circuit that causes the resistance to drastically decrease or the voltage to drastically increase, excessive current will flow. This can be caused by anything from a lightning strike to a short circuit. Since power is equal to $V \times I$, intensive heat is generated when excessive current flows. This heat can melt conductors, destroy components, and sometimes cause fires. For this reason, it is imperative that some means be used to detect and protect against an excessive current condition. The devices most often used to provide this protection are fuses and circuit breakers.

### I.3.1 Fuses

**Fuses** come in all shapes and sizes. No matter the shape or size of the fuse, they all operate by the same principle. Most fuses used in electronics consist of a short length of wire or metal ribbon that is enclosed in a cylinder-shaped container. The fuse is placed in one of the conductors that feed the load. All of the current flowing to the load also flows through the fuse. The wire or ribbon in the fuse is designed to melt at a certain current. When the current flowing through the fuse exceeds its rated current, the wire in the fuse melts, or blows. When the fuse blows, it will open the current path to the load, protecting it from damage.

Along with the current rating, the fuse is also rated with a voltage rating. This is the voltage that the fuse can open safely without arching. For this reason, the voltage rating of the fuse should always be higher than the circuit's maximum voltage. The fuse voltage and current ratings are usually stamped on one of the fuse end caps.

The metal ribbon is replaceable on some large fuses, so the fuse cylinder is reusable. However, the small fuses used in automobiles and personal computers are not reusable. If a fuse blows in your car or in your computer, it must be replaced with a fuse of the same physical size, current rating, and voltage rating. If the fuse is replaced with a fuse of the same type and blows once again, there is a problem in the equipment that must be repaired. Never replace a blown fuse with one with a higher current or lower voltage rating. Figure I.5 shows a typical fuse used in the compatible PC.

Fuses are a great and inexpensive means to protect a circuit from over-current. Their only drawback is that after a fuse blows, the metal ribbon or fuse itself must be replaced. The circuit breaker is another over-current protection device that does not have to be replaced after the device senses an over-current condition.

# Basic Electricity and Electronics

**Figure I.5** Fuses and Circuit Breakers

*Source:* Floyd, *Electronic Fundamentals, Circuits, Devices, and Applications,* 4th ed. (Upper Saddle River, NJ: Prentice Hall, 1998) p. 37.

## I.3.2 Circuit Breakers

**Circuit breakers** perform the same function as a fuse, but with one important difference. A circuit breaker is designed to open a circuit, without damaging itself when an over-current condition is detected. Circuit breakers can be reset and used over and over after being tripped because of an over-current condition.

Most circuit breakers have the same basic appearance as a light switch. In fact, the circuit breaker usually utilizes a toggle lever, labeled On and Off, that is often used as a switch to turn loads off and on. When the circuit breaker senses an over-current condition, it will automatically switch to a position between on and off. This is usually referred to as a *trip condition*. When the circuit breaker trips, it opens the circuit to the load, preventing current flow. Once tripped, the circuit breaker can be reset by manually moving the toggle lever to Off, and then to On.

Circuit breakers are not used in the compatible PC itself. However, they are used in the circuit breaker boxes found in most houses, surge suppressors, and UPS units. Surge suppressors and UPS will be discussed in greater detail in section I.11.

Protecting a circuit from over-current is one thing, but we also need to know a few things about protecting ourselves from the dangers of electricity. For this reason, I have decided to include the following section—not to scare you, just to make sure you understand the dangers.

## I.4 SOME SHOCKING FACTS

The human body will conduct current if exposed to the two conductors that carry and return the current. As little as 0.01 amperes of current-flow through the body can interfere with the electrical pulses from the brain that control the beat of the heart. As you have probably guessed, this interference can be lethal.

Although 0.01 amps of current seems like a small amount, there are several factors that have to be in place before even this small amount of current can flow through the body. We already know from section I.2.3 (Problem I.2) that the amount of current flow through the body is determined by the amount of voltage the body comes in contact with divided by the amount of resistance the body has at the time of the contact. Remember the formula given earlier, $I = V/R$. Most of the time, the human body has more than 200,000 ohms of resistance. This amount of resistance requires approximately 200 volts or more to cause 0.01 amperes of current to flow through the body.

This does not mean that lower voltages are not dangerous. To the contrary, the 120 vac available from a common wall outlet has probably killed or injured more people than any other. The reason is that the body's resistance varies from minute to minute, due to many factors. For this reason, lower voltages can definitely be fatal under the right circumstances.

If you are not a trained technician, never work on electrical or electronic equipment with power applied. In fact, it is highly recommended that you unplug all equipment any time you are working on installing or removing any component.

The most dangerous path for current flow through the body is from hand to hand. This is because this path will take the current through the heart. If you have to work on live circuits, make it a habit to work with one hand behind you. Also, because moisture and sweat cause body resistance to drop dramatically, avoid working on any live circuit in hot, humid conditions.

Electricity should not be feared, but it should definitely be respected. The best rule to follow is if you do not know what you are doing, find someone who does. Section I.12.1 provides more safety guidelines to follow when working with electrical and electronics circuits.

Now that we know a little about the basics of a circuit and some of the units, values, and calculations used in electricity and electronics, it is time to discuss the direction of the flow of electrical current.

## I.5  DIRECTION OF CURRENT FLOW

The two terminals (connections) on the electrical voltage source that produce the volts that push the current have a polarity of negative and positive. For reasons that will be explained later, some voltage sources constantly swap the polarity of these terminals, while others do not.

We learned early in our education that an atom's electrons are negatively charged. We also learned that *like charges repel each other* and *unlike charges attract.* Because of these facts, the terminal of the voltage source that carries the negative charge pushes the electrons out of the terminal through the conductor to the load. After the electrons flow through the load, the positive terminal of the voltage source attracts the electrons through the return conductor back to the voltage source. This means that the electrons that make the electrical current travel through the circuit from negative to positive.

There are two types of current flow we need to discuss—alternating current (ac) and direct current (DC). The ac current gets its name from the fact that the current constantly changes direction. On the other hand, DC current flows in a single direction. The formulas given in sections 0.4 and 0.6 are the same for both ac and DC. Notice that it is standard practice to abbreviate ac with lowercase letters and DC with uppercase letters.

### I.5.1  Alternating Current (ac)

**Alternating current** is produced when the voltage source constantly swaps the voltage polarity of the two terminals at a precise rate. This will cause the current to start at zero, reach a maximum in one direction, decrease to zero, then reverse itself and reach a maximum in the opposite direction, and then return to zero. This cycle is repeated continuously. Visually, this will create a voltage image similar to the one illustrated in Figure I.6.

An ac **cycle** is basically the completion of a polarity swap. In other words, when one of the terminals swings from zero to positive to zero to negative and then back to zero, one cycle is completed. To help visualize this, two cycles are labeled in Figure I.6. The term **frequency** is used to define the rate of the cycles. As an example, power companies in North America generate ac at a frequency of 60 cycles per second, whereas power companies in Europe generate 50 cycles per second. Instead of using the expression "cycles per second," the term **hertz** is usually used. Hertz stands for cycles per second. So, a frequency of 60 cycles per second can also be expressed as 60 hertz (60 Hz).

You may ask, "Why do power companies produce ac instead of DC?" There are a couple of answers to this question. First, ac generators are a lot cheaper to produce than DC generators. Second, ac is easier and cheaper to send over great distances with less power loss.

**Figure I.6** AC Cycles

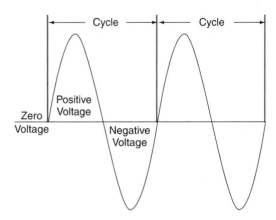

Customers of the power company pay for power (watts), not for voltage or current. In section I.2.4 (Problem I.3), we were introduced to the fact that power is the product of voltage and current ($P = V \times I$). Because all present conductors have some resistance, the more current the conductor has to carry to produce the required power, the more voltage the conductor will lose. The more voltage that is lost by the conductor, the less power will be available to the load. If the current can be reduced in the conductors, the conductors that carry the power will have less voltage drop and more power will be delivered to the load.

The method used by power companies to reduce the voltage loss of the conductors is to increase the value of the voltage for transmission and decrease the voltage value when it comes time to actually run loads. Power is the product of current and voltage, so if the voltage goes up, the current required to produce the same power will go down. This may be confusing to some, but simply stated, if the voltage is raised to extremely high levels for transmission, the current required to produce the same power will drop to an extremely low level. Less current flow through the transmission conductors will mean less voltage will be lost due to the small resistance of the conductors themselves. This is illustrated mathematically in Problem I.7.

**Problem I.7:** How much current is required to produce 1,200 watts of power when the voltage is 120 volts ac (vac) and 1,200 vac?

**Solution:** Referring to the equation in section I.2.4, Problem I.5, $I = P / V$. Using this formula, the current required to produce 1,200 watts with 120 vac will be:

$$I = 1{,}200 \text{ W}/120 \text{ vac} = 10 \text{ amps}$$

The current required to produced 1,200 watts with 1,200 vac will be:

$$I = 1{,}200 \text{ W}/1{,}200 \text{ vac} = 1 \text{ amp}$$

Because ac voltage can be easily stepped up (increased) for transmission and stepped down (decreased) to safer levels for residential and industrial use, ac is cheaper to send over great distances. Devices called *transformers* perform the process of stepping the voltage up and down. Transformers are discussed further in section I.6.4.

After generation by the power company, the ac voltage is typically stepped up to 275,000 volts ac for transmission. This extremely high voltage level will cause less current to flow through the transmission conductors to produce the power used by the many loads. As stated previously, less current through the conductors will cause less voltage to be lost due to the small resistance of the conductors themselves. After transmission, large transformer banks in installations called *substations* then usually step down the voltage to 2,400 volts. Once stepped down, the ac is transmitted over the utility poles that are located throughout our neighborhoods. Last, the voltage is once again stepped down to 240 and 120 volts ac for residential use. The transformers you see mounted on the utility poles in

**Figure I.7** A Typical Power Distribution System

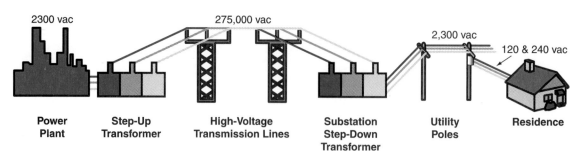

your neighborhood accomplish this feat. A basic drawing of this power distribution scheme is shown in Figure I.7.

Electrical devices such as refrigerators, heaters, dryers, and washing machines work directly off the ac that is provided by the wall outlets in your residence. Most electronic devices such as computers, televisions, and stereos require DC to operate.

### I.5.2 Direct Current (DC)

**Direct current** is electrical current that flows in a single direction. This is the type of current produced by batteries, solar cells, DC generators, and DC power supplies. These DC voltage sources usually have terminals marked positive (+) and negative (−). As with ac, direct electrical current flows out of the negative terminal, through the load, and back to the positive terminal. Unlike ac, however, the terminals never swap polarity.

The advantage DC has over ac is that it is very easy to control electronically with devices such as diodes, inductors, capacitors, and transistors. These few devices form the backbone of a technology called electronics. These few components are the main devices used in all electronic devices. This includes the computers about which this book is written.

You are probably thinking, "If electronic equipment, such as computers, operate off of DC, how can we plug these devices into an ac wall outlet?" The answer to the question is, "Inside the cover of the electronic equipment is a group of electronic circuits that form a device called a *DC power supply,* or simply *power supply.* The purpose of the power supply is to convert the ac supplied by the wall outlet into the DC required by all of the electronic circuits."

Since the DC power supply provides the current that makes all of the electronic devices in the computer operate, it is very important to have a basic understanding of its operations. It is also important to have a basic understanding of the electronic devices that make up the computer. Before we continue our discussion on power supplies, let us first discuss some of the electronic devices, plus another very important component of a computer called the digital clock.

### I.5.3 The Digital Clock

The information transferred inside the computer is synchronous. **Synchronous** is a term that means the information transfer is timed with a clock to make sure it is at the right place at the right time. Since DC powers the circuits in the computer, the clock that synchronizes the information transfer must also be DC. The DC clock in the compatible computer usually switches back and forth between +5 VDC and 0 VDC at a constant rate similar to ac. However, the major difference is the DC clock voltage never changes polarity, so the current flow is in a single direction. As we will find out later, when something is switched between two states, this is referred to as a *digital signal.* Because the computer uses a clock that switches between only two states, it is called a **digital clock.** Figure I.8 identifies two cycles of a typical digital clock. The use of the terms *cycle, frequency,* and *hertz* have the same meaning relative to a digital clock as they do in ac.

Now that we have a little basic knowledge under our belts about electricity and the terms used to describe it, let us begin our discussion on the main devices that are used to electronically control the flow of current.

Basic Electricity and Electronics 13

**Figure I.8** Digital Clock Showing Two Cycles

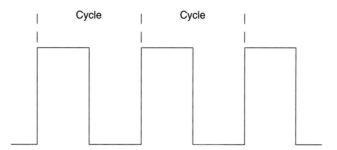

## I.6 ELECTRONIC DEVICES

As stated previously, when current flows through a load it uses the energy of the electron flow to produce work. The form of work produced can be anything from the illumination produced by a light bulb through providing the current needed to make all of the electronic circuits in a computer operate.

No matter what the load is, the flow of electrical current must be controlled by some means. In electronic circuits, the control process is performed by many devices with names such as switch, relay, resistor, capacitor, inductor, triac, diac, SCR, UJT, transformer, vacuum tube, diode, and transistor, just to name a few. These electronic devices are grouped together to form complex electronic circuits that form the loads of the power supply.

Electronic circuits are formed by the aforementioned electronic devices, powered by a voltage source, and connected together electrically by conductors. Electronic circuits perform all of the amazing feats you have come to expect from today's electronic equipment, including personal computers.

All of the electronic devices just mentioned control the flow of current by using resistance. Of those listed, the devices used most often in computers are the switch, resistor, capacitor, inductor, transformer, diode, and transistor. Because this book is about personal computers, the function of these devices is explained briefly in the following sections.

### I.6.1 Switches

A **switch** is a device that breaks (opens) or makes (closes) the current path to the load. When the switch opens the current path, current will not flow through the load. No current through the load means no work is performed. We all use switches daily to control the flow of electrical current.

A switch is usually placed in series (in line) with the source conductor and provides some kind of mechanical lever that the operator can use to control its operation. The open and close positions of the switch are usually labeled Off and On. On a computer, however, the position of the switch that turns the computer on is usually labeled *1* and the position of the switch that turns the computer off is labeled *0*.

Switches are categorized as either a toggle switch or a push button. Toggle switches use a rocker arm that moves to turn the switch either off or on. The wall switches we use daily to control lights are toggle switches.

Push buttons are either momentary contact or on/off. An on/off push button is usually pressed once to turn a device on and pressed again to turn it off. A momentary contact push button is on when it is pressed on, off when it is released, or vice versa depending on the application. Examples of momentary push buttons are doorbell and telephone buttons. Computers use all three of these types to turn off and on, reset, and select one of two operating speeds. A photo of a few typical switches is shown in Figure I.9.

### I.6.2 Resistors

**Resistors** are passive devices whose resistance stays constant. The number of ohms a resistor has is manufactured into the device. Resistors fall into two categories—fixed and variable. A fixed resistor's ohm value cannot be changed by the user, whereas a variable resistor's ohm value can. A good example of a variable resistor is a volume control on a radio. Computer CD-ROM drives and sound cards usually utilize a variable resistor. A fixed resistor and a variable resistor are shown in Figure I.10.

**Figure I.9** Switches

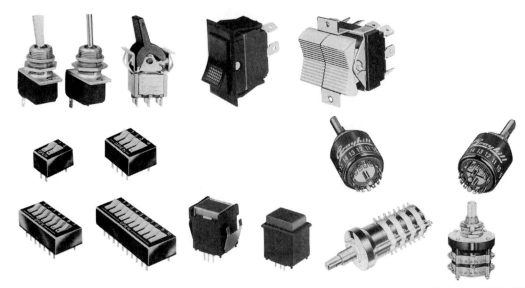

*Source:* Floyd, *Electronic Fundamentals, Circuits, Devices, and Applications,* 4th ed. (Upper Saddle River, NJ: Prentice Hall, 1998) p. 36.

**Figure I.10** Fixed and Variable Resistor

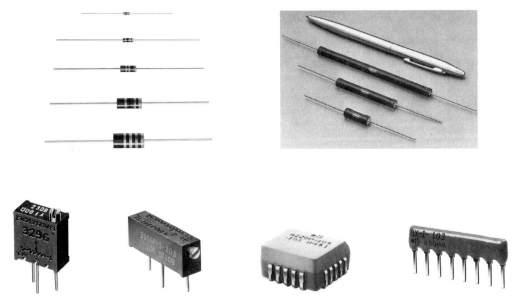

*Source:* Floyd, *Electronic Fundamentals, Circuits, Devices, and Applications,* 4th ed. (Upper Saddle River, NJ: Prentice Hall, 1998) p. 4.

### I.6.3 Inductors and Capacitors

**Inductors** and **capacitors** are electronic devices that vary their resistance depending on the current's rate of change. This rate of change is most often produced by alternating current. In personal computers, inductors are used as electrical noise filters and in devices called transformers.

# Basic Electricity and Electronics

**Figure I.11** Inductors and Capacitors

Source: Floyd, *Electronic Fundamentals, Circuits, Devices, and Applications,* 4th ed. (Upper Saddle River, NJ: Prentice Hall, 1998) p. 4.

Capacitors are also used as an electrical noise filter. However, the most important use of capacitors in the personal computer is as a storage element. This is because capacitors can store electrical charges, similar to a rechargeable battery, but for a very brief period of time. An inductor and a capacitor are shown in Figure I.11.

## I.6.4 Transformers

The primary use of **transformers** is to change the level of ac voltages. There are several types of transformers but only two are discussed in this book. One is called a step-up transformer because it is used to increase the value of the ac voltage. The other is called a step-down transformer because it is used to decrease the value of the ac voltage. These step-up and step-down characteristics make transformers very useful in controlling ac voltages.

The transformers used in electronics have two sides. One side is called the primary and the other is the secondary. The primary of a transformer is where the input ac voltage is connected, whereas the secondary is where the output ac is taken from. Transformers usually have only two connection points of the primary side, but they may have multiple connection points on the secondary side to provide different levels of output voltages.

A very important feature of a transformer is that it does not change the amount of power being transferred from primary to secondary, just the amount of voltage. For example, if there are 100 watts of power supplied to the primary, 100 watts of power are delivered on the secondary. This is an important characteristic because power is the product of voltage and current (section I.2.4). This means if the transformer steps the voltage up, the current will have to decrease by the same ratio in order to keep the power the same on both sides of the transformer. Also, if the transformer steps the voltage down, the current will have to increase in the secondary to keep the power the same on both sides of the transformer.

## I.6.5 Diodes

**Diodes** are electronic devices that allow current to flow in only one direction. This characteristic gives diodes many uses, but its primary use is to convert ac into DC. Diodes are used in both the linear and

**Figure I.12** Electronic Devices

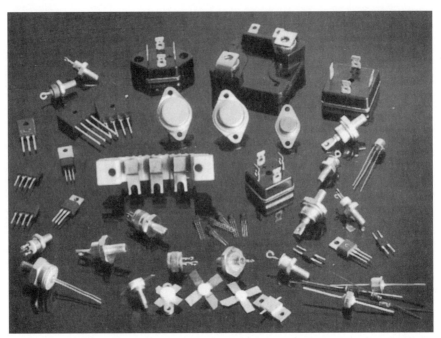

*Source:* Floyd, *Electronic Fundamentals, Circuits, Devices, and Applications,* 4th ed. (Upper Saddle River, NJ: Prentice Hall, 1998) p. 5.

switching power supplies, which are discussed in sections I.7.1 and I.7.2. When diodes are used to convert ac into DC, they are usually called *rectifiers*.

### I.6.6 Transistors

Probably the most important device in electronics and personal computers is a device called a **transistor**. Transistors are the current valves of electronics. A transistor's resistance can be electrically changed to precisely control the flow of current, similar to the valves used in a water system. Most transistors are made from silicon, which is a semiconductor. *Semiconductors* are elements that get their name from the fact that their electrical characteristics place them directly in the center between conductors and insulators. For this reason, transistor circuits are often called *semiconductor circuits*.

Some transistor circuits are designed to amplify small electrical signals, such as those coming from a compact disk player to a point where the signal can drive speakers. Other transistor circuits are designed to act as simple switches to electrically switch current on and off, similar to a wall switch. The big difference between a transistor switch and a wall switch is that a person operates a wall switch, whereas a transistor switch is operated by another electrical signal. It is this operation as a switch that is used most often in personal computers.

Several transistors are shown in Figure I.12. The appearance of the transistor varies depending on its applications. One characteristic that will change is the transistor's physical size. The less electrical current the transistor has to control, the smaller the size. The size of most of the transistors used in today's computer is less than 1 micron (1 millionth of a meter). Transistors of this small size are an important characteristic of a device in the compatible PC called a *microprocessor,* which is discussed in chapter 2.

## I.7 DC POWER SUPPLIES

DC power supplies currently fall into two categories—linear and switching. Linear DC power supplies are by far the easiest to design, troubleshoot, and repair. Switching DC power supplies are far more efficient than linear DC power supplies but are harder to troubleshoot and repair. Both categories are found in PCs, so both are explained.

**Figure I.13** Block Diagram of a Typical Linear DC Supply

**Figure I.14** Block Diagram of a Typical Switching Power Supply

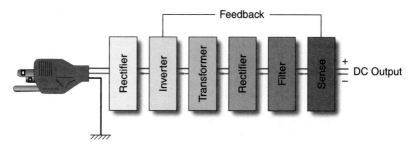

## I.7.1    The Linear DC Supply

The basic linear DC power supply usually uses transformers to step the ac voltage from the wall outlet down to the levels needed by the electronic circuits. After this is accomplished, diodes—called *rectifiers*—convert the ac into DC. Once the rectifiers produce DC, large capacitors are used to filter and smooth the DC voltage to take out the noise produced by the rectification process. Next, the DC flows through a complex electronic circuit in the power supply called a **regulator.** The function of the regulator is to produce the exact values of DC voltage and polarity needed to operate all of the electronic circuits in the equipment. A block diagram of a typical linear DC supply is shown in Figure I.13.

## I.7.2    The Switching DC Power Supply

Due to reasons that are too complicated to explain in this book, the 60 Hz supplied by the power company requires the transformers and electronic devices in a linear DC supply to be large, bulky, and expensive. If the frequency of the ac to be converted to DC can be increased, then the size of the transformers and electronic devices can be reduced and the DC power supply becomes more efficient. This is basically how switching power supplies operate.

When the ac is brought into a switching power supply, it is converted to DC by diode rectifiers. This DC is applied to an electronic circuit called an **inverter.** The function of an inverter is to convert DC into ac. In the switching power supply, the ac produced by the inverter will be at a much higher frequency than that provided by the ac power outlet. This high-frequency ac is applied to a high-frequency transformer that drops the ac voltage levels down to usable levels. The ac voltages produced by the high-frequency transformer are once again applied to diode rectifiers to produce the DC voltages required by the power supply loads. The output of the rectifier is supplied to filter circuits that remove the electrical noise generated by the inverter and rectifier circuits. The final DC voltages are applied to a sense circuit that detects the required DC voltages. This sense circuit provides a feedback signal that controls the frequency of the inverter, which determines the value of the final DC voltages.

As you can see, just explaining the basic operation of a switching DC power supply is a complex task. Even though these DC power supplies are complex, they are the preferred DC supply in computers because of their efficient operation. However, this complex operation makes switching DC power supplies very difficult to troubleshoot and repair if they fail. Figure I.14 shows a block diagram of a typical switching power supply.

This discussion on linear and switching power supplies was not in-depth. The reason is that the supplies used in the compatible computer are relatively inexpensive and are simply replaced when one fails. Before we continue our discussion on power supplies, let us talk a little about a few terms and

**Figure I.15** Sharing a Common Conductor between Power Supplies

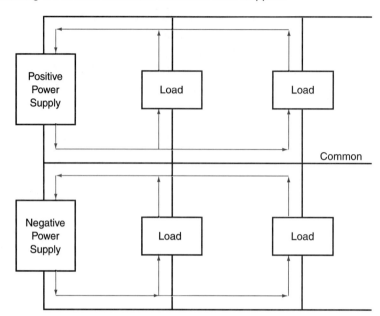

**Figure I.16** A Power Plug Providing Earth Ground

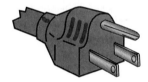

expressions that are used in electricity, electronics, and power distribution that will make the remainder of our discussion easier to follow.

## I.7.3 Common and Ground

Because there is no electrical voltage (push) on the conductor that returns the current to the voltage source, it can be shared by all of the electrical or electronic circuits in a piece of equipment. This characteristic is true for both ac and DC. The return conductor is often shared by all of the electronic circuits in a piece of equipment, so it is usually called **circuit common, common,** or **electrical ground.** Sharing a common conductor for the return path for all of the circuits saves a considerable amount of money in the design of electrical and electronic equipment. This means one large trace on a PCB can be used as a common instead of having a smaller, individual trace for each circuit. Also, most metals like steel are good conductors of electricity. One popular method used to reduce the cost of equipment design is to use the steel case of electrical appliances as common.

Another important characteristic of electrical current is that it will always return to the supply by which it was produced. This characteristic makes it possible for DC supplies that produce different voltage levels, even at different polarities, to share the same common. Figure I.15 illustrates negative and positive DC power supplies that share the same common.

For safety reasons and to help reduce electrical interference, sometimes the common conductor is also connected to earth ground. Earth ground is most often provided through the ac wall outlet. The offset round conductor protruding from the computer's power plug identifies the earth ground conductor. It is very important that the ground lead on the power plug not be cut or broken off, which happens quite often. If an extension cord is used to power the computer, it must provide an earth ground conductor. Running an ungrounded computer can cause safety problems and can make the computer more susceptible to static discharge and lightning strikes. Figure I.16 shows a grounded power plug.

## I.8 THE COMPATIBLE COMPUTER'S DC POWER SUPPLY

This book is written around computers that dominate the world market. These computers are referred to as the IBM-compatible personal computers (PCs) or simply compatible PCs. Just like most electronic equipment, the electronic circuits inside the compatible PC require DC to operate. Most compatible PCs receive their power from an ac wall outlet, like other electronic devices. To convert the ac coming from the wall outlet into the DC voltages required by the electronic circuits in the compatible PC, the PC uses a switching DC power supply.

There are currently two power supply categories used for the compatible computer. One is referred to as the *AT power supply;* the newer is called the *ATX power supply*. These two categories of supplies are not interchangeable and use different style connectors to connect to the computer's motherboard. However, both power supply categories have the same physical appearance and are of the switching design.

The AT power supply in the compatible PC actually consists of four individual DC supplies. These include +5 volt DC (VDC), −5 VDC, +12 VDC, and −12 VDC supplies. Each supply is rated according to the maximum current it can supply to the loads to which it is connected. The ATX power supply provides the same DC voltages provided by the AT supply plus an additional +3.3 VDC and +5 VDC standby supply.

In both the AT and ATX supplies, the +5 VDC supply is capable of delivering the most electrical current because it supplies the power to most of the computer's electronic circuits. With the AT power supply, the +5 VDC supply also provides power to the microprocessor. Because current microprocessors require voltage levels below +5 VDC, additional circuitry is required on the motherboard to provide the correct microprocessor voltage requirements.

The +3.3 VDC supplied by the ATX power supply is also a high-current supply that is rapidly gaining popularity among design engineers. The +3.3 VDC supply currently provides power to the microprocessor and to the PCI expansion bus. Both of these devices are discussed throughout this book.

The +12 VDC supply provides the next highest current level. It is used to power three different devices in the compatible computer. First, it supplies the power that runs the motors on the disk drives. Second, the +12 VDC also provides the power that runs the cooling fans in the computer. Third, the +12 VDC supply provides one-half of the power needed to drive the output section of special electronic circuits called serial ports.

The −12 VDC and −5 VDC supplies have very little use in the compatible computer. The −12 VDC supply powers the other half of the output power required by the serial ports. The −5 VDC supply is not used to power any of the standard circuits in the compatible computer. It was originally used because it was a standard voltage level used on computers when the compatible computer was designed. It is also available to engineers that may need this power supply for circuits that might be designed for future compatible computers. All of the devices mentioned here are discussed later in this book.

As the name implies, the ATX's +5 VDC standby supply provides power even when the power switch on the front panel of the computer is in the off (power-down) position. Of course, the computer has to be plugged into an ac outlet that provides +5 VDC standby power all of the time. Use of this supply will add instant on capabilities and will also give devices the ability to hold information when the computer is powered down.

Overall, the AT and ATX switching DC power supplies are rated in watts, similar to a light bulb. The number of watts is calculated by adding together the maximum watts that can be produced by the individual supplies. The maximum wattage of each supply is calculated using the formula given in Problem I.8 ($P = V \times I$). This formula tells us to multiply the power supply's voltage times the supply's maximum current. The maximum current capability of each individual supply is usually listed on a label attached to the top of the main DC power supply. Figure I.17 shows a typical power supply label. Problem I.8 demonstrates how the information obtained from the power supply's label can be used to determine the maximum wattage rating of the DC supply.

**Problem I.8** What will be the maximum rated power output of a compatible computer's main power supply if the individual supplies are rated as follows?

+5 VDC @ 25 A; +12 @ 10 A; −12 @ 0.5 A; and −5 @ 0.5 A

**Solution:** First: Calculate the individual power for each supply using the formula $P = V \times I$ (power = volts × amps).

$$5 \times 25 = 125 \text{ W}; 12 \times 10 = 120 \text{ W}; 12 \times 0.5 = 6 \text{ W}; \text{ and } 5 \times 0.5 = 2.5 \text{ W}$$

Second: To calculate the maximum wattage rating, add the power available from each supply. Total power equals = $125 + 120 + 6 + 2.5 = 253.5$ or 250 W.

**Figure I.17** A Power Supply's Label

**Figure I.18** Different-Sized Power Supplies

# Basic Electricity and Electronics

It should be noted that the maximum wattage rating of a compatible computer's power supply is not the actual power being consumed. This rating is the maximum power available if the power supply is fully loaded, with each supply providing its maximum rated current.

## I.8.1 Power Supply's Physical Size

The switching DC power supplies in compatible PCs are not only categorized by their wattage rating, but also by their physical size. There are currently four physical size categories: PC/XT, AT, Baby AT, and PS/2. The power supply is mounted inside a case with the computer's electronic circuitry. Each case is designed to accommodate a power supply of a particular physical size. If you purchase a power supply, it is very important that you buy the right wattage rating and physical size for the computer for which the power supply will be installed. Supplies of different sizes are shown in Figure I.18.

**Figure I.18** (Continued)

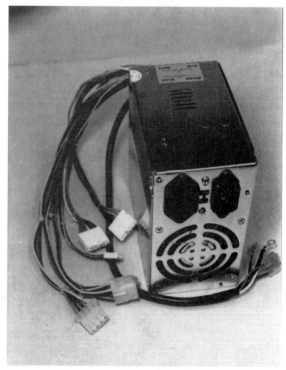

*Source:* Dan L. Beeson, *Assembling and Repairing Personal Computers* (Upper Saddle River, NJ: Prentice Hall, 1997) p. 33.

Table I.2  Color of Conductors for AT and ATX Power Supplies

| | AT-DC Supply Conductor Color Codes |
|---|---|
| Red | + 5 VDC |
| Yellow | + 12 VDC |
| White | −5 VDC |
| Blue | −12 VDC |
| Orange | Power good—Output used to initialize the computer when power is turned on |
| Black | Ground (return) |
| | ATX Supply Conductor Color Codes |
| Orange | +3.3 VDC |
| Red | + 5 VDC |
| Yellow | +12 VDC |
| Blue | −12 VDC |
| White | −5 VDC |
| Purple | +5 VDC standby |
| Brown | +3.3 VDC sense—Input that senses the presence of the 3.3 VDC supply |
| Green | Power-On—Input used to electronically turn the PSU on |
| Gray | Power OK—Output that initializes the computer when power is turned on |
| | ATX Optional Power Supply Conductors |
| White | FanM—Monitors operation of CPU cooling fan |
| White and blue striped | FanC—Controls the speed of the CPU cooling fan |
| White and brown striped | +3.3 VDC sense—Accurately senses the 3.3 VDC at the load |
| White and black striped | 1394R—Supports the IEEE-1394 standard |
| White and red striped | 1394V—Supports the IEEE-1394 standard |

*Note:* Color codes may vary slightly among power supply manufacturers.

### I.8.2 Power Supply Color Codes

The wire insulators used to connect the compatible PC switching power supply to the different electronic devices in the computer are usually a color that indicates the DC voltage level or power supply signal they supply. The color codes commonly used for the AT and ATX supplies are shown in Table I.2. Some of the power supply's wires do not supply voltage but are used as signal wires. The function of the signal wires for both the AT and ATX supply is listed to the right of the signal's name in Table I.2.

### I.8.3 Power Supply Connectors

All power supply conductors are terminated with connectors whose shape and size are determined by the electronic device to which the conductor is intended to supply power. These different connectors

**Figure I.19** ATX Power Supply Motherboard Connectors and Pin-out

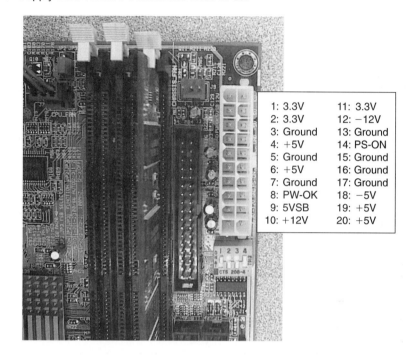

provide power to the motherboard and all of the other internal peripheral devices required by the computer. Identifying which connector goes with which device is very important for anyone who wishes to build or upgrade compatible computers. The connectors for the AT and ATX power supplies are categorized as motherboard connector(s), ATX optional motherboard connectors, 5¼″ disk drive connectors, or 3½″ floppy disk drive connectors.

An ATX power supply motherboard connector is shown in Figure I.19. An AT power supply motherboard connector is shown in the upper left-hand corner of Figure I.20. The motherboard connectors for both supplies are used to provide electrical power to all electronic circuits on the motherboard, including the expansion bus. The expansion bus will be discussed in greater detail in chapter 4, but basically, the expansion bus allows the user to add devices to the computer.

As you can see in Figures I.19 and I.20, the motherboard power connectors for the AT and ATX power supplies are completely different. This means the motherboard has to be designed for either an AT or ATX power supply.

As shown in Figure I.19, the AT power supply requires two motherboard connectors that are commonly referred to as P8 and P9. The power supply voltage levels are assigned to the different pins on the connector. The +5 volt supply uses three conductors on P9 and one conductor on P8. The −5, +12, and −12 VDC supplies are all provided by P8. All of the conductors labeled GND on P8 and P9 are used as the return conductor for all four of the voltage levels provided by the AT power supply. Even though the AT power supply served it purpose for several years, it did have some major problems, which brought about the introduction of the ATX supply.

The first problem is that the AT power supply uses two connectors that can be plugged into the motherboard connector backwards or offset by one or more pins. Even though the AT power supply is supposed to sense this condition and instantly shut itself off, there is no guarantee that the motherboard will not be damaged. The ATX power supply alleviates this problem by making the motherboard power connector in the shape of a rectangle. Along with the shape of the connector itself, the pins and sockets of both the power supply and motherboard connectors vary between a square and "D" shape to make it impossible to plug the connector in backwards or offset without being deliberately modified.

The second problem with the AT power supply is the current popularity of the +3.3 VDC supply. This is because in most computer applications, lowering the voltage gives the designer the abil-

**Figure I.20** AT Power, Standard and Mini-drive and Connector and Pin-out

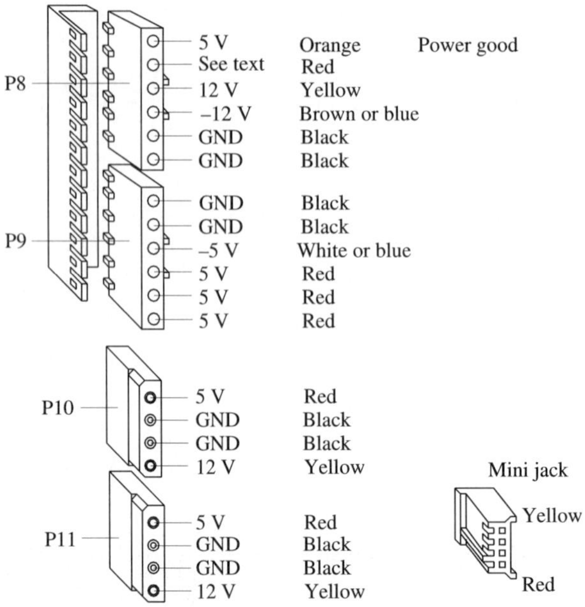

```
P8 ─┬─ 5 V         Orange        Power good
    ├─ See text    Red
    ├─ 12 V        Yellow
    ├─ −12 V       Brown or blue
    ├─ GND         Black
    └─ GND         Black

P9 ─┬─ GND         Black
    ├─ GND         Black
    ├─ −5 V        White or blue
    ├─ 5 V         Red
    ├─ 5 V         Red
    └─ 5 V         Red

P10 ─┬─ 5 V        Red
     ├─ GND        Black
     ├─ GND        Black
     └─ 12 V       Yellow        Mini jack

P11 ─┬─ 5 V        Red                        Yellow
     ├─ GND        Black
     ├─ GND        Black
     └─ 12 V       Yellow        Red
```

Note: Usual position on board is P9 front; P8 rear.

*Source:* Dan L. Beeson, *Assembling and Repairing Personal Computers* (Upper Saddle River, NJ: Prentice Hall, 1997) p. 25.

ity to increase the speed at which a device operates. Due to this fact, all current computers have major components that operate at voltage levels below +5 VDC. To obtain the lower voltages, motherboards that use an AT power supply have additional voltage regulators. On the other hand, the ATX supplies the +3.3 VDC as a standard output, so an ATX motherboard will not have the additional +3.3 VDC regulator.

Another advantage of the ATX power supply is that it provides +5 VDC standby supply for devices that need power when the rest of the devices in the computer are turned off. This feature is not available with the AT power supply.

Along with the motherboard connector(s), the power supply has additional connectors that are used to provide power to the other devices installed inside the computer. The bottom left illustration in Figure I.20 shows the 5¼″ disk drive power connector, including the pin-out. Even though this connector is named after the first device it was designed to provide power for, it is the workhorse connector in the compatible PC. It is used to supply power to devices such as 5¼″ disk drives, hard-drives, CD-ROM drives, CD-R drives, CD-R/W drives, DVD drives, Zip drives, and tape drives, just to name a few. All of these devices are discussed in chapters 11 and 12.

The 5¼″ disk drive power connectors are so popular that there are often not enough to go around. However, devices called *splitters* are available that will provide two connectors for one. Using splitters is acceptable as long as you do not overload the rated current of the supply. A splitter is shown in Figure I.21.

Currently, the 3½″ floppy drive connector is only used to provide power to the devices for which it is named. A drawing of this connector along with the pin-out is shown in the bottom right-hand corner of Figure I.20. Since the 3½″ floppy drive is the only device that uses this connector, there are usually two of the connectors provided by both the AT and ATX power supplies. The 3½″ floppy drive is discussed in chapter 11.

Now that we know the color codes and pin assignments for the various power supply connectors used in the compatible computer, let us move our discussion to a topic that has been argued ever since the personal computer was introduced.

**Figure I.21** Standard Drive Power Splitter

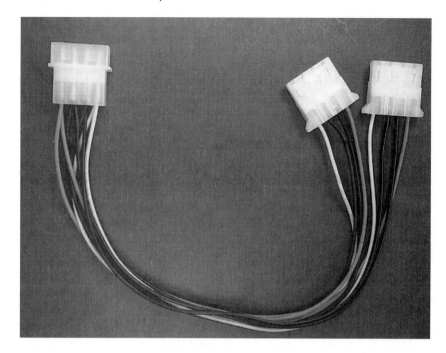

## I.9 ON OR OFF, THAT IS THE QUESTION

There has been lots of discussion over the years on whether it is better to leave a computer on 24 hours a day or turn it off any time it is not in use. Although leaving a computer on 24 hours a day might not be best for you, constantly turning the computer off and on is definitely bad for the computer.

In addition to the shock effect created by the in-rush of current when power is first applied to the computer, there is another consideration. When electrical current starts flowing through any electronic device, heat is generated. When power is removed, the electronic devices cool down. This heating up and cooling down of the electronic devices causes a problem much greater than the shock created by the in-rush current.

When electronic devices heat up, they expand (get bigger). When they cool down, they contract (get smaller). This constant expansion and contraction created when the computer is turned on and off can literally tear the electronic devices in the computer apart over time.

Computers are designed to run 24 hours a day and will probably last longer if they are allowed to do so. In some installations, the computer is required to be on 24 hours a day, so the decision is out of the operator's hands. In most home and office installations, the operator has to decide whether to leave the computer on 24 hours a day.

When the computer is turned on, it is consuming power for which you have to pay. There are additional pieces of equipment that are part of the computer that consume power, other than for the computer itself. These include devices such as a video monitor, printer, and speakers. With these devices, the computer can consume up to 400 watts of power. If the computer in this example is left on 24 hours a day, it will require approximately 288 KWH per month. I currently pay approximately 7.1 cents per KWH. At this rate, it costs me approximately $20.45 per month to leave my computer and its peripherals on 24 hours a day.

If you cannot afford this expense, the best practice for the home computer is once you turn it on, leave it on until you are ready to retire for the day. At work, the best practice is to turn the computer on when you arrive for work in the morning and leave it on until you leave for the day.

For those who leave their computer turned on most of the time, there are alternatives that will help reduce the overall power consumption. These features are referred to as *power management* and are configured in both the computer CMOS setup and operating system. Both are covered in later chapters. In the following section, we discuss the basics of the compatible computer's power management standards.

> ### Procedure I.1: Configuring Window 9X APM settings
>
> 1. Click on the Start button on the Task Bar.
> 2. Move the mouse pointer up to Settings, then over to Control Panel.
> 3. Click on Control Panel.
> 4. Locate, then double-click on the Power Management icon.
> 5. Make the desired selection and entries.
> 6. To exit without making any changes, click on the X in the window's upper right-hand corner or the Cancel button.
> 7. To accept changes and close the window, click on the OK button.

## I.10 POWER MANAGEMENT

To save money and reduce the effects that thousands of computers consuming power in the United States have on the environment, there is an effort underway to adopt two current standards for saving energy. One standard is a voluntary standard called *Energy Star* and the other is called *Advance Power Management* (APM).

### I.10.1 Energy Star

The Environmental Protection Agency (EPA) and the U.S. Department of Energy have established a voluntary standard called **Energy Star** for users of computer and peripherals to follow to save power. Basically, the Energy Star standard establishes a low-power state that the computer or peripheral switches to during states of prolonged inactivity. This state of inactivity is usually referred to as the *sleep,* or *standby, mode.* When the user attempts to use the device, it will wake up and consume its standard amount of power. If a computer or peripheral meets the Energy Star criteria, the device is allowed to display the Energy Star symbol.

Although compliance with the Energy Star criteria is voluntary, it is rapidly gaining the support of many manufacturers. Displaying the Energy Star label is an added selling feature because the vendor can advertise its products as being energy efficient.

### I.10.2 Advanced Configuration and Power Interface (ACPI)

ACPI is a specification created jointly by Intel, Microsoft, and Toshiba. It defines specifications of both hardware and software that are specifically designed to work together to reduce the power consumption of the computer and its peripherals. To take advantage of ACPI, the computer's BIOS, chipset, peripherals, drivers, operating system, and applications software must all conform to the specification. If you don't know the meaning of all of the terms in the preceding sentence, don't worry—they are all covered in later chapters in this text. The ACPI settings for the BIOS are covered in chapter 9, while Procedure I.1 lists the steps to configure the *advanced power management (APM)* features of Window 9X.

Configuring the computer to consume less power is one thing, but we also need to know some things we can do to protect our computer from the damage that can be caused by the wall outlet that provides the power in the first place. The following sections in this chapter cover this topic.

## I.11 PROTECTING YOUR INVESTMENT

Computers, like all electronic equipment, can be damaged beyond repair if exposed to the extremely high voltage spikes that occur when lightning strikes. Turning the equipment off during a thunderstorm offers very little protection from lightning. Lightning has enough energy to jump several miles through the air, so it will easily jump across a power switch. The only sure way to protect your electronic equipment from direct damage due to lightning strikes is to unplug it from the wall outlet. For some, this may be possible, but for most this is not an alternative.

To protect electronic equipment as much as possible and still leave it plugged into the power outlet, there are two alternatives. The first is to use a surge suppressor and the second is to use an uninterruptible power supply (UPS).

**Figure I.22** Typical Surge Suppressor

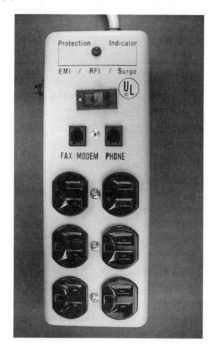

## I.11.1 Surge Suppressor

The surge suppressor is a special power outlet that has electronic devices that adsorb the energy of voltage spikes that are often caused by lightning strikes. When a voltage spike occurs, the surge suppressor will usually protect the devices that are plugged into its outlets even though the suppressor itself may be destroyed. Surge suppressors are available in several price ranges from the very cheap to the very expensive. The following list shows three important features you should look for in a surge suppressor. If the surge suppressor you currently own does not have these features, I highly recommend you replace it with one that does.

1. The surge suppressor should have an indicator lamp that indicates if it has been tripped.
2. The surge suppressor should conform to the Underwriters Laboratories standard UL 1449.[2]
3. The surge suppressor should have some kind of equipment replacement policy to replace equipment that is damaged when plugged into the surge suppressor.

Surge suppressors that protect your computer from voltage spikes that come in over the phone line or television cable are also available. If your computer connects to the phone line or television cable, you should have an appropriate surge suppressor between the incoming line and your computer. It is also a good idea to have surge suppressors on all electronic devices in your home and business. Figure I.22 shows a typical surge suppressor.

The UPS is another device that can protect your computer against voltage spikes and lower-than-normal ac voltage levels. Although they are more expensive than surge suppressors, UPS devices have features that may make them the protection device of choice for your particular needs.

## I.11.2 UPS

An **uninterruptible power supply (UPS)** provides power to the computer during a power outage or power brownout. The basic function of the UPS is to use a DC power supply to charge batteries when the ac is functioning correctly. An inverter is used to convert the DC provided from the battery into

---

[2]For more information on the UL 1449 standard, visit the web site http://www.ul.com/about/newsrel/nr1449.htm.

120 vac when power fails. There are currently two UPS types available commercially. One is called a *standby*, or *off-line, UPS* and the other is called an *on-line UPS*.

### I.11.3  Standby (off-line) UPS

A standby UPS powers the computer directly from the ac wall outlet until there is a power failure or brownout. When this condition occurs, an electronic sense circuit inside the UPS switches on an inverter to produce ac from the DC batteries. The computer is then switched over to the ac produced by the inverter. Even though this switch occurs in milliseconds, it can cause problems for some computers and devices. The main advantage of a standby UPS over an online UPS is that it is cheaper.

Because the equipment plugged into a standby UPS is powered directly from the ac line, voltage spikes can still cause equipment damage unless surge suppression is provided. If you purchase a standby UPS, make sure it has internal surge suppression. Also, make sure the standby UPS has an equipment replacement policy.

### I.11.4  On-line UPS

An on-line UPS always supplies power to the computer from the inverter. With this feature, there is no switch time when the ac from the wall outlet fails. Also, because the inverter is powering the computer, the computer is isolated from the ac wall outlet. This feature gives the computer excellent protection from the voltage spikes, power failures, and power brownouts.

Along with the type of UPS, there are other considerations to take into account when deciding which UPS to purchase. To understand these considerations, we must first discuss some of the ratings used for UPS devices.

### I.11.5  UPS Ratings

UPS devices are rated in volt-amperes (VA), whereas computer power supplies are usually rated in watts. In ac, a wattage rating is usually referred to as *true power* because this is the actual work being performed. Unfortunately, the UPS has to provide what is called *apparent power*. Apparent power is a rating used when ac provides power to loads that contain resistance and inductive or capacitive components. Apparent power is rated in VA and is usually higher than watts.

When transformers, inductors, and capacitors are encountered in an ac circuit, strange things happen between the current and voltage that cause the ac voltage source to have to provide more power than necessary. All of these aforementioned devices are found in the power supply used in the PC. Remember from section I.2.4 that the power being used is calculated by multiplying the voltage times the current ($P = V \times I$). In ac, if you multiply the voltage times the current, you are calculating VA, not watts. The actual work in watts being performed is usually less than VA.

To calculate the actual amount of work in watts, a fractional multiplier called a *power factor* is used. To come up with the actual power factor (PF) value requires the use of some expensive test equipment along with basic trigonometry. Needless to say, describing this method is beyond the scope of this book. However, another method that will give an approximate PF value works well for our purpose and provides a safety margin to boot. To convert the wattage rating of the PC power supply into an approximate VA value, simply multiply the watts by a PF of 1.43. Problem I.9 is an example of converting watts into an approximate VA value using this method.

---

**Problem I.9:** The computer's power supply is rated at 230 watts. What will be the calculated approximate VA of the power supply?

**Solution:** Since approximate VA = watts · 1.43, then VA = 230 · 1.43 ≈ 328.9 VA

---

Along with the VA rating of the UPS, there is also a time rating. The UPS time rating indicates the length of time the UPS can safely provide the rated VA to the loads plugged into it when the main ac fails. The time rating is usually given in minutes for full load and half load. The reason for giving these two ratings is that the batteries that supply power to the inverters inside the UPS do not have a linear discharge rate. This simply means that the half-load time will be greater than the full-load time divided by two.

Basic Electricity and Electronics

---

**Procedure I.2: Calculating UPS size**

1. List all of the devices you intend to be powered by the UPS on a worksheet.
2. Read the model label on each piece of equipment and write down voltage-ampere or wattage ratings next to the names of each piece of equipment on the list. This label is usually located on the rear panel of most equipment.
3. Multiply the voltage and current for each piece of equipment listed on the worksheet and write the answer beside each. If the equipment is rated in watts, multiply the wattage rating by 1.43 to convert to VA, and then write this calculation on the worksheet beside the name of the piece of equipment.
4. Add all of the VA calculations together and multiply the subtotal by 0.25 to compensate for future growth. Add this growth factor to the subtotal.
5. Select a UPS that has a VA rating at least as large as the one calculated.

---

Now that we have the technical stuff out of the way, let's see how to calculate the UPS size that you will need for your computer.

### I.11.6 Calculating UPS Size

Choosing the appropriate VA rating for a UPS is very important. Procedure I.2 outlines the steps to follow to calculate UPS size for your computer system. This procedure can also be used if you want to use a UPS to power devices other than a computer.

Well I guess that's about it as far as an introduction to electricity, electronics, and the power systems used in personal computers. I know the discussions in the chapter were not as in-depth as some people might like. If you are still thirsting for knowledge, I highly recommend you obtain additional textbooks or take courses in electronics. Now it is time to discuss basic troubleshooting procedures to use when a problem arises with the power distribution systems used in the PC.

## I.12 POWER SUPPLY TROUBLESHOOTING

Because the main DC power supply is what furnishes the currents required by all of the devices and circuits in the compatible PC, it can definitely cause all kinds of major problems. These problems can be anything from erratic behavior to complete failure.

Better DC supplies usually have all kinds of protection circuits that should shut all individual supplies down if the voltage or current levels on any of the individual DC supplies drift outside of set tolerances. Some of the less expensive DC supplies will not have these protection circuits, however. In any case, if the main DC power supply shuts down, the faults could be any of the individual supplies or any of the electronic devices that are connected to the supply.

If the power supply is determined to be defective, it is usually simply replaced. This is because the power supply itself is fairly inexpensive and is considered a consumable item. A consumable item is anything that will cost more to repair than to replace. If you are not a trained electronics technician, you should never remove the metal cover from the power supply itself because of the dangerous voltage levels inside. If your computer is in a critical application, you should consider purchasing a spare power supply of the correct wattage and size so it will be on hand if the power supply in the computer should fail. Also, having a spare power supply available is a valuable troubleshooting tool.

### I.12.1 Safety First

Before we get started with any troubleshooting procedures for the compatible PC switching power supply, we must first cover some basic safety precautions. To access the switching DC supply in a compatible computer, the computer's cover must be removed. Once the cover is removed, electrical circuits will be exposed that can easily be damaged or cause bodily harm if you are careless and do not follow basic safety procedures. If these basic safety procedures are followed, the danger to the computer and to your health will be minimal. If you do not follow basic safety procedures, the computer can be damaged or you can be hurt.

Even though the main DC supply in the compatible PC does not put out any DC voltages that are considered lethal, the DC power supply receives its power from an ac wall outlet. The 120 vac that is provided by the wall outlet can easily be fatal under the right circumstances. Also, the +5 VDC and +12 VDC supplies provided by both the AT and ATX power supplies contain enough current to create a lot of heat if they are allowed to short to ground or to another one of the DC supplies. This heat can quickly damage equipment or cause severe burns. To avoid these potential problems, here are a few safety notes:

- Learn the location of the circuit breaker that controls the power used by the computer or test equipment.
- Use tools and test equipment only for the purpose for which they were designed.
- Make sure the computer chassis is grounded. Never unground any electrical equipment or appliance by clipping the ground prong on the power cord, using a two-wire extension cord, or using a three-wire to two-wire plug adapter.
- Replace fuses only with those that have the same physical shape, voltage, and current ratings.
- Have the phone numbers available for local fire and rescue services.
- Replace all frayed or defective power cords. Do not attempt to repair them.
- Keep yourself dry. Never work on electrical equipment in wet conditions.
- Never remove the computer cover with power applied.
- Never work on any electrical circuit with power applied, if possible.
- Never disconnect or connect any electrical connectors with power applied.
- Never remove or install a printed circuit board with power applied.
- Remove all jewelry.
- Remove all items from your shirt pocket.
- Work with one hand in your pocket, if you have to work on live equipment.
- Avoid working alone on live electrical equipment.
- Do not remove the power supply's cover unless you are a trained technician.
- Use a static wristband when working inside the computer.
- Pay attention to what you are doing.

### I.12.2 Erratic Behavior

Troubleshooting a computer that exhibits erratic behavior is usually harder than troubleshooting one that completely fails. Some of the symptoms of a defective power supply that can cause erratic behavior are:

- Random memory failures that change locations;
- Random computer resets;
- Computer resets when there is a small power brownout;
- Static discharges to the computer case or keyboard that cause the computer to reset;
- Disk drive motors or cooling fans do not operate;
- Toggling the power switch several times to get the computer to operate; and
- Constant failure of equipment that receives power from the computer's main supply.

These are just a list of the symptoms I have seen that were caused by erratic power supply problems. Because the power supply provides current to all devices and circuits in the computer, there can be many more symptoms than those listed.

The computer's power cord and the ac wall outlet in which the computer is plugged are the first things that should be checked when a power supply is suspected of causing the computer to act erratically. The DC power supply cannot produce good DC if it is not getting good ac. The wall outlet should be a grounded receptacle with no loose connection.

Inspect the computer's power cord for any fraying or breaks. Also check any extension cords that the computer may be plugged into. If at all possible, swap the computer's power cord and extension cords with cords that are known to be good. Try plugging the computer directly into the wall outlet without the extension cord. If an extension cord is used, purchase one of high quality that supplies an earth ground connection. Never run the extension or computer power cord under a carpet that is walked on without enclosing

# Basic Electricity and Electronics

**Table I.3** DC Power Supply Voltage Measurements

| Supply | Tolerance | Maximum | Minimum |
|---|---|---|---|
| +3.3 VDC | ± 5% | +3.47 VDC | +3.14 VDC |
| +5 VDC | ± 5% | +5.25 VDC | +4.75 VDC |
| −5 VDC | ± 10% | −5.5 VDC | −4.5 VDC |
| +12 VDC | ± 5% | +12.6 VDC | +11.4 VDC |
| −12 VDC | ± 10% | −13.2 VDC | −10.8 VDC |

the power cord in a cable guard. Cable guards can be purchased at most office supply stores. Never run devices such as electric heaters or coffee pots on the same extension cord or power outlet with the computer.

You should feel resistance when plugging and unplugging the computer's power cord from the wall outlet. The power plug should not be warm to the touch after the computer has operated for several hours. Heat generation is an indication of resistance in the power outlet, which can be dangerous.

If at all possible, the ac power outlet should be checked with a voltmeter. In the United States, the reading should be between 105 vac and 130 vac. If you do not live in the United States, consult your regional power company for its ac voltage parameters. If the reading is not within the specified ranges, call your local power company.

Once the wall outlet has been eliminated as the source of the problem, the simplest and most often used method to determine if the power supply is causing the computer's erratic behavior is to simply try a known good power supply. If possible, replace the power supply with a supply that has a higher wattage rating because the problem might be that the original supply is being overloaded. If the erratic behavior ceases, then the original power supply is the problem. Of course, if the erratic behavior continues, the problem is not being caused by power supply.

## I.12.3 For the Technician

If you know how to use a DIGITAL Multimeter (DMM) and oscilloscope, there are a few checks you can make to help determine if the power supply is defective. The most common power supply problem that will cause the computer to act erratically is incorrect voltage levels. It is very important that the power supply be under normal loads when voltage measurements are made. The DC supplies in the compatible PC should all measure within a ± 5% tolerance. Table I.3 lists the minimum and maximum voltage levels for each of the supplies in the compatible computer.

Another power supply problem that can cause a computer to operate erratically is excessive DC ripple voltages or electrical noise. If the power supply is working correctly, the power supply's regulators should filter out excessive ripple and noise voltages. An oscilloscope is needed to find this problem. Once again, the power supply must be under normal loads when the electrical noise levels are measured. Also, the oscilloscope's ground lead must be connected to one of the ground leads on the power supply connector. Table I.4 lists the recommended maximum noise/ripple voltages on the indicated DC supplies.

## I.12.4 No Power at All

A complete power supply failure is indicated when the computer exhibits no activity at all. The most obvious indications are that no lights are illuminated on the computer's front panel, the computer's display is blank, and the power supply's cooling fan is not operating. If the computer's power supply does not operate at all, it does not necessarily mean the power supply is defective. The better power supplies have protection circuits that will shut down the supply in the case of a fault that exceeds factory-set limitations. This shutdown may be caused by the power supply itself or the circuits that are connected to the power supply.

The first thing to do if the power supply does not work is to check that the computer is plugged in and turned on. Also make sure that the power cord is plugged all the way into the back of the computer. Even though this may seem obvious, I have seen it happen several times over the years. It is also a good

Table I.4  Ripple Noise/Ripple Voltages

| Power Supply Voltage | Maximum Ripple Voltage |
|---|---|
| +12 VDC | 120 mVpp |
| −12 VDC | 120 mVpp |
| +5 VDC | 50 mVpp |
| +5 VSB | 50 mVpp |
| −5 VDC | 100 mVpp |
| +3.3 VDC | 50 mVpp |

idea to plug some other device, such as a lamp, into the same ac wall outlet the computer was plugged into. This will determine if the outlet is working. If you have another power cord that will fit the computer, try swapping cords. Sometimes one or more of the conductors in the power cable breaks, causing the power supply to not work. If the computer is plugged into an extension cord, try plugging the computer directly into a wall outlet or an extension cord that is known to be good. If these steps do not correct the problem, the computer's cover must be removed.

It might seem that the easiest way to determine if the power supply or circuits that the power supply feeds are defective is to simply unplug all of the power supply connectors from the loads. Although this is a common procedure used with **linear power supplies,** it will not work with **switching power supplies.** A switching DC power supply will not operate correctly unless it is connected to a load. For the switching power supply used in compatible computers, the +5 VDC and +12 VDC must be loaded for the main DC supply to operate. Because of this fact, all of the power connectors going to the different devices cannot be disconnected at the same time or the power supply will not operate. Procedure I.3 lists the steps to follow to determine if the power supplies or loads are defective.

If the power supply does not start working, the power supply is probably defective and will have to be replaced. Remember the defective power supply must be replaced with one of the same physical size and

### Procedure I.3: Determining if power supplies or loads are defective

1. Turn the computer off and unplug the computer's power cord.
2. Disconnect the power connectors going to the motherboard but leave the power connected to all of the disk drives.
3. Plug the computer into an ac outlet and turn it on.
4. If the power supply starts to work, which is indicated when the power supply's cooling fan begins to rotate, you have found your problem. If the power supply still does not operate, go to step 5.
5. Turn the computer off and unplug the computer's power cord.
6. Reinstall the motherboard's power connectors.
7. Remove all of the adapter cards plugged into the motherboard's expansion bus.
8. Plug the computer in and turn it on.
9. If the power supply starts to operate, the problem is with one of the adapter cards. If the power supply still does not operate, go to step 14.
10. Turn the computer off and unplug the power cord.
11. Reinsert one of the adapter cards.
12. Plug the computer in and turn it on.
13. If the power supply continues to operate, repeat steps 10 through 13 until the defective adapter card is located.
14. If the motherboard is not the problem, turn the computer off and unplug the power cord.
15. Reconnect the motherboard power connector(s) and unplug the connectors going to each disk drive.
16. Plug in the computer and turn it on.
17. If the power supply starts working, reconnect each disk drive connector individually until you find the drive that is causing the problem. You should turn the computer off and unplug the power cord before you reconnect any drive connector.

# Basic Electricity and Electronics

wattage rating. In fact, it is a good idea to replace the defective supply with one that has a higher wattage rating. The procedure to use to remove and install a power supply is explained in the next section.

## I.13  POWER SUPPLY REMOVAL AND INSTALLATION

Processes for removing and installing AT or ATX power supplies are very similar. The only difference is that the AT power supply will have its power switch mounted on the computer's front panel. The removal and installation procedures are given in Procedure I.4.

### Procedure I.4:  Power supply removal and installation

**Removal**

1. Shut down the computer's operating system and turn the computer off. On the AT power supply, the power switch is located on the front panel of the computer. On the ATX power supply, the switch on the front panel is not the true power switch. The true power switch is located on the back of the power supply that protrudes through the computer's back panel.
2. Remove the power cord from the back of the computer.
3. Remove the computer's chassis cover and store the screws in a safe location.
4. Remove the power connectors from the motherboard and all of the internal peripherals.
5. For a computer with an AT power supply, remove the electrical connector from the remote power switch located on the computer's front panel. Make a note of which color wire is connected to each power switch connector.
6. Remove the four screws that hold the power supply in the chassis and store them safely. The locations of these screws are shown in Figure I.23.
7. Remove the power supply and store it in a safe location.

**Installation**

1. Make sure the computer's power switch is in the off position.
2. Make sure the computer power cord is not plugged into the wall receptacle.
3. If in place, remove the computer's cover and store the screws in a safe location.
4. Position the power supply in the chassis so that its mounting screw holes align with those in the chassis. Once positioned, secure the power supply with the four screws.
5. Make sure the motherboard and peripheral power cables are routed inside the chassis so they will not interfere with the installation of the computer's cover.
6. On an AT power supply, connect the remote power switch conductors to the correct terminal on the computer's power switch located on the front panel. There should be four male terminals on the switch. If you are not sure of the connection, the terminals can be identified with an ohmmeter. Low resistance should be measured between terminal pairs when the switch is in the on position. The white and blue conductors should be connected to one terminal pair that measured low resistance and the black and brown should be connected to the other.
7. Insert the correct power supply connector into the motherboard connector. On AT power supplies, make sure the black conductors on both the P8 and P9 connectors are positioned toward the center of the connector. Also make sure the connectors from the power supply are not offset, leaving one or more pins exposed on the motherboard connector when both power supply connectors are inserted. The correct orientation of P8 and P9 is shown in Figure I.20.
8. Insert the correct power connector into each of the internal peripherals. Make sure the connector on the 3 1/2" drive is not inserted upside down.
9. Reinstall any data cables that had to be removed to install the power supply.
10. Insert the power cord into the receptacle located on the back of the power supply, and then plug the power cord into a operational wall outlet.
11. Turn the computer on using the power switch. On computers that use the ATX power supply, the power switch located on the power supply must be turned on by the soft power switch located on the computer's front panel.
12. If the power supply is working correctly, the power supply and CPU fan should be running and the computer should boot. If the power supply is not working properly, follow the troubleshooting procedures outline in the sections that begin with section I.12.
13. If the computer is working properly, shut down the operating system, turn the computer off, and reinstall the cover.

**Figure I.23** Power Supply Mounting Screw Locations

## I.14 CONCLUSION

It is not really necessary to fully understand the concepts of electricity and electronics to use a compatible computer. However, it is imperative that an individual who wants to be proficient in the upgrade and repair of the PC has a basic understanding of the terms and expressions used in the electrical and electronic fields. In this chapter, I have attempted to provide an explanation of the components, terms, and basic operation of devices that are most prevalent in the field of personal computers.

I have also made an attempt to give you enough knowledge about electrical and electronic devices such as transformers, resistors, capacitors, inductors, diodes, switches, and transistors so you will have a better understanding of the terms and expressions used daily in the field of computers. The majority of the basic computer books I have read use these terms without explanation.

Also, having a basic insight into the operation of the computer power supplies is very important for anyone who wishes to build, upgrade, and repair computers. Because power supplies are simply replaced when they are upgraded or defective, I did not get into great detail about the inner-workings of the power supply. Hopefully, enough information is supplied to determine if a power supply is defective. Also, I provided information that should help you to select the correct size and wattage rating for the supply if it requires replacement.

Basic explanations were given in the chapter about other computer power-related devices such as fuses, circuit breakers, surge suppressors, UPS, and power management. Hopefully you realize these are very important concepts for a computer-savvy person to know.

## Review Questions

1. What do the following acronyms stand for?
   a. APM
   b. ACPI
   c. KWH
   d. UPS

### Fill in the Blanks

1. The unit used to measure the amount of electrical push is the _____ .
2. One ampere is equal to _____ electrons moving past a point in _____ second.

# Basic Electricity and Electronics

3. Electrical current consists of the flow of _____ .
4. _____ is the best known conductor of electrical current.
5. The most often used conductors of electrical current are _____ and _____ .
6. _____ is mainly used as a plating because it does not rust.
7. A group of conductors contained within a common jacket is called a _____ .
8. The rate of doing work is measured in a unit called the _____ .
9. The unit of opposition to current flow is the _____ that can also be represented by the Greek symbol _____ .
10. Power can be calculated by multiplying the amount of _____ by the amount of _____ .
11. The _____ protection device melts a small ribbon conductor when an overcurrent condition is sensed.
12. A safety precaution to follow when working on live electrical circuits is to work with one _____ behind your back.
13. Current that flows in only one direction is called _____ _____ .
14. Current that changes direction at a fixed rate is called _____ _____ .
15. The rate at which an ac signal changes directions is called _____ .
16. A device that can step ac voltages up or down is called a _____ .
17. A unit that converts ac into DC is called a _____ _____ _____ .
18. The electronic device that works as an electronically controlled switch is the _____ .
19. The _____ is the electronic device most often used as a storage element within the computer.
20. A _____ power is the type most often used in the compatible computer due to its efficiency.
21. _____ _____ usually supplies the return path for all of the power supplied in the computer.
22. The compatible computer power supply whose motherboard connection cannot be inserted backwards is the _____ .
23. The insulator color used for the +5 VDC supply in the computer is _____ .
24. A(n) _____ _____ helps limit the damage caused by lightning strikes.
25. A(n) _____ power supply provides power to the computer during a power failure.
26. Another name for electronic devices is _____ .

List the color wire for each of the following power supplies in the compatible computer.

1. +5 VDC
2. −5 VDC
3. +12 VDC
4. −12 VDC
5. +3.3 VDC
6. +5 VDC standby

## Multiple Choice

1. The protection device that opens the circuit by melting a piece of metal into is called a:
   a. fuse.
   b. circuit breaker.
   c. UPS.
   d. surge suppressor.
2. A device that maintains power to the computer even when the power to the ac wall outlet fails is called a:
   a. power supply.
   b. UPS.
   c. circuit breaker.
   d. surge suppressor.
   e. upper pulsating supply.
3. What is the function of a transformer?
   a. increases and decreases power
   b. changes the frequency of the ac
   c. Converts ac into DC
   d. changes the value of the voltage
4. If a device has 500 ohms of resistance and 120 volts is applied to it,
   a. 500 amperes of current will flow.
   b. 120 amperes of current will flow.
   c. 0.24 amperes will flow.
   d. 4.17 amperes will flow.
5. If 5 amperes is flowing through a device with 15 volts applied, the power will be:
   a. 5 watts of power will be consumed.
   b. 15 watts of power will be consumed.
   c. 3 watts of power will be consumed.
   d. 75 watts of power will be consumed.
6. If a device that uses 500 watts is left on for 24 hours, how many KWH will it consume?
   a. 12 KWH
   b. 12,000 KWH
   c. 500 KWH
   d. 24 KWH
7. Which device will usually blow when current flow through it exceeds the rated value?
   a. circuit breaker
   b. fuse
   c. surge suppressor
   d. UPS
8. Which device carries the current throughout the circuit?
   a. ampere
   b. voltage source
   c. conductor
   d. resistor
9. A thin board that appears to have the conductors for the circuit printed on its surface is called a(n):
   a. electronic circuit card.
   b. plastic circuit board.
   c. engraved circuit board.
   d. printed circuit board.
10. The over-current detection device that does not destroy itself when an over-current condition is sensed is the:
    a. circuit breaker.
    b. fuse.
    c. surge suppressor.
    d. UPS.

11. DC is used most often in electronic circuits because:
    a. it is cheapest to produce.
    b. its voltage levels can easily be changed.
    c. it is faster than ac.
    d. it is the easiest to control.
12. Which device is usually used to mechanically turn current off and on?
    a. transistor
    b. transformer
    c. fuse
    d. capacitor
    e. switch
13. A transformer can be used to increase or decrease the amount of power.
    a. True
    b. False
14. Which unit in the power supply produces the exact values of DC needed by the loads?
    a. inverter
    b. regulator
    c. transformer
    d. rectifier
    e. filter
15. Which unit in the power supply converts the ac to DC?
    a. inverter
    b. transformer
    c. rectifier
    d. filter
16. The ATX and AT power supplies use the same motherboard connector.
    a. True
    b. False
17. All power supplies are the same physical size.
    a. True
    b. False
18. Which insulator color is used for the 3.3 VDC on the ATX power supply?
    a. blue
    b. red
    c. yellow
    d. green
    e. brown
19. A computer should be turned off anytime it is not in use.
    a. True
    b. False
20. All electronic equipment should be plugged into a(n):
    a. standard wall outlet.
    b. surge suppressor.
    c. UPS.
    d. battery.
21. The UPS that always powers the computer from the inverter is the:
    a. offline UPS.
    b. standby UPS
    c. on-line UPS
    d. enhanced UPS
22. Before working on the power supply, all jewelry should be removed.
    a. True
    b. False
23. A defective power supply can cause the computer to reset randomly.
    a. True
    b. False
24. What is the best known conductor of electricity?
    a. gold
    b. silver
    c. copper
    d. tin
25. Why are some electronics devices gold plated?
    a. to lower the resistance of the device
    b. because gold does not rust
    c. because gold is the best known conductor of electricity
    d. because it looks really pretty.
26. The approximate resistance of a 500-ft section of 22-gauge copper wire is:
    a. 20 ohms.
    b. 17 ohms.
    c. 8 ohms.
    d. 10 ohms.

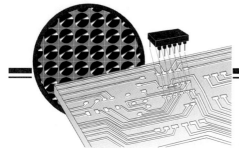

# CHAPTER 1

# Computer Numbering Systems and Terminology

## OBJECTIVES

- Convert between decimal and binary numbers.
- Convert between hexadecimal and decimal numbers.
- Convert between binary and hexadecimal numbers.
- Use ASCII and Enhanced ASCII charts.
- Describe the method of error detection called parity.
- Determine the parity of a binary code.
- Define kilo, meg, gig, milli, micro, nano, and cycle.

## KEY TERMS

| | | |
|---|---|---|
| binary | parity | cycle |
| hexadecimal | kilo | milli |
| ASCII | mega | micro |
| Enhanced ASCII | giga | nano |

One of the major problems with understanding how personal computers operate is that they are digital devices. The term *digital* indicates the computer does not use the base-10 (decimal) numbering system that we know and love so well. In fact, the only numbering system digital computers understand is **binary.** All numbers and information must be converted or coded to the binary system before the computer can actually process the data. This task is usually performed by a program and is transparent to the user and sometimes even the programmer.

You do not have to learn the binary numbering system to use today's personal digital computer. Programs inside the computer allow the average user to use the decimal system to enter numbers into the computer and a standard QWERTY keyboard to enter text. However, if you ever need to upgrade your computer, change hardware configurations, or diagnose computer malfunctions, you will probably need to have some understanding of the binary and hexadecimal numbering systems.

## 1.1 NUMBERING SYSTEM BASICS

The *base* of a numbering system indicates the number of combinations each digit can represent. For example, the decimal numbering system is also referred to as *base-10*. The term base-10 indicates each digit can have 10 numerical combinations—0 through 9. The binary numbering system is called *base-2*. The 2 indicates each digit can only have two combinations—0 and 1.

Both base-10 and base-2 are weighted numbering systems. The term *weighted* indicates the value (weight) each digit holds to the overall weight (value) of the number varies depending on its physical location. In every weighted numbering system, each digit holds a weight of the digit times the base raised to some power. The actual power the base is raised to depends on the digit's position in the number.

The number farthest to the right is called the *least significant digit (LSD)* because it holds the smallest weight to the overall weight of the number. The digit farthermost to the left is called the *most significant digit (MSD)* because it holds the greatest weight to the overall weight of the number.

Understanding binary fractions and negative numbers is a lot harder than understanding unsigned whole binary numbers and is not really necessary to comprehend the basic concepts of the binary numbering system. To avoid more confusion, the remainder of this discussion will deal only with unsigned whole numbers.

### 1.1.1 (Base-10) Basics

As stated previously, the decimal numbering system is a weighted system. The decimal numbering system is also referred to as base-10 because each digit can have 10 combinations—0 through 9. An example of a decimal number is $78,123_{10}$. You probably noticed the small 10 to the right of the number. This is called a subscript. It is standard practice to use subscripts to define the base when several bases are being used. I use this practice to help avoid confusion when we start representing numbers in bases other than 10.

We already know that each digit of the number $78,123_{10}$ holds a weight of the digit times 10 raised to some power. We also know that the power of 10 the digit is raised to depends on the digit's position in the number. To help with the explanation, refer to Example 1.1 as we examine the decimal number $78,123_{10}$.

**Example 1.1:** The number 78,123 showing weights of the power of 10s

| $10^4$ | $10^3$ | $10^2$ | $10^1$ | $10^0$ |
|---|---|---|---|---|
| 7 | 8 | 1 | 2 | 3 |

In Example 1.1, the first digit on the right (3) (LSD) is in the $10^0$ position. Ten raised to the zero power is one. The actual weight the 3 holds to the overall weight of the number is $3 \times 10^0$ or $3 \times 1 = 3$. The next digit to the left is the $10^1$ position. Ten raised to the first power is 10. So the weight 2 holds in the number is $2 \times 10 = 20$. The next digit to the left holds a weight of the number in that position times $10^2$, or 100. So this means the 1 holds a weight of $1 \times 100 = 100$. The next digit to the left holds a weight of the number in that position times $10^3$, or 1,000. So the digit 8 in the example number holds a weight of $8 \times 1,000 = 8,000$. The digit farthermost to the left (MSD) holds a weight of the number in that position times $10^4$, or 10,000. So the digit 7 holds a total weight of $7 \times 10,000 = 70,000$. If the number were larger, the power of 10s would keep increasing by 1 until the last digit of the number is reached.

The digits in the decimal number $78,123_{10}$ tell us that the total weight of the number is equal to 7 ten thousands + 8 thousands + 1 hundred + 2 tens + 3 ones. This may seem obvious because we have dealt with the base-10 system for so long. However, because this is the basic concept used by all weighted numbering systems, it is nice to start with something you already know and then go from there. To calculate the actual weight of the decimal number, we perform the process illustrated in Example 1.2.

**Example 1.2:** Calculating weight of the decimal number by adding each digit's weight

$$(7 \times 10,000) + (8 \times 1,000) + (1 \times 100) + (2 \times 10) + (3 \times 1) = 78,123_{10}$$

## 1.1.2 Binary (base-2) Basics

As the names *binary* and *base-2* imply, each digit of the binary numbering system can represent only two states—0 or 1. This may seem like a simple concept, but people have been using the decimal numbering system all of their lives, so learning the binary system is similar to learning a foreign language.

Just like base-10, base-2 is a weighted numbering system. Each binary digit, called a *bit* in binary, in the base-2 number holds a weight of the digit (0 or 1) times 2 raised to some power. An example of a binary number is 1 1 0 1 1 0 1$_2$. To help clear up this explanation, let us examine this binary number a little closer. To do this, we will refer to Example 1.3.

**Example 1.3:** Binary number showing positional weights

| 1 | 1 | 0 | 1 | 1 | 0 | 1 |
|---|---|---|---|---|---|---|
| $2^6$ | $2^5$ | $2^4$ | $2^3$ | $2^2$ | $2^1$ | $2^0$ |

Notice in Example 1.3 that we use the same technique that was used for the decimal number in Example 1.1. The LSb (least significant bit) actually holds a weight of 2 raised to the 0 power. As bits are added, the power of 2 increases by one until the last bit is reached (MSb).

**Problem 1.1:** If one more digit is added to the left of the binary number in Example 1.3, what power of 2 will its bit position hold?

**Solution:** Because the last position holds a weight of 2 raised to the 6th power, the new bit position will hold a weight of 2 raised to the 7th power.

This may seem fine and dandy, but if you were paid in binary you would like to know how much money you were making. If we were taught binary all of our lives instead of decimal, this would be second nature. However, this is not the case. Because the computer only knows binary, it has no problem with binary numbers, but humans need to convert the binary number into decimal before it makes any sense. Let us look at one simple way to perform this task in Example 1.4.

**Example 1.4:** Decimal equivalent of the power of 2s

| 1 | 1 | 0 | 1 | 1 | 0 | 1 |
|---|---|---|---|---|---|---|
| 64 | 32 | 16 | 8 | 4 | 2 | 1 |

The simplest method to convert a binary number into decimal, besides using a calculator that converts for you, is to convert each power of 2 into its decimal equivalent. Notice in Example 1.4 that the decimal weight of each power of 2 is written below each binary bit. To find the decimal weight of the binary number, simply add the decimal weights that have a 1 in the bit position above them. This process is illustrated in Example 1.5.

**Example 1.5:** Adding the decimal weights that have a 1 from Example 1.4

$$64 + 32 + 8 + 4 + 1 = 109_{10}$$

Hopefully you have noticed that the decimal weight each power of 2 holds doubles each time you move to the left. This means a calculator should not be necessary to convert binary numbers into decimal. Besides, a calculator that does base conversion may not always be available.

In Example 1.6, we will look at another binary to decimal example. The binary number is on top and the decimal weight of each power of 2 is on the bottom.

**Example 1.6:** Binary number with decimal weights for each bit position

| 1 | 0 | 1 | 1 | 0 | 1 | 0 | 1 | 0 | 1 | 0 | 0 |
|---|---|---|---|---|---|---|---|---|---|---|---|
| 2,048 | 1,024 | 512 | 256 | 128 | 64 | 32 | 16 | 8 | 4 | 2 | 1 |

As stated previously, to convert to decimal, simply add the decimal weights of the binary bit positions that contain a 1. Therefore, the process shown in Example 1.7 finds the decimal weight of the binary number in Example 1.6.

**Example 1.7:** Computing the decimal weight of the binary number in Example 1.6

$$2{,}048 + 512 + 256 + 64 + 16 + 4 = 2{,}900_{10}$$

### 1.1.3 The Reason for Binary

No matter how complex it may be, basically all a computer can do is input, store, process, and output information. The computers discussed in this book use binary to represent the information the computer inputs, processes, and outputs. You may wonder why binary is used when decimal seems so easy to us. Binary is used mainly because it is easy to transfer, process, and store.

To perform the task of inputing and outputing information, computers must have the ability to move this information around electrically. Computers transfer information over a group of electrical conductors called a *bus*. Information that goes to or from the different devices that make up the computer is placed on these conductors. If decimal were used, 10 voltage or current combinations would have to be represented on each wire to represent the 10 decimal combinations. This would be comparable to trying to distinguish between 10 brightness levels emitted from a light bulb. When binary is used, only two voltages or current levels are required to represent numbers and codes. This is similar to trying to tell if a light is turned on or off. It is very hard to tell how bright a flashlight is over distance, but it is very easy to tell when the flashlight is turned on or off.

The computer must have a place to hold the information until it is ready to be processed. It is very difficult to store information that has to be represented in 10 different levels for each digit. If base-10 were used, the electronic circuits that accomplish this feat would be very expensive to design and manufacture. When binary is used, the storage elements that make up the computer's memory can consist of nothing more then a lot of electronic switches called *flip-flops*. Each switch can be turned on or off easily to store a single binary digit (bit). These flip-flop storage elements are manufactured cheaply and occupy very little space on the material used to construct these computer memory circuits.

Finally, the computer has the ability to process information. Remember that the processing of information has to be performed by electronic circuitry. Designing circuitry to deal with the decimal numbering system is very complex and expensive because it has to be designed to handle the 10 combinations required by each decimal digit. Circuitry designed to process binary information is far less complex and, therefore, less expensive.

This section has hopefully given you some insight as to why binary was chosen by the engineers who first developed computers. Simply put, binary is easy to input, output, and process electrically, and easy to store electrically.

### 1.1.4 Decimal-to-Binary Conversion

Because the computer is binary and we understand decimal, we need to be able to convert a decimal number into binary. There are several methods that can be used to convert from the decimal numbering system to the binary numbering system. To help avoid confusion that may already exist, let us cover only one of the simpler methods. The method we use here involves the steps outlined in Procedure 1.1.

---

**Procedure 1.1:   Decimal-to-binary conversion**

1. Subtract the largest possible decimal weight of a power of 2 from the decimal number to be converted.
2. Write a 1 in the bit position of the power of 2 subtracted.
3. Repeat step 1 with the answer to each subtraction until the subtraction produces a difference of 0.
4. Place a 0 in all remaining bit positions to the right of the most significant 1.

# Computer Numbering Systems and Terminology

To help with the explanation, Examples 1.8 through 1.13 will be used to illustrate this process. The process usually begins by constructing a chart similar to the one shown in Example 1.8. The chart shows the decimal equivalent of binary powers of 2s that is one position greater than the decimal number to be converted. The number we are converting is 598, so the largest power of 2 used in Example 1.8 holds a weight of 1,024 ($2^{10}$).

Example 1.8:   Typical chart used to convert decimal to binary

| $2^{10}$ | $2^9$ | $2^8$ | $2^7$ | $2^6$ | $2^5$ | $2^4$ | $2^3$ | $2^2$ | $2^1$ | $2^0$ |
|---|---|---|---|---|---|---|---|---|---|---|
| 1,024 | 512 | 256 | 128 | 64 | 32 | 16 | 8 | 4 | 2 | 1 |
|  |  |  |  |  |  |  |  |  |  |  |

Notice in Example 1.8 that the largest power of 2's decimal weight that we can subtract from 598 is 512 ($2^9$). Following the procedure steps outlined previously, we subtract 512 from 598 and write a 1 in the $2^9$ (512) bit position. Example 1.9 shows the subtraction process.

Example 1.9:   Step 1, converting 598 into binary

$$\begin{array}{r} 598 \\ -\ 512 \\ \hline 86 \end{array}$$

| $2^{10}$ | $2^9$ | $2^8$ | $2^7$ | $2^6$ | $2^5$ | $2^4$ | $2^3$ | $2^2$ | $2^1$ | $2^0$ |
|---|---|---|---|---|---|---|---|---|---|---|
| 1,024 | 512 | 256 | 128 | 64 | 32 | 16 | 8 | 4 | 2 | 1 |
|  | 1 |  |  |  |  |  |  |  |  |  |

Next, we subtract the highest possible power of 2's decimal weight (64) from the results of the previous subtraction (86) and write a 1 in the $2^6$ (64) bit position. This is shown in Example 1.10. Notice that the 1 in the $2^9$ bit position that was created in Example 1.9 is still in place.

Example 1.10:   Step 2, converting 598 into binary

$$\begin{array}{r} 86 \\ -\ 64 \\ \hline 22 \end{array}$$

| $2^{10}$ | $2^9$ | $2^8$ | $2^7$ | $2^6$ | $2^5$ | $2^4$ | $2^3$ | $2^2$ | $2^1$ | $2^0$ |
|---|---|---|---|---|---|---|---|---|---|---|
| 1,024 | 512 | 256 | 128 | 64 | 32 | 16 | 8 | 4 | 2 | 1 |
|  | 1 |  |  | 1 |  |  |  |  |  |  |

Next, subtract the highest possible power of 2's decimal weight (16) from the results of the previous subtraction (22) and write a 1 in the $2^4$ (16) bit position. This step is illustrated in Example 1.11.

Example 1.11:   Step 3, converting 5

$$\begin{array}{r} 22 \\ -\ 16 \\ \hline 6 \end{array}$$

| $2^{10}$ | $2^9$ | $2^8$ | $2^7$ | $2^6$ | $2^5$ | $2^4$ | $2^3$ | $2^2$ | $2^1$ | $2^0$ |
|---|---|---|---|---|---|---|---|---|---|---|
| 1,024 | 512 | 256 | 128 | 64 | 32 | 16 | 8 | 4 | 2 | 1 |
|  | 1 |  |  | 1 |  | 1 |  |  |  |  |

Next, we speed up the process by performing the next two subtractions in Example 1.12. Notice in Example 1.12 that when 2 and 2 are subtracted, the result is 0. This indicates the subtraction process is complete.

**Example 1.12:** Steps 4 and 5, converting 598 to binary

$$\begin{array}{c} 6 \\ -4 \\ \hline 2 \end{array} \qquad \begin{array}{c} 2 \\ -2 \\ \hline 0 \end{array}$$

| $2^{10}$ | $2^9$ | $2^8$ | $2^7$ | $2^6$ | $2^5$ | $2^4$ | $2^3$ | $2^2$ | $2^1$ | $2^0$ |
|---|---|---|---|---|---|---|---|---|---|---|
| 1,024 | 512 | 256 | 128 | 64 | 32 | 16 | 8 | 4 | 2 | 1 |
|  | 1 |  | 1 |  | 1 |  | 1 | 1 |  |  |

Last, 0's are placed in all remaining bit positions to the right of the first power of 2 subtracted. This final process is shown in Example 1.13.

**Example 1.13:** Step 6, converting 598 to binary

| $2^{10}$ | $2^9$ | $2^8$ | $2^7$ | $2^6$ | $2^5$ | $2^4$ | $2^3$ | $2^2$ | $2^1$ | $2^0$ |
|---|---|---|---|---|---|---|---|---|---|---|
| 1,024 | 512 | 256 | 128 | 64 | 32 | 16 | 8 | 4 | 2 | 1 |
|  | 1 | 0 | 0 | 1 | 0 | 1 | 0 | 1 | 1 | 0 |

As indicated by the previous examples, by following this very simple process, we find that the decimal number 598 equals 1001010110 in binary.

**Problem 1.2:** Convert the decimal number 205 to binary.

**Solution:** Using the chart in Example 1.8, the following power of 2 decimal equivalent can be subtracted.

$$205 - 128 = 77 : 77 - 64 = 13 : 13 - 8 = 5 : 5 - 4 = 1 : 1 - 1 = 0$$

After the 1s are placed in the chart in Example 1.8, where powers of 2 are subtracted and 0s are placed in the other spots, we come up with the binary number 1 1 0 0 1 1 0 1.

### 1.1.5 Hexadecimal

The binary numbering system is relatively easy when dealing with a small number of bits. The more bits that are used the more likely it becomes for a person to make a mistake when converting base-2 to decimal. Even copying the binary number to a different location becomes more difficult as the binary number become larger. To get some feel of how cumbersome a large binary number is to people, Example 1.14 shows an example of a 32-bit binary number.

**Example 1.14:**

$$01101001101101001000111111010101_2$$

As you can see, it is very easy to make mistakes when converting large binary numbers to decimal numbers. Most computer hardware and software manufacturers understand this problem. To help reduce some of the common mistakes people make with the binary numbering system, the **hexadecimal** numbering system is usually used to represent large binary numbers.

# Computer Numbering Systems and Terminology

**Table 1.1** Decimal, Hexadecimal, and Binary Comparison

| Decimal | Binary | Hexadecimal |
|---------|--------|-------------|
| 0 | 0000 | 0 |
| 1 | 0001 | 1 |
| 2 | 0010 | 2 |
| 3 | 0011 | 3 |
| 4 | 0100 | 4 |
| 5 | 0101 | 5 |
| 6 | 0110 | 6 |
| 7 | 0111 | 7 |
| 8 | 1000 | 8 |
| 9 | 1001 | 9 |
| 10 | 1010 | A |
| 11 | 1011 | B |
| 12 | 1100 | C |
| 13 | 1101 | D |
| 14 | 1110 | E |
| 15 | 1111 | F |

The *hex* in hexadecimal means 6 and the *decimal* means 10. Add them together and you get 16, which is the base of the hexadecimal numbering system. In the *base-16* (hexadecimal) numbering system, each digit can have 16 combinations. The decimal numbers 0 through 9 are used to represent the first 10 combinations in hexadecimal. Because the decimal numbers 10 through 15 require two digits, they must be represented in some other way in hexadecimal. In hexadecimal, the numbers 10 through 15 are represented with the letters *A* through *F*. Table 1.1 shows the decimal numbers 0 through 15 with their hexadecimal and binary equivalents.

The hexadecimal numbering system is a weighted system, just like decimal. This indicates that each digit of a hexadecimal number holds a weight of the digit times 16 raised to some power. $AB9C2_{16}$ is an example of a hexadecimal number. This may seem great and wonderful, but if you were paid this much money every year you would like to know how much you were making. To get a feel for the weight of the number, we humans must somehow get the number into the only numbering system we fully understand—decimal. Besides using a base calculator, the simplest method to convert from hexadecimal to decimal is to follow the steps outlined in Procedure 1.2.

## Procedure 1.2: Converting hexadecimal to decimal

1. Convert each digit of the base-16 number into its base-10 equivalent.
2. Convert the powers of 16 into their decimal equivalents.
3. Multiply the digit in that power's position by the decimal weight it holds.
4. Add the decimal products together to get the actual decimal number.

Procedure 1.2 is used in Example 1.15 to convert $AB9C2_{16}$ into decimal. The results of the procedure show that the decimal weight of the hexadecimal number $AB9C2_{16}$ equals 702,914 decimal. Not bad pay for a year's work.

**Example 1.15:** Converting the hexadecimal number $AB9C2_{16}$ to decimal (Refer to Table 1.1)

| Hex | $(A \times 16^4) + (B \times 16^3) + (9 \times 16^2) + (C \times 16^1) + (2 \times 16^0)$ |
|---|---|
| Decimal | $(10 \times 65{,}536) + (11 \times 4{,}096) + (9 \times 256) + (12 \times 16) + (2 \times 1) = 702{,}914_{10}$ |

**Problem 1.3:** Converting the hexadecimal number $A43B_{16}$ to decimal

**Solution:** $10 \times 4{,}096 = 40{,}960: 4 \times 256 = 1{,}024 : 3 \times 16 = 48 : 11 \times 1 = 11$

$$40{,}960 + 1{,}024 + 48 + 11 = 42{,}043$$

The binary representations in Table 1.1 are deliberately shown using 4-bit groups. Notice that exactly 16 combinations can be represented with 4 bits of binary—no more and no less. This is the exact number of combinations that a single digit of hexadecimal represents. What all of this boils down to is any binary number can be converted to hexadecimal by simply splitting the binary number into groups of 4 bits starting from the right. Once this is done, use the hexadecimal number that corresponds to the weight of each group of 4 binary bits (see Example 1.16).

**Example 1.16:**

$$1\,0\,1\,0\,0\,1\,0\,1\,1\,1\,0\,1 = \underset{A}{1010} \quad \underset{5}{0101} \quad \underset{D}{1101}$$

This means that $1\,0\,1\,0\,0\,1\,0\,1\,1\,1\,0\,1_2 = A5D_{16} = 2{,}653_{10}$

Notice in Example 1.16 how the binary number is split into groups of 4 on the right. Extra 0s can be added to the left of the binary number to complete a 4-bit group, but never add 0s to the right or the weight of the number will be changed. Now let us convert the 32-bit binary number used earlier into hexadecimal in Problem 1.4.

**Problem 1.4:** Convert the binary number $01101001101101001000111111010101_2$ from Example 1.14 into hexadecimal.

**Solution:** Set the binary number off 4 bit groups starting on the right and write the base-16 number that each 4-bit pattern represents below the binary number.

| 0110 | 1001 | 1011 | 0100 | 1000 | 1111 | 1101 | 0101 |
|---|---|---|---|---|---|---|---|
| 6 | 9 | B | 4 | 8 | F | D | 5 |

The weight of the hexadecimal number $69B48FD5_{16}$ that is converted in Problem 1.4 and the 32-bit binary number shown in Example 1.14 are the same. Hopefully, you can see how much easier and accurate it is for the computer to convey the large binary number as $69B48FD5_{16}$ to a human than its binary equivalent. By the way, the weight of this number in decimal is $1{,}773{,}440{,}981_{10}$.

To go from hexadecimal to binary, just reverse the process. Below the hexadecimal number write the 4-bit binary pattern that equals the weight of each hexadecimal digit. Example 1.17 shows this technique.

**Example 1.17:** Converting hexadecimal to binary

$$A9C28_{16} = \underset{1010}{A} \quad \underset{1001}{9} \quad \underset{1100}{C} \quad \underset{0010}{2} \quad \underset{1000_2}{8}$$

There are a couple of good reasons for using hexadecimal to represent binary. First, hexadecimal is easier to communicate large binary numbers to people without making errors. Second, very large binary numbers can be represented with fewer hexadecimal digits.

It must be understood that the computer itself only understands binary. If hexadecimal is used to enter information or establish hardware configurations, it must somehow be converted to binary before the computer can process the information.

### 1.1.6 Calculating Numerical Combinations

Throughout this book we will need to determine the number of numerical combinations that can be represented given the number of digits a number has. For example, if you start counting at 0, how many numbers can be written using three decimal digits? Before you answer 999, you have to realize that 0 is one of the numerical combinations of a 3-digit decimal number. If you start counting at 0 and go through 999, there are 1,000 numerical combinations.

This seems easy enough when decimal is used because we have dealt with decimal all of our lives. However, if I asked how many numerical combinations there are when six binary bits are used, few people would be able to answer this question without writing the combinations down. A simple formula can be used to calculate the number of numerical combinations given the number of digits used to count, no matter the base.

To calculate the number of numerical combinations given the number of digits and the base, raise the base to the number of digits being used to count. The actual formula is $B^n$, where $B$ is the base and $n$ is the number of digits. For example, 6 digits of binary will have $2^6$, or $64_{10}$, combinations.

**Problem 1.5:** How many numerical combinations will a 24-bit binary number have?

**Solution:** $2^{24} = 16,777,216_{10}$

**Problem 1.6:** How many numerical combinations will an 8-digit hexadecimal number have?

**Solution:** $16^8 = 4,294,967,296_{10}$

### 1.1.7 BINARY COMPUTER TERMINOLOGY

The terms used most often when dealing with binary and computers are bit, nibble, byte, byte-word (8-bit word), word, long-word, and quad-word. A *bit (b)* refers to a single binary digit. A *nibble* is the term used to refer to the group of 4 bits that represents a single hexadecimal digit. A *byte (B)* refers to a group of eight consecutive bits, starting from the LSb.

A computer *word* usually varies in its number of bits depending on the communication size of the computer. But in the so-called *compatible computers,* the word sizes are fixed values that depend on the size of the data being processed. The term *byte-word* indicates an 8-bit word. The term *word* refers to a group of 16 consecutive bits. A *double word* is a group of 32 bits. The term *quad-word* is used in reference to a group of 64 consecutive bits. The largest word sizes used by your computer to communicate with other devices will depend on the microprocessor installed inside the computer. The relative differences among bit, nibble, byte, word, double-word, and quad-word are shown in Figure 1.1.

Figure 1.1  Terms to Represent Groups of Binary Bits

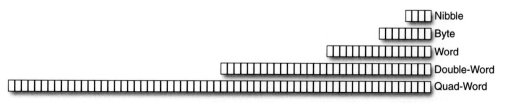

**Table 1.2**

| MSD → | 0 | 1 | 2 | 3 | 4 | 5 | 6 | 7 |
|---|---|---|---|---|---|---|---|---|
| 0 | NUL | DLE | SP | 0 | @ | P | ` | p |
| 1 | SOH | DC1 | ! | 1 | A | Q | a | q |
| 2 | STX | DC2 | " | 2 | B | R | b | r |
| 3 | ETX | DC3 | # | 3 | C | S | c | s |
| 4 | EOT | DC4 | $ | 4 | D | T | d | t |
| 5 | ENQ | NAK | % | 5 | E | U | e | u |
| 6 | ACK | SYN | & | 6 | F | V | f | v |
| 7 | BEL | ETB | ' | 7 | G | W | g | w |
| 8 | BS | CAN | ( | 8 | H | X | h | x |
| 9 | HT | EM | ) | 9 | I | Y | i | y |
| A | LF | SUB | * | : | J | Z | j | z |
| B | VT | ESC | + | ; | K | [ | k | { |
| C | FF | FS | , | < | L | \ | l | | |
| D | CR | GS | - | = | M | ] | m | } |
| E | SO | RS | . | > | N | ^ | n | ~ |
| F | SI | US | / | ? | O | _ | o | DEL |

## 1.2 ASCII

As previously stated, all data transfer within a computer must consist only of 1s and 0s. The question then arises, "How does the computer deal with letters and characters to communicate with devices that eventually have to send information to people?" The answer is, "The computer uses binary codes to represent characters and letters." The most popular code for text transmission in microcomputers is the *American Standard Code for Information Interchange* (**ASCII,** pronounced "*ask-key*"). Table 1.2 shows an ASCII chart that can be used to find the ASCII representation of different letters, numbers, characters, and control codes.

Programmers usually send text information to the computer's display or printer in ASCII. The electronic circuits of the printer or video controller translate the code and cause the correct character to be printed or displayed. Displays and printers are covered in more detail later in chapters 9 and 10. Note that current computers usually represent characters in the form of pictures. This is referred to as *graphics mode*. This does not mean that ASCII is obsolete—ASCII is still the preferred method of transferring information over the global Internet, which is discussed in greater detail in chapter 14.

There are actually two versions of ASCII. One is the original 7-bit code and the other is an 8-bit code called **Enhanced ASCII.** We will cover the 7-bit code first because it is the internationally accepted standard for transferring text information.

Using 7 bits, 128 combinations can be represented ($2^7$). To use the 7-bit ASCII chart provided in Table 1.2, find the character in the matrix and read the number of the column followed by the row number. For example, let us find the ASCII representation of the letter *R*. First, find *R* in the ASCII chart's

# Computer Numbering Systems and Terminology

**Table 1.3** ASCII Control Codes

| \multicolumn{4}{c}{ASCII Control Codes} | | | |
|---|---|---|---|
| NUL | Null | DC1 | Device Control 1 |
| SOH | Start of header | DC2 | Device Control 2 |
| STX | Start of text | DC3 | Device Control 3 |
| ETX | End of text | DC4 | Device Control 4 |
| EOT | End of transmission | NAK | Negative Acknowledge |
| ENQ | Enquiry | SYN | Synchronous idle |
| ACK | Acknowledge | ETB | End of transmission block |
| BEL | Bell | CAN | Cancel |
| BS | Backspace | EM | End of medium |
| HT | Horizontal tab | SUB | Substitute |
| LF | Line Feed | ESC | Escape |
| FF | Form Feed | FS | File separator |
| CR | Carriage return | GS | Group separator |
| SO | Shift out | RS | Record separator |
| SI | Shift in | US | Unit separator |
| DLE | Data link escape | DEL | Delete |

matrix. Second, follow the column up to the hexadecimal digit that will be used as the most significant digit (MSD) of ASCII representation and write it down (5). Third, follow the row over to the left for the hexadecimal digit that will be used for the least significant digit (LSD) of the ASCII representation (2). Fourth, combine the two. So, the ASCII representation of the letter $R$ is $52_{ASCII}$.

You must realize that ASCII is a binary code. The chart is laid out in hexadecimal to make it easier to follow. The 7-bit ASCII for the letter $R$ is actually 1010010.

Along with characters and punctuation, codes have to be available to control the operation of devices, for example, telling a printer or display to do a carriage return and a line feed. This is where the first two columns of the ASCII chart in Table 1.2 come into play. These two columns contain codes that are referred to as *device control codes*. The code for carriage return (CR) is $0D_{ASCII}$ and the code for line feed (LF) is $0A_{ASCII}$. I am not going to discuss any of the other control codes because this is just a basic introduction to ASCII. The function of these codes is usually listed in the manual for your printer or other device that uses ASCII. The meaning of the control code abbreviations is shown in Table 1.3. As you can see, most of the functions performed by the control codes are obvious.

**Problem 1.7:** What is the ASCII representation for the letter $L$?

**Solution:** Because the letter $L$ is at the intersection of column 4 and row C, the ASCII representation is 4C hexadecimal or 1001100 binary.

**Problem 1.8:** What character does an ASCII 3F represent?

**Solution:** At the intersection of column 3 and row F is the character "?."

## 1.2.1 Parity

Errors often occur when data are transmitted. To help alleviate this problem, a method of error detection called *parity* is sometimes used. When parity is used, an extra bit is added to the left of the most significant bit to indicate the number's parity. The *P* in the binary example shown in Example 1.18 indicates the parity bit's position.

**Example 1.18:**

| P | | | 7-bit ASCII | | | | | *Rich* |
|---|---|---|---|---|---|---|---|---|
| 1 | 1 | 0 | 1 | 0 | 0 | 1 | 0 | R |
| 1 | 1 | 0 | 0 | 1 | 0 | 0 | 1 | I |
| 1 | 1 | 0 | 0 | 0 | 0 | 1 | 1 | C |
| 0 | 1 | 0 | 0 | 1 | 0 | 0 | 0 | H |

Two types of parity may be used—*even* or *odd*. When even parity is used, the parity bit will be a 1 or a 0 to keep the total number of 1s, including the parity bit, in the 8-bit code always even. If odd parity is used, the parity bit will be a 1 or a 0 to keep the total number of 1s in the 8-bit code always odd.

The device transmitting the code and the device receiving the code must be set up to send and receive the same parity. Example 1.18 is an example of 7-bit ASCII using even parity to represent the name *Rich*. Notice when you count the number of 1s in every ASCII character in Example 1.18, including the parity bit, the number of 1s always come out even.

**Problem 1.9:** What will be the state of the parity bit for an ASCII-5A if odd parity is used?

**Solution:** First 5A must be converted to 7-bit ASCII. 5A = 1 0 1 1 0 1 0.

Next the number of 1s are counted in the 7-bit ASCII. The number of 1s in 1011010 is four. Because 4 is an even number and we are using odd parity in this example, a 1 needs to be in the parity bit to make the total number of 1s in the 8-bit code odd.

The final 8-bit code (parity + ASCII) will be $1\ 1\ 0\ 1\ 1\ 0\ 1\ 0_2$.

**Problem 1.10:** If the letter *F* is transmitted using 7-bit ASCII, what 8-bit code will be transmitted if even parity is used?

**Solution:** Referring to the 7-bit ASCII chart in Table 1.2, we find that the ASCII representation of the letter *F* is $46_{16}$. After $46_{16}$ is converted to 7-bit binary, we get $1000110_2$. There are three 1s in this binary number. This means the parity bit will have to be a 1 to make the number of 1s in the 8-bit code even. So the code $11000110_2$ will be transmitted for this example using even parity.

Parity is not that popular these days when transmitting large blocks (multiple bytes) of binary information. This is because if 2, 4, 6, or 8 bits change during transmission, the receiving digital equipment will not be able to detect the error. Currently, the most popular error-checking technique for large blocks of binary data transfer is called *cyclic redundancy checking* (CRC). CRC will be discussed in more detail in section 1.2.3.

Parity is not usually used in the transmission of data over distance, so you may be wondering why we are discussing it at all. Parity is still an option for devices that use serial communication and in the memory of the computer. Serial communication is discussed in chapter 9 and memory is discussed in chapter 7. We are discussing parity at this time because it was originally developed for sending information using 7-bit ASCII.

## 1.2.2 Enhanced ASCII

IBM® chose not to use parity when sending information to its display and printer. Instead it chose to use the 8th bit to add more character combinations. Using all 8 bits for characters means that 256 combinations can be represented instead of the 128 combinations of standard 7-bit ASCII. This 8-bit code is usually re-

Table 1.4

| MSD → | 8 | 9 | A | B | C | D | E | F |
|---|---|---|---|---|---|---|---|---|
| 0 | Ç | É | á | ░ | L | ⊥ | α | ≡ |
| 1 | ü | æ | í | ▓ | ⊥ | ┬ | ß | ± |
| 2 | é | Æ | ó | █ | ┬ | ┬ | Γ | ≥ |
| 3 | â | ô | ú | │ | ├ | ╚ | π | ≤ |
| 4 | ä | ö | Ñ | ┤ | ─ | ╘ | Σ | ⌠ |
| 5 | à | ò | Ñ | ╡ | ┼ | ╒ | σ | ⌡ |
| 6 | å | û | ª | ╢ | ╞ | ╓ | μ | ÷ |
| 7 | ç | ù | º | ╖ | ╟ | ╫ | τ | ≈ |
| 8 | ê | ÿ | ¿ | ╕ | ╚ | ╪ | Φ | ° |
| 9 | ë | Ö | ⌐ | ╣ | ╔ | ┘ | Θ | • |
| A | è | Ü | ¬ | ║ | ╩ | ┌ | Ω | · |
| B | ï | ¢ | ½ | ╗ | ╦ | ■ | δ | √ |
| C | î | £ | ¼ | ╝ | ╠ | ■ | ∞ | η |
| D | ì | ¥ | ¡ | ╜ | = | ▌ | φ | ² |
| E | Ä | ₧ | « | ╛ | ╬ | ▐ | ε | ■ |
| F | Å | ƒ | » | ┐ | ⊥ | ▀ | ∩ | |

ferred to as Enhanced ASCII. Enhanced ASCII is simply an extension of 7-bit ASCII. Referring to Table 1.4, notice that the column numbers of Enhanced ASCII begins with the number 8. This is because the first eight columns (0 through 7) are the same as those in the 7-bit ASCII chart shown in Table 1.2.

The Enhanced ASCII chart shown in Table 1.4 is used the same way as the 7-bit ASCII chart shown in Table 1.2. For example, to find the ASCII representation of the fraction 1/2, find 1/2 in the chart. Then write down the column number and then the row number. So, the ASCII for 1/2 is $AB_{ASCII}$.

**Problem 1.11:** What is the Enhanced ASCII representation for the character "¢"?

**Solution:** This character is at the intersection of column 9 and row B in the enhanced ASCII chart, so the code is *9B*.

### 1.2.3 Checksum and CRC

There are two popular methods in addition to parity that are used to detect errors in transmitted binary information. One is called *checksum* and the other is called a *cyclic redundancy code* (CRC). Whereas parity works with single bytes of binary information, both checksum and CRC work with multi-byte transfers of binary information.

Checksum adds all of the binary bytes to be transferred and attaches the sum to the end of the information. The receiving piece of equipment adds the bytes as they are received and compares the sum it generates to the sum that is attached to the end when it was transmitted. If the two sums match, there is approximately a 99.29% chance that the received information is error free. Of course, if the sum is different, there was an error in transmission.

CRC is a more reliable, yet complicated, method. This is because binary math is involved with its calculation. Basically, CRC works on the principle of shifting and dividing by a predefined number. The remainder is then attached to the original information and transmitted. Once the transmitted binary number is received, including the CRC, it is once again divided by the same predefined binary number that was used during transmission. If there is no remainder generated at the receiver, there is over a 99.99% chance that there was no error during transmission. The bigger the divisor, the larger the remainder and the more accurate the CRC. The CRCs currently used in the compatible computers are called *CRC-16* and *CRC-32*. The numbers indicate the number of remainder bits that are attached to the transmitted binary information.

## 1.3 COMPUTER NUMBERING SYSTEM TERMINOLOGY

This unit deals with computer numbering systems, so we also need to discuss other numerical terms that are used in computers when talking about speed and storage capacity. As if computers were not confusing enough, we use some of the same terms when talking about speed and storage capacity, but they mean different things.

### 1.3.1 Storage

The computer uses memory to store information until it is ready to be processed. The computer also uses mass storage devices called *disk drives* to hold the information before it is loaded into the computer's memory. The memory and mass storage devices store binary information in 8-bit groups, or 1 byte. You might have heard someone say, "My computer has 16 megabytes of memory." Just exactly what does this mean?

The metric prefixes **kilo, mega,** and **giga** indicate the number of 8-bit locations in a computer's memory and mass storage devices. Because a computer uses the binary numbering system, the meanings of these prefixes, relative to computer storage, are not the same as those you probably learned in high school. The meanings of these prefixes, relative to computer storage, are as follows:

$$1 \text{ kilo (k)} = 2^{10} = 1{,}024_{10}$$
$$1 \text{ meg (M)} = 2^{20} = 1{,}048{,}576_{10} = 1{,}024 \text{ K}$$
$$1 \text{ gig (G)} = 2^{30} = 1{,}073{,}741{,}824_{10} = 1{,}024 \text{ M}$$

When someone tells you his or her computer has 16 megabytes (MB) of memory installed, it does not mean the computer has 16,000,000 bytes of memory. The actual number of bytes of memory in the computer must be calculated. The calculation can be performed several different ways. First, if you have a calculator that provides exponents, the number of megs can be multiplied by $2^{20}$ ($16 \times 2^{20} = 16{,}777{,}216$ bytes). Second, the actual number of bytes can be determined by multiplying the number of megs by $1{,}048{,}576_{10}$ ($16 \times 1{,}048{,}576 = 16{,}777{,}216$ bytes). Third, a meg is equal to 1,024 K and a K equals 1,024, so the actual amount of bytes of memory can be calculated by multiplying the number of megs by 1,024 times 1,024 ($16 \times 1{,}024 \times 1{,}024 = 16{,}777{,}216$ bytes). Notice that no matter what method is used, the answer is the same. Let us work a few problems to help explain this further.

---

**Problem 1.12:** How much is 640 kilobytes (KB) of memory?

**Solution:** One K, relative to computer storage, equals 1,024. Therefore, 640 K equals $640 \times 1{,}024$, or 655,360 bytes of memory.

**Problem 1.13:** You hear someone say, "My computer has 48 megabytes of memory installed." How much memory is actually in the computer?

**Solution:** One meg, relative to computer storage, equals $1{,}024 \times 1{,}024$. Therefore, 48 megabytes equals $48 \times 1{,}024 \times 1{,}024 = 50{,}331{,}648$ bytes.

# Computer Numbering Systems and Terminology

**Problem 1.14:** The computer has 4 megabytes (MB) of memory. How many bytes of memory are available?

**Solution:** Relative to computer storage, one meg equals 1,024 K and a K equals 1,024. Therefore, 4 MB equals 4 × 1,024 × 1,024 or 4,194,304 MB of memory.

Most of the computers we are discussing in this text have memory in the megabyte range. It is becoming very popular to represent the number of megs using the kilo unit. As an example, 32 MB can also be expressed as 32,768 kilobytes (KB). To convert a number from MB to KB, multiply by 1,024 (32 MB × 1,024 = 32,768 KB). I know this might not make much sense to you, but this is one of those situations in which you have to take things the way they are and not the way you think they should be.

**Problem 1.15:** Convert 48 MB into KB.

**Solution:** 48 MB × 1,024 = 49,152 KB

**Problem 1.16:** How many bytes do 49,152 KB equal?

**Solution:** 49,152 KB × 1,024 = 50,331,648 bytes.

## 1.3.2 Speed

Computers are synchronous devices. As stated in section I.5.2, the term *synchronous* means that everything in the computer is timed with a clock. Remember, computers are binary devices, so the clock that times the computer is a digital clock. A digital clock is a signal that switches back and forth between a 1 and a 0 at a constant rate. The digital clock signal is produced with DC voltage levels. If a voltage is present, it represents a binary 1, and 0 volts (ground) represents a binary 0.

## 1.3.3 The Hertz

The digital clock in the computer moves in **cycles** similar to ac, as discussed in the Introduction. A cycle for a digital clock is shown in Figure 1.2. Similar to ac, the rate at which the cycles occur is given in hertz (Hz). Remember, one hertz is simply defined as one cycle in a second. The digital clock that times the computer is usually given in megahertz (MHz). When referring to the computer's speed, the metric prefixes kilo, mega, and giga do not mean the same thing as they do when talking about the computer's storage capacity. Following are the meanings of these terms in reference to the computer's speed:

$$k = kilo = 10^3 = 1,000$$
$$M = meg = 10^6 = 1,000,000$$
$$G = gig = 10^9 = 1,000,000,000$$

Notice that the metric prefixes jump powers of 10 in multiples of 3. This is referred to as *engineering notation*. Engineering notation is widely used in all technical fields, including computers, and is used throughout this book.

**Problem 1.17:** If a computer is operating at a speed of 40 MHz, what is the actual speed?

**Solution:** Because 40 MHz deals with speed, one meg equals 1,000,000. Therefore, the computer is operating at 40,000,000 hertz.

## 1.3.4 On the Other Side of the Decimal Point

If a computer is timed by a digital clock running at 40 MHz, how long does each cycle take? Believe it or not, this is something a person who is knowledgeable in computers needs to be able to calculate. One hertz equals one cycle per second, so 40 MHz must be 40 million cycles in 1 second. This also means

**Figure 1.2** Digital Clock

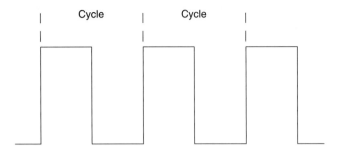

that the time for each cycle must take one 40 millionth of a second (1/40,000,000). In computers, we do not use fractions but we do use the decimal fraction. Therefore, 1/40,000,000 of a second equals 0.000000025 seconds. As you can see, leaving these large decimal fractions in this form is very error-prone and is hard to convey verbally. When we work with decimal fractions, the fractions are usually converted to the metric prefix **milli, micro,** or **nano.** Following are the meanings of these terms:

$$1 \text{ milli (m)} = 10^{-3} = 0.001$$
$$1 \text{ micro } (\mu) = 10^{-6} = 0.000001$$
$$1 \text{ nano (n)} = 10^{-9} = 0.000000001$$

The ultimate goal when converting decimal fractions into milli, micro, or nano is to start moving the decimal point to the right in groups of threes (engineering notation) until a whole number is reached. For example, 0.0004 seconds will take six moves to the right to get whole numbers (two groups of three). Six moves to the right is the same as $10^{-6}$ and is referred to as microseconds ($\mu$s). This means that 0.0004 seconds can also be expressed as 400 $\mu$s. Let us work a few problems to help understand this concept.

**Problem 1.18:** Convert 0.000000025 seconds into the correct metric unit.

**Solution:** The decimal point will have to be moved nine times to the right to achieve the whole number 25. Nine moves are required, so 0.000000025 seconds equals 25 nanoseconds (ns).

**Problem 1.19:** Convert 0.003 into the correct metric prefix.

**Solution:** 0.003 equals 3 milliseconds.

There is also a need to convert fractions given in metric prefixes into seconds. To do this, move the decimal point to the left the number of times indicated by the prefix. For example, to convert 3 $\mu$s into seconds, move the decimal point to the left six times. So, 3 $\mu$s equals 0.000003 seconds.

**Problem 1.20:** Convert 15 $\mu$s into seconds.

**Solution:** 15 $\mu$s equals 0.000015 seconds.

**Problem 1.21:** Convert 66.76 ns into seconds.

**Solution:** 66.76 ns equals 0.00000006676 seconds.

## 1.4 CONCLUSION

Binary is the numbering system used by microcomputers because it is easy to transfer, process, and store without errors. All other numbering systems have to be converted or coded into the 1s and 0s of the binary numbering system before the computer can actually process the information.

Digital devices often use hexadecimal to represent large binary numbers. The computer itself does not use hexadecimal. Because the computer has the ability to process binary information, it can use this processing ability to make binary numbers appear as if they are hexadecimal. The computer does this because hexadecimal helps relay large binary numbers to humans without error.

The most predominant code used by microcomputers to transfer text information is ASCII. ASCII gives digital devices the ability to process and store text. There are two ASCII codes available. One is a 7-bit code that gives 128 combinations, whereas the other is an 8-bit code that gives 256 combinations.

Parity was originally developed to detect errors in the transmission of binary information. Today, the most popular method used to detect binary transmission errors is cyclic redundancy checking, or CRC. Parity is still used to detect errors in information stored in the computer's memory.

Relative to computer storage, 1 K equals 1,024, 1 M equals 1,024 K, and 1 G equals 1,024 M. Relative to speed, 1 K equals 1,000, 1 M equals 1,000 K, and 1 G equals 1,000 M. The metric units used to represent decimal fractions are milli, micro, and nano.

# Review Questions

1. Convert the following binary numbers into decimal.
   a. 1 0 1 0 1 1 1 0 1
   b. 1 1 1 1 0 0 1
   c. 1 1 0 0 0 1 1 1 0 0 1 0 1 0 1 0
2. Convert the following decimal numbers into binary.
   a. 56
   b. 981
   c. 5,891
3. Convert the following binary numbers into hexadecimal.
   a. 1 1 0 0 1 1 1 0 1 0 1 0 0
   b. 1 1 1 0 1 0 1 0 1
   c. 1 0 0 1 0 0 1 0 1 0 1 1

## Multiple Choice

1. The only numbering system a computer understands is:
   a. decimal.
   b. hexadecimal.
   c. binary.
   d. base-10.
2. The only digits used in the binary numbering system are:
   a. 0 and 1.
   b. 0 through 9.
   c. 0 through A.
   d. two.
3. What are the digits in the binary numbering system called?
   a. bytes
   b. bites
   c. bits
   d. digits
4. In the decimal number 78945, the 8 holds a weight of.
   a. 8,945.
   b. 8,000.
   c. 8.
   d. none of the above.
5. In the binary numbering system, as the digits move to the left:
   a. the weight of the digits doubles.
   b. the weight of the digits is cut in half.
   c. the weight of the digits remains the same.
   d. the weight of the digits increases by 10.
6. The binary number 1 0 0 1 0 1 0 0 1 0 0 1 would equal _____ in decimal.
   a. 2,377
   b. 949
   c. 4,681
   d. none of the above
7. The decimal number 894 would equal _____ in binary.
   a. 1 0 0 0 1 0 0 1 0 1 0 0
   b. 1 1 0 1 1 1 1 1 0 1
   c. 1 1 0 1 1 1 1 1 1 0
   d. 1 1 1 1 1 1 1 1 0
8. Digital computers use binary because:
   a. binary is easy to transmit from one place to another.
   b. memory devices that store binary are easy to manufacture.
   c. decimal values would be almost impossible to store without it.
   d. all of the above.
9. Hexadecimal is:
   a. the only numbering system the computer understands.
   b. used by humans to represent large binary numbers.
   c. not often used when dealing with computers.
   d. handled by the computer faster than it handles binary.
10. The binary number 1 0 0 1 0 0 1 1 1 0 1 1 0 1 0 1 1 1 1 0 1 would equal _____ in hexadecimal.
    a. 93B5E1$_{16}$
    b. 1210045$_{16}$
    c. 9680353$_{16}$
    d. 1276BD$_{16}$
11. The hexadecimal number A9CD4 would equal _____ in binary.
    a. 695,508
    b. 1 0 1 0 1 0 0 1 1 1 1 0 1 1 0 0
    c. 1 0 1 0 1 0 0 1 1 1 0 0 1 1 0 1 0 0
    d. 1 0 0 1 1 0 1 1 1 0 0 1 0 0 1 1 0 1 0

12. The hexadecimal number 9B6 would equal _____ in decimal.
    a. 1 0 0 1 1 0 1 1 0 1 1 0$_2$
    b. 2,596
    c. 9,116
    d. none of the above
13. The ASCII representation for the letter "V" would equal:
    a. 1 0 1 0 1 1 0.
    b. 1 1 1 0 1 1 0.
    c. column 5 row 6.
    d. 1 1 0 0 1 0 0.
14. The character represented by 1 1 0 1 0 1 1 in ASCII would be:
    a. ?
    b. k.
    c. ¢
    d. j
15. If the ASCII from the previous question is transmitted using *even* parity, the parity bit will be:
    a. one (1)
    b. zero (0)
    c. zero (0) or (1)
    d. none of the above
16. A group of four consecutive binary bits is called a:
    a. byte.
    b. nibble.
    c. word
    d. bit.
17. A group of eight consecutive binary bits is called a:
    a. byte.
    b. nibble.
    c. word.
    d. bit.
18. If a computer has 2MBs of RAM installed, it will actually have:
    a. 2,000,000 bytes.
    b. 1,048,576 bytes.
    c. 2,097,152 bytes.
    d. 16,777,216 bytes.
    e. none of the above answers are correct.
19. If a microprocessor operates at 66 MHZ, it will be actually be operating at:
    a. 66,000,000 HZ
    b. 69,206,016 HZ
    c. 67,584 HZ
    d. 66,000 HZ
    e. none of the above.
20. Which of the following is equal to 28 milliseconds?
    a. 0.028 seconds
    b. 0.28 seconds
    c. 0.0028 seconds
    d. 28 seconds

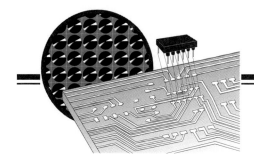

CHAPTER 2

# Microcomputer Basics

## OBJECTIVES

- List and describe the basic unit of any computer.
- List and describe the main units that make up any microcomputer.
- Describe the functions of the microprocessor's address bus, data bus, and control bus.
- Describe the characteristics of RAM and ROM.
- Define terms that are used relative to microprocessors.
- Explain the differences among machine language, assembly language, and high-level computer languages.
- List the microprocessors used in compatible computer chronologically.

## KEY TERMS

computer
arithmetic logic unit (ALU)
control/timing
input
output
memory
central processing unit (CPU)
wafer
integrated circuits (ICs)
pin(s)
microprocessor unit (MPU or μP)
microcomputer (MCU or μC)
program
programmer
bus
address bus
data bus
control bus
word
bus cycle
interrupt
write cycle
read cycle
stored program concept
random-access read/write memory (RAM)
read-only memory (ROM)
volatile
nonvolatile
static RAM (SRAM)
dynamic RAM (DRAM)
operating system
firmware
basic input-output system (BIOS)
power-on self-test (POST)
disk operating system (DOS)
application program
RESET pin
computer language
hardware
hardware drivers
machine language
assembly language
high-level language
operational code (Opcode)
object code
mnemonic
assembler
addressing mode
interpreter
compiler
syntax

Several definitions can be used to define the term *computer*. A **computer** is a device that inputs, stores, processes, and sooner or later, outputs information is one definition. There are two categories of computers—*analog* and *digital*. Analog signals have the ability to change an infinite number of combinations between two points, whereas digital signals change a fixed number of combinations. The digital computer is by far the most widely used and is the only category of computer covered in this book.

# Chapter 2

## 2.1 DIGITAL COMPUTER BASICS

Digital computers, no matter what size, consist of the same basic units. These units include the **arithmetic logic unit (ALU), control/timing unit, input port unit(s), output port unit(s),** and **memory** units, as illustrated in Figure 2.1. The ALU and control/timing units are often referred to as the computer's **central processing unit (CPU).** The ALU is responsible for all of the arithmetic and logic operations the computer performs. The control/timing unit provides all of the necessary signals to control and time the communication of the CPU with the computer's memory and input/output devices. The input port unit is where information is brought into the computer from the outside world, usually into the computer's memory. The output port unit is where information is sent from inside the computer to the outside world. Because the input and output units are our portal into and out of the computer, they are simply called *ports*. We communicate with the computer through its input and output ports. Any device that communicates with the computer through a port is called a *peripheral*.

Of all the units that the computer is comprised of, the CPU is considered by most to be the brain of the computer because the CPU controls all of the other computer units. Three decades ago, the electronic circuits that incorporated the computer's CPU consisted of several very large printed circuit boards (PCBs). But that is not the case for today's computers.

The semiconductor technology that most electronic devices are made from has improved immensely over the years. The biggest improvement has been in the reduction of the size of the electronic devices that all electronic circuits consist of. In fact, electronic devices that used to be measured in inches are now measured at less than a quarter micron (one thousandth of a millimeter). This ability to make the size of the electronic devices so small has made it possible to integrate (combine) an entire electronic circuit on a single piece of semiconductor material that had previously held only one electronic device.

The semiconductor material on which the entire circuits are etched is in the shape of a circle and is called a **wafer.** The individual circuits are etched on the wafer into small, rectangular patches called *dies*. The difference between a wafer and die is shown in Figure 2.2. Each individual die contains an entire electronic circuit, so it is also known as an **integrated circuit (IC).**

The individual dies are cut from the wafer and are so small and fragile that it is impossible for a person to handle them without damaging them. To make handling possible, the die is mounted in

**Figure 2.1** Block Diagram of a Basic Digital Computer

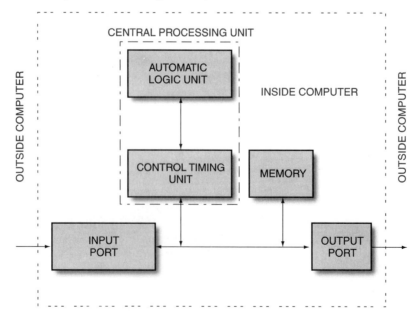

# Microcomputer Basics

a package made of ceramic or plastic. Electrical connections are made to the die through metal tabs, called **pins,** that protrude from the sides or bottom of the package. Figure 2.3 shows a cutaway view of the basic construction of an IC package. The package shown in this figure is called a *dual-in-line-package,* or simply *DIP.* Some of the most popular methods of packaging ICs are illustrated in Figure 2.4.

As semiconductor manufacturers continued to produce smaller and smaller electronic devices, it became possible to integrate the ALU and control/timing units—along with internal memory called *registers*—of the computer onto a single semiconductor die. The result of combining all of the units of the CPU onto a single integrated circuit is an IC called a **microprocessor unit** (**MPU** or **MP**). Figure 2.5 shows a fundamental block diagram of a typical microprocessor. A microprocessor is also referred to as a CPU, MPU, μP, or, simply, processor. All of these different names are used throughout this book.

The functions of the ALU and control/timing units of the microprocessor are the same as those already discussed. The primary function of a *register* is to hold the binary data while they are actually being processed inside the microprocessor. It is beyond the scope of this book to explain the functions of a microprocessor's registers, but to truly understand how a microprocessor works, one must understand these functions.

**Figure 2.2**   Semiconductor Wafer Showing an Individual Die

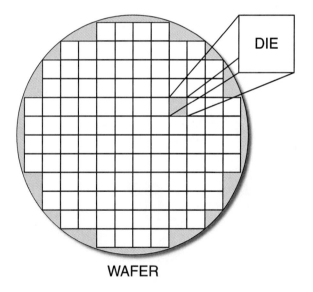

**Figure 2.3**   DIP

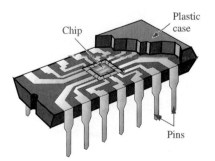

*Source:* Floyd, *Digital Fundamentals,* 7th ed. (Upper Saddle River, NJ: Prentice Hall, 2000) p. 19.

**Figure 2.4** Popular Methods of Packing ICs

**Figure 2.5** Microprocessor Block Diagram

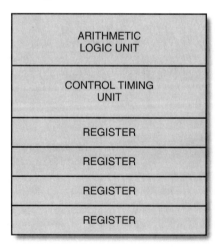

Any computer that uses a microprocessor as its central processing unit is correctly referred to as a **microcomputer** (**MCU** or **μC**). All microcomputers incorporate four basic units that include the microprocessor, memory, input, and output. Figure 2.6 shows a block diagram of a typical microcomputer.

### 2.1.1  Just What Can a Microprocessor Do?

Most people believe a microprocessor is totally amazing and beyond their grasp. From the television shows and movies we watch, we get the impression that a microprocessor is some mysterious device that thinks and acts on its own. It is true that the microprocessor controls the computer's overall operation. It is also true that current microprocessors can execute millions of instructions per second (mips).

**Figure 2.6** Basic Microcomputer Block Diagram

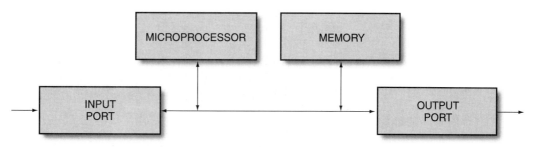

However, microprocessors are not the beast that people think they are. In fact, a μP would be totally useless if there were not a person to tell it exactly what to do.

All that a microprocessor can do is *fetch* (get) and *execute* (carry out) a set of individual instructions. A human arranges the individual instructions to make the microprocessor perform a desired task. This arrangement of instructions to perform a task is called a **program** and the person who does the arranging is called a **programmer.** The microprocessor's individual instructions allow it to perform basic arithmetic, logic, and data manipulation operations. The programmer writes programs using these instructions to control the computer's RAM, ROM, and input and output ports through the microprocessor.

All of the amazing tasks you hear or see a computer perform are actually accomplished by a program written by humans just like you and me. The programmers who write these programs are the real brains of the computer and are really what is amazing.

## 2.2 THE MICROPROCESSOR BUSES

To communicate with the microcomputer's memory or input and output devices, the microprocessor uses what is referred to as a **bus.** A computer bus is a group of electrical conductors that carries similar information between the microprocessor and the memory and input/output (I/O) ports. Memory and I/O ports have to be connected electrically to the microprocessor's buses if the microprocessor is to communicate with them. There are three buses used by all microprocessors—the **address bus,** the **data bus,** and the **control bus.** Figure 2.7 shows the interconnection of the microprocessor buses.

### 2.2.1 The Address Bus

The address bus is used by the microprocessor to select each device it wishes to communicate with. The engineers who design the microcomputer assign a binary address to every memory location or I/O port connected to the microprocessor. To select the device for communication, the microprocessor must place the device's binary address onto the address bus. Therefore, the size of the microprocessor's address bus will determine the total number of devices it can select for communication. The bigger the address bus, the more devices it can select. To calculate the number of storage locations a microprocessor can address, raise 2 to the power of the width of the address bus.

**Problem 2.1:** If a microprocessor has a 16-bit address bus, how many address combinations are available?

**Solution:** To calculate the number of address combinations, raise 2 to the 16th power ($2^{16}$). This gives an answer of 65,536 address combinations.

The 8088 μP used in the PC and XT computers has a 20-bit address bus. To calculate the number of addresses this processor can select, raise 2 to the 20th power ($2^{20}$). So, the 8088 μP can select $1,048,576_{10}$ (1 M) different memory devices. The 80286 has a 24-bit address bus, so it can select $16,777,216_{10}$ (16 M) different memory devices. The 80386, 80486, and Pentium μPs have a 32-bit address bus, so they can address $2^{32}$ different memory devices, or $4,294,967,296_{10}$ (4 G).

**Figure 2.7** Block Diagram of a Basic Microcomputer Showing the Address, Data, and Control Buses

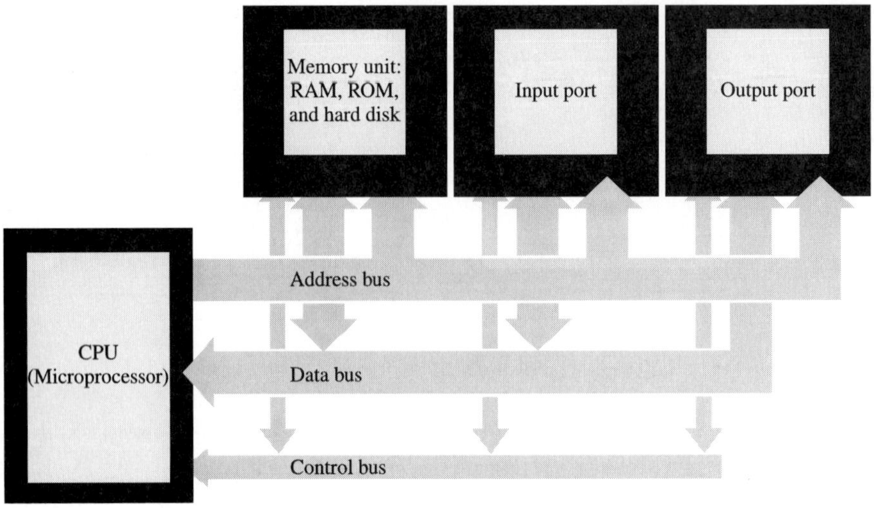

*Source:* Floyd, *Digital Fundamentals,* 7th ed. (Upper Saddle River, NJ: Prentice Hall, 2000) p. 19.

For reasons that will be discussed later, the number of input/output addresses for all Intel compatible μP since the 8086 is limited to $65,536_{10}$ (64 K) address combinations.

### 2.2.2 The Data Bus

The data bus is where the communication actually takes place between the selected device and the microprocessor. Every microprocessor communicates in a group of binary bits called a **word.** The larger the word size, the faster the communications. If you hear the term *8-bit microprocessor,* it is referring to the largest word size of the microprocessor, not the addressing range of the microprocessor. An 8-bit microprocessor communicates in 8-bit words but might have a 16-bit address bus. This means the microprocessor can communicate with 65,536 different 8-bit devices. Current microprocessors are available with 4-bit, 8-bit, 16-bit, 32-bit, and 64-bit word sizes. Word sizes and data bus sizes differ among microprocessors. For example, the 8088 is a 16-bit microprocessor that communicates over an 8-bit data bus. This means the 8088 will usually take two **bus cycles** for each word. A bus cycle is a term used to describe the time required to transfer data over the data bus. The 80386SX is another example of a microprocessor with a different word size and data bus size. This is a 32-bit microprocessor that communicates over a 16-bit data bus.

### 2.2.3 The Control Bus

The control bus controls the communication between the selected device and the microprocessor. Some of the control bus signals provide the timing needed to ensure that the address and data information are at the right place at the right time. Some other control bus signals are used to control the direction of the communication. Another signal on the control bus indicates whether a memory address or I/O address is currently on the address bus. Other control signals are used to **interrupt** the microprocessor to make it skip to a predefined program. On a hardware level, interrupts are a very important concept to understand and will be discussed later in chapters 8 and 15.

### 2.2.4 Let's Move Some Data

To aid in further explaining the function of the microprocessor's address, data, and control buses, we will use a hypothetical microprocessor with a 3-bit address bus, an 8-bit data bus, and a single control line to indicate whether the microprocessor is reading or writing. The drawing of the bus structure of this hypothetical microprocessor is shown in Figure 2.8. With a 3-bit address bus width, the micro-

**Figure 2.8** Typical Write Cycle

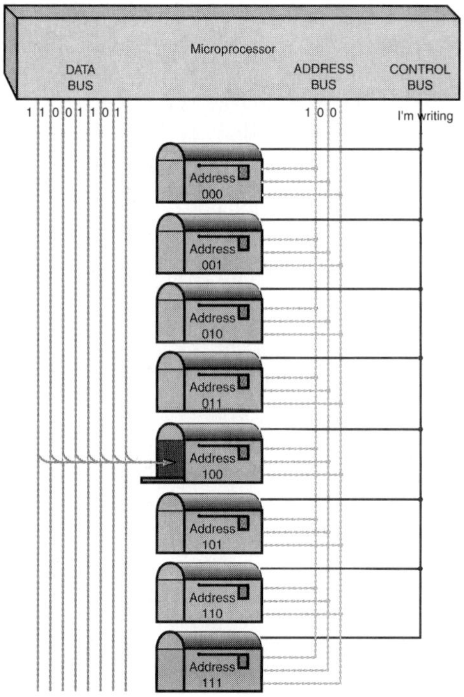

processor can address a maximum of eight devices ($2^3$). Each of these eight devices (either memory or I/O) is assigned an address by the engineer who designs the computer.

The way that the microprocessor selects devices is similar to the way the mail system works, so the devices in Figure 2.8 are drawn as mailboxes. Hopefully this will make it easier for you to understand the process the microprocessor uses to transfer information.

We will start out by moving data from the microprocessor to a device. This direction of data movement is called a **write cycle** (refer to Figure 2.8 during this discussion). The sequence of events the microprocessor will follow for a write cycle is:

1. The μP places the device's address out of its address bus. In the example in Figure 2.8, the binary address 100 is being selected by the microprocessor. The address will open the mailbox door that is assigned address 100 so the microprocessor can transfer data. Notice that all other mailbox doors are closed. The other devices that are not being selected by the current content of the address bus cannot transfer data with the μP.
2. A signal on the control bus called *read/write control* (or some similar name) informs the selected device that the microprocessor wishes to store (write) data.
3. The microprocessor places the data to be stored out of its data bus. In the example in Figure 2.8, the binary data being written is 11001101.
4. The device selected by address 100 accepts the data being placed on the data bus by the microprocessor. This concludes the write cycle.

Next let us discuss a read cycle. During a **read cycle,** data flows from the selected device to the microprocessor. In other words, the selected device places information on the data bus and the microprocessor reads the information off the data bus (refer to Figure 2.9 during this discussion of the read cycle). The sequence for a read cycle is:

1. The microprocessor places the device's binary address out of its address bus. In the example illustrated in Figure 2.9, this is the binary address 110. When the device assigned the address 110 sees its address on the address bus, it will open its door and allow the microprocessor to transfer data.
2. The control bus generates a signal telling the selected device "I'm reading."

**Figure 2.9** Hypothetical Read Cycle

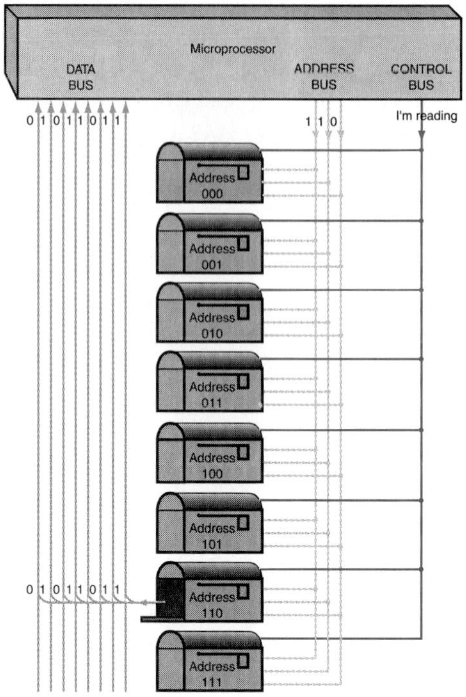

3. The device places its information on the data bus.
4. The μP reads the data bus. This concludes the read cycle.

Even though this explanation has been simplified quite a bit, this discussion has hopefully demonstrated the function performed by each of the microprocessor's buses. Remember that both the read and write cycles are completed in a few nanoseconds ($10^{-9}$) by the microprocessors used in today's microcomputers.

## 2.3 MEMORY

Digital computers operate by what is called the **stored program concept.** Simply stated, the stored program concept means that before a computer can run any program or use any information, it must somehow be loaded into the memory unit of the computer. There are two basic types of memory found in computers—**read-only memory (ROM)** and **random-access read/write memory (RAM).**

All memory further falls into two categories called **volatile** and **nonvolatile.** Volatile memory will lose its contents when power is removed from the ICs, whereas nonvolatile memory will maintain its contents even when power is removed. ROM is always nonvolatile; RAM used in the compatible PC is volatile.

RAM is even further categorized into **static RAM (SRAM)** or **dynamic RAM (DRAM),** which are discussed in greater detail in chapter 7. In the discussion that follows, I give only brief explanations and the basic use of ROM and RAM in the computer.

### 2.3.1 ROM

All living creatures are born with a basic operating system that makes certain systems in the body work before anything is learned. At birth, you can already breathe and your heart works although you were not taught these functions. Every species is born with an internal operating system, called the *automated nervous system,* that is the same for each member of a given species. Not being able to access the automated nervous system is fatal. Computers follow the same basic concept.

The only thing that a computer can actually do is run programs. Simply stated, if a program is not running, the computer is dead. When a computer is first turned on (born), the CPU follows a set of steps that were built in to the CPU or microprocessor when it was designed. The steps it follows are inherent to the CPU and cannot be controlled by the computer operator. These steps will always force the CPU to go to the same address in the computer's memory looking for the first instruction of a program. If a microprocessor instruction is not at this address, the computer will never come alive.

Once this first program instruction is located, the CPU is a slave to the program. This program will control the computer's input, output, and memory units, which cause the computer to operate. For this reason, the group of programs that make a computer operate is called the computer's **operating system.** Similar to the human automated nervous system, every computer must have a basic operating system that gets it going when power is applied. These programs have to be in a nonvolatile memory and must be difficult for the average person to change. Just like each species of living things is born with the same basic operating system, so is each species (type) of computer. This basic computer operating system is located in integrated circuits called *read-only memory (ROM)*.

As the name implies, ROMs can be read by the microprocessor but not written to. The programs and information stored in the ROM ICs are usually placed there when the ROM IC is manufactured.[1] In the old days, these programs could be changed only by physically changing the ROM ICs installed in the computer. Most of today's computers use a type of ROM called *Flash-ROM*. Flash-ROMs allow a trained individual to change the programs in the ROM without changing the ROM IC. ROMs and Flash-ROMs are covered in more detail in chapter 7. Because the programs in the ROM are not usually changed by the end user, these programs are often called the computer's **firmware.** If a computer has problems running its firmware from the ROMs, it will die.

Some computers run their entire operating system from ROM and do not require any additional memory. These computers usually perform the same basic, dedicated tasks such as the control of traffic signals, home appliances, or other things. To be flexible, today's personal computers must have the ability to add to the operating system contained in the ROM. For this reason, the entire operating system is not stored in ROM.

### 2.3.2 The Main Operating System

When IBM® developed the PC, it chose not to place its main operating system in the computer's ROM. IBM® learned from its dealings with mainframes and minicomputers that this made it very difficult to upgrade the operating system as programming techniques and I/O devices advanced. Instead, IBM® chose to place just a very basic operating system in the ROM to get the computer up and running. This ROM-based operating system is called the computer's **basic input/output system (BIOS).**

In compatible computers, the BIOS contains the programs that set up the I/O devices in the computer. The BIOS also performs a **power-on self-test (POST)** each time the power is turned on. The BIOS will eventually tell the computer to attempt to load the main operating system through ports from a mass storage device called a disk drive (disk drives are covered in more detail in chapters 11 and 12).

The main operating system (OS) is loaded from a disk drive, so it is called a **disk operating system (DOS).** The most widely used DOSs for the IBM-compatible computers are written by the Microsoft® Corporation and are called *MS-DOS*®, Windows 3.X®, Windows 95®, Windows 98®, Windows NT®, Windows 2000®, and Windows Me®. Some other disk operating systems currently available for the IBM®-compatible computers are Unix, Linux, BeOS, Netware, OS/2, and OS/2 Warp.

### 2.3.3 RAM

The IBM computer loads its main operating system from a disk drive into the computer's memory, so there needs to be a section of memory in the computer that the microprocessor can not only read from but also write to. ROMs are nonvolatile read-only devices, so there is no way that programs loaded from a disk drive can be stored there. Therefore, the computer contains a section of **random-access read/write memory (RAM)** where it can load (write) and execute (read) the disk operating system. In addition to holding the main operating system, there is another very important need for RAM in a computer.

Once a computer is made operational by the operating system, RAM is also needed in the computer to hold programs other than the operating system. These programs are loaded by the OS through an

---

[1] There are several types of ROMs that are user-programmable, but these devices require special programming equipment.

input device (usually a disk drive or keyboard) into the computer's RAM and executed by the microprocessor. Having a computer operate is great and wonderful, but if we could not use the computer to perform other applications, it would be a useless device.

The name given to the category of programs that apply the computer once it is made operational by the operating system is **application programs.** An application program is a program written for a specific application, such as word processing, database management, games, and spreadsheet programs. There has to be enough RAM in the computer to not only hold the OS, but also to hold one or more application programs at the same time.

The RAM used in the compatible computers is volatile memory. Remember that volatile memory loses its contents each time the computer is turned off. This is not a problem because the operating system automatically reloads from a disk drive into the computer's RAM each time it performs the start-up sequence. The start-up sequence is called *booting* the computer, and it will be discussed in greater detail in section 2.5.

Since RAM is read/write memory, the computer operator can change its contents easily. Applications can be loaded from a disk, used, and then terminated at will by the computer operator through the computer's OS. Because the contents of RAM can be changed without changing hardware, the information loaded into RAM is usually called *software*.

As stated previously, there are currently two categories of RAM used in the compatible computers—SRAM and DRAM. Basically speaking, SRAM is fast and expensive and DRAM is slow (compared to SRAM) and cheap. For this reason, DRAM is the main memory in the computer. However, because of its speed, SRAM is used in a much smaller special memory called a *cache*.

Remember that this has been only a brief discussion of RAM. Understanding how the RAM is organized in a computer is extremely important for any person who wants to build, upgrade, or repair a computer. For this reason, RAM is covered in more detail in chapter 7.

## 2.4 RESET

You may wonder how the microprocessor knows to go to ROM when power is first applied. To answer this question, let us discuss a very important pin that belongs to the microprocessor's control bus.

Microprocessors have a pin called the **RESET pin** that belongs to the control bus. When this pin is pulsed (brought from a 0, to a 1, back to a 0) the microprocessor is reset. When a microprocessor receives a reset pulse, it is forced back into programs that are stored in the computer's ROM. This occurs no matter what the microprocessor is doing at the time.

In most of the computers covered in this book, a reset pulse is generated in only two ways. First, each time the computer is turned on, an electronic circuit inside the computer's power supply automatically generates a reset pulse. Second, a reset is generated when a button mounted on the front of the computer labeled *RESET* is pressed and released.

All of the microprocessors covered in this book go to the same memory address, $FFFF0_{16}$, each time the RESET pin is pulsed. This address is always located in the computer's ROM and contains the first machine code instruction of the program that initializes the computer. The pin assigned to the reset signal of several of the microprocessors used in the IBM and compatible computers is discussed in chapter 3.

## 2.5 BOOTING THE COMPUTER

When a computer is told to load its main operating system, this is referred to as *booting* the computer. There are two types of boots used by compatible computers. The first is referred to as a *hardware reset, hard-boot,* or *cold-boot*. The second is referred to as a *soft-reset, soft-boot,* or *warm-boot*.

The hard-boot is initiated when the microprocessor's RESET pin is pulsed. As stated previously, this occurs automatically each time power is applied to the computer. Most compatible computers also supply a RESET button, located on the front panel, that will hard-boot the computer when pressed and released. The operating system or application program cannot ignore the hard-boot sequence.

The soft-boot is initiated in several ways depending on the OS controlling the computer. One method used by all of the operating systems discussed in this book is initiated by pressing a sequence of three keys on the computer's keyboard. Two keys, labeled *Ctrl* (Control) and *Alt* (Alternate), are pressed and held down together. Once this is accomplished, the third key, labeled *Del* (or Delete), is pressed and then all three keys are released. The other methods will be discussed in the individual section on each operating system.

A soft-boot is controlled by the OS running the computer and does not pulse the microprocessor's RESET pin. Because this boot sequence is controlled by software, it can be disabled by the operating system or application program. Also, because a soft-boot can be disabled, it may or may not actually boot the computer. In fact, the Ctrl-Alt-Del sequence usually has to be repeated twice to soft-boot a computer running Windows 3.X or Windows 95/98.

The first time the soft-boot sequence is initiated in Windows, a blue screen with white letters appears informing you that you must repeat the soft-boot sequence in order to restart the computer. The message also informs you that all current data in the computer's memory will be lost.

When the Ctrl-Alt-Del soft-boot sequence is initiated in Windows 95/98, a task window appears on the computer's display. One of the buttons on the task window is Restart the computer. If this button is selected, the computer will soft-boot. Windows 95/98 has several ways to soft-boot the computer that will be discussed in chapter 6.

## 2.6 PROGRAMMING TERMINOLOGY

Writing programs for a computer requires the use of a **computer language.** Each programming language has a group of instructions or statements that can be combined to make the computer perform a desired task. As stated previously, a group of individual instructions put together to perform a task is called a *program*. The person who writes program is called a *programmer*.

Why are programs so important? Well, the answer is that none of the units that make up a computer (micro, mini, or mainframe) will operate unless there is a program that makes them work. The physical parts of a computer are called **hardware.** Hardware includes the microprocessor, memory ICs, and all input and output devices. It is very important to understand that a hardware device will not operate unless there is a program that makes it work. If you add a new printer to your computer and there is not a program that uses it, the printer will not work. This is not because the hardware is defective, but because there is no software to make the hardware work. The programs that control the computer's hardware are often called **hardware drivers,** or simply *drivers*.

## 2.7 PROGRAMMING BASICS

It is beyond the scope of this book to actually try to teach how to write programs for the compatible computer. In fact, it is seldom necessary for an individual who plans to simply build, upgrade, or repair a computer to write a program. However, because programs are such an integral part of the computer, you need to have a basic insight into the structure of a program.

A programmer can write programs for a computer in three different levels of complexity. These levels are called **machine language** or *machine code*, **assembly language,** and **high-level language.** This is just an introduction to the basics of programming, so the following sections and examples give only a basic explanation of each level of computer programming. If you need more information, visit your local library or bookstore.

### 2.7.1 Machine Language

As stated earlier, the microprocessor is a binary device. The only thing it understands is the 1s and 0s that make up the base-2 numbering system. Everything the microprocessor is given must be represented in 1s and 0s. Each microprocessor has a group of instructions that the programmer uses to tell the microprocessor the tasks he or she wishes it to perform. Each instruction tells the microprocessor to perform a specific step such as math (add, subtract, multiply, divide), logic (and, or, invert, exclusive or), bit manipulation (shifts and rotates), and data movement (move data between the μP and memory or I/O).

Each individual instruction that a microprocessor understands is represented by a code of 1s and 0s. These codes are called **operational codes,** or **opcodes.** Programs written using opcodes are referred to as *machine code,* **object code,** or *machine language programs.* To communicate with a microprocessor on its level, one must first learn the microprocessor's machine language.

Example 2.1 shows a segment of a machine code program written for Intel's 8086 microprocessor (80X86). This program segment will move the contents of a block of memory addresses to other locations. Each row is a separate instruction.

**Example 2.1:** 8086 machine code program example

```
1 0 1 1 1 0 0 0 0 0 1 0 0 0 0 0 0 0 0 1 0 0 0 0
1 0 0 0 1 1 1 0 1 1 0 1 1 0 0 0
1 0 1 1 1 1 1 0 0 0 0 0 0 0 0 0 0 0 0 0 0 0 0 1
1 0 1 1 1 0 0 1 0 0 0 1 0 0 0 0 0 0 0 0 0 0 0 0
1 0 0 0 1 0 1 0 0 0 1 0 0 1 0 0
1 0 0 0 1 0 0 0 0 0 1 0 0 1 0 1
0 1 0 0 0 1 1 0
0 1 0 0 0 1 1 1
0 1 0 0 1 0 0 1
0 1 1 1 0 1 0 1 1 1 1 1 0 1 1 1
1 0 0 1 0 0 0 0
```

It should be noted that Example 2.1 shows only the data and not the memory addresses at which the data will be stored. For reasons that will be discussed in chapter 7, each group of 8 bits will require one memory address. To store the program segment shown in Example 2.1, 21 memory addresses will be required.

The 8086 microprocessor has over 100 machine code instructions. We can tell the 8086 microprocessor to do most of these instructions in several different ways, so there are almost 250 different opcodes. With this many instructions and this many opcodes, it is almost impossible for an untrained individual to actually communicate with the microprocessor on the machine code level. The majority of the time, programs are written in another language level that is easier for people to understand and then *translated* into the microprocessor's machine language.

### 2.7.2 Assembly Language

Each machine language instruction is assigned a **mnemonic** (memory jogger) that is usually three to four characters in length. Simply stated, assembly language programs are written using the instruction's mnemonic instead of the binary opcode. For example, the machine code program shown in Example 2.1 appears as the program shown in Example 2.2 when written in 80X86 assembly language. The *X* in the number 80X86 means that this program is the same for all Intel and compatible 80 series microprocessors from the 8086 to current models.

**Example 2.2:** 80X86 assembly language program

```
                MOV AX,1020H
                MOV DS,AX
                MOV SI,100H
                MOV CX,10H
       NXTPT:   MOV AH,[SI]
                MOV [DI],AH
                INC SI
                INC DI
                DEC CX
                JNZ NXTPT
                NOP
```

The assembly language program shown in Example 2.2 is still difficult for the untrained individual to understand. It should be obvious, however, that assembly language would be a lot easier to learn than machine code.

Even though assembly language is easier to learn and write programs for, the microprocessor cannot run this program because it only understands machine language. In order for the μP to execute this program, it must first be *translated* into the microprocessor's machine language. The process of translating an assembly language program into machine language is called *Assembling*. The translation from assembly language to machine language is performed by another machine language program called an **assembler.**

### 2.7.3 Some Facts about Machine and Assembly Languages

When programs are written in machine code or assembly language, the programmer is in total control of the microcomputer. He or she must know how to use all of the internal registers of the microprocessor (the insides of the μP). The programmer must understand the microprocessor's **addressing modes** (the way the μP addresses data from memory and I/O). He or she has total control of the computer's μP, memory, and I/O devices, and must also know all of the addresses assigned to the different memory and I/O devices connected to the μPs to make them work.

It is very important to know that companies such as Intel and Motorola, which manufactures front-end microprocessors, use their own machine language. The machine language programs written for one of these manufacturer's microprocessor will not run on the other. For example, Apple® uses microprocessors manufactured by Motorola, whereas compatible computers use microprocessors manufactured by Intel or other companies that make Intel-compatible microprocessors. The opcodes used by Motorola's microprocessors are not the same as the opcodes used by Intel or Intel-compatible microprocessors. Machine language programs are nontransportable among computers that do not run the same or compatible microprocessor.

### 2.7.4 High-Level Language

Programs written in a high-level language are usually written using terms and expressions that are common to the English language. *COBOL, Fortran, BASIC, Pascal,* and *C* are just a few of the high-level languages used by computer programmers. Although this text does not teach programming, a basic explanation of each of these high-level languages follows.

Fortran is an acronym for *Formula Translator.* Fortran is a language that uses terms and expressions that are most often used in the scientific and engineering community. COBOL is an acronym for *COmmon Business Oriented Language.* As this name implies, COBOL commands use terms and expressions that are commonly used in the business community. BASIC is the acronym for *Beginner's All-purpose Symbolic Instruction Code.* BASIC was the first high-level language written strictly for the microcomputer. Pascal is a computer language that was named after Blaise Pascal, the inventor of the first mechanical adding machine. Pascal and C are structured languages that are used principally to solve logic problems.

No matter which high-level language is used, each language has a set of commands the programmer can use to write programs. Example 2.3 illustrates a program written in BASIC that will calculate the numbers of days old a person will be on his or her next birthday.

**Example 2.3:** Program written in BASIC

```
CLS
INPUT "How old are you today";AGE
D = AGE * 365 + 365
PRINT "You will be";D;"days old on your next birthday"
```

When high-level languages are used, the programmer does not usually need to know what type of microprocessor is being used. The programmer does not need to know the codes that make the computer's keyboard or display work. The programmer does not need to know the address of the computer's memory, keyboard, or screen. Programming in a high-level language takes the task of actually controlling the computer's hardware away from the programmer.

**Figure 2.10** Block Diagram of Interpreter

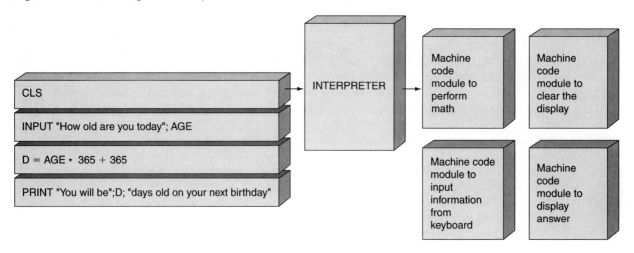

The microprocessor still needs the program to be in machine language before it can actually execute it. This means the program must be translated to machine language before the microprocessor can actually do anything with it. High-level languages are translated into machine language using two methods—**interpreter** or **compiler**.

Before we discuss interpreters and compilers, there are two programming terms that need to be introduced. A program written in any language besides machine language is called the *source code*. If the translation of the high-level language program generates a machine code program, the machine code program is called the *object code*.

### 2.7.4 Interpreter and Compiler Basics

To help explain interpreters and compilers, I will use an analogy. When the president of Russia makes a speech on television, he uses his native language. The only language I understand is English. Before I can understand what he is trying to say, his speech must be translated into English. If the speech is made on television, an interpreter is often used.

As the Russian president makes each statement, he will pause and the interpreter will translate the statement into English. The problem with this method is it slows down the overall speed of the speech. A much faster method to get his message across would be for the Russian president to give his speech to a translator beforehand. The translator would take the entire speech and translate it to English. I would get this compiled version in my native tongue, which I would read as fast as I could.

A computer language *interpreter* consists of machine code modules written to translate each statement used in the high-level language as it is being executed. As the source code program is executing, the interpreter examines each statement and causes the correct machine code module that performs each source code statement to run. As each statement is completed, the interpreter will go to the next statement and so on. If a **syntax** error occurs, the interpreter will display an error message and give some indication of where the syntax error is located. A syntax error indicates that the statement is written incorrectly. There is no object code generated by the interpreter. However, the source code is always available and is edited easily if there are any errors. A basic flowchart for an interpreter is shown in Figure 2.10.

A computer language *compiler* will take the entire source code program and convert it into machine language. Once this is accomplished, the machine language program will be the actual program that is executed by the CPU. The compiler works basically the same as an interpreter; however, the machine code module needed for each source code statement is combined in a single machine code program. The machine code program is usually stored on a disk as an object code file. If a syntax error occurs as the program is being compiled, the compiler will give some kind of error and indicate the location of the syntax error. Figure 2.11 shows a basic flowchart of a compiler.

Interpreters and compilers are written for each family of microprocessors. The high-level language program itself is usually stored or written in an ASCII format. This means the high-level language program can be transported to different computers and then compiled or interpreted for that computer's microprocessor.

**Figure 2.11** Basic Flowchart of a Compiler

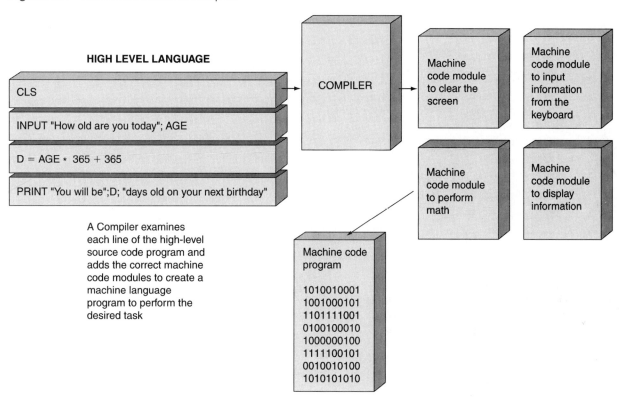

## 2.10 CONCLUSION

All microcomputers consist of four main units—microprocessor, memory, input, and output. The memory is where programs and information are stored as they are being processed. Input is the place where information is brought from the outside world into the computer. Output is where information is sent from the computer to the outside world.

The memory unit of a microcomputer is usually divided between two types of memory called RAM and ROM. The ROM contains the programs that initially get the computer started when power is applied or when a special button called RESET is pressed. The RAM holds the main operating systems along with application programs.

Microprocessors use three buses to select, communicate with, and control devices (memory, input, or output). These three buses are called the address bus, the data bus, and the control bus.

The only thing a microprocessor (or computer) can do is run programs written in a language called machine language. Programs written in any other programming language must first be translated into the microprocessor's machine language before they can be executed. The three forms of translation are called assembling, interpreting, and compiling.

## Review Questions

1. What do each of the following acronyms and abbreviations stand for?
    a. ALU
    b. CPU
    c. IC
    d. MCU
    e. RAM
    f. ROM
    g. SRAM
    h. DRAM
    i. POST
    j. BASIC
    k. COBOL
    l. FORTRAN

## Fill in the Blanks

1. The _____ _____ _____ performs the arithmetic in the microprocessor.
2. The _____ bus determines the word size of the microprocessor.
3. The _____ bus determines the maximum amount of memory that can be selected by the computer.
4. A _____ is a group of microprocessor instructions put together to perform a task.
5. When the microprocessor is running a program, it is either in the _____ or _____ phase.
6. A _____ cycle brings information into the microprocessor from the data bus.
7. _____ _____ _____ can be read from but not written to.
8. The _____ _____ is responsible for making the computer operate.
9. The programs stored in ROM are called _____.
10. _____ _____ memory can be read from and written to.
11. When the computer is _____, the primary operating system is loaded.
12. Programs that can be changed without changing hardware are called _____.
13. The _____ pin forces the microprocessor to run programs in ROM when pulsed.
14. To soft-boot the computer, the _____, _____, and _____ keys are used.
15. The binary language of the microprocessor is called _____ language.
16. The language that uses instruction mnemonics to represent instructions is called _____ language.
17. _____, _____, and _____ are used to translate programs written in other languages into machine language.

## Multiple Choice

1. The four main units a microcomputer consist of are the:
    a. fetch, execute, idle, and wait.
    b. CPU, ALU, MPU, and memory.
    c. ALU, control timing, I/O, and memory.
    d. input, output, memory, and microprocessor.
2. The microprocessor's address bus:
    a. selects a device for communication
    b. is where the communication takes place between the μP and the selected device.
    c. controls the communication between the selected device and the Mp.
    d. all of the above.
3. The microprocessor's data bus:
    a. selects a device for communication
    b. is where the communication takes place between the μP and the selected device.
    c. controls the communication between the selected device and the μP.
    d. all of the above.
4. The microprocessor's control bus:
    a. selects a device for communication.
    b. is where the communication takes place between the μP and the selected device.
    c. controls the communication between the selected device and the μP.
    d. all of the above.
5. If a microprocessor has a 20-bit address bus:
    a. it can select up to 20 devices to communicate with.
    b. it can communicate in 20-bit words.
    c. it can control up to 20 different devices.
    d. it can select up to 1,048,576 different memory devices.
6. If a microprocessor has a 32-bit data bus:
    a. it can select up to 4,294,967,296 different devices.
    b. it can communicate in 32-bit words (long words).
    c. it can select up to 32 different devices.
    d. it can control up to 32 different devices.
7. What is the primary difference between volatile and nonvolatile memory?
    a. Volatile memory will only be used in computers in which power will never fail.
    b. Volatile memory will lose its contents when power is removed, whereas nonvolatile memory will not.
    c. Nonvolatile memory will lose its contents when power is removed, whereas volatile memory will not.
    d. Volatile memory is less stable than nonvolatile memory.
8. A mnemonic is:
    a. another name for machine language program.
    b. a male programmer.
    c. three of four characters that are used to represent an instruction.
    d. another name for a opcode.
9. During a read cycle:
    a. information is flowing from memory to the CPU.
    b. information is flowing from the CPU to memory.
    c. information is flowing from a peripheral to memory.
    d. information is flowing from memory to a peripheral.
10. The store program concept means:
    a. information is stored on a disk.
    b. information is stored on a hard drive before it can be processed.
    c. all programs and information must be in the computer's memory before they can be processed.
    d. all programs and data must be properly stored or information will be lost.
11. The programs stored in ROM:
    a. can be changed only by changing hardware.
    b. are often called the computer's firmware.
    c. will not disappear when power is removed from the computer.
    d. contain the programs that get the computer going each time power is applied.
    e. all of the above.
12. The programs and information stored in RAM:
    a. cannot be changed without changing hardware.
    b. can be changed easily without changing computer hardware.
    c. is always accessed faster than the information stored in ROM.
    d. will not disappear when power is removed from the computer.

13. If a microprocessor's RESET pin is pulsed, the microprocessor
    a. will start running programs stored in the computer's ROM.
    b. will start running programs stored in the computer's RAM.
    c. will start running programs from a mass storage device called a disk drive.
    d. will stand up in its socket and sing "God Bless America."
14. A machine language program:
    a. is the only language the microprocessor understands.
    b. has to be translated before it can be used by the microprocessor.
    c. written for one microprocessor will run on all microprocessors.
    d. is written for a specific machine.
15. The basic operating system that the microprocessor will start running when power is first applied will be located in:
    a. an input unit.
    b. an output unit
    c. in the computer's ROM.
    d. in the computer's RAM.
16. A group of instructions put together to perform a task is a(an):
    a. compiler.
    b. program.
    c. interpreter.
    d. assembler.
    e. programmer.
17. Assembly language programs:
    a. are the only language the microprocessor understands.
    b. are executed faster than machine code.
    c. use mnemonics to represent opcode.
    d. have to be compiled before the computer can use them.
18. High-level language programs:
    a. are the only language the microprocessor understands.
    b. have to be translated into machine language.
    c. are executed faster than any other language level.
    d. use mnemonics to represent opcodes.
19. Multitasking is:
    a. time sharing programs between two microprocessors.
    b. the ability to run more than one program at the same time.
    c. the name of a computer with multiple microprocessors installed.
    d. a computer running multiple operating systems.
20. The term *object code compatible* means microprocessors made by different companies:
    a. speak the same machine language.
    b. have the same pin-out.
    c. speak the same high-level language.
    d. require the same voltage levels to operate.

# CHAPTER 3

# Microprocessors and the Compatible Computer

## OBJECTIVES

- Define terms, expressions, and acronyms used in relation to microprocessors and compatible computers.
- List the major manufacturers that make the microprocessors used in compatible computers.
- Describe the basic characteristics of the microprocessors currently used in compatible PCs.
- Explain microprocessor voltage characteristics.
- Explain the differences among Socket5, 6, 7 and Slot-1 and Slot-2 microprocessors.
- Describe the differences between CISC and RISC microprocessors.
- List the basic components of a compatible PC.
- List the major chipset manufacturers.
- Define terms, expressions, and acronyms used in relation to chipsets.
- Explain the basic characteristics of chipsets.
- Describe the four current motherboard form factors.

## KEY TERMS

compatible computer
downwardly compatible
Intel-compatible
fetch phase
execute phase
queue
pipeline
object code-compatible
die
pin-out
multiplexing
multitasking
protected mode

real mode
virtual memory
virtual 86 mode
cache
local bus
ZIF
heat sink
Pentium
superscaler
Socket4
Socket5
Socket7
MMX®

symmetric multi-
 processing (SMP)
over-clocking
complex instruction set
 computing (CISC)
reduced instruction set
 computing (RISC)
single-edge contact
 cartridge (SECC)
slot 1
SC242
dual independent bus
 (DIB)

frontside bus (FSB)
single-edge processor
 package (SEPP)
SC330
Pentium rating (PR)
motherboard
chipset
form factor
mini-AT
ATX
LPX
NLX

The term **compatible computer** refers to computers manufactured by different companies that use object code-compatible microprocessors and compatible programmable ports, and have everything located at the same address(es). If computers share these features, then machine code programs written for one compatible computer should run on another compatible computer without modification. IBM originally designed the compatible computers that dominate the microcomputer market. Because IBM no longer sets the standards for these types of computers, they are most often simply called *compatibles*.

73

The compatibles are split into several generations that were introduced in the following order: PC, XT, AT, 386, 486, Pentium (P5, P54C, P55C), Pentium Pro (P6), Pentium II, and the current Pentium III microprocessors. Compatible computers might also be referred to as *Socket7, Super Socket7* and *Slot-A* computers. All of the computers within each generation have object code-compatible microprocessors, compatible programmable ports, and the same input/output ports and memory address assignments. Programs written for a certain generation computer should run on other compatibles within that generation with no problems.

The designers of each new generation of compatible have strived to stay **downwardly compatible** with the previous generations of compatible computers. *Downwardly compatible* means that programs that are written around lower-generation compatible computers should run on the upper-generation machines without modification. However, programs written *specifically* for upper-generation machines will not run on lower-generation machines. For example, a program written for the XT-generation computer will run on a 386-generation computer but not the other way around. For this reason, software manufacturers list the generations of compatibles the software can run on the product's box label.

The IBM-compatible computers dominate the personal computer market all over the world. Compatible computers have many things in common that will be discussed throughout this chapter. The most important thing a compatible computer must have is a microprocessor that is machine code-compatible with those manufactured by Intel. This is the case, so this seems like a good starting point.

## 3.1 INTEL

All of the microprocessors currently installed in the compatible computers—no matter who is the manufacturer—are referred to as **Intel-compatible,** which means that microprocessors made by companies other than Intel recognize the same machine language instructions that are recognized by microprocessors made by Intel Corporation. Until recently, the packages that compatible microprocessors were mounted in were also compatible with Intel's microprocessors. This feature meant that a compatible microprocessor could be used in the same socket used by older Intel processors. This is one feature that Intel is currently trying to change.

Before we get too deep into our discussion on microprocessors, as usual, we need to first get some basic stuff out of the way.

### 3.1.1 The Other Guys

To my knowledge there are currently two semiconductor manufacturers that make microprocessors that are completely object code-compatible with Intel microprocessors. These companies are Advanced Micro Devices (AMD) and VIA Technology, Inc. In the early days of the compatible microprocessors, there were a few compatibility problems. However, I have not heard of any compatibility problems with the current versions of these second-source devices at the time of this writing.

Because these microprocessors are totally object code-compatible with the Intel processors, the Intel processors will be discussed in more detail in the following sections. A basic description of each of the aforementioned compatible microprocessors begins in section 3.3.

### 3.1.2 Fetch and Execute

As stated previously, when a microprocessor is running a machine language program, it is constantly switching between two phases—fetch and execute. During the **fetch phase,** a machine language instruction is brought into the microprocessor over the data bus and decoded. During the **execute phase,** the microprocessor does exactly what the machine language instruction told it to do.

Microprocessors below the third generation cannot fetch and execute instructions at the same time. During the execution of instructions that do not require a data transfer, the microprocessor's address, data, and control buses are idle. When the buses are idle, no data are transferred between the microprocessor and the other units of the computer.

Microprocessors above the second generation add a small first-in-first-out (FIFO) memory called a **queue** inside the microprocessor. Queue is another name for a line of items waiting to be served. When an instruction is executed that does not require data transfer, these microprocessors will pre-fetch the next program instruction and place it in the queue. When it is time to fetch the next instruction, the microprocessor does not have to go to the computer's memory because the instruction is already in the queue.

This technique gives microprocessors above the second generation the ability to fetch and execute instructions at the same time. This trait greatly enhances the speed of program execution over that of second-generation processors. All microprocessors used in the IBM and compatible computers are above the second generation and have the ability to fetch and execute program instruction at the same time. A **pipeline** is the term used for the path an instruction takes through the microprocessor as it is being processed.

## 3.2 THE MICROPROCESSORS

Now that we have a little basic terminology out of the way, we begin our discussion of the microprocessor by covering the Intel processors. Except for the 8086, the Intel processors are covered chronologically in the order they were used in the compatible computers.

### 3.2.1 8086

The 8086 is a third-generation microprocessor that is the predecessor to the 8088. Even though this microprocessor was Intel's first 16-bit microprocessor, it was not used in the original IBM computers. It was used later in some compatibles and IBM's low-end PS/1 models, however. We will cover the 8086 first because all of Intel's processors and compatible processors, up to the present, have stayed downwardly object code-compatible with the 8086. The term **object code-compatible** means that a given microprocessor will recognize the same machine language instructions as recognized by another microprocessor.

Figure 3.1 shows an actual 8086 and 8088 microprocessor integrated circuit die that has been enlarged by several magnitudes. The 8086 MPU contains approximately 29,000 transistors on a **die** that is approximately 0.25″ square.

**Figure 3.1** Die of 8086/88 Microprocessor

*Source:* Courtesy Intel Corporation.

**Figure 3.2** Pin-out of 8086 and 8088 Microprocessors

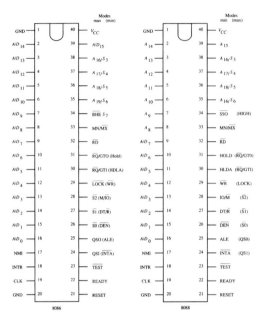

*Source:* Asser/Stigliano/Bahrenberg, *Microprocessor Theory and Servicing* (Upper Saddle River, NJ: Prentice Hall, 2001) p. 346.

The die itself is extremely small and fragile and, if left exposed, it would be impossible to handle without damaging it. Besides being fragile, there has to be some means to make electrical connections to the electronic circuitry on the die itself. To solve these problems, the 8086 die is placed in a package called a *dual in-line package* (DIP), so named because it has two rows of pins that are parallel to each other. The pin assignments of the 8086, called the **pin-out** of an IC, are shown on the left in Figure 3.2. Notice that each pin has a number and a function. The pin number is inside the drawing and the function of the pin is given outside the drawing. For example, pin number 21 is the RESET pin discussed previously in chapter 2.

The 8086 MPU requires 16 data bus pins, 20 address bus pins, 22 control bus pins, and 3 power pins. Using a little math, the 8086 microprocessor requires a total of 61 pins. When the 8086 was brought to market, the largest standard IC package available was the 40-pin DIP. To allow the 61 required pins to work in a 40-pin DIP, a technique called *multiplexing* is used.

**Multiplexing** is a widely used technique in which two or more signals share the same conductor at different times. The 8086's address information is needed before data is transferred, which allows the microprocessor's address and data bus to be multiplexed. The multiplexed address and data bus pins are labeled A/D 0 (pin 2) through A/D 15 (pin 39), as shown in Figure 3.2. Because the memory and I/O ports need to know when the microprocessor is providing address and data information, a pin on the control bus called *ALE* (pin 25) is provided. External circuitry uses the ALE pin to demultiplex the address, data, and control information for the rest of the computer. The remaining pins of the address bus are labeled A16 through A19 and are located at pin numbers 35 through 38.

With a 20-bit address bus, the 8086 can select 1,048,576 (1 M) different 8-bit memory devices, or 524,288 (512 K) different 16-bit memory devices. The control bus signal BHE (pin 34) indicates whether the microprocessor is transferring a 16-bit or 8-bit word. The word size used is set by each machine language instruction.

Even though the address bus is 20 bits, Intel chose to use only the lower 16 bits (A0 through A15) to select I/O devices. The use of 16 bits limits the I/O addresses to 65,536 combinations. Even though it may seem confusing to us, Intel had a good reason for doing this. There should be no way a computer will ever need more than 65,536 I/O devices. Because this is the case, limiting the size of the I/O addresses makes the machine language instructions that deal with I/O shorter. Shorter instructions execute faster and save memory.

The microprocessor indicates which type address is on the address bus—either memory or I/O—with a control bus signal called IO/$\overline{\text{M}}$ (pin 28). The logic state (0 or 1) of this control bus signal is controlled by each machine language instruction. An instruction that accesses information from a memory address will cause this pin to go to a 0, whereas an instruction that accesses information from an input or output port address will cause this pin to go to a 1.

Using a control bus signal to indicate whether the address bus contains an I/O or a memory address allows a memory storage location and an I/O device to have the same numerical address. This is because pin 28 tells all of the memory and input/output devices whether the microprocessor is trying to select an I/O device or a memory storage location. This technique is similar to having two different streets—one street where all of the I/O devices live and another where all of the memory devices live. Two houses can have the same numerical address because they are on different streets.

Since all of the other Intel processors have these same control pins, they will not be discussed in the sections for the other microprocessors. Only the pin number will be given. Even though there are more pins assigned to the control bus, these will not be discussed because they are beyond the scope of this book.

### 3.2.2    8088

The 8088 MPU is the microprocessor IBM chose to use in its IBM-PC and XT computers. This microprocessor is a 16-bit microprocessor with an 8-bit data bus. A drawing of the 8088 is shown along with its pin-out on the right in Figure 3.2. Similar to the 8086, the 8088 uses the same pins for the lower 8 bits of the address bus and the 8-bit data bus (pins 9 through 16).

IBM chose to use the 8088 over the 8086 because it was cheaper to build a computer with an 8-bit data bus than with a 16-bit data bus. Even though this microprocessor would run machine language programs slower than the 8086, it still out-performed all other microprocessors used in microcomputers at the time, with operational speeds available at 5 and 8 MHz.

From a programmer's point of view, the 8086 and 8088 are identical. However, with an 8-bit data bus, the 8088 usually requires two bus cycles to transfer information, whereas the 8086 requires only one.

The 8088 and 8086 both have a 20-bit memory address bus, so they can select 1,048,576 different 8-bit memory devices. The 8088 can address up to 65,536 different I/O devices, just like the 8086. If the 8088 and the 8086 were running the same program at the same speed, the 8086 would execute the program twice as fast because of its 16-bit data bus.

The pins used for the reset in the 8088 are the same as those for the 8086. The 8088 cannot transfer 16-bit data during one bus cycle, so the BHE pin is not available.

### 3.2.3    The Math Coprocessor

The math operations available to most of the 80X86 series of μPs (below the 80486) are addition, subtraction, multiplication, and division. This doesn't seem like a lot, but all other math functions can be performed using these four operations. The only problem is that a program must be written to perform the higher-level math operations. The more complicated the math operation, the longer the program has to be to perform that operation and, obviously, longer programs take more time to execute.

Intel developed a math coprocessor, also called a *floating-point unit* (FPU), for each microprocessor in the 80X86 family of microprocessors up to the 80486. The math coprocessor adds individual instructions to perform high-level math functions. Most programs do not need this higher level math, so the FPU is not usually installed in the compatible computer when purchased. Programs that need the higher math functions provided by the math coprocessor must be specifically written for them. This feature makes these special programs shorter, which means they are executed several magnitudes faster.

Most operating system and applications programs written for a computer that uses this microprocessor do not use, or need, the high level of mathematics provided by the math coprocessor. If a program is not written using the special math instructions recognized by the math coprocessor, the processor is just an extra expense.

For this reason, most compatible computers that use microprocessors below an 80486 are sold without the math coprocessor. However, a socket is usually available for adding a math coprocessor. If you do not have a coprocessor in your machine, make sure the software you purchase does not require one.

**Figure 3.3** 80286 Die

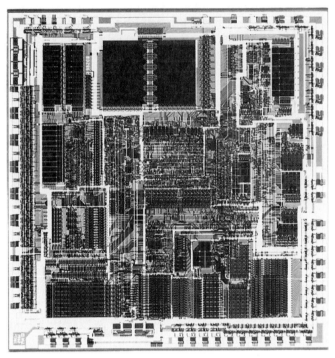

*Source:* Courtesy Intel Corporation.

If you install a coprocessor, choose one that is compatible with the microprocessor installed in your computer and operates at the same speed as the microprocessor.

Intel uses the digit 7 as the last digit for the matching math coprocessor. For example, the 8087 is the math coprocessor used with the 8086 and 8088 microprocessors.

### 3.2.4 80286

The 80286 (286) microprocessor is a 16-bit microprocessor with a 24-bit address bus and 16-bit data bus. This address bus size allows the 286 microprocessor to select 16,777,216 different 8-bit memory devices, or 8,388,608 different 16-bit memory devices. The I/O addresses for the 80286 are still limited to 65,536. A magnified photo of the actual 80286 die is shown in Figure 3.3. The 80286 die contains approximately 134,000 transistors.

When the 286 was introduced, Intel was trying to move the microprocessor into new realms of technology. Intel was trying to introduce the microcomputer community to the world of multitasking—a major feat. **Multitasking** is the ability to run more than one application program at the same time. To multitask, the microprocessor had to handle the computer's memory and I/O devices in a manner that was not compatible with the 8086 microprocessor. Losing compatibility with the 8086 was a drastic step for Intel because of the extremely large software base for the 8086 microprocessor at the time. To avoid failure from the start, Intel chose an approach that would allow it to market a totally new microprocessor architecture and yet keep it completely compatible with the 8086 microprocessor. To do this amazing feat, Intel gave the 286 microprocessors two modes of operation—**protected mode** and **real mode.**

In the real mode, the microprocessor thinks it is an 8086. When in the real mode, the 80286 can output only 1,048,576 memory address combinations even with its 24-bit address bus.[1] It is also totally object code-compatible with the 8086 microprocessor. The only difference one notices between the two

---

[1] There is a flaw in the 80286 chip that allows it to address slightly more memory than 1,048,576 memory addresses. This flaw will be discussed in greater detail in chapter 8.

Table 3.1  Microprocessor Bus and Core Speeds

| Bus | Core |
|---|---|
| 6 MHz | 6 MHz |
| 8 MHz | 8 MHz |
| 10 MHz | 10 MHz |
| 12.5 MHz | 12.5 MHz |

MPUs is an increase in instruction execution speed because the 286 can process instructions faster. The operation bus and core speeds available with the 80286 are listed in Table 3.1.

The protected mode adds additional instructions to the instruction set and manages the computer's memory in a way that is not compatible with the 8086 MPU. The protected mode also gives applications and operating systems written around this mode access to the entire addressing range of the microprocessor. And, on top of this, the protected mode allows multitasking. Note that multitasking is accomplished on a time-share basis, so the more applications that are running, the slower each will run.

The switch to 286's protected mode is made with a set of machine language instructions. Once in the protected mode, the only way to return the 286 MPU to the real mode, so that it can run 8086 programs, is with a machine reset using the microprocessor's RESET pin.

Intel believed that programmers would stop writing programs for the real mode and adopt the 286's protected mode. This would put an end to trying to stay compatible with the 8086. How wrong they were!

There were millions of IBM and compatible computers running the 8088/86 microprocessor when the 286 machines hit the market. To stop writing programs for the 8086 microprocessor and start writing protected mode programs would limit the number of potential customers for a program's applications. Needless to say, programmers continued writing programs for the 8086 because they would run on all of the IBM and compatible machines, including those running the 286 microprocessor.

Because the 286 does not have the ability to run the real mode program within the protected mode, all application programs (that I know of) run the 286 in the real mode to be compatible with the 8086. This means IBM or compatible computers that use the 286 microprocessor are nothing more than souped-up XT computers.

The 286 was the original microprocessor used in the IBM-AT and compatible computers. The 286 can run with either the 8087 or 80287 math coprocessor. The microprocessor was installed into three different packages. The one illustrated in Figure 3.4 is referred to as a *pin grid array* (PGA). The PGA has the pins protruding out the bottom of the package instead of the sides. This package allows for more pins than allowed by other standard IC packages.

Notice in Figure 3.4 that the RESET pin is numbered 29 on the 286 microprocessor. Pin 1 is the BHE pin that indicates an 8-bit or 16-bit data transfer. Pin 67 is the M/IO pin that indicates whether the microprocessor is trying to select either a memory or I/O device. Also notice that the data bus (D0 through D15) and address bus (A0 through A23) are not multiplexed because they use separate pins.

### 3.2.5  More Information on the Protected Mode and the Real Mode

To perform multitasking, each program must think it is the only one the microprocessor is running. The microprocessor accomplishes this feat by giving each program its own section of RAM. Each program's section of the computer's RAM is protected from being accessed by the other application programs the microprocessor is running; hence, the name *protected mode*.

The protected mode must be supported by the operating system running the computer in order to perform multitasking. Some of the operating systems currently available that support the protected mode are Windows 3.X, Windows NT, Windows 95, Windows 98, Windows 2000, Windows Mc, OS/2, OS/2 Warp, Linux, and UNIX.

**Figure 3.4** Pin-out of 80286 Microprocessor

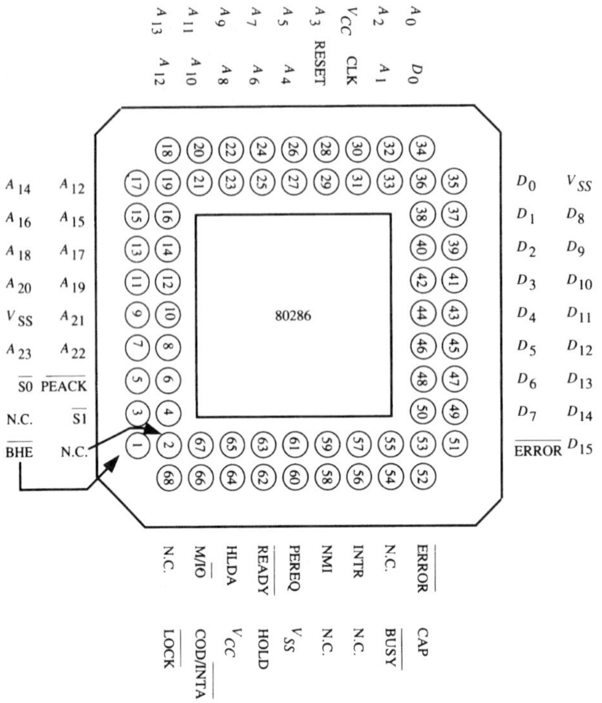

*Source:* Asser/Stigliano/Bahrenberg, *Microprocessor Theory and Servicing* (Upper Saddle River, NJ: Prentice Hall, 2001) p. 431.

Of these protected-mode operating systems, only Windows 3.0, 3.1, and 3.11 will run on an IBM and compatible computer using a 286 microprocessor. However, when the 286 is running one of these Windows operating systems, real-mode programs written for the 8086 will run but will not multitask.

Just like the protected mode, the computer's operating system must be written to support the microprocessor's real mode. Some of the most popular real-mode operating systems are MS-DOS, PC-DOS, and DR-DOS.

MS-DOS, Windows 3.X, and Windows 95 and 98 are the predominant operating systems for the IBM and compatible computers, so most applications are written for them. As a result, programs are not referred to as real-mode and protected-mode applications but as DOS, Windows, or Windows 95/98 applications. Applications that are referred to as DOS applications are usually written around the microprocessor's real mode, whereas applications referred to as Windows or Windows 95/98 applications are written around the microprocessor's protected mode. All four operating systems are covered in more detail in chapters 5 and 6.

Note that Windows 3.X is not a full-blown, stand-alone operating system but is a parasite, or add-on, operating system to MS-DOS. This characteristic of Windows 3.X is discussed in greater detail in chapter 5.

### 3.2.6 80386

The 80386DX (386) microprocessor was introduced in 1985 with a 32-bit address bus and a 32-bit data bus. This address bus size gives this microprocessor the ability to place a staggering 4,294,967,296 different physical memory addresses out of its address bus pins. Along with the physical addressing range, the 386DX also added a feature called *virtual memory*. With virtual memory, the operating system and applications can have access to up to $2^{46}$ bytes (64 terabytes [TB]) of virtual memory using a file on the drive called a *swap file*. The concept of virtual memory is discussed further in chapter 7. Similar to the 8086 and 80286, the 386DX has the ability to address 65,536 I/O devices. The magnified die of an 80386 microprocessor is shown in Figure 3.5. The 386 microprocessor consists of approximately 275,000 transistors.

**Figure 3.5** 80386 Die

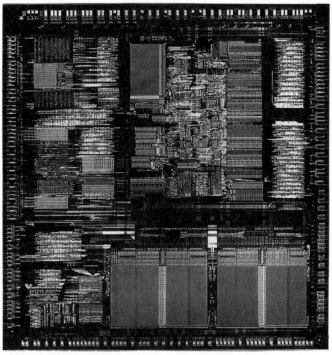

*Source:* Courtesy Intel Corporation.

The processor can be operated in real mode, protected mode, and **virtual 86 mode.** The switch to protected mode is made with software. Switching the processor back to the real mode requires a machine reset, similar to the 286 MPU. The virtual 86 mode is not available with the 80286 microprocessor, but this mode really added power to the 386 microprocessor.

The virtual 86 mode is available only with the microprocessor operating in the protected mode. The microprocessor can be switched between protected mode and virtual 86 mode with machine language instructions.

In the virtual 86 mode, the microprocessor can run machine language programs written for the 8086 microprocessor. This provides the ability to run DOS application within the protected environment. These last two sentences mean that the virtual 86 mode allows programs written for the 8086 microprocessor to be executed by the 386 MPU without switching back to the real mode.

When the correct operating system is used, the 80386 microprocessor allows multitasking of both DOS and Windows applications. Currently, the most popular protected mode of operating systems for the 386 microprocessor are Windows 95 and Windows 98. However, many computers are still running Window 3.X.

In 1988, Intel introduced the Intel 386SX microprocessor. The introduction of this microprocessor added quite a bit of confusion to the purchase of a 386 machine. The 386SX MPU is the same 32-bit microprocessor as the 386DX, but with two major differences. The 386SX MPU has a 16-bit data bus and a 24-bit address bus. If 32-bit words are used with the 386SX, two bus cycles are required to transfer data. This smaller address and data bus size cause a computer designed around this microprocessor to be almost $200 cheaper than one designed around the 386DX.

If two computers—one with a 386DX and the other with a 386SX—operating at the same speed are running a real-mode operating system, such as MS-DOS, you will not notice the difference between the two machines. However, if a protected-mode operating system is used, such as Windows 3.X or Window 9X, the 386SX will be noticeably slower. If you are planning on buying a used 386 computer to run Windows or Windows 95, watch out for the 386SX machines. The pin-out of the 386SX MPU is shown in Figure 3.6, and the available operational speeds are listed in Table 3.2.

**Figure 3.6** 386SX MPU Pin-out

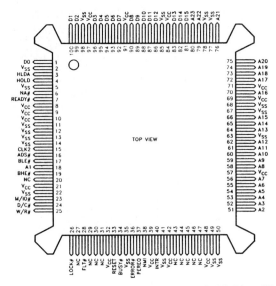

*Source:* Don L. Beeson, *Assembling and Repairing Personal Concepts* (Upper Saddle River, NJ: Prentice Hall, 1997) p. 64.

Table 3.2  The Available Bus and Core Operational Speeds for the 386DX and SX Microprocessors

| 386DX | | 386SX | |
|---|---|---|---|
| *Bus* | *Core* | *Bus* | *Core* |
| 16 MHz | 16 MHz | 16 MHz | 16 MHz |
| 25 MHz | 25 MHz | 20 MHz | 20 MHz |
| 33 MHz | 33 MHz | 25 MHz | 25 MHz |
| 40 MHz | 40 MHz | 33 MHz | 33 MHz |

Figure 3.7 shows the 386DX microprocessor along with the pin-out. Although there are other packages available, the standard 80386DX is mounted in a 132-pin programmable grid array package (PGA). Notice in Figure 3.7 that the pins are laid out in rows and columns. In computers, this is referred to as a *matrix*. The columns are lettered *A* through *P* and the rows are numbered *1* through *14*. If you look at the intersection of column C and row 9, you will see the RESET pin. During the remainder of this discussion, pin locations are referenced in the form of (column, row). For example, the RESET pin is located at (C,9).

Referring once again to Figure 3.7, notice that the address pins (A2 through A31) and data pins (D0 through D31) are located throughout the matrix. You may wonder, "Where are address bits A0 and A1?" The 80386DX microprocessor can access data in 8-bit, 16-bit, or 32-bit lengths, depending on what size word the machine language instruction calls for. To do this, the microprocessor must address 1, 2, or 4 memory addresses at one time depending on the word size. The control pins that are used to select 1, 2, or 4 memory address are the Byte Enable pins BE0 (E,12), BE1 (C,13), BE2 (B,13), and BE3 (A,13). These control signals replace address bits A0 and A1.

**Figure 3.7** 80386DX Pin-out

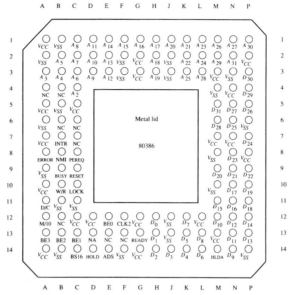

*Source:* Asser/Stigliano/Bahrenberg, *Microprocessor Theory and Servicing* (Upper Saddle River, NJ: Prentice Hall, 2001) p. 436.

The 80387 is the math coprocessor used with the 386DX μP, and the 80387SX is used with the 386SX μP.

---

**Problem 3.1:** What is the pin number of data bit D17 on the 386DX microprocessor?

**Solution:** The data bit D17 pin is at the intersection of column N and row 10, so its location is given as (N,10).

---

### 3.2.7 Cache Memory Basics

When the Intel 386DX-40 microprocessor hit the market, operating at a speed of 40 MHz, it brought about a major problem that had to be resolved. The DRAM technology being used at that time could not keep up with the bus speed of this version of the 386 microprocessor. Although using SRAM would have solved the problems, going to SRAM as the main memory meant adding thousands of dollars to the price of a machine. A computer costing that much would be out of the range of most individuals.

Another option was to deliberately add wait states to the bus cycle during memory access. This has the effect of stretching the bus cycle out, giving the memory more time to provide its data. This is akin to buying a sports car and constantly driving it with the brakes on. Why buy a high-speed microprocessor in the first place if you are not going to run it at full speed?

The final solution to the problem was to place a small section of high-speed SRAM called a **cache** on the motherboard. Cache is controlled by a special IC called a *cache controller*. The cache controller automatically copies the most used information from DRAM into this small section of SRAM.

Adding a cache to the motherboard allows the 386DX-40 microprocessor to run at full speed as long as the information it needs is in the cache. If the microprocessor has to make an access to DRAM, wait states are used. The technique of using a cache only added about a 5% average speed reduction, which is a small price to pay considering the alternatives.

**Figure 3.8** 80486 Die

*Source:* Courtesy Intel Corporation.

### 3.2.8  80486

The Intel 80486 (486) was introduced in 1989 in two versions—the 486DX and 486SX. Both processors run a 32-bit address and a 32-bit data bus. Both versions of the 486 microprocessor have the same physical, virtual, and I/O addressing ranges as the 386DX. Along with these features, both versions of the 486 have a memory management unit built inside the microprocessor. Both also have an internal 8 KB Level 1 (L1) cache memory.

An L1 cache is actually integrated onto the same silicon die as the 486 microprocessor. The L1 cache runs at the microprocessor's core speed. A Level 2 (L2) cache is the name given the cache that was developed for the 386DX-40 microprocessor. The L2 cache is not integrated onto the microprocessor die and is located outside the 486 microprocessor on the motherboard. The L2 cache on the 486 motherboard runs at the speed of the local bus. The differences between the L1 and L2 caches will become important as this discussion continues.

The magnified die for the 486 microprocessor is shown in Figure 3.8 and the pin-out is shown in Figure 3.9. Adding the L1 cache to the microprocessor plus some other major components made the transistor count of this processor jump up to approximately 1.2 million.

If both the 486DX and SX have all of these things in common, how are they different? The only difference between the 486DX and the 486SX is that the 486DX MPU has an internal math coprocessor and the 486SX MPU does not. Because the math coprocessor is part of the 486DX, it is not logical to call it a coprocessor. For this reason, the internal math processor is most often referred to as a *floating-point unit (FPU)*.

The 386 and 486 are object code-compatible and have the same memory and I/O addressing range. The 486 microprocessor also supports the same modes of operation—real, protected, and virtual 86—as the 386. All programs written for the 386 processor will run unmodified on the 486. However, the 486 will handle instructions slightly faster than the 386DX. If a 386DX and 486-compatible are operating at the same speed and running the same program, the 486 will execute the program faster. The operational speeds available with 486DX and SX microprocessors are listed in Table 3.3.

**Figure 3.9** 80486 Pin-out

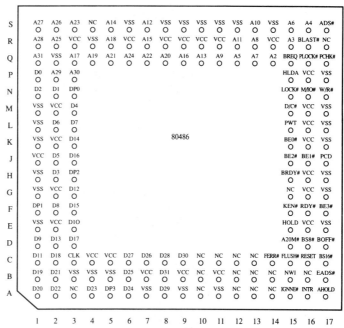

*Source:* Asser/Stigliano/Bahrenberg, *Microprocessor Theory and Servicing* (Upper Saddle River, NJ: Prentice Hall, 2001) p. 438.

**Table 3.3** 486DX and SX Microprocessor Bus and Core Speeds

| Bus | Core |
|---|---|
| 16 MHz | 16 MHz |
| 25 MHz | 25 MHz |
| 33 MHz | 33 MHz |
| 40 MHz | 40 MHz |

Note that even though the 486 and the 386 have the same address and data bus sizes, the IC package and pin assignments for the two microprocessors are completely different. The upgrade from a 386 to a 486 microprocessor *cannot* be accomplished by simply removing the 386 microprocessor and installing a 486 microprocessor in its place.

### 3.2.9 486DX2 and DX4

In its quest for speed, Intel developed internal speed doubling (DX2) and tripling (DX4) technology for its 486 line of microprocessors. This speed increase is located inside the microprocessor and not on the microprocessor's address, data, and control buses (or local bus).

The address, data, and control buses work in unison with one another, so they are most often referred to under a single name called the **local bus.** The local bus is the microprocessor's address, data, and control buses, before they go through any other logic. Figure 3.10 illustrates the concept of the local bus.

To upgrade the speed of the computer before the introduction of the 486DX2 and DX4 Intel-compatible microprocessors, one had to completely change out the main computer circuit card, called the *motherboard.* Once the motherboard was replaced, a microprocessor that was capable of running at the speed of the motherboard's local bus was installed.

**Figure 3.10** Local Bus Diagram

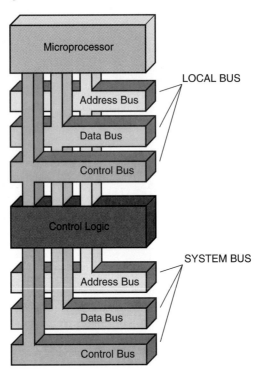

When DX2 and DX4 came along, it became possible to increase the speed of the computer by simply changing the microprocessor. If your computer is running a 486DX microprocessor with a **ZIF** (zero insertion force) called *Socket3,* and the motherboard also supports speed multiplication, the DX can be replaced easily with a DX2 microprocessor. Socket3 is the designator of the ZIF socket that has the same physical pin locations as the 486DX/DX2 microprocessors. An illustration of a typical ZIF socket, along with its basic operation, is shown in Figure 3.11. A ZIF socket allows for easy replacement of a microprocessor with little fear of damaging the pins.

I do not want to give you the impression that all you need to do to upgrade a 486DX to a DX2 or DX4 microprocessor is to simply change out the processor. You need to read the remainder of this section before making any attempts at this upgrade.

As stated previously, the DX2 and DX4 486 microprocessors have internal circuitry that increases the internal processing speed of the microprocessor, not the speed of the microprocessor's local bus. The 486DX2 MPU doubles the internal processing speed while the 486DX4 MPU triples the internal processing speed. This means that a 33-megahertz (MHz) 486DX can be replaced with a 486DX2/66 microprocessor that operates internally at 66 MHz. Note that the local bus speed of both of these microprocessors would be 33 MHz.

**Problem 3.2:** What is the local bus speed of a 486DX4-120 microprocessor?

**Solution:** The last digit of a microprocessor's part number indicates the microprocessor's internal speed of operation in megahertz. The part number ends with −120, so this microprocessor is operating at 120 MHz. The DX4 version of the 486 microprocessor is operating at three times the speed of the microprocessor's local bus. Thus, the microprocessor's local bus is operating at 40 MHz.

You may ask, "What is the advantage of going to a higher speed version of the 80486 microprocessor if the local bus is running at a slower speed?" The answer to this question is, "If the microprocessor is processing information faster, the overall processing speed of the computer will increase." This is similar to being able to think faster. However, the speed of the microprocessor's local bus is not increased, so the overall speed of the computer will not double with a DX2 chip or triple with a DX4 chip.

**Figure 3.11** Typical ZIF Socket

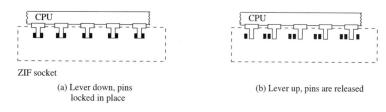

(a) Lever down, pins locked in place

(b) Lever up, pins are released

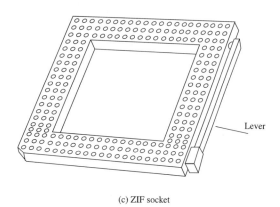

(c) ZIF socket

*Source:* Antonakos/Boylestad/Floyd/Mazidi/Mazidi/Miller, *The CEA Study Guide* (Upper Saddle River, NJ: Prentice Hall, 1999) p. 33.

**Table 3.4** 486DX4 Microprocessor's Local Bus (LB) Frequencies, Multiplier, and Core Frequencies

| 486DX2 | | | 486DX4 | | |
|---|---|---|---|---|---|
| LB | Mult | Core | LB | Mult | Core |
| 25 MHz | 2.0 | 50 MHz | 25 MHz | 3.0 | 75 MHz |
| 33 MHz | 2.0 | 66 MHz | 33 MHz | 3.0 | 100 MHz |
| 40 MHz | 2.0 | 80 MHz | 40 MHz | 3.0 | 120 MHz |
| 50 MHz | 3.0 | 150 MHz | | | |

Table 3.4 shows the relationship between the most popular 486DX2 and DX4 microprocessors and the local bus speed.

## 3.2.10 Upgrading a 486DX to a DX2 or DX4

It may seem that it is really easy to upgrade a 486 computer to a higher speed version by simply changing the microprocessor. Although I wish it was that easy, it is not. There are three very important requirements that have to be met to upgrade a 486DX to a DX2 or DX4.

The first thing to consider is the local bus speed. Each version of the 486 microprocessor, regardless of whether it is a DX, a DX2, or a DX4, requires a certain local bus speed. The available local bus speeds are listed in Table 3.4. If a motherboard supports only a 33.22 MHz local bus speed, the only processors that can be considered are the DX-33, DX2-66, or the DX4-100. The better 486 motherboards provide the ability to set the local bus speed to 25, 33, 40, or 50 MHz.

The second requirement for the motherboard to run a DX2 or DX4 is that the motherboard has to support speed multiplication. The speed multiplication feature gives the user the ability to set the 486 to a multiplier of times 1 for the 486DX, times 2 for the 486DX2, or times 3 for the 486DX4.

The third requirement deals with the problems that arise when trying to run microprocessors at such high speeds. This calls for a more in-depth explanation that will be covered in the next section.

**Figure 3.12** 486 with Heat Sink and Cooling Fan

## 3.2.11 Cooling It Down

As the internal (core) speed of the microprocessor increases, the heat it generates also increases. If the microprocessor die gets too hot, it will cease to operate and may even be damaged. This is why most microprocessors are mounted in a ceramic package instead of plastic—ceramic dissipates heat a lot better. But the ceramic package is often not enough.

Most high-speed microprocessors also have a grooved piece of metal called a **heat sink** that is mechanically attached to their tops. The purpose of the heat sink is to increase the surface area of the microprocessor to help dissipate the heat. A small fan is usually attached on top of the heat sink. The fan increases the efficiency of the heat sink by several magnitudes. A heat sink and cooling fan are illustrated in Figure 3.12. Even these features are not enough to decrease the heat produced by most high-speed microprocessor to levels that are acceptable. This means other methods will also have to be used.

The heat produced by any electronic circuit, including microprocessors, is proportional to the amount of voltage that is applied to it. Basically, *proportional* means that lowering the voltage will lower the heat.

The standard voltage level used for the logic circuits in a computer is +5 VDC. This voltage is approximately equal to three flashlight batteries connected together, one right after the other. Even though this does not seem like much voltage, it is more than enough for the microprocessor to provide correct binary logic levels inside the computer. In fact, to produce a legal one (1) logic level requires a voltage of approximately 2.4 VDC. This means the voltage powering the microprocessor can be lowered and still produce correct logic level. However, it can never be lowered all the way down to 2.4 VDC because at that level, there would be no safety margin. To reduce the heat produced by the microprocessor, the trick is lowering the voltage to the microprocessor and not all of the other computer logic.

The 486DX microprocessor operates at +5 VDC, just like all of the other computer logic. The 486DX4, which operates three times faster than its DX counterpart, would get way too hot if powered by +5 VDC. To avoid this problem, Intel decided to operate the DX4 at a voltage of around 3.45 VDC. In order for a 486DX4 to be installed in a motherboard, it must support the lower voltage for the microprocessor socket. If a 486DX microprocessor is replaced with a 486DX4 without lowering the voltage going to the microprocessor, the 486DX4 will cease to operate and may be damaged.

The amount of voltage required by the microprocessor is stamped on top of the microprocessor's package. Most of today's computers give the user the ability to set the voltage going to the micro-

**Figure 3.13** Pentium™ Die

*Source:* Courtesy Intel Corporation.

processor's mounting socket by moving electrical jumpers or changing the position of switches. Obviously, it is very important to have the correct voltage level going to the microprocessor.

### 3.2.12 Numbers and Names

The compatible chip manufacturers were using the same series of numbers for their chips that Intel used. Intel sued chipmaker AMD because of this fact and lost. The court ruled that a company could not copyright a series of numbers. Because of this very reason, Intel chose to start naming its microprocessors instead of giving them a sequence of numbers.

### 3.2.13 Pentium®

The Intel **Pentium**® microprocessor was first introduced in 1993. There are currently three versions available—P5, P54C, and P55C. All versions of the Pentium® are 32-bit microprocessors that have a 32-bit address bus and a 64-bit data bus. All have an internal FPU and two L1 caches that vary in size depending on the version. One L1 cache is used to cache data and the other is used for instructions.

The Pentium® microprocessor also utilizes what is referred to as a *dual-pipeline architecture,* which allows the Pentium processors to execute two 32-bit instructions during one bus cycle, one per pipeline. The term **superscaler** is used when referring to microprocessors that have more than one pipeline.

The 64-bit data bus allows the microprocessor to fetch two instructions at the same time to help keep the two pipeline instruction queues loaded. The 64-bit data bus also gives the Pentium® processor the ability to address data in 8-, 16-, 32-, or 64-bit groups.

Figure 3.13 shows a magnified photo of the P5. This version of the Pentium® processor requires approximately 3.1 million transistors. The Pentium® microprocessors support all addressing ranges and modes of operation that were introduced by the 386 and are totally machine language-compatible with the long line of X86 microprocessors. The original Pentium® processor does not support speed multiplication, as indicated in Table 3.5.

**Table 3.5** Original Pentium Local Bus (LB) and Core Frequencies

| LB | Core |
|---|---|
| 60 MHz | 60 MHz |
| 66 MHz | 66 MHz |

**Figure 3.14** First-generation Pentium

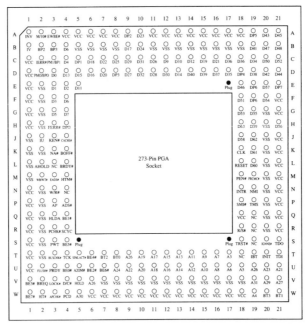

*Source:* Asser/Stigliano/Bahrenberg, *Microprocessor Theory and Servicing* (Upper Saddle River, NJ: Prentice Hall, 2001) p. 444.

### 3.2.14 Socket4, 5, and 7 and Other Pentium® Standards

The Pentium® and compatible microprocessors are currently the mainstay of the microcomputer industry. To allow the production of motherboards that support the Pentium® processor, Intel released standards for motherboard manufacturers to follow. There are currently three standards: first-generation (P5), P54C, and P55C.

The first-generation standard is the first version of the Pentium® processors sold by Intel, which was introduced in 1993. The first-generation Pentium® processor requires a 5 VDC supply and is available in two speeds—60 MHz and 66 MHz. The first-generation Pentium® microprocessor uses a times 1 (X1) multiplier, so it runs at the same speed as that of the local bus. The microprocessor is mounted in an industry-standard, 273-pin PGA package, as shown in Figure 3.14. The first-generation Pentium® processor is not pin-compatible with the Pentium microprocessors currently sold. Intel provided a 273-pin PGA industry-standard socket for the first-generation Pentium® processor called *Socket4*.

The P54C standard sets the operating voltage of the microprocessor at approximately 3.3 VDC, but this can vary slightly among manufacturers. The P54C uses an open technology, 320-pin SPGA (staggered pin grid array). This staggered pin arrangement is illustrated in Figure 3.15. It should be obvious that the SPGA package is not compatible with the 273-pin PGA used by the first-generation Pentium® processors. This means that a computer designed to run a first-generation Pentium® microprocessor cannot be upgraded to a P54C by simply changing microprocessors. Socket5 is the industry standard ZIF socket designed to match the pin-out for the P54C Pentium® and compatible microprocessors.

**Figure 3.15** P54C Pentium Pin-out

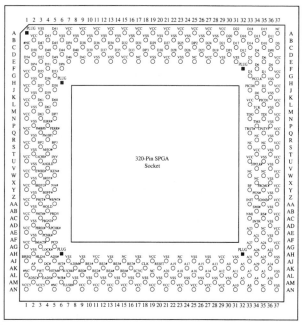

*Source:* Asser/Stigliano/Bahrenberg, *Microprocessor Theory and Servicing* (Upper Saddle River, NJ: Prentice Hall, 2001) p. 445.

**Table 3.6** Pentium P54C Microprocessor's Local Bus Frequencies (LB), Multiplier, and Core Frequencies

| LB | Mult | Core |
|---|---|---|
| 50 MHz | 1.5 | 75 MHz |
| 60 MHz | 1.5 | 90 MHz |
| 50 MHz | 2.0 | 100 MHz |
| 60 MHz | 2.0 | 120 MHz |
| 66 MHz | 2.0 | 133 MHz |
| 60 MHz | 2.5 | 150 MHz |
| 66 MHz | 2.5 | 166 MHz |
| 66 MHz | 3.0 | 200 MHz |

The P54C Pentium processors are currently available in versions that run at 75, 120, 133, 150, 166, 180, 200, and 233 MHz. All P54C Pentium processors use speed-multiplying technology. As with the 486DX2 and DX4, this means the microprocessor's core is operating at a speed faster than that of the computer's local bus. The current local bus speeds and multipliers for the Pentium P54C are listed in Table 3.6. Along with the P54C microprocessor, Intel also released an industry-standard, 320-pin ZIF socket called *Socket5*.

From the drawing in Figure 3.15, it is easy to see how the pin arrangement gets the name *SPGA*. This pin-out is very complex, so we are going to discuss only a few of the pins. Somewhere in the matrix you should find VCC, VSS, address bus (A3 through A31), data bus (D0 through D63), RESET, and

**Figure 3.16** Socket7

the INTR pin. The INTR pin is used by the computer's interrupt request signals. Interrupt requests are used by ports to get the microprocessor's attention and will be discussed in future chapters.

There are eight byte-enable (BE0 through BE7) pins that substitute for the A0, A1, and A2 address pins. These pins allow the microprocessor to address memory in 8- (byte), 16- (word), 32- (doubleword), and 64- (quad-word) bit groups.

To set the speed multiplier on the microprocessor, control bus pins with the designator *BF* are used. On the drawing in Figure 3.15, there is only one BF pin, which is located at X33. On later versions of the P54C and P55C Pentium® microprocessors, there are two additional BF pins: BF1 and BF2 added at X33 and W35. These pins are labeled *NC* (not connected) on this drawing. Jumpers on the motherboard usually set the combination of 1s and 0s that are applied to these pins to set the core speed multiplication factor used by the MPU.

The P55C Pentium® processors require two (dual) power supplies to operate. This is called *dual-voltage* because neither of the voltages is provided directly by the computer's main power supply. The dual voltage feature sets the internal voltage (core) lower than the voltage used to generate the binary level sent on the microprocessor's address, data, and control buses. This feature allows the core of the microprocessor to operate at lower temperatures, which translates to faster speeds. The two voltages required are usually called the *core* and *bus* voltages. The core voltage usually equals around 2.8 VDC but can vary between different versions of the chip and different manufacturers. The bus voltage on all versions of the P55C, both Intel and compatible, is 3.3 VDC. Current P55C-compatible microprocessors operate at local bus speeds between 66 MHz and 100 MHz and core speeds of between 166 MHz and 500 MHz.

The P55C microprocessors and P54C microprocessors are almost pin-compatible, but not quite. Additional dummy pins are added to the P55C microprocessor to prevent it from being inserted into Socket5, which does not provide the dual voltages needed by the P55C processor. The industry standard socket for the P55C processor is **Socket7.** Socket7 is unique in that it will mount both the P54C or P55C Pentium® microprocessor. A photo of Socket7 is shown in Figure 3.16. This standard calls for a 321-pin socket that matches the pin location of the P55C Pentium® microprocessor.

Socket7 is of a zero-insertion-force (ZIF) design. A handle is located to the side of the socket that, when raised, allows for easy removal and installation of the microprocessor. After the microprocessor is installed, the handle is lowered and latched into place. This firmly latches the microprocessor into the socket.

The Socket7 standard originally called for the motherboard to have another socket for an additional voltage regulator module (VRM) to provide the dual voltages to the microprocessor by regulating down

# Microprocessors and the Compatible Computer

**Table 3.7** Pentium P55C Local Bus Frequencies (LB), Multipliers, and Core Frequencies

| LB | Mult | Core |
|---|---|---|
| 66 MHz | 2.5 | 166 MHz |
| 66 MHz | 3.0 | 200 MHz |
| 66 MHz | 3.5 | 233 MHz |

the +5 VDC provided by the computer's main power supply unit. This feature allowed Intel to produce future-generation Pentiums with different voltage requirements and sell a different VRM that supplied the required voltages. This way, the user would not have to buy a new motherboard to upgrade processors.

Current Socket7 motherboards use on-board regulators whose output voltage can be programmed by the user with the arrangement of jumpers on the motherboard. This feature eliminates the need for the VRM.

## 3.2.15 Pentium® MMX™ (P55C)

The acronym **MMX™** stands for *multimedia extensions*. The Pentium® MMX™ microprocessor is simply a Pentium® microprocessor that has an additional set of 57 instructions designed specifically for multimedia applications. The Pentium® MMX™ chips also increase the L1 cache to 32 KB. To follow suit, all of the other companies that make MMX™-compatible microprocessors have also increased the L1 cache size above the Pentium®'s 16 KB. All Pentium® MMX™-compatible chips use the P55C standard.

MMX™ is designed to enable the latest multimedia software—from audio to video to 3D graphics—to run faster on your PC. Even though there are not currently many applications that take advantage of these new instructions, by the time you read this text, I am sure there will be.

Because it is very important, note again that the voltages required by the P54C and P55C microprocessors are not the same. In fact, some P54C microprocessors use other voltage levels besides 3.3 VDC. Also, the dual voltage required by all P55C microprocessors is not the same. It is very important to configure the motherboard, if possible, to provide the correct voltage levels to the microprocessor socket or the microprocessor will be damaged. If the required voltage is not available, try setting the motherboard to provide the next lower voltage to the microprocessor socket. The supported Local Bus, Multipliers, and Core speeds for Intel's P55C processors are listed in Table 3.7.

## 3.2.16 Symmetric Multiprocessing

One interesting fact about the Pentium® microprocessors is that they support multiprocessing. Multiprocessing means the computer is running more that one microprocessor on the motherboard. The feat of running more than one microprocessor on a single motherboard is referred to as **symmetric multiprocessing (SMP).** Two Pentium® microprocessors can be installed in a specially designed motherboard that has two Socket7s. To take advantage of the multiprocessing capabilities of the Pentium®, the operating system must be able to run in a multiprocessor environment. As far as I know, the only operating systems currently available that support SMP are Microsoft's Windows® NT and Windows® 2000, Be, Inc's BeOS, and Linux.

## 3.2.17 Over-clocking

With all of the local bus speeds and core multipliers available with the 486DX2 and DX4 microprocessors, it did not take long before someone tried to run a 486 at a faster local bus speed or a higher multiplier than that for which it was designed. When this happened, the technique of over-clocking was born. **Over-clocking** is a term used when a computer's microprocessor or local bus is run faster than its rated speed.

Even though I am not endorsing the practice, running a microprocessor faster than it is rated is a fairly safe practice. This is because the manufacturer usually rates the microprocessor's speed conservatively. If a microprocessor is run at a faster than designated speed, it will usually lock up (stop running programs) long before it is damaged. If you decide to attempt to over-clock a microprocessor, there are several things you should consider.

First, the microprocessor is running faster, so it is going to run hotter. To help increase the efficiency of the microprocessor's heat sink, heat-sink compound should be used between the microprocessor and the heat sink. Second, because you are running the microprocessor outside its design parameters, there is no guarantee that the microprocessor will not be damaged. And third, if the computer does not operate at all or starts to lock up occasionally, the local speed and core multiplier need to be returned to the settings for which it is rated.

There are two major concerns for both the consumer and the microprocessor manufacturer concerning over-clocking. There are computer vendors that over-clock the microprocessors in their computers and then sell the computer as if it has the faster microprocessor installed. There are also microprocessor vendors that relabel the microprocessor with the over-clock speed and then sell the microprocessor as the higher speed version. Either of these practices can lead to the microprocessor failing and the microprocessor manufacturer getting the blame.

Another obvious problem for microprocessor manufacturers is that they are losing money on the sale of their higher speed processors. For all of these reasons, microprocessor manufacturers are making every effort to prevent over-clocking their microprocessors. This includes bonding the multiplier speed setting inside the processor. Even though this helps, the processor can still be over-clocked by increasing the speed of the local bus. The current technique being implemented by Intel is to place a device called a *phase lock loop* (PLL) inside the microprocessor to sense local bus and core speeds. This technique should eliminate the ability to over-clock Intel's microprocessors.

### 3.2.18  CISC and RISC

When microprocessors were first developed, designers thought there had to be an individual instruction for each task that the microprocessor needed to perform. With the ability of the 80X86 microprocessors to process information in different lengths (32, 16, or 8 bits), not only are there different instructions for each task, but the instructions themselves are of variable lengths. In addition, Intel and other microprocessor manufacturers have maintained support for each previous generation of microprocessor, even their newest products. All of these factors result in microprocessors that have a very long instruction set.

Microprocessors that use this technology are referred to as **complex instruction set computing (CISC)** processors. CISC processors include the entire line of Intel 80X86 and Pentium® microprocessors.

As time passed, microprocessor designers developed single instructions that perform multiple tasks. The instructions are of a fixed length, so they are processed more efficiently. Using this technology also greatly enhances the performance of the microprocessor because the circuitry required is reduced by a large magnitude. Microprocessors that use this technology are referred to as **reduced instruction set computing (RISC)** processors.

To go to a total RISC design means scrapping the ability of the microprocessor to directly run programs written for previous-generation CISC microprocessors. With the millions of IBM compatibles running the 80X86 line of CISC processors, this is still not an option that Intel is willing to take.

Although RISC microprocessors will not run programs written for the 80X86 line of Intel processors, the computer can run these programs. A software interpreter, just like high-level interpreters, can translate CISC instructions into RISC instructions while the program is being executed.

The only problem with this technique is that using external software interpreters will decrease the overall execution speed of the program written using CISC instructions. To alleviate this major problem that arises with RISC processors, Intel gave its next-generation processor some unique features.

### 3.2.19  Pentium® Pro™ (P6)

In November of 1995, Intel released the Pentium® Pro microprocessor (sometimes called the *P6*). The Pentium® Pro is a 32-bit microprocessor with a 64-bit data bus. The 64-bit data bus gives the microprocessor the ability to transfer two 32-bit words during one bus cycle. The address bus size is 36 bits, which gives the microprocessor the ability to select an amazing $68,719,476,736_{10}$ bytes of physical

**Figure 3.17** Pentium Pro Die

*Source:* Courtesy Intel Corporation.

memory. The virtual memory and I/O addressing range are the same as those of the 386DX. A picture of the P6 die is shown in Figure 3.17.

Along with the larger address bus, the Pentium® Pro processor added out-of-order instruction execution, dynamic register renaming, multiple branch prediction, and data-flow analysis. These are all technical terms that the average person does not really need to understand, but, needless to say, they are major improvements to the superscaler design. Along with these improvements, which are used by most current-generation CPUs, the P6 offers another feature.

The 150 MHz, 166 MHz, 180 MHz, and 200 MHz Pentium® Pro microprocessors have a separate 256 KB L2 inside the same IC package as the microprocessor. There is also a 200-MHz version that has a 1 MB L2 cache in the same package as the microprocessor. The IC package that contains both the processor core and the L2 cache is called a 387-pin dual-cavity pin grid array (DCPGA) (Figure 3.18).

The location of the L2 cache inside the microprocessor package greatly enhances the Pentium® Pro's instruction execution time. This is because the P6 has two separate data buses. One 64-bit data bus is used to transfer information with devices located outside the microprocessor and the other 64-bit data is dedicated to the internal L2 cache. The DCPGA, including the Pentium® Pro processor and 1 MB L2 cache, contains approximately 62 million transistors. Of these, approximately 5.5 million belong to the P6 and the rest belong to the L2 cache.

The microprocessor is of superscaler design and has three instruction pipelines. This gives the microprocessor the ability to process up to three instructions per bus cycle.

The P6 is a RISC microprocessor, which seems to indicate that Intel has finally given up on support of the long line of 80X86 CISC processors. However, Intel realized that any attempt to do this would doom the microprocessor and leave it with no chance to survive.

To keep compatibility with the 80X86 line of microprocessors, Intel added the ability of the Pentium® Pro processor to perform on-the-fly translation of each CISC 80X86 machine language instruction into its RISC equivalent, which is used by the internal works of the Pentium® Pro processor. This process is transparent to the operating systems and application programs written around the 80X86 line of microprocessors.

There is a slight problem with the Pentium® Pro's translation of the 16-bit machine language operating systems and applications programs written for the 8086 and 80286 microprocessors. When the Pentium® Pro processor was being developed, Intel decided to optimize it to run 32-bit machine language programs. This means the Pentium® Pro processor does not execute 16-bit machine language programs as effectively as does the Pentium® microprocessor. In fact, if a Pentium® and Pentium® Pro processors

**Figure 3.18** P6 Dual Cavity Package

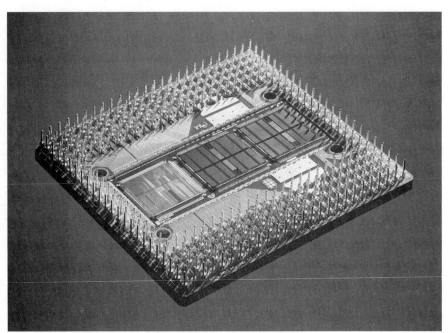

*Source:* Courtesy Intel Corporation.

**Table 3.8** Pentium Pro Local Bus Frequencies (LB), Multiplier, and Core Frequencies

| LB | Mult | Core |
|---|---|---|
| 60 MHz | 2.5 | 150 MHz |
| 66 MHz | 2.5 | 166 MHz |
| 60 MHz | 3.0 | 180 MHz |
| 66 MHz | 3.0 | 200 MHz |

are operating at the same speed while running 16-bit programs, the Pentium® processor will outperform the Pentium® Pro processor. However, when it comes to 32-bit programs, the Pentium® Pro processor will leave the Pentium® processor eating its dust.

Figure 3.18 shows the Pentium Pro microprocessor from the bottom. In this figure you can see the dies for both the L2 cache and the P6 microprocessor core. Also, it should be noted that the P6 uses a combination of both a PGA and SPGA. It should be obvious that the pin arrangement of this microprocessor is not compatible with any of the Pentium® microprocessors. This means that the only way to upgrade a computer from a Pentium® processor to a Pentium® Pro processor is to replace the motherboard.

Like the P5 microprocessor, the Pentium® Pro processor supports SMP, but with one major difference. Up to four Pentium® Pro microprocessors can be installed on a special motherboard. As with the P5, the motherboard will require an OS that supports multiprocessing.

The Pentium® Pro microprocessor is available at core frequencies of 150 MHz, 166 MHz, 180 MHz, 200 MHz, and 233 MHz. All of these frequencies run the local bus at 66 MHz and use core speed multiplying technology similar to that of the 486DX2/DX4, P54C, and P55C. The local bus, multipliers, and core speed available with the Pentium® Pro processor are listed in Table 3.8.

With all of the features available to the Pentium® Pro microprocessor, you may wonder why everybody did not go to this microprocessor. One reason is that the dual-cavity package used by the P6 made the price of the microprocessor quite a bit more than the price of the P5. Second, the location of the

**Figure 3.19** Pentium II Die

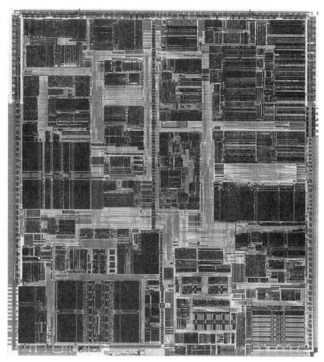

*Source:* Courtesy Intel Corporation.

MPU and L2 cache inside a single package made it very difficult to dissipate all of the heat. But neither of these reasons is why the Pentium® Pro processor did not dominate the market.

The socket standard for the Pentium® Pro microprocessor is called Socket8. The major difference with Socket8 and all other Intel socket standards is that Intel decided to make Socket8 a closed technology. Closing the technology means that no other chip manufacturer can produce a pin-compatible Pentium® Pro microprocessor.

### 3.2.20 Changing the Standards

Intel is currently in the process of trying to change a lot of the hardware standards used by today's compatible computers. One of the things Intel is trying to change is the use of the industry-standard Socket7 that is used by today's P54C and P55C microprocessors. Intel originally tried to change the standard with the Pentium® Pro processor. Now, it is taking additional steps with the Pentium® II and Pentium® III line of microprocessors.

### 3.2.21 Intel's Pentium® II™

The Pentium® II is available in three versions—the Pentium® II™, the Celeron™, and the Pentium® II Xeon™. All versions of the Pentium® II are totally machine language-compatible with the entire range of X86 microprocessors. They also support the MMX™ extensions. In fact, the Pentium® II CPUs are actually versions of the Pentium® Pro (P6) microprocessor's core, with a few major changes that will be discussed in the following paragraphs. Running a P6 core means that the Pentium® II microprocessors all have a 36-bit address bus and a 64-bit data bus and use the same address ranges and modes of operation as the P6. To help spread things out a little, I will continue to discuss the Pentium® II in this section and the Celeron™ and Pentium® Xeon™ in sections 3.2.22 and 3.2.23.

Intel's major change with the Pentium® II is that instead of using an industry-standard Socket7 or Socket8, the Pentium® II and its L2 cache are permanently mounted on a printed circuit card inside a cartridge. The microprocessor makes a connection from the cartridge to the rest of the computer through a single 242-pin, two-tier male edge card connector. The cartridge is called a **single-edge contact cartridge (SECC)** and looks similar to a video game cartridge. A Pentium® II microprocessor die is shown in Figure 3.19. Figure 3.20 shows the SEC cartridge.

**Figure 3.20** Pentium II SEC Cartridge

*Source:* Courtesy Intel Corporation.

The SEC cartridge installs into a special female edge card connector located on the motherboard. The connector is similar to an adapter card. This special motherboard connector is commonly called **Slot1,** but is technically referred to as the **SC242** (242-contact slot connector). Intel owns all of the patents to the SC242 connector and has chosen to make it closed technology. Closed technology means that Intel-compatible chip manufacturers such as AMD,and VIA cannot make Pentium® II pin-compatible microprocessors. Compatible manufacturers are taking steps to stay competitive with Intel, as discussed in later sections of this chapter.

The Pentium® II microprocessor has a 32 KB L1 cache that is divided equally between a data cache and an instruction cache. As with all of the other microprocessors discussed so far, the L1 cache runs at the core frequency of the microprocessor.

The Pentium® II processor has two data buses—the **dual independent bus (DIB)** and the **frontside bus (FSB).** The DIB connects directly to a 512 KB L2 cache located on the same circuit card as the microprocessor. The cache's DIB runs at half the speed of the microprocessor's core frequency. This means that the L2 cache will operate at approximately 117 MHz on a Pentium® II operating at 233 MHz. The DIB is sometimes referred to as a *backside bus.*

The FSB is the name Intel has given to the Pentium® II microprocessor's local bus. This bus is formed by the edge card pins coming out of the SEC cartridge and is where the computer's memory and communication ports reside. Similar to other speed-multiplying microprocessors, the actual speed of the FSC is a division of the microprocessor speed.

Two Pentium® II microprocessors can operate in a SMP arrangement on a specially designed motherboard running an OS that supports multiprocessing, similar to the P5s. Core frequencies of 233 MHz, 266 MHz, 300 MHz, 333 MHz, 350 MHz, 400 MHz, and 450 MHz are currently available. The FSB, Multiplier, and core frequencies available with the Pentium® II processor are listed in Table 3.9.

All versions of the Pentium® II processor below 400 MHz operate the FSB side bus at 66 MHz. The 400 MHz and 450 MHz operate the frontside bus at a blazing 100 MHz. Unlike the P5 and P6 microprocessors, the Pentium® II microprocessor's core multiplier is set automatically inside the SEC. This feature seems to rule out over-clocking this microprocessor. However, there is information on the Internet that gives specific instructions on how to open the SEC and place jumpers and resistors to change the multiplier and frontside bus speed to allow over-clocking. Note that I am not condoning this procedure.

# Microprocessors and the Compatible Computer

**Table 3.9** Pentium II Microprocessor's FSB, Multiplier, and Core Frequencies

| FSB | Mult | Core |
|---|---|---|
| 66 MHz | 3.5 | 233 MHz |
| 66 MHz | 4.0 | 266 MHz |
| 66 MHz | 4.5 | 300 MHz |
| 66 MHz | 5.0 | 333 MHz |
| 100 MHz | 3.5 | 350 MHz |
| 100 MHz | 4.0 | 400 MHz |
| 100 MHz | 4.5 | 450 MHz |

**Figure 3.21** Celeron SEPP and PPGA

*Source:* Courtesy Intel Corporation.

## 3.2.22 Intel® Celeron™

The Celeron™ microprocessor is Intel's attack on the comparatively low-priced compatible microprocessors made by AMD and VIA. These microprocessors are still designed around Socket7 technology but are more akin to a Pentium® Pro than a Pentium®. These microprocessor manufacturers are using techniques that allow them to compete with the Pentium® II processors at a reduced price. To counter this trend, Intel introduced the Celeron microprocessor.

The Celeron™ processor is available in the **single-edge processor package (SEPP),** which is basically a SEC without the cover. The SEPP uses the same pin-out as used in the Pentium® II and is totally compatible with the SC242 connector. The photo on the left in Figure 3.21 shows the Celeron™ processor's SEPP.

All Celeron™ microprocessors currently run the FSB at 66 MHz and have a 32 KB L1 cache. The Celeron™ is currently available at speeds from 266 MHz through 700 MHz, with faster speeds sure to

**Table 3.10** Current Pentium II Celeron Microprocessor's FSB, Multiplier, and Core Frequencies

| FSB | Mult | Core | L2 Cache |
|---|---|---|---|
| 66 MHz | 4.0 | 266 MHz | No |
| 66 MHz | 4.5 | 300 MHz | No |
| 66 MHz | 4.5 | 300A MHz | Yes |
| 66 MHz | 5.0 | 333 MHz | Yes |
| 66 MHz | 5.5 | 366 MHz | Yes |
| 66 MHz | 6.0 | 400 MHz | Yes |
| 66 MHz | 6.5 | 433 MHz | Yes |
| 66 MHz | 7.0 | 466 MHz | Yes |
| 66 MHz | 7.5 | 500 MHz | Yes |
| 66 MHz | 8.0 | 533 MHz | Yes |
| 66 MHz | 8.5 | 566 MHz | Yes |
| 66 MHz | 9.0 | 600 MHz | Yes |
| 66 MHz | 9.5 | 633 MHz | Yes |
| 66 MHz | 10.0 | 677 MHz | Yes |
| 66 MHz | 10.5 | 700 MHz | Yes |

follow. The 266 MHz and 300 MHz versions do not have an L2 cache on the SEPP but the other speed versions do. All Celeron™ processors above the 300 MHz version have a 128 KB L2 cache mounted on the SEPP with the microprocessor and use the same DIB used by the Pentium® II. The Celeron™ processor FSB, Multiplier, Core speed and L2 cache support current available are listed in Table 3.10.

One of the major reasons why Intel decided to go to the SEC/SEPP is because the technology did not exist to reliably mount the microprocessor and L2 cache inside the same IC without running into over-heating problems. However, with today's low core voltage levels and integration technology, this problem seems to have been overcome. This is because Intel is releasing the Celeron™ microprocessor in a new IC pin package called the *370-pin Plastic PGA (PPGA)*. The 370 PPGA is shown on the right in Figure 3.21.

Along with the 370 PPGA Celeron™ microprocessor, Intel has also released a new pin-compatible ZIF socket called *PGA370*. Like the SC242, the PGA370 is closed technology. You might wonder why a Celeron™ microprocessor can go from a 242-pin edge card connector to a 370-pin PPGA. The reason is that PPGA uses extra power and ground pins and there are 51 pins on the PPGA that are not used.

The cost of manufacturing an IC in a pin package is a lot less than manufacturing a microprocessor on a printed circuit card that mounts in a slot, so the future of the PGA370 socket looks bright. With 51 unassigned pins available, the PGA370 technology has carried over into the Intel Pentium® III microprocessors. Because the Celeron™ processor is designed around the home computer market, it does not support SMP.

### 3.2.23  Intel Pentium II Xeon™

The Intel® Xeon processor is the current high-end SEC cartridge microprocessor designed for network servers and high-speed work stations. People who have a lot of money and like to brag will probably buy this beast. The Intel® Xeon processor is currently available at speeds of 400 MHz and 450 MHz, both of which run the FSB at 100 MHz (Table 3.11). The Intel® Xeon processor has a 32 KB L1 cache split equally

Table 3.11 Current Pentium II Xeon Microprocessor's FSB, Multiplier, and Core Frequencies

| FSB | Mult | Core |
|---|---|---|
| 100 MHz | 4.0 | 400 MHz |
| 100 MHz | 4.5 | 450 MHz |

Figure 3.22  Xeon SEC Cartridge

*Source:* Courtesy Intel Corporation.

between a data cache and an instruction cache. The onboard L2 cache sizes available with the Xeon processor are 512 KB, 1 MB, and 2 MB. Unlike the Pentium® II, the L2 cache runs at the core speed of the microprocessor. Figure 3.22 shows the Xeon SEC cartridge.

The Intel® Xeon processor is totally machine language-compatible with the long line of X86 microprocessors. It has the same address ranges and modes of operating that the P6 microprocessor has.

The Intel® Xeon microprocessor's SEC installs on the motherboard into a 330-contact slot connector (**SC330**) that is also known as *Slot-2*. The SC330 is different than the SC242, so a Intel® Xeon processor motherboard is not compatible with a Pentium® II or Celeron™ processor. Similar to the P6, the Intel® Xeon and SC330 can run in an SMP arrangement of up to four CPUs using a special motherboard and OS that supports multiprocessing.

## 3.2.24  Intel® Pentium® III

Similar to the Pentium® II, the Pentium® III microprocessor is basically the P6 core that is currently available at 450 MHz, 500 MHz, 550 MHz, 600 MHz, 650 MHz, 700 MHz, 750 MHz, and 800 MHz core speeds. The Pentium® III processor also runs the FSB at 100 MHz, similar to current Pentium® IIs and Intel® Xeon processors. However, some of the newer versions of the Pentium® III processor will run on a 133 MHz FSB. The Pentium® III processor has a 32 KB L1 cache that is divided equally between

**Table 3.12** Pentium III and Pentium III Xeon Front Side Bus (FSB), Multiplier (Mult), and Core Frequencies

| FSB | Mult | Core |
|---|---|---|
| 100 MHz | 4.5 | 450 MHz |
| 100 MHz | 5.0 | 500 MHz |
| 100 MHz | 5.5 | 550 MHz |
| 100 MHz | 6.0 | 600 MHz |
| 100 MHz | 6.5 | 650 MHz |
| 100 MHz | 7.0 | 700 MHz |
| 100 MHz | 7.5 | 750 MHz |
| 100 MHz | 8.0 | 800 MHz |
| 133 MHz | 4.0 | 533 MHz |
| 133 MHz | 4.5 | 600 MHz |
| 133 MHz | 5.0 | 667 MHz |
| 133 MHz | 5.5 | 733 MHz |
| 133 MHz | 6.0 | 800 MHz |
| 133 MHz | 6.5 | 866 MHz |
| 133 MHz | 7.0 | 933 MHz |
| 133 MHz | 7.5 | 1.0 GHz |
| **Pentium III Xeon** | | |
| 133 MHz | 4.5 | 600 MHz |
| 133 MHz | 5.0 | 667 MHz |
| 133 MHz | 5.5 | 733 MHz |
| 133 MHz | 6.0 | 800 MHz |
| 133 MHz | 6.5 | 866 MHz |
| 133 MHz | 7.0 | 933 MHz |

instruction and data caches. A 512 KB L2 cache that is accessed at the microprocessor's core speed is connected to the DDB. The Pentium® III processor is currently available in SECC and SECC2 packages.

Along with the MMX™ instructions added by the Pentium® MMX™ microprocessor, Intel has added 70 additional instructions to the Pentium® III processors instruction set. These new instructions are designed specifically for enhanced 3D imaging using texturing and floating point operations. These instructions fall into a category of instruction called *single instruction multiple data (SIMD)*. SIMD is Intel's answer to the 3DNow!™ instructions currently available with AMD and Cyrix processors.

Along with the additional instructions, the Pentium® III processor adds a feature that is causing a lot of controversy in the microprocessor consumer market. This feature is the introduction of a unique serial number that is available to applications and networks to identify the individual Pentium® III microprocessor.

Fortunately, Intel added the ability to disable this feature with a program that can be executed from a diskette or CMOS.

Similar to the Pentium® II processor, the Pentium® III processor also has an Intel® Xeon version. This version has all the features of the Pentium® II Xeon™ processor plus those added with the Pentium® III processor. The FSB, multiplier, and core speeds for both the Pentium® III and Pentium® III Xeon™ processors are listed in Table 3.12.

### 3.2.25   The Intel® Itanium™

Intel's next-generation microprocessor, the *Itanium™ processor*, is under development and is expected to be released at about the same time as this book is being published. Intel will probably rename the processor when it becomes available for purchase commercially. The Itanium™ processor is going to use the Intel Architecture-64 (IA-64). The IA-64 architecture means the Itanium™ processor will be a true 64-bit microprocessor.

Unlike the Pentium® II and Pentium® III processors, the Itanium™ microprocessor is being designed from the ground up. In fact, the Itanium™ processors will support an entirely new 64-bit instruction set that Intel is developing jointly with Hewlett Packard. The Itanium™ processor is maintaining full X86 machine language-compatibility through on-the-fly translation, similar to the P6.

I wish I could tell you more about this microprocessor, but this is all the information I have. For the most current information on Intel microprocessors, visit the Intel Web site at developer.intel.com/design/processor/.

The Itanium™ winds up our discussion on the Intel processors used in the compatible PCs. I guess that means it's time we move on to the competition.

## 3.3   THE COMPATIBLE MICROPROCESSORS

Pentium®, Pentium® Pro, Celeron™, and Intel® Xeon™ are either trademarks or registered Intel® trademarks of Intel, which means that manufacturers of compatible microprocessors cannot use any of these names. However, there are manufacturers that make chips that offer the same, or more, features than those offered by Intel.

There are currently two major chip manufacturers of Intel compatible microprocessors: AMD and VIA Technology. Each of these companies' compatible microprocessors will be covered in the following sections. The compatible microprocessors through the 486 have almost identical characteristics as their equivalent Intel processors, so I will not be discussing compatible microprocessors that were made before the Intel Pentium®. Before we begin our discussion, however, we need to clear up the speed rating used by some compatible manufacturers.

### 3.3.1   Pentium Rating (PR)

Some Intel-compatible microprocessors do not operate at the same local bus speed and multiplier as the Pentium® processor they are trying to replace. These microprocessors use a rating called the **Pentium® rating (PR)** instead of a clock speed rating. Instead of comparing the CPU's core speed, these ratings attempt to compare the speed of program execution. For example, *PR200* indicates that the microprocessor runs machine language programs at a speed comparable to that of an Intel® Pentium® microprocessor operating at a core frequency of 200 MHz.

The PR of a compatible microprocessor compares the processing speed and not the microprocessor's actual core and local bus speeds. This statement means that the multiplier and local bus will usually be set lower than the speed and multiplier of the Intel® Pentium® processor to which it is being compared. The biggest problem that I have encountered is that individuals are setting the speed and multiplier to the PR instead of setting them according to the documentation that is provided with the microprocessor. This will lead to the MPU being over-clocked, which often causes the computer to lock up.

The important thing to remember is to consult the documentation that comes with any MPU to determine the correct voltage, multiplier, and local bus settings. Now that we have a basic understanding of the PR rating, let's start our discussion with Intel's biggest competitor—AMD.

Table 3.13  AMD-K5™ PR, Voltage, Local Bus (LB), Multipliers, and Actual MPU Core Frequencies

| PR | Voltage | LB | Mult | Core |
|---|---|---|---|---|
| 166 MHz | 3.45–3.60 | 66 MHz | 1.75 | 116.7 MHz |
| 133 MHz | 3.45–3.60 | 66 MHz | 1.5 | 100 MHz |
| 120 MHz | 3.45–3.60 | 60 MHz | 1.5 | 90 MHz |
| 100 MHz | 3.45–3.60 | 66 MHz | 1.5 | 100 MHz |
| 90 MHz | 3.45–3.60 | 60 MHz | 1.5 | 90 MHz |
| 75 MHz | 3.45–3.60 | 50 MHz | 1.5 | 75 MHz |

Table 3.14  AMD-K6™ Microprocessor Local Bus, Multiplier, and Core Frequencies

| Local Bus | Mult | Core | Core Voltage | Bus Voltage |
|---|---|---|---|---|
| 66 MHz | 2.5 | 166 MHz | 2.9 V | 3.3 V |
| 66 MHz | 3.0 | 200 MHz | 2.9 V | 3.3 V |
| 66 MHz | 3.5 | 233 MHz | 2.2 V | 3.3 V |
| 66 MHz | 4.0 | 266 MHz | 2.2 V | 3.3 V |
| 66 MHz | 4.5 | 300 MHz | 2.2 V | 3.3 V |

### 3.3.2   Advanced Micro Devices, Inc. (AMD)

For reasons that were discussed earlier, AMD calls its P54C Pentium-compatible microprocessor the *AMD-K5*™. All versions of the AMD-K5™ MPU come in a 296-pin PGA package that is pin-compatible with the Intel P54C microprocessors, whose pin-out was shown in Figure 3.15. Calling the AMD-K5™ processor a Pentium®-compatible microprocessor is not exactly accurate because the AMD-K5™ also supports some advanced features like dynamic branch prediction and out-of-order instruction execution, which are not available with the Pentium® microprocessor.

Table 3.13 shows the voltage levels, multipliers, and PR ratings used by the AMD-K5™ microprocessor. Notice on the first three listings of AMD-K5™ processors that the CPU frequency and processor speeds do not match. This is because this model of the AMD-K5™ processor uses the PR ratings that are discussed in section 3.3.1.

AMD's P55C Pentium®-compatible microprocessor is called the AMD-K6™ processor. The AMD-K6™ processor is actually more akin to the Intel Pentium® Pro microprocessor, containing nearly all of its features except the internal L2 cache and the 36-bit address bus. One major feature of the AMD-K6™ processor is that it runs in the industry-standard Socket7, which makes it a very inexpensive upgrade for all of the Socket7 motherboards currently installed in many compatible computers.

The AMD-K6™ processor is fully compatible with X86 machine language and also supports the 57 additional MMX instructions. The AMD-K6™ processor uses the same multipliers and local bus speeds and it is comparable with Intel's P55C microprocessor. In addition to the Intel multipliers, AMD added multipliers of 4.0 and 4.5, which added core speeds of 266 MHz and 300 MHz. Table 3.14 lists current AMD-K6™ microprocessors and their local bus speeds, multipliers, core frequencies, and core and bus voltages.

The AMD-K6-2™ processor, shown in Figure 3.23, is the first microprocessor to introduce 3DNow!™ technology. The 3DNow!™ technology is a new set of 3D instructions that AMD and VIA are trying to move into the mainstream PC community. A major boost in the support of 3DNow!™

**Figure 3.23** AMD-K6-2

*Source:* Reprinted by permission of Advanced Micro Devices, Inc.

**Table 3.15** AMD-K6-2™ Processor Local Bus, Multiplier, and Core Frequencies

| LB | Mult | Core |
|---|---|---|
| 66 MHz | 3.5 | 266 MHz |
| 66 MHz | 4.0 | 300 MHz |
| 100 MHz | 3.0 | 300 MHz |
| 66 MHz | 4.0 | 333 MHz |
| 95 MHz | 3.5 | 333 MHz |
| 100 MHz | 3.5 | 350 MHz |
| 66 MHz | 4.0 | 366 MHz |
| 95 MHz | 4.0 | 380 MHz |
| 100 MHz | 4.0 | 400 MHz |
| 100 MHz | 4.5 | 450 MHz |
| 100 MHz | 5.0 | 500 MHz |
| 100 MHz | 5.5 | 550 MHz |

occurred when Microsoft decided to support it with Direct Draw 6.0. 3DNow!™ includes an additional 21 instructions to the X86 instruction set that are specifically designed to bring a major enhancement to the speed of 3D imaging. These instructions are in addition to the standard 57 MMX™ instructions that were first introduced by Intel. Currently available local bus speeds (LB), multipliers (Mult), and core frequencies for the AMD-K6-2™ are listed in Table 3.15.

**Figure 3.24** AMD Athlon™ Microprocessor Cartridge

*Source:* Reprinted by permission of Advanced Micro Devices, Inc.

AMD's current mainstream microprocessor is the AMD-K6-III™ processor. The AMD-K6-III™ processor is fully compatible with X86, and MMX™, and 3D-Now! technologies. The AMD-K6-III™ processor is currently available at a core speed up to 450 MHz and installs in what is now referred to as a Super 7™ platform or motherboard. Super 7 platform is the terminology used to describe motherboards that use the P55C pin-compatible Socket7 supporting a 100 MHz local bus speed. Along with the 100 MHz local bus speed, the AMD-K6-III™ processor also has an integrated 256 KB L2 cache inside the same PGA package with the microprocessor, similar to the Pentium Pro. Unlike the Pentium® Pro, however, the L2 cache inside the AMD-K6-III™ package operates at the core speed of the microprocessor. Adding this to a large 64 KB L1 cache that is divided equally between instructions and data makes the AMD-K6-III™ processor a formidable opponent to Intel's Pentium® II.

Along with the L2 cache that is inside the microprocessor package of the AMD-K6-III™ processor, the Super 7 motherboard can also have an L2 cache. Under this arrangement, the motherboard L2 cache is referred to as the *L3 cache*.

At the time of this writing, AMD's newest entry into the ever-evolving X86 microprocessor market is the AMD Athlon™ processor with 3DNow! technology. Unlike all of the other microprocessors produced by AMD, the AMD Athlon™ microprocessor is definitely not Socket7-compatible. In fact, the AMD Athlon™ processor uses Digital Research's alpha bus protocol, which is referred to as *EV6*. This is a bus used by alpha RISC processors that will go by the name *Slot-A*. The Athlon processor is shown in Figure 3.24.

According to information provided by AMD, the AMD Athlon™ processor has the following features:

- Binary compatibility with the X86 line of microprocessors;
- Dynamic instruction execution;
- High-speed FPU;
- Support for MMX and AMD 3DNow!™ technologies;
- 128 KB L1 (64 KB data, 64 KB instruction) minimum;
- 72-bit backside data bus for an L2 cache from 512 KB to 8 MB;
- Core speed of 500 MHz and faster; and
- 72-bit FSB running at 200 MHz or faster.

As you can see from these features, the AMD Athlon™ is a high-end microprocessor with plenty of power for even the most powerful PCs. This processor should give Intel formidable competition for the high-end user, server, and work station.

For the most current information on AMD's microprocessors, visit its Web site at www.amd.com.

**Table 3.16** M1 Microprocessors

| PR  | Mult | Core    | LB     |
|-----|------|---------|--------|
| 120 | 2.0  | 100 MHz | 50 MHz |
| 133 | 2.0  | 110 MHz | 55 MHz |
| 150 | 2.0  | 120 MHz | 60 MHz |
| 166 | 2.0  | 133 MHz | 66 MHz |
| 200 | 2.0  | 150 MHz | 75 MHz |

**Table 3.17** Cyrix M2 Microprocessor PR, Core Frequency, Local Bus Frequency, and Multiplier

| PR  | Mult | Core | LB  |
|-----|------|------|-----|
| 300 | 3.0  | 225  | 75  |
| 300 | 3.5  | 233  | 66  |
| 333 | 3.5  | 262  | 75  |
| 366 | 2.5  | 250  | 100 |
| 400 | 3.0  | 285  | 95  |
| 433 | 100  | 300  | 3.0 |

### 3.3.3 Cyrix

In 1997, Cyrix was purchased by National Semiconductor and, after only 2 years, by VIA Technologies. Even though the company has changed hands, the processors still carry the Cyrix name.

The Cyrix-compatible P54C is called the 6x86, or M1. Like the AMD-K5, the M1 microprocessor is pin-compatible with the Intel® Pentium® P54Cs. The M1 microprocessor operates at voltage levels from 3.3 to 3.52 volts, depending on the version. Also, unlike the K5 microprocessor, all M1s use a PR, which makes its speed rating higher than the local bus speed times the multiplier. M1 microprocessors, along with their PR, local bus speed, and multiplier, are listed in Table 3.16.

The Cyrix-compatible P55C microprocessor is referred to as the M2, or 6x86L. Similar to the M1, the M2's PR is higher than the local bus speed times the multiplier. The current M2 microprocessor's PR, core frequency, local bus frequency, and multiplier are listed in Table 3.17.

In the year 2000, VIA Technologies introduced a Socket370-compatible microprocessor called the VIA Cyrix III. As far as I know, this is the only Socket370-compatible microprocessor currently available.

The VIA Cyrix III (Figure 3.25) is currently available at FSB speeds of 500 and 600 MHz, with faster speeds sure to follow. Unlike the Intel's Socket370 microprocessors, the VIA Cyrix III microprocessors can be run at FSB speeds of 66 MHz, 100 MHz, and 133 MHz, using the correct multipliers. Along with this feature, the VIA Cyrix III supports both MMX™ and 3DNow! multimedia extensions. The VIA Cyrix III contains a large 128-KB L1 cache, but does not contain an L2 cache.

For the most current information on the VIA Cyrix microprocessor, visit its Web site at www.cyrix.com.

### 3.3.4 Centaur Technology Inc.

Up until 1999, Centaur was a division of Integrated Device Technology. Like Cyrix, the Centaur division of IDT was purchased by VIA Technologies in 1999. Centaur currently offers two entries in the Intel-compatible field of microprocessors: the IDT WinChip™ C6™ and the WinChip™ 2. Both Centaur

**Figure 3.25** VIA/Cyrix III

*Source:* Courtesy VIA Technologies, Inc.

processors are packaged in a 296-pin PGA that is Socket7-compatible. Both processors are also X86 machine language- and MMX™-compatible. Along with MMX™, the WinChip2 also supports AMD's 3DNow!™ technology instructions. Neither chip is capable of multiprocessing.

Unlike all other P54C- and P55C-compatible microprocessors, the Centaur chips are not superscaler in design. This feature makes the IDT/Centaur dies very small and also lowers the power consumption. Centaur compensates for the lack of multiple pipelines with a large 64-KB split L1 cache that is divided equally between instructions and data. The small die size allows Centaur to sell the WinChips at a price that is far below that of its cousins.

The WinChip C6 is a P54C- and P55C-compatible microprocessor that is available at speeds of 180 MHz (60 MHz bus), 200 MHz (66 MHz bus), 225 MHz (75 MHz bus), and 240 MHz (60 MHz bus). The C6 is supported by two core voltages equal to 3.53 V and 3.3 V.

The WinChip2 is a P54C- and P55C-compatible microprocessor that is currently available at speeds of 225 MHz (75 MHz bus), 240 MHz (60 MHz bus), 250 MHz (83 MHz bus), 266 MHz (66 MHz bus), and 300 MHz (75 and 100 MHz bus). The WinChip2 runs at one of two core voltages equal to 3.52 V and 3.3 V.

## 3.4 MOTHERBOARDS

All of the aforementioned microprocessors are mounted on a large printed circuit board (PCB) that is mounted in the bottom of the desktop case or against one of the sidewalls of a tower case. This PCB is called the **motherboard,** or *system board*. Most of today's motherboards contain the following:

- Microprocessor(s);
- BIOS ROM;
- System RAM;
- L2 cache;
- CMOS;
- Real-time clock;
- Chipset;
- Keyboard port;
- Mouse port;
- Internal speaker port;
- Two EIDE ports;
- Two programmable serial ports (COM), one with infrared capabilities;
- One programmable parallel port (LPT);

# Microprocessors and the Compatible Computer

**Table 3.18** Motherboard Manufacturers and Internet Addresses

| Abit Computer Corp. | www.abit.com.tx |
| Acer America Corp | www.acer.com |
| AcerOpen | www.aopen.com.tw |
| American Megatrends | www.megatrends.com |
| ASUS | www.asus.com |
| First International Computer Inc. | www.fica.com |
| Giga-Byte Technology Co., Ltd. | www.giga-byte.com |
| Intel Corporation | www.intel.com |
| Ocean Office Automation, Ltd. | www.oceanhk.com |
| Supermicro Computer Inc. | www.supermicro.com |
| Tyan Computer | www.tyan.com |

- Two USB ports;
- PCI and ISA expansion bus; and
- AGP slot.

The motherboard runs a close second (or may actually be first) to the microprocessor in determining the speed and reliability of the compatible PC. The most important circuits that control the motherboard are a group of one or more ICs mounted on the motherboard called the **chipset.** The chipset controls almost every aspect of the motherboard, from supporting the microprocessor to keeping the correct time.

Because a motherboard is designed around a certain microprocessor and chipset, most motherboards that use chipsets that are made by reputable manufacturers are basically the same. In fact, I have purchased a few low-priced motherboards that were not made by any of the top 10 manufacturers and they have all functioned reliably. However, I recommend that you buy a name-brand motherboard because the manufacturers have better support for the end user. The Internet address of a Web site that I visit most often for references on motherboard manufacturers is www.motherboards.org. Table 3.18 lists several of the best-known motherboard manufacturers along with their Internet addresses.

In addition to a reputable manufacturer, some of the things that you should look for when purchasing a motherboard are the microprocessor it supports, BIOS ROM, chipset, form factor, FSB speeds, MPU voltage regulators, integrated peripherals, and the number and style of the expansion slots. The MPU, FSB, and MPU voltage regulators have already been discussed in previous sections of this chapter and will not be discussed further. All of the other major motherboard components will be covered in the following section. Some of the sections are brief because the featured component will be discussed in detail in later chapters.

## 3.4.1 Chipset

As stated previously, the chipset is probably one of the most important, yet overlooked, features of the compatible PC. All of the information that is transferred between the microprocessor and the computer's RAM and peripherals must pass through the chipset. The chipset controls the overall features of the computer, so it is very important that you have a basic understanding of its features. Some of the features controlled by the chipset are:

- Type of microprocessor;
- Memory type, amount, timing, and error detection/correction;
- Type and amount of L2 cache;

**Table 3.19** Chipset Manufacturers

| | |
|---|---|
| Advance Micro Devices (AMD) | www.amd.com |
| Intel Corporation | www.intel.com/design/chipsets/ |
| Silicon Integrated Systems Corporation (SiS) | www.sis.com.tw |
| VIA Technologies Inc. (VIA) | www.viatech.com |
| Acer Laboratories Inc. (ALi) | www.ali.com.tw |

- Amount of system RAM that is routed through the cache;
- Available local bus or frontside bus speed;
- The number of microprocessor speed multipliers;
- PCI bus type, bus mastering, and synchronization method;
- Symmetric multiprocessing capability;
- Available EIDE programmable input/output (PIO) and direct memory access modes;
- Built-in PS/2 mouse, keyboard controller and CMOS, and real-time clock circuitry;
- The number of peripheral components interface (PCI) and industry standard architecture (ISA) expansion slots; and
- Power-saving features.

There are currently five manufacturers whose chipsets I recommend for the compatible computer. These companies are reliable and offer full product support and technical documentation on their Web sites. The company names and Internet addresses of the recommended chipset manufacturers are listed in Table 3.19.

Chipsets are currently grouped into two categories—those that are designed to run Socket7-compatible microprocessors and those that are designed to run the three versions of the Pentium® II and the two versions of the Pentium® III. Due to the closed technology, Intel is the predominant manufacturer of chipsets for the Pentium® II and Pentium® III. The other manufacturers are just now releasing chipsets for these processors. However, Intel is no longer designing new chipsets for Socket7 microprocessors, so the other chipset manufacturers now dominate this market. This does not mean that the features offered by the Socket7 chipsets have degraded. In fact, most of the features offered by current Socket7 chipsets meet or exceed those offered by Intel for its line of Pentium® II and Pentium® III microprocessors.

### 3.4.2 Socket5 and Socket7 Chipsets

Tables 3.20 through 3.23 list basic specifications of chipsets that are currently installed on Socket5 and Socket7 motherboards. Data books that contain all of the technical information can be downloaded from the respective manufacturer's Web site.

Table 3.20  Intel P54C and P55C Chipsets

| | | \multicolumn{4}{c}{Intel P54C and P55C Chipsets} | | | |
|---|---|---|---|---|---|
| **Chipset** | | **430FX** | **430HX** | **430VX** | **430TX** |
| | | 82437FX<br>82438FX<br>83271SB | 82439HX<br>82371SB | 82437VX<br>82438VX<br>82371SB | 82439TX<br>82371SB |
| MPU | Bus Speeds MHx | 66 | 66 | 66 | 66 |
| | SMP | No | 2 | No | No |
| RAM | RAM Types | FPM | FPM<br>EDO | FPM<br>EDO<br>SDRAM | FPM<br>EDO<br>SDRAM |
| | Parity/ECC | No | Yes | No | No |
| | Maximum RAM | 128 MB | 512 MB | 128 MB | 256MB |
| | Cacheable | 64 MB | 512 MB | 64 MB | 64 MB |
| L2 Cache | L2 Cache Size | 512 K | 512 K | 512 K | 512 K |
| | Cache Type | PB | PB | PB | PB |
| | Cache Policy | W/B | W/B | W/B | W/B |
| I/O | PCI | 2.0 | 2.1 | 2.1 | 2.1 |
| | USB | No | Yes | Yes | Yes |
| | IDE | | UDMA | | ATA-33 |
| | APM | Yes | Yes | Yes | Yes |
| | AGP | No | No | No | No |

**Table 3.21** VIA P54C and P55C Chipsets

| Chipset | | VPX/97 | VP2/97 | VP3 | MVP3 | MVP4 |
|---|---|---|---|---|---|---|
| | | \multicolumn{5}{c}{VIA Chipset} | | | | |
| | Part Number | 585VPX<br>587<br>586B | 590VP<br>595<br>586B | 597<br>586B | 598AT<br>586B | 8501<br>686A |
| MPU | Max. Bus Speeds | 75 MHz | 66 MHz | 66 MHz | 100 MHz | 100 MHz |
| | SMP | 1 | 1 | 1 | 1 | |
| RAM | RAM Types | FPM<br>EDO<br>BEDO<br>SDRAM | FPM<br>EDO<br>BEDO<br>SDRAM | FPM<br>EDO<br>SDRAM | FPM<br>EDO<br>SDRAM<br>DDR SDRAM | FPM<br>EDO<br>SDRAM |
| | Parity/ECC | No | Yes | Yes | Yes | Yes |
| | Maximum RAM | 512 MB | 512 MB | 1 GB | 1 GB | 768 MB |
| | Cacheable | 512 MB | 512 MB | 1 GB | 1 GB | |
| L2 Cache | L2 Cache Size | 2 MB | 2 MB | 2 MB | 2 MB | 2 MB |
| | Cache Type | PB | PB | PB | PB | PB |
| | Cache Policy | W/B | W/B | W/B | W/B | W/B |
| I/O | PCI | 2.1 | 2.1 | 2.1 | 2.1 | 2.2 |
| | USB | Yes | Yes | Yes | Yes | Yes |
| | UDMA/33 | Yes | Yes | Yes | Yes | Yes |
| | ATA/66 | No | No | No | No | Yes |
| | Pwr. Man. | Yes | Yes | Yes | Yes | Yes |
| | AGP | No | No | Yes | Yes | Yes |

Table 3.22  SiS P54C and P55C Chipsets

| | | SiS P54C and P55C Chipsets | | | | | | |
|---|---|---|---|---|---|---|---|---|
| **Chipset** | | | | | | | | |
| | Part Number | 530 5595 | 5591 5595 | 5597 or 5598 | 5581 or 5582 | 5571 | 5596 5513 | 5511 5512 5513 |
| MPU | Max. Bus Speed | 75 MHz | 83.3 MHz | 75 MHz | 75 MHz | 75 MHz | 66 MHz | 66 MHz |
| | SMP | 1 | 1 | 1 | 1 | 1 | 1 | 1 |
| RAM | RAM Types | FPM EDO SDRAM | FPM EDO SDRAM | FPM EDO SDRAM | FPM EDO SDRAM | FPM EDO SDRAM | FPM EDO | FPM EDO |
| | Parity/ECC | No | No | No | No | No | No | No |
| | Maximum RAM | 384 MB | 768 MB | 384 MB | 384 MB | 384 MB | 512 MB | 512 MB |
| | Cacheable | 128 MB | ? MB | 128 MB | 128 MB | ? MB | ? MB | ? MB |
| L2 Cache | Maximum L2 Cache Size | 512 KB | 1 GB | 512 KB | 512 KB | 512 KB | 1 MB | 1 MB |
| | Cache Type | PB | PB | PB | PB | PB | PB | PB |
| | Cache Policy | WB | WB | WB | WB | WB/WT | WB/WT | WB/WT |
| I/O | PCI | 2.1 | ? | ? | 2.1 | 2.1 | 2.1 | ? |
| | USB | Yes | Yes | Yes | Yes | Yes | No | No |
| | UDMA/33 | Yes | Yes | Yes | Yes | No | No | No |
| | ACPI | Yes | ? | Yes | Yes | No | No | No |
| | AGP | Yes Integrated | Yes Integrated | No, VGA Integrated | No, VGA Integrated | No | No, FGA Integrated | No |

**Table 3.23** Ali P54C and P55C Chips

| | | Ali P54C and P55C Chips | |
|---|---|---|---|
| *Chipset* | | *Aladdin V* | *Aladdin IV+* |
| | Part Number | M1541<br>M1543 | M1531<br>M1543 |
| MPU | Max. Bus Speed | 100 MHz | 83.3 MHz |
| | SMP | 1 | 1 |
| RAM | RAM Types | FPM<br>EDO<br>SDRAM | FPM<br>EDO<br>SDRAM |
| | Parity/ECC | No | No |
| | Maximum RAM | 1 GB | 1 GB |
| | Cacheable | 512 MB | 64 MB<br>512 MB |
| L2 Cache | Maximum L2 Cache Size | 1 MB | 1 MB |
| | Cache Type | PB | PB |
| | Cache Policy | WB | WB |
| I/O | PCI | 2.1 | 2.1 |
| | USB | Yes | Yes |
| | UDMA/33 | Yes | Yes |
| | ACPI | Yes | Yes |
| | AGP | Yes | No |

### 3.4.3 Pentium® Pro, Pentium® II, and Pentium® III Chipsets

Tables 3.24 and 3.25 list basic specifications of Pentium® Pro, Pentium® II, and Pentium® III chipsets. If you need more detailed information, visit the manufacturer's Web site to download complete data books. The L2 cache is contained inside the microprocessor packages, so this specification is no longer controlled by the chipset.

**Table 3.24** Intel® Pentium® Pro, Pentium® II, and Pentium® III Chipsets

| | Intel® Pentium® Pro, Slot 1 and Socket370 | | | | |
|---|---|---|---|---|---|
| **Chipset** | | 440FX | 440BX | 440LX | 440EX |
| | Part Numbers | 82441FX<br>82442FM<br>82371SB | 82443BX<br>81371AB | 82443LX<br>81371AB | 82443EX<br>81371AB |
| MPU | Max. Bus Speed | 66 MHz | 100 MHz | 66 MHz | 66 MHz |
| | SMP | 2 | 2 | 2 | No |
| RAM | RAM Types | FPM<br>EDO | SDRAM | EDO<br>SDRAM | EDO<br>SDRAM |
| | | BEDO | | | |
| | Parity/ECC | Yes | | Yes | No |
| | Maximum RAM | EDO<br>1 GB<br>SDRAM<br>512 MB | 1 GB | EDO<br>1 GB<br>SDRAM<br>512 MB | 256 MB |
| I/O | PCI | 2.1 | 2.1 | 2.2 | NA |
| | PCI Bus Masters | 5 | 5 | 4 | NA |
| | USB | No | Yes | Yes | NA |
| | UDMA/33 | ? | Yes | NA | Yes |
| | ACPI | No | Yes | Yes | Yes |
| | AGP | No | Yes | Yes | Yes |

**Table 3.25** Compatible Slot 1 and Socket370 Chipsets

| Slot 1 & Socket370 | | VIA | | SiS | | | | Ali |
|---|---|---|---|---|---|---|---|---|
| **Chipset** | | *Apollo Pro Plus* | *Apollo Pro* | *5600* | *600* | *620* | *630* | *Aladdin Pro II* |
| | Part Numbers | 82C691 82C596 | 82C639 82C595A | 5600 5595 | 600 5595 | 620 5595 | 630 5595 | M1621 M1543 |
| MPU | Max. Bus Speed | 100 MHz | 100 MHz | 100 MHz | 100 MHz | 100 MHz | 100 MHz | 100 MHz |
| | SMP | No | No | No | No | No | No | No |
| RAM | RAM Types | FPM EDO SDRAM | FPM EDO SDRAM | FPM EDO SDRAM | FPM EDO SDRAM | NA NA SDRAM | NA NA SDRAM | FPM EDO SDRAM |
| | Parity/ECC | No | No | Yes | Yes | NA | NA | Yes |
| | Maximum RAM | 1 GB | 1 GB | 1.5 GB | | 1.5 GB | 1.5 GB | 1 GB |
| I/O | PCI | 2.1 | 2.1 | 2.1 | 2.1 | 2.2 | 2.2 | 2.1 |
| | PCI Bus Masters | 5 | 5 | 4 | 2.1 | 4 | 4 | 5 |
| | USB | Yes | Yes | Yes | Yes | Yes | Yes | Yes |
| | UDMA/33 | Yes | Yes | Yes | Yes | Yes | Yes | Yes |
| | ACPI | Yes | Yes | Yes | Yes | Yes | Yes | Yes |
| | AGP | Yes | Yes | Integrated | Integrated | Integrated | Integrated | Yes |
| | Integrated Audio | No | No | No | No | No | Yes | No |
| | Integrated NIC | No | No | No | No | No | Yes | No |
| | Integrated Modem | No | No | No | No | No | Yes | No |

# Microprocessors and the Compatible Computer

## 3.4.4 Chipset Summary

I hope the tables shown in the preceding sections are not too technical in nature. I know that a lot of this information can be overwhelming at times. The main things to consider, relative the chipset, when purchasing a motherboard are the MPUs it supports, the local bus speed (FSB), the number of MPU multipliers available, and the total amount of cacheable RAM.

The actual chipset used in most compatible PCs is not listed as part of the specifications of a pre-assembled computer. However, most of the features offered by the chipset are listed as part of the computer's specifications. You need to read these specifications carefully to determine if the computer has all of the options you desire.

One major chipset specification that is often overlooked is the amount of cacheable memory. The maximum amount of memory that can be installed in the motherboard is often higher than the maximum number of addresses that can be routed through the L2 cache. For example, the Intel 430VX and 430TX chipsets can support up to 128 MB and 256 MB of RAM respectively, but both only cache the first 64 MB of RAM. If more than the cacheable amount of RAM is installed in the computer, the computer's performance will be degraded greatly when running operating systems such as Microsoft Windows 95, 98, 2000, and NT.

Along with the cacheable memory a chipset supports, there are a few features you should look for in a chipset if you are planning on purchasing a new motherboard or computer. These include support for 133 MHz FSB (local bus), USB, AGP X2, IEEE 1394 (FireWire), PCI 2.1, UDMA 66 (ATA/66), IrDA (Infrared), and PC100 SDRAM. All of these features will be discussed in future chapters.

## 3.4.5 Identifying the Chipset

The motherboard's chipset can be determined using two methods. The first method is to locate the chipset's main IC on the motherboard itself. This is a large square IC that is usually located close to the microprocessor, then toward the center of the motherboard. The IC will have the manufacturer's name or logo printed on it along with the IC's part number. The part number will identify the chipset installed on the motherboard. Intel's part numbers are fairly easy to figure out because the part number has two letters toward the end such as TX, FX, or GX. These two letters identify the chipset installed on the motherboard. For example, the number 82439TX identifies the chipset as the Intel 430TX.

To use the second method, the computer has to be running either Windows 95 or 98. The steps to follow are listed in Procedure 3.1. Once you determine the chipset, visit the manufacturer's Web site to determine the features offered by the chipset.

---

### Procedure 3.1: Using Windows 95 or 98 to determine the chipset installed in the computer

1. Click the left mouse button on the Start button located on the Task Bar.
2. Move the mouse pointer up to Settings and then over to Control Panel.
3. Click on Control Panel.
4. After the Control Panel window is displayed, locate and double-click the left mouse button on the System Icon.
5. After the System Properties window displays, click the left mouse button on the Device Manager tab.
6. Locate then double-click the left mouse button on the System Devices icon.
7. Look for the listing in the table with the Processor (CPU) to PCI Bridge or PCI to ISA Bridge. This listing identifies the computer's chipset. Figure 3.26 shows an example of this listing with the chipset identification listing highlighted. In Figure 3.27, the chipset is listed as an 82439TX, which is an Intel 430TX chipset.
8. Close all open windows by clicking the left mouse button on the X in the upper right-hand corner of each window or by clicking the left mouse button or the Cancel buttons.

**Figure 3.26** Identifying the Chipset Using Windows 95 or 98

*Source:* Courtesy Intel Corporation.

**Figure 3.27** Typical Mini-AT Form Factor Motherboard

## 3.4.6 Motherboard Form Factors

The **form factor** is the physical shape of the motherboard. There are currently four form factors used in compatible computers: the Mini-AT (Baby-AT), ATX, LPX, and NLX.

The shape of the **Mini-AT** form factor has been basically the same since it was first introduced with the 286 compatible computers. A picture of the Mini-AT form factor is shown in Figure 3.27.

**Figure 3.28** Typical ATX Form Factor

Even though the microprocessor, chipset, and expansion slots on motherboards have evolved tremendously, the shape of the Mini-AT motherboard has stayed practically the same. This feature has made computer upgrades quite easy by simply changing out the computer's motherboard. This is because any Mini-AT form factor motherboard will mount into any chassis (case) that is designed around this form factor.

Even though the Mini-AT form factor has served the microcomputer industry well, it has a few problems that makes its replacement inevitable. First, the microprocessor socket is placed directly behind the expansion bus slots. This feature limits the length of adapter cards that can be inserted into the slots that are in front of the MPU socket. Second, the microprocessor's voltage regulators are usually located in a position that places them below the microprocessor when the motherboard is installed in a tower chassis. This feature causes the heat generated by the voltage regulators to pass over the microprocessor. Third, the power supply connectors on the Mini-AT motherboard can be inserted backwards or offset, which can cause damage to the motherboard or power supply. Fourth, ribbon cables are used to connect the motherboard connectors for the serial, parallel, and USB ports to the actual port connectors that are usually mounted on the back panel of the chassis. These additional cables clutter the inside of the case and also make it very difficult to get to the memory sockets.

The **ATX** form factor motherboard is designed to eliminate the shortcomings of the Mini-AT form factor by placing the microprocessor socket to one side of the expansion bus so it will not interfere with the adapter cards. It also mounts the connectors for the integrated parallel, serial, and USB ports on a metal plate that is located on the back of the motherboard. This feature eliminates many of the ribbon cables that are required by the Mini-AT form factor. Also, the ATX power supply connector cannot be inserted backwards or offset. All of these features make the ATX motherboard and chassis form factor preferred for new installations. A typical ATX form factor is shown in Figure 3.28.

The **LPX** and **NLX** form factors are used in most name-brand preassembled computers. This is because their features dramatically reduce the overall production cost of the computer. The biggest feature of the LPX and NLX motherboards is that practically all of the standard peripheral ports are integrated onto the motherboard itself. In some installations, this feature eliminates the need for an expansion bus altogether. Both of these motherboards have the ability to add expansion slots, if needed.

**Figure 3.29** LPX Motherboard Form Factor

**Figure 3.30** NLX Motherboard Form Factor

*Source:* Courtesy Intel Corporation.

The LPX motherboard has one large expansion slot located in the center of the PCB. If expansion slots are needed, a smaller PCB, called a *riser,* is inserted into the slot on the motherboard. A typical LPX form factor is shown in Figure 3.29.

The NLX motherboard uses a similar feature except this motherboard has a male edge card connector on one side of the motherboard. The riser PCB is plugged into this edge card connector on the motherboard and provides the additional expansion slots. Figure 3.30 shows a typical NLX form factor.

### 3.4.7 The Motherboard's ROM

One of the main concepts this book tries to convey is that almost everything in the computer requires a program to make it operate. The programs that make a computer operate are called the *operating system*. As discussed previously in chapter 2, the first operating system the compatible sees, when power is first applied, is stored in the computer's ROM. These programs are nonvolatile and cannot normally be changed by the user. The name given to this operating system is the *BIOS*. ROMs and BIOS are discussed in chapter 8.

### 3.4.8 The Motherboard's RAM

The motherboard's RAM is the random access read/write memory where programs and data are stored during work sessions. Information can be loaded and saved to or from this memory. The contents of this memory will be lost if power is removed from the computer. If you do not wish to lose the information in the RAM when you turn your computer off, it should be saved on a mass storage device called a *disk*.

There are three categories of user RAM installed on the motherboard in compatible computers: *conventional RAM, expanded RAM,* and *extended RAM*. Each category will be discussed in detail later in chapter 7. Regardless of the category, the amount of RAM in the machine is given in kilobytes or megabytes of storage. Remember, in computers 1 K of storage actually equals 1,024 storage locations and 1M of storage equals 1,024 K storage locations. Therefore, if your system has 640 K bytes of conventional RAM installed, there are actually 655,360 bytes (640x1024). Each application program will require a minimum amount of RAM to work properly. The RAM requirements are usually listed on the side of the box in which the application program is purchased.

### 3.4.9 Expansion Bus

A group of female edge card connectors called the *expansion bus* is located on the motherboard. The expansion bus is where printed circuit boards (PCBs), called *adapters,* are inserted. There will usually be an adapter card for the computer's video, sound, and modem. There may also be adapter cards that contain additional parallel or serial ports.

Several types of expansion buses are used on the different generations of compatible computers. They include PC-Bus (8-bit), ISA, MCA, EISA, VL-Bus, and PCI. The meanings and characteristics of each of these expansion buses are covered in chapter 4.

### 3.4.10 Integrated Ports

Integrated input/output ports are located on the motherboard as separate ICs or as part of the chipset. The most often included are EIDE, FDD, keyboard, PS/2 mouse, parallel, serial, and USB. Some chipsets provide additional integrated ports such as video, sound, game port, and network ports. These integrated ports and their configurations are discussed in later chapters.

## 3.5 MOTHERBOARD REMOVAL AND INSTALLATION

Removal and installation of the motherboard is probably the most complicated task in the assembly of the computer. This is because the motherboard almost always has to be configured correctly before it is installed. Also, if the motherboard is not installed correctly, there is a good chance that it will be damaged beyond repair the first time power is applied. Also, if the motherboard MPU, DIMM, and Flash-ROM voltages are not set correctly, these devices can also be damaged.

Procedure 3.2 lists a generic removal and installation procedure that should apply to most motherboards. As usual, if the motherboard is supplied with mounting instructions, those should be followed instead.

## Procedure 3.2: Motherboard removal and installation

### Removal

1. Make sure the computer's power switch is in the off position and the power cord is unplugged from the back of the computer.
2. If applicable, unplug the monitor's power cable from the rear of the chassis.
3. Remove the computer chassis cover and store the screws in a safe location.
4. Wear an anti-static wristband attached to a bare metal surface on the computer chassis any time you are in contact with motherboard. If a wristband is not available, touch a bare metal surface on the chassis often to avoid static buildup.
5. Attach an anti-static mat to a bare metal surface on the computer's chassis.
6. Remove all data cables from the computer's back panel.
7. If the motherboard is in a tower chassis, lay the chassis on its side with the motherboard cavity facing up.
8. Remove all of the adapter cards and place them in an anti-static bag or on an anti-static mat. Store the screws in a safe location. (The procedure for removing and installing adapter cards is found in chapter 9.)
9. Disconnect all data, power, and front panel connectors from the motherboard. If the same motherboard is going to be reinstalled and you are not sure of their placement, label each connector with a piece of masking tape and marker as it is removed.
10. Remove all screws securing the motherboard and store them in a safe location. These screws can be located at various locations on the motherboard.
11. Removing the motherboard with an AT form factor:
    a. If the motherboard is mounted in a tower chassis, slide the motherboard toward the bottom of the chassis until the nipples on the nylon standoffs disengage from the slots on the chassis mounting panel.
    b. For a desktop chassis, slide the motherboard toward the right, looking from the front of the computer, until the nipples on the nylon standoffs disengage from the slots on the chassis mounting panel. If the motherboard will not slide, it usually means one or more mounting screws has not been removed.
    c. Gently tilt the motherboard up from the edge that is at the bottom of a tower chassis or the right edge on a desktop. If the motherboard will not lift, it usually means one or more of the nipples on the nylon standoff are not disengaged from the slots in the chassis.
    d. Lift the motherboard out of the chassis and place on an anti-static mat or store it in an anti-static bag.
12. Removing a motherboard with an ATX form factor:
    a. Gently tilt the motherboard from the edge that is toward the front of the chassis. If the motherboard will not tilt, it usually means one or more of the mounting screws is still in place.
    b. Remove the motherboard from the chassis and place it on an anti-static mat or store it in an anti-static bag.

### Installation

### Preparing the chassis

1. Make sure the computer power switch is in the off position and the power cord is unplugged from the back of the computer.
2. Remove the computer's cover and store the screws in a safe location.
3. Determine if the chassis and power supply connector(s) are compatible with the motherboard form factor to be installed. Do not attempt to install the motherboard into a chassis for which it is not designed.
4. If a tower chassis is being used, lay it on its side with the motherboard cavity facing up and the bottom of the chassis toward you.
5. Attach an anti-static mat to a bare metal surface on the computer chassis.
6. Wear an anti-static wristband attached to a bare metal surface on the computer chassis.
7. Remove the motherboard from the anti-static bag and place it on the anti-static mat.
8. Make a note of the location of the mounting holes through the motherboard.
9. Gently lower the motherboard into the computer's chassis with the expansion connectors toward the back of the computer.
10. Make sure the keyboard connector is aligned with the correct hole, located on the back of the computer.
11. On an ATX chassis, make sure all of the integrated port connectors align with the appropriate knockout on the back of the chassis. On some ATX chassis, the metal panel through which the integrated port connectors protrude can be loosened to allow easy alignment. On some ATX chassis, different panels may be provided to match the port connectors.

**Figure 3.31** Metal Standoff

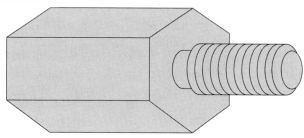

**Figure 3.32** Mounting Nylon Standoffs

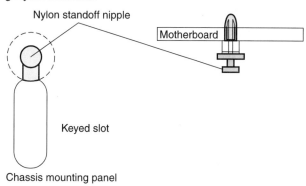

12. Once the motherboard is aligned, use a felt-tipped pen to make a mark through each hole in the motherboard that aligns with a threaded screw hole in the chassis mounting panel. On an AT form factor motherboard and chassis, also mark each of the keyed slots of the mounting panel that align with a mounting hole through the motherboard.
13. Remove the motherboard from the chassis and place it on an anti-static mat.
14. Screw the metal standoff into each of the marked threaded holes. Do not over-tighten the screws. These metal standoffs look similar to the one shown in Figure 3.31 and are provided with the chassis.
15. Insert nylon standoffs into the holes in the AT form factor motherboard that are aligned with the marked keyed slots on the chassis mounting panel, as shown in Figure 3.32.

### Installing the MPU
16. Installing the MPU in a Slot-1 motherboard:
    a. Install the MPU cooling fan on the MPU using the documentation provided with the cooling fan (sometimes the fan is pre-installed).
    b. If not already done, raise the MPU rails located on each side of the Slot-1 connector.
    c. Raise the clips on each side of the rails to allow the MPU to be installed.
    d. Slide the outside edges along the longest sides of the MPU module down the channels on the inside of each rail. Make sure the keys in the male MPU edge card connector align with the keys in the female Slot-1 connector.
    e. Once aligned, push down evenly on the top of both sides of the MPU module until it is seated into the Slot-1 connector. This is indicated when the clips on each rail snap into the sides of the MPU module, locking it into position.
17. Installing the MPU in a Socket7 motherboard
    a. Raise the handle on the ZIF socket by pushing it away from the side of the socket and then lifting up.
    b. Insert the MPU into Socket7, making sure pin 1 of the microprocessor aligns with pin 1 of the Socket7. Pin 1 of the microprocessor is identified by pins aligned at an angle on one of the microprocessor's corners. This set of angled pins must align with angled holes in Socket7. If aligned correctly, and no MPU pins are bent, the MPU should insert into the socket with very little pressure.

c. Once the microprocessor is inserted in the ZIF socket, press down on the microprocessor gently while pushing the ZIF socket's handle down. You will feel some resistance on the handle as the pressure starts to secure the processor in the socket. This resistance is normal and will not damage the processor. The processor is locked securely in the socket when the handle locks into place in the down position.
d. If the microprocessor does not already have a heat sink/cooling fan installed, install the cooling fan assembly on top of the microprocessor with the short retainer clip on the short side of the socket.
e. Connect the short retainer clip on the cooling fan assembly by hooking the clip under the tab on the short side of the socket. The short side of the socket is the side away from the handle hinge.
f. Connect the long retainer clip on the cooling fan assembly by pressing down with your finger on the clip until it snaps onto the tab on the long side of the socket. This takes quite a bit of pressure. Be careful not to break the clip on the socket.

18. Install the memory using the procedure given in chapter 7.
19. Use the documentation provided with the motherboard to configure the jumpers on the motherboard for the MPU to be installed. If no documentation is provided, it is often painted on the motherboard's component side.
20. Installing the ATX motherboard:
    a. Lower the ATX motherboard into the chassis cavity with the expansion slots toward the rear of the computer.
    b. Align the integrated port connectors with the connector holes provided on the chassis.
    c. Once the connectors are aligned, position the motherboard until all of the mounting holes are aligned with the thread holes in the metal standoffs. Make sure there are no extra standoffs that do not align with a hole in the motherboard.
    d. Secure the motherboard to the metal standoffs with the correct size screw. Do not over-tighten the screws. Also, make sure the head of the screw does not press into any trace on the motherboard.
21. Installing an AT form factor motherboard:
    a. Lower the motherboard into the chassis with the connectors toward the back of the computer.
    b. Align the nipples on the nylon standoffs that were previously installed in the holes in the motherboard with the larger opening of the keyed slots in the chassis mounting panel. This is illustrated in Figure 3.32.
    c. Slide the motherboard in the direction required to move the nipples of the nylon standoff into the narrower opening in the keyed slot. The nipples should move all the way to the end of the narrower opening.
    d. Make sure the appropriate holes in the motherboard align with the threaded holes in the metal standoffs installed previously. Make sure there are not extra standoffs that do not align with a hole in the motherboard.
    e. Secure the motherboard in the chassis by screwing the correct size screw into the threaded holes in the metal standoff. Do not over-tighten the screws. Make sure the heads of the screws do not press against a trace of the motherboard.
22. Install the chassis front panel connectors onto the appropriate pin connectors on the motherboard. The front panel connectors are usually labeled with the name of the signal they control. To determine the correct connector on the motherboard for each front panel signal, consult the documentation that is provided with the motherboard.
23. Install the power supply connector into the correct connector on the motherboard. If the motherboard is an AT, make sure the black conductors on both the P8 and P9 connectors meet at the center of the connector. For more information on these connectors, refer to the Introduction.
24. If any internal peripherals need to be installed, refer to the installation procedure in the appropriate chapters.
    a. Power supply installation, Introduction
    b. Floppy drive installation, chapter 10
    c. CD-ROM installation, chapter 10
    d. Hard-drive installation, chapter 11
25. Install the data connectors for the internal peripherals and external port connectors (AT only) into the correct pin connectors on the motherboard. The location of the connectors is provided in the documentation that comes with the motherboard.
26. Install the adapter cards using the procedure outlined in chapter 9. If an adapter metal mounting plate does not align correctly with the hole in the back panel, loosen the screws that mount the motherboard, and then move the motherboard slightly until the plate aligns. Once aligned, re-tighten the motherboard mounting screws, but avoid over-tightening.
27. Make the following checks before applying power:
    a. The motherboard is not touching the metal chassis.
    b. The motherboard power connector(s) is inserted correctly.

c. The power and data connectors are correctly inserted and oriented in each internal peripheral.
28. After making the check in step 27, disconnect the anti-static wristband and mat.
29. Make sure the computer's power switch is in the off position.
30. Connect the power cord to the receptacle in the back of the computer.
31. Turn the power switch to the on position. If the cooling fans on the power supply or MPU do not rotate, turn the power off immediately.
32. The computer should only beep once, indicating the completion of the POST. If the computer beeps more than once, turn the computer off and refer to chapter 15.

## 3.6 CONCLUSION

In this chapter I have attempted to explain some of the basic features of the microprocessors, motherboards, and chipsets used in the compatible computer. Even though the information in this chapter is not presented in great depth, it should be plenty for an individual who plans to build, upgrade, and repair compatible computers. Personally, I believe you cannot know too much about anything, especially computers. For this reason, I highly recommend that you read all of the technical information you can get your hands on relative to microprocessors and chipsets.

In order for computers to be compatible, machine code programs that run on one computer must run on all computers that claim to be compatible. For this to be possible, each generation of compatible computer, no matter the manufacturer, must use object code-compatible microprocessors, compatible programmable ports, and everything in the computers has to be at the same address. Each new generation of the PC has attempted to stay downwardly compatible with previous-generation machines. This means that machine code programs written for the 8088 computer that came out in 1981 will run on the compatible computers that are designed and sold today.

Along with technical information, this chapter has tried to give a little history into the development of the microprocessors used in the compatible PC and the different technologies that are currently available. These include Super Socket7 and Slot-1 and Slot-2 motherboards and microprocessors. Hopefully, enough information is given to aid in the decision on which technology to choose.

The motherboard's chipset is another important topic that is covered in this chapter. The chipset is probably one of the most important, yet overlooked, features on the motherboard. It is so important because the chipset controls almost all of the features available on the motherboard, down to the microprocessor it supports. Manufacturers are constantly introducing new chipsets. For this reason, it is impossible to cover the features of the newest chipsets in this textbook. To get the most up-to-date chipset information, visit the manufacturer's Web site.

## Review Questions

1. Write the meaning of each of the following acronyms.
   a. ZIF
   b. MMX
   c. CISC
   d. RISC
   e. SECC
   f. DIB
   g. FSB
   h. SEPP
   i. PR

## Fill in the Blanks

1. Another name for the address, data, and control buses is the _____ bus.

2. The drawing that gives the name of the signal that is provided by each pin of the microprocessor is called the _____ out.

3. _____ compatible is a term that indicates a computer supports the technologies used by previous-generation computers.

4. During the _____ phase, the microprocessor gets an instruction from memory.

5. An instruction moves though the microprocessor over a _____ .

6. A single microprocessor running multiple programs at the same time is called a(n) _____ .

7. Simulated memory on a hard drive is called _____ memory.

8. A _____ is a high-speed memory where the most used data are placed.
9. The Pentium II renamed the local bus the _____ _____ bus.
10. In order to multitask, the microprocessor must be running in the _____ mode.
11. A device that helps dissipate the heat generated by the microprocessor is a _____ _____.
12. To increase the efficiency of a heat sink, a _____ _____ is often used.
13. _____ _____ _____ refers to the ability to run multiple microprocessors.
14. The _____ determines most of the features offered by a motherboard.
15. The _____ form factor corrected most of the shortcomings of the Baby-AT.

## Multiple Choice

1. The Pentium® II microprocessor:
   a. inserts into a socket.
   b. mounts in a slot.
   c. is the most current Intel MPU.
   d. is also made by AMD and Cyrix.
2. During which phase is the microprocessor looking for an instruction?
   a. execute
   b. run
   c. lap
   d. fetch
3. All P54C-compatible microprocessors:
   a. run at the same speed.
   b. operate at the same voltage level.
   c. use speed multiplying.
   d. cannot be installed in Socket7.
4. The Pentium® Pro:
   a. has an internal L2 cache.
   b. is the fastest microprocessor Intel makes.
   c. mounts in an industry-standard Socket7.
   d. does not utilize an L1 cache.
5. Socket6:
   a. is open technology.
   b. is closed technology.
   c. is also used by compatible microprocessors.
   d. was never sold.
6. The local bus:
   a. is a combination of the microprocessor's address, data, and control buses.
   b. is used to select the location of the device the computer needs to communicate with.
   c. runs at a faster speed than the microprocessor.
   d. always runs at the same speed as the microprocessor.
7. Socket3:
   a. is designed to mount a P5-compatible microprocessor.
   b. is designed to mount a 486-compatible microprocessor.
   c. is designed to mount a P54C-microprocessor.
   d. will only mount AMD microprocessors.
8. The 386DX:
   a. does not support multitasking.
   b. was the first to provide the Virtual 86 mode.
   c. is not downwardly compatible with other Intel processors.
   d. has a 16 KB L1 cache.
9. The 286:
   a. can be switched back and forth between the real mode and the protected mode without reseating the microprocessor.
   b. automatically switches to the protected mode after reset.
   c. was the first Intel microprocessor to offer the protected mode.
   d. is only available in a PGA package.
10. The Pentium Pro is classified as a CISC.
    a. True
    b. False
11. A math coprocessor is required in all compatible computers.
    a. True
    b. False
12. The 486DX4 runs the MPU's core speed at:
    a. the same speed as that of the local bus.
    b. twice the speed of the local bus.
    c. three times the speed of the local bus.
    d. four times the speed of the local bus.
13. The 486SX:
    a. has a different address bus size than the 486DX.
    b. has a different data bus size than the 486DX.
    c. has a different control bus size than the 486DX.
    d. does not have an internal math coprocessor.
14. An instruction pipeline:
    a. Is the path an instruction takes through the computer.
    b. Is in the shape of a tube and is the unit in the computer that all instructions have to pass through on their way to the microprocessor.
    c. Is the path an instruction takes through the microprocessor.
    d. allows the microprocessor to execute more than one instruction at a time.
15. A dual pipeline:
    a. gives the microprocessor the ability to operate on two program instructions at the same time.
    b. carries water a lot faster than a single pipeline.
    c. is another name given to the Windows operating system.
    d. is a term used to indicate that there are two data paths on the computer's motherboard.
16. Which Intel microprocessor was the first to use the L1 cache?
    a. 80286
    b. 80386
    c. 80486
    d. Pentium®
    e. Pentium® II
17. First-generation P5 microprocessors are pin-compatible with all other Pentiums®.
    a. True
    b. False
18. Socket7 is capable of running the:
    a. first-generation P5.
    b. P54C.
    c. P55C.
    d. P6.
    e. Both answers b and c are correct.

19. Which Intel processor was the first to use an L2 cache mounted in the same package as the microprocessor?
    a. Pentium® Pro
    b. Pentium® II
    c. Pentium® III
    d. 80486
20. A microprocessor that has multiple instruction pipeline is called a:
    a. fast microprocessor.
    b. multi-pipeline microprocessor.
    c. superscaler microprocessor.
    d. third-generation microprocessor.
21. Which of the following motherboard form factors has one large expansion slot located toward the center of the card in which a smaller PCB called a riser installs?
    a. Mini-AT
    b. ATX
    c. LPX
    d. NLX
22. A 486DX4-120 operates the local bus at a speed of:
    a. 120 MHz.
    b. 60 MHz.
    c. 30 MHz.
    d. 40 MHz
23. A P55C Pentium® compatible microprocessor:
    a. requires a single power supply.
    b. operates at the same speed as that of the local bus.
    c. needs core and bus power supplies to operate.
    d. is not object code-compatible with the other Intel-compatible microprocessors.
24. A P5-200 runs the local bus at a speed of:
    a. 50 MHz.
    b. 60 MHz.
    c. 66 MHz.
    d. 200 MHz.
25. The Pentium® II microprocessor:
    a. is pin-compatible with the P5 microprocessors.
    b. is pin-compatible with the P6 microprocessors.
    c. is not pin-compatible with older-generation Intel-compatible microprocessors.
    d. operates at the same speed as that of the local bus.
26. The term *downwardly compatible* means:
    a. the computer will run programs written for future-generation machines.
    b. the computer will run programs written for other computers within its generation.
    c. the computer will run programs written for previous-generation machines.
    d. the computer will not run programs written for previous-generation machines.
27. The 386's ability to use disk storage to supplement RAM is called.
    a. Virtual memory.
    b. RAM drive.
    c. Disk RAM.
    d. Play RAM.
28. The PR rating used by some Intel-compatible computers:
    a. specifies the core speed of the microprocessor.
    b. specifies the processing speed of the microprocessor.
    c. is not currently used.
    d. is a rating that was adopted by Intel for its microprocessors.
29. In order to multitask, the microprocessor must be:
    a. in the fetch phase.
    b. in the execute phase.
    c. in the real mode.
    d. in the protected mode.
30. The motherboard form factor that uses an edge card connector in which another PCB that contains the expansion bus installs into is the:
    a. Mini-AT.
    b. ATX.
    c. LPX.
    d. NLX.
31. The virtual 86 mode:
    a. allows the microprocessor to act like an 8086 microprocessor.
    b. allows MS-DOS applications to run in the protected mode.
    c. allows an 8086 microprocessor to emulate a 386DX microprocessor.
    d. allows the microprocessor to run at a speed faster than that of the local bus.
32. What is the technical name for multiprocessing?
    a. multitasking
    b. multi-microprocessor
    c. symmetric multiprocessing
    d. protected mode
33. The chipset runs the motherboard.
    a. True
    b. False
34. Cache memory:
    a. is physically part of the computer main memory.
    b. is made from an expensive memory.
    c. keeps often-used data.
    d. is considered an I/O port.
35. Which of the following is not located on the motherboard?
    a. chipset
    b. RAM
    c. ROM
    d. hard drives
36. The motherboard form factor that mounts the microprocessor directly behind the expansion slots is the:
    a. Mini-AT
    b. ATX
    c. NLX
    d. LPX

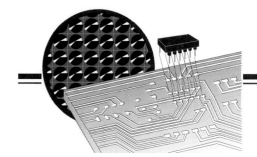

# CHAPTER 4

# Evolution of the PC and the Expansion Bus

## OBJECTIVES

- Define the meaning of terms, expressions, and acronyms used in relation to the evolution of the PC and the different expansion buses.
- List the characteristics of each generation of compatible computer through the Pentium III.
- Name the microprocessors used in each generation of compatible computer.
- Name the expansion buses used in each generation of compatible computer.
- Explain the characteristics of each expansion bus.

## KEY TERMS

IBM personal computer
expansion bus
peripheral
8-bit PC bus
industry-standard architecture
open technology
reverse-engineer
IBM-extended technology (IBM-XT)

turbo
advanced technology (AT)
ISA bus
personal system/2 (PS/2)
micro-channel architecture (MCA)
enhanced industry-standard architecture (EISA)

VL-Bus
peripheral component interface (PCI)
personal computer memory card international association (PCMCIA)
PC-card

The International Business and Machine Corporation (IBM) first introduced the **IBM personal computer** (IBM-PC) on Tuesday, August 11, 1981. At the time of its introduction, there were only a few personal computer manufacturers that IBM had to compete with in the consumer group in which it was planning to market its computer. IBM wanted to develop a computer aimed at small businesses that did not need the processing power and speed offered by very expensive mini- and mainframe computers that were predominant at the time. Most of the other personal computers on the market at the time were aimed at engineers and individual home use and were considered toys by the business community. When a major mainframe computer manufacturer like IBM released its PC, it gave legitimacy to the small personal computer. Little did IBM know that the introduction of this computer would revolutionize the personal computer market.

129

## 4.1 THE IBM-PC

Due to the cost of construction and the large support for the 8-bit data bus structure, the IBM-PC was built around the 5 MHz Intel 16/8-bit 8088 microprocessor. Color monitors were very expensive at the time, so IBM decided to run the PC's MPU at 4.77 MHz to generate the frequencies required to run an optional radio frequency (RF) modulator. The RF modulator allowed a standard color TV set to be used as a display monitor, similar to most of the current dedicated video game computers.

The PC came with five expansion slots for optional adapter controller cards and 64 KB of base RAM expandable to 256 KB on the motherboard. The computer could be upgraded to a total of 640 KB with an additional RAM adapter card that was installed into one of the computer's expansion slots. The maximum RAM memory available on most personal computers at the time was around 48 KB, so 640 KB of RAM seemed like an unlimited source for software writers to develop application programs for the IBM-PC.

### 4.1.1 The PC Bus

From the start, the IBM-PC was unique among personal computers because it was designed to be very easy to upgrade. This was because all of the input and output (I/O) ports were not designed into the motherboard, like most personal computers at the time, but an **expansion bus** was developed to allow the use of optional **peripheral** adapter cards (ports). As previously stated, a *peripheral* is any device that communicates with the computer through an input or output port. Devices such as monitors, keyboards, disk drives, joysticks, mice, and modems are examples of peripherals.

The original expansion bus consists of an industry-standard 62-pin female edge card connector, as shown in Figure 4.1. IBM's engineers assigned the pins of this connector for 8 data lines, 20 address lines, 24 control lines, and 8 power supply lines. Of the 62 pins on the connector, only one was left unused. This expansion bus became known as the **8-bit PC bus, or 8-bit industry standard architecture** (8-bit ISA). The pin assignments for the 8-bit PC bus are shown in Table 4.1.

Notice in Table 4.1 that the connector has two columns of pins with 31 pins on each side. One side is called the *A* side and has pins numbered A1 through A31. The other side is called the *B* side and has pins numbered B1 through B31. This makes a total of 62 pins.

This gets confusing sometimes because the 20 address lines use the names A0 through A19. However, they are located on the right-hand side of the connector from connector pins A31 through A12. The pins named D0 through D7 constitute the 8-bit data bus. The data bus is also located on the right side of the connector from pin numbers A9 through A2. The power supply pins are labeled +5 (B3 and B29), −5 (B5), +12 (B9), −12 (B7), and GND (B1, B10, and B31). Of the 62 pins, all are assigned to signals except B8, which is not electrically connected (N/C). All of the other pins belong to the control bus.

Explaining the function of each signal on the control bus is beyond the scope of this book. However, some of the control signals that will be discussed in chapter 8 are the IRQ signals (B4 and B25 through B21) and DMA signals (B6, B15, B16, B17, B18, and B26).

An adapter card that is designed to plug into the PC bus is called an *8-bit card* because there are only 8 data bits available on the 8-bit ISA. Figure 4.2 shows an illustration of a typical 8-bit card.

IBM made some interesting decisions when it began to design the IBM-PC. At the time, those decisions made the company tons of money. However, in the long run, these decisions proved to be IBM's demise in the PC market. For more insight, let us continue our discussion.

**Figure 4.1** PC-bus

**Table 4.1** PC bus or 8-bit ISA

| Signal | A | B | Signal | Signal | A | B | Signal |
|---|---|---|---|---|---|---|---|
| Ground | 1 | 1 | I/O Check | DACK1 | 17 | 17 | A14 |
| Reset | 2 | 2 | D7 | DRQ1 | 18 | 18 | A13 |
| +5 V | 3 | 3 | D6 | REFRESH | 19 | 19 | A12 |
| IRQ2 | 4 | 4 | D5 | $CLK_{4.77 MHz}$ | 20 | 20 | A11 |
| −5 V | 5 | 5 | D4 | IRQ7 | 21 | 21 | A10 |
| DRQ2 | 6 | 6 | D3 | IRQ6 | 22 | 22 | A9 |
| −12 V | 7 | 7 | D2 | IRQ5 | 23 | 23 | A8 |
| N/C | 8 | 8 | D1 | IRQ4 | 24 | 24 | A7 |
| +12 V | 9 | 9 | D0 | IRQ3 | 25 | 25 | A6 |
| Ground | 10 | 10 | I/O CH Ready | DACK2 | 26 | 26 | A5 |
| MEMW | 11 | 11 | AEN | T/C | 27 | 27 | A4 |
| MEMR | 12 | 12 | A19 | ALE | 28 | 28 | A3 |
| IOW | 13 | 13 | A18 | +5 V | 29 | 29 | A2 |
| IOR | 14 | 14 | A17 | $OSC_{14.3MHz}$ | 30 | 30 | A1 |
| DACK3 | 15 | 15 | A16 | Ground | 31 | 31 | A0 |
| DRQ3 | 16 | 16 | A15 | | | | |

**Figure 4.2** Typical 8-bit PC-bus Card

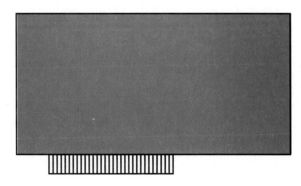

## 4.2 IBM-COMPATIBLE COMPUTERS

When IBM decided to enter the PC market, it wanted to do so in the shortest time possible. In fact, the IBM-PC went from idea to delivery in only 1 year. This short time period meant there was no time for IBM to develop unique ICs, peripheral ports, and operating systems for the IBM-PC. Instead, IBM used off-the-shelf ICs and peripheral port standards that were already developed and available.

IBM even chose a small company (at the time) called *Microsoft* to provide the disk operating system (DOS). Another interesting wonder is that IBM chose to make the 8-bit PC bus pin assignments open technology. The term **open technology** means that companies can use the technology developed by another company without paying royalties. All of the IBM-PC's hardware was off-the-shelf technology and the 8-bit PC bus was open technology, so the hardware was very easy for any manufacturer to **reverse-engineer** and copy legally.

Microsoft sold exclusive rights to IBM for an operating system that IBM named *PC-DOS* for approximately 80,000 dollars. Although this seems like a small sum of money, Microsoft desperately wanted IBM's business for its application programs. One of the smartest decisions made by Microsoft was to retain the rights to sell the operating system to other computer manufacturers. Microsoft was gambling that other manufacturers would copy the IBM-PC and that these companies would need an operating system for their computers.

With all of the components in the computer available to anyone for purchase and a readily available operating system, there was only one major problem that a manufacturer had to overcome to produce a copy of the IBM-PC. IBM had a copyright on the basic input/output system (BIOS) ROM programs that made all of the hardware in the computer work. This stranglehold on the PC was short-lived, however.

Compaq Computers Corporation took on the task of reverse-engineering the ROMs in the IBM-PC. It analyzed the IBM-BIOS programs to figure out the task performed by each module. After the tasks were determined, new programs to perform the same tasks were written differently. Because these programs were different, copyright infringements were avoided, even though they performed the same task. After the firmware problem was solved, fully IBM-compatible computers started to appear on the market.

At first IBM did not worry about the compatible computer manufacturers because it could not keep up with consumer demand for the computer anyway. IBM and Microsoft were the only major software writers for the computer, so they made money on the sale of software no matter who made the computer. Also, IBM was the only company making adapter cards for the PC. So, no matter who made the computer, IBM still made money from software, hardware, and any subsequent upgrade. Besides, by the time the other companies had time to reverse-engineer the IBM-PC, IBM released its next generation of the personal computer—the IBM-XT.

## 4.3　IBM-XT

The **IBM-extended technology** (IBM-XT) computer was introduced in March of 1983. By this time, new versions of the 8088 microprocessor had almost doubled in speed and the price of color monitors had dropped considerably. This meant that there was no longer a need to run the computer at a speed compatible with the frequencies required by a color TV set. IBM first ran the XT computer at 4.77 MHz, but later raised the speed to 8 MHz.

IBM kept the same PC bus as that of the IBM-PC computers. Unfortunately for IBM, because the bus structure and hardware were almost exactly the same as those of the PC, compatibles appeared on the market a short time later. However, coming out with a computer after someone else does is not always a disadvantage.

Because of problems that arose with running older software at higher speeds, the compatible manufacturers gave their computers the ability to run at the old 4.77 MHz and a second, much higher, speed. The second, faster speed (8, 10, or 12 MHz) became known as the **turbo** mode. This feature allowed the computer to be switched to the slower bus speed to run software that ran too fast on the higher bus speed.

The basic IBM-XT came with 256 KB of user RAM that could be upgraded to 512 KB on the motherboard. Most compatible computers allow 640 KB on the motherboard. With the XT, the number of expansion slots was increased from six to eight. To do this, the spacing between the connectors was narrowed. This spacing is still in use today.

Another feature added by the XT that is still in use is an indication that the power-on self-test (POST) is in progress by counting the memory off in the upper left-hand corner of the display as it is being tested. Also, the XT was the first computer that added BIOS support for hard disk drives and came with a hard drive installed.

## 4.4 IBM-AT

Next, IBM came out with the **advanced technology** (AT) computer in August of 1984. The IBM-AT is built around the 16-bit Intel 80286 microprocessor (286). The 286 MPU has a 16-bit data bus and a 24-bit address bus. There is only one spare pin on the 8-bit ISA bus, so there is no way it could be modified to handle the additional lines needed for 286's data, address, and control buses. To support the new microprocessor, IBM developed a new expansion bus for this computer that is currently known as *ISA*.

### 4.4.1 The ISA Bus

When IBM came out with the IBM-AT, it did not want to make all of the 8-bit expansion cards it was selling on the market obsolete. This meant that IBM had to develop an expansion slot with 16-bit data lines, plus additional address and control lines, that was totally compatible with the 8-bit ISA expansion bus. To keep this compatibility, the pin assignments on the 62-pin edge card connector were kept the same. An additional 36-pin connector was added in front of the 62-pin connector. IBM assigned the pins on the short connector to the additional data, address, and control lines needed by the AT computer. Figure 4.3 shows a single expansion slot of the AT that includes both the upper and lower connectors. The combination of these two connectors is known as the *16-bit ISA* bus, or simply, the **ISA bus.**

Figure 4.4 shows a typical ISA adapter card, which is usually referred to as a *16-bit adapter*. Notice the *key* (notch) that allows the card to be inserted into both the long connector originally used by the PC bus and the additional 36-pin connector added to form the ISA bus.

What's nice about the ISA is that an 8-bit card can still be plugged into the back 62-pin connector. With this type of architecture, upgrading from a PC or XT to an AT did not make all of the 8-bit peripheral cards obsolete. The pin-out for the ISA expansion bus is listed in Table 4.2.

One of the problems with the PC bus was that there was really no set standard for the data transfer speed even though it is often rated at around 2 megabytes per second (MBps). This means that all

**Figure 4.3** ISA Expansion Bus

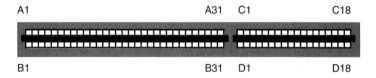

**Figure 4.4** Typical ISA Adapter

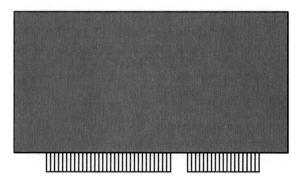

**Table 4.2** ISA Pin-out (16-bit)

| Signal | B | A | Signal | Signal | B | A | Signal | |
|---|---|---|---|---|---|---|---|---|
| Ground | 1 | 1 | I/O Check | DACK2 | 26 | 26 | A5 | |
| Reset | 2 | 2 | D7 | T/C | 27 | 27 | A4 | |
| +5 V | 3 | 3 | D6 | ALE | 28 | 28 | A3 | |
| IRQ2 | 4 | 4 | D5 | +5 V | 29 | 29 | A2 | |
| −5 V | 5 | 5 | D4 | $OSC_{14.4MHz}$ | 30 | 30 | A1 | |
| DRQ2 | 6 | 6 | D3 | Ground | 31 | 31 | A0 | |
| −12 V | 7 | 7 | D2 | Signal | D | C | Signal | Second connector |
| N/C | 8 | 8 | D1 | MEM CS16 | 1 | 1 | SBHE | |
| +12 V | 9 | 9 | D0 | I/O CS16 | 2 | 2 | A23 | |
| Ground | 10 | 10 | I/O CH Ready | IRQ10 | 3 | 3 | A22 | |
| MEMW | 11 | 11 | AEN | IRQ11 | 4 | 4 | A21 | |
| MEMR | 12 | 12 | A19 | IRQ12 | 5 | 5 | A20 | |
| IOW | 13 | 13 | A18 | IRQ15 | 6 | 6 | A19 | |
| IOR | 14 | 14 | A17 | IRQ14 | 7 | 7 | A18 | |
| DACK3 | 15 | 15 | A16 | DACK0 | 8 | 8 | A17 | |
| DRQ3 | 16 | 16 | A15 | DRQ0 | 9 | 9 | MEMR | |
| DACK1 | 17 | 17 | A14 | DACK5 | 10 | 10 | MEMW | |
| DRQ1 | 18 | 18 | A13 | DRQ5 | 11 | 11 | D8 | |
| REFRESH | 19 | 19 | A12 | DACK6 | 12 | 12 | D9 | |
| $CLK_{8.33MHz}$ | 20 | 20 | A11 | DRQ6 | 13 | 13 | D10 | |
| IRQ7 | 21 | 21 | A10 | DACK7 | 14 | 14 | D11 | |
| IRQ6 | 22 | 22 | A9 | DRQ7 | 15 | 15 | D12 | |
| IRQ5 | 23 | 23 | A8 | +5 V | 16 | 16 | D13 | |
| IRQ4 | 24 | 24 | A7 | Master | 17 | 17 | D14 | |
| IRQ3 | 25 | 25 | A6 | Ground | 18 | 18 | D15 | |

8-bit adapters cannot be guaranteed to operate in all 16-bit ISA or 8-bit PC bus expansion slots. To give engineers the ability to design reliable adapter cards around the ISA bus, IBM decided to set the speed of the ISA at 8.33 MHz. This speed allowed reliable cards to be developed and set the data transfer rate at approximately 8.33 MBps. This speed is still the speed of the ISA slots installed on today's computers.

Along with the speed, IBM also started another trend with the PC bus that moved into ISA that is still a major design factor in today's computer. This is the maximum number of allowable I/O addresses available on an expansion bus.

### 4.4.2 ISA Addressing Range

As discussed previously in chapter 2, Intel and the compatible microprocessor manufacturers gave their microprocessors the ability to select up to 65,536 I/O devices, regardless of the size of the address bus. Even though this range is available on the ISA bus, IBM decided to limit the number of I/O addresses used by its adapter cards even further, to 1,024 addresses. Limiting the number of the I/O address range reduces the cost of adding the electronic circuitry needed to select the additional addresses.

For this reason, the address range for the ISA bus is limited from address $000_{16}$ through $3FF_{16}$. IBM thought that 1,024 addresses should be a sufficient number to not only meet the demands of the AT computer, but also the needs of any future versions.

## 4.5 IBM-PS/2 AND MICRO-CHANNEL

By the time the IBM-AT computer came out, the compatible manufacturers were very proficient in producing copies. This allowed them to introduce their version of the AT shortly after IBM started selling it. The compatible manufacturers had the advantage of seeing what IBM did wrong with its version of the computer, so most of the compatibles came out better, cheaper, and more powerful than the originals sold by IBM. The PC-style computer had been out for some time, so there were many software houses writing programs for the computer and manufacturers making adapter cards. With better computers available and more software and hardware, IBM started losing a major portion of the PC market. These facts forced IBM to make major changes in the path it envisioned for the personal computer.

When IBM introduced its **Personal System/2** (PS/2) computer in 1987, it chose to completely change the expansion bus connector and also reassign the pins. The PS/2 expansion bus was designed to support up to 32 data lines running at 8.33 MHz. This speed and this number of data bits allowed the PS/2 to transfer information at speeds of approximately 32 MBps, which is four times the speed of ISA. IBM received a copyright on this new expansion interface and called it the **Micro-Channel Architecture (MCA).**

IBM's Micro-Channel is actually available in three versions: a 32-bit and 16-bit version, plus a slot specifically designed for a high-speed video adapter. IBM chose this arrangement because all I/O devices do not need to transfer data in 32-bit groups. When designing the connector for Micro-Channel, IBM realized that the contacts used in the ISA connector were a lot bigger than they had to be to carry address, data, and control information. Using smaller pins allows the contacts to be spaced closer, which makes the connector smaller. In fact, the physical size of the 32-bit MCA connector is about the same as that of the 16-bit ISA. A 16-bit MCA adapter card will work in a 32-bit slot, but a 32-bit MCA card will not work in a 16-bit MCA slot. A 32 bit and a 16-bit MCA expansion slot are shown in Figure 4.5.

IBM has a copyright on the MCA connector and pin assignments, so it decided to make them closed technology. Closed technology means the production of a PS/2-compatible computer or MCA adapter cards is possible only if royalties are paid to Big Blue (IBM).

IBM changed the connector that adapter cards are designed around, so an adapter designed around the ISA expansion bus no longer fits into an MCA expansion bus slot, or vice versa. This can be seen in Figure 4.6, which shows both a 16-bit ISA and 16-bit MCA adapter for comparison purposes. The pin assignments for both the 16-bit and 32-bit MCA connectors are listed in Table 4.3.

**Figure 4.5**  16-bit and 32-bit Micro-Channel

**Figure 4.6**  16-bit ISA and 16-bit Micro-Channel

Table 4.3  Pin-out for the 16- and 32-bit MCA Expansion Bus*

| Signal | B | A | Signal | Signal | B | A | Signal |
|---|---|---|---|---|---|---|---|
| Audio Gnd | 1 | 1 | CD Setup | Check | 32 | 32 | Status 0 |
| Audio | 2 | 2 | Made 24 | Ground | 33 | 33 | Status 1 |
| Ground | 3 | 3 | Ground | CMD | 34 | 34 | M/IO |
| $OSC_{14.3MHz}$ | 4 | 4 | A11 | CHRDYRTN | 35 | 35 | +12 V |
| Ground | 5 | 5 | A10 | CD SFDBK | 36 | 36 | CD CHRDY |
| A23 | 6 | 6 | A9 | Ground | 37 | 37 | D0 |
| A22 | 7 | 7 | +5 V | D1 | 38 | 38 | D2 |
| A21 | 8 | 8 | A8 | D3 | 39 | 39 | +5 V |
| Ground | 9 | 9 | A7 | D4 | 40 | 40 | D5 |
| A20 | 10 | 10 | A6 | Ground | 41 | 41 | D6 |
| A19 | 11 | 11 | +5 V | CHRESET | 42 | 42 | D7 |
| A18 | 12 | 12 | A5 | Reserved | 43 | 43 | Ground |
| Ground | 13 | 13 | A4 | Reserved | 44 | 44 | DS 16 RTN |
| A17 | 14 | 14 | A3 | Ground | 45 | 45 | Refresh |
| A16 | 15 | 15 | +5 V | Key | 46 | 46 | Key |
| A15 | 16 | 16 | A2 | Key | 47 | 47 | Key |
| Ground | 17 | 17 | A1 | D8 | 48 | 48 | +5 V |
| A14 | 18 | 18 | A0 | D9 | 49 | 49 | D10 |
| A13 | 19 | 19 | +12 V | Ground | 50 | 50 | D11 |
| A12 | 20 | 20 | ADL | D12 | 51 | 51 | D13 |
| Ground | 21 | 21 | Preempt | D14 | 52 | 52 | +12 |
| IRQ9 | 22 | 22 | Burst | D15 | 53 | 53 | Reserved |
| IRQ3 | 23 | 23 | −12 V | Ground | 54 | 54 | SBHE |
| IRQ4 | 24 | 24 | ARB0 | IRQ10 | 55 | 55 | CD DS 16 |
| Ground | 25 | 25 | ARB1 | IRQ11 | 56 | 56 | +5 V |
| IRQ5 | 26 | 26 | ARB2 | IRQ12 | 57 | 57 | IRQ14 |
| IRQ6 | 27 | 27 | −12 V | Ground | 58 | 58 | IRQ15 |
| IRQ7 | 28 | 28 | ARB3 | Reserved | 59 | 59 | Reserved |
| Ground | 29 | 29 | ARB/Grant | Reserved | 60 | 60 | Reserved    16-bit MCA ends |
| Reserved | 30 | 30 | Terminal Count | Reserved | 61 | 61 | Ground |
| Reserved | 31 | 31 | +5 V | Reserved | 62 | 62 | Reserved |

(Continued)

**Table 4.3** (Continued)

| Signal | B | A | Signal | Signal | B | A | Signal |
|---|---|---|---|---|---|---|---|
| Ground | 63 | 63 | Reserved | BE1 | 77 | 77 | +12 V |
| D16 | 64 | 64 | Reserved | BE2 | 78 | 78 | BE3 |
| D17 | 65 | 65 | +12 V | Ground | 79 | 79 | DS 32 RTN |
| D18 | 66 | 66 | D19 | TR 32 | 80 | 80 | CD CS 32 |
| Ground | 67 | 67 | D20 | A24 | 81 | 81 | +5 V |
| D22 | 68 | 68 | D21 | A25 | 82 | 82 | A26 |
| D23 | 69 | 69 | +5 V | Ground | 83 | 83 | A26 |
| Reserved | 70 | 70 | D24 | A29 | 84 | 84 | A28 |
| Ground | 71 | 71 | D25 | A30 | 85 | 85 | +5 V |
| D27 | 72 | 72 | D26 | A31 | 86 | 86 | Reserved |
| D28 | 73 | 73 | +5 V | Ground | 87 | 87 | Reserved |
| D29 | 74 | 74 | D30 | Reserved | 88 | 88 | Reserved |
| Ground | 75 | 75 | D31 | Reserved | 89 | 89 | Ground |
| BE0 | 76 | 76 | Reserved | | | | |

*The 16-bit connector ends at A60 and B60.

## 4.6 THE SPLIT

IBM made a major miscalculation when it assumed that all compatible manufacturers would pay royalties to produce a PS/2-compatible computer. IBM thought it was the leader and that all of the compatible manufacturers would just follow. Of course, up to the PS/2, IBM was right about that.

Upgrading from a PC/XT/AT to the PS/2 had several drawbacks. First, all new expansion cards would have to be purchased because the MCA and ISA pin assignments and spacing are not compatible. Second, MCA expansion cards cost more than ISA because there are fewer manufacturers. Third, the selection of optional expansion cards for the 8/16-bit ISA seems unlimited, whereas cards for the MCA are harder to find.

Because there is so much hardware and software support available for the ISA bus, the compatible manufacturers chose to stay with the 16-bit ISA. Even though they did go with the Intel 80386 MPU, with its 32-bit address and data buses, the compatibles stayed with the 16-bit ISA. This forced the compatible motherboards to support only 32-bit communication between the microprocessor and the computer's RAM. It was obvious, however, that to remain competitive, compatible manufacturers had to develop their own 32-bit expansion bus.

## 4.7 EISA

A consortium of compatible manufacturers was formed to take on the task of developing a new 32-bit expansion bus, independent of IBM's MCA. This consortium chose to make its 32-bit expansion bus totally compatible with the 16-bit ISA bus. To do this, a two-tier (two-story) connector was developed with the top tier wired as the 16-bit ISA. The bottom tier added the additional pins for the 32-data bus structure. This expansion bus is known as the **enhanced industry-standard architecture** (EISA, pronounced "ee-sa"). Figure 4.7 shows a motherboard that uses EISA connectors.

**Figure 4.7** EISA Expansion Bus

**Figure 4.8** EISA and ISA Adapter Cards

Plastic keys in the EISA connector prevent an ISA card from descending far enough into the slot to access the EISA signals in the bottom tier. Keys (notches) cut in the EISA card allow it to slide smoothly into the bottom of the EISA connector to pick up the signals on both tiers. For comparison purposes, both an ISA card and an EISA adapter card are shown in Figure 4.8. Notice that the edge card contacts on the EISA card are much thinner than those on the ISA. This allows the EISA signal to be interleaved with those that belong to the ISA. This arrangement almost doubles the number of pins on the EISA bus when compared to the ISA bus.

**Table 4.4** Pin-out of the EISA Expansion Bus Connector*

| Signal | F | E | Signal | Signal | F | E | Signal |
|---|---|---|---|---|---|---|---|
| Ground | 1 | 1 | CMD | Reserved | 14 | 14 | Reserved |
| +5 V | 2 | 2 | Start | BE3 | 15 | 15 | Ground |
| +5 V | 3 | 3 | EXRDY | Key | 16 | 16 | Key |
| Reserved | 4 | 4 | EX32 | BE2 | 17 | 17 | BE1 |
| Reserved | 5 | 5 | Ground | BE0 | 18 | 18 | A31 |
| Key | 6 | 6 | Key | Ground | 19 | 19 | Ground |
| Reserved | 7 | 7 | EX16 | +5 V | 20 | 20 | A30 |
| Reserve | 8 | 8 | SLBURST | A29 | 21 | 21 | A28 |
| +12 V | 9 | 9 | MSBURST | Ground | 22 | 22 | A27 |
| M/IO | 10 | 10 | W/R | A26 | 23 | 23 | A25 |
| LOCK | 11 | 11 | Ground | A24 | 24 | 24 | Ground |
| Reserved | 12 | 12 | Reserved | Key | 25 | 25 | Key |
| Ground | 13 | 13 | Reserved | A16 | 26 | 26 | A15 |
| A14 | 27 | 27 | A13 | D18 | 8 | 8 | D19 |
| +5 V | 28 | 28 | A12 | Ground | 9 | 9 | D20 |
| +5 V | 29 | 29 | A11 | D21 | 10 | 10 | D22 |
| Ground | 30 | 30 | Ground | D23 | 11 | 11 | Ground |
| A10 | 31 | 31 | A9 | D24 | 12 | 12 | D25 |
| Signal | H | G | Signal  Second connector | Ground | 13 | 13 | D26 |
| A8 | 1 | 1 | A7 | D27 | 14 | 14 | D28 |
| A6 | 2 | 2 | Ground | Key | 15 | 15 | Key |
| A5 | 3 | 3 | A4 | D29 | 16 | 16 | Ground |
| +5 V | 4 | 4 | A3 | +5 V | 17 | 17 | D30 |
| A2 | 5 | 5 | Ground | +5 V | 18 | 18 | D31 |
| Key | 6 | 6 | Key | MAKx | 19 | 19 | MREQx |
| D16 | 7 | 7 | D17 | | | | |

*The upper pins are the same as those of the 16-bit ISA, so only the additional 32-bit lower pin assignments are given in this table.

As with the IBM-PC/XT/AT computers, this consortium of compatible manufacturers could look at all of the shortcomings of IBM's new MCA, and also the old ISA bus, and improve on both designs. Some of the features of EISA are:

- No copyright royalties have to be paid to use EISA.
- MCA has a maximum throughput of 20 MBps, while the EISA bus can support a maximum burst transfer rate of 33 MBps.

**Figure 4.9** VL-bus Motherboard Block Diagram

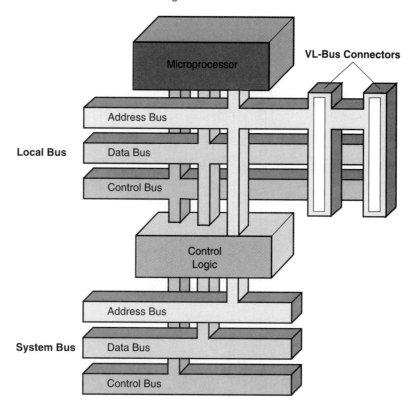

- EISA users can install cards designed for ISA machines while waiting for fast EISA cards to become available.

You are probably wondering, "Why hasn't EISA caught on in the PC community?" After all, it seems to have everything going for it. Of course, what everything usually boils down to is how much something costs and, as you can probably guess, EISA is very expensive. The high price is due mainly to the cost of EISA's two-tier connectors and also the cost of the ICs required to make EISA work.

To keep EISA compatible with ISA, the maximum rated speed for EISA is 8.33 MHz. However, because EISA can transfer 4 bytes of information at this speed, it has a transfer speed of over 33 MBps (4 × 8.33 MHz). The pin assignments for the EISA connector are listed in Table 4.4.

## 4.8 VL BUS

As more and more applications used the video adapter's graphics mode to be attractive visually, the video circuitry became, and still is, one of the major bottlenecks in the compatible computer. To increase the speed at which the application can update graphics video information, a cheap 32-bit interface had to be developed. The **Video Electronics Standards Association** (VESA) developed a local bus expansion slot that is used on the late-production 386DX and most all 486 machines.

As discussed previously in chapter 2, the local bus connects directly to the microprocessor address, data, and control lines without going through the chipset that controls the ISA or EISA expansion buses. Figure 4.9 shows a basic block diagram of a computer that supports the VL bus. The solid black line represents the address, data, and control buses.

VESA local bus (**VL bus**) is a 32-bit I/O interface that dominates 486-compatible machines. To keep the VL bus expansion slot totally compatible with the PC bus and ISA slots, VESA chose to add a third connector in front of the 16-bit ISA. Figure 4.10 shows a motherboard with a VL bus connector. Figure 4.11 shows a typical 32-bit VL bus adapter card. Notice the three keys in the VL bus card that allow it to insert into all three connectors.

**Figure 4.10** VL-bus Motherboard with 2 VL-bus Slots

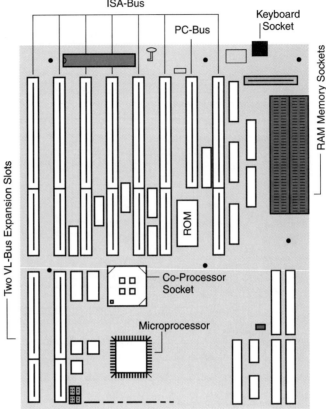

**Figure 4.11** Typical VL-bus Expansion Card

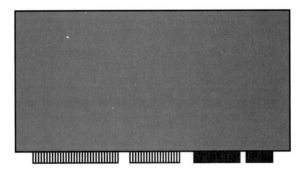

The trend is for the 486 motherboard to use the ISA expansion slots to run the slow-speed floppy port, serial port(s), printer port(s), and game port(s). VL bus adapter cards are usually used for the video adapter and hard drive adapters.

The maximum rated speed for the VL bus was originally set at 50 MHz. However, this high bus speed causes problems because of the loading effect on the microprocessor's local bus. For this reason, the accepted speed limit of the VL bus is 33 MHz. If the VL bus is used on a motherboard with a local bus speed above 33 MHz, buffers are required and wait states are usually used so the microprocessor will not run off and leave a VL bus adapter card.

At 33 MHz, a transfer rate of up to 132 MBps can be theoretically achieved (4 bytes $\times$ 33 MHz). The pin assignments for the VL bus connector are listed in Table 4.5.

**Table 4.5** Pin-out for the VL-bus Expansion Slot

| Signal | B | A | Signal | Signal | B | A | Signal |
|---|---|---|---|---|---|---|---|
| D0 | 1 | 1 | D1 | A15/D47 | 31 | 31 | A14/D46 |
| D2 | 2 | 2 | D3 | +5 V | 32 | 32 | A12/D44 |
| D4 | 3 | 3 | Ground | A13/D45 | 33 | 33 | A10/D42 |
| D6 | 4 | 4 | D5 | A11/D43 | 34 | 34 | A8/D40 |
| D8 | 5 | 5 | D7 | A9/D41 | 35 | 35 | Ground |
| Ground | 6 | 6 | D9 | A7/D39 | 36 | 36 | A6/D38 |
| D10 | 7 | 7 | D11 | A5/D37 | 37 | 37 | A4/D36 |
| D12 | 8 | 8 | D13 | Ground | 38 | 38 | Write Back |
| +5 V | 9 | 9 | D15 | A3/D35 | 39 | 39 | BE0/4 |
| D14 | 10 | 10 | Ground | A2/D34 | 40 | 40 | +5 V |
| D16 | 11 | 11 | D17 | Unused/LBS64 | 41 | 41 | BE1/5 |
| D18 | 12 | 12 | +5 V | Reset | 42 | 42 | BE2/6 |
| D20 | 13 | 13 | D19 | Data/Code Status | 43 | 43 | Ground |
| Ground | 14 | 14 | D21 | Mem I/O Status/D33 | 44 | 44 | BE3/7 |
| D22 | 15 | 15 | D23 | W/R Status/D32 | 45 | 45 | Address Data Strobe |
| D24 | 16 | 16 | D25 | Key | 46 | 46 | Key |
| D26 | 17 | 17 | Ground | Key | 47 | 47 | Key |
| D28 | 18 | 18 | D27 | Ready Return | 48 | 48 | Local Ready |
| D30 | 19 | 19 | D29 | Ground | 49 | 49 | Local Device |
| +5 V | 20 | 20 | D31 | IRQ9 | 50 | 50 | Local Request |
| A31/D63 | 21 | 21 | A30/D62 | Burst Ready | 51 | 51 | Ground |
| Ground | 22 | 22 | A28/D60 | Burst Late | 52 | 52 | Local Bus Grant |
| A29/D61 | 23 | 23 | A26/D58 | ID0 | 53 | 53 | +5 V |
| A27/D59 | 24 | 24 | Ground | ID1 | 54 | 54 | ID2 |
| A25/D57 | 25 | 25 | A24/D56 | Ground | 55 | 55 | ID3 |
| A23/D55 | 26 | 26 | A22/D54 | Local Clock | 56 | 56 | ID4/ACK64 |
| A21/D53 | 27 | 27 | +5 V | +5 V | 57 | 57 | Unused |
| A19/D51 | 28 | 28 | A20/D52 | Local Bus Size 16 | 58 | 58 | Local/External Address Data Strobe |
| Ground | 29 | 29 | A18/D50 | | | | |
| A17/D49 | 30 | 30 | A16/D48 | | | | |

**Figure 4.12** PCI Expansion Card

The VL bus does speed up I/O transfer on the 486 DX/SX machines by taking some of the I/O devices off of the 8-MHz 16-bit ISA bus and connecting them directly to the microprocessor's 32-bit data bus. As stated, this allows I/O devices to run at system clock speeds of up to 33 MHz. The major disadvantage of the VL bus is that it operates synchronously with the microprocessor. Because most peripherals run considerably slower than the 486DX/SX microprocessors, wait states are often inserted into the computer's bus cycle while the microprocessor waits for slow expansion cards to finish their data transfer. The **peripheral component interface (PCI)** is designed to alleviate this major VL bus shortcoming.

## 4.9 PERIPHERAL COMPONENT INTERFACE (PCI)

Currently, the most widely used version of the PCI expansion bus gives I/O ports full 32-bit data transfer capabilities. A 64-bit version is also available, but I have never seen it used. A 32-bit PCI expansion bus connector is shown in Figure 4.12.

PCI allows peripherals on the PCI bus to operate asynchronously from the microprocessor. *Asynchronous* means that the PCI bus can switch to a slower speed independent of the CPU's speed to avoid peripheral wait states. This dual-speed technique is referred to as a *mezzanine*. The CPU can send out information to an adapter on the PCI bus through the bus control logic and then access the system memory without waiting for the peripherals to respond. This feature is available because memory (called *buffers*) in the PCI control logic holds the data going to the expansion slot until the port can respond.

Even though PCI adapter cards do not run any faster than VL bus cards, they will not tie up the computer because the microprocessor will not be waiting for slow I/O devices during data transfer. This one fact greatly enhances the overall speed of the computer because it is not constantly waiting on slow I/O devices. A typical PCI adapter card is shown in Figure 4.13.

The PCI bus originally operated at half the speed of the computer's local bus. At a 66-MHz local bus speed, the PCI bus runs at 33 MHz (66/2). At a 60-MHz local bus speed, the PCI bus runs at 30 MHz.

**Figure 4.13** PCI Adapter Card

At a 50-MHz local bus speed, the PCI bus runs at 25 MHz. Transferring 32 bits (4 bytes) while running at 33 MHz translates to a maximum speed of 132 MBps.

Current versions of the Pentium II, III, and P55C-compatible microprocessors run at local bus speeds faster that 66 MHz. Motherboards are currently being produced with local bus speeds of 75, 83, 100, 133, and 200 MHz. These local bus speeds will create PCI bus speeds of 37.5 MHz, 41.5 MHz, 50 MHz, 66.5 MHz, and 100 MHz, respectively. At speeds above 33 MHz, PCI adapter cards can become unstable or cease to operate entirely. For this reason, motherboards that support local bus speeds above 66 MHz limit the PCI bus to 33 MHz.

Besides practically eliminating I/O wait states, two more major advantages added by PCI expansion bus standard are its Plug-and-Play (PnP) and bus-mastering capabilities. PnP allows a PnP operating system such as Windows 95/98/Millennium/2000 to assign the resources used by a PnP adapter so they will not conflict with adapters that were installed previously in the computer. This concept is covered in greater detail in chapter 9. Bus-mastering allows the peripheral to use direct memory access (DMA). DMA is a technique that allows the peripheral to transfer information directly with the computer's memory without going through the microprocessor. DMA is covered in greater detail in chapters 9 and 15.

Note that the pin spacing and pin assignments of the PCI expansion bus are not compatible with ISA. This means an ISA adapter will not fit into a PCI slot, or vice versa. This is no big problem because PCI is an open technology and all kinds of PCI adapters are available. All Pentium- and Pentium II-compatible motherboards have at least three PCI expansion slots available and are now the preferred interface for video adapter cards, hard drive ports, and high-speed network adapters. The pin assignments for each PCI bus connector are listed in Table 4.6.

## 4.9.1 Addressing Range of PCI

By the time PCI was introduced, the 1,024 I/O address limit of ISA was starting to get cramped. Realizing that something had to be done, Intel decided to give PCI the ability to support the full I/O addressing range of its microprocessors. This means that PCI can support 65,536 I/O addresses. This makes the address range of PCI from $0000_{16}$ through $FFFF_{16}$.

Table 4.6  PCI Pin-outs*

| Signal | B | A | Signal | Signal | B | A | Signal |
|---|---|---|---|---|---|---|---|
| −12 V | 1 | 1 | Test Reset | Address/Data 17 | 32 | 32 | Address/Data 16 |
| Test Clock | 2 | 2 | +12 V | C/BE 2 | 33 | 33 | +3.3 V |
| Ground | 3 | 3 | Test Mode Select | Ground | 34 | 34 | Cycle Frame |
| Test Data Output | 4 | 4 | Test Data Input | Initiator Ready | 35 | 35 | Ground |
| +5 V | 5 | 5 | +5 V | +3.3 V | 36 | 36 | Target Ready |
| +5 V | 6 | 6 | Interrupt A | Device Select | 37 | 37 | Ground |
| Interrupt B | 7 | 7 | Interrupt C | Ground | 38 | 38 | Stop |
| Interrupt D | 8 | 8 | +5 V | Lock | 39 | 39 | +3.3 V |
| PRSNT1 | 9 | 9 | Reserved | Parity Error | 40 | 40 | Snoop Done |
| Reserved | 10 | 10 | +5 V I/O (+3.3 V) | +3.3 V | 41 | 41 | Snoop Backoff |
| PRSNT2 | 11 | 11 | Reserved | System Error | 42 | 42 | Ground |
| Ground (Access key) | 12 | 12 | Ground (Access key) | +3.3 v | 43 | 43 | PAR |
| Ground (Access key) | 13 | 13 | Ground (Access key) | C/BE 1 | 44 | 44 | Address/Data 15 |
| Reserved | 14 | 14 | Reserved | Address/Data 14 | 45 | 45 | +3.3 V |
| Ground | 15 | 15 | Reset | Ground | 46 | 46 | Address/Data 13 |
| Clock | 16 | 16 | +5 V I/O (+3.3 V) | Address/Data 12 | 47 | 47 | Address/Data 11 |
| Ground | 17 | 17 | Grant | Address/Data 10 | 48 | 48 | Ground |
| Request | 18 | 18 | Ground | Ground | 49 | 49 | Address/Data 9 |
| +5 V (+3.3 V) | 19 | 19 | Reserved | Access key (Ground) | 50 | 50 | Access key (Ground) |
| Address 31 | 20 | 20 | Address/Data 30 | Access key (Ground) | 51 | 51 | Access key (Ground) |
| Address 29 | 21 | 21 | +3.3 V | Address/Data 8 | 52 | 52 | C/BE 0 |
| Ground | 22 | 22 | Address/Data 28 | Address/Data 7 | 53 | 53 | +3.3 V |
| Address/Data 27 | 23 | 23 | Address/Data 26 | +3.3 V | 54 | 54 | Address/Data 6 |
| Address/Data 25 | 24 | 24 | Ground | Address/Data 5 | 55 | 55 | Address/Data 4 |
| +3.3 V | 25 | 25 | Address/Data 24 | Address/Data 3 | 56 | 56 | Ground |
| C/BE 3 | 26 | 26 | Init Device Select | Ground | 57 | 57 | Address/Data 2 |
| Address/Data 23 | 27 | 27 | +3.3 V | Address/Data 1 | 58 | 58 | Address/Data 0 |
| Ground | 28 | 28 | Address/Data 22 | +5 V I/O (+3.3 V) | 59 | 59 | +5 V I/O (+3.3 V) |
| Address/Data 21 | 29 | 29 | Address/Data 20 | Acknowledge 64-bit | 60 | 60 | Request 64-bit |
| Address/Data 19 | 30 | 30 | Ground | +5 V | 61 | 61 | +5 V |
| +3.3 V | 31 | 31 | Address/Data 18 | +5 V | 62 | 62 | +5 V |

(Continued)

# Evolution of the PC and the Expansion Bus

**Table 4.6** (Continued)

| Signal | B | A | Signal | Signal | B | A | Signal |
|---|---|---|---|---|---|---|---|
| (64-bit) | | | | +5 V I/O (+3.3 V) | 79 | 79 | Address/Data 48 |
| Reserved | 63 | 63 | Ground | Address/Data 47 | 80 | 80 | Address/Data 46 |
| Ground | 64 | 64 | C/BE 7 | Address/Data 45 | 81 | 81 | Ground |
| C/BE 6 | 65 | 65 | C/BE 5 | Ground | 82 | 82 | Address/Data 44 |
| C/BE 4 | 66 | 66 | +5 V I/O (+3.3 V) | Address/Data 43 | 83 | 83 | Address/Data 42 |
| Ground | 67 | 67 | Parity 64-bit | Address/Data 41 | 84 | 84 | +5 V I/O (+3.3 V) |
| Address/Data 63 | 68 | 68 | Address/Data 62 | Ground | 85 | 85 | Address/Data 40 |
| Address/Data 61 | 69 | 69 | Ground | Address/Data 39 | 86 | 86 | Address/Data 38 |
| +5 V I/O (+3.3 V) | 70 | 70 | Address/Data 60 | Address/Data 37 | 87 | 87 | Ground |
| Address/Data 59 | 71 | 71 | Address/Data 58 | +5 V I/O (+3.3 V) | 88 | 88 | Address/Data 36 |
| Address/Data 57 | 72 | 72 | Ground | Address/Data 35 | 89 | 89 | Address/Data 34 |
| Ground | 73 | 73 | Address/Data 56 | Address/Data 33 | 90 | 90 | Ground |
| Address/Data 55 | 74 | 74 | Address/Data 54 | Ground | 91 | 91 | Address/Data 32 |
| Address/Data 53 | 75 | 75 | +5 V I/O (+3.3 V) | Reserved | 92 | 92 | Reserved |
| Ground | 76 | 76 | Address/Data 52 | Reserved | 93 | 93 | Ground |
| Address/Data 51 | 77 | 77 | Address/Data 50 | Ground | 94 | 94 | Reserved |
| Address/Data 49 | 78 | 78 | Ground | | | | |

*The pin-out for the 32-bit PCI bus ends at pin 62.

## 4.10 PCMCIA (PC-CARD)

Even though this text focuses primarily on the desktop and tower personal computers, there is one expansion slot used on the smaller notebook and palm computers that bears mentioning. The name of the slot is the **Personal Computer Memory Card International Association (PCMCIA).**

PCMCIA was originally developed to make adding more memory to the notebook computer a fairly easy process. However, the current version of PCMCIA has evolved into a very powerful expansion slot used by both memory and peripherals from modems to hard drives. PCMCIA nonvolatile, flash-memory cards have also made a huge impact in the digital camera market.

PCMCIA accomplishes its assigned tasks through slots of three sizes called *Type I, II,* and *III.* All three slot types consist of a small, male 68-pin connector into which very small cards called PC-cards insert. All slot types accommodate PC-cards with a length of 85.6 mm (3.34") and a width of 54.0 mm (2.11"). In fact, the only difference among the card types is their physical height. A Type I PC-card has a height of 3.3 mm (0.13"), whereas Type II and III cards have heights of 5.0 mm (0.195") and 20.5 mm (0.8"), respectively. Because the types differ only in height, the thinner cards can be used in the thicker slots, but not the other way around. Figure 4.14 shows the relative size differences.

PCMCIA is a Plug-and-Play standard that allows PCMCIA memory and peripherals to be plugged into a PCMCIA slot only when they are needed. PCMCIA also allows "hot swapping." This is a technique that allows a PC-card to be removed and inserted into a PCMCIA slot without turning the computer off.

**Figure 4.14** Relative Size Difference of Type I, II, and III PC-cards

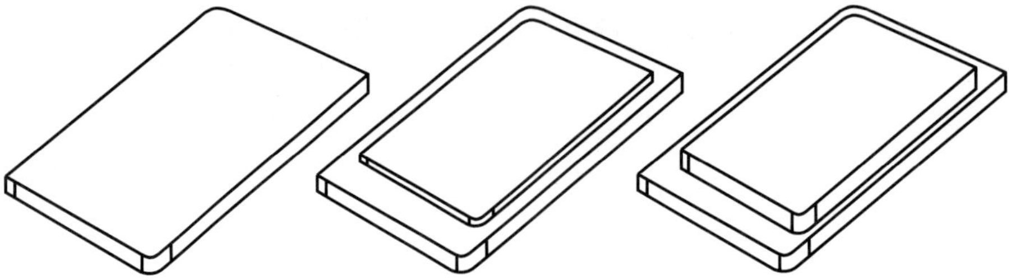

Type I          Type II          Type III

**Table 4.7** PCMCIA Expansion Slot Pin-out

| Pin | Signal | Pin | Signal | Pin | Signal | Pin | Signal |
|---|---|---|---|---|---|---|---|
| 1 | GND | 35 | GND | 18 | Vpp1 | 52 | Vpp2 |
| 2 | D3 | 36 | CD1# | 19 | A16 | 53 | A22 |
| 3 | D4 | 37 | D11 | 20 | A25 | 54 | A23 |
| 4 | D5 | 38 | D12 | 21 | A12 | 55 | A24 |
| 5 | D6 | 39 | D13 | 22 | A7 | 56 | A25 |
| 6 | D7 | 40 | D14 | 23 | A6 | 57 | VS2# |
| 7 | CE1# | 41 | D15 | 24 | A5 | 58 | RESET |
| 8 | A10 | 42 | CE2# | 25 | A4 | 59 | WAIT# |
| 9 | OE# | 43 | VS1# | 26 | A3 | 60 | RSRVD |
| 10 | A11 | 44 | RESRVD | 27 | A2 | 61 | REG# |
| 11 | A9 | 45 | RSRVD | 28 | A1 | 62 | BVD2 |
| 12 | A8 | 46 | A17 | 29 | A0 | 63 | BVD1 |
| 13 | A13 | 47 | A18 | 30 | D0 | 64 | D8 |
| 14 | A14 | 48 | A19 | 31 | D1 | 65 | D9 |
| 15 | WE# | 49 | A20 | 32 | D2 | 66 | D10 |
| 16 | READY | 50 | A21 | 33 | WP | 67 | CD2# |
| 17 | Vcc | 51 | Vcc | 34 | GND | 68 | GND |

Most computers that utilize this expansion slot will have two PCMCIA connectors, but the opening will be the size of a Type III PC-card. This technique allows two Type I or Type II cards to be inserted, or one Type III. The pin-out for the PCMCIA slot is listed in Table 4.7. For more information on the PCMCIA standard, visit www.pc-card.com/pccard.htm.

## 4.11 ISA LIVES ON

As stated previously, ISA was introduced in 1984 as the expansion bus for the IBM-AT-compatible computers. In the world of the constantly evolving personal computer, it seems like ISA should be dead and buried by now. As a matter of fact, ISA is still very much alive and well. Even though PCI is the current de facto standard in the world of high-speed I/O communication, you must realize that all ports do not need to run at the speeds offered by PCI. This is especially true for peripherals that are controlled directly by humans. ISA adapters are cheaper to design and build and still work very well with relatively slow peripherals such as mice, floppy drives, printers and modems, and the 10-Mb network adapter, just to name a few.

For this reason, today's motherboards still have one or two ISA expansion bus slots to support these slower adapters. However, it should be mentioned that even on today's thoroughbreds that run at local bus speeds up to 200 MHz, the ISA bus is still clipping along at 8.33 MHz.

## 4.12 WHAT IS THE RIGHT EXPANSION BUS FOR ME?

With motherboards that use an 8088 microprocessor, the only expansion bus available is the 8-bit expansion bus. The 286 motherboards will probably have two 8-bit slots and the rest ISA. However, an 8-bit card will operate in an ISA slot, so the 8-bit slots are not really required.

The 386 motherboards will probably have all ISA slots or no more than two 8-bit slots. Some late-production 386DX-40 motherboards have up to three VL bus slots. This is the preferred configuration for 386DX motherboards, but these styles are very hard to find.

If you are going to buy an older computer with a motherboard that uses a 486 microprocessor, make sure it has at least two VL bus expansion slots and the rest ISA. The VL bus expansion slots should be used for the video and hard disk adapters. The newer 486 motherboards will have PCI expansion slots. If available, PCI is the preferred expansion bus for hard drives and video because of its performance advantages over VL bus.

The current expansion bus configuration for SC242, SC230, PGA370, and Super Socket7 motherboards is to have at least four PCI expansion slots, three ISAs, and one AGP slot. AGP stands for *accelerated graphics port*. As the name implies, AGP is not an expansion slot but a port. AGP is discussed in greater detail in chapter 12.

## 4.13 GROWING PAINS

Now, a few words for the wise. Just because a new-style computer hits the market with a new microprocessor and expansion bus, it doesn't mean you have to go out and buy one. The techniques used when a new-style computer is first introduced may not become the established standards when the computer finally becomes accepted in the PC community. If you are planning to buy a computer that runs a new processor or expansion bus, you should wait until the new-style computers get over their "growing pains" before you purchase one of these beasts, or you may be stuck with an expensive machine with the wrong technology.

## 4.14 CONCLUSION

The IBM-compatible computer has evolved from being just a copy of computers built by IBM to no longer depending on IBM to set the standards. In fact, IBM has found itself following the lead set by the two primary standard setters—Intel and Microsoft.

IBM's decision to make the IBM-PC through AT computers an open technology is probably the main reason why these types of computers are the most popular and reasonably priced computers in the world. This open technology is also one of the major problems with the machines. There are so many manufacturers, styles, and options available that the consumers' task of choosing the correct computer for their needs is a nightmare. This chapter has hopefully cleared the air on some of the confusion associated with compatible computers and expansion buses.

To help tie things together, Table 4.8 lists the standard expansion slots covered in this chapter along with their data bus widths and maximum data transfer speeds.

Table 4.8  Expansion Bus Comparison

| Name | Data Bit | Maximum Transfer Speed |
|---|---|---|
| PC-bus | 8 | 2 MBps |
| ISA | 16 | 8.33 MBps |
| MCA | 32/16 | 33 MBps |
| EISA | 32/16 | 33 MBps |
| VL-bus | 32 | 132 MBps |
| PCI | 32/64 | 132 MBps/264 MBps |

# Review Questions

1. What do each of the following acronyms and abbreviations stand for?
   a. IBM
   b. ISA
   c. BIOS
   d. XT
   e. AT
   f. PS/2
   g. MCA
   h. EISA
   i. VESA
   j. VL
   k. PCI
   l. PCMCIA

## Fill in the Blanks

1. The first expansion bus used on the IBM-compatible computer was called the _____ _____ .
2. _____ technology makes the technology available to other manufacturers at no charge.
3. The 32-bit version of the _____ expansion bus has a maximum transfer rate of 132 MBps.
4. The _____ bus is the name of the first 32-bit expansion bus used on the compatible computer.
5. The ability for companies to use technology developed by another without paying royalties is called _____ _____ .
6. The _____ bus is where adapter cards are inserted to add additional ports to the compatible computer.
7. The term _____ is used to describe motherboards that have the ability to run at multiple speeds.
8. The hexadecimal addressing range of the ISA bus is from _____ through _____ .
9. The hexadecimal addressing range for the PCI bus is from _____ through _____ .
10. The PCI expansion bus is both a _____ -bit and _____ -bit standard.
11. The predominant expansion bus for the 486-compatible computer is the _____ , which is _____ technology.
12. The IBM PS/2 computers use the _____ expansion bus, which is _____ technology.
13. The technology that allows the operating system to configure the adapter card is called _____ -and- _____ .
14. For a peripheral to transfer information directly with the computer's memory, the expansion bus must support _____ _____ .
15. PCI was originally designed to run at a speed that is _____ _____ the speed of the computer's local bus.

## Multiple Choice

1. The term *32-bit expansion bus* indicates that the connector used by the bus has:
   a. 32 address lines.
   b. 32 data lines.
   c. 32 control lines.
   d. All of these answers are correct.
   e. None of these answers is correct.
2. The ISA expansion bus has:
   a. 20 data lines.
   b. 24 data lines.
   c. 16 data lines.
   d. 8 data lines.
3. Why did the original IBM-PC run at 4.77 MHz?
   a. so it would not exceed the speed of the computer's memory devices
   b. to allow the use of a standard color TV as a monitor
   c. because this was the speed rating of the microprocessor
   d. because of no particular reason
4. The ISA expansion bus is designed for the:
   a. ISA adapter card.
   b. PC-bus adapter card.
   c. PCI adapter card.
   d. All of these answers are correct.
   e. Both answers (a) and (b) are correct.
   f. Both answers (a) and (c) are correct.

5. ISA is an open standard.
   a. True
   b. False
6. The term *turbo-motherboard* is used to indicate:
   a. the microprocessor is turbo-charged.
   b. the motherboard is turbo-charged.
   c. the microprocessor can operate at different speeds.
   d. there is greater air flow over the motherboard.
7. Micro-Channel is:
   a. a 16-bit expansion bus.
   b. smaller than 16 bits.
   c. a 32-bit expansion interface.
   d. a 64-bit expansion interface.
   e. both (a) and (c) are correct.
8. Micro-Channel is an open standard.
   a. True
   b. False
9. EISA is a 32-bit expansion bus. This indicates that there are:
   a. 32 address lines.
   b. 32 data lines.
   c. 32 control lines.
   d. 32 total pins on the connector.
10. Which of the following adapters can be used in an EISA expansion bus?
    a. PC-bus
    b. ISA
    c. EISA
    d. All of these answers are correct.
11. VL-bus:
    a. is used to interface additional memory to the AT-style computer.
    b. was developed by IBM.
    c. is a 32-bit expansion bus that connects directly the MPU's local bus.
    d. was not used on the 386DX motherboard.
12. VL-bus is an open technology expansion bus.
    a. True
    b. False
13. IBM designed the VL-bus expansion bus.
    a. True
    b. False
14. PCI is a mezzanine interface bus. The term *mezzanine* means:
    a. the computer can run at more than one speed.
    b. the peripheral interface can run at a speed independent of the MPU.
    c. the motherboard has more than one microprocessor.
    d. the peripheral interface bus actually runs faster than the MPU.
15. The speed of the ISA expansion bus:
    a. depends on the speed of the computer's local bus.
    b. is half the speed of the computer's local bus.
    c. is approximately 8 MHz regardless of the computer's local bus.
    d. operates at twice the speed of the computer's local bus.
16. VL-bus operates at the speed of the local bus.
    a. True
    b. False
17. The PCI expansion bus operates at the speed of the computer's local bus.
    a. True
    b. False
18. The name of the expansion bus that connects directly to the microprocessor's buses is called the:
    a. ISA.
    b. EISA.
    c. PCI.
    d. VL-bus.
    e. MCA.
19. Bus mastering:
    a. is the name of the next-generation expansion bus.
    b. was originally used with the ISA expansion bus.
    c. is the technology that supports PnP adapters.
    d. allows a peripheral to transfer information directly with the computer's memory.
20. Different expansion bus technology cannot be used on the same motherboard.
    a. True
    b. False
21. PCI is an open technology.
    a. True
    b. False
22. One of the advantages that PCI has over VL-bus is:
    a. PCI practically eliminates I/O wait states.
    b. PCI operates at the speed of the local bus.
    c. PCI accepts 8-, 16-, and 32-bit adapter cards.
    d. PCI operates at 1/2 the speed of the local bus.
23. Which of the following is the 32-bit I/O interface bus that operates asynchronously with the computer's local bus?
    a. EISA
    b. ISA
    c. PCI
    d. VL-bus
24. Which of the following is the proprietary 32-bit I/O expansion bus that was developed by IBM for the PS/2 computer?
    a. EISA
    b. ISA
    c. PCI
    d. VL-bus
    e. MCA
25. Which of the following is the 16-bit I/O expansion bus that is capable of running an 8-bit or 16-bit adapter card?
    a. EISA
    b. ISA
    c. PCI
    d. VL-bus
26. Which of the following is the 32-bit I/O bus that does not connect directly to the microprocessor's local bus?
    a. EISA
    b. ISA
    c. PCI
    d. VL-bus
27. The PCI expansion bus practically eliminated:
    a. multiple clock cycles per I/O transfer.
    b. the ISA bus.
    c. I/O wait states.
    d. IBM and the predominant compatible computer manufacturers.

# CHAPTER 5

# Operating System Basics, Including MS-DOS and Windows 3.X

## OBJECTIVES

- Identify several operating systems used with the compatible computers.
- Define terms and expressions relative to operating systems.
- Describe DOS and Windows 3.X filename syntax.
- Describe DOS and Windows directory structure.
- Explain the difference between a hard-boot and soft-boot.
- Explain the meaning of version numbers.
- List device names used by DOS.
- Describe the purpose of an operating system's shell.
- Explain the main differences between DOS and Windows 3.X operating systems.
- Explain the expression *parasite operating system*.
- List and explain the characteristics of the Windows 3.X operating modes.
- Explain the basic DOS boot sequence for the compatible computers.
- List the three required boot files and two optional boot files.
- Explain the function of the files IO.SYS, MSDOS.SYS, and COMMAND.COM.
- Explain the purpose of DOS's CONFIG.SYS and AUTOEXEC.BAT files.
- Explain the purpose of the WINDOWS INI files.
- Explain the purpose of TSRs.

## KEY TERMS

application program interface (API)
DOS applications
Windows applications
End-User License Agreement (EULA)
freeware
shareware
PC-DOS
MS-DOS
boot
version number
filename
extension
8.3 format
attribute
directory
address
root directory
subdirectory
parent directory
command line
graphical user interface (GUI)
default
cursor
DOS shell
double-click
parasite operating system
standard mode
enhanced mode
program manager
file association
foreground program
background program
cooperative multitasking
program information file (PIF)
boot sequence
configuration
batch file
AUTOEXEC.BAT
CONFIG.SYS
terminate and stay resident (TSR)
information files (INI)

153

As stated in chapter 2, the main purpose of the operating system is to make the computer operate so it can run application programs. For this reason, the main portion of the operating system is also referred to as the **application program interface (API),** which provides the connection between the operating system and application. Because the application program uses the operating system to perform its amazing feats, application programs are linked to a certain operating system. For this reason, application programs are usually categorized according to the operating system for which they are written. For example, application programs written to run on the MS-DOS operating system are called **DOS applications.** Application programs written for the Windows operating system are called **Windows applications.**

Just like the compatible computers for which this book is written, the operating systems for the compatible computer have strived to maintain downward compatibility. *Downward compatibility* means that application programs written for previous-generation operating systems will usually run on a newer-generation operating system without modification. For example, programs written for MS-DOS will run under Windows 98.

Along with running applications, the operating system also has to give the user the ability to control the basic functions required to maintain the application programs and their data. The operating system must also give the user the ability to configure the operating system for his or her specific needs. These two requirements of the operating system will be discussed in this chapter and in chapter 6.

We learned in chapter 2 that the ROMs in a compatible computer contain only a basic OS called the *BIOS* to get the computer started. The main OS is usually loaded into the RAM of the machine from a disk drive; hence the name *disk operating system (DOS)*. There are several DOSs available for the different families of compatible computers. These include MS-DOS, Windows, Windows 95, Windows 98, Windows for Workgroups, Windows NT, Windows 2000, Windows Millennium, OS/2, OS/2 Warp, UNIX, Linux, BeDOS, and Netware, just to name a few. Of those listed, the four that are currently most popular are MS-DOS, Windows 3.X, Windows 95, and Windows 98. Microsoft writes all four of these disk operating systems along with the newest versions—Windows 2000 and Windows Millennium (Me).

Windows 2000 Profession came out in February 2000, which is about the time I was finishing up the final version of this text. Like all of the other Microsoft operating systems, I am sure it will become the driving force for the compatible PC in the future. For this reason, I have included a section on this operating system in the next chapter. Unfortunately, at the time of this writing, Windows Me's anticipated release date is fall 2000, so information on this operating system is not included in this text.

Having a basic understanding of the computer's disk operating system is so important that I have divided this topic into two chapters. This chapter covers disk operating systems in general, along with the basics of MS-DOS and Windows 3.X. Windows 95, 98, NT, and 2000 are covered in chapter 6.

Before we get started, note that it is not the intent of this text to teach all there is to know about the aforementioned disk operating systems. The following discussions are just a basic DOS overview that is needed to understand the remainder of this book. If you need more detailed information, there are hundreds of books available on all of the disk operating systems mentioned. Also, before we get started with our discussion about operating systems, we need to clear up one major misconception about software that people have had for years. This is, "Who actually owns the software?"

## 5.1    WHO OWNS SOFTWARE?

Most people believe that when they buy a new OS or application program, it becomes their property. In almost all circumstances, this is not the case. Either the programmer or the company that employs the programmer retains full ownership of the software.

When you pay your money, you are not actually buying the software. You are paying for a right to *use* the software. This right is given in the form of a license agreement between the purchaser and the owner of the software. The license contains the terms the user has to follow to legally use or copy the software. Included with the software is a document called an **End-User License Agreement (EULA).** By installing the software, the user accepts the terms outlined in the EULA. The EULA is sometimes part of the documents that come packaged with the software, but it is currently popular to display the EULA during the software's installation process. When this is the case, the user has to accept the license agreement before the installation process will continue. The EULA outlines your legal rights for using the software, so it is a very important document that you should always read.

Not all software requires the user to pay for the EULA in order to install, use, and copy. **Freeware** is the name given to software that the user can install, use, copy, and distribute freely for as many times as he or she desires. **Shareware** is the name given to software that the user can install and use freely for a predefined evaluation period. If the user keeps using the shareware software beyond the evaluation period, he or she should purchase a license.

### 5.1.1 A Little DOS History

Contrary to popular belief, Microsoft did not fully write the first DOS that was used with the original IBM-PC computer when it came out in 1981. Believe it or not, Microsoft bought the first DOS along with full rights from Seattle Computer Products (SCP) for $50,000. Microsoft sold the rights for this same DOS to IBM under the stipulation that Microsoft would retain ownership and could license the DOS to anyone it desired.

IBM called this first and all subsequent DOSs it supplied with the IBM-PC computers **PC-DOS**. Microsoft licensed basically the same DOS to compatible computer manufacturers under the name **MS-DOS**. Other companies, such as Digital Research, wrote DOSs that had the same identical look and feel of MS-DOS and PC-DOS. Digital Research called its DOS *DR-DOS*. All of these versions of DOS are basically the same from a user's point of view, so they are all generically known as simply *DOS*. To this day, even though all of the main operating systems are a DOS, the term DOS is used to generically refer to all of the DOSs written for the IBM-PC/XT/AT computers. All of the other disk operating systems are referred to by names such as Unix, Windows, Windows 95, and OS/2.

To avoid confusion, I refer to disk operating systems in general as simply *OS*, or *operating system*. I use the term *DOS* when referring to the original DOS written for the of IBM PC/XT/AT-compatible computers. I will refer to all of the other disk operating systems by name.

### 5.1.2 Why Load from Disk?

The primary reason why almost all compatible computers load their main OS from disk is so the operating system can be easily changed with the development of new programming techniques and advances in hardware technologies. If the main OS was contained entirely in the computer's ROM, the ROM ICs themselves would have to be changed every time the main operating system changed. Changing ROM requires technical expertise that is beyond the range of most individuals.

When a new OS is purchased, it will be supplied on some type of transportable media such as a floppy diskette or a CD-ROM. The new OS is not intended to be loaded and run from the diskettes or CD-ROM. Instead, a program called SETUP or INSTALL will be loaded and executed from the first diskette or CD-ROM that will install the new OS onto the computer's hard drive. It will reside on the hard drive until it is loaded and executed from the computer's RAM memory when the computer is booted.

## 5.2 SIMILARITIES BETWEEN DOS AND WINDOWS 3.X

DOS operates the microprocessor in real mode by itself whereas Windows 3.X operates the microprocessor in protected mode. However, in most current computers, the microprocessor is actually in the virtual 86 mode when DOS is running. This switch is not performed by DOS itself but by special programs called *memory managers*. Memory managers will be discussed in greater detail in chapter 7.

Even though the operation of the microprocessor in the protected mode and the real-mode is drastically different between DOS and Windows 3.X, these operating systems have a lot in common. For this reason, we will start our discussion with the similarities and then move to the differences between the two operating systems.

### 5.2.1 Version Numbers

Operating systems and application programs usually change to keep up with the changing trends. To help keep track of which OS or application you are currently using, the company that writes and releases the new software assigns a group of numbers to each new release. This group of numbers is called the **version number,** or *release number.* At the time of writing, the current stand-alone version of MS-DOS is 6.22, the current version of Windows is 3.11, the current version of Windows 95 is Operating System Release 3.0 (OSR3) and the current version of Windows 98's Second Edition (SE). As you probably guessed, the higher the version number, the more recent the release of the software. Unfortunately, comparing version numbers between two companies' comparable software is not a valid indication of how recently the software was released.

You may wonder, "What is the purpose of the fraction in the version number?" First, to answer this question you need to know that the numbers to the right of the period in the version number are not really a decimal fraction but an indication of the number of times the version has been revised. Sometimes problems appear after a version number is released, which happened with Windows 3.0. To solve these problems, Microsoft released Windows 3.1 to indicate the problem had been corrected. The number(s) to the right of the period is called the *revision number.* A revision indicates a small change has been made in the current version.

Sometimes only minor revisions are needed to correct problems. A number in the second place after the period indicates a minor revision. An example is the MS-DOS version 4.01 release. The 01 to the right of the period indicates this is the first minor revision to MS-DOS release 4.0. Another example is the MS-DOS 6.22 release. The numbers to the right of the period indicate there have been two revisions and two minor revisions to MS-DOS release 6.0. Applications written for the first version should run on the revisions without any required modifications.

**Problem 5.1:** What is indicated by the version number 3.11?

**Solution:** The version number 3.11 indicates that version 3 has been revised once and also has had one minor revision.

### 5.2.2 What Is Meant by "Windows 3.X"?

Windows versions below 3.0 are written for the 286-compatible microprocessors. Because the 286 does not support the virtual 86 mode, Windows versions 1.0 and 2.0 cannot multitask DOS applications. For this reason, very few of these versions of Windows were purchased. In fact, I have never seen either of these versions.

Windows versions 3.0, 3.1, and 3.11 are the last releases of the Windows operating system before Windows 95. There is one very important difference between these releases of Windows and the previous releases. Even though they were written primarily for the 286, all three of the 3.0 and above releases support the 386 microprocessor's virtual 86 mode. This allows the multitasking of Windows applications with DOS applications. Of course, this only holds true when Windows is used on a compatible microprocessor above a 286. For this reason, Windows versions 3.0 and above are still very popular and are still being used on thousands of compatible computers. For obvious reasons, all of these releases of Windows are referred to generically as Windows 3.X. I will use the *Windows 3.X* reference for these combined releases throughout the remainder of this book.

### 5.2.3 DOS and Windows 3.X Filenames

Remember that a computer uses disk drives in a manner similar to the way we use books. A disk drive is a place where data is stored until it can be loaded through an input port into the computer's RAM where it is actually processes. There are really only two types of data stored on a disk drive: programs, and data used by a program. No matter the type, the OS you are using has to organize the data on the disk using a format that makes its storage and retrieval fast and accurate. To do this, the operating systems used by the compatible computers all use a method similar to the filing system that has been used in businesses for years.

The basic office filing system consists of a file cabinet, file drawers, and individual file folders, as shown in Figure 5.1. In a typical disk file system used by compatible computers, the file cabinet is comparable to the diskette or hard drive platter. The file cabinet drawers are comparable to file directories and the file folders are comparable to filenames. Each of these concepts is labeled in Figure 5.1 and will be discussed in the following sections.

**Figure 5.1** DOS Office File Cabinet

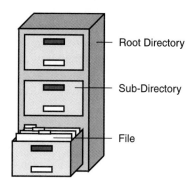

**Table 5.1** Reserved Punctuation Symbols and Their Meanings

| |
|---|
| : = colon = identifies a drive letter |
| = space = separates sections of a DOS command line |
| \ = back slash = tells DOS to take a path |
| / = slash = identifies a command switch |
| * = asterisk = replaces all letter combinations |
| ? = question mark = replaces a single letter |
| . = period = separates the file name from the extension |
| > = greater than = redirects output |
| < = less than = redirects input |
| \| = vertical bar = pipes (routes) the output of one DOS command into another DOS command |

Just as a file folder is used in the business office to hold like information, similar information is stored on a disk in units called *files*. The file folders used in a business office have a raised tab on which a name is written so that the file folder can be located easily in a file drawer. Similarly, a file stored on a disk is given a **filename** so that it can be located easily on the disk. A typical filename is given in Example 5.1.

**Example 5.1:** Example of a typical filename. Case does not matter.

> Progress.exe

To understand the syntax used in Example 5.1, we need to discuss several rules that define the parameters used by Windows 3.X and DOS for filenames. First, the filename is split into two parts: a name and an **extension.** The name portion is required but the extension is optional.

The name portion of the filename can be from one to eight characters in length. Any combination of letters and numbers can be used. Some punctuation symbols are legal as part of the name and some are not. Most punctuation symbols have a special meaning to DOS and Windows and are not allowed as part of the filename or extension. A brief explanation of the reserved punctuation symbols appears in Table 5.1. To be on the safe side, the novice DOS user should avoid punctuation as part of the name and extension.

The optional filename extension can be from one to three characters in length. As in the name, punctuation should be avoided as part of the extension to avoid accidentally using one of the reserved punctuation symbols. Just because the extension is not required to store information on a disk does not mean it is not required to actually access the information stored in the file. In fact, most applications use the extension to identify data files. When this is the case, the application will not be able to access the information stored in the data file without the required extension.

The filename extension usually identifies the type of information stored in the file. Table 5.2 lists some commonly used filename extensions along with the type of file they identify.

EXE and COM are two very important extensions that are required to use the information stored in the file. If a file has one of these extensions, it identifies the file as being a machine language program that can usually be executed by the user through the OS.

---

**Problem 5.2:** What file extension is usually used by Windows 95 and 98 program information files?

**Solution:** According to the common extensions listed in Table 5.2, INF is the extension used by Windows 9X program information files.

---

If a filename extension is used, it becomes part of the uniqueness of the filename. For example, the filename CARDS with the extension DOC will be considered a different file than the filename CARDS with the extension TXT. With 8 characters available for the filename and 3 characters for the extension, there is a total of 11 characters. If we used only letters and numbers, each filename could have up to $11^{36}$ combinations. This equals approximately 309,126,805,328,706,726,356,733,529,368,874,530 decimal combinations, which is almost an unlimited number.

When a filename is entered from the keyboard, DOS and Windows need to know where the filename's name ends and the extension begins. To separate the two—name and extension—a period (.) is used. For example, the files listed in the previous paragraph will be accessed by the complete filenames CARDS.DOC and CARDS.TXT. The period is for separation purposes only and is not part of the 11 characters. Because DOS and Windows allow up to 8 characters for the file's name and up to 3 characters for the extension, with the two separated with a period, this is referred to as the **8.3 format.**

### 5.2.4 File Attributes

The file system used by the operating systems covered in this book gives the ability to assign each filename one or more **attributes.** Currently there are six attributes that a file can be given. These include hidden, read-only, system, archive, label, and subdirectory. A file's attributes are usually assigned automatically to the file when the user creates it or when the file is installed as part of an application or OS. As you have probably guessed, all of the operating systems covered in this book give the user the ability to change some of these attributes.

If a file is assigned the hidden attribute, it will not be visible in the directory when DOS is being used. Hidden files will only be visible in Windows if the option to display hidden files is enabled, which is covered in the next chapter. The read-only attribute means the file cannot be altered, copied, or erased. The system attribute is a combination of the hidden and read-only attributes. System files are used exclusively by the operating system. An archive attribute indicates that the file has never been backed up. Backing up is a method of saving information on another media in case the original becomes corrupted. The archive attribute is the default attribute given to a file when it is created or changed.

If a file is given the label attribute, it indicates the directory entry is not the name of a file, but is instead a software label that can be written on the disk. This feature is handy if the printed label on the disk jacket is lost or becomes unreadable. If the filename is assigned the subdirectory attribute, it indicates it is not the name of a file, but the name of a subdirectory. A filename that is given this attribute contains information that points to the location of the subdirectory on the disk. Subdirectories are covered in more detail in section 5.2.7.

The procedure used to change and display a file's attributes is different for each OS. For this reason, consult the OS user's manual or use the Windows 9X/2000 Help option for the correct procedure. The syntax for DOS command ATTRIB is covered in Appendix C.

---

**Problem 5.3:** What file attribute would be used to prevent a file from being copied, altered, or erased?

**Solution:** The file should be given the read-only attribute.

**Table 5.2** Commonly Used File Extensions

| | |
|---|---|
| ARC | Archived file that is compressed to save disk space |
| ASC | ASCII file |
| ASM | Assembly language source code program |
| BAK | MS-DOS and Windows 3.X backup file |
| BAS | BASIC program source code |
| BAT | Batch file (ASCII file that contains MS-DOS commands, can be executed for the OS shell) |
| BIN | Binary file |
| DAT | Data file |
| DLL | Dynamic link library |
| DOC | Document file used by Microsoft Word and WordPad |
| DRV | Device driver program |
| C | C language program source code |
| COM | Command program file (machine code program that can be launched for the OS shell) |
| CPI | Code page information (DOS) |
| CRF | Cross-reference file created by compilers and assemblers |
| ERR | Error file |
| EXE | Executable program file (machine code program that can usually be launched from the OS shell) |
| FNT | Printer font file |
| GIF | Graphic image format file |
| INF | Windows 95 and 98 program information file |
| INI | Windows 3.X and MS-DOS information file |
| INC | Include file |
| LIB | Library file (usually contains macro used by compilers) |
| LST | Listing file created by assemblers and compilers |
| OBJ | Object file (a machine code program that has not been linked to an operating system) |
| PIF | Program information file (contains information about MS-DOS applications used by Windows) |
| SCR | Screen saver program file |
| SYS | System file (files used to boot the OS and also Terminate and Stay Resident (TSR) programs loaded with the CONFIG.SYS) |
| TMP | Temporary data files |
| TXT | Text file (usually ASCII) |
| VXD | Windows 95 and 98 virtual hardware driver |
| WPG | WordPerfect graphics file |
| ZIP | Compressed file format |

## 5.2.5 Directories

No matter what OS is running your computer, it has to be able to locate information stored on a disk rapidly and accurately. To do this, all of the aforementioned operating systems used by the compatible computers utilize **directories.** The disk's directory is similar to the table of contents in a textbook. The table of contents shows the different topic headings along with the page number at which the starting point of the indicated topic is located. Similarly, the disk's directory holds the names of each file along with its starting location on the disk. Along with the file's name and starting location, the directory also holds the file's attributes, extension (if used), size, and the creation date and time.

When the computer's OS is told by the user to load a file from the disk into the computer's RAM, the OS first checks the directory to determine if the filename is present and, if so, the starting location of the file. If the filename is not listed in the directory, the OS will usually display an error message indicating that the file was not found.

When a file is saved to a disk, the OS will first check the directory to see if the filename already exist. If it does, the new information will replace the old. In DOS versions before 5.0, this replacement is done without warning. If the name does not exist, once the new file is saved, its filename, extension, starting location, attributes, and creation date and time will be added to the listings in the directory.

The procedure used to display the list of the filenames contained in the directories of a floppy diskette, hard drive, or CD-ROM varies depending on the OS being used. Because this is a very important command, I highly recommend that you consult your operating system's user manual or Help files for this procedure.

Just as a file cabinet has multiple drawers to more efficiently manage the files in an office, a disk can have multiple directories to divide the files into groups according to their function. These directories fall in one of two categories. One category is called the **root directory** and all of the others are called **subdirectories.** The differences between these two directories will be covered in the following sections.

## 5.2.6 Root Directory

Even though a disk can have multiple directories, there can be only one root directory on each disk or hard disk partition. The root directory is the most important directory on the disk and can never be moved or deleted by the user. The root directory is created automatically when the disk is formatted, which is required to make a disk compatible with the OS that is running the computer. Formatting is an important process that will be discussed in the section on each OS.

The root directory is always located at the very front of the disk or hard drive partition and the OS always knows its exact location. For this reason, the root directory is the only directory on the disk that does not have a name. Figure 5.2 shows an example of the root directory listing on my computer, which uses DOS. In fact, the directory listing illustrated in Figure 5.2 is the basic appearance of all directories on the disk, whether root or sub.

The first column on the left contains all of the filenames. The second column lists the file extensions. If the third column contains <DIR>, it indicates the name is that of a subdirectory and not a file. The fourth column shows the size of the file in bytes. The fifth column indicates the date that was in the computer when the file was created or modified. The sixth column shows the time that was in the computer when the file was created or modified.

There can be multiple subdirectories located throughout the disk. In fact, subdirectories can contain other subdirectories. No matter how many subdirectories are contained on a disk, all of them originally evolved from the root directory.

## 5.2.7 Subdirectories

A hard drive can contain hundreds of files and each file has a filename listed in a directory. For example, the hard drive I am presently using to save and load the files for this book contains over 10,000 files. Several problems occur if an attempt is made to place all of the filenames in the same directory.

First, it would be very difficult to locate a single file within the hundreds or thousands of listed files. Second, the individual files that belong to an application, or even this book, would be almost

**Figure 5.2** Typical Root Directory

```
Volume in drive C is STACVOL_DSK
Directory of C:\

BASIC        <DIR>      09-16-92   8:09p
DBASEIV      <DIR>      07-23-92   1:34p
DOS          <DIR>      05-15-91  12:33p
GRADE        <DIR>      05-15-91  12:21p
GRAPHICS     <DIR>      07-10-91  10:34a
MSPUB        <DIR>      10-29-92   9:48p
MSWORKS      <DIR>      10-22-92   7:33a
PCTOOLS      <DIR>      01-20-93  10:51a
QUICK        <DIR>      08-06-91   9:09a
SCANGAL5     <DIR>      07-21-92   2:14p
SCHEMA       <DIR>      05-15-91  12:18p
STACKER      <DIR>      02-23-93   7:35p
WINDOWS      <DIR>      07-20-92   2:06p
WP51         <DIR>      05-15-91  12:21p
WINSLTH      <DIR>      07-07-93   6:09a
RICH2    BAT         26 04-18-91   5:34p
COMMAND  COM      47845 04-09-91   5:00a
WINA20   386       9349 04-09-91   5:00a
TREEINFO NCD       1451 01-13-92   7:50a
QFINDEX      <DIR>      02-25-93  10:57a
DM           <DIR>      07-08-93   6:27a
AUTOEXEC QDK        471 04-05-93  12:36p
CONFIG   SYK        507 07-12-93   6:15a
CONFIG   STC        371 02-23-93   8:11p
AUTOEXEC STC        434 02-18-93   7:06p
WPWIN        <DIR>      02-24-93   7:33a
WPC          <DIR>      02-24-93   7:35a
GMKW         <DIR>      02-24-93   8:06a
AUTOEXEC BAK        462 03-02-93  11:08a
AUTOEXEC OLD        434 02-18-93   7:06p
EASYFLOW     <DIR>      03-23-93   7:16a
QEMM         <DIR>      04-05-93  12:35p
CONFIG   SYS        507 08-16-93   1:29p
CONFIG   QDK        569 04-05-93  12:51p
OPT3     BAT        820 04-05-93  12:51p
AUTOEXEC BAT        491 04-05-93  12:51p
123          <DIR>      08-26-93   9:32a
CONFIG   BAK        511 08-16-93   1:15p
       38 file(s)      64248 bytes
                    29245440 bytes free

War Eagle C:\>
```

impossible to find and separate from all of the other files. Third, every filename within a directory has to be unique. As new application programs and data are saved on the disk, odds are that one will eventually have the same filename as one that already exists within the directory. As stated previously, when this happens the old file will be wiped out and replaced with the new. This would make the original application that used this filename unusable. These reasons alone make it necessary to have a better way to organize the files stored on a disk so that all application programs can be separated.

Just like most file cabinets have several drawers that are used to separate the file folders in an organized manner, the operating systems used on the compatible computers use multiple directories to separate the many files according to their application and function. These different directories are called *subdirectories.* You can think of each disk's root directory as a file cabinet and each subdirectory as one of the individual drawers in the file cabinet.

Subdirectories can be located anywhere on the surface of the disk. Just like program and data files, each subdirectory must have a unique name within the directory in which it is located. Subdirectories created under DOS and Windows 3.X are named using the same 8.3 format used by any other file. Subdirectories are usually not given extensions to make them easier to locate using the DOS file search commands, however.

A subdirectory is nothing more than a file, and directories hold filenames, so subdirectories can be created and listed within other subdirectories. This is similar to having file cabinet drawers located within file cabinet drawers. For example, the root directory can contain a listing for a subdirectory named YEAR. The subdirectory YEAR can contain a listing for the subdirectory MONTH. And the subdirectory MONTH can contain a listing for a subdirectory named DAY.

A root directory that has subdirectories, which themselves have subdirectories, creates a structure similar to a tree. The root directory is considered the trunk of the tree and each subdirectory is a branch. To help visualize this, DOS provides a TREE command that displays the directory structure of the disk. An example of the TREE command is shown in Figure 5.3.

Example 5.2 shows the listing of a subdirectory named PLAY. The first two directory filenames, dot (.) and dot-dot (..), are the first filenames in all subdirectories. In fact, dot and dot-dot will be listed in the subdirectory even when the subdirectory is empty. The reason is that these files are not part of the application or data filenames listed in the directory, but are used to keep track of the location of the subdirectory on the disk.

**Figure 5.3** Typical Directory Tree Displayed Using the DOS Tree Command

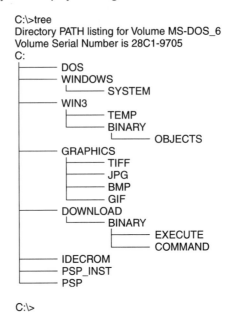

**Example 5.2:** Typical subdirectory listing

```
C:\play>dir

Volume in drive C has no label
Volume Serial Number is 3E55-1BF4
Directory of C:\play

.                <DIR>         02-15-99    9:53p
..               <DIR>         02-15-99    9:53p
SETUP    EXE      92,672       09-30-96   11:58p
ENGINE   DLL     166,744       09-30-96   11:58p
PLAY     HLP      35,841       09-30-96   11:56p
LICENSE  TXT       5,919       04-09-96   11:16p
PROBLEM  TXT       5,310       02-18-96    7:34p
README   TXT       9,495       04-09-96   10:13p
PLAY     EXE     191,488       09-30-96    7:44p
COM      EXE      10,000       09-30-96    8:34p
PLAY1    VXD      77,966       09-30-96   11:56p
PLAY2    VXD      55,598       09-30-96   11:57p
        10 file(s)             651,033 bytes
         2 dir(s)        1,930,489,856 bytes free

C:\play>
```

**Problem 5.4:** Referring to the directory listing shown in Example 5.2, what is the size of the file named LICENSE.TXT?

**Solution:** According to the directory listing shown in Example 5.2, LICENSE.TXT is 5,919 bytes in length.

Because a disk's root directory can contain listings for subdirectories, and subdirectories can contain listings for other subdirectories, there has to be a method of giving direction to the operating system about how to locate an individual file. The method used by the operating systems covered in this text is referred to as the *path*.

## 5.2.8 Path

Every operating system discussed in this book gives the user the ability to specify a file's location by typing the location on the computer keyboard. Like most computer commands, the information has to be given using the correct syntax. Before we cover the actual syntax used, we need to first get some basic information.

As stated previously, files are located in the root directory or in a subdirectory. If located in a subdirectory, the subdirectory itself can originate from another subdirectory. When the user tells the operating system how to locate a filename on the disk, it is referred to as a *path*. Subdirectories use the same 8.3 format as that used for naming files, so there has to be a way to tell the operating system which names are directory names and which are filenames when specifying the location of a file. To indicate the name of a directory to the operating system, each directory name is preceded by a backslash (\) with the name of the file following the last backslash. Example 5.3 shows two file path examples.

**Example 5.3:** File path examples

---

The file path for the file RR.COM located in the SEC directory is:

\sec\rr.com

The path for the file TOT.DAT is located in the TIP subdirectory. The TIP subdirectory is located within the TOP directory.

\top\tip\tot.dat

**Note:** Although case does not matter, most file paths are typed on the keyboard using lowercase letters.

---

As you have probably guessed, it is very important to know how to tell the operating system the location of a file. It is also very important to be able to identify machine language programs within a directory.

## 5.2.9 Identifying Programs in a Directory

When you look at a directory listing that contains an application program or operating system, you will see at least one file with the extension **EXE** or **COM.** If a file has either one of these extensions, it indicates the file is a machine language program that can usually be executed by the user through the operating system's shell. The extension **COM** identifies a *command file* and **EXE** identifies an *executable file*. The only real difference between a COM and an EXE program is the way the operating system handles the computer's RAM. Because of this, a COM file can never be over 64 KB in length, whereas an EXE can be any size.

Just because a file does not have an EXE or COM extension does not mean it is not a machine code program. In fact, there are lots of machine code programs that do not end with either of these extensions. If a machine code program does not have an EXE or COM extension, the user cannot execute it through the operating system's shell, however. These other programs are loaded and executed either by the operating system itself or through other machine code programs.

Most people believe that a program consists of a single file. This is true in a few cases, but most of the time a program will consist of many files. For example, the subdirectory shown in Example 5.2 is a listing of a hypothetical application program named PLAY. This single application program actually consists of 10 files. Some of the 10 files are data files and some are machine language program modules. Some of the machine language programs end with an EXE extension and some do not.

An example of a machine language program that does not end with EXE or COM extension are those files shown in Example 5.2 that end with DLL extensions. Programs with this extension are called a *dynamic link library*. A dynamic link library is a machine language program that is shared between multiple Windows applications.

Even though programs can have multiple machine language modules that are saved as individual files within the same directory, at least one of the modules must end with either an EXE or COM extension. This program will launch the other machine language modules as they are needed. The program that launches the other modules is called the *kernel*. The kernel program that launches the hypothetical PLAY application that is illustrated in Example 5.2 is PLAY.EXE.

Along with files, the operating system also needs some method that allows the user to locate the most commonly used peripherals without knowing their actual binary address. To do this, all of the operating systems used by compatible computers allow the user to access the most commonly used peripherals by name instead of by address. This technique is covered in the next section.

## 5.2.10 DOS and Windows 3.X Device Names

We already know from previous chapters that the engineers who design a computer assign addresses to everything in the computer. We also know that a programmer writes a program that tells the microprocessor to use a port's assigned address to select the port for communication. To control the smart ports used in the compatible computer also requires the use of binary codes that are specific to each port. These codes are sent to the port address by the program to tell the port how to control its attached peripheral.

Writing machine language programs to control the computer's hardware requires the programmer to not only know all of the port addresses, but also all of the codes that are required to tell each port the task that needs to be performed. As you can probably guess, this feat is beyond the capabilities of most computer users.

One of the primary functions of an operating system is to provide a user-friendly interface by taking over the machine language control of the computer's ports. Instead of learning addresses and codes, the operating systems used on the compatible computers allow the user to control the most common ports by using command and device names. The command will be discussed in later sections. For now, let us concentrate on the device name. Table 5.3 contains a list of the device names assigned to the most commonly used compatible computer ports.

Notice in Table 5.3 that the floppies are named *A* and *B*. This seems to cause a problem in that A and B are also legal filenames. How does DOS know when A and B are files and when A and B are drives? Once again, punctuation comes into play. To indicate to DOS that A and B are not files, a colon (:) follows the drive name. So, to identify the two floppies, the names A: and B: are used. The name C: is used for the first hard drive (if one is installed). To access information on any drive, the name of the drive is used instead of the actual drive's address (which most DOS users do not know in the first place). For

**Table 5.3** Commonly Used Peripheral Names

| COM1 | Program serial port #1 |
| --- | --- |
| COM2 | Program serial port #2 |
| AUX | Default serial port |
| LPT1 | First parallel port |
| LPT2 | Second parallel port |
| CON | Keyboard |
| A: | Floppy drive one |
| B: | Floppy drive two |
| C: | Hard drive one |

example, to display the directory on the B: drive, the command and device name will be DIR B: when the user selects the device using MS-DOS. Writing machine language programs to retrieve the directory off the drive is a complex task, to say the least. However, using device names and the OS to access these ports makes life a lot easier on the user.

Now that we know a little about operating systems in general, let us move our discussion to controlling the operating system itself.

### 5.2.11 The Shell

The main purpose of an operating system is to allow the user to manage the overall operation of the computer's hardware. To do this, the OS contains one program that functions as the link among the user, the computer hardware, the operating system, and other applications programs. This program is called the operating system's *shell,* so named because this program starts out in control of the computer and shells out the computer's time to other programs. Each operating system has at least one shell program that will take control of the computer each time the computer is booted (reset). The different shells utilized by the operating systems used on the compatible computer will use either a typed **command line** (text) or a **graphical user interface (GUI).** The command line method requires the operator to learn the commands of the shell and type them in from the keyboard. A GUI shell usually uses a mouse as a pointer to select the different applications and commands. Mice are covered in greater detail in chapter 9.

The default shell for DOS is called *COMMAND.COM,* whereas the default shell for Windows 3.X is called *Program Manager.* COMMAND.COM uses a command line typed on the keyboard while Program Manager uses a GUI. As you have probably guessed, using the two shells is entirely different.

Learning how to use an operating system's shell efficiently is the key to controlling the computer. It is not the intent of this book to teach DOS, so we will only cover enough information on shells to allow you to perform basic operations. However, there is a brief discussion on using shells in the following sections.

This concludes our discussion of the similarities between DOS and Windows 3.X. Now let us discuss the differences between the two.

## 5.3 DIFFERENCES BETWEEN MS-DOS AND WINDOWS

From the previous discussions, you have probably noticed that DOS and Windows 3.X have many similarities. This is due primarily to the fact that Windows 3.X is basically an add-on to the DOS operating system. This statement will be clarified later, but first let us discuss the primary differences between the two operating systems. We will discuss DOS in the next few sections and Windows 3.X starting in section 5.5.

## 5.4 DOS

As stated previously, DOS is a real-mode operating system. In most of today's computers, however, the microprocessor is actually running in the virtual 86 mode if DOS is used. This is because DOS is usually used along with a high-level OS, memory manager, or DOS application that uses the protected mode to execute. Operating in virtual 86 has no effect on the use and operation of DOS itself. However, it should be mentioned that some DOS applications may have problems running in the virtual 86 mode. When this is the case, the documentation that is supplied with the application usually contains the steps needed to remedy this situation.

Because DOS is a real-mode operating system, it cannot multitask applications. Not being able to multitask means that only one DOS application can be running in the computer's RAM at a time. Although this might seem like a disadvantage to some, it allows DOS to require very few of the computer's resources. This important feature of DOS will be discussed in later sections. For now, however, let us discuss the very important shells used by DOS.

### 5.4.1 The Shells Used by DOS

DOS versions below 4.0 have only one shell available. This shell is usually located on the disk in the root directory under the filename COMMAND.COM. DOS versions 4.0 and above have a second shell available called DOS-shell. These shells are different from each other, so each will be discussed separately.

### 5.4.2 COMMAND.COM

COMMAND.COM is the default shell used by all versions of DOS. Due to reasons that, hopefully, should become clear during this discussion, this is the shell you should learn. In computers, the term **default** means this is what the computer will use if not told differently. COMMAND.COM is a very cryptic shell that requires the user to learn a bunch of DOS commands that give instruction to the computer in a form of a command line. If COMMAND.COM is the shell running your computer, a prompt appears on the screen similar to the one shown in Example 5.4. The prompt indicates that the computer is waiting for the operator to tell it what to do in the form of a DOS command line or the name of an application program the operator wishes the computer to execute.

**Example 5.4:** DOS prompt

```
C:\>_
```

Referring to Example 5.4, the DOS prompt consists of a letter that indicates the default drive. In this case, the default drive is C. This can be any letter from A to Z depending on the number of drive letters available on the computer. The backslash (\) precedes the name of the working (default) directory. There is no name following the backslash, so the root directory is the default directory. Following the backslash is the greater-than sign (>). The greater-than sign is simply a command separator. To the immediate right of the greater-than character is a blinking underscore (_) that is called the **cursor.** The cursor indicates where information will be displayed on the screen as the user types on the computer keyboard. Another example of a DOS prompt is shown in Example 5.5.

**Example 5.5:** Another DOS prompt

```
D:\DOCS\IMPORT>_
```

In Example 5.5, D is the default drive and the default directory is a subdirectory named IMPORT. The subdirectory IMPORT is located in another subdirectory named DOCS. The subdirectory DOCS is located in the root directory on the default drive D.

If you have never used DOS, you are probably wondering, "Now that I know about the prompt, what do I do next?" To answer this question, you need to read on.

### 5.4.3 Internal and External DOS Commands

At the prompt, the user types the name of the application program or a DOS command line. All DOS command lines begin with a command and all commands fall into one of two categories that are referred to as *internal* and *external.*

When COMMAND.COM loads, it does not contain all of the program modules to execute all of the available DOS commands. In fact, COMMAND.COM contains only those DOS commands that are most frequently used. The other DOS commands that are used less often are left on the disk and loaded into the RAM by the user through COMMAND.COM as needed. For this reason, the DOS commands loaded as part of COMMAND.COM are called *internal commands,* whereas the DOS commands left on the disk are called *external commands.*

Table 5.4 shows a list of commonly used DOS commands with a brief explanation of each. The internal and external commands are indicated by an "I" or an "E" in the "Type" column. This list is by no means the entire list of DOS commands available under the DOS prompt. For more information on how

**Table 5.4** Basic DOS Commands

| Command | Function | Type |
|---|---|---|
| ATTRIB | Sets, changes, or displays a file's attributes | E |
| BACKUP | Backs up one or more files from one disk to another | E |
| CHDIR (CD) | Changes the working (default) directory | I |
| CHKDSK | Checks for file allocation errors | E |
| CLS | Clears the computer screen and returns the cursor to the upper right-hand corner of the display | I |
| COPY | Copies the specified file(s) to the specified location(s) | I |
| DEL | Deletes the specified file(s) | I |
| DIR | Lists the contents of the specified directory | I |
| DISKCOPY | Copies the entire contents of one diskette to another of the same density | E |
| FORMAT | Formats a floppy diskette for hard drive to receive files | E |
| LABEL | Places a software label on the diskette or hard drive | E |
| MEM | Displays the computer's RAM configuration | E |
| MKDIR (MD) | Creates subdirectories or folders | I |
| PATH | Sets a directory search path | I |
| PRINT | Prints the contents of a text file to a printer | E |
| PROMPT | Changes the appearance of the MS-DOS prompt | I |
| RENAME (REN) | Changes the name of a file | I |
| RESTORE | Restores files that were previously backed up with the BACKUP command | E |
| RMDIR (RD) | Removes and empties directory | I |

to actually use the DOS commands and the command line, refer to the DOS user's manual or any of the dozens of books that have been written on DOS. A DOS quick reference is supplied in Appendix C.

What one has to realize is that even though the command prompt is visible on the computer's screen, it does not mean that all of the DOS commands are available. The only commands that can truly be counted on are the internal commands. When DOS is installed and used from a hard drive, all of the external commands available with the version of DOS installed will probably be available. If DOS is booted from a diskette, however, most of the external commands will not be available because the DOS boot files will require most of the diskette's storage capacity. This is the reason why DOS is provided on several diskettes. Also, DOS versions above 3.3 are all designed to be installed on a hard drive. To limit the number of diskettes required, most of the external commands are saved on the installation diskettes in compressed form and are uncompressed by the installation utility.

It will, hopefully, become obvious during the remaining discussions that having a diskette available that will boot the computer to MS-DOS is very important, even if the computer is running one of today's advanced operating systems. It is also important to have the external commands that will most likely be needed on this boot disk. The procedure for creating a bootable DOS diskette is given in section 5.4.7. For now, however, let us get into our discussion on basic DOS command lines.

### 5.4.4 Basic DOS Command-line Syntax

Basically, **syntax** is the rules that govern the way something is written. Proper syntax includes both correct spelling and correct punctuation. Current operating system shells do not have the ability to reason, so the user must learn and use the correct DOS command-line syntax or the operating system will not carry out the command, or will carry it out incorrectly. This can cause major problems.

Because the user supplies most of the information for each command, the syntax in most books uses symbols to represent different user-defined options. The user provides the optional information to tailor each command for his or her specific needs. For example, the syntax from Appendix C for the COPY command is shown in Example 5.6. The COPY command will copy the contents of a source file to a target location. If desired, the COPY command can also be used to give the target file a different name than the one used by the source file. Because it is not the intent of this book to teach DOS, COPY will be the only DOS command line we will discuss in detail in this section. Any time a DOS command line is given during the remainder of this textbook, its entire syntax will be given.

**Example 5.6:** Syntax of the copy command

> COPY [s:]sfilespec [t:][tfilespec]

Before we discuss the COPY command, let us review some general information about the command-line syntax in Example 5.6. First, the placement of spaces is very important because they are used to separate the different sections of the DOS command line. The placement of spaces is probably the one thing that gives people the most problems with command-line syntax. Second, brackets are used in the command-line syntax to indicate optional, user-defined information. If the user omits optional information, default information will be used. For example, if an optional drive letter is omitted, the default drive letter that is indicated by the DOS prompt will be used. Also, if the user omits the optional path, the one indicated by the DOS prompt will be used.

Every DOS command line begins with a DOS command. In the syntax shown in Example 5.6, the command line begins with the DOS command COPY. A space that separates the command from the other components of the command line follows the DOS command.

All of the information that pertains to the source file follows the COPY command. The source file is the file that is being copied. In the syntax shown in Example 5.6, the source file information includes "[s:]sfilespec." As shown, there are no spaces within the source information. The source file information starts with the combination of symbols [s:]. Because "s:" is enclosed within brackets, it is optional information that may or may not be supplied by the user. In this instance, the [s:] represents the letter of the drive where the source file is located. If a letter is not supplied, the default drive letter is used.

The letters "sfilespec" follow the optional source drive letter. These letters ask the user to supply the path to the filename that is going to be copied. Because "sfilespec" is not enclosed within brackets, this information is required and must be supplied by the user. A space follows the source file information to separate it from the rest of the command.

The sequence of symbols "[t:][tfilespec]" is next in the syntax. The [t:] is the optional entry where the user specifies the target drive letter. The target drive is where the file is to be copied. Once again, if this is omitted, the default drive indicated by the DOS prompt will be used.

The symbols [tfilespec] are last in the syntax example. This is the path to the target directory's locations. If this is omitted, the default directory path indicated by the DOS prompt will be used. If a filename is not specified, the target file will be given the same name as the source file. To help explain the COPY command-line syntax further, let us look at an example of a command line that will copy the contents of the file ILT.COM from the subdirectory ELECT into the subdirectory TOT under the same name (see Example 5.7).

**Example 5.7:** Copy command

> C:\>copy_\elect\ilt.com_\tot
>
> **Note:** The underscores show the location of spaces in the command line and are not typed.

In Example 5.7, the optional source and target drive letters are not given, so the default C, indicated by the DOS prompt, will be used. Because no filename is given as part of the target information, the target file will be given the same name as the source file.

Example 5.8 is another example of a COPY command line that will copy the file named PROG.TXT from the root directory on the A: drive into the CATALOG directory on the D: drive, changing the name to PROG1.TXT.

**Example 5.8:** COPY command example

> C:\>copy_a:\prog.txt_d:\catalog\prog1.txt
>
> Note: The underscores show the location of spaces in the command line and are not typed.

As stated previously, it is the intent of this textbook to teach hardware, not operating systems. However, it is very important for the technical professional to be proficient in using an operating system's shell. The default shell for MS-DOS is COMMAND.COM and is the shell you should learn. For this reason, Appendix C contains a list of the most often used DOS commands and their syntax.

In addition to COMMAND.COM, there is an optional shell included with DOS versions 4, 5, and 6 called *DOS shell*. Even though DOS shell is not used often by computer technicians, I have included a brief discussion of it in the next section.

### 5.4.5 DOS Shell

DOS versions 4.0 and above provide an optional shell called **DOS shell.** The DOS shell program is usually located in the DOS subdirectory under the filename DOSSHELL.EXE. To use DOS shell, it must first be loaded into the computer's RAM by typing "dosshell" at the DOS prompt. DOS shell can also be loaded automatically during the boot process by placing DOSSHELL as the last line in the AUTOEXEC.BAT. The AUTOEXEC.BAT file will be discussed in greater detail in section 5.8.4.

When DOS shell is used, the computer's display is more graphics-oriented, which makes it easier to use. DOS shell also gives the user the ability to use a mouse as a pointer if a mouse drive has been loaded. When using DOS shell, the application programs and OS commands available are listed in series of menus. To choose an application to run or a command to carry out, the user can use the cursor control keys on the edit keypad to move the highlighted area over the name of the desired application or command, and then press the Enter key. If a mouse is available, the mouse is moved to position the pointer on top of the application or command the user wishes to run. Then the left mouse button is pressed and released quickly two times in succession. This procedure is called a **double-click.**

If the DOS shell is set up correctly, it is very user-friendly and easy for the novice computer user to use. However, to set up the options in the different menus used by DOS shell requires knowledge of using the command line and directory organization. If you do not have this knowledge, someone will have to set the computer up for you. Figure 5.4 shows a typical view of DOS shell's main menu, which contains a list of different program and command groups.

This concludes our introduction to using DOS. If you need more information, consult one of the hundreds of texts that are available. Now we need to talk about the reasons why you need to learn how to use DOS even though you are using a newer and more advanced operating system.

### 5.4.6 Reasons for Learning DOS

If you are currently using Windows 95, Windows 98, Windows NT, or Windows 2000, you may think there is no need to learn how to use the DOS command line. Although this might be true for individuals who simply use the compatible computer, it is not true for anyone who wishes to assemble, upgrade, install, or repair compatible computers. It is also not true for anyone who installs and customizes operating systems.

**Figure 5.4** DOS Shell Main Menu

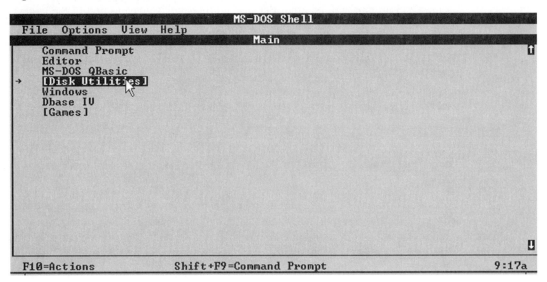

Advanced operating systems such as Windows 3.X/9598/NT/2000 are very complex and require many computer resources. As a result, these operating systems have a tendency to crash occasionally. When this occurs, the OS or the operating system's configuration files can become corrupted and the OS will no longer boot the computer. If the OS no longer boots, the computer is dead. When this happens, you need an OS that can boot the computer from a floppy and requires very few computer resources. DOS is such an operating system. Once the computer is operational, the advanced OS can often be repaired or reinstalled.

Another reason why you will need to learn to use DOS is that when Microsoft releases a new OS, it has to be installed from a previous OS. If the new OS can be installed from the previous-generation Microsoft OSs, it is called an *upgrade*. For example, a Windows 95 Upgrade is usually installed from Windows 3.X. All current upgrades of Microsoft operating systems can also be installed from DOS, however. To install from DOS, the original installed diskettes or CD-ROM from the previous-generation OS that you are upgrading from will be needed during the installation process.

If the new operating system can only be installed from DOS, it is called a *full install*. Full-install versions of the OS cannot be installed from any other operating system except DOS. Full installs are usually used when the OS is first installed on the computer after it is assembled.

I hope this discussion has shown you why all of the people who believe that DOS is dead are terribly mistaken. Even though there are no new major applications currently being written for DOS, the basic use of the DOS operating system is required by any person who plans to repair, upgrade, or install OS on compatible computers.

It is seldom necessary to have a complete version of DOS with all of its external commands when it is simply used to boot the computer so that the main operating system can be repaired. For this reason, a simple DOS boot disk is often created with a few of the external commands that are needed for the repair job. The procedure used to create a DOS boot disk is explained in the following section.

### 5.4.7 Making a DOS Bootable Floppy

Every so often it becomes necessary to boot the computer from a floppy diskette instead of the hard drive. When this situation arises, you will need to have a bootable floppy available. When a bootable disk is created, it contains only the two DOS system files and the shell COMMAND.COM. It does not contain any external DOS commands.

A DOS boot diskette can be created in two ways. First, if the diskette is already formatted, the SYS command can be used. Second, a DOS boot diskette can be created as part of the DOS format command using the /S switch. Both of these methods are shown in Example 5.9.

**Example 5.9:** Creating a boot diskette on a previously formatted blank diskette

> C:\>SYS_A:
>
> Creating a boot diskette during the format process
>
> C:\>format_a:_/s
>
> **Note:** Underscores indicate the location of spaces and are not typed.

---

### Procedure 5.1: Copying additional files to the DOS boot diskette

1. Place the boot diskette in the A: drive.
2. At the C:\> DOS prompt, type the following command lines. The computer's Enter key must be pressed at the end of each line.

Copy_c:\dos\format.com_a:
Copy_c:\dos\fdisk.com_a:
Copy_c:\dos\attrib.com_a:

**Note:** Underscores show the location of spaces and are not typed.

---

### Procedure 5.2: Shutting down DOS

1. If possible, always turn the computer off with the DOS prompt display. Never turn the computer off with an application running. Always exit any application before shutting down the computer.
2. Before you exit the application, save any data to a diskette or hard drive, if required. The procedure for exiting an application and saving data should be in the documentation that came with the application.
3. Remove any diskette from the floppy disk drives and store them properly. Finally, turn the computer off by following the procedure outlined in the computer's documentation.

---

Along with the DOS system files and COMMAND.COM, you will probably want to include the external DOS commands FORMAT, FDISK, and ATTRIB. An example procedure that will copy these files from the hard drive to the boot diskette is shown in Procedure 5.1. The actual procedure used on your computer may vary depending on the computer's directory structure.

### 5.4.8 Shutting Down DOS

To avoid corruption of the files used by DOS applications, the steps outlined in Procedure 5.2 should be followed before a computer running DOS is turned off. Sometimes the computer will lock up and this procedure cannot be followed. If this happens, the computer must be rebooted using either the soft- or hard-boot sequence covered in section 5.7. Note, however, that if the computer is booted with an application running, all unsaved data will be lost.

## 5.5   WINDOWS 3.X

Even though no new version of Windows 3.X has been released in several years, it is still a very popular operating system for compatible computers above a 286. One reason for its popularity is that Windows 3.X added the ability to run 386 and above microprocessors in the protected mode. The protected mode gives the ability to tap some of the potential of the 386 and above microprocessors. Another reason is that Windows 3.X uses a GUI shell. If set up correctly, the GUI shell makes Windows

3.X very user-friendly. If you are considering using Windows 3.X, you need at least a 386DX machine with a minimum of 4 MB of RAM installed. However, 8 MB is recommended to allow more efficient multitasking.

### 5.5.1 Some Things You Need to Know about Windows 3.X

Windows 3.X is a **parasite operating system** that runs on top of DOS. This statement means several things. First, two operating systems—both DOS and Windows 3.X—have to be installed on your hard drive in order to boot Windows 3.X. Second, DOS must first be loaded into the computer's RAM, and then Windows 3.X is loaded on top of DOS by typing WIN at the DOS prompt or by a line in the AUTOEXEC.BAT file. Third, because Windows 3.X runs on top of DOS, most of the hardware drivers are written around the microprocessor's real mode.

Windows 3.X can actually be operated in two modes: **standard mode** and **enhanced mode.** Standard mode will run on a 286-based computer or a 386DX computer with less than 4 MB of RAM. The standard mode does not allow multitasking of DOS applications and will not be covered further in this textbook.

The enhanced mode requires at least a 386DX microprocessor-based machine that has a minimum of 4 MB of RAM. The enhanced mode provides multitasking of Windows and DOS applications as well as advanced hard drive and memory management. If the computer on which Windows 3.X is installed meets or exceeds these hardware recommendations, the enhanced mode is automatically selected during the installation process of Windows 3.X.

There is another thing you should know about Windows that prevents it from tapping the full potential of the 386DX and above microprocessors, even when operating in the enhanced mode. Because Windows 3.X is written to operate on computers that use either a 286 or 386 microprocessor, most of the operating system's machine code is written using 16-bit instructions. This is because 16-bit instructions can be executed by either the 286- or 386DX-compatible microprocessors. You may ask, "Why write 16-bit code for a 32-bit microprocessor?" The answer is, "If Microsoft wrote 16-bit code for the standard mode and a 32-bit code for the enhanced mode, the operating system would be so large it would probably occupy an entire hard drive."

**Problem 5.5:** In which mode must Windows 3.X be running in order to multitask?

**Solution:** Windows 3.X must be running in the enhanced mode.

### 5.5.2 Why the Name *Windows*?

Instead of using the command lines required by DOS, Windows 3.X uses a graphical user interface (GUI). The GUI consists of the entire computer's screen, small pictures called *icons,* and rectangular boxes called *windows*. All are located at various positions on the display.

The entire screen is referred to as the *desktop*. The desktop itself can be a graphics image or a solid color that is set by the user. A graphic image that is displayed for the desktop is called *wallpaper*. Rectangular images on the desktop are referred to as *windows* because of their appearance. The default window that is displayed when Windows loads contains small pictures called *icons*. An icon is a picture that is used to launch a program or execute an operating system command.

When an application program is running, it must be able to display information on the computer's display. In a multitasking operating system, multiple application programs can be running at the same time. To keep things from getting jumbled up, each application must have its own section of the display. Windows performs the separation by creating a rectangular box on the display for each application that is running. It is in this dedicated rectangle—a window—that each application displays its unique information. I guess you have figured it out by now that this is how Windows got its name.

An example of a typical application window is shown in Figure 5.5. The different components that make up this typical window are labeled. Most application windows are basically the same, so we will discuss the basic functions of each labeled component shown in Figure 5.5.

**Figure 5.5** Typical Windows 3.X Application Window

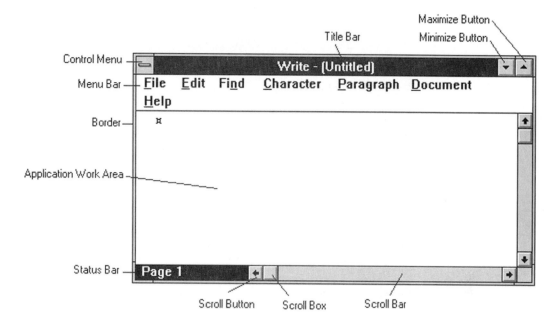

The *Title Bar* runs across the very top of the window. The function of the Title Bar is to display the name of the application running within the window. This information is included in the program's kernel, and not the name of the file that is loaded and running in the computer's RAM.

On the left and right sides of the title bar are squares that look like a button. In fact, because of their appearance and use, these images are called *buttons*. Positioning the mouse pointer over the button, and then clicking the left mouse button once simulates pressing the button.

The button on the left-hand side of the Title Bar that is labeled with a minus (−) sign is used to close (exit) an application along with its window. On the right-hand side of the Title Bar there are two buttons: one with an up arrow and the other with a down arrow. These buttons are used to minimize and maximize the program's window. When the program is *maximized,* its window will be displayed on the desktop. When the program is *minimized,* it is reduced to an icon on the desktop and is switched to the background. Background programs will be discussed in greater detail in section 5.5.7. But basically, when a program is running in the background, information cannot be entered into the application by using either the keyboard or the mouse.

The *Menu Bar* is immediately under the Title Bar. Each word on the Menu Bar is the title of a menu. A menu title is selected by clicking on it with the mouse. This will cause a menu to pop down from the Menu Bar, listing the items available with the selected menu.

The large box directly below the Menu Bar is called the application's *work area.* It is within this area that the application displays its information and the user inputs information. The size of the entire window can be adjusted by clicking and holding the left mouse button on the window's borders and dragging the border to the desired location. This procedure can only be done when the program is running in the foreground, however. Running a program in the foreground is discussed in section 5.5.7.

Now that we have examined the components of a basic window, we move our discussion to the amazing little mouse.

### 5.5.3 The Mouse

Unlike MS-DOS, Windows 3.X uses a mouse as the primary input device to control the computer's operation. The technical aspect of the operation of the mouse itself will be covered in chapter 9. In this section we will discuss only its use.

A typical two-button mouse is shown in Figure 5.6. As the mouse is moved across a flat surface with the user's hand, a pointer will move across the Windows 3.X desktop. Using the mouse, the pointer is

**Figure 5.6** Typical Two-button Mouse

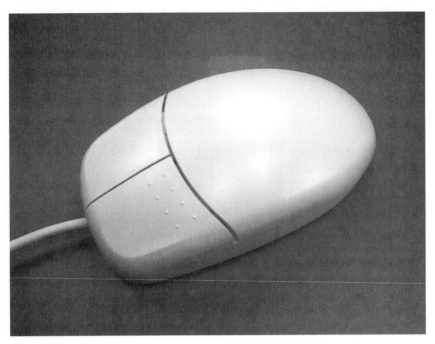

positioned over an icon or a word on a window's Menu Bar. Once the pointer is in position, the user has three options.

First, the user can click on the item. A *click* is performed when the user rapidly presses and releases the left mouse button one time with the pointer positioned over the item. A click is used to select an item.

Pressing and releasing the left mouse button twice in succession is called a *double-click*. A double-click is performed with the mouse pointer positioned over an icon and is used to launch an application or execute a command.

Third, with the mouse pointer positioned over an icon or filename, the left mouse button is pressed and held. With the left mouse button held in the down position, the mouse is moved. As the mouse moves, the selected item moves with the mouse pointer. When the item is at the desired location, the left mouse button is released. This procedure is called *dragging*.

### 5.5.4 Windows 3.X's Shell

The default shell for Windows 3.X is called **Program Manager.** Program Manager is loaded as part of the Windows 3.X boot sequence and is located in the Windows subdirectory under the name PROGMAN.EXE. Like other Windows 3.X applications, Program Manager is displayed as a window on the desktop when it is maximized. An example of Windows 3.X Program Manager's window is shown in Figure 5.7.

Windows 3.X's Program Manager displays icons in units called *program groups*. The icon used to identify a program group is labeled in Figure 5.7. To open a program group, simply double-click on the group's icon. When the window for the selected program group opens, there will be additional icons that represent the applications within the program group. An example of the contents of the open program group GAMES is shown in Figure 5.7.

Programs can be launched from Program Manager in three different ways. First, an icon can be created for the application in a program group. Double-clicking on the icon launches the application. Second, the program that launches the application can be located by using File Manager, and then double-clicking on the program's name. Third, the Run window can be opened and the path and name of the programs can be typed in. The procedure to follow to launch an application is covered in the Windows 3.X user's manual.

**Figure 5.7** Windows 3.X Program Manager

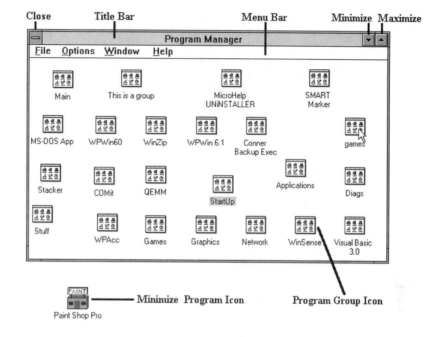

## 5.5.5 Windows File Manager

Windows 3.X uses a GUI application called *Windows File Manager* to allow the user to manage the files and directories. The name of the program that launches File Manager is WINFILE.EXE and it is located in the Windows subdirectory. It is nice to know the actual name of the program, but not really necessary because File Manager is launched with an icon in the Main Program group.

To start Windows File Manager, double-click the MAIN program group in Program Manager. This will cause the MAIN program group window to display. An example of a typical MAIN group is shown in Figure 5.8. Once the MAIN program group window opens, double-click on the File Manager icon (small file cabinet). The typical File Manager icon is shown in Figure 5.8. A File Manager window is shown in Figure 5.9.

Managing files in Windows is quite simple. The process of copying, deleting, and moving files is mostly done by using the mouse to position the pointer over the file or file folder you are interested in, then clicking the left mouse button either once or twice. To find the procedures for the many functions of File Manager, consult the Windows 3.X user's manual.

## 5.5.6 File Association

As stated previously, files stored on a disk fall into one of two categories—program files or data files used by a program. Data files are usually unique to a certain applications. As a way of identifying what data files belong to what applications, file extensions are usually used. For example, Microsoft Word® data files use the DOC extension whereas Corel WordPerfect® data files use the WPD extension. Most people do not know the difference between a program file and data file. Because this is the case, there may be a problem if someone attempts to run a data file by double-clicking on the name of a data file using Windows File Manager. To take care of this problem, Windows 3.X uses what is called *file associations*.

**File association** is a method in which data file extensions are associated with an application program. If an attempt is made to run a data file, the application program whose data files use the extension will be launched automatically and the data from the selected data file will be loaded automatically. Associations are usually established automatically when an application is installed on the computer's hard drive.

Problems can arise when associations that are assigned automatically by the installation of applications sometimes change an association that was previously set by an application already installed on the computer. This has happened to me a few times with programs that display graphics data files. For

**Figure 5.8** Main Program Group

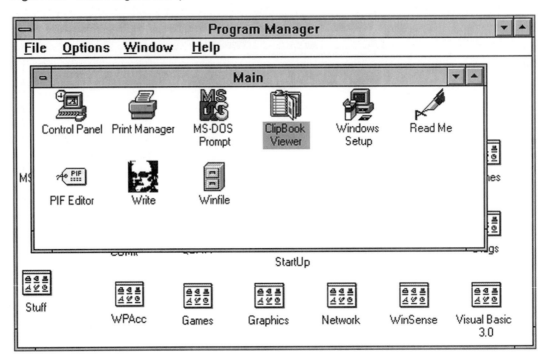

**Figure 5.9** File Folder in Windows File Manager

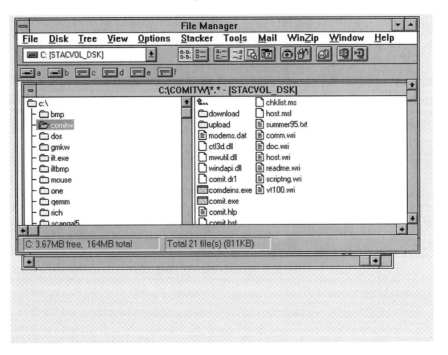

example, when I installed Microsoft's Internet Explorer®, it changed the associations of the graphics files with the JPG and GIF that were originally associated with a drawing program I use. To solve this problem, I had to change the association back to the original program manually. The procedure used to set and modify file associations is not covered in the Window's 3.X user's manual, so I will cover the procedure in a little greater detail.

> **Procedure 5.3: Display, create, or edit Windows 3.X data file associations**
>
> 1. Start Windows File Manager.
> 2. On File Manager's Menu Bar, click on File.
> 3. On the File menu, click on Associate.
> 4. The Associate window will appear. In the File Extension text box, type the extension for which you want to set or edit the association.
> 5. If an association has been created previously, the application's name will appear in the "Associate with" text window. If no association has been made, the word "None" will appear in this text box.
> 6. If you do not wish to change the association, click on the Cancel button.
> 7. The association can be changed three different ways:
>    a. Type the path and the name of the file that launches the application in the Associate With text box.
>    b. Use the scroll bar to locate the Windows 3.X applications that have been documented through installation.
>    c. Use the Browse button to find the program that launches the application. Once located, double-click on the filename and it will automatically be entered in the Associate window.
> 8. Once completed, click on the OK button to establish associations.

File associations are set and edited through Windows File Manager. To open Windows File Manager, double-click on the Main Program Group, then double-click on the File Manager icon. Once File Manager is open, follow Procedure 5.3 to display, create, or edit data file associations.

### 5.5.7 Foreground and Background Programs

Programs that are multitasking under Windows 3.X run in either the foreground or the background. There can only be one foreground program; all other programs are background programs. The **foreground program** is the program the user is currently inputting information into through the computer's keyboard or mouse. The foreground program is also the program that is getting the most execution time.

An application program is brought to the foreground by simply clicking the left mouse button on the application's window. Programs can also be brought to the foreground by pressing and holding the Alt key, then pressing and releasing the Tab key on the computer's keyboard. This key sequence will cause a small window that lists the name of the current foreground program to pop up in the middle of the screen. By repeatedly pressing the Tab key while holding down the Alt key, the name of each application that is being multitasked will be listed one at a time. When the name of the application you wish to bring to the foreground appears in the window, simply release the keys. This procedure is quite useful when there are many program windows on the desktop.

Programs that are not currently selected are called **background programs.** Applications running in the background get less computer execution time. Background programs are still running, but the user cannot enter information into a background program from the keyboard or mouse. In order for the user to enter information into the application, the program must be brought to the foreground using one of the procedures outlined in the previous paragraph.

### 5.5.8 Multitasking and Windows 3.X

Windows 3.X uses what is referred to as **cooperative multitasking.** As the name implies, cooperative multitasking requires each program that is running on the computer to cooperate with all other programs. If this does not happen, the computer will usually lock up. When it is a program's turn to execute some of its instructions, Windows 3.X's cooperative multitasking allows the program to take control of the computer. The program executes a predetermined number of instructions and then turns control of the computer back over to Windows 3.X. Windows 3.X then turns the computer over to another program being multitasked. This process continues until each program has its turn to execute some of its instructions. The process is repeated over and over.

The cooperative multitasking used by Windows 3.X can cause a few problems. First, if the program does not return control back over to Windows 3.X, the other applications running will not get

> ### Procedure 5.4: Loading a DOS application's PIF file
>
> 1. With the PIF Editor running, click on the File option in the Menu Bar.
> 2. Select the Open option. The Open window will display the current PIFs in the Windows directory.
> 3. Click on the name of the PIF file you wish to display or edit.
> 4. Click on the OK button.
> 5. The PIF file will load into the PIF editor.

any of the microprocessor's time. Second, if the program that is controlling the computer crashes, it will usually lock up the computer, requiring the user to reboot. Any time the computer is rebooted, all of the information in the computer's RAM will be lost. Third, if an application program crashes and Windows 3.X does not, Windows 3.X cannot recover the resources that were being used by that faulty application. This usually means that the computer will have to be rebooted and Windows 3.X reloaded. If this is the case, make sure the information from programs that did not crash is saved before the computer is rebooted.

Believe me, cooperative multitasking can cause a lot of heartbreak. Even with these problems, Windows 3.X's ability to multitask applications is far superior to DOS's ability to run only one application at a time.

### 5.5.9 Program Information Files

Because DOS applications are not written for the Windows 3.X operating system, Windows 3.X needs information about the DOS application so it will know how to handle it when it is being multitasked. For this purpose, Windows 3.X uses a **program information file (PIF)** to hold information about each DOS application.

A PIF file for each DOS application is located in either the Windows directory or the DOS application's directory. The PIF file uses the same name as the program that launches the DOS application except it has a PIF extension instead of the EXE or COM. When Windows 3.X is told to launch the DOS application, Windows 3.X first searches for the PIF file that belongs to the DOS program. If Windows 3.X does not find a PIF file, it automatically creates a generic one from a set of defaults that should work with most DOS applications. Note that the PIF file can also be used to launch the DOS application by double-clicking on the PIF's name using File Manager.

To create and edit the PIF files used by Windows 3.X, a PIF editor is provided. To launch the Windows 3.X PIF editor, double-click on the Main program group in Program Manager. Once the Main group window opens, double-click on the PIF Editor icon. Procedure 5.4 lists the steps required to load a DOS application's PIF file.

Figure 5.10 shows the first of two windows used by the PIF Editor. This window shows the basic setting, and Figure 5.11 shows the advanced setting window of the PIF file for a DOS program called *Microsoft Advanced Basic*. Table 5.5 gives a basic explanation of the function performed by each option provided by the PIF editor.

### 5.5.10 Information Files

Windows 3.X also gives the ability for Windows 3.X applications to supply information to the Windows 3.X operating system each time that an application is launched. Text files (ASCII) called *information files* perform this task. If needed, the Windows information file for the Windows 3.X application will be located in the \Windows\System subdirectory. Windows information files can be identified by the INI extension. Because of the extension, these files are often referred to as "any" files.

In addition to supplying configuration information about each Windows 3.X application, INI files are also used to configure the Windows 3.X operating system itself. The Windows 3.X configuration files are named SYSTEM.INI, WIN.INI, PROGMAN.INI, WINFILE.INI, and CONTROL.INI. These Windows 3.X configuration files are discussed in greater detail in section 5.8.9.

**Figure 5.10** PIF Editor's First Window

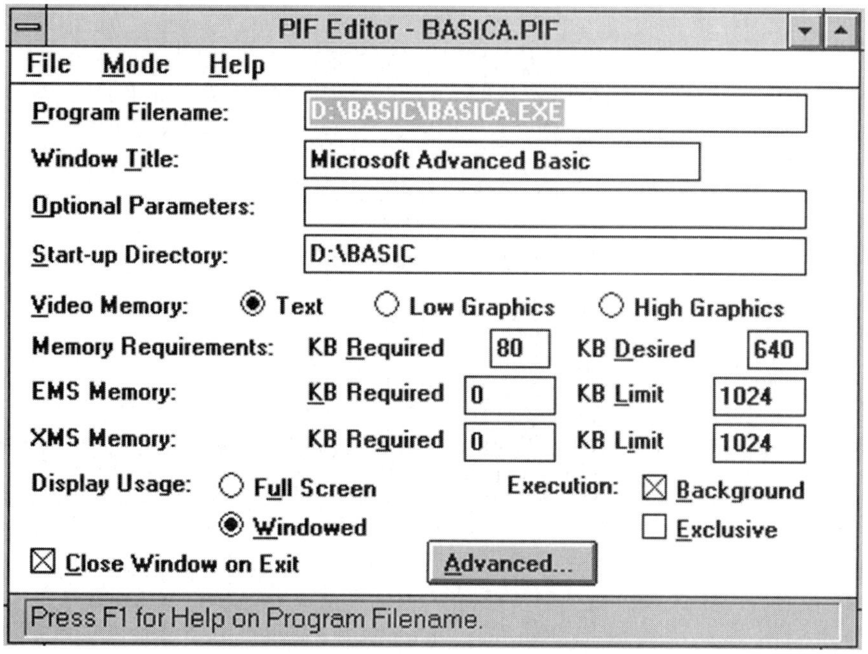

**Figure 5.11** The PIF Editor's Advanced Window

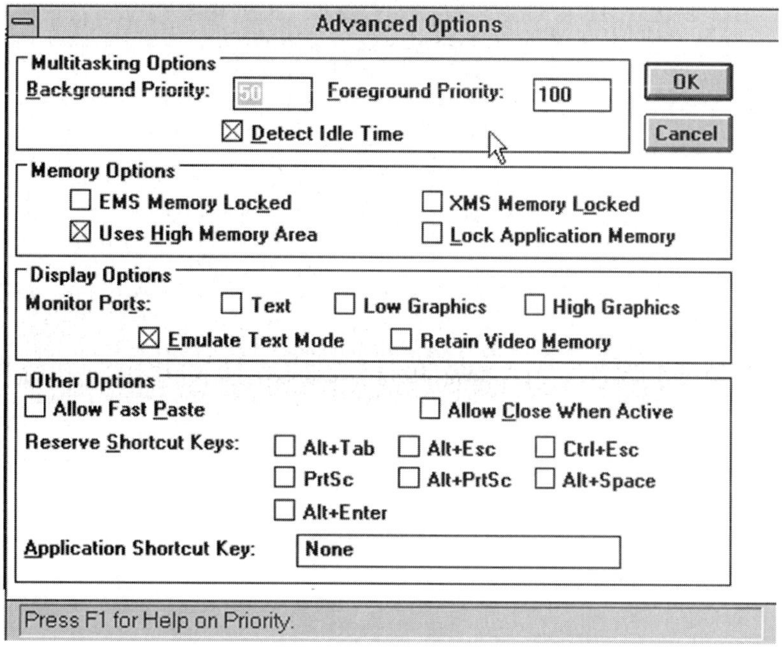

## 5.5.11 Shutting Down Windows 3.X

If at all possible, the computer should never be turned off with Windows 3.X running. This can cause the Windows 3.X operating system or Windows 3.X applications to become corrupted. When Windows 3.X is shut down, it will automatically close all of the open files that are being used by the OS. It will also update the Windows 3.X configuration files that will be discussed in later sections. Procedure 5.5 lists the steps that should be followed to shut down Windows 3.X.

**Table 5.5** PIF Editor Options

| Option Title | Description |
|---|---|
| Program Name | The name of the actual executable (EXE) or command file (COM) that launches the DOS application |
| Windows Title | The description of the program that will appear in the program's window Title Bar |
| Optional Parameters | If the DOS program requires any switches or options after the name of the program, they should be entered here. |
| Start Up Directory | The DOS path to the location of the file |
| Video Memory | A check in one of these options determines the default video mode Windows 3.X will use when the program is launched. |
| Memory | Allows the user to enter a minimum value (required) and maximum (desired) values of conventional memory that will be assigned to the program |
| EMS Memory | Allows the user to enter the minimum (required) and maximum amount of expanded memory used by the application. If none is needed, enter 0. |
| XMS Memory | Allows the user to enter the minimum (required) and maximum amount of extended memory used by the application. If none is needed, enter 0. |
| Display Usage | Selects whether the application will run within a window or full screen. |
| Execution | Background is selected to allow the program to run in the background. Exclusive is selected to disable multitasking when the program is run. |
| Close Window or Exit | Automatically closes the window the program is running in when the DOS application is terminated |
| ADVANCED WINDOW | Entered by clicking on the Advanced button |
| Multitasking Options | The Background and Foreground options are used to configure the priority the program has when it is running one of the indicated modes. The larger the number, the higher the priority. Detect Idle Time check box allows Windows 3.X to give a higher priority to other programs when this program is waiting for entry from a keyboard. |
| Memory Options | All of the lock options in this frame prevent the information in the indicated memory category from using virtual memory. The Use High Memory Area allows the program to have access to the HMA. For more information, refer to chapter 7. |
| Display Memory | Tells Windows 3.X to make sure the DOS application uses the video mode selected by the check boxes. |
| Other Options | Fast Paste allows the fast pasting of information into the program. Some DOS applications do not support this feature. Reserve Shortcut Keys is used to reserve shortcut key sequences normally used by Windows 3.X for the DOS application. Allows Close When Active terminates the program as soon as the window it is running is closed. The program will be terminated regardless of the normal shutdown procedure and any data being used by the program will be lost. |

## Procedure 5.5: Shutting down Windows 3.X

1. Exit all Windows 3.X applications using the procedure outlined in the application's documentation.
2. Click on the minus sign in the upper left-hand corner of the Program Manager window.
3. Click on the Exit Windows option in the pop-down menu.
4. Click on the OK button on the Exit Windows information box.

## 5.6 SUMMARY OF THE DIFFERENCES

This concludes our discussion on the differences between the DOS and Window's 3.X operating systems. Now that we have a basic understanding of the operating system itself, let us now discuss how the operating system is actually loaded from the disk into memory. We will also discuss how the user can configure the operating system for the computer it is operating on and for the user's individual needs.

## 5.7 BASIC BOOT SEQUENCE

To *boot* a computer means to tell the computer to load its main operating system. In the compatible computer, this task is accomplished in three ways: (1) when the power is turned on; (2) when the RESET button, usually located on the front of an AT or above computer, is pressed and released; and (3) when the keys labeled Ctrl and Alt are pressed together and held, then the Del key is pressed. For obvious reasons, the last method is also referred to as simply Ctrl+Alt+Del.

As discussed previously in chapter 2, section 2.4, when power is turned on or the RESET button is pressed and released, this will create a pulse on the microprocessor's RESET pin. When this happens, the microprocessor will be forced to go to the computer's ROM looking for the first instruction of the reset routine. For this reason, this process is often referred to as a *cold-boot, hard-boot,* or *hardware reset.*

The keyboard is a peripheral attached to a port that is controlled by the operating system, so when the Ctrl+Alt+Del key sequence is pressed, the response of the computer will vary depending on the operating system running. These differences will be discussed in the section on each operating system. The Ctrl+Alt+Del sequence is controlled by software, so it is referred to as a *soft-boot* or *warm-boot.*

To fully understand the compatible computers, it is necessary to understand the sequence that compatibles follow each time they are booted. The basic **boot sequence** used to load MS-DOS into the computer RAM is covered in Table 5.6. The microprocessor actually goes through several steps before it attempts to load the main operating system from the disk. These steps will be the same no matter which operating system is loaded. The steps that are common to booting all operating systems are listed in Part I of Table 5.6.

As indicated by step 5 in Table 5.6, the BIOS attempts to load the operating system's boot files from the default disk drive. Some of the boot files are required and some of the boot files are optional. If any of the required files are not found, the boot process will terminate and an error message will be displayed. If any of the optional boot files are not found, the computer will usually continue the boot process. However, the performance of the computer will probably be greatly degraded.

The required files and optional files required to boot MS-DOS and Windows 95 and 98 have the same names but are significantly different. The files loaded during the boot sequence for DOS are shown in Table 5.6, whereas the boot sequence for Windows 95 and Windows 98 will be discussed in chapter 6. Because Windows 3.X is a parasite operating system, it never boots the computer from power-up, reset, or Ctrl+Alt+Del. All of the files in Table 5.6 are listed in the order that they are loaded.

### 5.7.1 Booting DOS

Table 5.6 lists an abbreviated version of the actual boot sequence followed by the compatible computers. I give an abbreviated version because the unabridged version is very complicated and is not really necessary to understand. In Table 5.6, the DOS boot sequence is divided into two parts. The first part of the boot sequence is controlled by the BIOS and will be the same no matter what operating system is loaded. The second part of the boot sequence is controlled by the operating systems installed on the computer's disk drive and will vary depending on the boot filenames required by the operating system.

Referring to Part II of Table 5.6, the files IO.SYS and MSDOS.SYS comprise the main portion of the DOS operating system. IO.SYS is loaded by the Boot Record first and contains all of the low-level drivers that interact with the drivers in the BIOS.

MSDOS.SYS is loaded by the IO.SYS and contains the programs that handle the operating system's disk drives and file access. The file named COMMAND.COM is DOS's default shell, which is discussed in section 5.4.2. COMMAND.COM contains all of the internal DOS commands along with the ability to load external commands and DOS applications from the disk drives.

Table 5.6  Basic DOS Boot Sequence

| | Part I: Basic Boot Sequence Controlled by the BIOS for the IBM-compatible Computer |
|---|---|
| 1. | The BIOS gets its configuration information from the CMOS. |
| 2. | The BIOS initializes the motherboard's programmable ports and chipset. |
| 3. | The BIOS checks for a warm- or cold-boot. If a warm-boot is indicated, a large portion of the POST (power-on-self-test) will be skipped. |
| 4. | The BIOS performs a central hardware test called the POST to verify that the computer's basic functions are operational. If the POST completes without errors, the computer will usually beep once. Any error will result in a multiple beep code and an error message will usually be displayed. |
| 5. | The BIOS performs a scan of addresses C0000 through DFFFF looking for video and other adapter ROMs. This is covered in greater detail in chapter 8. |
| 6. | The BIOS loads a program from the very front of the diskette called the *Volume Boot Record* if booting from a floppy or the *Master Partition Boot Record* if booting from a hard drive. The hard drive's Master Partition Boot Record will point the boot process to the location of the Volume Boot Record on the bootable partition. The Master Partition Boot Record, Volume Boot Record, and bootable partition for hard drives will be discussed further in chapter 11. No matter the source—floppy or hard-drive—the Volume Boot Record program is loaded into the computer's RAM and control of the computer is turned over to the Boot Record. |
| | Part II: Portion of the Boot Sequence Controlled by the Boot Record |
| 7. | The required file IO.SYS is loaded into the computer's RAM. |
| 8. | Control of the boot sequence is turned over to the IO.SYS. |
| 9. | The required file MSDOS.SYS is loaded into the computer's RAM. |
| 10. | If present, the commands in an optional file named CONFIG.SYS are executed. |
| 11. | The required file COMMAND.COM is loaded into the computer's RAM and control of the computer is turned over to COMMAND.COM. |
| 12. | If present, COMMAND.COM executes the commands in an optional file named AUTOEXEC.BAT. |

The AUTOEXEC.BAT and CONFIG.SYS files are optional files that are used to load drivers and change the default hardware configuration used by DOS and Windows. These files will be discussed in detail later in sections 5.8.4 and 5.8.5, respectively.

The files IO.SYS and MSDOS.SYS cannot be seen when the directory of the drive is listed using DOS's DIR command. This is because these files are automatically given the attributes hidden, system, and read-only. If the DOS command CHKDSK is used on the drive the computer booted from, it will indicate that there are at least two hidden files on this disk. These two hidden files are IO.SYS and MSDOS.SYS. They are visible using the FILE LIST command under DOSSHELL or Windows 3.X File Manager when the View Hidden Files option is enabled. For more information, consult an operating system's user manual on setting up the OS to display hidden files.

### 5.7.2  Loading Windows 3.X

Windows 3.X is loaded into the RAM of the computer just like any other application. Before any attempt is made to load Windows 3.X, the computer must first be booted to MS-DOS. Also, an extended memory manager (XMM), similar to HIMEM.SYS (Chapter 7), must be loaded by a command in the CONFIG.SYS file (section 5.8.5). To load Windows 3.X, type WIN at the DOS prompt or enter WIN as the last entry in the AUTOEXEC.BAT file (section 5.8.4).

**Figure 5.12** Typical Configuration Switches for PC- and XT-compatible Computers

## 5.8 CUSTOMIZING YOUR COMPUTER

To *customize* your computer means to change it to your personal needs and the needs of the computer's hardware. This might include loading drivers or setting up directories, programs, or data to accommodate your task and suit your style. One way that the computer is customized is by changing its **configuration.** A computer's configuration is basically how the operating system handles the computer's hardware. Each time a compatible computer boots, it encounters at least two operating systems. The first is the one contained in ROM—the BIOS. The second is the one loaded from disk—the DOS. If Windows is being used, the computer encounters BIOS, DOS, and then Windows 3.X. The configuration for each of these operating systems is established in different ways.

### 5.8.1 BIOS Configuration

There are certain things the BIOS needs to know about the hardware installed in a computer so the correct software drivers in the BIOS will be used. In the IBM-PC- and XT-compatibles, this configuration information is supplied by small switches located on the motherboard. An example of these switches is shown in Figure 5.12. In the PC/XT-compatible computers, these switches are at I/O address 060H. On the compatibles above the AT, switches are replaced by the CMOS, which is discussed later. The BIOS reads the position of these switches or CMOS each time the computer boots to determine how to treat the indicated hardware devices.

Figure 5.12 shows the hardware options provided by each switch on the IBM-PC/XT computer. Knowing the information supplied by each switch is one thing, but knowing at which position to set the switch—ON or OFF—is another. To set the switches in the correct position, other charts (not shown), which are very confusing, are referenced.

When IBM introduced the AT, it decided to change the method used to establish hardware configurations for the BIOS. IBM chose to use a special type of RAM, called *CMOS,* to hold the 1s and 0s provided by the configuration switches on the PCs and XTs. CMOS is a special semiconductor that consumes very little power when not in use. This gives the ability to install a small battery on the motherboard to keep power on the CMOS-RAM when the computer is turned off. The 1s and 0s stored in the CMOS-RAM are usually placed there by running a program called **SETUP.** The location of the CMOS and use of the CMOS Setup program are covered in more detail in chapter 8.

## 5.8.2 DOS Configuration

When DOS loads, it uses a set of default configurations that will probably not meet the needs of your applications. To change the default hardware configurations used by DOS, an optional file named CONFIG.SYS is primarily used. This file is loaded after the two system files (IO.SYS and MSDOS.SYS), but before COMMAND.COM.

The AUTOEXEC.BAT file is another optional file that is used primarily to allow the execution of a set of DOS commands each time the computer is booted, but it can also be used to establish basic configurations such as loading drivers. The AUTOEXEC file's BAT extension indicates that it is a batch file. Before we start our discussion of the AUTOEXEC.BAT and CONFIG.SYS files, let's discuss batch files in general.

## 5.8.3 Batch Files

A **batch file** is basically defined as an ASCII (text) file that contains DOS commands. Batch files can be created to replace a set of commonly used DOS commands. Example 5.10 illustrates of a set of DOS command lines that will copy two files, GALLUPS.TXT and CHEROKEE.DOC, from a subdirectory named DOCUMENT, located on the A: drive, into a subdirectory named BLISTER on the B: drive.

Example 5.10: Command lines that will copy the two files GALLUPS.TXT and CHEROKEE.DOC from the A: drive subdirectory DOCUMENT to the B: drive subdirectory BLISTER

```
C:\>COPY A:\DOCUMENT\GALLUPS.TXT B:\BLISTER
C:\>COPY A:\DOCUMENT\CHEROKEE.DOC B:\BLISTER
```

Suppose the same set of command lines shown in Example 5.10 had to be used many times during the course of the day. It would be nice if there was a way to store the commands in a file and tell the computer to execute the commands in the file instead of having to retype the command line each time. Guess what? There is.

Batch files are created with any text editor or word processor that gives one the ability to save information in ASCII. A batch file can also be created by using the COPY command to copy information directly from the computer keyboard into a file. No matter the method, the file *has to* have the extension .BAT before it can be executed from the DOS prompt or with Windows Program Manager.

What if the command lines listed previously were saved in a batch file named MOVE.BAT? Then DOS would perform the list of commands by simply entering the name of the batch at the command prompt. An example of creating the batch file MOVE.BAT and then launching it through a command line is given in Example 5.11.

Example 5.11: Creating a batch directly for the keyboard using the copy CON command. Press the Enter key after each line.

```
COPY CON MOVE.BAT
C:\>COPY A:\DOCUMENT\GALLUPS.TXT B:\BLISTER
C:\>COPY A:\DOCUMENT\CHEROKEE.DOC B:\BLISTER
```

After the last command line entered, press the F8 key to close the file and return to the DOS prompt. Once entered, the command lines in the batch file will be executed automatically by typing in the following command line at the DOS prompt.

```
C:\>MOVE
```

The last command line in Example 5.11 will cause the commands in the batch file MOVE.BAT to be executed as if they were typed as individual command lines at the DOS prompt. One advantage of using batch files is that it saves time by reducing the number of keystrokes needed to enter long DOS commands. Another advantage is if someone sets up batch files to perform all of the tasks the computer operator needs, the operator can make the computer perform complex tasks without knowing anything about DOS.

### 5.8.4 AUTOEXEC.BAT

The **AUTOEXEC.BAT** file is a batch file with one unique characteristic: DOS will execute the commands in this batch file automatically each time that the computer is booted. The DOS commands in the AUTOEXEC.BAT are executed after COMMAND.COM is loaded into the computer's RAM. The AUTOEXEC.BAT file is not required to boot the computer. In fact, if an AUTOEXEC.BAT file does not exist on the disk from which the computer boots, the boot will usually continue without any errors. Even though an AUTOEXEC.BAT file is optional, it greatly enhances the overall operation of the computer. Being able to create and modify this file is very beneficial to those who have chosen a career that requires knowledge of the compatible computers. Example 5.12 is an example of an AUTOEXEC.BAT file and a basic explanation of each command.

**Example 5.12:** Example of an AUTOEXEC.BAT file

```
@ECHO OFF
CLS
PATH=C:\DOS\;C:\WINDOWS
C:MOUSE
PROMPT $P$G
WIN
```

Refer to Example 5.12 during this discussion. The first line will prevent the commands in the batch file from being displayed on the computer's screen as they are being executed. The second line will clear the computer's display. The third line will tell DOS to search the indicated directories if it cannot find the desired file in the default directory. The fourth line loads the mouse driver into the computer's RAM. The fifth line changes the DOS prompt to indicate the default directory and the default drive as part of the prompt. The sixth line will load the Windows operating system.

Teaching all of the uses of the AUTOEXEC.BAT file is beyond the scope of this book. However, I hope I have indicated the need to learn how to create and edit this file. As stated previously, if compatible computers are a part of your chosen career, make sure you obtain additional information on creating and editing the AUTOEXEC.BAT file.

One major reason for learning to edit the AUTOEXEC.BAT is that most DOS application programs will automatically update the AUTOEXEC.BAT and CONFIG.SYS files when they are installed on the computer. This is great and wonderful, but if the application is deleted from the disk, the user will have to know what statements to remove from the AUTOEXEC.BAT file because this is not done for you when the file is deleted.

### 5.8.5 CONFIG.SYS

The **CONFIG.SYS** is an ASCII file that you can create and edit with a text editor or word processor. The CONFIG.SYS file's main function is to allow the user to customize DOS default configurations. It performs this function by letting the user specify additional hardware drivers, the number of files that can be open at one time, the amount of memory used to transfer information with the disk drives (buffers), and the file name of the command shell.

Each time the computer is booted, DOS executes the commands in the CONFIG.SYS file before it loads the operating system's shell and before it executes the AUTOEXEC.BAT. Like the AUTOEXEC.BAT file, a CONFIG.SYS file is optional. However, just like the AUTOEXEC.BAT file, it will greatly enhance the overall operation of the computer. Table 5.7 contains a list of the directives, with basic explanations, that can be used in the CONFIG.SYS running MS-DOS 5.0 or above.

Example 5.13 is an example of a CONFIG.SYS. This is a stripped-down version of the actual CONFIG.SYS file used on one of the computers on which this book was generated. Just as with the AUTOEXEC.BAT file, it is beyond the scope of the textbook to teach all of the uses of the CONFIG.SYS

**Table 5.7** CONFIG.SYS Directives

| Directive | Action |
|---|---|
| BREAK | Determines when DOS recognizes the Ctrl-Break sequence |
| BUFFERS | Sets the number of file buffers DOS uses |
| COUNTRY | Sets country-dependent information |
| DEVICE | Loads and installs device drivers into the operating system |
| DEVICEHIGH | Loads device drivers in the upper memory block |
| DOS | Sets the area of RAM where DOS will be located, and specifies whether to use the upper memory block |
| DRIVPARM | Sets characteristics of a disk drive |
| FCBS | Sets the number of file control blocks (FCBs) that DOS can open concurrently |
| FILES | Sets the number of file handles used at one time |
| INSTALL | Runs a TSR program while DOS reads the CONFIG.SYS file |
| LASTDRIVE | Sets the highest disk drive letter available on the computer |
| REM | Labels the line as a comment |
| SHELL | Informs DOS what command interpreter should be used and where the interpreter is located |
| STACKS | Sets the number of stacks that DOS uses |
| SWITCHES | Disables extended keyboard functions |

file. However, it is very important to learn how to create and edit this file. A basic explanation of the CONFIG.SYS file shown in Example 5.13 is given following the example.

**Example 5.13:** CONFIG.SYS example

```
FILES = 40
BUFFERS=20
BREAK=OFF
DEVICE=C:\DOS\HIMEM.SYS
DOS=HIGH
DEVICE=C:\DOS\ANSI.SYS
DEVICE=C:\DOS\SETVER.EXE
DEVICE=C:\SCANGAL5\SJDRIVER.SYS
DEVICE=C:\WINDOWS\MOUSE.SYS /Y
```

Refer to Example 5.13 during this discussion. The first line (FILES = 40) sets the maximum number of filenames and device names that an application can use during a session. The second line (BUFFERS = 20) sets the amount of RAM the OS uses when transferring information with the disk drives. Line 3 (BREAK=OFF) only allows the DOS break function (Ctrl+C) to be except during certain times. A DOS break allows the user to abort a DOS command. The fourth line (DEVICE=C:\DOS\HIMEM.SYS) loads the memory manager HIMEM.SYS located on the C: drive in the subdirectory DOS into the computer's RAM (chapter 7). Line 5 (DOS=HIGH) tells DOS to load a portion of DOS into a section of memory called the High Memory Area (chapter 7). Line 6 (DEVICE=C:\DOS\ANSI.SYS) loads a driver program named ANSI.SYS into the RAM that allows the display circuitry to recognize commands that set colors when the computer is operating under DOS. The seventh line (DEVICE=C:\DOS\SETVER.EXE) loads a program into RAM that allows DOS to return the correct DOS version number to applications to request a certain version. Line 8 (DEVICE=C:\SCANGAL5\SJDRIVER.SYS) loads the driver for a scanner into the computer's RAM. The final line (C:\WINDOWS\MOUSE.SYS /Y) loads the mouse driver into RAM.

### 5.8.6 TSRs

All DOS and Windows programs, commands, drivers, and data that are needed by the computer *cannot* be loaded into the RAM of the computer at once. Programs and drivers that are loaded off disks into the computer's RAM fall into one of two categories: *transient* or *nontransient.*

When a transient program is loaded into the RAM of the computer, control of the computer is turned over to the program. Once the program's task is completed it will terminate and return control of the computer back over to the operating system's shell. The memory occupied by the program is released for other transient programs. If the task performed by the transient program needs to be executed again, it must once again be loaded into the RAM. All of the external DOS commands, utility programs, and application programs are transient programs.

A nontransient program is called a **terminate and stay resident (TSR)** program. Once terminated, these programs will stay resident in the computer's RAM so they can be used repeatedly without reloading them from the disk. In the CONFIG.SYS file, TSRs are loaded with the DEVICE= command. In the CONFIG.SYS example shown in Example 5.13, HIMEM.SYS, ANSI.SYS, SETVER.SYS, SJDRIVER.SYS, and MOUSE.SYS are all TSRs.

### 5.8.7 Why TSRs?

Hopefully, you understand by now that programs control all of the hardware devices in the computer. The different programs that control the computer's hardware are called *device drivers,* or, simply, *drivers.* Some of the device drivers are located in the computer's BIOS, some are loaded into the computer with the operating system, and some are in ROM BIOS located on expansion cards.

As computers become more and more complex, the list of device drivers seems endless. With the ability to install several types of displays, mice, CD-ROMs, scanners, and so on, there is no way that all of the drivers required could possibly be supported by the BIOS and OS. To alleviate this problem, DOS adds the ability to load drivers as TSRs into the computer's RAM from disk. Once loaded as a TSR, the driver is automatically added to the list of drivers already supported by the BIOS and the internal portion of DOS. The driver TSR can be loaded with the AUTOEXEC.BAT file if it ends with a COM or EXE extension. However, if the driver ends with the SYS extension, it must be loaded with the CONFIG.SYS file using the DEVICE command.

### 5.8.8 What Is the Difference in the Way a TSR is Loaded?

You may wonder, "How is loading TSRs different with the AUTOEXEC.BAT and CONFIG.SYS files?" The answer is, "Some TSRs have to be in the computer's RAM before control of the computer is turned over to the operating system's shell." Some examples include memory managers, file compression utilities, and CD-ROM drivers. These TSRs must be loaded using the DEVICE command in the CONFIG.SYS file. Other TSRs depend on the functions supplied by the operating system's shell. If this is the case, a line in the AUTOEXEC.BAT file must load the TSR.

How do you know whether to load the TSR with the AUTOEXEC.BAT or the CONFIG.SYS? If the TSR ends with the SYS extension, it will have to be loaded with the CONFIG.SYS file. If the TSR ends with an EXE or COM extension, its documentation will have to the checked to determine which method to use.

Most of the time, the device will include a diskette or CD-ROM that contains the driver for the device along with a program that is designed to install the driver on the hard drive. This installation program will usually add the required entries to the appropriate configuration file automatically to load the device's TSR correctly.

## 5.8.9 Windows Configuration

As stated previously, Windows 3.X is a parasite operating system because it requires DOS to work. Before Windows is loaded into the computer's RAM, the computer must have already been booted into DOS. Windows also requires that an extended memory manager TSR (usually HIMEM.SYS) be resident in the computer's RAM before any attempt is made to load Windows. Memory managers are covered in chapter 7. The HIMEM.SYS TSR has to be loaded with DOS's DEVICE command in the CONFIG.SYS file. Another TSR used by Windows that will greatly enhance its performance is SMARTDRV.EXE. This TSR is usually loaded with the AUTOEXEC.BAT file, but might be loaded in the CONFIG.SYS file depending on the DOS or Windows version being used.

After the computer is booted into DOS and the indicated TSRs are loaded into the computer's RAM, DOS is the operating system running the computer. Windows is loaded by entering the command WIN at the DOS prompt or through a menu item in the DOS shell. For people who do not use DOS, Windows is usually loaded automatically with the WIN statement as the last line in the AUTOEXEC.BAT file.

WINDOWS 3.X uses text files (ASCII) that are referred to as **information files (INI)** to establish the Windows 3.X configuration information. These files are usually located in the WINDOWS subdirectory and can be identified by the INI extension. Due to the extension name, these configuration files are often simply referred to as *"any"* files. The five INI files used to configure Windows 3.X are named SYSTEM.INI, WIN.INI, PROGMAN.INI, CONTROL.INI, and WINFILE.INI. These INI files are similar to the configuration files used by DOS only by the fact that they all establish hardware configurations and they all are text files. Except for these two similarities, the Windows 3.X configuration files are entirely different than those used by DOS.

An entire book can be written on just these five INI files, so I will only give a basic overview of these files' contents. Of the five mentioned, the SYSTEM.INI and WIN.INI are the most important and will be the only ones covered in any detail. Due to the length of these two files, complete examples will not be given in this discussion. However, if you have Windows 3.X running your computer, the contents of these files can be viewed and edited using the DOS editor or Windows 3.X notepad. Also, Windows 9X provides a SYSEDIT program to edit the Windows 3.X configuration files. SYSEDIT is discussed in section 6.0.0.

An example of the actual DOS command line to view the contents of the SYSTEM.INI file is shown in Example 5.14. Words to the wise: If you do not know what you are doing, do not attempt to change the contents of any Windows' files while viewing them.

**Example 5.14:** DOS command line to display the contents of the SYSTEM.INI file

> C:\>EDIT_C:\WINDOWS\SYSTEM.INI
>
> Note: Underscores indicate the location of spaces and are not typed.

## 5.8.10 The SYSTEM.INI File

The SYSTEM.INI file is used by Windows 3.X to establish the interface (connection) to the computer's hardware. This includes identifying all of the Windows 3.X hardware drivers. When Windows 3.X is first installed on the computer, the SYSTEM.INI file is divided into eight sections. Each of the eight sections contains information pertaining to hardware drivers and their setup. The contents of these sections are usually modified automatically as new Windows 3.X applications are installed. The eight section names and a basic description of the contents of each are listed in Table 5.8.

To give you a better feel of how complicated and in-depth the SYSTEM.INI file is, the [boot] section of the SYSTEM.INI file on the computer on which this text was produced is listed in Example 5.15. Remember that this is only one of the eight sections.

Table 5.8  Explanation of the Eight Original Sections of the SYSTEM.INI

| [boot] | Lists the device drivers used by Windows and its own program components |
|---|---|
| [boot description] | Provides a description for each device driver listed in the [boot] section |
| [driver] | Lists ALIAS names for each Windows driver |
| [keyboard] | Gives the setting for keyboard control and driver |
| [mci] | Lists each Media Control driver |
| [non-windows app] | Lists the options to control DOS applications running under Windows |
| [standard] | Lists the options to control Windows running in standard mode; standard mode is for a 286 μP |
| [386 enh] | Lists options to control Windows running in 386 enhanced mode (protected mode) |

**Example 5.15:** Example of the [boot] section of the SYSTEM.INI

```
[boot]
386grabber=TVGA.GR3
oemfonts.fon=vgaoem.fon
286grabber=VGACOLOR.2GR
fixedfon.fon=vgafix.fon
fonts.fon=vgasys.fon
display.drv=T800C.DRV
shell=progman.exe
mouse.drv=mouse.drv
language.dll=
sound.drv=mmsound.drv
keyboard.drv=keyboard.drv
system.drv=system.drv
drivers=mmsystem.dll
SCRNSAVE.EXE=C:\WINDOWS\SSSTARS.SCR
network.drv=;
comm.drv=comm.drv
```

### 5.8.11  The WIN.INI File

The WIN.INI file contains user-selectable options that affect the visible appearance of Windows 3.X and user-customized hardware interaction. The options in this file basically define the display and printing abilities of Windows 3.X. The WIN.INI also contains the user-defined color schemes, fonts, wallpaper, screen saver, and mouse setup, among other things. When Windows 3.X is first installed, the WIN.INI contains 18 sections. As other Windows 3.X applications are added, they will usually automatically add sections to the WIN.INI to add needed features.

For example, the WIN.INI file on one of the computers on which this book was created contains 50 sections. Only the name and basic description of the original 18 sections are listed in Table 5.9. Some of the options listed in the WIN.INI file are defined by the user through the Control Panel. The Control Panel is opened under the MAIN program group that is shown in Figure 5.8.

As with the SYSTEM.INI file, the WIN.INI file is very complex for the average user and even most technicians to understand. As an example of the WIN.INI files' complexity, Example 5.16 contains a listing of the [windows] section of the WIN.INI file on one of the computers on which this book was created.

**Table 5.9** Basic Sections of the WIN.INI File

| | |
|---|---|
| [windows] | Controls the elements of the Windows environment |
| [desktop] | Controls the appearance and position of the different elements of the desktop (screen) |
| [extensions] | Associates file extensions with certain applications |
| [intl] | Defines how to display items inside windows for countries other than the United States |
| [ports] | Lists available output ports |
| [fonts] | Lists character fonts loaded by Windows |
| [fontSubstitutes] | Lists interchangeable fonts |
| [TrueType] | Sets options for using TrueType fonts |
| [mci extension] | Associates media files with their driver |
| [network] | Sets options relating to Windows and the Local Area Network (LAN) |
| [embedding] | Sets options that control Object Linking and Embedding (OLE); OLE is the ability to pass information among different applications |
| [Windows Help] | Sets the options for window size and placement, and object color in the Windows on-line Help |
| [sounds] | Sets sounds to different Windows events; must have a sound card installed in the computer |
| [printerPorts] | Contains a list of the printers available to Windows applications |
| [devices] | Provides a list of output devices that are used for compatibility with Windows versions before 3.0 |
| [programs] | Lists the directory paths for locations of Windows applications |
| [colors] | Defines the colors used for each element of the Windows display |
| [Compatibility] | Provides Windows 3.1 with compatibility to certain Windows 3.0 applications |

**Example 5.16:** The [windows] section of Windows WIN.INI file

```
[windows]
spooler=yes
load=C:\TRIDENT.CXI\trackwin.exe D:\MHUNI2\remind.exe
run=
Beep=yes
BorderWidth=3
CursorBlinkRate=530
DoubleClickSpeed=700
Programs=com exe bat pif
Documents=
DeviceNotSelectedTimeout=15
TransmissionRetryTimeout=45
KeyboardDelay=0
KeyboardSpeed=31
ScreenSaveActive=1
ScreenSaveTimeOut=300
device=Epson FX-85,EPSON9,LPT1:
CoolSwitch=1
```

(Continued)

**Example 5.16:** (Continued)

```
DoubleClickHeight=8
DoubleClickWidth=8
DefaultQueueSize=16
NetWarn=0
```

**Figure 5.13**  Control Panel

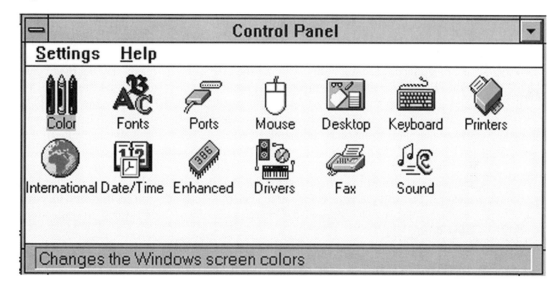

Because the WIN.INI contains most of the settings that define user preferences, it only makes sense that Windows 3.X provides a method of modifying the settings in the WIN.INI using some kind of GUI. The GUI that provides this service is called Control Panel. To open Control Panel, double-click on the Main program group in Program Manager. In the Main program group window, double-click on the Control Panel icon. Figure 5.13 shows the options available in Control Panel.

### 5.8.12 The PROGMAN.INI, CONTROL.INI, and WINFILE.INI

The other INI files used by Windows 3.X to establish hardware configurations are the PROGMAN.INI, CONTROL.INI, and WINFILE.INI. I will give only a basic explanation of these INI files.

The PROGMAN.INI file contains the options that control how Windows 3.X's Program Manager functions. As stated previously in the chapter, the Program Manager is Windows 3.X's default shell. An example of the appearance of a typical Program Manager display is shown in Figure 5.7.

The CONTROL.INI file contains information that governs the color schemes used by Windows 3.X. The user can select color schemes specified by the CONTROL.INI through the options in the Control Panel located in the MAIN program group. Once in the Control Panel, double-click on the Color icon, which is a picture of crayons. This icon can be seen in Figure 5.13.

The WINFILE.INI file contains the information that controls the Windows File Manager. The File Manager can be accessed through the MAIN program group by double-clicking the left mouse button on the File Manager icon, which is a picture of a two-drawer file cabinet. This can be seen in Figure 5.9. The File Manager allows the user to manage the files on all disk drives.

### 5.8.13 CONFIG.SYS, AUTOEXEC.BAT, and Windows 3.X INI Problems

The operating system's default configuration will often not meet the configuration needed by all applications. When this is the case, these applications will not run or their execution will be degraded. As you have probably seen, the configuration files for the DOS and Windows 3.X are so complex that the average person would have a very difficult time creating and modifying them to meet the demands of each application. Because of this reason, when a new application is installed on the computer's hard drive, it will usually modify the appropriate configuration files automatically so it can execute without errors.

This seems great, but the configuration for one application may not be the configuration for another. This is especially true for DOS applications that do not have individual configuration files that are available with Windows 3.X applications. This means when a new DOS application is installed, an old DOS application could stop functioning.

Before DOS 6.0 was released, the solution to this problem was to place the configuration files on the hard drive that met most DOS applications' configuration needs. Then separate boot diskettes were created that contained the correct configuration files for each of the other DOS applications. This process ended up with several boot disks to keep track of.

MS-DOS 6.0 gives the user the ability to create multiple configurations in a single CONFIG.SYS file. Each time the computer is booted, a menu appears from which the operator can select the correct configuration needed for the application to be executed. This makes it easier to use multiple configurations. However, if another configuration is needed, the computer has to be rebooted. As you have probably guessed, creating these multi-configuration CONFIG.SYS files requires a lot of knowledge.

Windows 3.X helps a little by allowing each Windows application to have its own unique INI file. However, most of the time, the application still makes changes automatically to the SYSTEM.INI, WIN.INI, and PROGMAN.INI in order to function correctly.

One major problem with an application modifying the operating system's configuration files automatically when they are installed is that the modification stays there even after the application itself is deleted. This will cause the computer to attempt to load information that is no longer there or load information that is no longer needed. This can cause the operating system to crash (stop functioning) or slow down.

Most DOS applications, which modify the CONFIG.SYS or AUTOEXEC.BAT, will create backup copies of these files under another extension such as CONFIG.UMB. This gives the ability to return to the original configuration if the application is deleted or if other problems pop up when the application is executed. Again, this seems like a good idea, but you end up with so many configuration files on the disk that you don't know which is which.

Because Windows 3.X's INI files are so large, most Windows 3.X applications do not make backup copies of these files when they are installed. After the Windows 3.X INI files have been modified by the installation of an application, there is no way of going back unless you edit out the statements in the INI file. From the examples given in this chapter, you can see that editing the Windows 3.X INI files is a major task, to say the least. For this reason alone, it might be a good idea to copy the SYSTEM.INI and WIN.INI files to a floppy disk before any Windows application is installed.

All is not lost if you have been using Windows 3.X for some time without knowing all of this scary information. Most better-written Windows 3.X applications have an uninstall program available that will not only delete the application from your hard drive, but will also remove modifications the program made to the Windows 3.X configuration files. Also, to help manage the Windows 3.X INI files, there are several second-party software packages available that will help keep the INI files optimized and will help uninstall unwanted Windows applications. Ask your local software retailer for information on these products.

## 5.8.14 Backing Up, Restoring, and Printing DOS and Windows Configuration Files

The Windows 3.X and DOS configuration files should be backed up regularly in case the originals become corrupted. This can happen during the normal operation of DOS and Windows 3.X or after the installation of an application that fails to install correctly. The steps outlined in Procedure 5.6 will back up and restore the Windows 3.X configuration files to and from a floppy diskette. Procedure 5.7 lists the steps to back up and restore the DOS configuration files.

Along with backing up and restoring the MS-DOS and Windows 3.X configuration files, the need often arises to create a hardcopy print-out in order to closely examine these files' contents. Also, it is nice to be able to show a hard copy to other individuals when they ask for or give advice. To create hard copies of the configuration files, follow the steps outlined in Procedure 5.8.

## Procedure 5.6: Backing up the Windows 3.X configuration files to a floppy diskette

General information: The Enter key must be pressed at the end of all command lines. Replace all underscores ( _ ) with spaces. Case does not matter.

**BACKUP:**
1. Label a blank, formatted diskette "Windows 3.X Configuration."
2. Place the diskette in the A: drive.
3. Type the following command lines at the DOS prompt:

   copy_c:\windows\system.ini_a:
   copy_c:\windows\win.ini_a:
   copy_c:\windows\progman.ini_a:
   copy_c:\windows\winfile.ini_a:
   copy_c:\windows\control.ini_a:

4. Remove the diskette from the A: drive and store it in a safe location.

**RESTORE**
1. Exit Windows 3.X if it is running.
2. Place the diskette that contains the Windows 3.X backup files into the A: drive.
3. At the DOS prompt, type the following command line. The Enter key must be pressed at the end of the line. Replace all underscores with spaces. Case does not matter.

   copy_a:\*.ini_c:\windows

4. Load Windows 3.X by typing WIN at the DOS prompt.

## Procedure 5.7: DOS configuration files backup and restore

General procedures: The Enter key must be pressed at the end of each command line. Replace all underscores ( _ ) with spaces. Case does not matter.

**BACKUP:**
1. Label a blank, formatted diskette "DOS Configuration Backup."
2. Place the diskette in the A: drive.
3. At the DOS prompt, type the following command lines. The Enter key must be pressed at the end of each line. Replace all underscores with a space. Case does not matter.

   copy_c:\config.sys_a:\
   copy_c:\config.sys_a:\

4. Remove the diskette from the A: drive and store it in a safe location.

**RESTORE:**
1. Place the diskette that contains the backup files in the A: drive.
2. At the DOS prompt, type the following command lines:

   copy_a:\config.sys_c:\
   copy_a:\autoexec.bat_c:\

3. Soft-boot the computer.

## Procedure 5.8: Creating hard copies of the DOS and Windows 3.X configuration files

Enter one or more of the following command lines. The underscores ( _ ) are used in the syntax to show the location of spaces. Instead of the underscores, type a space.

copy_c:\config.sys_prn
copy_c:\autoexec.bat_prn
copy_c:\windows\system.ini_prn
copy_c:\windows\win.ini_prn
copy_c:\windows\winfile.ini_prn
copy_c:\windows\progman.ini_prn
copy_c:\windows\control.ini_prn

## 5.9 CONCLUSION

I realize that this was not an in-depth coverage of DOS and Windows 3.X. As mentioned several times in this chapter, the intent of this book is not to teach operating systems, but to describe the basics of how knowledge of the operating system is required to control the computer's hardware. However, it is imperative to have a basic understanding of the operating system running on your computer to understand the discussions throughout the remainder of this book. Also, a basic knowledge of the OS is required if you are planning to build, upgrade, or repair PCs.

Hopefully, I have instilled in you the need to learn how to use DOS's command line even if the main operating system on your computer is Windows 3.X/95/98/NT/2000. You also need to understand the directory structure of the operating system on your computer. And, finally, any person who intends to upgrade and repair computers needs to dig deeper into how to create and edit the configuration files used by all of the operating systems discussed in this book.

## Review Questions

1. Write what each acronym stands for.
   a. API
   b. DOS
   c. EULA
   d. GUI
   e. TSR

### Fill in the Blanks

1. An MS-DOS filename can use up to _____ characters for the name and _____ characters for the extension.
2. The portion of the operating system that supports the running of an application is called the _____ _____ _____.
3. The _____ _____ _____ has to be accepted before most software can be installed and used.
4. Software that can be copied and distributed freely is called _____.
5. Each release of an operating system or application is usually assigned a _____ number.
6. A disk file can either be _____ or _____ used by a program.
7. Read-only can be one of a file's _____.
8. The first directory on any diskette is called the _____ directory.
9. The user uses a _____ to tell the operating system how to locate a file on the disk.
10. The two categories of machine language programs that can usually be run from the operating system's shell are called _____ files and _____ files.
11. The MS-DOS commands that are loaded into the computer's RAM along with COMMAND.COM are called _____ commands.
12. When Windows 3.X is multitasking, only one program can be running in the _____.
13. The file that tells Windows 3.X how to multitask a DOS application is called a _____ _____ file.
14. The Windows 3.X configuration file that loads the hardware drivers is called the _____ file.
15. The _____ Windows 3.X configuration file holds most of the user settings.
16. The first portion of a compatible computer's boot sequence is controlled by the _____.
17. Additional DOS device drivers that must be in place before COMMAND.COM is loaded must be loaded with the _____ file.
18. A program that stays resident in the computer's memory even after it has been terminated is called a _____ _____ _____.
19. The DOS configuration file that will execute a set of DOS commands automatically each time the computer is booted is called the _____ file.

### Multiple Choice

1. A DOS and Windows 3.X file can have a name up to (not including the extension):
   a. 26 characters long.
   b. 8 characters long.
   c. 11 characters long.
   d. 13 characters long.
2. CON is the name DOS uses for the:
   a. serial port.
   b. parallel port.
   c. keyboard.
   d. display.
3. MS-DOS is a stand-alone operating system.
   a. True
   b. False
4. Windows 3.X is a stand-alone operating system.
   a. True
   b. False
5. The names given to the two floppy drives are:
   a. Jack and Jill.
   b. Drive 1 and Drive 2.
   c. Drive 0 and Drive 1.
   d. Drive A and Drive B.
6. Which punctuation mark indicates a drive name?
   a. colon

b. period
  c. backslash
  d. space
7. An operating system's shell:
  a. is the operator's interface with the computer's operating system.
  b. is the computer's cover.
  c. is always a program called COMMAND.COM.
  d. is the same for all operating systems.
8. The main purpose of the AUTOEXEC.BAT file is to:
  a. load TSRs.
  b. execute a series of DOS commands each time the computer is booted.
  c. automatically execute the POST each time the computer is reset.
  d. establish hardware configurations.
9. The main purpose of the CONFIG.SYS file is to:
  a. load TSRs.
  b. execute a series of DOS commands each time the computer is booted.
  c. automatically execute the POST each time the computer is reset.
  d. establish hardware configurations.
10. What is the CONFIG.SYS command that loads a TSR?
  a. BUFFERS
  b. STACKS
  c. DEVICE
  d. REM
11. A batch file:
  a. is another name for a machine code program.
  b. executes programs in a batch.
  c. contains DOS commands.
  d. is not used by WINDOWS.
12. A TSR will:
  a. not be removed from the computer memory once its task is completed.
  b. be removed from the computer memory once its task is completed.
  c. automatically reset the computer each time it is executed.
  d. solve all of the problems in the world if given enough time.
13. BIOS is the acronym for:
  a. beginner's input/output system.
  b. basic I/O operating system.
  c. blue institute of organized smerfs.
  d. basic input/output system.
14. DOS is the acronym for:
  a. dumb output system.
  b. display operating scheme.
  c. disk operating system.
  d. disk output system.
15. POST is the acronym for:
  a. power-on self-test.
  b. powerful operating system terminal.
  c. power-off self-test.
  d. premature operating system test.
16. Relative to computers, an icon is:
  a. used by all operating systems.
  b. an object of uncritical devotion.
  c. a pictorial representation.
  d. another name for a program.
17. Relative to computers, a shell is:
  a. the operator interface (connection) to the computer's operating system.
  b. the metal cover in which the motherboard is mounted.
  c. the programs that protect the operating system.
  d. the name given to a computer that is designed to operate at the beach.
18. Including the extension, a DOS filename can be:
  a. up to 8 characters in length.
  b. up to 12 characters in length.
  c. up to 11 characters in length.
  d. up to 24 characters in length.
19. Windows 3.X uses:
  a. fast multitasking.
  b. slow multitasking.
  c. 386 multitasking.
  d. cooperative multitasking.
20. Files that end with the extensions EXE or COM:
  a. are assembly language programs that have to be assembled before an attempt is made to run them.
  b. are ASCII files that contain DOS commands.
  c. are the system files used to boot the computers.
  d. indicate the files are machine language programs that can usually be launched from the DOS prompt.
21. The Windows 3.X configuration file that determines what driver programs are used by Windows 3.X is the:
  a. CONFIG.SYS file.
  b. WIN.INI file.
  c. SYSTEM.INI file.
  d. PROGMAN.INI file.
22. Windows 3.X uses the 8.3 filename format.
  a. True
  b. False
23. The technique of using graphics instead of text is called:
  a. a graphics operating system.
  b. a graphical user interface.
  c. a mouse interface.
  d. a user-graphical interface.

# CHAPTER 6

# Windows 95/98/2000

## OBJECTIVES

- List the basic features provided by the Windows 9X/2000 shell.
- Explain the options provided by the Start menu.
- Describe the basic function of each Control Panel icon.
- List the reserved characters that cannot be used in Windows 9X/2000 filenames.
- Explain the use of the Windows 9X/2000 Help feature.
- List the Windows 9X/2000 shortcut keys.
- Explain the Windows 9X support for MS-DOS and Windows 3.X applications.
- Explain the difference between cooperative and preemptive multitasking.
- Describe the differences between DOS mode and DOS prompt.
- List the basic Windows 9X boot sequence.
- Describe the function of the system and user registry files.
- Describe the different sections of the Regedit display.
- Use the correct procedure to back up and restore the Windows 9X/2000 registry files.
- Describe the procedure for installing Windows 9X.
- Describe Windows 9X/2000 filename syntax.

## KEY TERMS

Windows Explorer
desktop
Task Bar
Start button
System Tray
Title Bar
Menu Bar
Windows Help

folder
shortcut
Start menu
My Computer
Control Panel
long filenames
reserved characters
DOS prompt

DOS mode
preemptive multitasking
Safe mode
registry files
USER.DAT
SYSTEM.DAT
regedit
System Policy Editor

Microsoft realized that it needed to move its operating systems into the realm of the 32-bit microprocessors. It did this by introducing an operating system named *Windows NT* (New Technology) in May of 1993. Windows NT is written for the business, networking, and power user communities and offers only limited support for Windows 3.X and DOS applications. Due to its limited support for these applications, very few personal computer users have purchased this operating system. Windows NT has evolved over the years into two distinct versions. One is called *Windows NT Workstation,* which is written for the power user and business computer. The other is called *Windows NT Server,* which is written to run a client server network.[1]

---

[1]Networks are covered in chapter 7.

In August 1995, Microsoft released the Windows 95 operating system, which is now known as *operating system release 1* (*OSR1* or "*a*"). This OS is written to meet the needs of the average computer user. Since then, Microsoft has released two revisions of this operating system: OSR2 (or "95b") and OSR3 (or "95c"). In 1998, Microsoft released the upgrade to Windows 95, called *Windows 98,* followed in 1999 by Windows 98 Second Edition (SE).

All releases of Windows 95 and 98 offer full support for both 32-bit and 16-bit applications. The 32-bit applications are written specifically for Windows 95 or 98 and take full advantage of most of the features offered by 32-bit compatible microprocessors. The support for 16-bit applications includes those written for Windows 3.X and DOS operating systems. These include DOS applications that are not written to run in a multitasking environment.

In February 2000, Microsoft added more operating systems to its arsenal with the release of Windows 2000. Similar to Windows NT, Windows 2000 is available in two versions. One is aimed at the business and power user community and is called *Windows 2000 Professional.* The other is written to run a network and is called *Windows 2000 Server.* Both versions of Windows 2000 are upgrades of Windows NT Workstation and Windows NT Server. Like Windows NT, they are both true 32-bit operating systems and will not run 16-bit programs that are not written for a multitasking environment.

In the Fall of 2000, Microsoft is planning the release of Windows Millennium (Me), which is the next logical progression of the popular Windows 98 operating system. This operating system will be written for the average personal computer user and will probably be the system installed on most pre-assembled computers. This OS is supposed to be the final step in Microsoft's plan to combine its 32-bit operating systems into one. From some of the advanced information I have read, this operating system will limit some of the support for DOS and Windows 3.X applications. This operating system is not out at the time of this writing, so it will not be discussed further in this text.

Windows 95 and 98 require at least a 386 DX microprocessor with a minimum of 8 MB of RAM to run. However, it is recommended that the minimum platform be a 486 SX- or DX-compatible microprocessor with 16 MB of RAM. Of course, the faster and more advanced the microprocessor and the more RAM installed, the faster Windows 9X will run. Along with the memory requirements, Windows 95 and 98 require approximately 30 MB of free hard drive storage space once installed.

Windows 2000 requires at least a Pentium®-compatible microprocessor with 32 MB of RAM. However, it is recommended that the minimum clock rate for the Pentium® be 233 MHz with a minimum of 64 MB of RAM. Just like Windows 9X, the faster the microprocessor and the more RAM in the computer, the faster and more efficiently Windows 2000 will run. Windows 2000 requires a minimum of 650 MB of free disk space to install and also a minimum total drive capacity of 6.5 GB.

Windows 95 and 98 are currently the most popular operating systems for personal computer use. Because this is the case, the major portion of this chapter discusses these two operating systems. Windows NT and 2000 will be discussed only when similarities or differences exist. Even though Windows 95, 98, and 2000 are different operating systems, they have quite a few similarities, however. In fact, similar to our discussion of MS-DOS and Windows 3.X, we will cover the similarities first and then move on to the differences in later sections. When referring to features that are the same for Windows 95, 98, and 2000, I refer to them collectively as *Windows 9X/2000.* The shared features offered by Windows 9X/2000 include the following:

- Improved shell;
- Use of filenames up to 255 characters;
- Continued support of 8.3 filename format;
- Support of both MS-DOS and Windows 3.X applications (limited support with Windows NT/2000);
- Support for 32-bit applications;
- Support for Windows 3.X cooperative multitasking;
- Preemptive multitasking to provide a stable multitasking platform;
- Uninstall wizard for Windows 9X/2000 applications;
- Support for power-management features;
- New Hardware Wizard to aid in the installation of non-PnP devices;
- Device Manager to manage peripherals, drivers, and ports;
- Better handling of protected mode program crashes;

- Better multitasking environment;
- Better environment for multimedia applications;
- Network-ready with full support for peer-to-peer networking;
- Dial-up networking;
- Dial-up server (Windows 95a and b require the Microsoft Plus! upgrade);
- Direct connection of two computers using either the serial or parallel ports;
- Support for Plug-and-Play adapter cards;
- Support for FAT32 using Windows 95b, 98, or 2000;
- Active right mouse button; and
- Much more.

## 6.1 THE SHELL

The default shell for Windows 9X/2000 is called **Windows Explorer.** The shell program is located in the Windows folder (subdirectory) on the C: drive under the name EXPLORER.EXE. Like any operating system's shell, Windows Explorer is loaded automatically as part of the operating system's boot sequence, which will be covered in section 6.9.

Even though Windows 95, Windows 98, and Windows 2000 use the same shell name, the shells do have a few differences that will be discussed in later sections. In this section, we discuss the similarities shared by the shells. Note that the Windows 95 shell can be upgraded to the Windows 98 and 2000 shells by installing Microsoft Internet Explorer™ 4.0. Because this is the case, the illustrations shown in this section are of the shell used by Windows 98/2000 and the upgraded Windows 95.

Windows 9X/2000 use a GUI shell that is somewhat different from the shell used by Windows 3.X. This is evident in the example of a typical Windows 98 GUI illustrated in Figure 6.1. As in Windows 3.X, the entire displayed image is referred to as the **desktop.** Unlike Windows 3.X, icons that launch applications and execute commands can be placed directly on the desktop.

Windows Explorer's main feature is a bar called the **Task Bar** that usually runs across the bottom of the display screen. The actual position of the Task Bar may vary because the user can reposition it to any display edge by dragging it with the mouse. In Figure 6.1, the Task Bar is the gray bar that runs across the bottom of the desktop.

**Figure 6.1** Windows 98 Desktop

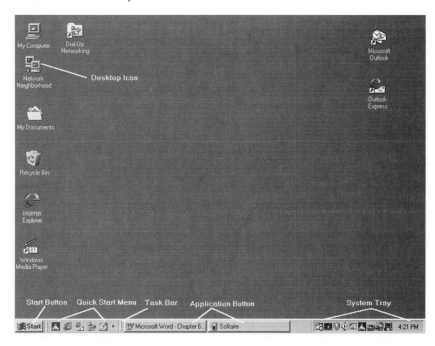

Graphics images that look like buttons appear on the Task Bar. The button on the left-hand side is called the **Start button.** Pressing this button is the primary method used to launch application programs and will be discussed in greater detail in section 6.1.6. The other buttons represent each application program running in the computer's memory. On the top of each button is the name of the application program the button represents. The button that appears to be pressed shows the name of the application program that is running in the foreground. All of the other buttons that do not appear to be pressed indicate the names of the application programs that are running in the background. In Figure 6.1, the program that is running in the foreground is Microsoft Word®, whereas Microsoft Solitaire is running in the background. Microsoft Word is the word-processing application program I used to write this book. Solitaire is a card game that can be installed from the Windows 9X/2000 CD-ROM.

The background programs can be brought to the foreground easily by a click of the left mouse button on the application's button on the Task Bar. This feature allows foreground programs to be run full screen. A switch to background programs can be made easily without minimizing the current foreground program.

A group of icons, including a small speaker and the time of day, appears on the right-hand side of the Task Bar. This part of the Task Bar is called the **System Tray** menu. The program icons in the System Tray menu represent programs that are running in the background that are activated by some event, such as a certain time of day. Some other means, such as pressing a sequence of keys on the keyboard or interrupts, may also activate these programs. Interrupts are discussed in greater detail in chapters 9 and 15. Programs are automatically added to the System Tray when they are installed on the hard drive. The procedure for removing programs from the System Tray will be covered in section 6.10.6.

If you are using Windows 98 or 2000 or have installed Internet Explorer 4.0 on a computer running Windows 95, another group of icons will appear immediately to the right of the Start button. This group of icons is called the *Quick Start* menu. To launch a program, simply click once on the application's icon in the Quick Start menu. To add items to the Quick Start menu, simply locate the name of the application you wish to add using Windows Explorer or locate the application's shortcut on the desktop, and then drag the program title or icon to the Quick Start menu section of the Task Bar. To remove an item from the Quick Start menu, drag the icon from the Quick Start menu onto the desktop.

### 6.1.1 Using the Mouse

The click, double-click, and drag functions of the mouse in Windows 9X/2000 are exactly the same as those in Windows 3.X, which are covered in chapter 5. One additional feature has been added to the mouse in Windows 9X/2000, however. This is called a *right-click*. As the name implies, to perform a right-click, use the mouse to position the mouse pointer over the desired object. Once the mouse pointer is in position, click the right mouse button once. This procedure will cause a menu window with options to appear.

### 6.1.2 Typical Windows 9X/2000 Application Window

A typical Windows 9X/2000 application window is shown in Figure 6.2. As you have probably noticed, the window used by Windows 9X/2000 is similar in appearance to a Windows 3.X application window. The **Title Bar** appears across the top of the window and is used to display the name of the application that is running within the window. On the right side of the Title Bar are three buttons. The button labeled with the minus sign (−) is used to move the program to the background and to minimize the program to a button on the Task Bar. The button labeled with the $X$ is used to close the program and all of its open files. The button with the two small windows is used to switch the program back and forth between full screen and running in a window.

The **Menu Bar** is directly beneath the title bar. The Menu Bar is similar to Windows 3.X's Menu Bar, which was introduced in chapter 5. The options that are available on the Menu Bar vary from application to application.

Additional bars called *Tool Bars* are below the Menu Bar. If available, Tool Bars are unique to the application running within the window. To use and make changes to the options available on the application's Tool Bar, consult the documentation or Help files that are included with the application.

# Windows 95/98/2000

**Figure 6.2** Typical Application Window

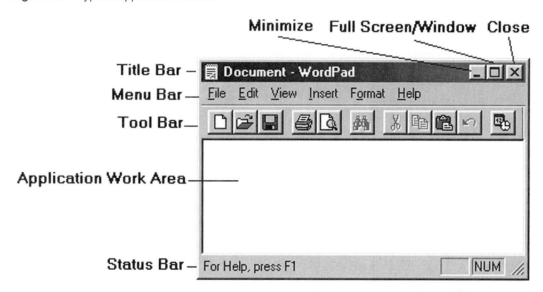

---

### Procedure 6.1: Opening the Windows 9X/2000 Help file

1. Position the mouse pointer over the Start button on the Task Bar.
2. Click on the Start button.
3. Move the mouse pointer up to Help.
4. Click on Help.

---

## 6.1.3 Windows Help

Windows 9X/2000 is supplied with a manual called "Getting Started." This manual covers the operating system's installation and basic use. The most extensive information on the procedures available with Window 9X/2000 is contained in a group of large Help files stored on the hard drive. The information contained in these Help files is accessed by using the Help option in the Start menu. Procedure 6.1 lists the steps to follow to open **Windows Help.**

When opening Windows Help, a window similar to the one shown in Figure 6.3 will display. The Help window contains three tabs that are labeled *Contents, Index,* and *Search.* To select one of the options, click on the tab. Clicking on the Contents tab will display a list of basic topics in the text box below the tab. The user can click on any of the listed topics for more information.

Clicking on the Index tab will cause a display similar to the one shown in Figure 6.3. In the small text box, the user types the information for which help is needed. As the user types, Index will automatically scroll through topics that most closely match the information the user is typing. Once the topic is displayed in the text window, the user double-clicks on the topic to open up the Help information.

Clicking on the Search tab allows the user to search through the Help topic for certain words or phrases. To use the Search function of Help, type the word or phrase in the text box, and then click on the List Topics button. Windows 9X/2000 will list all Help topics that contain the entered word or phrase.

Windows Help is a valuable tool that everyone should learn to use. In fact, Windows Help will be referenced throughout this chapter and in various chapters throughout the remainder of this text.

## 6.1.4 Running Applications

Applications can be launched in several different ways with Windows Explorer. First, double-click on the application's icon, if one is located on the desktop. Second, double-click on the application program's icon, if one is available in the Start menu. Third, double-click on the program name that launches

**Figure 6.3** Typical Help Window

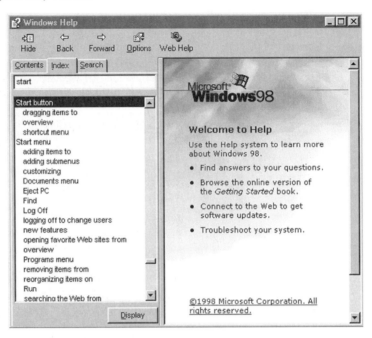

the application using Windows Explorer. Fourth, if available, click on the application's icon in the Quick Start menu. And fifth, type the path and name of the program in the text box using the Start menu's Run option. To display the procedures used to run an application, type "running applications" in the Index section of Windows Help.

### 6.1.5 Icons on the Desktop

Icons on the desktop are used to either open a program group or run an application program. If the icon opens a program group, it is referred to as a **folder.** If the icon launches an application, it is called a **shortcut.** The user can create folder and shortcut icons on the desktop, but they are most often created automatically when the Windows 9X/2000 operating system or an application is installed on the computer's hard drive.

To open the folder or launch the application, simply double-click on the appropriate icon. To list the procedure used to add icons to the desktop, type "icons" in the Index option of Windows Help.

### 6.1.6 The Start Menu

Like the icons on the desktop, most applications will automatically add folders or shortcuts to the Start menu when they are installed. To open the Start menu—to run an application—the Task Bar's Start button is usually used. The Start button is located to the left on the Task Bar (see Figure 6.1). When the left mouse button is clicked on the Start button, a second menu will pop up, similar to the one shown on the left in Figure 6.4. This first menu that pops up is called the **Start menu.** One of the listings in the Start menu is Programs. If you move the mouse pointer up to the Programs option, yet another menu will pop up, similar to the menu shown in the center of Figure 6.4. Simply move through the menus until you find the icon for the application you wish to run. Once the application program's icon is located, click the left mouse button on the icon.

When Windows 9X/2000 is installed, several standard options are available on the Start menu that usually cannot be removed by the user without changing entries in the Windows 9X configuration files. These options are Shut Down, Run, Help, Find, Settings, Documents, and Programs. The options Log Off and Favorite will usually be available if you have Windows 98, 2000, or Internet Explorer 4.0 on a computer running Windows 95.

**Figure 6.4** Start Menu

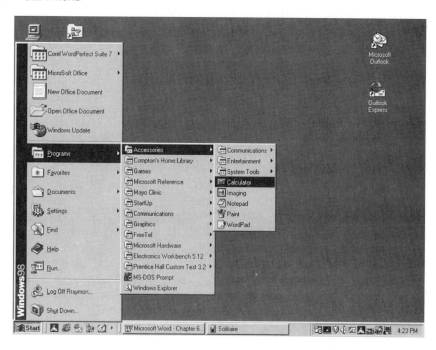

---

### Procedure 6.2: Displaying the contents of the Startup folder

1. Click on the Start button.
2. Move the mouse pointer up to Programs, and then over to Startup.

---

Anyone who has basic knowledge of Windows 9X/2000 should know how to add folders and shortcuts to the Start menu. The procedure for adding program shortcuts to the Start menu can be found by typing "start menu" in the Index option of Windows Help.

### 6.1.7 The Startup Folder in the Start Menu

The Startup folder is similar to the AUTOEXEC.BAT file used by DOS. Application can be launched automatically each time Windows 9X/2000 is started. This is accomplished by placing the application program's shortcut icon in the Startup folder in the Start menu. Procedure 6.2 lists the steps to follow to display the program shortcuts in the Startup folder.

Because these programs are started automatically each time Windows 9X/2000 boots, this can be a major source of boot problems. For this reason alone, it is imperative that one learns how to add and remove program shortcuts in this folder. The procedures to follow can be found by entering "startup folder" in the Index option of Windows Help.

## 6.2 MY COMPUTER

The Windows 9X or 2000 desktop includes an icon labeled **My Computer.** My Computer is used to control and set most of the user-defined functions and features of the Windows 9X/2000 operating systems. To open My Computer, double-click on its icon on the desktop. When the My Computer window opens, an icon for each of the computer's hard drive(s), floppy drive(s), CD-ROM drive(s), and network drives (if applicable) should be displayed. Along with these icons will be folders for the **Control**

## Procedure 6.3: Entering Control Panel

1. Use the mouse to position the pointer over the Start button and then click on the left mouse button.
2. Move the mouse pointer up to Settings and then over to Control Panel.
3. Click the left mouse button on Control Panel.

Figure 6.5  Typical Control Panel

**Panel** and, in Windows 9X, there will be one for printers. There may be additional icons that are placed in My Computer by applications when they are installed.

One or more of these icons may not be in the My Computer section of the computer you use because these icons can be deliberately removed from My Computer by changing settings in the Window 9X configuration files. The configuration files are covered in section 6.10.

### 6.2.1 Control Panel

Almost all of the user options and computer controls are established through Control Panel. Because Control Panel is so important, it can be entered in several different ways. The steps we will use most often to enter Control Panel are listed in Procedure 6.3.

A typical Control Panel display is shown in Figure 6.5. The options shown in this figure are the options available when Windows 9X/2000 is initially installed. It should be noted that there might be additional options available in the Control Panel on your computer. Table 6.1 shows a brief explanation of each option along with the keyword to type in the Index option of Windows Help. Most of the mentioned Control Panel options will be discussed in greater detail throughout this text.

## 6.3 WINDOWS 9X/2000 FILENAMES

Windows 9X/NT/2000 allows for **long filenames**—filenames and directory names (called *folders*) that are up to 255 characters in length. The 255 characters is the total length including the filename and the filename extension. Unlike DOS and Windows 3.X, spaces are allowed as part of the filename and filename extension. Similar to DOS and Windows 3.X, a period (.) is used to separate the filename from

**Table 6.1** Control Panel Options

| Option | Brief Description | Key word |
|---|---|---|
| Accessibility options | Settings for individuals with disabilities | accessibility |
| Add new hardware | Adds new PnP hardware | add |
| Add/remove programs | Adds/removes Windows applications | add/ |
| Date/time | Changes computer's date, time | date |
| Display | Changes the display settings | display |
| Font | Displays and installs available fonts | font |
| Game controller | Calibrates and sets up game controllers | game controller |
| Keyboard | Sets user keyboard options | keyboard |
| Mouse | Sets user mouse options | mouse |
| Passwords | Sets and edits Windows passwords | passwords |
| Power management | Sets the power management features | power |
| Printers | Adds/removes and sets up printer | printer |
| Regional settings | Selects regional settings | region settings |
| Sound | Assigns sounds to events | sound schemes |
| System | Views/changes hardware settings | device manager |
| Users | Sets user profile options | users, profiles |

the extension. Unlike the 8.3 format, multiple periods can be used, with the last period to the right in the filename indicating the extension. The uses of long filenames offered by Windows 9X/2000 makes keeping track of files and file folders much easier than with DOS and Windows 3.X.

A Windows 9X/2000 file can have a name such as *Letter to Mr Jones written on Feb 2_1998.doc*. In DOS and Windows 3.X, this would have to be represented in the 8.3 format. So a name such as *LETJON98.DOC* might be used. As you can see, the long filename gives a lot more insight into the information contained in a file.

As one can imagine, a problem arises when a long filename created in Windows 9X/2000 is displayed or accessed by DOS or Windows 3.X applications, which expects the filename to be the 8.3 format. To alleviate this problem, Windows 9X/NT/2000 stores filenames under both the long format and the 8.3 format in a directory structure called *VFAT*.

When long filenames are stored, they are automatically truncated by the OS into names that are within the limitations of MS-DOS and Windows 3.X. This is done by using the first six characters of the long filename, and then adding a tilde (~) followed by the number of occurrences of the first six characters. For example, the long filename *Letter to Mr Jones written on Feb 2_1998.doc* will appear as *LETTER~1.DOC* to a Windows 3.X or MS-DOS application. Also, if the long filename *Letter to Mr. Smith written Nov 8_1995.doc* is encountered later in the directory by a MS-DOS or Windows 3.X application, it will appear as *LETTER~2.DOC*. As you can see, it is almost impossible to tell which file contained what information.

### 6.3.1 Reserved Characters

Like MS-DOS and Windows 3.X, some punctuation symbols are not legal as part of the Windows 9X/2000 filename because they are used for other things. The only **reserved characters** in Windows 9X are the asterisk (*), slash (/), back slash (\), and colon (:).

> **Procedure 6.4: Opening Windows Explorer**
>
> 1. Click the left mouse button on the Start button on the Task Bar.
> 2. Move the mouse pointer up to Programs.
> 3. In Windows 9X, move the mouse pointer down to the Windows Explorer option.
> 4. In Windows 2000, move up to the Accessories option and then down to the Windows Explorer option.
> 5. Click the left mouse button on Windows Explorer.

Table 6.2  Often-used Keyboard Shortcuts*

| | |
|---|---|
| Ctrl + Esc | Opens the Start menu |
| Ctrl + C | Copies |
| Ctrl + V | Pastes |
| Ctrl + X | Cuts (used along with Paste to move a file) |
| Ctrl + F4 | Closes the current window (used in applications that use a multiple document interface (MDI) such as word processors, spreadsheets, and database applications) |
| Ctrl + Z | Undo |
| F10 | Activates the applications Menu Bar |
| Alt + underlined letter in the menu | Selects the menu or menu option |
| Alt + Tab | Brings the last-used application's window to the foreground (by continuously pressing the Tab key while holding the Alt key, cycles through all applications being multitasked) |
| Alt + F4 | Closes the current application |
| Delete | Deletes highlighted selections |

*To use, press and hold the key to the left of the plus sign, then press the key to the right of the plus sign.

## 6.4 WINDOWS EXPLORER

Windows Explorer allows the user to manage the files and folders (directories) on the computer's mass storage devices such as floppy disks, hard drives, and CD-ROMs. The steps to open Windows Explorer are listed in Procedure 6.4.

Once opened, Windows Explorer allows the user to manipulate files and folders by simply dragging the object to a new location. Along with controlling the Windows 9X/2000 shell with the mouse, Windows Explorer also offers an extensive set of shortcuts that often make the manipulation of files and folders faster than using the mouse.

### 6.4.1 Windows Explorer Shortcut Keys

Along with using the mouse, Windows 9X supports an extensive set of keyboard commands. Learning these keyboard functions is imperative for a couple of reasons. First, the mouse sometimes fails due to hardware or software conflicts. Second, using the keyboard commands is often faster than using the mouse. For this second reason, most of the key sequences used to perform commands are often called shortcuts. The key sequences for the most often used shortcuts are listed in Table 6.2.

Learning how to effectively use Windows Explorer is very important for anyone who plans to use Windows 95, 98, NT, or 2000. Luckily, Windows Explorer functions basically the same in all of these

Windows 95/98/2000

operating systems. This means if you learn how to use it proficiently in one of the mentioned operating systems, it will be very easy to move to one of the others. For more information, type "Windows Explorer" in the Index option of the Help menu.

### 6.4.2 File Associations

Like Windows 3.X, Windows 9X/2000 allows a data file to be associated with an application. If an attempt is made to run a data file, Windows 9X/2000 will automatically launch the application program to which the data file has been associated. If there has been no previous association setup, Windows 9X/2000 will display a window listing the applications installed on the hard drive. The user can then click on the appropriate application and set a permanent association, if desired.

As with Windows 3.X, associations are usually established automatically when the application is installed on the computer's hard drive. However, sometimes it becomes necessary for the user to change data file associations. To change or edit a data file association, use the procedure outlined in the Index option of Help by typing "associating file" in the text box.

### 6.4.3 Controlling Files

To be proficient in any aspect of the compatible PC, it is imperative to know how to control the computer's file system. The most important file features, in my opinion, are formatting, finding, copying, deleting, renaming, moving, backing up, and changing attributes. The only way you can learn these functions is through practice.

As with most other commands, the procedure to follow to perform these tasks can be accessed using the Windows 9X/2000 Help feature. To access these procedures, simply enter "File" in the Index section of the Help menu.

## 6.5 SUPPORT FOR WINDOWS 3.X

Windows 9X/2000 offers support for Windows 3.X's 16-bit applications. To provide this support, some of the Windows 3.X information (INI) files that were introduced in chapter 5 remain a part of these operating systems. The INI files included with Windows 9X are PROGMAN.INI, WIN.INI, and SYSTEM.INI. Windows 2000 includes only the WIN.INI and SYSTEM.INI files. On computers that run Windows 9X, these files are found in the Windows folder (directory) on the C: drive, similar to Windows 3.X. On computers that run Windows NT/2000, the Windows 3.X configuration files are located in the WINNT folder.

To provide support for a Windows 3.X application means that Windows 9X/2000 must support cooperative multitasking. Remember from chapter 5 that when cooperative multitasking is used, the operating system turns control of the computer over to the application when it is the application's turn to execute instructions. For this reason, if a Windows 3.X application crashes when running, the computer will usually have to be rebooted.

### 6.5.1 Windows 3.X and Windows 9X/2000 Information Files

Like Windows 3.X, Windows 9X/2000 applications also have the ability to specify unique configurations using information files. These files will be stored on the drive with the same name as the applications. To help differentiate among Windows applications, which use the INI extension, Windows 9X/2000 information files end with the *INF* extension.

## 6.6 SUPPORT FOR DOS APPLICATIONS

One advantage (and curse) of the DOS operating system is the ability of the programmer to access hardware directly without going through the OS. To overcome the shortcomings of DOS, there are lots of DOS applications that contain their own drivers, especially video drivers. Needless to say, this cannot be allowed in a multitasking operating system because one application cannot be allowed to take control of any peripheral.

When Microsoft introduced Windows 95 and 98, one of its main objectives was to maintain full support for DOS applications. To provide this support, Windows 9X provides two modes of operation that offer two environments for a DOS application to run. One is called **DOS prompt,** or *DOS window,* and the other is called **DOS mode.** DOS prompt is provided for DOS applications that can use Windows 9X drivers, while DOS mode is provided for DOS applications that contain their own drivers.

When Windows NT came out, Microsoft decided not to support DOS applications that contain their own drivers. This feature has been carried over to Windows 2000. Instead of DOS prompt, these operating systems provide a feature called *Command prompt,* which emulates the DOS shell COMMAND.COM. Command prompt allows the use of DOS commands and DOS applications that will run in a multitasking environment.

### 6.6.1 Windows 9X/2000 DOS Prompt (DOS Window) and Command Prompt

DOS applications that will run in DOS or Command prompt are actually operating under the Windows 9X/2000 operating system. Running under the Windows 9X/2000 operating system provides two important features: first, the DOS application is using all of Windows 9X/2000's 32-bit drivers; and second, the DOS applications can be multitasked with Windows 9X/2000 applications.

When a DOS application is launched, Windows 9X/2000 looks for a PIF file for the applications. If a PIF file is available, it will usually tell Windows 9X/2000 how to handle the DOS application. If a PIF is not available, Windows 9X/2000 will try to determine if the program will function correctly within a DOS or Command prompt. If it will not, Windows 2000 will terminate the program, while Windows 9X will suggest that the program be run in DOS mode.

### 6.6.2 Windows 9X DOS Mode

As mentioned previously, due to the limitations of the DOS operating system, some DOS applications contain their own drivers. This is especially true for many DOS games that contain their own driver to run the computer's display. Some DOS applications also use their own memory managers. When this is the case, these applications will not run within a DOS or Command prompt and cannot be run under the Windows NT or 2000 operating systems.

For DOS applications that will not run in a DOS window (DOS prompt), Windows 9X provides what is referred to as *DOS mode.* DOS mode can be selected by the user through the DOS application PIF or recommended by Windows 9X when the program is launched through the Windows 9X shell.

In DOS mode, the computer is running a 16-bit DOS real-mode operating system that is part of Windows 9X. When DOS mode is selected, all Windows 9X applications will be closed and no multitasking will take place. Also, none of the Windows 9X 32-bit drivers will work in DOS mode, so any driver that is not supported by DOS (mouse, CD-ROM, and sound card) must be loaded in the CONFIG.SYS and AUTOEXEC.BAT files, which were discussed in chapter 5.

DOS mode can be entered in several different ways: (1) by selecting DOS mode as one of the Shut Down options in the start menu; (2) DOS mode can be selected as one of the options in the PIF file for the DOS applications; and (3) Windows 9X will automatically suggest DOS mode if an attempt is made to run a DOS application that does not have a PIF but it detects that the program might have problems running in a DOS window.

### 6.6.3 Program Information Files (PIFs)

Like Windows 3.X, Windows 9X/2000 uses a PIF (program information file) to determine how it should handle DOS applications that will run within a DOS or Command prompt. Also similar to Windows 3.X, Windows 9X/2000 has a PIF editor that allows the user to display and edit the PIF file for any DOS application. To launch the PIF editor, use the steps outlined in Procedure 6.5. For more information on how to use PIF editor, type "PIF" into the Index option of Windows 9X/2000 Help menu.

---

### Procedure 6.5: Starting the PIF Editor

1. Find the name of the DOS PIF using Windows File Manager.
2. Once the filename is located, right-click on the file's icon. This will cause a menu window to display with options that are available with the file.
3. Click on the Properties option. This will cause the PIF editor to load automatically.

---

### Procedure 6.6: Setting up unique DOS configuration files for DOS applications that run in DOS mode

1. Locate the name of the file that launches the DOS application using Windows Explorer.
2. Right-click on the filename.
3. Click on the Properties option in the pop-up menu.
4. In the Properties window, click on the Program tab.
5. Click on the Advanced button.
6. The DOS mode option box should already be checked.
7. Click on the white option dot beside "Specify a new MS-DOS configuration."
8. Type the desired CONFIG.SYS and AUTOEXEC.BAT contents in the appropriate text boxes.

---

#### 6.6.4 Windows 9X DOS Mode and the CONFIG.SYS and AUTOEXEC.BAT Files

One of the major problems with DOS applications is that the configuration specified in the CONFIG.SYS and AUTOEXEC.BAT files used by one DOS application will probably not be good for all DOS applications. Before Windows 9X, to overcome this problem the user placed a CONFIG.SYS and AUTOEXEC.BAT file that met most of the configuration needs of the DOS applications that were in use on the hard drive. For the other DOS applications, the user usually placed a unique CONFIG.SYS and AUTOEXEC.BAT file on a bootable floppy disk and booted the computer from the floppy. Once the computer was booted with the unique DOS configuration files, the application program would be run. This led to having to keep up with different floppies to boot the computer for different DOS applications.

Windows 9X avoided this problem by allowing the user to specify unique CONFIG.SYS and AUTOEXEC.BAT files for each DOS program that executes in DOS mode. This unique CONFIG.SYS and AUTOEXEC.BAT file is specified as part of the DOS application's PIF. Procedure 6.6 lists the steps to follow to establish a unique CONFIG.SYS and AUTOEXEC.BAT file in a DOS application that requires DOS mode.

#### 6.6.5 System Editor (SysEdit)

Windows 9X/2000 gives the user the ability to display and edit all of the DOS and Windows 3.X configuration files using a single application called *System Editor*. This application is located in the Windows folder under the name SYSEDIT.EXE. System Editor can be launched from the Run window or a shortcut can be added to the Start menu. To launch System Editor, follow the steps outlined in Procedure 6.7.

---

### Procedure 6.7: Launching System Editor

**Warning:** Avoid making changes to the configuration files unless absolutely necessary.

1. Click on the Start button on the Task Bar.
2. Click on the Run options.
3. In the text box on the Run window, type the following command: SysEdit
4. Click on the OK button.

---

**Figure 6.6** Windows 9X Close Program Window

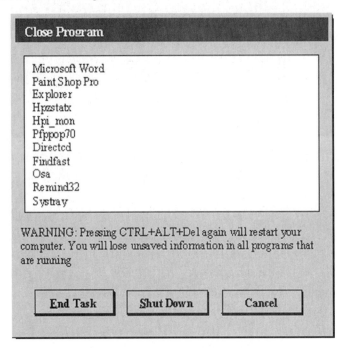

## 6.7 WINDOWS 95/98/2000 APPLICATIONS (PREEMPTIVE MULTITASKING)

The Windows 9X/2000 operating systems and their applications use **preemptive multitasking.** Preemptive multitasking is quite different from Windows 3.X's cooperative multitasking in the fact that the operating system always maintains control over the computer. This feature means that a Windows 9X/2000 application can crash and can usually be terminated without rebooting the computer or without affecting other applications that are running at the time of the crash.

To shut down a Windows 9X application that is not responding, "Close Program" is used. To start Close Program, press the Ctrl key plus the Alt key then press the Del key. Only do this once because if the sequence is performed a second time, the computer will soft-boot. The Ctrl+Alt+Del sequence will display a window similar to the one shown in Figure 6.6.

If Windows 2000 is being used, the Ctrl+Alt+Del sequence will open "Windows Security." This window is shown in Figure 6.7. From this window, click on the Task Manager button, then click on the Applications tab to display the task currently running in the computer's memory.

As seen in Figure 6.6, Task Manager or Close Program displays a windowpane that contains a list of all of the applications that Windows 9X/2000 is currently running. Any program that is having a problem will usually be labeled as "not responding." Any program that is not responding can be shut down by clicking on the program to highlight it, and then clicking on the End Task button. Note that closing some programs may cause the system to become unstable and may require a hard-boot to return the system to normal operation.

## 6.8 THE MAJOR DIFFERENCES AMONG WINDOWS 95, 98, AND 2000

The main visible difference among Windows 95, 98, and 2000 is the operating system's shell. Even though Windows 98 and 2000 use a shell called Windows Explorer, it is actually a version of Microsoft's Web browser called *Internet Explorer.* Using a Web browser as the shell adds some unique

**Figure 6.7**  2000 Windows Security Window

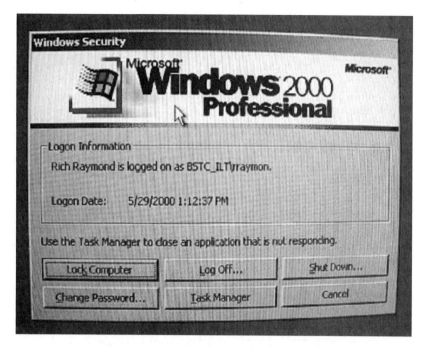

features that are not available with the default shell used by Windows 95. These features include, but are not limited to, the following:

- Forward and Back buttons appear on the Windows Explorer Tool Bar to go back and forth between opened windows;
- Quick edit buttons can cut, copy, past, undo, delete, and display properties on the Windows Explorer Tool Bar;
- Internet URLs (address) can be typed in the Address text box of Windows Explorer to view Web pages; and
- Internet favorite sites can be added to the Start menu.

Note that Windows 95c also uses the Internet Explorer shell. Windows 95a and 95b can be upgraded to this shell by installing Internet Explorer releases greater than 4.

In addition to the improved shell, Windows 98 also offers many more improvements over Windows 95 that make upgrading to this operating system advisable. Some, but not all, of these improvements are contained in the following list:

- Support for FAT 32 (also available on Windows 95b and 95c);
- Improved support for device drivers;
- Continued support for FAT 16;
- Conversion from FAT 16 to FAT 32 without reformatting the hard drive;
- Maintenance Wizard that schedules certain hard-drive preventive maintenance tools to run automatically;
- Advanced power management;
- System information for an overall look at a computer's hardware and software environments;
- Full support for the universal serial bus (USB) (Windows 95c); and
- More troubleshooting features.

Comparing Windows 2000 to Windows 98 is really not a valid comparison because Windows 2000 is an upgrade to Windows NT, not Windows 98. As previously stated, Windows 2000 is not written for the home computer user but for the business and power-user community. If you are into playing a lot of games, Windows 2000 is probably not the route you want to travel.

Another problem with upgrading to Windows 2000 is that just because a driver or peripheral works with Windows 9X does not mean that it will work with Windows 2000. I found this out the hard way after I installed Windows 2000 and the software for my scanner, digital camera, and backup would no longer run. When I visited the Web sites for these devices, drivers for Windows 2000 were not yet available or were not even planned. For this reason, before you buy or upgrade to Windows 2000, I highly recommend a visit to the Microsoft upgrade site at *www.microsoft.com/windows2000/upgrade/compat/* for help in determining the problems that will be encountered.

Similar to Windows 98, 2000 uses Internet Explorer as its default shell. All of the features provided in the previous list of enhancements offered by Windows 98 are included. Some of the additional features offered by Windows 2000 are provided in the following list:

- Very stable platform (few lockups);
- Multiple operating system boot-up without changing active partition;
- Continued support for FAT 16 and FAT 32;
- The New Technology File System (NTFS);
- Support for extremely large-capacity drives using NTFS;
- Allows combining several drives into a single volume using NTFS;
- High security using NTFS;
- FAT to NTFS conversion utility;
- Support for multiprocessing;
- Support for Unicode filenames;
- Troubleshooter to aid in fixing hardware problems;
- Support for uninterruptable power supplies (UPS); and
- More.

### 6.8.1 Service Packs

Very seldom are operating systems made available commercially without undetected bugs. Also, new hardware technology is often introduced after the operating system's release. In the old days of operating systems, bugs in the software were patched and support for new technologies was added by releasing the operating system under a new revision number. For example, the problems in MS-DOS 6.0 were repaired with 6.1, which were repaired by 6.11, and so on.

Currently, the most popular way for Microsoft to provide software bug patches, minor upgrades, and new hardware support is to bundle them together in one package called a *service pack*. Service packs for Microsoft's operating systems and applications can be downloaded free from Microsoft's Web site at www.microsoft.com/msdownload, or provided for a minimal charge on CD. I highly recommend regular visits to this Web site.

## 6.9 BOOTING WINDOWS 9X

The basic boot sequence for Windows 9X is listed in Table 6.3. The first steps of the Windows 9X boot are identical to the boot sequence used by MS-DOS. The difference occurs when the boot sequence goes to the master boot record on the hard drive. This is where the actual boot files are identified and loaded from the hard drive. Similar to the boot sequence used by MS-DOS, the Windows 9X boot sequence is divided into two parts. The first part is controlled by the BIOS and is exactly the same as the MS-DOS boot sequence. Windows 9X controls the second part of the boot sequence, however.

To make the boot sequence more efficient, the functions performed by MS-DOS's IO.SYS and MSDOS.SYS are combined into the single Windows 9X's IO.SYS file. In some installations of Windows 95, this file might be named WINBOOT.SYS.

Although it is still part of the Windows 9X boot sequence, MSDOS.SYS has been changed completely. Under Windows 9X, the MSDOS.SYS file is now a text file (ASCII) that is very similar to a Windows 3.X configuration INI file. An example of a typical MSDOS.SYS file is shown in Example 6.1.

# Windows 95/98/2000

**Table 6.3** Basic Windows 9X Boot Sequence*

| |
|---|
| 1. The Volume Boot Record program loads IO.SYS. The file might be called WINBOOT.SYS if Windows 9X was installed as an upgraded version. Control of the computer is turned over to IO.SYS. |
| 2. IO.SYS loads and takes control of the boot sequence. |
| 3. IO.SYS loads and executes the commands in MSDOS.SYS. |
| 4. If present, the commands in the optional CONFIG.SYS file are executed. |
| 5. If an AUTOEXEC.BAT file is present, COMMAND.COM loads and executes the optional AUTOEXEC.BAT file. |
| 6. WIN.COM is loaded and takes control of the boot sequence. |
| 7. VMM32.VXD is loaded, which loads the drivers indicated by the SYSTEM.DAT and SYSTEM.INI configuration files. VMM32.VXD also switches the microprocessor into the protected mode. |
| 8. The Windows 9X kernel is loaded. This includes the files KERNEL32.DLL, GDI.EXE, CDI32.DLL, USER.EXE, and USER32.EXE. |
| 9. The configuration settings in the WIN.INI are established. |
| 10. The Window 9X shell (EXPLORER.EXE) is loaded and takes control of the computer. |
| 11. EXPLORER.EXE loads the configuration settings in the USER.DAT file. |

*The BIOS portion of the boot sequence is covered in chapter 5, section 5.7.1.

**Example 6.1:** Typical MSDOS.SYS used by Windows 9X

```
;FORMAT
[Paths]
WinDir=C:\WINDOWS
WinBootDir=C:\WINDOWS
HostWinBootDrv=C

[Options]
BootMulti=1
BootGUI=1
;
;The following lines are required for compatibility with other programs.
;Do not remove them (MSDOS.SYS needs to be >1024 bytes).
;xxxxxxxxxxxxxxxxxxxxxxxxxxxxxxxxxxxxxxxxxxxxxxxxxxxxxxxxxxxxa
;xxxxxxxxxxxxxxxxxxxxxxxxxxxxxxxxxxxxxxxxxxxxxxxxxxxxxxxxxxxxb
;xxxxxxxxxxxxxxxxxxxxxxxxxxxxxxxxxxxxxxxxxxxxxxxxxxxxxxxxxxxxc
;xxxxxxxxxxxxxxxxxxxxxxxxxxxxxxxxxxxxxxxxxxxxxxxxxxxxxxxxxxxxd
;xxxxxxxxxxxxxxxxxxxxxxxxxxxxxxxxxxxxxxxxxxxxxxxxxxxxxxxxxxxxe
;xxxxxxxxxxxxxxxxxxxxxxxxxxxxxxxxxxxxxxxxxxxxxxxxxxxxxxxxxxxxf
;xxxxxxxxxxxxxxxxxxxxxxxxxxxxxxxxxxxxxxxxxxxxxxxxxxxxxxxxxxxxg
;xxxxxxxxxxxxxxxxxxxxxxxxxxxxxxxxxxxxxxxxxxxxxxxxxxxxxxxxxxxxh
;xxxxxxxxxxxxxxxxxxxxxxxxxxxxxxxxxxxxxxxxxxxxxxxxxxxxxxxxxxxxi
;xxxxxxxxxxxxxxxxxxxxxxxxxxxxxxxxxxxxxxxxxxxxxxxxxxxxxxxxxxxxj
;xxxxxxxxxxxxxxxxxxxxxxxxxxxxxxxxxxxxxxxxxxxxxxxxxxxxxxxxxxxxk
;xxxxxxxxxxxxxxxxxxxxxxxxxxxxxxxxxxxxxxxxxxxxxxxxxxxxxxxxxxxxl
;xxxxxxxxxxxxxxxxxxxxxxxxxxxxxxxxxxxxxxxxxxxxxxxxxxxxxxxxxxxxm
```

(Continued)

**Example 6.1:** (Continued)

```
;xxxxxxxxxxxxxxxxxxxxxxxxxxxxxxxxxxxxxxxxxxxxxxxxxxxxxxxxxxn
;xxxxxxxxxxxxxxxxxxxxxxxxxxxxxxxxxxxxxxxxxxxxxxxxxxxxxxxxxxo
;xxxxxxxxxxxxxxxxxxxxxxxxxxxxxxxxxxxxxxxxxxxxxxxxxxxxxxxxxxp
;xxxxxxxxxxxxxxxxxxxxxxxxxxxxxxxxxxxxxxxxxxxxxxxxxxxxxxxxxxq
;xxxxxxxxxxxxxxxxxxxxxxxxxxxxxxxxxxxxxxxxxxxxxxxxxxxxxxxxxxr
;xxxxxxxxxxxxxxxxxxxxxxxxxxxxxxxxxxxxxxxxxxxxxxxxxxxxxxxxxxs
DoubleBuffer=1
AutoScan=1
WinVer=4.10.1998
```

Under Windows 9X, the MSDOS.SYS file is used to tailor the Windows 9X startup sequence. Like the Windows 3.X configuration files, MSDOS.SYS is divided into sections, with each section label contained within brackets. In Example 6.1, two sections are identified. One is labeled [Paths] and the other is labeled [Options]. Table 6.4 lists and gives a basic explanation of the options that can be used in MSDOS.SYS.

The required file SYSTEM.DAT replaces the CONFIG.SYS file that was optional booting MS-DOS. The SYSTEM.DAT file will be explained in greater detail in section 6.10. The CONFIG.SYS, COMMAND.COM, and AUTOEXEC.BAT are optional files and come into play if Windows 9X is forced into the DOS mode. The user can do this during the boot sequence by pressing Shift+F5 when the "Loading Windows 95" message appears on the screen. Also the DOS mode can be selected as one of the Shut Down options offered in the Start menu. WIN.COM is the actual Windows 9X operating system kernel that takes care of loading the default shell EXPLORER.EXE.

### 6.9.1 Windows 9X Boot Sequence Options

There are several options available to the user during the Window 9X boot sequence that are very useful in troubleshooting problems. As the computer boots, press and hold the Ctrl key on the keyboard. Eventually a menu will appear similar to the one shown in Example 6.2.

**Example 6.2:** Windows 9X boot sequence Option menu

1. Normal
2. Logged (BOOTLOG.TXT)
3. Safe mode
4. Step-by-step confirmation
5. Command prompt only
6. Safe mode command prompt only

If the Normal option is selected from this menu, Windows 9X will attempt to boot normally. If the Logged option is selected, Windows 9X will attempt to boot normally but will keep a record of the boot sequence in a file named BOOTLOG.TXT.

If the **Safe mode** option is selected, Windows 9X will boot in its most basic form. If Windows 9X will not boot in Safe mode, it usually means it will have to be reinstalled or a known good set of registry files will have to be restored. If Windows 9X detects that a wrong driver is being used during a normal boot sequence, it will automatically boot into Safe mode. Once in Safe mode, the user can change drivers. The Safe mode is discussed further in section 6.9.3.

If Step-by-step confirmation is selected, the boot sequence will pause between each major section of the boot. At each pause, information will be displayed indicating the next section of the boot sequence that will take place. The user is asked to press the Y key (yes) or N key (no) to determine if the indicated step will be processed or skipped. Skipping steps in the boot sequence is very useful when Windows 9X fails to boot at all. One of the drivers or configuration files might be the problem.

The Command Prompt Only option will take the computer to a real-mode version of MS-DOS. COMMAND.COM will be loaded and will control the computer. If any drivers that are not supported

**Table 6.4** MSDOS.SYS Options*

| Entry | Description |
|---|---|
| [Paths] section: | |
| HostWinBootDrv=c | Defines the location of the boot drive root directory |
| WinBootDir= | Defines the location of the necessary startup files. The default is the directory specified during Setup; for example, C:\WINDOWS. |
| WinDir= | Defines the location of the Windows 95 directory as specified during Setup |
| [Options] section: | |
| BootDelay=n | Sets the initial startup delay to *n* seconds. The default is 2. BootKeys=0 disables the delay. The only purpose of the delay is to give the user sufficient time to press F8 after the Starting Windows message appears. |
| BootFailSafe= | Enables Safe mode for system startup. The default is 0. (This setting is enabled typically by equipment manufacturers for installation.) |
| BootGUI= | Enables automatic graphical startup into Windows 95. The default is 1. |
| BootKeys= | Enables the startup option keys (that is, F5, F6, and F8). The default is 1. Setting this value to 0 overrides the value of BootDelay=n and prevents any startup keys from functioning. This setting allows system administrators to configure more secure systems. (These startup keys are described in General Troubleshooting.) |
| BootMenu= | Enables automatic display of the Windows 95 Startup menu. The default is 0. Setting this value to 1 eliminates the need to press F8 to see the menu. |
| BootMenuDefault=# | Sets the default menu item on the Windows Startup menu; the default is 3 for a computer with no networking components, and 4 for a networked computer. |
| BootMenuDelay=# | Sets the number of seconds to display the Windows Startup menu before running the default menu item. The default is 30. |
| BootMulti= | Enables dual-boot capabilities. The default is 0. Setting this value to 1 enables the ability to start MS-DOS by pressing F4 or by pressing F8 to use the Windows Startup menu. |
| BootWarn= | Enables the Safe mode startup warning. The default is 1. |
| BootWin= | Enables Windows 95 as the default operating system. Setting this value to 0 disables Windows 95 as the default; this is useful only with MS-DOS version 5 or 6.x on the computer. The default is 1. |
| DblSpace= | Enables automatic loading of DBLSPACE.BIN. The default is 1. |
| DoubleBuffer= | Enables loading of a double-buffering driver for a SCSI controller. The default is 0. Setting this value to 1 enables double-buffering, if required by the SCSI controller. |
| DrvSpace= | Enables automatic loading of DRVSPACE.BIN. The default is 1. |
| LoadTop= | Enables loading of COMMAND.COM or DRVSPACE.BIN at the top of 640K memory. The default is 1. Set this value to 0 with Novell,® NetWare,® or any software that makes assumptions about what is used in specific memory areas. |
| Logo= | Enables display of the animated logo. The default is 1. Setting this value to 0 also avoids hooking a variety of interrupts that can create incompatibilities with certain memory managers from other vendors. |
| Network= | Enables Safe mode with Networking as a menu option. The default is 1 for computers with networking installed. This value should be 0 if network software components are not installed. |

*The information in this table was obtained from the Windows 95 Resource Kit.

by MS-DOS are needed, such as those for the CD-ROM, mouse, or sound card, they must be specified in the CONFIG.SYS or AUTOEXEC.BAT file. Command prompt is very useful for executing DOS applications in the real mode because the DOS mode available through the Windows 9X Start menu actually operates the microprocessor in the Virtual 86 mode.

If the Safe Mode Command Prompt Only option is selected, the computer will drop to a real-mode operating system without processing the CONFIG.SYS file and AUTOEXEC.BAT file.

### 6.9.2 Windows 2000 Boot-up Options

Like Windows 9X, Windows 2000 also provides a few startup options to aid in troubleshooting a boot problem. To display the Setup options, press the F8 key when the "Starting Windows" message appears at the bottom of the screen during the boot sequence. This will display the boot options listed in Example 6.3.

**Example 6.3:** Windows 2000 boot options

> Safe Mode
> Safe Mode with Networking
> Safe Mode with Command Prompt
> Enable Boot Logging
> Enable VGA Mode
> Last Known Good Configuration
> Directory Services Restore Mode (Windows 2000 Server)
> Debugging Mode
> Boot Normally
> Note: Use the cursor control arrow keys to select an option, then press Enter.

If Safe mode is selected, it will start Windows 2000 using the most basic drivers available (PS/2 mouse, basic video, keyboard, hard drives, floppy drives, and default system services). If the computer does not start in Safe mode, the Emergency Repair disk may be needed to repair the system.

As the name implies, Safe Mode with Networking starts Windows 2000 with the basic services available with Safe mode, along with network support.

Safe Mode with Command Prompt starts Windows 2000 with the same basic services that are available with Safe mode. After logging onto the system, the Command prompt is displayed instead of the Windows 2000 GUI desktop.

Enable Boot Logging is similar to the Logged (BOOTLOG.TXT) option in the Windows 9X startup. If this option is selected, Windows 2000 will attempt to boot in Normal mode while creating a log named NTBTLOG.TXT, in the WINNT directory, listing all of the drivers and services that were loaded along with those that were not.

Enable VGA will boot Windows 2000 in Normal mode except the basic VGA video driver will be used. This is very useful because a major cause of system problems is selecting the wrong video driver or video mode, which prevents the computer from booting or makes the displayed image unreadable. From this mode, the original display driver can be restored or the correct display mode can be selected.

Last Known Good Configuration is selected during the boot to restore the Windows 2000 configuration (registry) files that were saved after the last shutdown. This option will not solve problems created by corrupted or missing drivers, however.

Debugging mode starts Windows 2000 while sending debug information out one of the computer's serial ports to another computer.

### 6.9.3 Windows 98/2000 Safe Mode

When Window 3.X is told to load a defective or incompatible driver with the SYSTEM.INI file during the boot sequence, Windows 3.X simply aborts the boot without even displaying an error message. When this occurs, it is up to the user to find the faulty driver in the SYSTEM.INI and correct the problem. This occurs quite often when attempting to change the video driver. If a display mode is chosen that is not compatible with the display adapter, Windows 3.X will no longer boot.

To solve this major problem with Windows 3.X, Microsoft added the Safe mode to Windows 9X/2000. If Windows 9X/2000 detects an error with a driver during the boot sequence, Windows 9X will automatically continue the boot into the Safe mode. Also, Safe mode can be entered manually through the startup options discussed in the previous section.

As stated, Safe mode loads the most basic drivers available to the operating system. These usually include basic drivers for the video adapter, keyboard, mouse, hard drives, floppy drives, and basic system services. These are drivers that are almost certain to work unless there is a major problem with the Windows 9X/2000 operating system itself.

It is obvious when the computer boots into Safe mode because a message window will pop up letting you know this fact. Also, the words "Safe Mode" will be displayed in all four corners of the display. In Safe mode, the user can change hardware drives or display settings to get Windows 9X/2000 to boot in Normal mode.

### 6.9.4 Soft-booting Windows 9X/2000

The most often used method for soft-booting Windows 9X/2000 is to use the Restart option in the Shut Down menu. To select this option, click on the Start button, then move the mouse pointer up to Shut Down, and then click on this option. When the Shut Down Windows Options menu appears, click on the small white dot beside Restart, and then click on the OK button, if using Windows 9X. For Windows 2000, use the scroll button to locate the Restart option, and then click on OK.

Another useful method to initiate the soft-boot sequence for Windows 9X/2000 is the Ctrl+Alt+Del sequence discussed in chapter 5. When this key sequence is used with MS-DOS, the computer will initiate a soft-boot sequence immediately. This is not the case with Windows 9X/2000, however. The first time the Ctrl+Alt+Del sequence is pressed, a window will appear in the middle of the screen called the Windows 9X Close Program or Windows 2000 Windows Security window. These windows have several functions that will be discussed later; in this section, only the soft-boot options will be discussed. Displayed on the bottom of both of these windows is a warning message informing you that pressing the Crtl+Alt+Del sequence again will restart the computer, which initiates a soft-boot. If this option is used, a backup will not be performed on the system files and files may even be corrupted. This procedure may also corrupt any application that may be running at the time. For this reason, I highly recommend that you do not use this procedure if at all possible. A Shut Down button is also displayed on the window. Clicking on this button will also soft-boot the computer.

### 6.9.5 Close Program and Task Manager Windows

Windows and DOS applications often lock up and stop responding. When this happens, the program cannot be closed using the application's normal Shut Down procedure. In Windows 3.X, the computer usually has to be rebooted to resolve the problem. When the computer is rebooted, all of the data being processed by the other programs being multitasked will also be lost and there is a chance of corrupting all open applications. Windows 9X allows the user to close nonresponding applications without rebooting the computer. To shut down nonresponding applications, the Close Program window is used.

To open the Close Program window, use the Ctrl+Alt+Del sequence once. In Windows 9X, the Close Program window, shown in Figure 6.6, will display immediately showing a list of all of the applications currently running in the computer's RAM. The scroll button to the right of the window can be used to list programs that are not currently shown. Programs that have the "not responding" message beside their name are programs that are causing problems. To close each program that is not responding, click on its name once and then click on the End Task button.

In Windows 2000, Task Manager is used to terminate nonresponding programs. When the Ctrl+Alt+Del sequence is initiated under Windows 2000, the Windows Security window will display. From this window, click on the Task Manager button, and then click on the Applications tab on the next window that displays. This window will display all of the applications that are currently running. To close a nonresponding application, use the same procedure as that used with Windows 9X.

## 6.10 WINDOWS 9X/2000 CONFIGURATION

All of the configuration files used by DOS and Windows 3.X are text files. Text files are saved on the disk in ASCII. The problem with ASCII is that each ASCII character requires a whole byte of storage. This means that to put only the term [boot] in Window's SYSTEM.INI file will require 6 bytes of storage. Using ASCII is very inefficient because of the amount of storage required for each configuration file. If a binary combination was used instead, each byte of memory could contain 256 configurations ($2^8$). This is why Microsoft decided to use a combination of ASCII and binary in the Windows 9X's configuration files.

The configuration files used by Windows 9X are called **registry files.** Actually there are only two registry files—SYSTEM.DAT and USER.DAT. These files are located in the Windows folder and contain configuration information using a combination of binary and ASCII. They are automatically given the file attributes hidden, system, and read-only so they will not display using Windows Explorer unless the Show All Files option is enabled. Under Windows 2000, the Display Protected System Files option must also be selected.

The registry file **USER.DAT** contains the setup information defined by the user. This includes such things as the icons that appear on the desktop, the screen saver selected, the background image, saved passwords, the location of the Task Bar, and much more.

The registry file **SYSTEM.DAT** contains all of the hardware configuration information that controls the computer. This includes, but is not limited to, hardware driver information, hardware configuration settings, and network configuration.

Unlike Windows 9X, Windows 2000s registry consists of several files that are too numerous to mention. All of the Windows 2000 registry files are located in the \WINNT\System32\config directory, which is located on the Windows 2000 boot drive.

No matter the number of files, Windows 9X/2000 provides two means to edit the contents of the registry. One is called Regedit, which provides access to the contents of the entire registry. The second is called System Policy Editor, which provides limited access to the registry.

### 6.10.1 REGEDIT

Using binary as part of the configuration information makes these files impossible to edit using DOS's text editor. Windows 9X/2000 does have an editor, called **Regedit,** that allows an extremely knowledgeable person to view and even change the contents of the registries.[2] This program is located in the Windows folder (WINNT) under the name REGEDIT.EXE. It is very important that a backup be made of the registry files before using Regedit.

To run REGEDIT, type "regedit" without the quotes in the text box under the Run option in the Start menu, and then click on the OK button. An example of the window displayed by the REGEDIT program is shown in Figure 6.8. As you can see from this figure, REGEDIT's display looks similar to the Windows Explorer display, with a left and right pane. In the left-hand pane are the main registry keys called *HKEY*s, which stand for *host key.*

Even though there are only two or more registry files, there are six HKEYs displayed in the left windowpane by REGEDIT. Of the six listed, the keys labeled HKEY_LOCAL_MACHINE (HKLM) and HKEY_USERS (HKU) are the master keys and are the actual registry files saved on the disk. All of the other keys are derived from these two main keys.

When the small plus (+) sign beside each key is clicked on, the left windowpane view expands in hierarchical form, similar to Windows Explorer. When a key (folder) is highlighted with a left-click of the mouse button, the values associated with the highlighted key are displayed in the right windowpane. The three categories that a key value will fall into are binary, double-word (DWORD, or 32-bit), and ASCII. Binary values are displayed in 1s and 0s, DWORDs are displayed in hexadecimal, and ASCII is displayed as text.

Unfortunately, it is beyond the scope of this book to teach the meanings of all of the entries and key values that can be displayed and modified with REGEDIT. Table 6.5 lists the major HKEY categories and a brief explanation of their contents.

---

[2]**Warning!** Changing entries in the Windows 9X/2000 registry can cause the system to become unstable and may prevent Windows 9X/2000 from booting. Changes should never be made to the registry unless absolutely necessary. Be extremely careful and always back up all registry files before making any changes.

**Figure 6.8** Typical REGEDIT Window

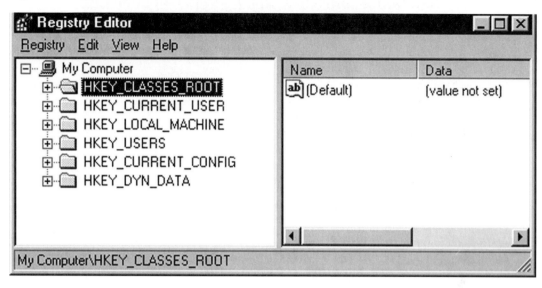

**Table 6.5** HKEY Functions in the Order They Appear When REGEDIT Is Used

| HKEY_CLASSES_ROOT (HKCR) | Points to the information contained in HKLM\Software\classes. The values in this HKLM key include file association information along with information on object linking and embedding (OLE). The purpose of HKCR is to maintain compatibility with Windows 3.X |
|---|---|
| HKEY_CURRENT_USER (HKCU) | Contains the settings for the current user who is logged onto the computer. HKU\DEFAULT contains the default key values used by HKCU so that a user can log on without prior user profile information. |
| HKEY_LOCAL_MACHINE (HKLM) | Contains information on all of the software installed on the computer, including hardware drivers, file associations, and OLE information. The settings in this key apply to all users. |
| HKEY_USERS | Contains information on all computer users and also the default user configuration. The entries in this key are used to create HKCU. |
| HKEY_CURRENT_CONFIG (HKCF) | Points to the collection of information contained in HKLM\Config\0001. |
| HKEY_DYN_DATA (HKDD) | Contains dynamic configuration information that remains in RAM when Windows 9X is running. This key keeps a dynamic entry of the computer's current configuration. HKDD also contains various network statistics in HKDD\PrefStats. |

## 6.10.2 Windows 9X System Policy Editor

Another utility program called **System Policy Editor** is available on the Windows 9X CD-ROM that allows the user to change some of the settings in the registry files using menu options. This program is a lot easier to understand than REGEDIT. To install System Policy Editor, follow the steps in Procedure 6.8. As with REGEDIT, it is imperative that the registry files be backed up before making any changes to the registry using System Policy Editor.

## Procedure 6.8: Installing System Policy Editor

1. Place the Windows 9X CD-ROM in the CD-ROM drive.
2. Click on the Start button, move up to Settings, and then over to Control Panel.
3. Click on Control Panel.
4. In the Control Panel window, double-click on the Add/Remove Programs icon.
5. Click on the Windows Setup tab.
6. Click on the Have Disk button.
7. On the Install From Disk window, click on the Browse button.
8. On the Open window, click on the scroll button beside the Drives box and find the drive letter of the computer's CD-ROM drive that contains the Windows 9X CD. Once located, click on the letter.
9. If the computer is running Windows 95, double-click on the folders in the right-hand window in the following sequence:

   ADMIN
   APPTOOLS
   POLEDIT

10. If running Windows 98, double-click on the folders in the right-hand window in the following sequence:

    TOOLS
    RESKIT
    NETADMIN
    POLEDIT

11. In the right window, click on the filename "poledit.inf" and then click on the OK button. Click on the OK button on the Install From Disk window.
12. In the Have Disk window, click on the small white check box beside System Policy Editor, and then click on the Install button.

**To launch system policy editor:**
1. Click on the Start button and then move the mouse pointer through the Start menus in the following sequence:

   Programs
   Accessories
   System Tools

2. Click on System Policy Editor.

**Figure 6.9** System Policy Editor Display

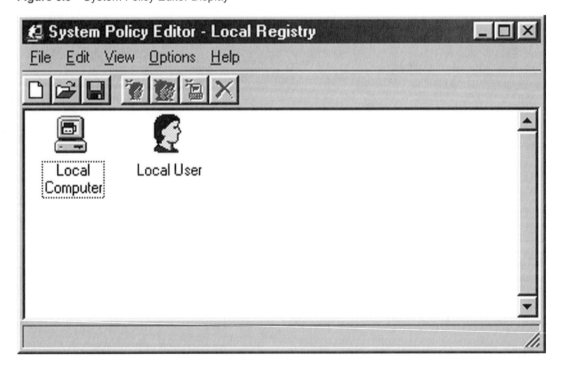

After the System Policy Editor loads, open the computer's registry by clicking on File in the Menu Bar, and then click on the Open Registry option. The display will change to the one shown in Figure 6.9. The changes available to SYSTEM.DAT are located under the Local Computer icon. USER.DAT is represented by the Local User icon. To open the menus to change or display the available settings in the registry file, double-click on the appropriate icon.

> **Procedure 6.9: Downloading and installing Tweak UI**
>
> **Warning!** Making changes to the registry using Tweak UI can cause the computer to become unstable and may prevent it from booting. For this reason, always back up the registry before making any changes.
>
> 1. Connect to the Internet through your network or internet service provider (ISP).
> 2. Start your Web browser and enter the following Internet address:
>
>    www.microsoft.com/windows95/downloads
>
> 3. Under the POWER AND KERNAL TOYS heading, select and download Windows 95 Power Toys Set.
> 4. Follow the download and installation instructions provided by Microsoft.
>
> **Installing Tweak UI from the Windows 98 CD:**
> 1. Insert the Windows 98 CD into the CD-ROM drive.
> 2. Click on the Start button.
> 3. Move up to Settings, and then over to the Control Panel.
> 4. Click on the Control Panel listing.
> 5. After the Control Panel window opens, locate and then double-click on the Add/Remove Programs icon.
> 6. After the Add/Remove Properties window opens, click on the Windows Setup tab.
> 7. Click on the Have Disk button.
> 8. On the Install From Disk window, click on the Browse button.
> 9. Select the correct drive letter for the computer's CD-ROM from the Drive scroll pane.
> 10. Once the CD-ROM drive has been selected, select the folder on the Windows 98 CD-ROM in the following sequence:
>
>     Tools
>     reskit
>     powertools
>
> 11. Highlight the tweakui.inf file with a left-click of the mouse button, and then click on the OK button.
> 12. Click on the OK button on the Install From Disk window.
> 13. In the pane displayed in the Have Disk window, place a checkmark in the small check box beside Tweak UI with a left-click of the mouse button, and then click on the Install button.
> 14. After the installation, close all open windows.
> 15. To launch Tweak UI, double-click on the Tweak UI icon in the Control Panel.

As with the other operating system configuration files, it is far beyond the scope of this book to teach the Windows 9X/2000 registry file with any great detail. There are entire books written on how registry files work and how to make changes to them. If you are going to fool with the registry, I highly recommend that you purchase one of these books before you make any attempts. The procedures for modifying and displaying the contents of the registry that are covered in this book will be given in the sections as needed.

### 6.10.3 Windows 9X Tweak UI (User Interface)

Tweak UI is a utility program provided by Microsoft that allows the user to tweak registry settings that affect appearance and performance. This utility can be installed directly from the Windows 98 CD, but must be downloaded from Microsoft's Web site for users of Windows 95. The Internet address to visit to download Tweak UI for Windows 95 users and the steps to follow to install Tweak UI from the Windows 98 CD are listed in Procedure 6.9. Once again, back up the registry before making any changes using the Tweak UI utility program.

It should be noted that Microsoft does not support Tweak UI. This means do not call Microsoft for help if any problems occur due to the use of this program.

### 6.10.4 Backing Up and Restoring the Registry Files

Because of the importance of registry files, Windows 9X/2000 automatically makes a backup copy of both registry files each time the operating system is shut down properly. There are a few differences in the number of backups kept and the number of files saved, however.

Windows 98 automatically keeps up to five backup copies of SYSTEM.DAT, USER.DAT, WIN.INI, and SYSTEM.INI in compressed files with the CAB extension. These files are stored in

## Procedure 6.10: Manual backup of the Windows 98 configuration files

**While Windows 98 is running:**
1. Click on the Start button.
2. Move up the mouse pointer through the following Start menu options:

   Programs, Accessories, System Tools

3. Click on the System Information option.
4. Click on the Tools option in the Menu Bar.
5. Click on Registry Checker.
6. Click on the Yes button when asked if you would like to back up the registry.
7. Close all open windows.

**Backing up the registry manually during the Boot Sequence:**
1. During the soft- or cold-boot sequence, press and hold the Ctrl key.
2. When the Boot Options menu displays, choose the Command Prompt Only option.
3. At the DOS prompt, type "scanreg" (without the quotes), and then press and release Enter.
4. When the Check Your Registry text window displays, select Start, followed by Enter.
5. After the check is completed, the Good Registry text window should display. From the window, choose the Create Backup option.
6. Once the backup is completed, a text window informing you that your registry is backed up will display. Select OK.
7. Soft- or hard-boot the computer.

## Procedure 6.11: Restoring any of the five Windows 98 configuration backups

1. During the soft- or cold-boot sequence, press and hold the Ctrl key.
2. When the Boot Options menu displays, choose Command Prompt Only.
3. At the DOS prompt, type "scanreg" (without quotes), followed by Enter.
4. When the Check Your Registry text window displays, select the Start option.
5. After the successful completion of the check, choose the View Backups option.
6. Use the up and down cursor control keys to select the desired backup.
7. Use the left and right cursor control keys to select Restore or Cancel to abort without restoring.

the \WINDOWS\SYSBCKUP\ folder under the filenames RB000.CAB through RB004.CAB, with RB000.CAB being the most recent.

Each time Windows 98 boots, a program called Registry Editor automatically runs before Windows 98 loads. If there is no problem with the registry, Windows 98 will make a new backup the next time it is shut down properly. If five backups already exist, the oldest is tossed. If an error is encountered, an error message will inform you of the situation and will restore the most recent backup automatically.

It is a good practice to get into the habit of backing up the Windows 98 registry manually before any new application or driver is installed. This can be accomplished by following the steps outlined in Procedure 6.10.

When Windows 98 detects an error in the current registry, it will display an error message informing you of the fact and will restore the last good backup automatically. If the user prefers to use one of the other backups, it will have to be selected manually using SCANREG. The steps to follow to select any of the five backups manually are listed in Procedure 6.11.

Windows 95 makes a single backup copy of SYSTEM.DAT and USER.DAT but does not back up any of the Windows 3.X INI files. These registry backups are stored in the Windows folder under the names SYSTEM.DA0 and USER.DA0. Like the originals, these files are given the hidden, read-only, and system attributes automatically and will not display using Windows Explorer unless the Show All Hidden Files option is enabled.

Because Windows 95 makes only a single backup and the Windows 98 backups are compressed, it is a very good idea to have additional backups of SYSTEM.DAT and USER.DAT of your own. Since I have been using Windows 95, I have lost both the original and the backup twice.

## Windows 95/98/2000

---

**Procedure 6.12: Backing up the Windows 9X registry files using DOS**

1. Exit to DOS mode by:
   a. Clicking on the Start Button.
   b. Selecting Shut Down.
   c. Clicking on Restart in DOS Mode.
   d. Clicking on OK.

**Note:** The Enter key must be pressed at the end of all DOS command lines. The locations of spaces in the command line are critical. For this reason, underscores are used in all of the command lines to indicate where spaces must be placed. For each underscore (_), press the space key on the computer's keyboard.

2. The following DOS prompt should appear:

   C:\Windows>

3. Perform this procedure step only during the first time you back up the registry files. Enter the following command lines:

   C:\Windows>md_\regback

4. Enter the following DOS command lines. Underscores are used to show the location of spaces in the command.

   C:\Windows>attrib_-s_-r_-h_system.dat
   C:\Windows>attrib_-s_-r_-h_user.dat
   C:\Windows>copy_system.dat_c:\regback
   C:\Windows>copy_user.dat_c:\regback

**Restoring the registry files**

1. Soft-boot or hard-boot the computer.
2. If booting from the Startup diskette, go to step 6.
3. If the computer is running Windows 98, press and hold the Ctrl key.
4. If the computer is running Windows 95, press the F8 key when the Loading Windows 95 message is displayed.
5. When the Boot Options menu displays, select the Command Prompt Only option.
6. If booting from the Startup diskette, at the A: prompt, type C: and then press the Enter key.
7. At the C: prompt, type the following command lines to restore the registry:

   Cd_/windows
   Attrib_-r_-s_-h_system.dat
   Attrib_-r_-s_-h_user.dat
   Ren_system.dat_*.bak
   Ren_user.dat_*.bak
   Copy_c:/regback/system.dat
   Copy_/regback/user.dat

8. If inserted, remove the Windows 9X Startup diskette from the A: drive.
9. Soft-boot the computer.

---

There is an option in Regedit that allows you to export (save) and import (load) the registry files to and from a floppy diskette. Selecting the File option from the Menu Bar of Regedit performs this feat. The problem with using this option is that Windows 9X has to be operational for it to work.

Windows 9X will no longer load when the registry files become corrupted, so they will have to be backed up and restored in a DOS environment. For this reason, the steps in Procedure 6.12 for backing up and restoring the registry files use DOS command lines with the computer operating in DOS mode. This procedure will also work if the computer is booted from the Windows 9X Startup diskette.

Because DOS is not available with Windows 2000, Procedure 6.12 will not work. Also, booting the computer from a bootable DOS floppy disk to make repairs is usually not an option with Windows 2000. To repair Windows 2000 when it will not boot and back up the registry, use the Emergency Repair disk (ERD). To create an ERD and back up the registry, use the algorithm given in Procedure 6.13. The procedure for using the ERD to restore the registry backup and repair Windows 2000 when it will not boot is also included.

### 6.10.5 Creating a Windows 9X Startup Disk

Hopefully, by now you realize that it is imperative that you have a DOS boot diskette available at all times when running Windows 9X. Because Windows 9X runs its own version of DOS, the computer must be booted using this version. The Windows 9X boot diskette is referred to as the *Startup disk*. Along with the DOS boot files, the Startup diskette will also contain the DOS utilities FDISK, FORMAT, and HIMEM.SYS. In addition, the Startup disk created by Windows 98 will also provide support for a CD-ROM. Use the steps outlined in Procedure 6.14 to create a Startup diskette.

### Procedure 6.13: Creating an ERD and backing up the registry

1. Insert a blank, formatted diskette into the A: drive.
2. Click on the Start button on the Task Bar.
3. Move the mouse pointer up to Programs, over to Accessories, over to System Tools, and then click on Backup.
4. From the Backup window, click on the Emergency Repair Disk button.
5. Click on the white check box beside the message "Also backup the registry to the repair directory."
6. Click on the OK button.

**To restore the Registry and Repair Windows 2000 if it will not boot:**

1. First, use the Last Known Good Configuration option available in the Startup options discussed in section 6.9.2.
2. If step 1 does not work, boot the computer from the Windows 2000 CD or Startup diskette.
3. Select the option to run the Windows 2000 setup program.
4. When asked, "Do you want to continue installing Windows 2000?", press the Enter key to continue.
5. A menu will appear asking if you want to install a fresh version of Windows 2000 or if you want to repair an existing installation. Press the R key to repair.
6. Insert the ERD into the A: drive.
7. This will display the Recovery Console. From this menu, press R to repair the system using the ERD.
8. Choose the Fast Repair option to restore the registry. Of course, any system changes made since the last backup will be lost.
9. The Manual Repair option, available in the Recovery Console, should only be used by advanced users.
10. For more information on using the ERD and repairing Windows 2000, type "Emergency" in the Index option of the Help menu.

### Procedure 6.14: Creating a startup disk

1. Place a formatted diskette in the A: drive. **Warning:** All data on this diskette will be wiped out when the Startup diskette is created.
2. Click on the Start button on the Task Bar.
3. Move up to Settings and then over to Control Panel.
4. Click on the Control Panel option.
5. When the Control Panel window opens, double-click on the Add/Remove Programs icon.
6. Click on the Startup Disk tab.
7. Click on the Create Disk button.
8. Follow the on-screen instructions.

### 6.10.6 Removing Items for the System Tray (SysTray)

As discussed previously, the icons displayed in the SysTray portion of the Task Bar usually represent programs that are running in the background. Sometimes this is desirable and sometimes it is not. If the computer is randomly locking up for no apparent reason, the problem might be in one of the SysTray programs.

Icons are usually added to SysTray when the program is launched through either the Startup folder in the Start menu or through the Run, Run Once, or Run Service entries in the SYSTEM.DAT registry file. To remove programs from the Startup folder, follow the steps outlined in Procedure 6.15.

### Procedure 6.15: Removing programs from the Startup menu

1. Right click on the Start button.
2. Click on Open.
3. Double-click on Programs.
4. Double-click on Startup.
5. Use the mouse to drag the icon for the program you do not want to start automatically onto the desktop.

> **Procedure 6.16: Removing programs from the Run, Run Once, and Run Service entries in the SYSTEM.DAT registry file using System Policy Editor**
>
> **Warning:** Making changes to the registry can cause the system to become unstable and prevent it from booting. Make sure the programs being removed for the Run, Run Once, and Run Service registry keys are not vital programs.
>
> 1. Back up the registry files.
> 2. Click on the Start button, then move up to Programs, then over to Accessories, then over to System Tools, and then over to the System Policy Editor.
> 3. Click on System Policy Editor.
> 4. After the System Policy Editor window opens, click on the File option in the Menu Bar.
> 5. From the File menu, click on the Open Registry option.
> 6. Double-click on the Local Computer icon.
> 7. Click on the plus (+) sign beside the Windows 9X System icon.
> 8. Click on the plus (+) sign beside Program Run.
> 9. Click on the words Run, Run Once, or Run Service (not on the check box).
> 10. Click on the Show button located toward the bottom of the window.
> 11. Click on the name of the file you wish to remove.
> 12. If you plan to add the entry back at a later time, write the Value name and Value entry down on paper and store it in a safe place.
> 13. Click on the Remove button.

Programs can be removed from the Run, Run Once, or Run Service entries in the SYSTEM.DAT registry file by using either Regedit or System Policy Editor. The steps covered in Procedure 6.16 utilize System Policy Editor. Before System Policy Editor can be used, it must first be installed on the computer.

## 6.11 INSTALLING WINDOWS 95

Installing Windows 9X is a smooth process if correct preparations are made. To install the upgraded version from Windows 3.X or from Windows 95, simply follow the instructions provided with the CDs or diskettes.

The installation procedure covered in this section is for an installation from MS-DOS. It is necessary to know this procedure or have it available when the upgrade installation fails or when Windows 9X has to be reinstalled on a new hard drive or after the hard drive has been formatted. Even though versions are called an upgrade, they can be installed from MS-DOS without installing the previous version of Windows, if you know the correct procedure.

If you are going to install Windows 9X on a new hard drive or format and reinstall, it is imperative that you make sure you have a copy of the MS-DOS CD-ROM driver. This driver is usually located on a floppy diskette that comes with the CD-ROM drive. If the CD-ROM drive was already installed in the computer at the time of purchase, there is probably a copy of the driver on the hard drive installed in the computer. The location of the CD-ROM driver can usually be obtained by examining the contents of the CONFIG.SYS file. The hard copy of this file can be obtained by following the steps in Procedure 5.8 in chapter 5. The line in the CONFIG.SYS file that loads the CD-ROM driver will begin with the DEVICE command.

Once a copy of the CD-ROM driver is made, the file MSCDEX.EXE must also be copied to a floppy disk before the old hard drive is removed or before the format process is started. This file is explained in greater detail in chapter 10. If you do not have this file, a copy can be obtained from any computer currently running Windows 9X.

Once the DOS CD-ROM driver and MSCDEX.EXE are safely copied on floppy, the installation procedure can begin. A generic procedure is covered in Procedure 6.17. The exact procedure will vary from computer to computer, depending on the hardware installed.

## Procedure 6.17: Installing Windows 9X

**Materials Needed:**
- Windows 9X Upgrade CD-ROM along with book or case with install number;
- Windows 95 CD-ROM, if upgrading to Windows 98;
- Windows 3.X diskette that contains the SETUP program;
- One blank, formatted 3½″ diskette;
- Copy of the DOS CD-ROM driver for the CD-ROM installed in the computer; and
- Copy of MSCDEX.EXE.

**Note:** This procedure will vary depending on the number of partitions you wish to place on your hard drive. Most compatible computers use one partition.

It is very important that you not only follow the procedure steps, but that you also read and understand the information displayed on the computer's screen after each step. If a computer displays information that contradicts or does not follow the installation procedure listed here, follow the procedure displayed by the computer.

### CMOS changes:

1. Hard-boot the computer by pressing and then releasing the Reset button.
2. When a message similar to "Press Del to run setup" appears on the screen, press the indicated key. If you miss the message, go back to step 1. This key may vary from BIOS to BIOS.
3. Use the ↑↓ keys on the keyboard to select the items in the setup program. To open each menu, use the Enter key.
4. Use the procedure in step 3 to locate and change the following options in the CMOS. The name given for the items may not be exactly the same as those listed.

   | Virus Warning | Disable |
   | Boot Sequence | A, C, ..... |

5. Use the procedure shown on the Setup Programs display to exit and save the changes made.
6. If not already done, partition and format the hard drive using the "/s" switch. These procedures are covered in chapter 11.
7. Install the DOS CD-ROM driver and MSCDEX.EXE.

### Windows 95 installation:

8. During the installation process, a message will appear informing you to insert the media that contains the original operating system being upgraded. When this message appears, follow the displayed instructions.
9. Press the button on the CD-ROM drive to open the door.
10. Insert the Windows 9X CD-ROM into the CD-ROM driver drawer with the print side up, and then press the button on the CD drive to close the drawer.
11. Switch to the CD-ROM by typing its drive number at the C: > prompt and pressing Enter.
12. At the prompt for the CD, type SETUP, and then press the Enter key. This should start the installation process.
13. The Setup program for Windows 95 will automatically run SCAN-DISK to check the integrity of the hard drives before the actual installation process begins. When the prompt appears, choose the option that allows Setup to run a routine check on your computer.
14. The Windows 95 Setup window should appear next. Move the mouse to position the pointer on the display screen over the button labeled Continue. With the pointer positioned over this button, press the left mouse button once.
15. A Setup License Agreement window appears asking if you accept the license agreement. Move the mouse until the pointer is over the Yes button, then press the left mouse button.
16. Next a window will appear that shows the steps the computer will follow during the Windows 95 installation process. This window will display three different times during the installation process. Each time this screen appears, click on the button labeled Next.
17. Next the Choose Directory window appears. If not already selected, choose the C:\WINDOWS directory, and then click on the Next button.
18. The Setup Options window appears next, if not already selected (indicated by a black dot within the circle next to the option). Click on the small circle next to Typical installation, and then click on Next.
19. The Certificate of Authenticity window appears next. In the text boxes available (white rectangles), enter the series of numbers on the label attached to the Windows 9X CD-ROM box, and then click on Next.
20. A window appears with two blank text boxes asking for a NAME and COMPANY. Enter the desired information then click on Next.

21. When a window appears that asks if you have a network adapter and sound card, use the mouse and left mouse button to check the appropriate boxes, depending on your computer. Then click on Next.
22. The Windows Components window appears next. Here, choose Install the Most Common Components, and then click on Next.
23. When a window appears that asks if you want to make a Startup disk, answer Yes. Have a blank diskette available. Insert this diskette into the A: drive when told to do so.
24. The installation program now copies Windows 95 files into the correct folder on the hard drive. This process will take a few minutes, so this is a good time to take a break.
25. The Finish Setup window eventually appears. Click Finish. The computer should now reboot into the Windows 95 operating system. This first-time boot usually takes longer than usual.
26. The Date/Time Properties window appears next. Click the left mouse button on the arrow next to the text box that displays the current time zone. A pop-down menu appears that shows all of the time zones available. Position the mouse pointer over Central Time Zone and click the left mouse button on it. Next click the left mouse button on the Date/Time tab (similar to a file folder). Enter the correct date and time. When you have finished, click on the OK button.
27. Next a window displays that allows you to select the printer attached to your computer. Select the correct printer or install a driver from a diskette. Once a printer driver has been installed, click on the OK button. If a printer is not attached to the computer, click on the Cancel button.
28. A window is now displayed that informs you that the computer must be restarted. Click on the OK button.
29. The Welcome to Windows 9X window appears when the computer reboots. Click on the Close button.

This concludes the installation procedure for Windows 9X.

## 6.12 CONCLUSION

Currently, Windows 9X is the predominant operating system used on the compatible PCs. The intent of this chapter is to give an individual an overview of this powerful operating system. However, it is not the intent of this chapter to teach everything there is to know about the Windows 9X operating systems. As you probably guessed, because this operating system is so widely used, there are dozens of books that can be purchased at your local book store that will give you more in-depth information.

Due to the limited time between the release of Windows 2000 and the completion of this text, I have included as much information on this OS as I have available. The expected release of Windows Me is after the completion of this text, so it is not included.

It is imperative that any individual who plans to use, build, upgrade, and repair a computer learns how to manipulate the computer's operating system. When dealing with the compatible PC's, the primary operating systems you need to learn in-depth are MS-DOS and Windows 9X. You might encounter other operating systems such as Windows NT, Windows 2000, OS/2, Unix, and Linux, but the main point is to become very familiar with the operating system used on the computers you are attempting to maintain.

Understanding the computer's operating systems is an integral part of fully understanding the computer's hardware. Without the operating system, hardware will not work. For this reason, operating systems will be discussed throughout the remainder of this textbook.

Along with understanding the operating system, it is also imperative that an individual who deals with the compatible PC on the hardware level has the ability to examine and modify the operating system's configuration files. Unfortunately, due to all of the other material covered in this book, only a basic insight into the complexity of these files was given. I highly recommend that you obtain additional information from books, individuals, and the Internet relating to the configuration files used on the computers you will be servicing.

# Review Questions

## Fill in the Blanks

1. The name of the Windows 9X shell is _____ .
2. Windows 9X is a combination of a(n) _____-bit and _____-bit operating system.
3. The entire surface of the Windows 9X display is called the _____ and the small pictures that represent programs or commands are called _____ .
4. The Start button is located on the _____ .
5. Windows supplies aid to the user through the _____ menu.
6. Windows 9X supports filenames up to _____ characters in length.
7. Setting the program name to be launched automatically with data files is called _____ _____ .
8. DOS programs that will not run within a DOS window will probably run in _____ _____ .
9. The Windows 9X _____ _____ can be used to display and edit all of the DOS and Windows 3.X configuration files.
10. To aid in troubleshooting problems with booting, Windows 9X provides the _____ mode.
11. The configuration files used by Windows 9X are called _____ files.
12. The Windows 9X configuration file that is responsible for loading all of the device drivers is called _____ .DAT
13. The Windows configuration file that is responsible for loading the user settings is called _____ .DAT.

## Multiple Choice

1. The Windows 9X operating systems are strictly 16-bit.
   a. True
   b. False
2. Windows 9X is strictly a 32-bit operating system.
   a. True
   b. False
3. Windows 2000 is strictly a 32-bit operating system.
   a. True
   b. False
4. The computer should be turned off with Windows 9X running.
   a. True
   b. False
5. Windows 9X operates the computer's microprocessor in the:
   a. real mode.
   b. protected mode.
   c. fast mode.
   d. preemptive mode.
6. Windows 9X fully supports Windows 3.X applications.
   a. True
   b. False
7. The first section of all boot sequences is controlled by:
   a. Windows 9X.
   b. the BIOS.
   c. the disk drives.
   d. the volume boot record.
8. Programs represented by icons on the Quick Start menu:
   a. are running in the background.
   b. are running in the foreground.
   c. are launched with a single-click of the mouse button.
   d. are launched with a double-click of the mouse button.
9. Windows 2000 is a combination of a 16- and a 32-bit operating system.
   a. True
   b. False
10. To cause a program to run automatically every time Windows 9X is booted:
    a. place its icon on the desktop.
    b. place its icon in the Start menu.
    c. place its icon in the Quick Start menu.
    d. place its icon in the Startup folder.
11. Most of the user-defined settings are found in the:
    a. Start menu.
    b. Control Panel.
    c. Network Neighborhood.
    d. System Tray.
12. Programs represented by icons in the System Tray:
    a. are running in the background.
    b. are running in the foreground.
    c. are launched by a single-click of the mouse button.
    d. are launched by a double-click of the mouse button.
13. The filenames used by Windows 9X:
    a. strictly follow the 8.3 format used by MS-DOS and Windows 3.X.
    b. use the same reserved characters as those used by MS-DOS and Windows 3.X.
    c. can be up to 255 characters long.
    d. can only have extensions that are up to three characters in length.
14. The Windows Explorer shortcut key sequence to open the Start menu is:
    a. Ctrl+Esc.
    b. Ctrl+Copy.
    c. Ctrl+C.
    d. Ctrl+V.
15. One method of placing a programs icon in the System Tray is:
    a. with the Startup disk.
    b. with the RunOnce key in the registry.
    c. by placing the application program's icon in the Start menu.
    d. by booting the computer to DOS.
16. A PIF file:
    a. contains multitasking information for DOS applications.
    b. contains memory management information.
    c. contains display information.
    d. all of these are correct.

17. DOS mode:
    a. must be used by all DOS applications.
    b. uses all of the hardware drivers supplied with Windows 9X.
    c. supports DOS programs that will not run within the DOS prompt.
    d. switches the microprocessor to the real mode.
18. Windows 9X/2000 applications specify configurations through:
    a. INI files.
    b. CAB files.
    c. EXE files.
    d. INF.
    e. PIF.
19. DOS and Command prompt:
    a. must be used by all DOS applications.
    b. use all of the hardware drivers supplied by Windows 9X.
    c. switch the microprocessor to the real mode.
    d. cannot multitask.
20. A utility program provided with Windows 9X/2000 that will display all of the DOS and Windows 3.X configuration files is called:
    a. System Editor.
    b. System Policy Editor.
    c. Regedit.
    d. Windows Explorer.
21. The configuration files used by Windows 9X/2000 are:
    a. strictly binary files.
    b. strictly text files.
    c. a combination of binary and text.
    d. similar to the configuration files used by Windows 3.X
    e. similar to the configuration files used by DOS.
22. The name of the Windows 9X configuration file that holds the individual user's setting is:
    a. SYSTEM.EXE.
    b. USER.DAT.
    c. USER.EXE.
    d. SYSTEM.DAT.
    e. WIN.INI.
23. The name of the Windows 9X configuration file that holds the computer's configuration is called:
    a. SYSTEM.DAT.
    b. SYSTEM.INI.
    c. WIN.INI.
    d. SYSTEM.EXE.
24. The Windows 9X/2000 configuration files are called:
    a. INI files.
    b. CONFIG files.
    c. Registry files.
    d. configuration files.
25. It is not necessary to understand the computer's operating system in order to understand the hardware in the computer.
    a. True
    b. False
26. Windows 9X/2000 automatically makes backups of its configuration files.
    a. True
    b. False
27. There is no need to make additional backups of Windows 9X/2000 configuration files.
    a. True
    b. False
28. DOS programs that can use the drivers provided by Windows 9X/2000 can run in:
    a. DOS prompt or Command prompt.
    b. only in DOS mode.
    c. only in DOS command.
    d. DOS speed.
29. A computer running Windows 9X/2000 can be turned off safely with the operating system running.
    a. True
    b. False
30. Windows 95 offers automatic updates over the Internet.
    a. True
    b. False
31. Preemptive multitasking:
    a. Allows the operating system to maintain control of the computer as applications are multitasked.
    b. Allows the application to gain control of the computer when it is its turn to multitask.
    c. Is used by Windows 3.X applications.
    d. Is the only multitasking available with Windows 9X.
32. The automatic backup configuration files created by Windows 98 have the _____ extension.
    a. BAT
    b. SYS
    c. CAB
    d. BAK
    e. DAO
33. The automatic backup files created by Windows 95 during the boot process have the _____ extension.
    a. BAT
    b. SYS
    c. CAB
    d. BAK
    e. DA0
34. The program that allows the user to make changes to all of the registry entries is called:
    a. REGMODIFY
    b. System Editor
    c. RegEdit
    d. Progman
35. To manually back up the registry under Windows 2000:
    a. requires the creation of a Startup disk.
    b. Perform SCANREG.
    c. Perform REGEDIT.
    d. requires the creation of an Emergency Repair Disk.

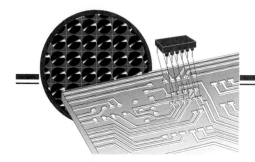

# CHAPTER 7

# PC Random-Access Memory

## OBJECTIVES

- Define terms and expressions used relative to computer memory.
- Draw the memory map for the IBM-compatible computer.
- Describe the characteristics of conventional, expanded, extended, and virtual memory.
- Describe the procedure used to set up MS-DOS's memory managers.
- Explain the procedure used to change Windows virtual memory settings.
- State the different characteristics of DIPs, SIMMs, and DIMMs.

- List the basic characteristics of the different types of DRAM used in the compatible computers.
- Describe the function of the computer's CMOS.
- Explain the differences among nonparity, parity, and ECC memory modules.
- State the procedure for the installation of memory modules in the compatible computer.
- Describe the procedure used to troubleshoot memory problems.

## KEY TERMS

byte address
page address
memory map
conventional memory
overlaying
kernel
upper memory area (UMA)
Beginner's All-purpose Symbolic Instruction Code (BASIC)
upper memory block (UMB)
memory paging
bank switching
expanded memory

Lotus-Intel-Microsoft (LIM)
LIM-expanded memory specification (LIM-EMS)
expanded memory manager (EMM)
extended memory
extended memory specification (XMS)
high-memory area (HMA)
HIMEM.SYS
EMM386.EXE
memory manager
MEMMAKER

actual memory
reserved memory
microprocessor bus cycle
access time
cache controller
single-inline memory module (SIMM)
configuration switches
complementary metal oxide semiconductor (CMOS)
swap file
virtual memory

When IBM chose the 8086/88 (86/88) microprocessor (μP), with its ability to address 1,048,576 bytes (1 MB) of memory, for its line of personal computers (PC), it moved the PCs into new realms of software development. Up until this time most microprocessors could only address 65,536 bytes (64 KB) of memory or less. This 64-KB addressing range had to be shared between ROM and RAM, and sometimes I/O. After the memory addresses were divided on most of these computers, approximately 48 KB of user RAM remained.

The 1 MB-memory addressing range of the PC's 86/88 microprocessor is split between ROM and RAM. The PC's I/O devices have their own address range that is separate from those used by memory, and will be covered in chapter 9. After the PC's 1 MB-address range is divided between RAM and ROM, there is room for up to 655,360 bytes (640 KB) of user RAM. This means the IBM-PC can have over 13 times the RAM available in most other PCs. With all of this memory available, software programmers started writing application programs for the PC that, up until the time, had only been able to run on mini- and mainframe computers. As these programs became more and more complex, even the 640 KB of RAM was not enough memory. To help alleviate the memory problem, methods were developed to give the programmer access to more RAM memory.

Before we start our discussion on the actual types of memory, we need to first get some terminology out of the way. Remember from chapter 1 that in computer storage, 1 K = 1,024 and 1 M = 1,024 K. These values will be used throughout this chapter, so don't forget them.

## 7.1 FUNDAMENTAL STORAGE UNIT

To allow the access of data in variable lengths, Intel and the compatible microprocessors use an 8-bit fundamental storage unit. This means that each memory address holds 8 bits of information. You might wonder, "How do we have microprocessors with 8-, 16-, 32-, and 64-bit data buses if each memory address can only hold 8 bits?" The answer to this question is that microprocessors that have a data bus greater than 8 bits have the ability to transfer data with more than one memory address at the same time. This feat is accomplished with signals on the microprocessor's control bus called *byte enable (BE) pins*.

Using a fundamental storage unit equal to 8 bits gives the microprocessors the ability to address information using different word sizes. For example, a 32-bit microprocessor can address information in 8-, 16-, or 32-bit word sizes by accessing memory in one, two, or four memory addresses at a time.

## 7.2 PAGES AND BYTES

To help explain the memory in compatible computers, we are going to divide the memory addresses into two units called *pages* and *bytes*. When pages and bytes are used to represent addresses, the first four hexadecimal digits from the right are called the **byte address,** or *byte number,* and all other digits are called the **page address,** or *page number*. When this method of indicating memory addresses is used, a byte address equals a fundamental storage unit of 8 binary bits.

For example, the hexadecimal address $82F9A_{16}$ is page $8_{16}$ byte $2F9A_{16}$. Address $10B9C4_{16}$ is page $10_{16}$ byte $B9C4_{16}$. Notice that the number of digits used to specify the byte address is always four, whereas the number of digits available to page numbers varies depending on the size of the microprocessor's address bus.

Remember that the number of combinations that can be represented by a given number of digits in any base can be calculated by raising the base to the number of digits. Using four hexadecimal digits to indicate the word address means that there are $65,536_{10}$ words on each page ($16^4$). The **word addresses** on any page will range from $0000_{16}$ through $FFFF_{16}$. When only the page number is given, it represents a 65,536-byte range of addresses.

**Problem 7.1:** What is the range of addresses for page $C_{16}$?

**Solution:** The range of addresses will be from $C0000_{16}$ through $CFFFF_{16}$.

Figure 7.1 shows the memory address assignments in the IBM-PC and XT-compatible computers. This type of illustration is usually referred to as a **memory map** because it shows what memory device is assigned to what address. Figure 7.1 is actually the memory map for the first 1,048,576 (1 M) addresses for all compatible computers using MS-DOS operating in the real mode. Each circle in Figure 7.1 contains a hexadecimal number that represents the page number. The number within the circle farthermost to the left is a 0, indicating page 0. This circle actually represents memory addresses between the range of $00000_{16}$ through $0FFFF_{16}$.

**Figure 7.1** Conventional RAM

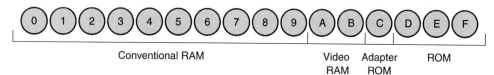

**Problem 7.2:** Referring to Figure 7.1, what address range does the "B" circle represent?

**Solution:** Address $B0000_{16}$ through $BFFFF_{16}$.

## 7.3 MEMORY IN THE PC

The memory address in the compatible computers is grouped into types that depend basically on the address location. The names of these memory types are conventional memory (base), upper memory, expanded memory, extended memory, and high memory. Each of these is discussed in the following sections.

### 7.3.1 Conventional Memory (Base Memory)

As stated previously, the IBM-PC computer splits the 8088 microprocessor's 1,048,576-byte addressing range into two types of memory: RAM and ROM. Referring to the memory map in Figure 7.1, ROM is used for the main BIOS, hard-drive BIOS, video adapter card BIOS, and some FDC adapter card BIOSs. The RAM is split into addresses for user RAM and video RAM. The user RAM is placed from page $0_{16}$ through page $9_{16}$, which is a total of 655,360 (640 K) address combinations. This 640-KB addressing range is sometimes called *base memory*, but is most often known as the computer's **conventional memory.**

**Problem 7.3:** What is the address range, in hexadecimal, for the compatible computer's conventional memory?

**Solution:** The conventional RAM occupies pages 0 through 9, so the address range is from 00000 through 9FFFF.

### 7.3.2 Never Enough Conventional Memory

As stated earlier, as the power of the PC began to be realized, more and more programmers started writing DOS application programs to perform tasks that had previously only been available on mini- and mainframe computers. Even though the computer can have a maximum of 640 KB of conventional memory installed, an application program does not get access to this entire range. The computer's DOS, interrupt tables (chapter 15), ROM BIOS tables, and TSRs require a large portion of the first few pages of addresses assigned to conventional memory. The more conventional memory needed by these items, the less is available for DOS application programs.

### 7.3.3 The Overlay

One way that a programmer compensates for the shortage of conventional RAM is through a technique called program **overlaying.** In this process, the program is divided into modules. A control module, called the **kernel,** is loaded into the computer's RAM. The kernel loads other modules from the disk as needed. Modules are loaded off the disk and discarded from RAM to keep the program's size within the computer's memory limitations. This is why most large application programs consist of several files. One file is the kernel and the other files are the individual program and data modules.

This technique theoretically gives the ability for the program's size to be unlimited. This is not the case, however, because the module's size cannot be variable. Some of the larger application programs can require over 600 K of conventional memory to handle the kernel and different modules as they are loaded off the disk. This means if 600 K of conventional memory is not free when the program attempts to load, an "insufficient memory" error message will be displayed.

Even though using overlays helped reduce the memory requirements of large programs, two problems still exist. First, with the computer constantly making disk access, the speed of program execution is greatly reduced. Second, if a new operating system is installed that requires more conventional memory or if new TSRs are loaded, even overlaid programs may not have enough memory available to load. This brought about some interesting techniques that are used to reduce the amount of conventional memory needed by the operating system to free up conventional memory for applications.

### 7.3.4 The Upper Memory Area (UMA)

Above the 640 K range of addresses used by conventional memory is an additional 384 K range of addresses (pages $A_{16}$ through $F_{16}$). This 384 K was originally used for video RAM, adapter ROMs, and the BIOS. This 384 K range of addresses is called the computer's **upper memory area (UMA)**. Referring to Figure 7.1, we see the video RAM is assigned UMA pages $A_{16}$ and $B_{16}$. This is a total of 128 KB of addresses. The video RAM is where images are stored while the video display controller (VDC) is displaying them. VDCs and video RAM are discussed in more detail in chapter 12.

As shown in Figure 7.1, page $C_{16}$ of the UMA is assigned to adapter ROMS. When IBM developed the address map for the compatible computers, it knew the BIOS would never be able to support all of the video and other adapters that would become available for the computer. To make sure the computer's BIOS would not become obsolete simply by the introduction of a new video standard, IBM assigned the addresses between $C0000_{16}$ and $CFFFF_{16}$ to *adapter BIOS*. The computer's main BIOS scans this area each time the computer is booted. If a new video adapter card is installed that is not supported by the BIOS, the manufacturer will place its driver program in the ROM installed on the adapter card. The video adapter ROM is assigned addresses in the range of C0000 to C7FFF. When the computer's BIOS detects the adapter ROM, the driver programs in the adapter ROM will supersede (take the place of) the driver programs in the computer's BIOS.

Hard-drive controllers that need additional BIOS support usually use the upper half of UMA page $C_{16}$, located in addresses C8000 through CFFFF. Because all main BIOS writers now place full support for hard drives in the computer's main BIOS, this area in the UMA usually goes unused.

Pages $D_{16}$ through $F_{16}$ of the UMA are assigned to ROMs. In the original IBM-PC/XT/ATs, pages $D_{16}$ and $E_{16}$ were used for ROMs that contained a BASIC interpreter, which is discussed in the next section. The main BIOS in compatible computers is located in page $F_{16}$. You might wonder, "With the need for memory, what are the unused pages of the UMA used for?"

### 7.3.5 Unused Addresses in the UMA

When IBM developed the memory map for the PC, the primary language used to program this computer was called the **Beginner's All-purpose Symbolic Instruction Code (BASIC)**. To facilitate this language, a BASIC interpreter program was located in the ROMs at pages $D_{16}$ and $E_{16}$. As the computers evolved, fewer and fewer people were writing programs because someone had already written them. Because the BASIC interpreter is very seldom needed, the ROMs that contained this programming language were eliminated from the computer. However, if BASIC programs are required, they can still be purchased or loaded from disk into RAM.

Eliminating the BASIC ROMs leaves the addresses in pages $D_{16}$ and $E_{16}$ unused on most compatible computers. Also, depending on the type of video adapter used, there may be some unused addresses in pages $A_{16}$ and $B_{16}$. As previously stated, full firmware support was added to the computer's main BIOS for hard drives. This eliminates the need to have adapter ROMs located on the hard drive adapter. Eliminating the hard drive adapter ROMs leaves the upper half of page $C_{16}$ unused on most compatible computers. An unused address in the UMA is illustrated in Figure 7.2.

**Figure 7.2** Unused Addresses

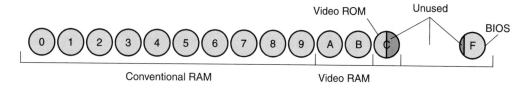

**Figure 7.3** Expanded Memory

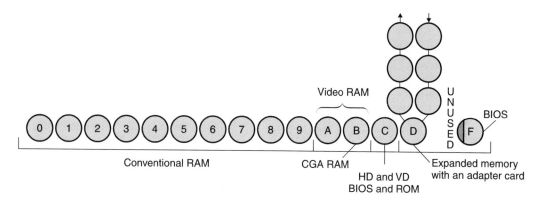

As shown in Figure 7.2, this leaves a lot of unused addresses that were originally assigned to ROM, video RAM, and adapter ROM in the IBM-PC. These unused blocks of memory in the UMA are called **upper memory blocks (UMBs)**.

Leaving all of these unused addresses in a machine that does not have enough RAM is not very smart. To help alleviate this problem, methods were developed to use the UMBs. Along with these methods came new terms and confusion. One method that probably created the most confusion is a type of memory called *expanded memory*.

### 7.3.6 Expanded Memory

One method used to exceed the 640-KB conventional RAM limitations of the 86/88 IBM-PC/XT is a technique called **memory paging**, or **bank switching**. In this process, additional memory, called memory windows, is switched into a 64-KB UMB. Memory access using this technique is called **expanded memory**. The operation of expanded memory is illustrated in Figure 7.3.

A group of companies—**Lotus, Intel, and Microsoft (LIM)**—established a standard for the hardware and software drivers that were required to run an expanded memory adapter card. The name of this standard is called the **LIM-expanded memory specification (LIM-EMS)**. EMS established the means to access up to 32 MB of expanded memory installed on an EMS expansion card using a TSR called an **expanded memory manager (EMM)**.

The expanded memory is divided in small groups of 64 K called *windows*. The individual windows are switched into the UMB, usually on page $E_{16}$ or $D_{16}$, in the UMA one window at a time by the EMM. When a 64-K window is filled, it is switched out of the UMA and another 64-K window is switched in. The microprocessor can only access the information in the window currently switched into the UMA. If the information needed by the microprocessor is not in the current window, the current window is switched out and the desired window is switched in.

Adding memory using this technique has several drawbacks. First, because the memory windows are switched into the memory map by the EMM software, they are not directly accessible to the microprocessor for program execution. This means that the additional memory can only be used to store and retrieve large blocks of data, not to run programs.

Second, to keep up with all of the locations where the data are stored within the expanded memory pages, the EMM has to establish a database in the conventional memory of the computer. This reduces the amount of conventional memory available for application programs.

Third, the expanded memory is available only to application programs that are written around EMS. When a program stores or retrieves data, it uses the EMM to specify what memory page to use. If the program is not written to use EMM, the expanded memory will go unused.

Fourth, because the memory is switched into the memory map, it slows down the program's access to the data stored in the expanded memory. However, this is still several times faster than constantly accessing the data from the disk.

As you can tell from this discussion, the introduction of expanded memory brought lots of confusion into the world of the IBM-PC/XT computer. Consumers thought that by simply adding an expanded memory card to their computer, they would have direct access to the extra memory available. This was, and still is, a major misconception because MS-DOS and most DOS applications programs do not have the ability to access the expanded memory. People could not (and still can't) understand how they could run out of memory with 8 MB or more of expanded memory in their computer.

For this reason, expanded memory is very seldom used on today's applications. This is because the memory addressing range of today's MPUs is far beyond the 1-MB limit of the 8088 MPU when operating in the protected mode. However, it is still very important to understand the concept when setting up these computers to run older DOS applications written around this technology.

### 7.3.7 Real Mode and Protected Mode Revisited

When IBM introduced the AT microcomputers, it used Intel's 80286 MPU (286). The introduction of these machines, based on this microprocessor, only added more confusion to the memory game. Before we go any further, let us review what we already know about the 286.

The 286 has a 16-bit data bus and 24-bit address bus. This address bus size gives the MPU the ability to place 16,777,216 ($2^{24}$) memory address combinations out of its address bus. To keep this microprocessor totally compatible with the 86/88, Intel gave the 286 two modes of operation—real mode and protected mode. In the real mode, the microprocessor thinks it is an 8086 and can access only 1,048,576 (1 M) memory addresses. Although this addressing range is characteristic of MS-DOS applications, there is a flaw in the chip that gives the 286 the ability to slightly exceed the 1-MB addressing range of the 8088. This flaw will be discussed later in this chapter.

Only in the protected mode can the 286 place 16,777,216 address combinations out of its address bus. This address bus size added additional addresses above the page $F_{16}$ limitation of the 8088. Just like all of the other memory address ranges, the range above the limits of the 8088 also has a name.

### 7.3.8 Extended Memory

**Extended memory** is the name given to the RAM addressed above the 1-MB limit of the 86/88 microprocessor. The memory map shown in Figure 7.4 illustrates this. This type of memory is accessible only to computers that run at least a 286 processor operating in the protected mode. MS-DOS is written around the microprocessor's real mode, so extended memory is almost impossible to access as true extended memory using this operating system and a 286 microprocessor. The **extended memory specification (XMS)** states the procedure and hardware used to address extended memory. Similar to expanded memory, a memory manager called the *extended memory manager (XMM)* is also needed.

---

**Problem 7.4:** What is the first page number of extended memory?

**Solution:** The addressing range of the 88/86 microprocessor ends at page F, so the first page of extended memory is page 10.

---

Because MS-DOS does not need extended memory to operate, it does not contain the drives that are needed to access it. If a DOS application needs to use extended memory, an XMM must be loaded with

**Figure 7.4** Extended Memory

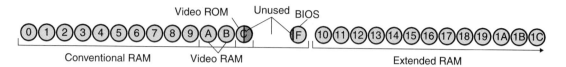

the CONFIG.SYS file because the memory management needs to be in place before the command interpreter is loaded. MS-DOS 3.3 and above come with a TSR called HIMEM.SYS, which is the default XMM. Along with managing the extended memory, HIMEM.SYS also has another important memory management function that is discussed next.

### 7.3.9 The 80286 Flaw

In the real mode, the 80286 is supposed to act just like a 8086 microprocessor. Remember, the 8086 has a 20-bit address bus that allows it to place $2^{20}$ memory address ($1,048,576_{10}$) combinations out of its address pins. In hexadecimal, this consists of addresses between the range of $00000_{16}$ to $FFFFF_{16}$. We already know that each digit of hexadecimal represents four consecutive binary bits, so a 5-digit hexadecimal number represents 20 bits of binary. The 20 pins of the microprocessor that carry the signals that belong to the address bus are called *A0*, for the least significant bit, up through *A19*, for the most significant bit.

Intel microprocessors calculate memory addresses by adding the contents of a register called a *segment register* inside the microprocessor to a 16-bit offset (byte address). Even though it is not important to know the exact process used, it is important to know that sometimes this addition can create a sum greater that $FFFFF_{16}$, but no larger then $10FFEF_{16}$. This does not create a problem with the 8086/88 microprocessor because there is not an A20 bit on the address bus that can handle the extra digit, so it is simply dropped. For example, if the sum creates an answer equal to $1045B9_{16}$, $045B9_{16}$ would be the actual address that goes out of the 8088 microprocessor's address bus on pins A0 through A19. When the maximum address ($FFFFF_{16}$) is reached, the address rolls back around to 0 ($00000_{16}$), similar to the odometer on an automobile or the counter on a tape player.

The 80286-compatible microprocessor has a 24-bit address bus. This means the address pins go from A0 through A23. Intel should have disabled address bits A20 through A23 when the microprocessor is in the real mode. This would have forced the addresses to roll around just like the 8086 microprocessor does, but this tiny mistake slipped by. The A20 bit is active when the 80286 microprocessor is in the real mode. This means that the real-mode addressing range for the 80286 microprocessor does not stop at page $F_{16}$ as the 8086 does, but extends up to $10FFEF_{16}$.

Even though it was a mistake in the 80286 microprocessor, it was one of those mistakes that turned out for the good. Microsoft discovered the flaw and decided to use page $10_{16}$ in the real mode to help free up some conventional memory addresses for use by applications. Similar to conventional memory, expanded memory, and extended memory, this page of memory also has a name.

### 7.3.10 High-Memory Area

When an 80286 or above microprocessor is operating in the real mode, page $10_{16}$ of extended memory is called the **high-memory area (HMA)**. MS-DOS does not directly support addressing of page $10_{16}$ in the real mode, so a driver is required. Microsoft decided to use the XMM driver HIMEM.SYS to control access of the HMA in the real mode.

You might wonder, "Why does MS-DOS need to address page $10_{16}$ in the real mode?" The reason is that with **HIMEM.SYS** (or similar) and at least a 286 processor, a large portion of MS-DOS can be loaded into the HMA. After this is accomplished, more of the precious conventional memory for application programs and TSRs remains.

To perform this amazing feat, HIMEM.SYS is loaded using the DEVICE= command in the CONFIG.SYS file. After this command, the DOS=HIGH command is placed in the CONFIG.SYS file. A typical example on how these lines will appear in the CONFIG.SYS file is shown in Example 7.1.

**Example 7.1:** Typical CONFIG.SYS lines to load HIMEM.SYS and MS-DOS into the HMA

```
DEVICE=C:\DOS\HIMEM.SYS
DOS = HIGH
```

At this time, we are going to put our discussion of HIMEM.SYS on hold. However, we are not quite finished with this memory manager, so look for its discussion in the coming sections of this chapter. We now need to move our discussion to the microprocessor that came along after the 286.

### 7.3.11 Beyond the 286

After the 80286, Intel introduced the 80386 (386), 80486 (486), and Pentium microprocessors. In addition to many other enhancements, these microprocessors have a 32-bit address bus. This address bus size gives these MPUs the ability to place 4,924,967,296 (4 GB) memory address combinations out of their address bus when in the protected mode. And let us not forget the Pentium Pro, Pentium II, and Pentium III microprocessors, with their 36-bit address bus that yields a memory addressing range of 68,719,476,736 (68 GB) when in the protected mode. However, in the real mode, they all operate as a very high-performance 86/88. In the real mode, the memory address range of these processors is limited to the 1-MB conventional memory addressing range plus the HMA.

These microprocessors automatically switch to the real mode at hardware reset. Like the 286, the 386/486/Pentium/Pentium Pro/Pentium II, and Pentium III MPUs are switched to the protected mode with software. Once in the protected mode, they can only be returned to the real mode by a hardware reset. These MPUs are different from the 286, however, in that a third mode of operation, called the *virtual 86 mode,* is added. The 386/486/Pentium/Pentium Pro MPUs can be switched back and forth between the protected mode and Virtual 86 mode with software. In fact, the virtual 86 mode operates within the protected mode. In the virtual 86 mode of operation, the microprocessor can directly run programs written for the 86/88 microprocessors. (Most MS-DOS programs are written around the microprocessor's real mode.) These MPUs have several other advantages that make them the microprocessor of choice in current versions of compatible computers.

First, new DOS versions, such as Microsoft's Windows 3.X, Windows NT, Windows 95, Windows 98, Windows 2000, Linux, and O/S2 support these MPUs' protected and virtual 86 modes. These operating systems will switch the MPU to the virtual 86 mode to run the thousands of MS-DOS application programs currently available. Second, these MPUs fully support multitasking. As stated previously, multitasking is the ability to run more than one application program (real or protected) simultaneously. Third, these microprocessors can address extended memory when operating in the virtual 86 mode. Being able to address extended memory in the virtual 86 mode opens up some features that are used by another MS-DOS memory manager known as **EMM386.EXE** (or similar).

### 7.3.12 More about MS-DOS and Windows Memory Managers

Practically all MS-DOS applications are written around the microprocessor's real mode and use the RAM addresses below 640 K (conventional memory). For this reason, it is imperative that as much conventional memory as possible be made available to applications.

One way that this is accomplished is by using HIMEM.SYS to move a large portion of MS-DOS into the HMA. But TSRs are also located in conventional memory and these programs often use more conventional memory space than MS-DOS uses before it is moved. It just does not make sense to have all of that extended memory going unused just because of the real-mode limitation of MS-DOS and MS-DOS applications.

Remember from previous discussions that HIMEM.SYS is not only the memory manager that gives access to the HMA, but it is also the extended **memory manager.** EMM386.EXE, however, is an expanded memory manager that allows the use of some of the extended memory for real-mode programs. The primary function of EMM386 is to allow extended memory to be used as if it were expanded memory.

Before this can be accomplished, EMM386.EXE must be loaded with a DEVICE= command in the CONFIG.SYS file. Because it must be able to access extended memory, EMM386.EXE is loaded after HIMEM.SYS. This sequence is shown in Example 7.2.

Once EMM386 is loaded, it monitors a program's call for expanded memory. When this occurs, EMM386 will route the address into a page in extended memory. This process is transparent to the applications and to the MS-DOS operating system. As far as MS-DOS and the application know, the computer has real expanded memory installed in one of the computer's UMBs.

Being able to use extended memory as expanded memory is great, but very few programs currently use expanded memory. What about freeing up more conventional memory for MS-DOS applications? As you might guess, like HIMEM.SYS, EMM386.EXE has a second function that aids in accomplishing this feat. It does this by allowing as many TSRs as possible to be moved out of the conventional memory. To do this, EMM386 translates memory addresses to make as much extended memory as possible appear to be addressed in the computer's UMBs. Once this is accomplished, TSR can be moved into the UMBs, which frees up more conventional memory.

Let's take time for a quick review before we continue. The programs HIMEM.SYS and EMM386.EXE are called *memory managers*. HIMEM has the job of managing the computer's extended memory and high-memory area (HMA). EMM386 is the expanded memory manager that allows some of the extended memory to be used as expanded memory. EMM386 is also the upper memory block (UMB) memory manager that allows TSRs to be loaded into the UMBs and not into conventional memory. If the computer's memory managers are set up correctly, there should be no spaces in the unused 1-M memory addressing range of the 86/88 microprocessor.

Example 7.2 shows how to load the memory managers and move DOS into the HMA. But what about loading TSRs into UMBs? Remember, TSRs can be loaded with either the CONFIG.SYS file or the AUTOEXEC.BAT file, depending on when they need to be in place.

**Example 7.2:** CONFIG.SYS lines that will load memory managers located in the DOS subdirectory and move a large portion of MS-DOS into the HMA

```
DEVICE=C:\DOS\HIMEM.SYS
DEVICE=C:\DOS\EMM386.EXE
DOS=HIGH
```

Once the memory managers are loaded with the CONFIG.SYS file, TSRs do not move magically to the UMBs. To accomplish this, each line in the CONFIG.SYS file and AUTOEXEC.BAT files that load TSRs have to be modified. In the CONFIG.SYS file, all of the DEVICE= commands, except the ones that load HIMEM.SYS and EMM386.EXE, must be changed to the DEVICEHIGH= commands. Also, LOADHIGH must be added in the front of each line in the AUTOEXEC.BAT file that loads a TSR.

When a TSR loads into a UMB, by default, it takes the entire block. This means that only a few TSRs will load high before all of the UMBs are used up. When a TSR has been told to load high and it will not fit, it will load automatically into conventional memory without displaying an error message. Like packing clothes in a suitcase, the more precisely things are packed a higher number of things will fit. To pack TSRs into UMBs precisely, one must know the number of UMBs available and the size, in bytes, of each TSR to be loaded high. Once this is known, additional information is added to the DEVICEHIGH and LOADHIGH commands to specify into which region the TSR should load and the amount of memory required. An example of the command modifications in the CONFIG.SYS file is shown in Example 7.3.

**Example 7.3:** Loading TSRs into specific regions

```
DEVICE=C:\WINDOWS\HIMEM.SYS
DEVICE=C:\WINDOWS\EMM386.EXE NOEMS
BUFFERS=21,0
FILES=30
DOS=UMB, HIGH
LASTDRIVE=F
FCBS=4,0
DEVICEHIGH /L:1,16976 =C:\WINDOWS\SETVER.EXE
DEVICEHIGH /L:1,22112 =C:\CDROM12X\CD1200.SYS /D:CD003
```

Line 8 in Example 7.3 indicates that the TSR is to be loaded into UMB region 1 and the length of the TSR is 16,976 bytes. Also notice that NOEMS has been added to the DEVICE= command that loads in the EMM386 memory manager. This is telling EMM386 that no extended memory is to be used as expanded memory. This should be added if Windows is used as the primary operating system.

You have probably noticed from the examples that installing and setting up the computer's DOS-based memory managers are major tasks even for individuals who are well versed in compatible computers. The computer's CONFIG.SYS and AUTOEXEC.BAT files have to be modified with special commands to load and set up the memory manager TSRs correctly. Special commands are required to move a portion of MS-DOS into the HMA and to load TSRs into the UMB. The good news is that if you are using MS-DOS 6.0, the task is a lot simpler.

**Example 7.4:** Command line used to run the MEMMAKER

> C:\>MEMMAKER

If you are using MS-DOS 6.0 or greater, a special utility called **MEMMAKER** can be executed by a command line similar to the one shown in Example 7.4. MEMMAKER will set up the computer's memory managers and modify the CONFIG.SYS and AUTOEXEC.BAT automatically to optimize the DOS memory environment. It will ask the user several questions during the installation that will allow the memory to be customized according to the DOS applications the user is running. This program can be executed at any time to keep the memory at its optimum configuration. This one feature alone makes upgrading to versions of MS-DOS above 6.0 worthwhile.

Even though this discussion has focused on the memory managers provided by MS-DOS, there is one major competitor in the memory manager market. We will discuss this competitor in our conclusion on memory managers.

### 7.3.13 Memory Manager Conclusion

There are several DOS memory managers available for purchase in addition to those supplied by MS-DOS. Two of the most widely used alternates are Symantec Corporation's (www.symantec.com) Quarterdeck's Expanded Memory Manager® called QEMM386® and QEMM97®. QEMM replaces both MS-DOS's HIMEM and EMM386. Once installed, a special utility called OPTIMIZE will set up the computer's memory. From my experience, this memory manager uses tricks and techniques that free up more conventional memory than that freed up by both of the MS-DOS memory managers combined. Also, QEMM97 aids in setting up the memory managers used by DOS mode programs in Windows 9X.

Configuring a computer's memory managers and keeping them optimized are very important, especially in a DOS environment. If any new DOS applications are installed that add new TSRs or change the computer's CONFIG.SYS or AUTOEXEC.BAT files, the memory-optimizing utility provided by the memory manager should be run.

Even though Windows 9X has its own extended memory managers, HIMEM and EMM386 are still needed by DOS applications that must run in DOS mode. If DOS memory managers are needed, they must be specified in the Advanced programs settings located under the Program option using PIF editor. If a DOS application will run in DOS prompt, its memory requirements are set using the Memory option in the PIF editor.

If you need more in-depth information on setting up DOS memory managers, it is available in many of the books written about DOS. Consult one of these books if you are having insufficient memory problems when running DOS applications. This is true even if the DOS application is running in Windows 9X's DOS mode.

We have discussed quite a bit of terminology used in reference to the memory in the compatible computer. Some of the terminology deals with the names given to specific memory address ranges. These names include conventional memory, UMA, HMA, expanded memory, and extended memory.

Also, we have discussed UMBs that reside within the UMA. Now that we know all of this memory terminology, we can discuss a command that will let us display the compatible computer's memory configuration.

### 7.3.14 DOS's MEM Command

It often becomes necessary to determine how much of each type of RAM exists within the computer. To access this information, MS-DOS's MEM command can be used. The MEM command is available only with version 5.0 and above. The MEM command is also available on computers running Windows 9X through DOS prompt. To use the command, simply type MEM at the MS-DOS prompt. Example 7.5 shows the results of this command.

**Example 7.5:** Using the MEM command

```
C:>MEM
```

| Memory Type | Total | Used | Free |
|---|---|---|---|
| Conventional | 640K | 36K | 604K |
| Upper | 155K | 155K | 0K |
| Reserved | 384K | 384K | 0K |
| Extended (XMS) | 31,589K | 113K | 31,476K |
| | | | |
| Total memory | 32,768K | 688K | 32,080K |
| Total under 1 MB | 795K | 191K | 604K |
| Largest executable program size | | 604K | (618,512 bytes) |
| Largest free upper memory block | | 0K | (0 bytes) |

MS-DOS is resident in the high memory area.

Refer to Example 7.5 for this discussion. The first column on the left indicates the memory type. The second column lists the total amount of each type of memory in the computer. The third column shows the amount of each type of memory used. The fourth column lists the quantity of each category of memory that is unused. According to Example 7.5, this particular computer has a total of 640 K of conventional RAM, of which 36 K is used. This leaves a total of 604 K of conventional memory to be used by MS-DOS applications.

**Problem 7.5:** For the computer illustrated in Example 7.5, how much extended memory is available to Windows applications?

**Solution:** According to Example 7.5, there are 31,476 KB of free extended memory.

Notice in the listing labeled "Largest executable program size," that memory is given in two forms. One states that there are 604 KB and the other, inside parentheses, states that there are 618,512 bytes of RAM available. You might wonder, "Which is correct?" The answer to this question is that they both are. The 618,512 is referred to as **actual memory** because it does not need to be multiplied by 1,024. Remember from chapter 1 that the rules to convert to the actual amount of memory are: (1) If the number ends in K, multiply by 1,024. (2) If the number ends in M, multiply by $1,024 \times 1,024$. (3) If the number is written out completely, do not multiply by anything because this is the actual number.

The /C (Clarify) switch can be added to the MEM command to display more in-depth information about the memory in the computer. An example of the information added by the /C switch is shown in Example 7.6.

**Example 7.6:** Example of the information added with the /C switch

```
C:>MEM /C

Modules using memory below 1 MB:
  Name           Total              Conventional        Upper Memory
  ---------      ---------------    ---------------     ---------------
  SYSTEM         18,128   (18K)     9,856    (10K)      8,272    (8K)
  HIMEM          1,168    (1K)      1,168    (1K)       0        (0K)
  EMM386         4,320    (4K)      4,320    (4K)       0        (0K)
  WIN            3,824    (4K)      3,824    (4K)       0        (0K)
  vmm32          99,056   (97K)     9,536    (9K)       89,520   (87K)
  COMMAND        7,664    (7K)      7,664    (7K)       0        (0K)
  SETVER         864      (1K)      0        (0K)       864      (1K)
  CD1200         20,592   (20K)     0        (0K)       20,592   (20K)
  IFSHLP         2,864    (3K)      0        (0K)       2,864    (3K)
  MSCDEX         36,224   (35K)     0        (0K)       36,224   (35K)
  Free           618,528  (604K)    618,528  (604K)     0        (0K)

Memory Summary:
  Type of Memory         Total          Used           Free
  ---------------        ---------      ---------      ---------
  Conventional           655,360        36,832         618,528
  Upper                  158,336        158,336        0
  Reserved               393,216        393,216        0
  Extended (XMS)         32,347,520     116,096        32,231,424
  ---------------        ---------      ---------      ---------
  Total memory           33,554,432     704,480        32,849,952
  Total under 1 MB       813,696        195,168        618,528

  Largest executable program size       618,512   (604K)
  Largest free upper memory block       0         (0K)
  MS-DOS is resident in the high memory area.
```

Notice in Example 7.6 that the first information given is under the heading, "Modules using memory below 1 MB." These entries give the name of each program loaded in the 1-MB DOS addressing range, including programs that belong to the operating system and TSRs. The first column shows each program name. The second column lists the total size in the program in bytes. The third column lists the amount of conventional memory used by the program. The fourth column shows the amount of UMB area that each program uses.

**Problem 7.6:** In the computer illustrated in Example 7.6, what is the size of the MS-DOS shell COMMAND.COM and what type of memory is it loaded into?

**Solution:** According to Example 7.6, COMMAND.COM uses 7,664 bytes of conventional memory.

**Problem 7.7:** In the computer illustrated in Example 7.6, what size is the TSR MSCDEX and what type of memory is it loaded into?

**Solution:** According to Example 7.6, MSCDEX requires 36,224 bytes of UMB.

Notice in Example 7.6 that along with other categories of memory, there is one called *reserved memory*. The discussion of this category of memory justifies a section of its own.

### 7.3.15 Reserved Memory

We already know that conventional memory occupies pages $0_{16}$ through $9_{16}$ and the UMA occupies pages $A_{16}$ through $F_{16}$ of the compatible computer's memory map. In decimal, conventional memory takes up the first 640 KB (655, $360_{10}$) memory address combinations and the UMA occupies the next 384 KB ($393,216_{10}$). When memory is added to today's compatible computers, the minimum unit of addition is in multiples of 256-KB groups. Note that this is the minimum unit of addition. The next higher unit of addition is in 1-MB units, then 4 MB, then 16 MB, and so on.

Using 256-KB units, to get the 640-KB required for conventional memory, at least three 256-KB units are needed. Three units of 256 KB will give a total of 768 KB of memory. This is the 640 KB used by conventional memory and an additional 128 KB more than is needed. What happens to the additional 128 KB of RAM? It cannot be split out and used somewhere else because it is part of the minimum addition unit. It cannot be used as conventional memory because its addresses go into pages $A_{16}$ and $B_{16}$, which are already being used by video RAM. Because it cannot be split out and it cannot be used as conventional memory, the only thing left to do is to disable it (turn it off) and not use it at all.

To get the mass amounts of RAM required for today's compatible computers, memory addition units of 1 MB or greater are used. The problem with units greater than 1 MB is that the first memory addition unit, which gives us the addressing range for conventional memory, will try to spill over into the entire UMA address range. Because two things cannot be at the same address, the 384-KB RAM addresses that try to occupy the same address range as the UMA are disabled. The memory that is disabled in both of these situations is called **reserved memory.**

Notice in both Examples 7.5 and 7.6 that there is a listing for reserved memory. This is the memory that is lost because its addresses are the same as those that belong to the UMA. However, all of this memory is not lost, as we will find out in the next discussion.

### 7.3.16 Shadow RAM (ROM Shadowing)

Just like RAM, each address in ROM can hold only 8 bits of data. This is fine for the PC- and XT-compatible computer because both use the 8088 microprocessor. The early ATs use the 80286 microprocessor, which has a 16-bit data bus. Each ROM holds only 8 bits per storage location, so two BIOS ROMs are used in computers with the 286 microprocessor. One is usually called the *LO ROM* because it supplies data bits D0 through D7. The other is usually called the *HI ROM* because it supplies data bits D8 through D15.

Computers that use an 80838DX or 80486SX/SX2/DX/DX2/DX4 microprocessor have 32-bit data buses. It seems that four BIOS ROM ICs will be needed to supply the 32 bits of data needed by these microprocessors. What about the Pentium and Pentium Pro, Pentium II, and Pentium III microprocessors that have 64-bit data buses? It seems that eight BIOS ROM ICs will be needed to supply the 64 bits of data needed by these microprocessors. However, if you look at any of the compatible computer's motherboards that run any of these microprocessors, you will find only one BIOS ROM. You are probably wondering, "How do they do that?"

Remember the reserved memory that is lost because it occupies the same addresses as the UMA. This RAM can be accessed by the size of the microprocessor's data bus, and it is going unused. This is where shadow RAM comes into play.

As soon as power is applied to the computer, the contents of the single BIOS ROM are copied by the chipset into the reserved memory that occupies the same address range. After all of the data are copied, the BIOS ROM is disabled and the RAM that contains a copy of the ROM's data is write-protected by the chipset, so it acts like ROM. Once this feat is accomplished, the BIOS is actually executed out of reserved RAM, and not ROM. This technique of copying the contents of ROM into RAM and running the code out of RAM is called *shadow RAM,* or ***ROM shadowing.***

It should be obvious that this technique saves a lot of money on production costs because only a single BIOS ROM is required and the data path to the ROM is only 8 bits wide. Also, a typical ROM requires between 150–400 ns to provide 8 bits of data to the microprocessor. This is relatively slow when compared to the 32 bits at 20–80 ns that are available from the computer's RAM. It also makes use of

> **Procedure 7.1: Displaying a computer's memory map using Windows 95 or 98**
>
> 1. Click on the Start button on the Task Bar.
> 2. Move up to Settings and then over to Control Panel.
> 3. Click on Control Panel.
> 4. After the Control Panel window displays, locate and then double-click on the System icon.
> 5. Click on the Device Manager tab.
> 6. Double-click on the Computer icon.
> 7. Click on the white dot beside Memory.
> 8. To exit, click on the X in the upper right-hand corner of each open window or click on the Cancel button.
>
> **Displaying a computer's memory map using Windows 98 and 2000**
>
> 1. Click on the Start button located on the Task Bar.
> 2. Move up to Programs, over to Accessories, then to System Tool, and then click on System Information.
> 3. Click on the plus (+) sign beside Hardware Resources.
> 4. Click on the Memory options.
> 5. To exit, click on the X in the upper right-hand corner of each open window or click on the Cancel button.

some of the reserved memory that would normally be lost. In addition, shadowing does not have to be limited to the system BIOS.

Remember, from the computer's memory map, ROMs may also be installed in UMA pages $C_{16}$, $D_{16}$, and $E_{16}$. The lower half of page $C_{16}$ will usually be the location of the video adapter ROMs. On most of today's motherboards, ROMs in these pages can also be shadowed. ROMs may or may not be installed in these pages, so the user has to inform the computer whether or not to shadow these pages. Shadows are a very important feature that can reduce the cost of the motherboard and adapter by allowing the use of only one 8-bit ROM. For more information on the ROM shadowing configuration, refer to chapter 8.

### 7.3.17 Viewing a Computer's Memory Map

Windows 95/98/2000 allows the user to view the actual memory map of a computer on which any one of these operating systems is installed. For the average user, this may be useful information to know. But from a technical point of view, the location of this information is very important. For this reason, the steps to follow to view the computer's memory map are listed in Procedure 7.1.

## 7.4 TYPES OF RAM

Two categories of RAM are used most often in microcomputers: *dynamic random-access read/write memory (DRAM)* or *static random-access read/write memory (SRAM)*. DRAM is slow and cheap while SRAM is fast and expensive. At the time of this writing, the cost of 1 MB of DRAM is about the same as that for 32 KB of SRAM. For this reason alone, DRAM is used almost exclusively in the compatible computers as the computer's main memory. Because SRAMs are so fast, they are used in the L1 and L2 cache memory.

Currently the DRAMs that are most often used in the compatible PCs are available in four categories: *fast page mode* (FPM), *extended data output* (EDO), *burst extended data output* (BEDO), and *synchronous dynamic random-access memory* (SDRAM). Two new categories called *double data rate SDRAM* (DDR SDRAM) and *rambus DRAM* are on the horizon and will probably be installed in some of the high-end network servers in the near feature. As always, before we discuss the different types of RAM, we need to cover some basics about DRAM.

### 7.4.1 DRAM Basics

DRAM is a very complex memory, so I am going to try to explain it without getting too technical. Each DRAM storage cell basically consists of one transistor and an electronic device called a *capacitor*. The capacitors used in DRAM are similar to a rechargeable battery that goes dead really fast (4 milliseconds or less). Just like rechargeable batteries, DRAMs have to be recharged constantly or their contents will be lost. The act of recharging the DRAM is called *refreshing*. This means DRAMs have to be refreshed constantly or they will lose their contents.

The memory cells in DRAM are set up in a matrix. Basically, a matrix contains cells that are aligned in rows and columns. To select a cell for data transfer, a cell's row number is selected, and then the cell's column number is selected. Circuits on the motherboard must split each single memory address that the microprocessor places on its address bus into separate row and column addresses for the DRAMs. Pins on the DRAM called *row address strobes (RASs)* and *column address strobes* (CASs) time the selection process.

After each cell selection using the RAS and CAS signals, there has to be a delay between read and write cycles to allow the circuitry in the DRAM to do what is called a *precharge*. Along with precharge, there are other DRAM specifications such as RAS to CAS delays, lead-off times, and burst read cycles, just to name a few.

Now, to the average person, all of this seems interesting, but who cares, right? You are probably saying about now, "There's circuitry on the motherboard that takes care of all of this stuff, isn't there?" Of course the answer to this question is, "Yes there is." However, there are some motherboards that allow the user to set the parameters of the computer's DRAM to optimize the computer's performance. Shaving one clock cycle off the RAS, CAS, or precharge times will greatly increase the overall performance of the computer. Of course, setting the memory parameters faster than the DRAM can handle will cause the computer to lock up.

You might wonder why DRAM is used if it is so much trouble. DRAM has been around for quite awhile, so there are all types of integrated circuits that take care of the refresh, RAS, CAS, and precharge. But this is not the reason why DRAM is used for the computer's main memory. The main reason is that DRAMs cost so much less than SRAMs, so the extra trouble is well worth the effort.

## 7.4.2 DRAM Read Timing

Not so long ago, information in computers was read from memory a single word at a time. This is similar to opening a book to a page, reading a word, and then closing the book. On the next read cycle, the book is opened to a page, a word is read, and then the book is closed again. This process is repeated for each read cycle. What you have to understand is that programs and data are usually stored in consecutive memory addresses. This means that the memory page will probably stay the same over thousands of read cycles.

Most of the time, the book in our example is turned to the exact same page that was used during the last read cycle. It just makes sense that if the book is already open to a page, why not go ahead and read several words before closing the book.

The main memory used in current compatible computers is accessed in a burst. The memory page has to be selected during the first read cycle of the burst, so this read cycle takes the longest. After the page is selected, the next read cycles in the burst do not have to select a page, so they will occur quite a bit faster. Over a four-word burst read, the time required to access the stored data is represented in the form *y-x-x-x*. The *y* represents the number of timing cycles for the first read, which is the longest because the memory page is selected. The *x*'s represent the number of timing cycles for the other consecutive reads, which is faster because they do not have to select a page.

This may be a little confusing, but one thing that should be obvious: The smaller these numbers are, the faster the microprocessor is getting information from memory. The faster the memory can present data to the microprocessor, the faster the computer will run.

Some computers give the user the ability to set the DRAM timing as part of the CMOS setup. This will be covered later in chapter 8.

## 7.4.3 Access Time

Computers operate at a certain speed. This speed is usually given in megahertz. One hertz is one cycle of a clock every second. One megahertz is 1,000,000 cycles of the computer's clock every second. If your computer is operating at 10 MHz, there are 10,000,000 clock cycles occurring every second. To find the actual time that each cycle of the clock takes, divide 1 by the speed of the clock. If the computer is operating at 10 MHz, one cycle of the clock will take 0.0000001 seconds, or 100 nanoseconds (1/10000000). The smaller the number of nanoseconds, the faster the clock speeds.

A **microprocessor's bus cycle** is the time it takes to transfer data over the computer's data bus. Sometimes a microprocessor's bus cycle consists of more than one clock cycle. For example, the 286 usually takes three clock cycles per bus cycle. In an 8-MHz 286-microcomputer system, each system clock cycle has a duration of 125 nanoseconds (1/8000000). Therefore, a bus cycle is 375 nanoseconds long (3 × 125).

Remember from chapter 2 that microprocessors above the 486 usually run faster than the computer's local bus. An example is the 486DX4-120. This microprocessor is running at 120 MHz internally, but the computer's local bus is running at 40 MHz. The bus cycle speed is set by the computer's local bus, not by the internal speed of the microprocessor. Because of this, the time for one bus cycle for the 486DX4-120 is 25 nanoseconds (1/40000000). The 486DX4 requires one clock cycle per bus cycle, so a bus cycle equals 25 nanoseconds.

The memory of the computer has to keep up with the speed of the microprocessor's bus cycles. To give some indication of the speed of the memory, **access time** is used. A memory's access time is the maximum time the memory should take to provide information from the selected location during a read cycle. In our 286 example, RAM with an access time of 375 nanoseconds or less can be used with no problem. If the RAM has an access time greater than 375 nanoseconds, it might not get the data to the microprocessor in time. This would be fatal for the microprocessor (causing the computer to lock up).

RAM for the 486DX4-120 example seems to need memory with an access time equal to 25 nanoseconds or less. The problem is that when this microprocessor came out, the fast DRAMs that were available had an access time of 70 ns. Of course there is the SRAM alternative, but using SRAM would have added thousands of dollars to the cost of the computer. As you have probably guessed, there had to be a better, less expensive way.

The faster the speed of the RAM, the more it costs. To allow the use of RAM with slower access times on the computer, wait states are sometime deliberately inserted into the microprocessor's bus cycle during RAM access. This may seem great and wonderful, but the more wait states, the longer it takes to transfer data with the computer's memory. Why buy a 486DX4-120 if you have to insert three wait states in the microprocessor's bus cycle to use 70-ns RAMs? RAM with an access time of 25 ns or less would have to be used to avoid wait states in a computer operating at this speed.

If 25-ns memories are required and they are not available in DRAM, it would seem like wait states is the only alternative. However, this is not the case, as we will find out when we discuss a type of memory called a *cache,* which is discussed in later sections. If your computer uses wait states, the number of wait states is selected either with jumpers on the motherboard or with the CMOS setup (see CMOS later in the unit). Consult the documentation that is provided with your computer to determine the correct settings.

It is time to put memory speed on hold while we discuss some more very important memory terms and technology. Computer technicians must understand the techniques used to detect and correct memory errors. Even though it seems logical that these techniques should be available on all PCs, this is not the case.

### 7.4.4 Parity and Error Correcting Code (ECC)

Due to the unreliability of DRAM at the time IBM designed the PC, IBM decided to include an additional parity bit to each storage address. This parity bit was standard equipment and was part of the ICs soldered onto the motherboard. On today's PCs, the use of parity is optional. If used, the parity bit used by the RAM functions is exactly the same as the parity bit discussed in chapter 1.

To provide support for parity, the motherboard contains special circuitry that handles storing the parity bit during a memory write operation and checking the parity bit during a memory read operation. If a parity error is detected during a memory read cycle, a nonmaskable interrupt (NMI) is generated to the microprocessor. NMI is discussed in greater detail in chapter 16. When the NMI occurs due to a RAM error, the OS will usually display an error message informing the user that a parity error has been detected.

When parity is used, each 8-bit storage location actually stores 9 bits of information, including the parity bit. Thus, if a computer contains 16 MB of memory (16,777,216 bytes), there will be an additional 16,777,216 bits of memory just to handle the parity bit. As you can guess, this adds quite a bit to the overall cost of the computer's memory.

Along with the additional cost, parity has a couple of minor problems. First, parity can only detect an error if an odd number of bits change. This means if 2, 4, or 8 bits change in the storage address, parity cannot detect the error. This is a small problem because multi-bit changes are practically nonexistent. The second problem with parity is that it cannot correct an error once it has been detected. This last problem with parity brings about a newer type of memory error detection that is currently available—*error correcting code (ECC)*.

The major difference between parity and ECC is that memory and motherboards that support ECC have the ability to not only detect, but also to correct single-bit errors. This feature occurs "on the fly" without any indication that an error occurred and with no disruptions in the processing of information.

For ECC to work, the computer has to access memory more than one storage location at a time. This is not a problem because current microprocessors rip through memory either 4 addresses (32 bits) or 8 addresses (64 bits) at a time. However, implementing ECC on a 32-bit system (4 addresses) requires 7 bits. Memory modules with parity only have four additional bits, so special ECC memory modules have to be purchased to support ECC on 32-bit systems. When information is accessed in 64-bit groups (8 addresses), 8 bits are required for ECC, which is exactly the same number of bits required for parity. This means standard memory modules that store parity can be used to store the ECC information. For this reason, ECC is more prevalent in 64-bit computers such as those using microprocessors above a 486.

ECC works on the principle of storing a code in the extra bits that were previously used by parity. The ECC is a mathematical representation of the information stored in the eight consecutive addresses. When the microprocessor writes to these eight addresses, special circuits on the motherboard or on the memory module itself store a code in the eight bits that are used for error detection.

When the information is read from memory, the error detection circuit recalculates the error code and compares it with the code stored in memory. If the two codes match, no error exists. If the codes do not match, a calculation is performed with the two error codes to determine which bit changed. Then the error is corrected. ECC can detect multiple-bit changes but cannot correct them. When a multiple-bit change occurs, an NMI is generated by the ECC circuitry, similar to parity.

Because ECC has the ability to detect and correct errors "on the fly," you may wonder why all of the memory used on the compatible PC is not ECC. The main reason is that memory that provides the extra bits to store ECC costs considerably more than memory that does not provide error detection. This extra cost is in addition to the cost of the ECC circuitry also required on the motherboard. As an example, in a computer I just put together installing 512 MB of RAM added $400 to its cost.

Most compatible computers do not use any form of memory error detection because of the added cost. This is usually not a problem because the DRAMs available today are very reliable and in most applications, error detection is not necessary. Memories that do not provide error detection are referred to as *nonparity*. However, computers running in critical applications usually need some way to detect memory failures while the computer is operating.

Because ECC is relatively new, there are thousands of motherboards that support parity but do not support ECC. For this reason, there are memory modules that have all of the ECC circuitry mounted on the memory module itself. The ECC is totally transparent to the motherboard. If an uncorrectable memory error occurs, it simply reports a parity error to the motherboard. Currently one of the most popular memory modules is called the *SIMM,* which will be discussed later. For this reason, memory modules that provide ECC are called *Error correction on SIMM (EOS)*. Due to the additional circuitry, the EOS memory modules cost considerably more than parity and nonparity modules.

Before ECC or parity can be used, the motherboard chipset must also support it. Also, all of the memory modules in the computer must support parity or ECC. Do not buy parity or ECC memory modules if any of the memory modules currently installed in your computer are nonparity. On the other hand, if parity/ECC memory is installed in your computer, all of the memory you buy for upgrades must provide parity/ECC, or error detection will have to be disabled for all the memory in the computer.

Now that we know about memory error and detection, let us start our discussion on the actual types of memory used in the compatible computers. The memory types that will be discussed are FPM, EDO, BEDO, SDRAM, RDRAM, DDR-SDRAM, and last but not least, SRAM.

### 7.4.5 Fast Page Mode (FPM)

Fast page mode has been around for a few years now. It is the type of DRAM memory used in almost all 386 computers, and most 486 and older P5 Pentiums. It is also the type of memory used on a lot of video adapter cards. FPM is currently available in two speeds: 70 ns and 60 ns.

The fastest access speed for FPM DRAM is 5-3-3-3 CPU cycles. This sequence of numbers indicates that the first read cycle will take five clock cycles and each additional read of the burst will take three clock cycles. These numbers hold true whether the microprocessor is transferring a byte (8 bits), a word (16 bits), or a double word (32 bits).

### 7.4.6 Extended Data Output (EDO)

EDO DRAM holds data on the data bus from one read cycle while the next memory address is being selected by the next read cycle. This, in effect, reduces the number of total clock cycles required to retrieve information. EDO is accessed differently than FPM, so the motherboard has to support EDO memory before its advantages can be realized.

EDO is currently available in three speeds: 70 ns, 60 ns, and 50 ns. If your computer is running at a local bus speed of 66 MHz, the 60, ns DRAMs are the slowest you should use. Motherboards that use Intel's Triton HX chipsets will support the 50-ns EDO memory. If your motherboard supports 50-ns EDOs, you should use these. The fastest access speed for EDO DRAM in CPU cycles is 5-2-2-2 for byte, word, or double-word (dword) burst reads.

The only problem with EDO is that it currently can only operate at a local bus speed up to 66 MHz. Even though this is one of the standard motherboard speeds still supported by Intel's processors, Intel, AMD, and VIA all currently have microprocessors that will operate at local bus speeds of 100 MHz and faster. Using EDOs on computers running this fast will slow them down considerably.

### 7.4.7 Burst Extended Data Output (BEDO) RAM

BEDO RAM is designed around the burst reads performed by most microprocessor **cache controllers.** This enables BEDO RAM to be accessed in a burst of a 5-1-1-1 timing pattern. Even with this speed, BEDO was short lived due to the introduction of the SDRAM, which is discussed in the next section. In fact, Intel supported BEDO only with the 440 FX chipset.

### 7.4.8 Synchronous Dynamic RAM (SDRAM)

SDRAM is the predominant DRAM available at the time of this writing. This RAM is able to handle all data transfer signals synchronized with the system clock. This is a feat that used to be accomplished only by the SRAM used in the computer's cache memory.

Due to its synchronous nature, SDRAM is not rated in access time but in clock speed. The three present speeds available are 66 MHz, 100 MHz, and 133 MHz. You should choose the speed that operates at the local bus (FSB) speed of your computer. The fastest access speeds of SDRAM are 5-1-1-1 CPU cycles for a byte, word, or dword burst read. Another note: SDRAM is available only in 168-pin DIMMs, which are discussed later.

### 7.4.9 Direct Rambus DRAM (DRDRAM)

Direct Rambus DRAM® (DRDRAM) is a technology developed by Rambus, Inc. and is currently sanctioned by Intel and AMD. In fact, it appears that DRDRAM will be incorporated into computers using AMD's Athalon® microprocessor. It will also be the RAM used in computers running Intel's future-generation 64-bit microprocessor, the Itanium, while it is in development.

DRDRAM is a completely new concept and design shift from the technique used by current PCs. It requires additional controllers and dedicated buses used strictly by the computer's RAM. The use of dedicated buses means that memory buses will not have to be shared with all of the other devices that need to transfer data. Unfortunately, this feature also adds considerably to the overall cost of the computer. Due to the high cost of a computer using DRDRAM chipset, motherboard, and compatible MPU, manufacturers are also pursuing a cheaper standard that will probably be adopted on most low- to medium-priced compatible PCs that will be discussed in the next section.

DRDRAM transfers information over multiple 16-bit channels. The smaller channel size and dedicated buses give DRDRAM the ability to transfer data at blazing speeds. The current clock rate for RDRAM is 800 MHz, which allows a transfer rate of 1.6 gigabytes per second. Faster speeds are currently being developed.

DRDRAM will use a new module called a *Rambus in-line memory module (RIMM)*. The RIMMs will have the same physical size as DIMMs but will be different electrically. They will not be interchangeable with the DIMMs because of the mechanical features of the RIMMs and RIMM sockets.

Even though DRDRAM allows data transfer at the amazing speeds, it does not mean it will be adopted by the entire PC community. It fact, due to the added cost, this technology will first be adopted in computers that function as network servers and high-end workstations. Of course, it will also be used in computers that are purchased by people who have to own the best and the fastest.

### 7.4.10   Double Data Rate SDRAM (DDR SDRAM)

DDR SDRAM is the next high-speed DRAM technology that will probably be adopted in the low- to medium-priced PC. In other words, this will more than likely be the memory technology that will be used by the average individual. Current SDRAM, although fast, only transfers data on the rising edge of each clock cycle. This means that a local bus running at 100 MHz can transfer data at less than 100 MB per second, not taking into account the time required to open the page.

DDR SDRAM transfers data on both the rising and falling edges of the clock. Theoretically, DDR SDRAM will be able to transfer data at a rate of 200 MB per second when timed with a 100-MHz clock. However, this does not take into consideration the number of clock cycles required to open the page during the start of a burst read.

The main advantage of DDR SDRAM is that it is an open standard that can be adapted readily to motherboard and chipset technology currently in use. These features make DDR SDRAM an attractive alternative to DRDRAM. At the time of this writing, Intel has not indicated if it will support DDR SDRAM with its future chipsets.

### 7.4.11   Static RAM (SRAM)

Currently, SRAM is the fastest memory available, with access times that are less then 2 ns. Access times at this level will support clock speeds of 500 MHz (1/2 ns). Static RAM is very expensive when compared to DRAM, so there is no way that all of the memory of the computer is going to be SRAM. SRAM is so much faster because it does not require refresh. Also, all of the memory cells are addressed directly, so no RAS or CAS is required. These features give the SRAMs the ability to run at blazing speeds. SRAM can be accessed as fast as 3-1-1-1 in a burst read sequence.

Because of its high cost per storage cell, it is highly unlikely that SRAM will ever be used for the main memory in the PC. However, there is a category of memory on the motherboard and microprocessor that can make use of all of the advantages offered by SRAM at a relatively low cost—the microprocessor and computer's L1 and L2 cache memories.

## 7.5   DIPS, SIMMS, AND DIMMS

Upgrading the memory on compatible computers can be intimidating and confusing. Installing additional memory incorrectly can damage the new memory or the computer itself. In the beginning, computer manufacturers intended to only have trained technicians upgrade the memory. As time passed, attempts have been made to make memory upgrade safe and easy for the user. There are three methods currently used to package the memory used by the compatible computers: DIPs, SIMMs, and DIMMs.

**Figure 7.5** Typical DIP

(a) Dual-in-line package (DIP)	(b) Small-outline IC (SOIC)

*Source:* Floyd, *Digital Fundamentals,* 7th ed. (Upper Saddle River, NJ: Prentice Hall, 2000) p. 19.

**Figure 7.6** 30-pin and 72-pin SIMM Comparison

## 7.5.1 The DIP

The RAM in PC-, XT-, and early AT-compatibles is upgraded by installing IC DIPs (dual-inline-packages). The DIP gets its name from the fact that it has two rows of pins that are parallel (inline) to each other. A DIP is shown in Figure 7.5. The RAM DIPs plug into IC sockets on the motherboard.

It is very easy to make mistakes when working with DIPs. They are easy to install backwards and the pins bend easily under the DIP or even break off when they are inserted into their sockets. Even though DIPs are not used as the computer's main memory in today's computers, they are still popular to use in a type of memory installed on video adapters.

## 7.5.2 The SIMM (30-pin and 72-pin)

**SIMM** is the acronym for **single-inline memory module.** SIMMs have preinstalled RAM ICs on a tiny printed circuit board (PCB). Instead of having pins that insert into sockets as DIPs do, the SIMMs have a male edge card connector on the bottom edge of the PCB. Even though pins (electrical connectors) are available on both sides of the SIMM's PCB, the pins on each side that are par-

**Figure 7.7** Motherboard with a 30-pin SIMM

allel to each other are connected together electrically so they act as a single pin. This is where the name *SIMM* comes from.

Because these devices are fairly easy to install without error, SIMMs are currently one of the most popular methods of memory upgrade for the compatible computers. SIMMs are currently available in two sizes: 30-pin and 72-pin. Figure 7.6 compares a 30-pin (top) SIMM with a 72-pin SIMM (bottom).

SIMMs install into special sockets on the motherboard called *SIMM sockets*. A motherboard with a 30-pin SIMM installed in the sockets is shown in Figure 7.7.

The data in a 30-pin SIMM are accessed in 8-bit groups. SIMMs can be purchased with or without parity. If parity is not used, the SIMM is called a "By eight ($\times$ 8)." For example, a 1-MB 30-pin SIMM without parity is called a "1 Meg by 8." If parity is used, it is called a "By nine ($\times$ 9)" because an extra bit is added to each 8-bit data group to store parity. The 30-pin SIMMs are available in three storage capacities: 256 KB, 1 MB, and 4 MB. Table 7.1 lists the pin-out for the standard 30-pin SIMM.

A 72-pin SIMM can access data in 8-, 16- or 32-bit groups. Similar to the 30-pin SIMMs, it can also be purchased with or without parity. If parity is not used, the 72-pin SIMM is called a "by 32" ($\times$ 32). If parity is used, it is called a "by 36" ($\times$ 36) because each fundamental storage unit has to have a parity bit. Remember that a fundamental storage unit is equal to 8 bits per memory address. A 32-bit data group contains four fundamental storage units, so four parity bits are needed. Currently, 72-pin SIMMs are available in four storage capacities: 4 MB, 8 MB, 16 MB, and 32 MB.

Like the 30-pin SIMMs, special sockets referred to as *72-pin SIMM sockets* are located on the motherboard. The number of sockets depends on the maximum amount of RAM the motherboard can hold. A motherboard with the SIMM sockets labeled is shown in Figure 7.8.

One noted feature of the 72-pin SIMM is its ability to indicate its storage capacity and access time by using *Presence Detected* (PD) pins. Table 7.2 lists the PD pins PD1 through 7 that are located at pins 67, 68, 69, 70, and 11, respectively. Having five PD pins give a total of 32 ($2^5$) size/speed combinations that can be represented. These combinations are shown at the bottom of Table 7.2. This feature allows auto configuration of the memory CMOS settings.

**Table 7.1** Pin-out of a Standard 30-pin SIMM

| Pin | Description | Pin | Description |
|---|---|---|---|
| 1 | + 5 VDC | 16 | Data Bit 4 |
| 2 | Column Address Strobe | 17 | Address Bit 8 |
| 3 | Data Bit 0 | 18 | Address Bit 9 |
| 4 | Address Bit 0 | 19 | Address Bit 10 |
| 5 | Address Bit 1 | 20 | Data Bit 5 |
| 6 | Data Bit 1 | 21 | Write Enable |
| 7 | Address Bit 2 | 22 | Ground |
| 8 | Address Bit 3 | 23 | Data Bit 6 |
| 9 | Ground | 24 | Not Connected |
| 10 | Data Bit 2 | 25 | Data Bit 7 |
| 11 | Address Bit 4 | 26 | Parity Bit Out (if used) |
| 12 | Address Bit 5 | 27 | Row Address Strobe |
| 13 | Data Bit 3 | 28 | Column Address Strobe Parity |
| 14 | Address Bit 6 | 29 | Parity In (if used) |
| 15 | Address Bit 7 | 30 | + 5 VDC |

**Figure 7.8** Motherboard with Both 72-pin SIMM and 168-pin DIMM Sockets

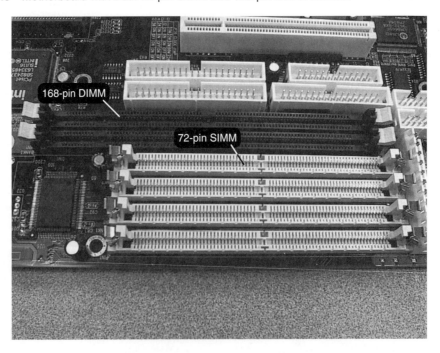

**Table 7.2** Pin-out of a Standard 72-pin SIMM

| | | | | | |
|---|---|---|---|---|---|
| 1 | Ground | 25 | Data Bit 22 | 49 | Data Bit 8 |
| 2 | Data Bit 0 | 26 | Data Bit 7 | 50 | Data Bit 24 |
| 3 | Data Bit 16 | 27 | Data Bit 23 | 51 | Data Bit 9 |
| 4 | Data Bit 1 | 28 | Address Bit 7 | 52 | Data Bit 25 |
| 5 | Data Bit 17 | 29 | Address Bit 11 | 53 | Data Bit 10 |
| 6 | Data Bit 2 | 30 | +5 VDC | 54 | Data Bit 26 |
| 7 | Data Bit 18 | 31 | Address Bit 8 | 55 | Data Bit 11 |
| 8 | Data Bit 3 | 32 | Address Bit 9 | 56 | Data Bit 27 |
| 9 | Data Bit 18 | 33 | Row Address Strobe 3 | 57 | Data Bit 12 |
| 10 | +5 VDC | 34 | Row Address Strobe 2 | 58 | Data Bit 28 |
| 11 | Presence Detect 5 | 35 | Parity Data Bit 2 | 59 | +5 VDC |
| 12 | Address Bit 0 | 36 | Parity Data Bit 0 | 60 | Data Bit 29 |
| 13 | Address Bit 1 | 37 | Parity Bit 1 | 61 | Data Bit 13 |
| 14 | Address Bit 2 | 38 | Parity Bit 3 | 62 | Data Bit 30 |
| 15 | Address Bit 3 | 39 | Ground | 63 | Data Bit 14 |
| 16 | Address Bit 4 | 40 | Column Address Strobe 0 | 64 | Data Bit 31 |
| 17 | Address Bit 5 | 41 | Column Address Strobe 2 | 65 | Data Bit 15 |
| 18 | Address Bit 6 | 42 | Column Address Strobe 3 | 66 | Not Connected |
| 19 | Address Bit 10 | 43 | Column Address Strobe 1 | 67 | Presence Detect 1 |
| 20 | Data Bit 4 | 44 | Row Address Strobe 0 | 68 | Presence Detect 2 |
| 21 | Data Bit 20 | 45 | Row Address Strobe 1 | 69 | Presence Detect 3 |
| 22 | Data Bit 5 | 46 | Reserved | 70 | Presence Detect 4 |
| 23 | Data Bit 21 | 47 | Write Enable | 71 | Not Connected |
| 24 | Data Bit 6 | 48 | Not Connected | 72 | Ground |

| Presence | Detect | Size | Decoding | Presence | Detect | Speed | Decoding |
|---|---|---|---|---|---|---|---|
| Gnd = Connected to Ground; — = Left Open | | | | PD3 | PD4 | PD5 | Speed |
| PD1 | PD2 | PD5 | Size | — | — | — | 60 ns |
| — | — | — | 8M | Gnd | — | — | 70 ns |
| Gnd | — | — | 1M | — | Gnd | — | 80 ns |
| — | Gnd | — | 2M | Gnd | Gnd | — | 100 ns |
| Gnd | Gnd | — | 4M | — | — | Gnd | 60 ns |
| — | — | Gnd | | Gnd | — | Gnd | 70 ns |
| Gnd | — | Gnd | 16M | Gnd | Gnd | Gnd | 50 ns |
| — | Gnd | Gnd | 32M | | | | |

**Note:** IBM 72-pin SIMMs use a different decoding scheme for the Presence Detect pins.

**Figure 7.9** Comparison of a Typical 168-pin DIMM and a 72-pin SIMM

### 7.5.3 DIMM (168-pin)

DIMM is the acronym for *dual-inline memory module*. The difference between DIMMs and SIMMs is that the DIMM pins on both sides of the memory module's edge card connector are used; hence, the name. A comparison of a typical 168-pin DIMM (top) to a 72-pin SIMM (bottom) is shown in Figure 7.9. 168-pin DIMM sockets are shown in Figure 7.8.

As seen in Figure 7.9, DIMMs are physically larger than the 72-pin SIMMs and have 168 pins. Because a DIMM uses the conductors on both sides of the circuit card, it gives it the ability to access data in 8-, 16-, 32-, and 64-bit groups (72 bits with ECC/parity). For this reason, the DIMM is the current memory technology used in Socket 7, Slot 1, Slot 2, Socket 370, and Slot A motherboards.

DIMMs are currently purchased as either EDO or SDRAM. Both EDO and SDRAM are available as nonparity, parity, or ECC. If used, the parity or ECC bits are available on the Check Bits pins listed in Table 7.3. The DIMM's Presence Detect is handled through serial data stored in circuits on the DIMMs. The PD information is accessed through the serial PD and serial clock pins listed in Table 7.3.

## 7.6 MEMORY BANKS

A *bank of memory* can be defined as the number of DIPs, SIMMs, or DIMMs needed to store the microprocessor's data bus. This does not include error detection. For example, a compatible computer that uses an 80286 will have banks capable of storing 16-bit groups of data. Memory banks on a 80286 motherboard use 30-pin SIMMs. A 30-pin SIMM is capable of holding only 8-bit groups, so two SIMM sockets are required for each bank.

Computers that use an 80386DX or 80486SX/SX2/DX/DX2/DX4 will have 32-bit data memory banks if parity is not used. Depending on the time of purchase, computers that use one of these microprocessors can have either 30-pin SIMM sockets, 72-pin SIMM sockets, or a combination of both. If 30-pin SIMMs are used, four SIMMs will be required for each bank of memory ($8 \times 4 = 32$). However, if the banks use 72-pin SIMMs, only one SIMM will be required per bank.

The P5, P54C, P55C, P6, Pentium II, and Pentium III computers have banks capable of storing 64 bits of data each, not including error detection. If the motherboard uses 30-pin SIMM sockets, eight

**Table 7.3** Pin-out of Standard 168-pin DIMM

| Pin | Signal Name | Pin | Signal Name | Pin | Signal Name |
| --- | --- | --- | --- | --- | --- |
| 1 | Ground | 29 | Byte Mask 1 | 57 | Data Bit 18 |
| 2 | Data Bit 0 | 30 | S0 | 58 | Data Bit 19 |
| 3 | Data Bit 1 | 31 | NC | 59 | + 3.3 or + 5 VDC |
| 4 | Data Bit 2 | 32 | Ground | 60 | Data Bit 20 |
| 5 | Data Bit 3 | 33 | Address Bit 0 | 61 | NC |
| 6 | + 3.3. or + 5 VDC | 34 | Address Bit 2 | 62 | Voltage Reference |
| 7 | Data Bit 4 | 35 | Address Bit 4 | 63 | Clock Enable 1 |
| 8 | Data Bit 5 | 36 | Address Bit 6 | 64 | Ground |
| 9 | Data Bit 6 | 37 | Address Bit 8 | 65 | Data Bit 21 |
| 10 | Data Bit 7 | 38 | Address Bit 10 | 66 | Data Bit 22 |
| 11 | Data Bit 8 | 39 | Bank Address 1 | 67 | Data Bit 23 |
| 12 | Ground | 40 | + 3.3 or + 5 VDC | 68 | Ground |
| 13 | Data Bit 9 | 41 | + 3.3 or + 5 VDC | 69 | Data Bit 24 |
| 14 | Data Bit 10 | 42 | Block 0 | 70 | Data Bit 25 |
| 15 | Data Bit 11 | 43 | Ground | 71 | Data Bit 26 |
| 16 | Data Bit 12 | 44 | NC | 72 | Data Bit 27 |
| 17 | Data Bit 13 | 45 | S2 | 73 | + 3.3 or + 5 VDC |
| 18 | + 3.3 or + 5 VDC | 46 | Byte Mask 2 | 74 | Data Bit 28 |
| 19 | Data Bit 14 | 47 | Byte Mask 3 | 75 | Data Bit 29 |
| 20 | Data Bit 15 | 48 | NC | 76 | Data Bit 30 |
| 21 | Check Bit 0 (if used) | 49 | + 3.3 or + 5 VDC | 77 | Data Bit 31 |
| 22 | Check Bit 1 (if used) | 50 | NC | 78 | Ground |
| 23 | Ground | 51 | NC | 79 | Clock 2 |
| 24 | NC | 52 | Check Bit 2 (if used) | 80 | NC |
| 25 | NC | 53 | Check Bit 3 (if used) | 81 | NC |
| 26 | + 3.3 or + 5 VCD | 54 | Ground | 82 | Serial Data I/O |
| 27 | Write Enable | 55 | Data Bit 16 | 83 | Serial Clock Input |
| 28 | Byte Mask 0 | 56 | Data Bit 17 | 84 | + 3.3 or + 5 VDC |

*(Continued)*

**Table 7.3** (Continued)

| Pin | Signal Name | Pin | Signal Name | Pin | Signal Name |
|---|---|---|---|---|---|
| 85 | Ground | 113 | Byte Mask 5 | 141 | Data Bit 50 |
| 86 | Data Bit 32 | 114 | S1 | 142 | Data Bit 51 |
| 87 | Data Bit 33 | 115 | NC | 143 | + 3.3 or + 5 VDC |
| 88 | Data Bit 34 | 116 | Ground | 144 | Data Bit 52 |
| 89 | Data Bit 35 | 117 | Address Bit 1 | 145 | NC |
| 90 | + 3.3 or + 5 VDC | 118 | Address Bit 3 | 146 | Voltage Reference |
| 91 | Data Bit 36 | 119 | Address Bit 5 | 147 | NC |
| 92 | Data Bit 37 | 120 | Address Bit 7 | 148 | Ground |
| 93 | Data Bit 38 | 121 | Address Bit 9 | 149 | Data Bit 53 |
| 94 | Data Bit 39 | 122 | Bank Address 0 | 150 | Data Bit 54 |
| 95 | Data Bit 40 | 123 | Address Bit 11 | 151 | Data Bit 55 |
| 96 | Ground | 124 | + 3.3 or + 5 VDC | 152 | Ground |
| 97 | Data Bit 41 | 125 | Clock 1 | 153 | Data Bit 56 |
| 98 | Data Bit 42 | 126 | Address Bit 12 | 154 | Data Bit 57 |
| 99 | Data Bit 43 | 127 | Ground | 155 | Data Bit 58 |
| 100 | Data Bit 44 | 128 | Clock Enable 0 | 156 | Data Bit 59 |
| 101 | Data Bit 45 | 129 | S3 | 157 | + 3.3 or +5 VDC |
| 102 | + 3.3 or + 5 VDC | 130 | Byte Mask 6 | 158 | Data Bit 60 |
| 103 | Data Bit 46 | 131 | Byte Mask 7 | 159 | Data Bit 61 |
| 104 | Data Bit 47 | 132 | Address Bit 13 | 160 | Data Bit 62 |
| 105 | Check Bit 4 (if used) | 133 | + 3.3 or + 5 VDC | 161 | Data Bit 63 |
| 106 | Check Bit 5 (if used) | 134 | NC | 162 | Ground |
| 107 | Ground | 135 | NC | 163 | Clock 3 |
| 108 | NC | 136 | Check Bit 6 (if used) | 164 | NC |
| 109 | NC | 137 | Check Bit 7 (if used) | 165 | Serial PD Address 0 |
| 110 | + 3.3 or + 5 VCD | 138 | Ground | 166 | Serial PD Address 1 |
| 111 | Column Address Strobe | 139 | Data Bit 48 | 167 | Serial PD Address 2 |
| 112 | Byte Mask 4 | 140 | Data Bit 49 | 168 | + 3.3 or + 5 VDC |

PC Random-Access Memory 257

SIMMs are required per bank. If the motherboard has 72-pin SIMM sockets, two SIMMs are required per bank. Currently, most of these microprocessors have 168-bit DIMM sockets. When this is the case, only one DIMM is required per bank.

Despite the number of memory modules required per bank, all of the memory modules in each bank must be the same storage capacity. Most motherboards and chipsets allow different storage capacity memory modules to be used in different banks. For example, one bank may contain 16-MB 72-pin SIMMs and another bank may contain 32-MB 72-pin SIMMs.

All memory modules in all of the banks must be the same type. In other words, all of the memory has to be FPM, EDO, or SDRAM. Different types of memory cannot be installed in the same computer. The motherboard must also support the type of memory modules installed. For example, if a motherboard supports only FPM and EDO memory modules, SDRAM modules will not operate.

If error detection or correction is to be used, all of the memory modules in all of the banks must support it. If any memory in the computer does not support error correction or detection, this feature must be turned off in the CMOS. On the other hand, if all of the memory modules support error detection or correction, this feature must be turned on in the CMOS or it will not be used.

Last, but not least, before you add memory to your computer, consult the documents that are supplied with the computer or motherboard to determine the correct size and style of the memory modules.

## 7.7 MEMORY INSTALLATION

Probably one of the cheapest and easiest upgrades that can be performed to increase the overall performance of the compatible PC is to increase the amount of RAM in the machine. Installing memory in today's computers is a fairly simple task, even for people with little technical knowledge.

The most complicated aspect of memory installation is making sure that the correct memory module size, storage capacity, and error detection are purchased. The information needed to upgrade the memory in a PC is usually contained in the documentation that is supplied with the computer. If you are purchasing a preassembled computer, make sure this information is provided.

### 7.7.1 Choosing the Right Memory Modules

Before any memory is purchased for the computer, there are a few very important preparation steps that should be followed. As stated previously, purchase memory modules that are the correct size, storage capacity, and error detection/correction needed by the computer in which the memory is to be installed.

The physical size can be 30-pin SIMMs, 72-pin SIMMs, or 168-pin DIMMs. All three of these sockets are shown in Figures 7.7 and 7.9. It is imperative that the right size memory modules be purchased for the computer in which the memory is to be installed. If this information is not available in the computer's documentation, Figures 7.7 and 7.9 can be used for reference.

Note that motherboards that are sold during the transition of memory module standards usually have memory sockets that support the older standard along with the newer. For example, the motherboard in one of my computers has four 72-pin SIMM sockets and two 168-pin DIMM sockets. This feature allows easy upgrade because the new motherboard can use memory modules from the previously installed motherboard. To help clarify this feature, a photo of a motherboard's memory sockets is shown in Figure 7.8.

Even though the motherboard shown in Figure 7.8 has six memory sockets, there are not six memory banks. Most Pentium motherboards have two banks of memory—Bank 0 and Bank 1. Two banks of memory require four 72-pin SIMM sockets, or two 168-pin DIMM sockets. The Socket 7 motherboard shown in Figure 7.8 has four 72-pin SIMM sockets and two 168-pin DIMM sockets. Two of the 72-pin SIMM sockets belong to Bank 0 and the other two belong to Bank 1. On the other hand, one 168-pin DIMM socket will belong to Bank 0 and the other DIMM socket will belong to Bank 1. It should be obvious that both SIMMs and DIMMs cannot occupy the addresses that belong to Bank 0 or 1 at the same time. This means if SIMMs are installed in their Bank 0 sockets, a DIMM cannot be installed in its Bank 0 socket. However, most motherboards that support both SIMMs and DIMMs will allow memory modules of different types to be installed in different banks. In other words, SIMMs can be installed in Bank 0 and DIMMs in Bank 1, or vice versa.

The second thing that needs to be considered before memory is purchased is the type of DRAM that the computer supports. The types currently available are FPM, EDO, and SDRAM. FPM and EDO are available in the 30-pin and 72-pin SIMMs, while the 168-pin DIMMs are available in EDO and SDRAM. Once again, consult the documents that are supplied with the computer to determine the type memory supported. If the documents are not available, the computer's cover must be removed to determine the sockets and the motherboard chipset. The memory socket determines the size and shape of the memory modules while the chipset determines the DRAM type: FPM EDO or SDRAM. For the procedure to determine a computer's chipset, along with the types of memory it supports, consult chapter 3.

The motherboard's chipset also determines the memory access times that are available. Installing memory modules with access times smaller than those supported by the chipset will not cause problems, but will be a waste of money. An example of this would be installing EDO memory modules with an access time of 50 ns when the fastest access time supported by the chipset is 60 ns. On the other hand, if memory modules are installed in a computer with an access time that is slower (higher numbers) than the highest access time supported by the chipset, the computer will not to operate at all. An example of this would be installing FPM memory modules with an access time of 80 ns when the slowest supported by the chipset is 70 ns.

The best practice is to consult the motherboard or computer documents to determine the correct speed and types of memory to purchase. If documents are not available when it is time to upgrade the computer's memory, buy memory modules that match the speed and type of the modules already installed. If memory is being purchased for a new motherboard, try to buy the fastest memory modules supported by the motherboard chipset. This will allow the computer to use minimum wait states, which translates to optimum speed.

The speeds for most DIMMs and 72-pin SIMMs are set automatically by the computer from information obtained from the modules themselves, and do not have to be set by the user. The access time of 30-pin SIMMs will have to be set by the user through the CMOS setup program, however. For more information on using the CMOS setup program, refer to chapter 8.

The maximum amount of memory that can be installed in a motherboard is another very important consideration. Once again, the motherboard's chipset determines these values. In addition to the chipset, the computer's operating system also determines the minimum amount of memory that must be installed in a computer. For example, Windows 98 requires a minimum of 8 MB of RAM. For information on the minimum requirements needed by an operating system, refer to the "Minimum Hardware Requirements" label on the box in which the OS was purchased. Also, technical information can be obtained from the operating system writer's Web site.

Besides the minimum and maximum storage capacity, there is yet another important specification of a chipset that determines the maximum amount of memory that should be installed in a computer, no matter how many banks the motherboard contains. This is a specification called the *maximum cacheable area*. This specification indicates the upper range of memory addresses whose data can be routed through the L2 cache. Surprisingly, the amount of memory the chipset allows to be installed is usually greater than the cacheable area. No matter how much memory can be installed in the motherboard, it is not a good idea to put more memory in the computer than can be routed through the L2 cache, especially when using operating systems such as Windows 95, 98, NT, or 2000. All of these operating systems load in the upper memory regions. If more memory is put in the computer than is cacheable, the operating system will load into the noncacheable RAM, which will cause the computer's performance to be degraded significantly. For the cacheable range specifications on the current popular chipsets, consult chapter 3.

Another important concept is that once all of the memory banks are filled, the only option to increase the amount of memory in the computer is to replace one or both banks with larger storage capacity memory modules. What do you do with the older memory modules that are being replaced? Some local vendor may allow you to trade in your older memory modules for the larger capacity modules. Vendors usually do not advertise this, so make it a point to ask. Also, ask around to see if anyone will purchase the older memory modules. I learned a long time ago that what you are getting rid of may be an upgrade to someone else.

We also need to discuss error detection and correction. Although this feature may not be an option on all of the compatible computers, I highly recommend some type of error detection bus on computers that are running in critical applications. Because of its ability to correct single bit errors, ECC is pre-

> **Procedure 7.2: Static safety precautions**
>
> 1. When working inside the chassis of the computer or handling PCBs, wear an anti-static wristband, if available. The wristband must be connected to an unpainted surface on the computer chassis. If a wristband is not available, always touch an unpainted surface on the computer's chassis before handling any of the computer's PCBs or components. Never wear an anti-static wristband when working with high voltages such as those in the main power supply and monitor.
> 2. If available, connect an anti-static mat to an unpainted surface on the computer chassis.
> 3. Place the IC or PCB to be installed in the computer on the anti-static mat before handling. If an anti-static mat is not available, touch an unpainted surface of the chassis with the anti-static bag in which the PCB or IC is packaged.
> 4. Any PCB or IC not installed in the computer should be stored in an anti-static media such as an anti-static bag or foam. Do not remove PCBs or ICs from their anti-static protective bag or tubes until you are ready to install them in the computer.
> 5. If possible, do not touch the pins of ICs or PCBs while they are being removed or installed in the computer.
> 6. If any cleaners are used, make sure they are anti-static.
> 7. Handle PCBs and ICs by the edges and avoid touching the pins.

ferred over parity, if it is available. Remember, to use ECC or parity, all of the memory modules installed in the computer must support it along with the motherboard's chipset. Also, the ECC parity option must be enabled in the computer's CMOS.

Finally, one important factor that is often overlooked is the metal used to plate the contacts on the end of the memory modules and the memory sockets located on the motherboard. Currently, the two metals most often used are tin and gold. Although both of these metals serve the function of preventing oxidation quite well, there is a problem with mixing the two, for example, using memory modules whose pins are plated with tin in sockets whose pins are plated with gold, or vice versa. Mixing dissimilar metals such as gold and tin creates a phenomenon known as electrolysis, which causes the metals to corrode. For this reason, I highly recommend that the plating used by the memory module and motherboard's memory module sockets match. If you start getting memory errors after the computer is in service for a few months, check the modules for corrosion caused by the mixing of the two metals. The module's contacts can be cleaned with oil-free electronics cleaning fluid and a simple pencil eraser.

In summation, to get the best performance from operating systems such as Windows 95, Windows 98, Windows NT, and Windows 2000, you should install as much memory as you can afford above the minimum required by the OS up to the maximum cacheable limits of the chipset. Also, install memory modules with the fastest access times or clock rate supported by the motherboard's chipset. And finally, install memory modules that match the plating used by the motherboard sockets.

### 7.7.2 Electrostatic Discharge

Most of the ICs in the computer, including those installed on memory modules, can be damaged if subjected to *electrostatic discharge (ESD)*. For this reason, it is imperative to follow a few static safety precautions. If you are planning on building, upgrading, or repairing personal computers, it is a good idea to invest in some reasonably priced anti-static safety devices. These include an anti-static wristband, anti-static mat, anti-static foam, and anti-static parts bags. Correct use of these devices will usually guarantee that the static-sensitive devices in the computer will not be damaged due to ESD. Procedure 7.2 list some basic anti-static steps to follow to prevent damaging devices due to ESD.

### 7.7.3 30-pin and 72-pin SIMM Installation

As discussed previously, SIMMs come in different storage capacities and access times. Before you install any new SIMMs in your computer, check the computer documents to determine what size, type, storage capacity, and error detection/correction the motherboard supports. If all of the banks in the

### Procedure 7.3: SIMM installation

Refer to Figure 7.10 when installing 30-pin SIMMs or Figure 7.11 when installing 72-pin SIMMs.

1. If upgrading the memory in the computer, back up important information contained on the computer's hard drive.
2. Follow the anti-static procedures outlined in Procedure 7.2.
3. Turn the computer off and unplug the ac power cord from the rear of the computer.
4. If not already done, remove the computer cover and store the screws in a safe location.
5. Locate the memory module sockets on the motherboard.
6. Pick up the SIMM modules by the edges. Avoid touching the components or edge card pin connectors.
7. Referring to the appropriate figure number, orient the key(s) on the SIMM with the key(s) on the socket.
8. Insert the SIMM's edge card connector into the channel that runs in the bottom of the socket at an angle away from the snaps (see appropriate figure). Do not force the SIMM into the socket because both the SIMM and socket can be damaged. If one side the SIMM will not go into the channel, it is probably being inserted backwards. Turn the SIMM around and try again.
9. Once the SIMM's edge connector is seated correctly in the channel, rotate the top of the SIMM toward the vertical position until it snaps behind the plastic or metal retainers.
10. Repeat the procedure for the remaining SIMMs.
11. Once all SIMMs have been installed, check all motherboard and internal peripheral data and power cables in case any were knocked loose.
12. Plug in the power cord and turn the computer power switch on.
13. Turn the computer off if it does not boot, if it displays a memory error message, or if it starts beeping continuously. Unplug the power cord and reseat all of the SIMM, taking electrostatic discharge precautions. Once all of the SIMMs have been reseated, repeat step 12.
14. Check the amount of memory displayed by the POST to make sure it matches the amount of memory installed. If not, return to step 13.
15. Allow the main operating system to load. Launch a few applications to make sure the computer is operating correctly.
16. If everything seems to be working correctly, turn the computer off and remove the ac power cord. Replace the computer's cover, plug in the ac power cord, and return the computer to service.

**Figure 7.10** 30-pin SIMM Installation

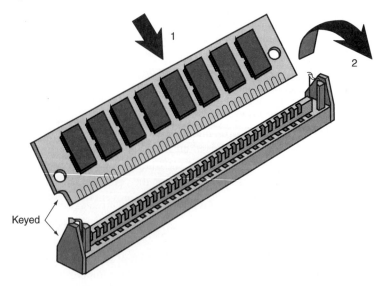

**Figure 7.11** 72-pin SIMM Installation

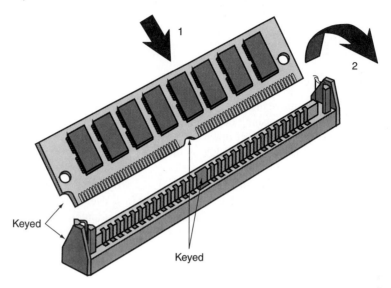

computer are full, currently installed memory modules will have to be removed and replaced with those of a higher storage capacity to increase the amount of RAM in the machine.

The installation procedures for 30-pin and 72-pin SIMMs are very similar. Both are covered in Procedure 7.3. When installing SIMMs, always handle them carefully. Do not remove the modules from their anti-static package until you are ready for installation. Do not flex or bend the SIMMs. Always use procedures to avoid static discharge and always hold the modules by the edges.

### 7.7.4 DIMM Installation

The installation procedure for DIMMs is slightly different from that for SIMMs. However, the handling precautions are the same. The installation steps for 168-pin DIMMs are listed in Procedure 7.4.

---

**Procedure 7.4: DIMM installation**

1. If upgrading the memory in the computer, back up the important information contained on the computer's hard drive.
2. Turn the computer off and disconnect the power cord from the rear of the computer.
3. Remove the computer's cover and store the screws in a safe location.
4. Use the anti-static measures outlined in Procedure 7.2.
5. Locate the memory expansion sockets on the motherboard. If all sockets are full, smaller capacity memory modules will have to be removed to make room for the newer, high-capacity modules.
6. As shown in Figure 7.12, push down both ejectors, which are located on each side of an empty DIMM socket. These tabs are also used to remove a previously installed DIMM by pressing down on both handles at the same time.
7. Insert the DIMM by aligning the keys (small notches) in the DIMM with the raised keys in the channel in the bottom of the DIMM socket (see Figure 7.12).
8. Firmly press down on the DIMM until it is seated in the socket. The ejector handles will rise and touch the side of the DIMM when it is seated. Some DIMM ejector tabs have tabs that will fit into the notch in each side of the DIMM.
9. Press on the sides of the ejector handle to force the tabs against the DIMM.
10. Repeat steps 4 through 9 for each DIMM to be installed.
11. Check all motherboard and peripheral data and power cables to see if any were accidentally

knocked loose while the DIMM(s) were being installed.
12. Plug in the ac power cord and apply power to the computer.
13. Turn the computer off if it does not boot, if it displays a memory failure error message, or if it starts beeping continuously. Remove the power cord and reseat all of the DIMMs by pushing down on the eject handles and then pressing firmly down on the DIMM. Plug in the ac power cord and once again apply power to the computer.
14. If all of the memory modules support ECC or parity, press the key sequence required to run the CMOS Setup program. Locate the Setup menu that allows the enabling of ECC or parity. Make sure this option is enabled. If any of the memory modules in the computer do not support ECC or parity, make sure the option in the Setup program is disabled.
15. Make sure the correct amount of memory is counted off during the POST. The POST displays the amount of memory in "K," so 64 MB will be displayed as 65,536 K. If the correct amount of memory is not counted during the POST, reseat the DIMMs using the procedure in step 13.
16. Allow the main operating system to load. Launch a few applications to make sure the computer is operating correctly.
17. If everything seems to working correctly, turn the computer off and remove the ac power cord. Replace the computer's cover, plug in the ac power cord, and return the computer to service.

**Figure 7.12**  168-pin DIMM Installation

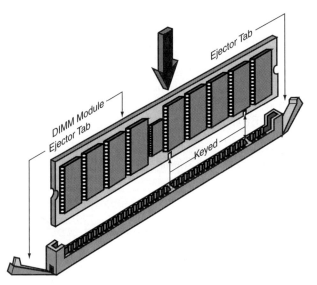

## 7.8  CACHE MEMORY

When a microprocessor runs a program, more often than not it is looping around in a small section of the program waiting for some form of input. If that section of the program in which the microprocessor is looping can be copied into high-speed SRAM, and the microprocessor switched into the SRAM, wait states can be eliminated or reduced. This is the method used by most compatibles running above 25 MHz. This high-speed SRAM is called a *cache*. A basic cache setup is shown in Figure 7.13. There are currently two types of caches used on today's compatible computers—*level one (L1)* and *level two (L2)*.

The L1 cache is located on the same silicon die as the microprocessor. This means the L1 is tightly coupled to the MPU and operates at the speed of the microprocessor's core. This is very important because current microprocessors operate at a speed faster than that of the motherboard (local bus). The L1 cache is usually small, somewhere between 8 K and 512 K, because there is not enough room on the die for a microprocessor and a bunch of SRAM.

The L2 cache is not part of the microprocessor die. On computers using Socket4, 5, or 7, the L2 cache is located outside the microprocessor on the local bus. An L2 cache that is connected to the local bus is

**Figure 7.13** Cache

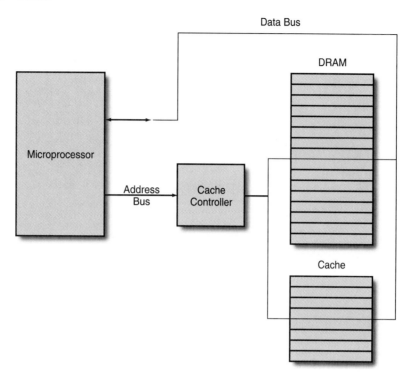

not operating at the speed of microprocessor, but at the speed of the local bus. Some of the newer microprocessor such as the Pentium® Pro, Pentium® II, 300A Celerons™ (and above), Pentium® II, Xeon, Pentium® III, AMDK6-3™, and AMD Athlon™ have the L2 cache mounted in the same package as the microprocessor. The L2 caches on these MPUs can run faster than the speed of the local bus. Because the L2 cache is not part of the microprocessor wafer, it can be a lot larger. L2 caches are usually 64 K, 256 K, 512 K, 1,024 K, or 2,048 K, depending on the microprocessor being used.

A cache uses a special circuit called a *cache controllers* and *TAG RAMs*. The cache controller anticipates the section of the computer's DRAM the microprocessor will be looping in while running the program. The cache controller will copy thousands of bytes around the looping area into the computer's high-speed SRAM cache. Once the information is copied into the cache, the cache controller will switch the CPU into the high-speed cache by way of the TAG RAMs. The TAG RAM chips intercept the memory address coming from the microprocessor and select the actual storage location within the cache array. This entire process is transparent to the CPU and the program running at the time.

As long as the CPU is running the program within the cache, wait states are reduced considerably or eliminated altogether. However, if the program sends the microprocessor outside the section of the program contained in the cache, the microprocessor is switched back into the DRAM, with wait states inserted into the bus cycle. The program will continue executing from the DRAMs until the cache controller can reload the cache.

This may seem very confusing, but the moral of the story is that caches allow slower and cheaper DRAMs to be used as the system's main memory without degrading the performance of a high-speed MPU a great deal.

The Pentium® Pro, Pentium® II, Pentium® III, AMD K6-3™, and AMD Athlon™ microprocessors actually have two data buses. One data bus is part of the computer's local bus and runs at that speed. This is the data bus that is used by the computer's main memory and I/O ports. The other data bus, called the *dedicated data bus (DDB)*, is a high-speed bus dedicated to the L2 cache. The L2 cache's DDB runs a lot faster than the computer's local bus. In fact, the DDB data bus runs at either one-half the speed or at the speed of the microprocessor's core. For example, a Pentium® II running at a core frequency of 233 MHz uses a multiplier of 3.5. With this multiplier, the local bus is running at 66 MHz. However, the DDB used by the L2 cache is running at one-half the core speed, which is 126 MHz.

Some computers give the user the ability to control some of the functions of the L1 and L2 cache using the CMOS setup. These configuration options are covered in chapter 8.

**Figure 7.14** Windows 3.X Control Panel

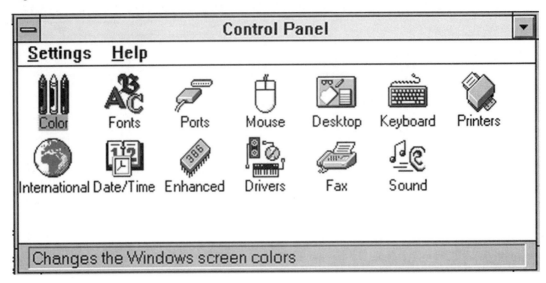

## 7.9 WINDOWS 3.X VIRTUAL MEMORY

Windows 3.X will load automatically in the enhanced mode if there is a 386 microprocessor and a minimum of 4 MB of RAM installed in the computer. If the Windows 3.X is running in enhanced mode, an image of an IC that is labeled *386* will appear in the Control Panel group. This icon is visible in the Control Panel window shown in Figure 7.14.

Windows will use expanded memory but prefers and operates faster with extended memory. Even though Windows will load in enhanced mode if a computer has only 4 MB of RAM, this is really not enough memory to make Windows run efficiently. Windows compensates for the shortage of RAM by using a **swap file** on the hard disk that is similar to expanded memory. The name given to the Windows swap file is **virtual memory.** The term *virtual memory* means it appears real to Windows and applications but is not.

Virtual memory is not really memory at all, but a file on the hard disk. As Windows starts running out of memory, it will swap information between the swap file and the computer's RAM. Hard drives are accessed in milliseconds, and RAM is accessed in nanoseconds, so the more often the swap file is used, the slower Windows will run. What this all boils down to is that Windows 3.X really needs at least 8 MB of RAM installed in the computer to run efficiently. Some Windows 3.X applications are so large and complex that even 8 MB of RAM is not enough. In fact, when using Windows 3.X, the more RAM the better.

Windows 3.X virtual memory can be configured as either a permanent or a temporary swap file. The size of a temporary swap file is determined by the amount of free hard drive space that is available each time Windows 3.X loads. When you exit from Windows, the temporary swap file is deleted automatically. The only control you have over the temporary swap file is which hard drive it is located on in a multiple-hard drive computer. A temporary virtual memory swap file is the default when Windows 3.X is installed.

Hard drive space assigned to a permanent virtual memory swap file will stay the same even after Windows 3.X is exited. The user not only gets to control the hard drive location, but also the size of the swap file. I suppose each virtual memory type has its advantages, but I prefer the permanent swap file because of the extra control available. To reconfigure Windows 3.X virtual memory settings, use the steps listed in Procedure 7.5.

**Problem 7.8:** What is the drive location, size, and type of the virtual memory swap file shown in the Virtual Memory control window shown in Figure 7.15?

**Solution:** The swap file is located on D: drive; the size of the file is 9,954 KB; and it is of the permanent type.

## PC Random-Access Memory

---

**Procedure 7.5: Windows 3.X virtual memory configuration**

1. Double-click on the Main program group icon in the Windows Program Manager.
2. Double-click the left mouse button on the Control Panel icon.

**Note:** After the Control Panel window appears, a 386-enhanced icon similar to the one in Figure 7.14 should be visible. If the 386-enhanced icon is not visible, Windows is operating in the standard mode and virtual memory is not supported.

3. If visible, double-click the left mouse button on the 386-enhanced icon. This should open up the Enhanced Mode control panel.

4. Click on the Virtual Memory button. A virtual memory control Window, similar to that shown in Figure 7.15, should appear. The Virtual Memory control window shows the drive's location, size, and type.
5. To change the setting for the virtual memory swap file, click the left mouse button on the Change button option in the Virtual Memory control window. This will bring up another window that will let you change the drive, type, and size of the swap file.

---

**Figure 7.15** Windows 3.X Virtual Memory Settings

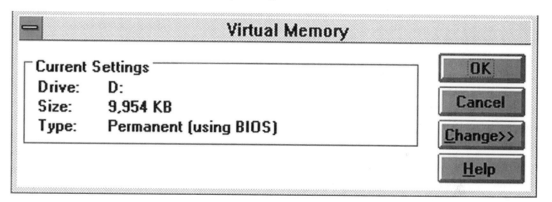

---

**Procedure 7.6: Windows 9X Virtual Memory settings**

1. Right-click on the My Computer icon on the desktop.
2. Left-click the mouse button on the Properties options.
3. Left-click the mouse button on Performance.
4. Left-click on the Virtual Memory button.

---

### 7.9.1 Setting Up Virtual Memory in Windows 9X

Windows 9X uses virtual memory just like Windows 3.X. The algorithm to follow to display and edit Windows 9X's virtual memory settings is given in Procedure 7.6. There are several ways to access the virtual memory control options in Windows 9X, but the procedure covered here is probably the most often used.

The steps outlined in Procedure 7.6 will open the Virtual Memory control window. Even though virtual memory options can be changed from this window, it is highly recommended that you select the option that lets Windows manage your virtual memory for you.

## 7.10 TROUBLESHOOTING MEMORY PROBLEMS

If the computer displays insufficient memory error messages, there are two options available to individuals running Windows 3.X or 9X. The first and most desirable method is to increase the memory capacity of the computer. The second and least desirable method is to increase the computer's virtual memory settings. The cost of using the second option is that the overall performance of the computer will decrease the more often virtual memory is used.

If the computer displays memory read/write or parity errors during the POST or normal operation, the first course of action is to reseat all of the memory modules in the computer. Also, when each module is removed from the socket, check the module pins for signs of corrosion and clean, if necessary. If this does not solve the memory problem, the next course of action is to try to determine which module is defective.

In memory systems that are currently available, the simplest method to determine the defective module is to swap each memory module in the computer with a known good module. The modules should be swapped one at a time with the computer unplugged. Take anti-static precautions. After swapping each module, test the computer. A second method is to swap the memory modules between banks. Eventually, this process should locate the defective module.

Note that diagnostic programs are available to test the memory in the computer. These programs are loaded from a disk and should be DOS based. This is because a DOS-based program will run in conventional memory, which allows the other memory in the computer to be tested extensively.

## 7.11 CONCLUSION

Fully understanding the memory layout in the compatible computers is probably the hardest thing to learn. The many terms, acronyms, and abbreviations make it similar to learning a foreign language. The only suggestion I can give is to read this chapter as many times as it takes to understand this very important subject.

When you purchase any application for the compatible computer, make sure you read all of the information on the product's box. A list of the memory requirements needed by the application for proper installation and operation is located somewhere on the box.

If the computer is running MS-DOS as the operating system, I highly recommend that you run at least version 6.0. This version of DOS has made it extremely easy to set up the computer's memory manager. Also, if you have not optimized the memory in your computer recently, it is probably about time you ran MEMMAKER.

If you are using Windows 95, 98, NT, or 2000, all of the memory managers are already in place and do not need to be optimized. This is because these operating systems have access to all of the memory in the computer, whereas MS-DOS and its applications operate in conventional memory.

If a computer is running any version of the Windows 3.X operating system, I suggest you upgrade the RAM in your computer to at least 8 MB, and keep it optimized. Even though Microsoft's Windows 9X will load into 8 MB of RAM, the computer will need at least 16 MB to run efficiently. The more memory in a Windows 9X computer, up to the cacheable limits of the chipset, the faster it will run when multitasking several applications. For this reason, I highly recommend that you have a least 32 MB, and preferably 64 MB, of RAM.

Windows NT and 2000 will run on a 32-MB machine but virtual memory is used extensively. This greatly degrades the performance of computers running one of these two operating systems. For this reason, 64 MB should be considered the minimum memory requirements for these operating systems.

## Review Questions

1. Write the meaning of the following acronyms:
    a. RAM
    b. ROM
    c. UMA
    d. UMB
    e. HMA
    f. XMM
    g. EMM
    h. EMS
    i. DRAM
    j. SRAM
    k. FPM
    l. EDO
    m. SDRAM
    n. DDRSDRAM
    o. DRDRAM
    p. LIM
    q. BASIC
    r. ECC
    s. DIMM
    t. SIMM
    u. DIP

### Fill in the Blanks

1. The first 640 KB of RAM is called _____ _____.

2. To access the computer's extended memory, the microprocessor must be running in the _____ mode.

3. SRAM is used for the _____ memory in the compatible computer.
4. Memory addresses are usually given in _____ and _____.
5. The _____ _____ illustrates how the memory addresses in a computer are assigned.
6. The ROM is located in the _____ _____ area.
7. A video adapter's ROM is usually addressed in page _____.
8. The term _____ _____ _____ is used to describe unused sections of the UMA.
9. Expanded memory is switched into a UMB _____ KB at a time.
10. The HMA can be addressed with the microprocessor operating in the _____ mode.
11. The name of the expanded memory manager provided with Microsoft operating systems is _____.
12. HIMEM.SYS manages the computer's _____ memory.
13. Reserved memory is the memory that is lost because it conflicts with the _____ _____ _____.
14. The speed rating of memory is called the _____ time.
15. The error-detecting scheme that can also correct single-bit errors is called _____ _____ _____.
16. The memory module that stores data in 64-bit groups is called the _____-pin _____.
17. The use of a swap file on a hard drive to supplement memory is called _____ memory.
18. Copying the contents of ROM into RAM is called _____.

## Multiple Choice

1. Conventional RAM is located between hexadecimal addresses:
   a. 0 and FFFFF.
   b. 0 and 3FFFF.
   c. 0 and 9FFFF.
   d. none of these answers is correct.
2. ROM:
   a. is volatile.
   b. is faster than RAM.
   c. can be read from but not written to.
   d. can be written to but not read from.
3. Expanded RAM:
   a. is only used on computers using at least an 80286 microprocessor.
   b. cannot be used in a computer running Windows 9X.
   c. is accessed faster than conventional RAM.
   d. can be installed in the IBM-PC/XT-style computer.
4. Extended RAM:
   a. is only available on computers running at least an 80286 microprocessor.
   b. is another name for expanded memory.
   c. is accessed faster than conventional RAM.
   d. can only be installed in the IBM-PC/XT-style computer.
5. Extended memory:
   a. begins at address 0.
   b. begins at address 100000H.
   c. ends at address FFFFFH.
   d. is an obsolete technology no longer used in the compatible computers.
6. When operating in the real mode, the 80486 microprocessor can place:
   a. 4,924,967,296 address combinations on its address bus.
   b. 1,114,096 address combinations on its address bus.
   c. 16,777,216 address combinations on its address bus.
   d. 655,360 address combinations on its address bus.
7. The virtual 86 mode:
   a. is the default mode of the 80386 microprocessor.
   b. is only available when the 80386 is operating in the protected mode.
   c. is only available when the 80386 is operating in the real mode.
   d. is only available with the 8086 microprocessor.
8. The upper memory block:
   a. is the unused location below 640 K.
   b. is the unused locations above 640 K but below 1,024 K.
   c. is another name for expanded memory.
   d. is another name for extended memory.
9. The upper memory area is the name:
   a. of the memory addresses between 640 K and 1,024 K.
   b. of the memory addresses between 0 K and 640 K.
   c. of the memory address above 1,024 K.
   d. used most often for expanded memory.
   e. used most often for extended memory.
10. The term *high-memory area* refers to the:
    a. memory addresses in the range of 0 K to 640 K.
    b. memory addresses in the range of 640 K to 1,024 K.
    c. first 64 K addresses of extended memory.
    d. first addresses of expanded memory.
11. What is the page number for the address $5B56A9C_{16}$?
    a. 5
    b. 6A9C
    c. 5B5
    d. 5B
12. How many byte addresses are contained within a memory page?
    a. 1,024 per K
    b. 64 K, or 65,536
    c. 1 M, or 1,048,576
    d. varies with the number of pages
13. DRAM:
    a. requires refreshing, but consumes less power and is inexpensive.
    b. does not require refreshing and is the fastest memory in the computer.
    c. is nonvolatile.
    d. is never cached.
14. In order for MS-DOS to access extended and expanded memory, the computer must:
    a. have any Intel microprocessor below an 80486.
    b. be running a version of MS-DOS above 4.0.
    c. load a memory manager TSR.
    d. have an expanded memory adapter card installed.
15. Adapter BIOS ROMs will usually be located between addresses:
    a. 00000 and FFFFF.
    b. C0000 and CFFFF.

c. 10000 and FFFFF.
d. 00000 and 9FFFF.
16. SRAM:
    a. has to be refreshed.
    b. is the fastest memory in the computer.
    c. is the type of memory used by the BIOS.
    d. is nonvolatile.
17. A caching motherboard:
    a. allows a slow microprocessor to be used with high-speed DRAMs.
    b. uses SRAM instead of DRAMs as the computer's main storage.
    c. allows slower and cheaper DRAMs to be used with a high-speed microprocessor.
    d. is a technique used on all IBM-compatible motherboards.
18. Cache memory occupies addresses within UMA.
    a. True
    b. False
19. An L1 cache is:
    a. located on the motherboard.
    b. located outside the microprocessor.
    c. is part of the microprocessor's integrated circuit.
    d. is usually DRAM.
20. Virtual memory:
    a. is part of extended memory.
    b. is part of expanded memory.
    c. is part of conventional memory.
    d. is not really memory at all.
21. The computer will operate faster when virtual memory is used.
    a. True
    b. False
22. A temporary virtual memory swap file:
    a. is erased from the hard disk each time Windows is exited.
    b. is left on the hard disk even when Windows is not running.
    c. will occupy the same disk space every time Windows is loaded.
    d. will allow the user to determine its size.
23. The acronym SIMM stands for:
    a. smiling insane muscular memory.
    b. several Inline memory module.
    c. system internal memory manager.
    d. single in-line memory module.
24. UMB stands for:
    a. unregulated memory bank.
    b. upper memory block.
    c. upper memory bank.
    d. upper memory barrier.
25. The memory access time specification is:
    a. the time the microprocessor takes to access the memory.
    b. the time the memory guarantees it will be able to be accessed.
    c. the speed of the microprocessor's local bus.
    d. not important.
26. UMB is:
    a. the name given to the unused addresses between 640 K and 1,024 K.
    b. the name given to addresses between the range 0 K and 640 K.
    c. All of the above answers are correct.
    d. Both answers *a* and *b* are correct.
27. CMOS is the acronym for:
    a. complementary motive of success.
    b. complete memory overload syndrome.
    c. complementary metal oxide semiconductor.
    d. complementary memory operating system.
28. A backup of important information of the computer's hard drive should be made before any memory upgrade.
    a. True
    b. False
29. CMOS:
    a. holds the basic configuration information supplied by switches in the PC/XT.
    b. replaces the CONFIG.SYS.
    c. replaces the AUTOEXEC.BAT.
    d. All of the above answers are correct.
30. All compatible computer memory modules have error-detecting capabilities.
    a. True
    b. False
31. How many 72-pin SIMMs will be required per bank in a computer running a Pentium-compatible microprocessor?
    a. 1
    b. 2
    c. 3
    d. 4
32. When 168-pin DIMMs are used in a computer running a Pentium-compatible microprocessor:
    a. two will be required in each bank.
    b. they must use ECC/parity.
    c. one will be required per bank.
    d. the computer will usually have more than two banks.
33. Which memory technology uses a dedicated data bus?
    a. DDRSDRAM
    b. CMOS
    c. SDRAM
    d. DRDRAM
    e. EDO
34. ECC:
    a. has the ability to detect and correct memory errors.
    b. can detect memory errors but cannot correct them.
    c. is utilized in the computer CMOS memory.
    d. does not require special memory modules to implement.
35. Reserved memory:
    a. is RAM that conflicts with the UMA.
    b. is RAM that conflicts with the HMA.
    c. is RAM that conflicts with conventional memory.
    d. is RAM that conflicts with EMS.

# CHAPTER 8

# BIOS and the CMOS

## OBJECTIVES

- Define terms, acronyms, and expressions used in relation to the computer's BIOS and CMOS.
- List the major manufacturers of the BIOSs used in the compatible PC.
- Describe how the contents of the CMOS are accessed.
- Define the typical settings in the CMOS Setup programs.
- Explain the characteristics of the types of ROMs used in the compatible computer.

## KEY TERMS

real-time clock (RTC)
offset
setup
extended configuration setup data (ECSD)
manufactured ROM
erasable programmable read-only memory (EPROM)
electrical erasable programmable read-only memory (EEPROM)
flash ROM

As you have probably guessed, the BIOS is a very important component of the compatible PC. Even though there are a number of companies that make motherboards for the PC, there are only four companies that write most of the BIOSs that are currently in use. These companies are American Megatrends, Inc. (AMI), Award Software, Inc. (Award), Phoenix Technology Ltd. (Phoenix), and Microid Research, Inc. (MR BIOS). These four became three in September 1998 when Award Software and Phoenix Technology Ltd. merged under the name Phoenix Technologies Ltd. However, Phoenix is currently maintaining both BIOSs as separate entities under their respective names. The Internet URLs (addresses) for these companies are given in Table 8.1.

Table 8.1  BIOS Manufacturers' URLs

| Company Name | URL (Internet Address) |
| --- | --- |
| American Megatrends, Inc. (AMI) | www.megatrends.com |
| Award Software, Inc (Award) | www.phoenix.com |
| Phoenix Technologies Ltd. (Phoenix) | www.phoenix.com |
| Microid Research, Inc. (MR BIOS) | www.mrbios.com |

These companies' BIOSs are all reliable and are basically the same because they have to be written around a given microprocessor and chipset. It stands to reason that if the microprocessor and chipset are the same, the BIOSs have to be similar, no matter who writes them. For this reason, choosing a BIOS is more of a user preference than a performance preference.

The features offered by the BIOS are usually visible to the user by way of the microprocessor and chipset it is written to control. The user can control most of the features of the BIOS through settings in the CMOS. The basics of the BIOS and CMOS are covered in the following sections.

## 8.1 BIOS CONFIGURATION USING THE CMOS

As stated previously in chapter 5, the computer's BIOS needs to have information provided about the hardware configuration as soon as the computer is turned on. This is necessary so the BIOS will know how to handle the computer's basic hardware correctly. This configuration information needs to be available to the BIOS before any attempt is made to load the main OS from disk.

The AT- (and above) compatible computers use a very small section of RAM that is part of the **real-time clock (RTC)** port to hold the 1s and 0s that define the computer's basic configuration. This configuration is used not only to establish hardware configurations for the BIOS, but also to define the hardware configuration used by the chipset.

There is a slight problem with using the system RAM in that it is volatile memory. To keep the configuration information from disappearing when the computer is turned off, a small lithium battery is mounted on the motherboard to maintain power to the RTC and special RAM contained within the port. To keep the batteries small and cheap, the RTC-RAM that holds the BIOS configuration is SRAM made from a semiconductor material that requires very little power. This semiconductor material is called *complementary metal oxide semiconductor* (*CMOS*). Do not confuse CMOS SRAM with the SRAM used for the cache. The SRAM used for the cache is constructed from a semiconductor technology that allows it to operate at a very high speed. The higher the speed at which the memory operates, the more electrical power it consumes.

Because the memory that holds the BIOS configuration information in the AT- (and above) compatible computers is constructed using CMOS technology, the BIOS and chipset configuration information is usually referred to simply as the computer's *CMOS*.

CMOS technology consumes far less power than do other technologies. However, there are two tradeoffs for this extremely low power consumption. First, CMOS costs more than other memories and, second, it is really slow. In fact, the CMOS used on compatible computers is so slow that it is accessed through a pair of ports, similar to any input/output device. The speed of this memory is not of great concern because the computer reads its contents only once during the boot process.

The port addresses used by the RTC and CMOS are $070_{16}$ and $071_{16}$. Through these two port addresses either 64 bytes of CMOS for older PCs, or 128 bytes of CMOS for current PCs, are selected. Contained within the 64 or 128 bytes of the CMOS are 14 bytes for the RTC. The remaining bytes hold BIOS and chipset configuration information.

You may wonder how 64 or 128 bytes of information can be stored at two port addresses. This feat is accomplished by accessing only a single byte of the CMOS information at a time through the two ports. An **offset** value is assigned to each byte of information contained in the CMOS. Storing the offset value into port address $070_{16}$ makes the selected information available at port $071_{16}$. Then port $071_{16}$ is read or written to get or change the contents of the desired CMOS offset. This is similar to going to one window (port $070_{16}$) to place your order and another window (port $071_{16}$) to pick up the order.

The configuration information originally assigned by the IBM engineers used only a small portion of the 64 bytes available in the original RTC, which included offsets 00H (hexadecimal) through 0DH. Offsets 10H, 12H, 14H–1AH, 2EH, and 2FH were also used. Because IBM assigned these values, all of the aforementioned BIOS writers use the same CMOS offsets to store the same configuration information. They are the same in all CMOS. Table 8.2 lists the information contained in these offsets.

The other undefined offsets in the original 64 bytes and additional 64 bytes of the 128-byte CMOS are assigned by the various BIOS writers to hold such things as memory timing, user-defined hard drive parameters, shadowing, passwords, boot sequence, and wait-state, just to name a few. Unfortunately, these CMOS offset assignments are different among the various BIOS writers.

# BIOS and the CMOS

**Table 8.2** CMOS Offsets and the Information Each Offset Contains.*

| Offset | Contents |
|---|---|
| 00H | RTC Current second |
| 01H | RTC Alarm second |
| 02H | RTC Current minute |
| 03H | RTC Alarm minute |
| 04H | RTC Current hour |
| 05H | RTC Alarm hour |
| 06H | RTC Current day of the week |
| 07H | RTC Current day of the month |
| 08H | RTC Current month of the year |
| 09H | RTC Current year |
| 0AH | RTC Status Register A |
| 0BH | RTC Status Register B |
| 0CH | RTC Status Register C |
| 0DH | RTC Status Register D |
| 10H | A and B Floppy Drive Type. The settings for this offset are: The code for drive A is set by bits 7–4 and the code for drive B is set by bits 3–0 using the following codes: <br><br> **Binary**     **Drive** <br> 0000     No drive installed <br> 0001     360-KB ¼″ drive <br> 0010     1.2-MB 5¼″ drive <br> 0011     720-KB 3½″ drive <br> 0100     1.44-MB 3½″ drive <br> 0101     2.88-MB 3½″ drive |
| 14H | Equipment Byte. The settings for this offset are: <br><br> Bits 7–6 denote the number of floppy drives installed <br> 00—One floppy drive <br> 01—Two floppy drives <br> Bits 5–4 denote the type display installed <br> 00—Not defined <br> 01—Color display using 40 columns <br> 10—Color display using 80 columns <br> 11—Monochrome <br> Bits 3–2 are not defined <br> Bit 1 denotes if a math coprocessor (FPU) is installed <br> 0—No math processor available <br> 1—Math processor available <br> Bit 0 denotes if floppy drives are installed <br> 0—No floppy drives installed <br> 1—Floppy drives are installed |

(Continued)

**Table 8.2** (Continued)

| | |
|---|---|
| 15H | Base Memory Size MSB (Conventional Memory) |
| 16H | Base Memory Size LSB (Conventional Memory) |
| 17H | Extended Memory Size LSB |
| 18H | Extended Memory Size MSB |
| 19H | Hard disk 0 extended drive type |
| 1AH | Hard disk 1 extended drive type |
| 2EH | Standard CMOS checksum MSB |
| 2FH | Standard CMOS checksum LSB |

*Offsets are given in hexadecimal.

The biggest problem from a technical point of view is that the various BIOS writers do not make the CMOS offset assignments used by their BIOS available to the general public. This makes it very difficult to determine what configuration information is used by the offsets in the CMOS that were not originally assigned by IBM. Luckily, there are individuals who have taken it upon themselves to hack the CMOS offset addresses for AMI, Award, and Phoenix in an attempt to determine the configuration information that each contains. One of the best Web sites that I have found lists all 128 bytes of CMOS for the three mentioned BIOSs. The Internet address is irb.cs.uni-magdeburg.de/~zbrog/asm.

Due to its complexity, it would be very hard to display and edit the contents of the CMOS directly. Fortunately, the BIOS allows us to display and change the contents of the CMOS indirectly using a program contained within the BIOS itself. The name of the program is called **Setup.**

One thing nice about using a program to edit the contents of the CMOS is that all of the different options can be displayed in menu form and can be changed easily. This one feature alone makes using a Setup program several magnitudes easier than setting binary 1s and 0s directly in the CMOS.

## 8.2 USING THE SETUP PROGRAM

As stated, the BIOS is written to control a certain microprocessor and chipset. For this reason, the settings available in the CMOS are similar among the various BIOS writers. Because the CMOS setups for these BIOSs are similar, only the BIOS written by Award Software, Inc. is covered in this text. I am not choosing Award's BIOS because it is better than the others. It simply happens to be the BIOS in the computer I am presently using. The Award's BIOS that I discuss is for a P55C motherboard running an Intel 430TX PCI chipset.

The compatible computers vary over a wide range, so the CMOS Setup options discussed in the following sections may contain more or fewer options than those available with the BIOS installed in your computer. Also, because some of the CMOS settings will lock up the computer if set incorrectly, it is becoming popular for motherboard manufacturers to hide these settings in their BIOS Setup menus. When CMOS settings are hidden, they cannot be altered directly by the Setup program supplied with the BIOS. However, there are a few Setup utility programs that can be downloaded from the Internet that will allow you to change these hidden options. The Setup utility program on most compatible computers is part of the programs that are stored in the computer's BIOS. If the Setup utility is located in the computer's ROM, you do not have to worry about keeping up with a diskette.

If the Setup program is contained in the BIOS, it is usually started by the user when the computer is first booted. To determine exactly how to invoke the Setup program, consult the documentation that came with your computer. When the video adapter's information is displayed, a message will usually appear somewhere on the screen that indicates the key you must press to run the Setup program. For example, Award and AMI BIOSs require the Delete (Del) key to be pressed, whereas Phoenix requires the F1 key to be pressed to run Setup. The Del key is located on the right side of the computer's keyboard on the edit keypad. It may also be labeled *Delete*. The F1 key is located to the left on the top row of keys, which are called the Function keys. Press the appropriate key when the message appears and it will invoke the Setup program.[1]

Before we go any further, there is one very important thing you should know. An incorrect CMOS entry can be fatal to the computer, causing it to lock up during the boot. Before you make any changes, document the current settings so that you can return the CMOS to its original configuration if problems occur. In fact, if you own or work with a compatible computer, it is a good idea to document the current CMOS settings for three reasons:

1. The lithium battery that powers the CMOS when the computer is turned off will eventually run down. When this happens, the contents of the CMOS will be lost. Once a new battery is installed, the Setup utility must be used to return the CMOS to the settings that were used prior to the battery failure.
2. The CMOS contents may be altered if a program crashes. A program crash can simply lock up the computer or it can cause the microprocessor to get out of step. When this occurs, the microprocessor may hit the addresses where the CMOS information is stored. If this happens, say goodbye to the contents of your CMOS.
3. If you make incorrect modifications to a CMOS entry, the computer may not get far enough into the boot sequence to allow you to run the Setup program again. When this happens, you will have to clear the contents of the CMOS and start from a set of defaults. Clearing the contents of the CMOS and loading defaults are discussed later in this chapter.

For these three reasons alone, it is nice to have a good copy of the contents of your CMOS so the computer can be returned to its original configuration.

### 8.2.1 Setup's Main Menu

After the Setup program loads by using the method outlined in the computer or motherboard's documentation, the Setup program's Main menu will display. The Main menu may be GUI or text, depending on the BIOS writer. No matter how the menu is displayed, the Main menu will have options similar to those listed in Table 8.3. It should be noted that the options listed in Table 8.3 may not have the same title as those listed in the BIOS in your computer.

Once again, note that the explanations in the following sections will be similar, but will probably not be exactly the same as the CMOS settings available in your computer due to several facts: Different BIOS writers place the same options in different menus; the options offered by one BIOS may be named something entirely different with another BIOS; and, the options offered by the BIOS will vary among chipsets. I highly recommend that you visit the Web site for the BIOS installed in your computer and download the Setup settings for the CMOS.

### 8.2.2 Standard Setup

The Standard Setup menu sets the options that were available in the original AT-compatible computers. If this option is selected from the listings in the Main menu, it provides options similar to the following:

> **Date:** Changes the date in the RTC.
> **Time:** Changes the time in the RTC.
> **Daylight Savings:** (may not be visible) Automatically adds (if enabled) one hour to the time when Daylight Savings Time begins.
> **Primary Master; Primary Slave; Secondary Master; and Secondary Slave:** Set the configuration parameters of the hard drive. The settings available are discussed in chapter 11. It is recommended that these be set to Auto.

---

[1] If Del or F1 does not start the Setup program, try the following. If these do not work, consult the writers of the BIOS. The setup program can be entered only during the boot sequence. These alternate keys start the setup program: All Function keys; Esc; Ctrl+Esc; Ctrl+S; and Ctrl+Alt+S.

Table 8.3  Typical Setup Program Main Menu Options

| Standard CMOS Setup |
| --- |
| BIOS Features Setup or Advanced CMOS Setup |
| Chipset Features Setup |
| Power Management |
| PCI/PnP Configuration |
| Load BIOS Defaults |
| Load Setup Defaults |
| Integrated Peripherals |
| IDE Auto Detection |
| Password Settings |
| HDD Low Level Format (may not be visible) |
| Save and Exit Setup |
| Exit without Saving |

**Floppy Drive A:** Defines the floppy drive installed as drive A: in the computer. Set to the appropriate setting depending on the drive installed.

**Floppy Drive B:** Defines the type floppy drive installed in the computer as drive B. Set to the appropriate setting depending on the drive installed.

**Video:** Selects the primary display device.

**Halt On:** Selects whether or not the boot sequence will stop if the BIOS detects non-fatal hardware failures during the POST.

**Memory:** This portion of the Standard Setup menu is automatically calculated during the boot POST and cannot be changed by the user. This memory display includes the amount of base memory (conventional), extended memory, and other memory detected.

### 8.2.3  CMOS Features

Most of the settings in Award's CMOS Features menu are found in the Advanced Setup section of AMI's BIOS. When the CMOS Features option is selected from the Main menu, it provides options similar to the following:

**Virus Warning:** Displays an error message (if enabled) if any program tries to change the contents of the master partition boot record of the hard drive or volume boot sector on a floppy. This does not detect other viruses in applications. You should install and run anti-virus software for this purpose. Enable except when running FDISK or installing or upgrading an operating system.

**Enable Internal Cache:** Enables or disables the L1 cache inside the microprocessor. It is highly recommended that this feature be enabled if the microprocessor in the computer has an L1 cache.

**Enable External Cache:** Enables or disables the L2 cache in the computer. It is highly recommended that this feature be enabled if the computer has an L2 cache available.

**Quick Boot:** Reduces (if enabled) the amount of time required to boot the computer because portions of the POST are skipped. It is recommended that this option be disabled.

**Boot Sequence:** Determines the order in which the BIOS will search the mass storage devices looking for the operating system boot files. It is recommended that this be set to attempt to boot from the C: drive or SCSI first, depending on the computer and location of the operating system boot files. If a need arises to boot from the floppy, this option can be changed by running the CMOS Setup program again.

**Swap Floppy Drive:** Allows the user to logically swap the A: and B: floppy drives without removing the cover and swapping the connections on the cable

**Memory Test above 1 MB:** Prevents the testing of memory above 1 MB. This option usually is contained within the quick-boot options. If available, it is recommended that this option be enabled to allow testing of memory above 1 MB.

**Boot Up Num Lock:** Determines if the small calculator keypad on the right of the keyboard is initially set to the number or cursor control feature. It is recommended that this be set to Num Lock because most keyboards already contain a separate cursor control keypad.

**Floppy Disk Seek:** If enabled, the BIOS will test communication with the floppy drives. To decrease the time for the boot sequence, it is recommended that this setting be disabled.

**Boot Up System Speed:** Most motherboards support two speeds of operation. If set at Low, the computer usually operates at 8 MHz. The High setting operates the computer at its rated speed. It is recommended that this option be set at High.

**Gate A20 Option:** Controls the A20 line for DOS applications. The selection determines if the keyboard controller, MPU, or the chipset controls A20. It is recommended that this option be set to Turbo to allow the microprocessor to control the A20 line.

**Typematic Rate:** On some BIOSs, the only selections available with the option are Fast and Slow. If this is the case, Fast is recommended for this setting. Other BIOSs will have Enable and Disable options. When this is the case, other Typematic settings will be available to select the typematic rate at which the characters repeat if the key is held down, and the typematic delay, which is the number of times a key has to be pressed before it starts to repeat.

**Security Option or Password Check:** Determines if the password set in the Main menu is required during the boot sequence or when the CMOS Setup program is launched.

**PS/2 Mouse Function Control:** If your computer uses a serial mouse, this option must be disabled. On the other hand, if you use a PS/S mouse, this option must be enabled.

**Parity/ECC Check:** Determines if enable, parity, or ECC. Do not enable this option unless all of the RAM in the computer supports parity or ECC.

**OS/2 Compatible Mode:** Allows the BIOS to load the OS/2 operating system if enabled. This should be set to disabled if the computer is running any other operating system besides OS/2.

**OS Select for DRAM > 64MB:** Enable this option only if the computer is running the OS/2 operating system.

**PCI/VGA Palette Snoop:** Supports ISA VGA adapter cards. If one of these cards is installed in the computer, consult the video card's documentation to determine if this option should be enabled. This option must be disabled for all other video cards.

**System BIOS Shadow Cacheable:** Routes (if enabled) the system BIOS through the cache memory. It is recommended that this option be enabled.

**Video BIOS Shadow Cacheable:** Allows (if enabled) the video shadow RAM to be routed through the computer's cache. It is recommended that this option be enabled.

**Assign IRQ to VGA:** Enable this option only if the display adapter installed in the computer requires an interrupt request (IRQ). This information should be in the documents that are provided with the video card.

**HDD SMART Capability:** Enable this option if it is supported by the hard drives installed in the computer. This information should be contained in the hard drive documents.

**Delay for HDD (Sec):** Selects a delay during the cold boot sequence to allow the hard drives to come up to speed. It is recommended that this be set as low as possible to reduce the boot time. However, if the computer freezes after the memory test, try higher values for this setting.

**Shadowing:** The remaining settings deal with shadowing ROMs into reserved memory. The system BIOS is often shadowed automatically without an option. If the option is available, make sure the system BIOS is shadowed. The video BIOS is also often automatically shadowed. If the option is available, make sure it is also enabled. The other Shadow options should be disabled unless shadowing is required by additional adapter BIOSs. This information should be contained in the documents for the adapter card. If shadow ROM is enabled when no adapter ROMs are present, the space available to the UMA will be reduced.

### 8.2.4 Chipset Setup

The settings in the chipset Setup menu predominantly control the RAM timing. If these settings are wrong, the computer will cease to function and may not allow the Setup program to run without clearing the CMOS. Because most of these settings are critical, the motherboard manufacturer often masks them. Some or most of the following options may not be visible in your CMOS Setup program:

- **Auto Configuration:** Allows (if enabled) the BIOS to select timing values for the system RAM. However, note that the settings chosen by the Auto Configuration option are not usually the optimal settings for the memory installed in the computer.
- **DRAM Timing:** This should be set to the access time of the memory modules installed in the computer. Lower values mean a faster system. Values set too low can cause the computer to malfunction. This setting only applies to 30-pin SIMMs. The timing setting for 72-pin SIMMs and 168-pin DIMMs are automatically set by the modules themselves.
- **DRAM Leadoff Timing:** Selects the CPU clock cycle combinations that are required before each memory read or write operation. Lower values mean a faster system. Values set too low can cause the computer to malfunction.
- **DRAM Read Burst (EDO/FP):** Sets the burst read clock sequence for both EDO or FPM memory. Lower values mean a faster system. Values set too low can cause the computer to malfunction.
- **DRAM Write Burst Timing:** Sets the burst write clock sequence. Lower values mean a faster system. Values set too low can cause the computer to malfunction.
- **Fast EDO Leadoff:** This feature should be enabled if EDO RAM is being used in the computer. Disable this feature if any of the memory banks in the computer contain FPM DRAM.
- **Refresh RAS# Assertion:** Selects the number of clock cycles the RAS signals are asserted during a memory refresh. A low setting means the computer will access memory at a faster speed. Values that are set too low can cause the computer to malfunction.
- **Fast RAS# to CAS# Delay:** Inserts a timing delay between the CAS and RAS signals. Disabling this feature means a faster system. Enabling this feature means a more stable system.
- **DRAM Page Idle Time:** Selects the number of clock cycles that the memory controller waits to close a DRAM page after the CPU becomes idle.
- **DRAM Enhanced Paging:** Keeps (if enabled) the memory page selected until a page/row miss. If disabled, the chipset keeps the memory page open in case the host may return.
- **SDRAM (CAS Lat/RAS-to-CAS):** Allows the user to set the CAS latency and RAS to CAS delay in clock cycles of 2/2 or 3/3. Lower settings mean faster system performance. Values set too low can cause the system to malfunction.
- **SDRAM Speculative Read:** Speculates (if enabled) the next DRAM read address. This allows the memory controller to issue an address slightly before the data has been read from the previous memory address. I recommend that this feature be enabled unless memory errors are encountered.
- **System BIOS Cacheable:** Allows (if enabled) the system BIOS to be routed through the cache, which means a faster system. I recommend that this feature be enabled. If system errors occur after enabling, disable this feature.
- **Video BIOS Cacheable:** Enables cacheing of the video BIOS, which results in better system performance. If system errors occur after enabling, disable this feature.
- **SDRAM RAS to CAS Delay:** Sets the number of clock cycles between a RAS and CAS. Lower values mean a faster system. Values set too low can cause the computer to cease to operate.
- **8/16 Bit I/O Recovery Time:** (may not be visible) Specifies the number of system clock cycles inserted between 8-bit or 16-bit I/O operations. Lower values mean faster speed. Values that are set too low can cause the adapter to malfunction.
- **Memory Hole 15 MB to 16 MB:** Allows support for older ISA cards. This should be set to Disable unless an ISA card is installed in the computer that specifies that this setting be enabled in the adapter's documentation.

### 8.2.5 Power Management

Power management is rapidly becoming a major issue in the compatible PCs. Both Intel and Microsoft are adding features to the hardware and software of the computer to reduce its power consumption when not in use. Following is a list of options available when Power Management is selected in the CMOS Setup program's Main menu:

> **Power Management:** Allows the user to select the degree of power saving used by the computer. If the disable option is selected, no power management is used by the BIOS, however. The Advanced Power Management (APM) feature for Windows 95 and 98 still may be used.
> **PM Control by APM:** If APM is available with the operating system installed on the computer, select the Yes option.
> **Video Off Method:** Determines the method use to blank (turn off) the screen on the computer's display. The option should be set according to the monitor's documentation.
> **Video Off After:** Selects the mode desired for the monitor to blank.

The user can only set the following options when Power Management is set to User-defined:

> **Dose Mode:** Sets the elapse time of system inactivity before the system clock's duty cycle is decreased. The percent of system clock activity is set by the Throttle Duty Cycle option.
> **Standby Mode:** Allows the user to select the time of system inactivity before the system's fixed drives and video are shut down. All other devices remain active.
> **Suspend Mode:** Shuts down all devices except the CPU after the selected time of system inactivity.
> **HDD Power Down:** Sets the time of inactivity before the hard drive(s) shut down. All other devices remain active.
> **Throttle Duty Cycle:** Sets the percent duty cycle for the system clock during the Doze mode.
> **ZZ Active in Suspend:** Enables the ZZ signal during Suspend mode.
> **VGA Active Monitor:** If enabled, any video activity will restart the power down timers for Standby mode.
> **CPUFAN Off in Suspend:** When enabled, the CPU fan will turn off during the Suspend mode if this option is supported by the motherboard.
> **IRQ8 Clock Event:** If turned off, the RTC is not monitored, so it will not wake the system from Suspend mode.

### 8.2.6  PnP/PCI Configuration

As the heading implies, the PnP/PCI Configuration option in the Main menu is used to configure the PnP (Plug-and-Play) capabilities of the BIOS. It is also used to select the user-defined options of the peripheral components interface (PCI) expansion bus. These terms, along with interrupts and DMA, will be covered in later chapters.

> **PnP OS Installed:** Enable this option if the computer is running a PnP operating system such as Windows 95, 98, or 2000.
> **Resource Controlled By:** If set to Auto, the BIOS configures all of the IRQs and DMAs used by the PnP adapters.
> **Reset Configuration Data:** When this option is enabled, the Extended System Configuration Data (ESCD) will reset. It is recommended that this option be set to disabled.
> **IRQ n Assigned to:** Allows the user to set IRQs manually as either Legacy ISA or PCI/ISA PnP.
> **DMA n Assigned to:** Allows the user to set DMAs manually to either Legacy ISA or PCI/ISA PnP.
> **PCI IRQ Activated by:** Allows the user to set either level or edge-triggered interrupt specification for the PCI expansion bus. It is recommended that this be set to Level unless instructed to set to Edge-triggered by the documentation supplied with the PCI adapter card.
> **PCI IDE IRQ Map to:** If the IDE hard drive interface is not on the motherboard, this option is used to select the location of the IDE adapter card—either PCI or ISA.
> **Primary/Secondary IDE INT#:** PCI provides one to four interrupts for PCI devices. These are called interrupts A, B, C, and D. The A interrupt is assigned to each device by default. If additional interrupts are required, they are selected with this option.
> **Used Mem base addr:** Allows the user to select the base address for any port that requires use of the HMA.

## 8.2.7 Integrated Peripherals

This option, in the main Setup menu, is used to configure the ports that are integrated on the motherboard.

- **IDE HDD Block Mode:** Set to Enable to automatically detect if the hard drives installed in the computer support block transfer; otherwise, disable.
- **IDE Primary/Secondary Master/Slave PIO:** Selects the PIO mode used by the BIOS to transfer information with the IDE device attached to the primary and secondary. This should be set according to the IDE devices documentation. If the Auto option is selected, the BIOS will automatically determine the best mode.
- **IDE Primary/Secondary Master/Slave UDMA:** Select Auto only if the installed hard drive supports UDMA/33 and the OS has the required drivers for UDMA.
- **On-Chip Primary/Secondary PCI IDE:** Enables or disables the motherboard integrated primary and secondary IDE ports. Disable this feature only if an IDE adapter is used instead of the onboard IDE.
- **USB Keyboard Support:** Enable this option if a USB keyboard is attached; otherwise, disable.
- **USB Latency Timer (PCI CLK):** Sets the number of PCI clock cycles the USB controller maintains control of the PCI bus.
- **Onboard FDC Controller:** Enables or disables the motherboard integrated floppy disk controller. Disable this option only if a FDC adapter is used or no floppy drives are installed in the computer.
- **Onboard UART 1 / 2:** Selects the address and interrupt used by the two integrated serial ports. This option can also be used to disable the motherboard's integrated serial ports.
- **IR Duplex Mode:** Select either full duplex (two directions at the same time) or half duplex (two directions but not at the same time) used by an infrared device connected to one of the serial ports. Set this to disabled if an IR device is not being used.
- **Use IR Pins:** Selects the pins used to transmit and receive IR information. Set this according to the documentation supplied with the IR device.
- **Onboard Parallel Port:** Selects the base port address and interrupt used by the motherboard's integrated parallel. This option can also be used to disable the integrated parallel port.
- **Parallel Port Mode:** Selects either the EPP or ECP modes for the parallel port.
- **ECP Mode Use DMA:** Selects the DMA channel to use if the ECP mode is selected.

## 8.2.8 Other Options in the Setup Program

The Password Setting option is used to set a password for the BIOS or boot sequence. When this option is selected, a message appears asking you to enter a password. Once a password is entered, another message will appear asking you to confirm the password. The process can be aborted without setting a password by striking the Esc key. Once a password is entered, the Security Option in BIOS Feature should be set to either System or Setup. If you forget the password, the CMOS will have to be cleared in order to once again gain access to the CMOS setup.

The IDE HDD Auto Detection option in the Setup program Main menu is used to automatically detect the configuration parameters of the hard drives installed in the computer. In order for this feature to work, the hard drive has to support the ATA-2 or EIDE standard. This information should be supplied with the drive documentation.

A low-level format prepares the surface of a hard drive to store files. This option should never be used with IDE or EIDE drives. In fact, this may not appear in the Main menu of all computers. To low-level format an IDE or EIDE drive, contact the manufacturer of the hard drive to obtain the correct procedure.

If the contents of the CMOS are cleared for one reason or another, the next boot sequence will abort with a CMOS CHECKSUM error. Along with this message will be another message informing you that Setup will have to be run. Once the Setup program launches, there are two options available that will return the computer to operation until the exact options in the CMOS can be configured. These options are Load BIOS Defaults and Load Setup Defaults.

The Load BIOS Defaults will load a set of settings from the BIOS ROMs into the CMOS that will provide the most stable operation. All of the CMOS settings provided by this option will be for minimal performances, but the system will more than likely boot unless other major problems exist.

> **Procedure 8.1: Determining the location of the ECSD**
>
> 1. Click on the Start button located on the Task Bar.
> 2. Move the mouse pointer up to Settings and then over to Control Panel.
> 3. Click on the Control Panel icon.
> 4. In the Control Panel window, locate and then double-click on the System icon.
> 5. On the System Properties window, click on the Device Manager tab.
> 6. In the Device Manager window, double-click on the Computer icon.
> 7. Click on the white dot beside "Memory."
> 8. All address ranges in the Memory option that are labeled "System board extension for PnP BIOS" are used for the ECSD.
> 9. Click on the X in the upper right-hand corner or on the Cancel button to close all open windows.

The Load Setup Defaults option will load optimal performance settings for the BIOS into the CMOS. These settings depend on the hardware installed in the computer to support these optimal standards. If the system locks up or memory failure error messages are displayed when the computer boots or is in operation, try the Load BIOS Default settings.

The next two options provided by the Setup program's Main menu are Save & Exit Setup and Exit Without Save.

### 8.2.9 Extended Configuration Setup Data (ECSD)

All current BIOSs support Plug-and-Play (PnP) technology. With this feature, peripheral ports are configured automatically by the operating system. To provide this option, additional memory addresses are needed than those provided by the CMOS in order to hold the PnP configuration information. The PnP information contained in the memory is the **extended configuration setup data (ECSD).** The address location of this memory varies from computer to computer and its contents are updated automatically by the BIOS. To find the location of the ECSD on a computer running Windows 9X, follow the steps outlined in Procedure 8.1.

## 8.3 CLEARING THE CMOS

Incorrect CMOS settings can cause the computer to lock up so badly that the boot sequence will not start. If the boot sequence will not start, the CMOS Setup program cannot be used to correct the problem. Because of this, most major motherboards provide a jumper to completely clear the contents of the CMOS. The location of this jumper varies from computer to computer. Its precise location is given in the documentation that is provided with the motherboard. If not provided, the jumper number that provides this service is usually written somewhere on the motherboard itself and is usually located near the real time clock battery. An example of a typical clear CMOS jumper is shown in Figure 8.1.

If the computer does not start the boot sequence due to incorrect settings in the CMOS, the computer should be turned off and the Clear CMOS jumper should be placed in the CLEAR position for approximately 10 seconds, and then moved back to the NORMAL position. When power is once again applied to the computer, a message will be displayed indicating that the CMOS has a CHECK-SUM failure. A message will also be displayed informing the user that the Setup program must be run to correct the problem. After the CMOS Setup program starts, the Load Bios Defaults or Load Setup Defaults should be selected from the Main menu. This will load parameters stored in the ROM that will allow the computer to once again boot. The Setup program can be used at a later time to enter the original configuration information.

If the Clear CMOS jumper cannot be located, turn the computer off and then remove the battery from the motherboard for approximately 10 seconds. This should be enough time to clear the computer's CMOS. When the computer is turned back on, the contents of the CMOS can be reestablished using the procedure given in the previous paragraph.

**Figure 8.1** Motherboard Clear CMOS Jumper

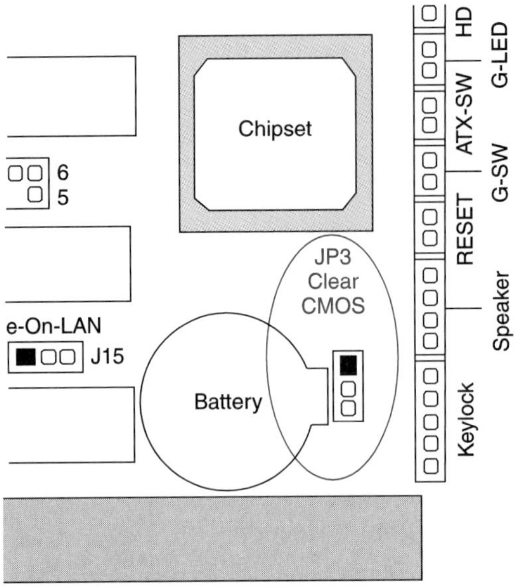

## 8.4 UPGRADING THE BIOS

Because the ROMs are read-only, you have probably wondered how all of the programs and data get into the ROMs in the first place. To understand this, we need to discuss the three categories of ROMs used in the compatible computer: *MROMs, EPROMs,* and *EEPROMs* (also known as *Flash ROMs*).

**Manufactured ROM (MROM)** is manufactured with the contents already contained within. Software engineers write the programs to be placed in the MROMs and then send the binary information to MROM manufacturers such as Motorola, Inc. The MROM manufacturer then takes the submitted binary information and produces the circuits in the MROM IC to provide the correct information for each ROM address. Once an MROM has been produced, it cannot be reprogrammed. MROMs are very inexpensive when produced in mass quantities.

**Erasable programmable read-only memory (EPROM)** is a unique ROM. Exposing the IC wafer to high-intensity, short-wavelength ultraviolet light will erase it. A device called an *EPROM eraser* provides the required ultraviolet light through a small window in the IC package that exposes the IC wafer. This window is shown in the illustration of a typical EPROM in Figure 8.2. After the EPROM is erased, its contents are programmed through the use of a special, and expensive, piece of equipment called an *EPROM programmer*. After an EPROM has been programmed, it is placed in the computer, where it operates just like an MROM. To reprogram the EPROM it must be removed from the computer, erased in the EPROM eraser, and reprogrammed in the EPROM programmer.

**Electrical erasable programmable read-only memory (EEPROM)** gets its name from the fact that it can be erased and programmed electrically. This feature allows the EEPROM to be erased and reprogrammed in the equipment in which it is intended to operate. This is currently the ROM used in most current compatible PCs and it is called **Flash ROM.**

### 8.4.1 Flash ROM

The original IBM-PC and XT used MROMs to hold the BIOS and BASIC interpreter. Most of the compatible PCs manufactured before the 486 used an EPROM. As stated, EPROMs can be erased and reprogrammed, but they must be removed from the computer.

Flash ROMs are unique in the fact that they can be erased and reprogrammed in the equipment in which they are intended to operate. Note that in order for a motherboard to use Flash ROMs, it must contain special circuitry that provides the erasing and programming services. Even though this added circuitry increases the overall cost of the motherboard, this computer programming feature

**Figure 8.2** Typical EPROM

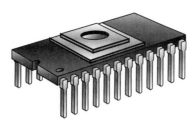

*Source:* Floyd, *Digital Fundamentals,* 7th ed. (Upper Saddle River, NJ: Prentice Hall, 2000) p. 636.

has made Flash ROMs very popular in all kind of applications, including the BIOS in the compatible PC.

The reprogramming of Flash ROMs requires the user to use software supplied by either the BIOS writer or motherboard manufacturer. This software is often available free over the Internet from the ROM or motherboard manufacturer's Web or FTP site.

Once the software is obtained, most motherboards require that a configuration jumper be moved in order for the programming to take place. After the motherboard is configured and the software is launched, the user follows the instructions displayed by the BIOS upgrade program.

Although this sounds like a simple task, upgrading the computer's BIOS is a major step and should be attempted only as a last resort. Sometimes the Flash ROM does not take the new programs or something happens that causes the upgrade program to abort. If this occurs, the information in the BIOS is usually corrupted, making the computer unusable. Flash ROMs will then have to be purchased from the manufacturer with the BIOS programs preinstalled.

## 8.5 CONCLUSION

The BIOS contains all of the basic drivers for the standard ports in the compatible PC. These drivers get the computer operational when power is first applied or when a reset or soft-boot sequence is initiated. For this reason, it is a very important component of any compatible computer.

The BIOS controls the basic functions of the computer and also initializes the motherboard chipset. The user configures the basic function of the BIOS and chipset by storing settings in a small memory called the CMOS that is part of the RTC assembly. Due to the complexity of the CMOS, the user usually sets its contents through the use of a Setup program that is contained within the BIOS. Unfortunately, motherboard manufacturers may mask some of the settings available with the CMOS from the user to avoid incorrect settings.

The contents of the CMOS will eventually have to be restored because it is a volatile R/W memory whose information is maintained by a battery that will eventually go dead. The contents of the CMOS can also disappear for no apparent reason. For these reasons, it is imperative that a hard copy be kept of the CMOS settings so they can be restored.

Along with the CMOS, additional configuration information is stored by the BIOS in a section of memory called the ESCD. The location of the ESCD varies from machine to machine.

The BIOS ROMs used in the compatible can include MROMs, EPROMs, or EEPROMs. MROMs cannot be reprogrammed and must be replaced in order to upgrade a computer's BIOS. EPROMs can be erased and reprogrammed, but this is done outside of the computer using special equipment. EEPROMs, also called Flash ROMs, are usually erased and reprogrammed in the equipment in which they are intended to operate. This feature has made Flash ROMs very popular in today's compatible computers.

## Review Questions

1. What do each of the following acronyms stand for?
   a. BIOS
   b. CMOS
   c. ESCD
   d. RAM
   e. ROM

f. EPROM
g. EEPROM
h. ESCD
i. RTC

## Fill in the Blanks

1. The program that is usually used to display and edit the contents of the CMOS is called _____ .
2. The CMOS is contained within the _____ _____ _____ port.
3. The CMOS is accessed through hexadecimal port addresses _____ and _____ .
4. In most CMOS memories used in the compatible computer, offset 09H contains the settings for the _____ _____ _____ .
5. On computers running AMI BIOS, pressing the _____ key during the _____ sequence enters the Setup program.
6. The settings for the hard drive are usually contained in the _____ CMOS Setup option of Setup options.
7. A _____ _____ jumper is usually provided on the motherboard to erase the contents of the CMOS.
8. PnP information is kept in _____ outside the CMOS and is called _____ _____ _____ data.
9. The technical name for Flash ROM is _____ _____ _____ _____ _____ _____ .
10. The configuration settings for the types of floppy drives are usually contained at offset _____ hexadecimal in the CMOS.
11. If enabled, the virus warning feature in the Setup menu will display an error message if any program tries to change the _____ _____ _____ of the boot hard drive.
12. If the computer locks up due to incorrect settings in the CMOS, the _____ _____ jumper should be used to erase the contents of the CMOS.

## Multiple Choice

1. The CMOS:
   a. holds configuration information used by the operating system.
   b. holds the configuration information used by the BIOS.
   c. is nonvolatile memory.
   d. is the fastest memory in the computer.
2. The contents of the CMOS:
   a. are maintained by the computer's power supply.
   b. are held in the computer's ROM and copied to RAM each time the computer is booted.
   c. are copied into ROM each time the computer is booted.
   d. are maintained by a small battery.
3. The contents of the CMOS will never disappear.
   a. True
   b. False
4. The CMOS is:
   a. part of the RAM of the computer
   b. part of the ROM of the computer.
   c. contained within the real-time clock module.
   d. similar to ROM.
5. All the offsets used by BIOS manufacturers are the same.
   a. True
   b. False
6. To obtain the optimal setting available for the RAM in the computer, the Auto Configuration option in the Setup menu should be enabled.
   a. True
   b. False
7. To get the most out of the RAM in the computer, the DRAM timing setting should be set:
   a. as high as possible without locking up the machine.
   b. as low as possible without locking up the machine.
   c. in the middle of the lowest and highest setting to ensure the memory operates in all conditions.
   d. to the Auto Configuration settings.
8. The ESCD memory:
   a. holds standard CMOS setup information.
   b. holds PnP information.
   c. is located with the CMOS.
   d. is set and maintained by the computer user.
9. If available, the low-level format option available in some Setup programs:
   a. should never be used to format IDE or EIDE drives.
   b. can be used to format floppy drives.
   c. should be used to format all hard drives.
   d. activates the low-level format program provided by the drive manufacturer.
10. The Load BIOS Defaults option in the Main menu of the Setup program:
    a. will load a set of CMOS settings contained in the computer's ROM.
    b. will load a set of CMOS from the computer's hard drive.
    c. should never be used.
    d. will load a set of configuration settings into the computer's CMOS that will be the optimum settings for the computer.
11. Most of today's compatible computers use:
    a. manufactured ROMs.
    b. EEPROMs.
    c. EPROMs.
    d. regular ROMs.
12. To program Flash ROMs:
    a. special erasing and programming equipment is required.
    b. the binary information must be sent to an IC manufacturer.
    c. the programming utility and new contents must be obtained from the manufacturer.
    d. ultraviolet light is required.

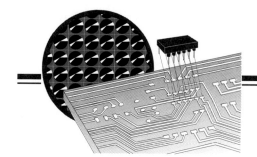

# CHAPTER 9

# Input/Output

## OBJECTIVES

- Define terms, expressions, and acronyms used in relation to I/O devices.
- List I/O address assignments for an AT-compatible computer.
- List the interrupt and DMA assignments for both the XT- and AT-style computers.
- Explain the characteristics of fixed, hardset, softset, and PnP adapters.
- State the function of the keys on an enhanced keyboard.
- Define the standard ports used in the compatible computer.
- Describe the different connectors and cables used in the compatible computers.
- State the characteristics of the compatible computer's programmable serial ports.
- Define terms used in serial communication.
- Define the flow-control pins used by the serial and parallel ports.
- Explain the characteristics of the USB and FireWire serial ports.
- State the characteristics of the compatible computer's programmable parallel ports.
- Explain the differences among SPP, EPP, and ECP parallel ports.
- Describe the characteristics of the keyboards used on the compatible computers.
- Explain the characteristics of the pointing devices used on the compatible computers.
- Explain the function of the sound card and game port.

## KEY TERMS

peripheral
input port
output port
input/output (I/O) port
I/O address map
direct memory access (DMA)
jumpers
fixed ports
hardset
DIP-switch
softset

Plug-and-Play (PnP)
cable
male connector
female connector
parallel
Electronic Industries Association (EIA)
RS232C
software flow control
hardware flow control
hot-swap

FireWire
isochronous
Institute of Electronic/Electrical Engineers (IEEE)
function keys
software defined
multimedia
musical instrument digital interface (MIDI)
sound blaster-compatible

We know that we communicate with a computer through its input and output ports. All input and output devices that are connected to these ports are located outside the computer, so they are often called **peripherals** because they are on the computer's periphery (outside). A computer port is a portal in which information is transferred between a peripheral and the computer. A computer's ports are very important because they are how we get information into and out of the computer.

283

An **input port** allows information to pass from an input peripheral, such as a keyboard, to the microprocessor or memory. An output peripheral, such as a printer, gets its information from the microprocessor or memory through an **output port.** An **input/output (I/O) port,** such as the port that the disk drives are connected to, allows information to flow between the computer and the peripheral in both directions. Even though all ports are not input/output ports, the term *I/O* is usually used generically to describe all of the ports in the compatible PC.

The peripheral connects to the I/O port and the I/O port connects to the computer's address, data, and control buses. The I/O port is assigned the addresses, not the peripheral itself. The I/O port consists of ICs that are specifically designed to handle the data transfer between the computer and the peripheral for which it was designed. The port consists of ICs that are mounted either on the computer's motherboard or on an adapter card that plugs into one of the computer's expansion slots.

When a computer system is purchased, it has several I/O ports and peripherals that are preinstalled. In addition to these preinstalled I/O devices, there are all kinds of additional I/O ports and peripherals that can be purchased for the different styles of compatible computers. The list of available peripherals and I/O ports keeps growing the longer that computers are on the market. It would be impossible to cover all of the available I/O ports and peripherals in this or any book. For this reason, this chapter focuses on keyboards, serial ports, mice, parallel ports, joysticks, sound cards, and modems.

The peripherals referred to as floppy drives, hard drives, video monitors, and printers are covered in chapters 10, 11, 12, and 13, respectively.

## 9.1 COMPUTER RESOURCES

Every I/O port that is installed in the compatible computer requires the use of computer resources. The I/O resources available in the compatible computer are limited and usually cannot be shared between ports. The I/O resources that are available in the compatible computer are addresses, interrupts, and direct memory access (DMA).

### 9.1.1 Input/Output Port Addresses

Just like the memory in the computer, every I/O port must have at least one address so that a program can select it for communication. The engineers who design the computer assign the I/O port addresses. The I/O addresses are usually listed in a table called the **I/O address map.** A partial I/O address map for the compatible computer is shown in Table 9.1. Notice in Table 9.1 that most of the defined ports require a range of addresses instead of just one. This is because most of the ports used in the compatible computer use programmable ICs that require several port addresses so the program can tell the port how to operate.

The engineers who designed the IBM-PC in 1981 originally developed the I/O address map for the compatible computers. The I/O address map was updated when the IBM-AT was introduced in 1984. The I/O address map developed for the IBM-AT computer is the base for the map in current use. This is the address map shown in Table 9.1.

You might wonder, "How can the computer have memory at address 0 and a port called the DMAC at the same address?" Addresses assigned to the computer's I/O ports can be the same as those assigned to memory because they are in different address maps. A good analogy is to think of the memory and I/O addresses in the computer as being on two separate streets—one street belongs entirely to memory devices and the other belongs entirely to I/O ports. Having the addresses in different address maps means the same numerical address can be assigned to both a memory device and an I/O port.

Table 9.1 is only a partial listing of the AT I/O addressing map. A complete listing contains I/O addresses that are not necessary to know and would require explanations that are beyond the scope of this text.

Windows 9X and Windows 2000 allow you to view the I/O address assignments in your computer using Device Manager or System Information. In fact, the I/O address map that belongs to each computer should always be the one referenced because some I/O addresses vary from computer to computer. Procedure 9.1 lists the steps you should follow to view a computer's I/O address map using Windows 9X or 2000.

Table 9.1  Partial I/O Address Map for the Compatible Computers*

| Address | Port |
|---|---|
| 000–00F | Direct memory access controller #1 (DMA) |
| 020–021 | Programmable interrupt controller #1 (PIC) |
| 040–05F | Programmable interval timer (PIT) |
| 060 & 064 | Keyboard |
| 061 | Motherboard speaker |
| 070–07F | Real-time clock & CMOS |
| 080 | POST code (some computers) |
| 081–083 | DMA page register |
| 0A0–0A1 | PIC #2 |
| 0B2–0B3 | Advance power management |
| 0C0–0DF | DMAC #2 |
| 0F8–0FF | Math coprocessor |
| 168–16F | 4th IDE interface |
| 170–177[1] | Secondary IDE |
| 1E8–1EF | 3rd IDE interface |
| 1F0–1F8 | Fixed disk (primary IDE) |
| 200–20F | Game port |
| 208–02F | Motherboard |
| 220–233 | Sound card (default) [1,2] |
| 238–23B | PS/2 or bus mouse adapter (alternate)[1] |
| 23C–23F | PS/2 or bus mouse adapter (default)[1] |
| 270–277 | PnP[1,5] |
| 278–27F | Parallel/printer port, checked 3RD (LPTX)[3] |
| 2B0–2DF | Enhanced graphics adapter (alternate) |
| 2E8–2EF | Programmable serial port 4 (COM4)[1] |
| 2F8–2FF | Programmable serial port 2 (COM2) |
| 330–331 | Musical instrument digital interface (MIDI) port[1,2] |
| 360–363 | Network (alternate) |
| 368–36B | Network (alternate) |
| 376 | Secondary IDE[1,5] |
| 378–37F | Parallel/printer port, checked 2nd (LPTX)[3] |
| 388–38B | FM synthesizer[1,2] |

(Continued)

Table 9.1 (Continued)

| Address | Port |
|---|---|
| 3B0–3BB | Monochrome display adapter |
| 3BC–3BF | Parallel/printer port, checked 1st (LPTX)[3,4] |
| 3C0–3CF | Enhanced graphics adapter |
| 3D0–3DF | Color graphics adapter |
| 3E8–3EF | Programmable serial port 3 (COM3) |
| 3F0–3F5 | Floppy disk controller (FDC) |
| 3F6 | Primary IDE |
| 3F8–3FF | Programmable serial port 1 (COM1) |
| **Added with PCI** | |
| 0480–048F | PCI bus[1,5] |
| 4D0–4D1 | PCI interrupt controller[1,5] |
| 678–67A | ECP parallel port[1,5] |
| 778–77A | ECP parallel port[1,5] |
| 7BC–7BE | ECP parallel port[1,5] |
| CF8–CFB | PCI configuration address register[1,5] |
| CFC–CFF | PCI configuration data register[1,5] |
| 4000–403F | PCI bus[1,5] |
| 5000–501F | PCI bus[1,5] |
| 6400–641F | Universal serial bus (USB)[1,2] |
| F000–F007 | Primary IDE controller[1,5] |
| F008–F00F | Secondary IDE controller[1,5] |
| FFA0–FFA7 | Primary bus master IDE register[1] |
| FFA8–FFAF | Secondary bus master IDE register[1] |

*All numbers are in hexadecimal.
[1] Not part of the original IBM-AT I/O addressing map
[2] May differ if another device already occupies this address
[3] LPT numbers are assigned by the order in which they are found
[4] All modes except EPP
[5] PCI computers only

## 9.1.2 A Little about Interrupts

To complete our discussion on I/O, I am going to disseminate just enough information about a topic called *interrupts* to get through this chapter. Interrupts are covered in detail in chapter 16.

Basically, interrupts provide a way for an I/O port to get the microprocessor's attention when the peripheral needs to transfer information with the computer. Once again, the IBM-AT-compatible computer established the interrupt standards still used on all current compatibles.

> **Procedure 9.1: Using Windows 95 and 98 Device Manager to display the I/O address map**
>
> 1. Click on the Start button.
> 2. Move up to Settings and then over to Control Panel.
> 3. Click on Control Panel.
> 4. Locate and double-click on the System icon.
> 5. Click on the Device Manager tab.
> 6. Click on the Computer icon at the very top of the list.
> 7. Click on the small white dot beside Input/Output (I/O).
> 8. Click on the Cancel buttons on all open windows to exit.
>
> **Using Windows 98 and 2000 System Information to display the I/O address map**
>
> 1. Click on the Start button.
> 2. Move up to Programs, over to Accessories, over to System Tools, and click on System Information.
> 3. Click on the plus (+) sign beside Hardware Resources.
> 4. Click on the listing for I/O.
> 5. Once completed, close the System Information window by clicking on the "X" in the upper right-hand corner.

Compatible computers above the XT have the ability to accept 16 interrupts using two ICs called *programmable interrupt controllers (PICs)* #1 & #2 that are located on the motherboard. Along with the PICs, control bus signals called *interrupt requests (IRQs)* are also used. Each IRQ is assigned a number from IRQ 0 through IRQ15. The I/O ports commonly assigned to these 16 IRQs are listed in Table 9.2.

Table 9.2  Interrupt Request (IRQ) Assignments

| 8-bit Interrupts | |
|---|---|
| IRQ0 | Timer |
| IRQ1 | Keyboard |
| IRQ2 | Interrupt controller |
| IRQ3 | COM2 and COM4 (serial ports 2 and 4) |
| IRQ4 | COM1 and COM3 (serial ports 1 and 3) |
| IRQ5 | LPT2 (parallel/printer port 2) (sound card[1]) |
| IRQ6 | Floppy disk |
| IRQ7 | LPT1 (parallel/printer port 1) |
| **16-bit Interrupts** | |
| IRQ8 | Real-time clock |
| IRQ9 | System (PCI = available) |
| IRQ10 | Available (network card[1]) |
| IRQ11 | Available (sound card[1]) |
| IRQ12 | Available (P/S2 mouse[1]) |
| IRQ13 | Math coprocessor |
| IRQ14 | Hard drive (primary EIDE) |
| IRQ15 | Available (secondary EIDE) |

[1] Not part of the original I/O address map

> **Procedure 9.2: Displaying the interrupt assignments using Windows 9X Device Manager**
>
> 1. Click on the Start button.
> 2. Move up to Settings and then over to Control Panel.
> 3. Click on Control Panel.
> 4. Locate and double-click on the System icon.
> 5. Click on the Device Manager tab.
> 6. Click on the Computer icon at the very top of the list.
> 7. Click on the small white dot beside Interrupt Request (IRQ).
> 8. To exit, click on the Cancel buttons.
>
> **Display the interrupt assignments using Windows 98 and Windows 2000 System Information**
>
> 1. Click on the Start button.
> 2. Move up to Programs, over to Accessories, over to System Tools, and then click on System Information.
> 3. Click on the plus (+) sign beside Hardware Resources.
> 4. Click on the IRQ listing.
> 5. To exit, click on the "X" in the upper right-hand corner of the System Information window.

Most of the IRQ control signals are assigned to pins on the expansion bus connector. Because these computers are downwardly compatible, the IRQ assignments started with the PC bus and then additions were made for the ISA bus. In chapter 4, we learned that the ISA connector contains the original connector and pin assignments used by the original 8-bit PC bus. A second connector was added to facilitate the additional pins needed for the 16-bit bus we called ISA. The pins used by the IRQs are split between the original 8-bit PC bus connector and 16-bit extension connector used by ISA. For this reason, the lower eight IRQs (IRQ 0 through IRQ 7) are called *8-bit interrupts* because they are assigned to the PC bus connector and are the only ones available to 8-bit ports. Even though a 16-bit port has access to all interrupts, the upper IRQs (IRQ 8 through IRQ 15) are called 16-bit interrupts because they are assigned to the 16-bit extension connector.

You will not have to worry about ports that are permanently assigned interrupts, such as the floppy disk, timer, keyboard, real-time clock, math coprocessor, and primary EIDE. The interrupts you will have to worry about are those assigned to multiple ports of the same type, such as the COM and LPT ports. Also, I/O ports that are not part of the original AT I/O address map require the use of an IRQ (note 1, Table 9.2). These will have to be configured so they will not conflict with an IRQ already in use.

Throughout this chapter, Table 9.2 will be referenced when discussing a device that may require the user to either assign or reassign an IRQ. Procedure 9.2 lists the steps you can follow to display the actual interrupt assignments in your computer using Device Manager or System Information.

### 9.1.3 Direct Memory Access

**Direct memory access (DMA)** is a method of transferring data to and from an I/O port directly with the computer's memory without going through the microprocessor. Needless to say, this technique speeds up the process of peripherals getting information in and out of the computer.

Two ICs located on the motherboard called *direct memory access controllers (DMAC) #1 and #2* handle DMA. Using the two DMACs, the AT-compatible and above can handle a total of eight devices that need DMA.

The device that needs DMA will inform the DMACs through control bus signals called *device request channels (DRQs)*, or *DMA channels*. The DRQ channels are numbered DRQ0 through DRQ7 on the expansion bus, but they are most often referenced as DMA0 through DMA7. The assignment of each of these DMA channels is listed in Table 9.3. The DMAs are split into two groups: 8-bit DMAs and 16-bit DMAs, for the same reason that the IRQs are split.

The DMACs currently used in the compatible computers allow data to be transferred between the computer's memory and the port only in 8-bit groups. For this reason, the only port that is permanently assigned a DMA is the floppy disk controller (FDC) port. Other ports, such as sound cards and ECP parallel ports, have to be assigned a DMA that will not conflict with other ports that need DMA. ECP parallel ports and sound cards will be discussed in sections 9.4.4 and 9.7, respectively. Procedure 9.3 lists the steps to follow to display the DMA assignments in your computer using Windows 9X Device Manager and Windows 2000 System Information.

Table 9.3  DMA Assignments

| 8-bit DMAs | |
|---|---|
| DMA0 | Available |
| DMA1 | Available (sound card) |
| DMA2 | Floppy disks |
| DMA3 | Available |
| **16-bit DMAs** | |
| DMA4 | First DMAC cascade |
| DMA5 | Available (sound card) |
| DMA6 | Available |
| DMA7 | Available |

---

**Procedure 9.3: Using Device Manager to display DMA assignments**

1. Click on the Start button.
2. Move up to Settings and then over to Control Panel.
3. Click on Control Panel.
4. Locate and double-click on the System icon.
5. Click on the Device Manager tab.
6. Click on the Computer icon at the very top of the list.
7. Click on the small white dot beside Input/Output (I/O).
8. To exit, click on the Cancel buttons.

**Using System Information to display DMA assignments**

1. Click on the Start button.
2. Move up to Programs, over to Accessories, over to System Tools, and then click on System Information.
3. Click on the plus (+) sign beside Hardware Resources.
4. Click on DMA.
5. To exit, click on the "X" in the upper right-hand corner of all open windows.

---

## 9.2 CONFIGURING AND I/O PORT RESOURCES

Sooner or later you will want to upgrade your computer by adding features that were not originally installed in your computer or were not available at the time of purchase. When this time comes, you can do the upgrade yourself and save a lot of money or you can pay someone a lot of money to do it for you.

When a new adapter card (I/O port) is purchased, it is usually preconfigured to an address, interrupt (if used), and DMA channel (if used) that are most likely to work. If the adapter card does not respond to its drivers or the computer locks up after the card is installed, the problem may be that the address, interrupt, or DMA used by the new adapter card is wrong or conflicts with an adapter card or port already installed in the computer. When this happens, the resources used by one of the ports will have to be changed to eliminate the conflict.

The act of setting the resources used by an I/O port is referred to as setting the port's configuration or configuring the port. The configuration of the ports in the compatible computer is either fixed, hardset, softset, Plug-and-Play (PnP), or a combination of these.

### 9.2.1 Fixed Ports

Most of the ports in the compatible computer are assigned fixed resources. If this is the case, the port's resources are set by the manufacturer and cannot be changed by the end user. Most of the ports that are

**Figure 9.1** Two- and Three-post Jumpers

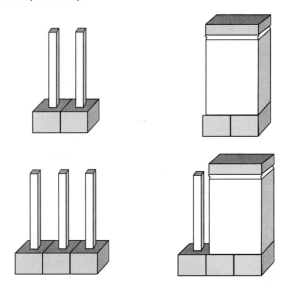

listed in the I/O address map shown in Table 9.1 use fixed resources. If the port's resources are fixed, only one of the ports can be installed in the computer. Some examples of **fixed ports** are monochrome or color video adapters; keyboard port, FDC port, hard drive port 1, hard drive port 2, game port; real-time clock and CMOS port; PIC #1 and #2; DMAC #1 and 2; and PS/2 mouse port, just to name a few.

Ports that were not part of the original AT I/O addressing map (identified by note 1 in Table 9.1) or ports that can have more than one of the same type installed, such as COMs and LPTs, usually must have their resources set by someone other than the port manufacturer. These resources have to be set so they will not conflict with the resources that are already in use by the ports installed previously in the computer.

### 9.2.2 Hardset Cards

The process of configuring a **hardset** adapter card (port) is accomplished by the placement of jumpers or the position of switches. Figure 9.1 shows a typical configuration of jumpers. A jumper consists of metal pins, called *posts,* that protrude from the adapter card or motherboard. The posts are positioned parallel to each other and are spaced precisely. The posts are usually arranged in two- or three-post groups. Each group of posts is responsible for setting an option such as an address, interrupt, or DMA channel. The options are selected by installing or removing a small plastic device called a *shorting plug,* or **jumper,** across a set of two posts.

If switches are used to configure an adapter card, they are contained in small blocks called a **DIP-switch.** A typical DIP-switch is shown in Figure 9.2. The DIP-switch get its name from the fact that its width and length are the same as the width and depth of the standard DIPs used by most ICs. A single DIP-switch can contain up to 10 individual switches. The switches are toggled either on or off to set the desired configurations.

When a computer is purchased as a complete system, the installed hardset adapter cards are configured by the computer manufacturer and should work. The user does not normally reconfigure these pre-installed, hardset adapter cards because the configuration information is not normally supplied with the computer.

When a new hardset adapter card is purchased by the end user, it is preconfigured to use resources that will most likely work. The new card is usually supplied with documentation that explains how to set the jumpers or switches on the card to reconfiguration it if its factory defaults do not work. This documentation should have an easy-to-follow, step-by-step process that will enable you to successfully configure the hardset adapter card to operate in your computer. It should also have a help number you can call if the installation process does not go smoothly.

**Figure 9.2** DIP-Switch

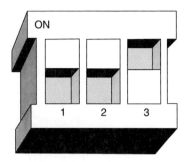

One of the things you might ask when the hardset adapter card is purchased is, "Is there someone I can call if I need help with this installation?" If the answer is no, shop somewhere else. Examples of a typical hardset adapter card configuration are covered in section 9.9.

### 9.2.3 Softset Cards

Hardset adapters are the cheapest and easiest to set up when the adapter contains only a single port, like a serial port or a parallel port. However, some hardset cards contain multiple ports such as sound cards or MI/O cards. When this is the case, the jumper and switch settings can become numerous and confusing. Another problem with hardset cards is that you have to keep up with the configuration sheet that comes with the card. Without the configuration information, reconfiguring the card becomes an almost impossible task. In addition, many people are intimidated by having to move jumpers around or change the position of switches.

As the name implies, a **softset** adapter's configuration is set by using software. A disk that contains a program that configures the resources of the card is supplied with a softset adapter. Once the adapter is installed in the computer, the program, with a name similar to SETUP or INSTALL, is run from the supplied diskette. This program will display the available resources in menu form. The user selects the desired addresses, IRQ (if required), and DMA (if required) from the list of resources to configure the port so it will not conflict with other ports already installed in the computer. Once the desired resources are selected, the user selects a menu option that stores the selected resources on the card in a special memory called *Flash ROM*.

Most of the integrated ports installed on the motherboard use the softset feature. The configuration of motherboard ports was covered in chapter 8.

### 9.2.4 Plug-and-Play (PnP)

In the never-ending quest to make the compatible computer easy to upgrade and use by the average person, adapters called **Plug-and-Play (PnP)** are now available. PnP adapters are configured by the computer's operating system so that its resources will not conflict with resources already being used by the computer. To take advantage of PnP technology, you must have three things: First, the adapter must be a PnP adapter; second, the computer must have a PnP-compatible BIOS; and last but not least, the computer must be running a PnP operating system. Currently, the only PnP operating systems available for the compatible computers are Windows 95, Windows 98, Windows Me, and Windows 2000.

Even though PnP seems like it is the answer to all of our configuration problems, there are still some catches that give it the name *Plug-and-Pray* by most computer technicians. First, the ports on a PnP adapter card can be assigned nonstandard resources. When this happens, the application that needs to use the port may not be able to find it or configure it correctly. This is especially true when Windows 3.X and MS-DOS applications try to use ports that have been assigned resources by Windows 95/98/Me/2000.

Second, if the PnP adapter is installed in the computer when the PnP operating system is installed or upgraded, the OS may place the adapter in a hardware category that will not allow it to work correctly.

**Figure 9.3** Motherboard with Integrated Peripherals

Most PnP cards also give the user the ability to softset or hardset them. This feature allows them to be used in computers that run non-PnP operating systems. Also, Windows 9X allows the user to softset most PnP adapters. This process is covered later in this chapter.

## 9.2.5 Standard Ports

Some of the fixed ports have always been permanently mounted on the motherboard as part of the chipset. These include the real-time clock and CMOS; keyboard port; PICs #1 and #2; DMACs #1 and #2; and motherboard speaker, just to name a few. Some ports are mounted on individual adapter cards that are installed in individual expansion slots on the motherboard. This technique allows for easy upgrade but requires the computer to have many expansion slots, which costs money and occupies space on the motherboard.

With the introduction of the AT computer, the port requirements became standardized, which brought about some major changes in the way that I/O ports are installed in the compatible computer. One of the major standards introduced with the AT is referred to as the *advanced technology attachment-1 (ATA-1)*. This standard is most often referred to as the *integrated drive electronics (IDE)* standard. This standard is covered in detail in chapter 11, but basically, the IDE standard moved the hard drive controller to the drive itself and established the way that two IDE hard drives are attached and accessed by the operating system. The two IDE hard drives are daisy-chained together (connected one after the other) on a single 40-pin ribbon cable.

The AT also established the standard ports that should be installed in the computer. Of course, this included the standard fixed ports (DMAC, IRQ, keyboard, and internal speaker) that have always been permanently mounted on the motherboard. Added to these fixed ports were one FDC port, two programmable serial ports, one parallel port, one game port, and one IDE port.

As the technology used to manufacture electronic circuitry advanced, it allowed all of the additional standard ports to be fabricated on a single IC and mounted on a single adapter card called an *AT-I/O,* or *multifunction I/O (MI/O),* card. A typical MI/O card is shown in Figure 9.3. The MI/O card occupies only one expansion slot and is usually a hardset card. An example of configuring one of these cards is given in section 9.9.

The ATA-2 and 3 standards which are collectively called the *enhanced integrated drive electronics (EIDE)* standard, followed the ATA-1 standard. Among other things, the EIDE standard increased the

**Figure 9.4** Typical VL Bus MI/O Card

number of IDE ports (HDD) to two. The first is called the *primary IDE* port and the second is called the *secondary IDE* port. Two IDE ports allow the installation of up to four IDE devices—two per port. Also, the EIDE standard added support for CD-ROM drives, which up to this time had been connected to the sound card. CD-ROMs and the EIDE standard are discussed in greater detail in chapters 10 and 11.

In their never-ending quest to make cheaper, user-friendly motherboards, motherboard manufacturers started placing the I/O chip from the MI/O card on the motherboard in order to reduce the number of expansion slots. These motherboards are referred to as *With I/O* motherboards and the ports are called *integrated peripherals.* The integrated ports are usually softset and are configured using the CMOS Setup program. The CMOS configuration for a typical motherboard's softset ports was covered in chapter 8. The integrated port connectors on a typical motherboard with I/O is shown in Figure 9.4.

### 9.2.6 Connection to the Port

One very important thing to remember is that no matter the location of the port, the peripheral has to be connected to the port electrically. The peripherals that are actually mounted on the motherboard, such as DMACs and PICs, use electrical conductors on the motherboard to establish this connection. Any peripheral that is not mounted on the motherboard, such as the keyboard, mouse, floppy drives, display monitor, hard drives, and CD-ROM, must use a cable to make the electrical connection to the port. A **cable** is a bundle of individual conductors that are insulated from one another inside a common jacket. As discussed in the Introduction, two types of cables are used in the compatible PC: flat and round.

The purpose of the cable is to transfer the information between the port and the peripheral. But we still have to make the electrical connection between the conductors in the cable and the port and peripherals themselves. The electrical connection is made on each end of the cable with a device called a *connector.*

In the compatible computers, there are five styles of connectors used to make the connection from the port, the peripheral, and the cable: the DB, Mini-DB, DIN, Mini-DIN, and pin connector. Each connector style is available in two genders. The connector that has the pins is called a **male connector,** while the connector that has the sockets into which the pins insert is called the **female connector.** Each connection requires two connectors of the same type but of a different gender. For example, if the port uses a male DB connector, the end of the cable that connects to the port must have a female DB connector.

**Figure 9.5** DB and Mini-DB Connectors

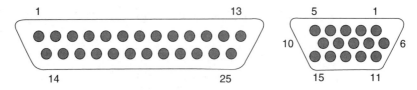

**Figure 9.6** Standard and Mini-DIN Connectors

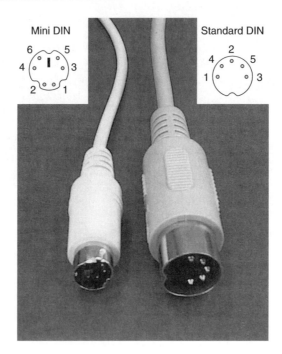

The DB and Mini-DB get their name from the fact that they are in the same basic shape of the letter D. The female connector has the sockets raised up on a plateau while the male connectors have the pins enclosed within a metal shell. This feature makes it impossible to insert the two connectors into each other backwards without making some modifications. Male and female DB and Mini-DB type connectors are shown in Figure 9.5. The Mini-DB uses the same features as the DB, except the pins are spaced closer together. This allows for more pins in a smaller connector.

The female DB connector is given a part number similar to DB25S. The DB specifies the DB style connector, the 25 indicates that there are 25 connections, and the S indicates that the connection is made with sockets (female). The male counterpart to the DB25S is the DB25P. The DB25 indicates (again) the DB style connector and 25 connections, and the P indicates the connection is made with pins (male).

The DIN and Mini-DIN connectors are round. These connectors are used to connect the keyboard and some mice. The major difference between the DIN and the Mini-DIN is the size of the connector. The Mini-DIN uses closer pin spacing so its connectors are quite a bit smaller. Both DIN and Mini-DIN connectors are shown in Figure 9.6.

Pin connectors are most often used to terminate ribbon cables. Just like the DB and DIN connectors, the pin connectors have a gender. A typical pin connector is shown in Figure 9.7.

Note that each end of the cable does not have to be terminated with the same style connector, for example, a printer cable may use a DB25P on one end and a male 36-pin Centronics connector on the other. The DB25P end of the cable plugs into a DB25S connector on the back of the computer and the male Centronics connector plugs into a female Centronics connector on the back of the printer.

**Figure 9.7** Pin Connectors

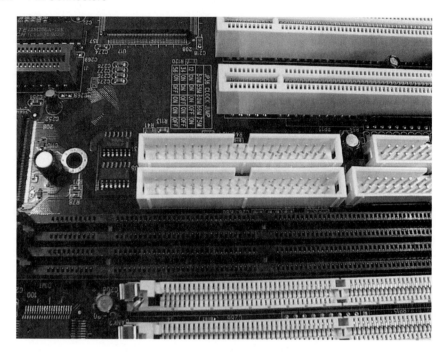

**Figure 9.8** Parallel Transfer of an 8-bit Character

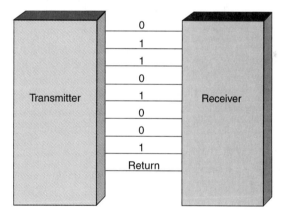

## 9.3   PROGRAMMABLE SERIAL PORTS

The data transfer over the microprocessor's address and data bus is in parallel. The term **parallel** means side-by-side paths. To accomplish parallel communication, each bit of information being transferred must have its own electrical conductor. A computer that has a 32-bit data bus and 32-bit address bus requires 64 electrical conductors just to transfer address and data information.

An analogy of parallel data transfer is a multi-lane freeway with all of the cars traveling abreast in predefined groups. This would be the fastest way to get the cars from one place to the next, but the more lanes, the higher the cost. Parallel data transfer is the fastest form of communication, but it requires a good deal of hardware to accomplish. The more hardware, the higher the cost. An example of transmitting an 8-bit number using parallel data transfer is shown in Figure 9.8. Remember from the Introduction that all signals have to have a common conductor, so to transfer 8 bits of information requires a minimum of nine conductors.

*Serial transmission* occurs when all information is transmitted over a single data path, one bit after another. This takes more time than does parallel transfer but requires far less hardware. Serial

**Figure 9.9** Serial Transfer of an 8-bit Character

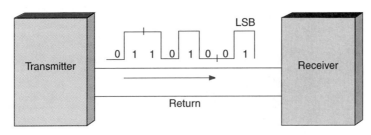

**Table 9.4** Interrupts Used by the Programmable Serial Ports*

| Port | Address | IRQ |
|---|---|---|
| COM1 | 3F8H | IRQ4 |
| COM2 | 2E8H | IRQ3 |
| COM3 | 3E8H | IRQ4 |
| COM4 | 2E8H | IRQ3 |

*The "H" at the end of each address indicates the number is hexadecimal.

transmission is comparable to a single-lane road where all of the cars have to travel one after another. All of the cars still get from one place to the next, but it takes a lot longer. Of course, a one-lane road is a lot cheaper to build than a multi-lane road. The serial transmission of the same 8-bit word shown in Figure 9.8 is illustrated in Figure 9.9. Even though all of the data are transferred on a single electrical conductor, an additional conductor is still required for the circuit common.

You may ask, "Why use serial when parallel is so much faster?" The answer to this question is, "Faster is not always better." There are a lot of input and output devices that do not need the speed offered by parallel transmission. In fact, the compatible computer uses serial to transfer data with the keyboard, floppy disk drives, mouse, and video monitor. Also, there are other times when parallel is impractical, such as when sending information across the United States or to another country. Here the cost would far outweigh the need for speed.

Current versions of MS-DOS and Windows support up to four programmable serial ports. Even though there are other fixed serial ports such as the keyboard, floppy disk, and video ports, these ports are designed to control a specific peripheral, these four serial ports are programmable serial ports. The term *programmable* means that these ports can be reconfigured by software, so many different types of serial peripherals can be attached to the same port at different times.

Even though MS-DOS and Windows support up to four programmable serial ports, the standard configuration is to have two programmable serial ports installed in the computer. One is usually used for a serial mouse and the other is usually used with an external modem (modem/fax). Mice and modems are covered in later sections of this chapter.

The programmable serial ports usually go by the names COM1, COM2, COM3, and COM4. Each serial port has its own set of unique addresses but must share the two interrupts. COM1 and COM3 share IRQ4; COM2 and COM4 share IRQ3. When a new COM port adapter card is purchased, it must be addressed so that it will not conflict with the other serial port(s) already installed in the computer. The interrupt assignment must also be set, depending on the port address used. Most COM port adapter cards have a group of switches or jumpers that allow the installer to set the port address and interrupt.

Some adapters allow the installer to set the serial port address by COM name, while others require that the actual COM port address be set. The address or COM number selected will determine whether IRQ3 or IRQ4 is used. Table 9.4 shows the COM port, address, and interrupt that go together.

Note that each address listed in Table 9.4 is referred to as the *base address,* which means the starting address. This is because each COM port actually occupies eight consecutive addresses. For exam-

> **Procedure 9.4: Displaying the COM port resources**
>
> 1. Click on the Start button.
> 2. Click on Setting and then click on Control Panel.
> 3. Locate and then click on the System icon.
> 4. In Windows 2000, click on the Hardware tab.
> 5. Click on the Device Manager tab or button.
> 6. Click on the plus (+) sign beside Port (COM & LPT).
> 7. Right-click on the COM port whose resource you wish to display and then click on Properties on the Options menu.
> 8. Click on the Resource tab.
> 9. To exit, click on the "X" located in the upper right-hand corner on all open windows.

**Figure 9.10** COM Port Connectors

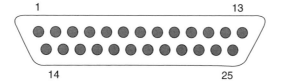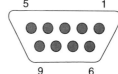

ple, COM1 is located at addresses 3F8H though 3FFH. Because each COM port always occupies eight consecutive addresses, the port is configured for the first (base) address and the other seven are understood. To display the resources used by the COM port on a computer running Windows 95, 98, or 2000, follow the steps outlined in Procedure 9.4.

Now that we know a little about COM ports, let's discuss a few of the serial port standards used by the compatible PCs.

### 9.3.1 Serial Port Standards

As the popularity of serial communication increased, some form of standardization had to be developed to make it possible to communicate with serial equipment sold by different manufacturers. A standardization committee of the **Electronic Industries Association (EIA)** established the **RS232C** serial port standard, which defines the electrical characteristics and pin assignments for a serial port. IBM chose to use the RS232C standard as the serial port on its line of microcomputers. Needless to say, all of the compatible manufactures also adopted the RS323C port standard.

The EIA-RS232C serial port uses either a 25-pin male connector called a DB25P or a 9-pin male connector called a DB9P. The DB9P connector is not standardized by EIA but is still used on all compatible computers above the XT because of its size. Figure 9.10 shows the layout of the DB9P and DB25P connectors. The pin assignments for both connectors are given in Table 9.5.

A newly purchased compatible has two programmable serial ports installed as standard equipment. On new computers that use the AT form factor, one of the serial ports usually uses a DB9P connector (usually COM1) and the other usually uses a DB25P connector (usually COM2). The move is on, however, to the ATX form factor that utilizes DB9P connectors for both COM ports.

### 9.3.2 Communication Terms

Several terms are used in reference to sending and receiving information. Some of these terms are only used in reference to serial transfer, but most are used in information communication in general. A few of the terms, with brief explanations, are listed in Table 9.6.

### 9.3.3 Flow Control

As discussed in chapter 2, the microprocessor uses its control bus to control the flow of data between the microprocessor and the selected memory or port. The need often arises to control the flow of information between the port and the peripheral. The control of the flow of information between the port and

Table 9.5  Pin-out for the DB9P and DB25P used by the COM Ports

| Signal Name | DB9P Pin | DB29P Pin |
| --- | --- | --- |
| Data Carrier Detect (DCD) | 1 | 11 |
| Receive Data (Rx) | 2 | 3 |
| Transmit Data (Tx) | 3 | 2 |
| Data Terminal Ready (DTR) | 4 | 20 |
| Signal Ground (SG) | 5 | 7 |
| Data Set Ready (DSR) | 6 | 6 |
| Request To Send (RTS) | 7 | 4 |
| Clear To Send (CTS) | 8 | 5 |
| Ring Indicator (RI) | 9 | 22 |

Table 9.6  Communication Terms

| | |
| --- | --- |
| Asynchronous | Data transmitted without a clock but at a precise rate. Even though this is a general communication term, the RS232C serial port is an asynchronous port. |
| Baud rate | A serial term used to signify the number of signal transitions per second. In asynchronous serial communication, baud rate is often synonymous with bit-per-second (bps). |
| Bits per second (bps) | A serial term that indicates the number of synchronous serial bits transmitted per second. With baud rates above 300, a modem will retransmit the asynchronous data coming from the serial port synchronously over the telephone lines. |
| Cyclic Redundancy Check (CRC) | General communication term for a popular method that detects transmission errors more efficiently than does parity. |
| Simplex | General communication term for data that is sent in one direction. Commercial television and radio stations use simplex. |
| Half duplex | General communication term for data sent in two directions but not at the same time. Most public radio systems such as HAM and CB use half duplex. |
| Full duplex | General communication term for data transmitted in two directions at the same time. The telephone system uses full duplex. |
| Parity | General communication term for a method of error checking. Parity can be selected as EVEN, ODD, or none. If used, transmitter and receiver must be set the same. |
| Start bit | A serial term for the bit used to indicate the start of a serial transmission (cannot be set by user). |
| Stop bit | A serial term for the bit(s) used to signal the end of each serial character. As with parity, transmitter and receiver should be set the same (number set by user). |
| Synchronous | General communication term for transmitted with a clock. |

**Table 9.7** RS232C Hardware Flow Control Pins

| Flow Control | Originator | Description |
| --- | --- | --- |
| Request To Send | Computer | Asks the peripheral if it is okay to send data |
| Clear To Send | Peripheral | Indicates if the peripheral is ready to receive data |
| Data Terminal Ready | Computer | Indicates if the computer's serial port is ready |
| Ring Indicator | Modem | Indicates an incoming call |
| Data Carrier Detect | Modem | Indicates that the modem is receiving a signal |

the peripheral is called *flow control,* or *handshake.* Flow control is accomplished in two ways: software flow control (sometimes called XON and XOFF) and hardware flow control.

As the name implies, **software flow control,** or *XON* and *XOFF,* uses binary codes transmitted between the port and the peripheral to help ensure that data flows smoothly. The electrical conductors used to send and receive data are the same as those used to send and receive the software flow control codes. Software control works fine for slow-speed communication that uses a transmission speed below 1,200 bps. And, because it requires fewer conductors in the interface cable, it is considerably cheaper than hardware flow control. In fact, full duplex communication, using the RS232C, requires only three conductors when software flow control is used. Referring to Table 9.5, these conductors are Tx, Rx, and SG.

When high-speed transmission is required, the overhead needed by software flow control greatly affects the speed of transmission. This is where hardware flow control comes into play. **Hardware flow control** uses conductors in the cable that connects the port to the peripheral to control the flow of information. The hardware flow control lines used by the RS232C serial port are RTS-, CTS-, DSR-, and DTR-ready. These are listed in Table 9.7 along with the location where the signal originates and a basic explanation. The procedure for setting the flow control options in Windows 95, 98, and 2000 will be given later in this chapter.

### 9.3.4 The Serial Port's UART

The integrated circuit that actually forms the programmable serial port is called a *universal asynchronous receiver transmitter,* or *UART.* This chip is responsible for converting the synchronous parallel information on the microprocessor's data bus into asynchronous digital information that travels out of the programmable serial port. It is also responsible for handling all of the hardware flow control signals. There are currently two UARTs used in the compatible computers. The first is called an 8250 UART, which supports bit speeds up to 9,600 bps. The other is the 16550, which supports bit speeds up to 115,200 bps. If you buy a new serial port, make sure it uses a 16550 UART. The procedure for determining the type of UART installed in a computer is discussed in later sections.

### 9.3.5 The Serial Character

Serial information is transferred in binary bit groups called *characters, frames,* or *packets.* The most popular term is *characters.* Each asynchronous serial character consists of one start bit, data bits (an optional parity bit), and stop bit(s). The start bit is required and is not user-definable. All of the other bits in the serial characters are under software and user control. This includes the number of data bits, the use and type of parity, and the number of stop bits. The size of the serial characters and the bps determine the actual speed of transmission. The most popular serial character size is 8-N-1 (or 8, none, and 1). This means 8 bits of data, no parity, and 1 stop bit. 8-N-1 creates a character that is 10 bits in length (don't forget the start bit). The rate at which serial characters are transferred is called the *character rate.* To find the character rate of the serial information being transferred, divide the bps rate by the character's size. For example, if a character's size of 8-N-1 is transmitted at 9,600 bps, the character's rate is

960 characters per second (9,600/10). It should be obvious that the smaller the character's size, the faster the characters will be transferred for a given bps rate.

**Problem 9.1:** If a character size of 8-E-2 (8 bits of data, even parity, and 2 stop bits) is transferred at a rate of 14,400 bps, what is the character rate?

**Solution:** The character size is 12 bits (1 start bit + 8 character bits + 1 parity bit + 2 stop bits). The character rate is approximately 1,200 characters per second.

**Problem 9.2:** If a character size of 8-N-1 (8 bits of data, no parity, and 1 stop bit) is transferred at a rate of 14,400 bps, what is the character rate?

**Solution:** The character size is 10 bits (1 start bit + 8 character bits + 0 parity bits + 1 stop bit). The character rate is approximately 1,440 characters per second.

### 9.3.6 Setting the COM Port Speed

A COM port's bit rate, character size, and type of flow control are configured by software. The user, through the operating system, sets the default configuration. A COM port can also be reconfigured at any time by the application that uses the port. It is becoming popular for Windows 95/98/2000 applications to use the default configuration settings in the operating system.

In MS-DOS, the COM port's configuration is set by means of the MODE command. An example of a typical command line used to configure the COM port is shown in Example 9.1. The syntax for the command is MODE COMX:baud, data bits, stop bits, then parity. The example in Example 9.1 will initialize COM1 to use 9,600 baud, 8 bits of data, 1 stop bit, and no parity.

**Example 9.1:** MS-DOS command line used to initialize COM1

```
C:>MODE COM1:9600,8,1,N
```

Windows 3.X's serial ports are set by double-clicking the left mouse button on the Main program group in the Program Manager window. This will display the Main program group window. In the Main program group, double-click on the Control Panel icon. This will open up the Control Panel window. From here, double-click on the Port icon. A typical Windows 3.X Control Panel window showing the COM port's settings is shown in Figure 9.11.

In Windows 9X/2000, the COM ports are configured using Device Manager, which is located in System option of the Control Panel. The algorithm to follow is listed in Procedure 9.5.

Even though the tasks required of the COM ports have been performed reliably for years, their one disadvantage is the relatively slow communication speed. To help quench the PC community's constant thirst for speed, some exciting additions in programmable serial communication have occurred within the last few years. Probably the most exciting additions is the introduction of two new PnP serial interfaces: USB and FireWire.

### 9.3.7 Universal Serial Bus (USB)

There is a new serial bus in town called the *Universal Serial Bus,* or *USB*. The USB is an open-technology, Plug-and-Play standard that is rapidly gaining the support of software writers and computer and peripheral manufacturers. In fact, Windows 95 OSR3, Windows 98, Windows 2000, and Windows Mc operating systems fully support the USB standard. With USB, users can connect as many as 127 USB devices to a single USB port on the back of the computer using USB hubs. Most USB devices themselves will serve as a hub.

USB devices do not have to be hardware configured. They are automatically detected and configured by the OS. One really nice feature of USB is that USB peripherals can be hot-swapped. The term **hot-swap** means devices can be installed and removed without shutting down and restarting the computer.

**Figure 9.11** Windows 8.X Control Panel Setting COM Port Speed

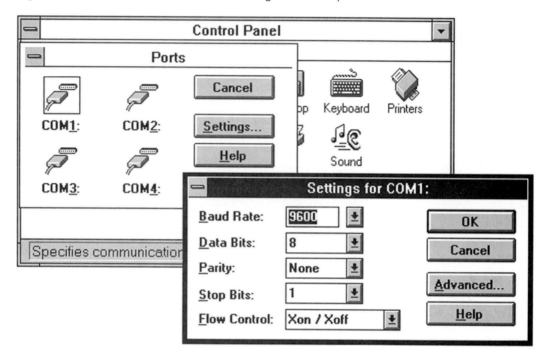

---

### Procedure 9.5: Configuring COM ports using Windows 9X/2000

1. Click on the Start button located in the Task Bar.
2. Move the mouse pointer up to Settings and then over to Control Panel.
3. Click the left mouse button on Control Panel.
4. Once the Control Panel window opens, double-click the left mouse button on the System icon.
5. If using Windows 2000, click on the Hardware tab.
6. Click on the Device Manager tab (on top) or button.
7. Click on the small plus (+) sign beside Ports (COM & LPT).
8. Right-click on the communications port you need to configure.
9. Once the desired port is selected, click the left mouse button on the Properties option displayed in the pop-up menu. The Communication (COM) Properties window should appear. Figure 9.12 illustrates a typical window showing the resources used by COM1. The bps, data bits, parity, stop bits, and flow control are set using the small scroll buttons to the right of each text box.
10. Once changes are made, click on the OK button located at the bottom of the window for the changes to take effect.
11. If do not want the changes to take effect, Click on the Cancel button or on the small X in the upper right-hand corner.

---

USB peripherals usually mount outside the system unit and attach to the USB port on the back of the computer's chassis.

USB is rated at transmission speeds up to 12 Mega-bits-per-second (Mbps) and is designed to support peripherals such as modems, printers, microphones, speakers, graphics tablets, game controls, joysticks, scanners, monitors, and digital cameras just to name a few.

Intel and compatible chipsets have supported the USB for a couple of years. In fact, a single male pin connector for two USB ports has been on motherboards for the same amount of time. Some older motherboards do not supply the actual USB connectors that mount on the computer's back panel. If your motherboard has the ports and does not supply the connectors, you will have to buy them at your

**Figure 9.12** COM Properties Using Windows 98

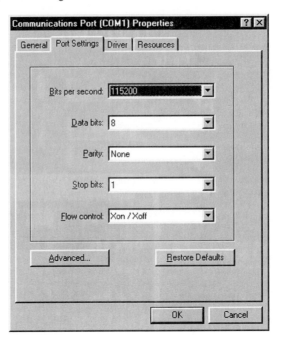

**Figure 9.13** Typical USB Setup

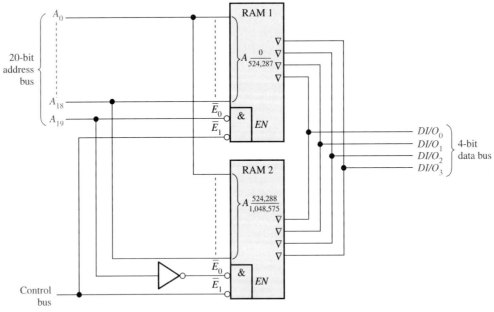

*Source:* Floyd, *Digital Fundamentals,* 7th edition (Upper Saddle River, NJ: Prentice Hall, 2000) p. 718.

Input/Output

Table 9.8   USB Pin-out

| Pin | Signal |
|---|---|
| 1 | + 5 VDC |
| 2 | Data Return |
| 3 | Data Signal |
| 4 | Ground |

---

**Procedure 9.6:   Displaying the resources used by the USB under Windows 98/2000 (USB is available only with Windows 95 OSR3)**

1. Click on the Start button.
2. Move the mouse pointer up to Settings and then over to Control Panel.
3. Click on Control Panel.
4. Locate and double-click on the System icon.
5. In Windows 2000, click on the Hardware tab.
6. Click on the Device Manager tab or button.
7. Click on the plus (+) sign beside the Universal Serial Controller.
8. Right-click on the listing that has "PCI to USB Universal Controller" as part of the heading.
9. Click on the Properties options.
10. Click on the Resources tab.
11. Once completed, close all open windows by clicking on the "X" in the upper right-hand corner or the Cancel button.

---

local computer parts store. A typical USB system setup, along with the connector, is shown in Figure 9.13. The USB pin-out is listed in Table 9.8. Procedure 9.6 lists the steps to follow to display the resources used by the USB under Windows 98/2000.

The USB port that is integrated on the motherboard is configured using the CMOS Setup program that was covered in chapter 8. The options available in the integrated peripheral section of the CMOS are listed in section 9.10.1.

## 9.3.8   IEEE 1394 FireWire

A new serial port that was originally developed by Apple Computer™ is on the horizon and is creating a lot of talk in the PC community. This serial port is referred to as the IEEE 1394 standard, more commonly known as **FireWire.** FireWire appears destined to move into the compatible PC community within the next few years.

Like USB, FireWire supports PnP and hot-swapping. FireWire can also connect up to 63 devices in either a daisy or treed arrangement. This connection is made with cables that are similar to phone cords. Unlike USB's 12 Mbps data transfer rate, FireWire currently supports blazing data transfer rates of 100 Mbps, 200 Mbps, and 400 Mbps.

Along with the higher transfer speeds, FireWire also supports asynchronous and **isochronous** data transfer timing schemes. The asynchronous transfer is similar to that used by the programmable serial port discussed previously. Isochronous is a method that sends data to the computer at a pre-set rate. This feature is required to stream audio and video information from a hard disk, CD-ROM, or DVD in real time.

Along with these features, FireWire allows devices to transfer information without even being connected to a computer. With all of these features and its high-speed transfer, it is hard to imagine that this port will not find its way into the compatible PCs.

**Table 9.9** Parallel Port Base Addresses and IRQ Assignments

| Name | Address | IRQ |
|------|---------|-----|
| LPT1 | 3BCH    | 7   |
| LPT2 | 378H    | 5   |
| LPT3 | 278H    | —   |

## 9.4 PARALLEL PORTS

The current operating systems used by the compatible computers can support up to three 8-bit parallel ports directly. The ports can be programmed to operate in one of three modes: Standard Parallel Port (SPP), Enhanced Parallel Port (EPP), or Enhanced Capabilities Port (ECP). Each of these modes will be discussed in detail in this section, but first we need to discuss the parallel port characteristics that apply to all three categories.

Most current compatible computers come with at least one parallel port as standard equipment. The compatible computer's OS uses the names LPT1, LPT2, and LPT3 for the parallel ports. If more than one parallel port is installed in the computer, each must be assigned a set of unique resources. The resource assignments can be either hardset, softset, or PnP, depending on the port and the operating system. Because of the way the operating system assigns LPT numbers, the resource assignments are usually made by address instead of by name. Table 9.9 lists the I/O address and IRQ assignment that correspond to each LPT name when all three parallel ports are installed.

The LPTs shown in Table 9.9 only hold true if all three parallel ports are installed in the computer. If only one parallel port exists, it will use LPT1 despite the address for which it is configured. If more than one parallel port exists, the ports will be assigned as LPT1, LPT2, or LPT3 according to the order in which they are found. The operating system looks for the printer port in the order listed in Table 9.9. For example, if one printer port, at address 278H, is installed in the computer, the operating system will use it as LPT1 (the first printer port). If a second port is added, configured for 378H, the operating system will used 378H as LPT1 and will automatically bump 278H to LPT2.

If a parallel port is simply used as a printer port, interrupts are not usually required. In any case, the printer's documentation should be consulted to determine the need for an IRQ. Because MS-DOS, Windows, and most application programs use the parallel ports as printer ports, leaving the interrupts disabled when installing a new parallel/printer port should cause no problems. If other peripherals such as a tape backup, scanner, or CD-ROM use the parallel port, consult the documents that accompany these devices to determine if an interrupt is required.

If the parallel port uses an interrupt on compatible computers above the XT, IRQ7 and IRQ5 are the IRQs assigned to the parallel ports (see Table 9.2). As a general rule, the port with the highest number address should be assigned IRQ7. For example, if two ports are installed at addresses 278H and 3BCH, 278H should be assigned IRQ5 and 3BCH should be assigned IRQ7. If the third parallel port is installed, it should be configured with the interrupt disabled.

To check the resources used by the parallel ports on a computer that runs Windows 95/98/2000, follow the steps outlined in Procedure 9.7.

### 9.4.1 Parallel Port Standards

The only standard that was used by the compatible computer's parallel port before 1994 was an open-technology standard established by the Centronics company. The Centronics standard only defined the connector to be used on the printer and the signals in the cable to support a printer port. The standard did not define whether the port was unidirectional (send only) or bi-directional (send and receive). It also did not define transfer speeds or specify which devices (other than a printer) could be connected to the port. In 1994, the **Institute of Electronic/Electrical Engineers (IEEE)** used the Centronics stan-

Input/Output

---

**Procedure 9.7: Display the resources used by the parallel ports**

1. Click on the Start button located on the Task Bar.
2. Move up to Settings.
3. Click on the Control Panel listing.
4. Once Control Panel opens, locate and then double-click on the System icon.
5. In Windows 2000, click on the Hardware tab.
6. Click on the Device Manager tab or button.
7. Click on the listing for "Port (COM & LPT)."
8. Right-click the listing for one of the LPT ports. Most computers show only show LPT1.
9. Click on the Properties listing.
10. Once completed, click on the "X" in the upper right-hand corner to close all open windows.

---

**Figure 9.14** Centronics-compatible Printer Connector

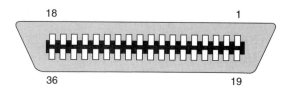

dard as a base for the IEEE 1284 standard, which defines five modes of operation for the compatible computer's parallel port. These modes are:

1. Computability mode (Centronics mode)
2. Nibble mode
3. Byte mode
4. EPP mode (Enhanced Parallel Port mode)
5. ECP mode (Extended Capabilities mode)

The IEEE 1284 standard gives manufacturers the ability to design peripherals and drivers so that devices other than printers can be connected to the parallel port and remain downwardly compatible with the standard parallel port (SPP). No matter which modes are available with a parallel port on your computer, they all use the same DB25S connector, which is located on the back panel of the computer. Figure 9.14 shows the connector's pin placement and numbering.

### 9.4.2 Standard Parallel Port (SPP)

The Standard Parallel Port mode (SPP) actually incorporates the first three modes specified by the IEEE 1284. These modes include the Compatibility mode (Centronics mode), Nibble mode, and Byte mode. To better understand the SPP modes, we need to first discuss a little printer history.

When printers were first attached to the microcomputer, there were no standards in place on how the printer should be connected. This led to each computer manufacturer developing its own connector and pin assignments for the printer port and printer. This caused a lot of confusion because only certain printers could be attached to certain computers.

Centronics, a printer manufacturer, developed an open-technology (anybody could use it) printer port to operate its line of printers. This open technology made it possible for computer manufacturers to use the Centronics printer port on their computers without paying royalties to Centronics. Needless to say, the Centronics printer port quickly gained wide acceptance in the microcomputer community.

With an open technology, other printer manufacturers started making their printers Centronics-compatible in order to connect to a microcomputer with a Centronics printer port. IBM, along with the compatible manufacturers, chose to adopt the Centronics printer port for its line of personal computers. The Centronics interface established the connector to be used on the printer side of the cable and the pin assignments for the data and hardware flow control signals for a unidirectional parallel port.

The connector developed by Centronics to plug into the printer goes by the same name and is still used on all printers that are designed to connect to the compatible's parallel port. The Centronics printer connector is shown in Figure 9.14. The pin-out for the connector is listed in Table 9.10.

Table 9.10  Pin-out of the Centronics Connector Located on the Printer

| Signal | Pin | Pin | Signal |
|---|---|---|---|
| Strobe# | 1 | 19 | Ground |
| Data Bit 0 | 2 | 20 | D0 Ground |
| Data Bit 1 | 3 | 21 | D1 Ground |
| Data Bit 2 | 4 | 22 | D2 Ground |
| Data Bit 3 | 5 | 23 | D3 Ground |
| Data Bit 4 | 6 | 24 | D4 Ground |
| Data Bit 5 | 7 | 25 | D5 Ground |
| Data Bit 6 | 8 | 26 | D6 Ground |
| Data Bit 7 | 9 | 27 | D7 Ground |
| Acknowledge | 10 | 28 | Ack Ground |
| Busy | 11 | 29 | Busy Ground |
| Paper Out | 12 | 30 | Reset Ground |
| Select | 13 | 31 | Reset |
| Auto Feed | 14 | 32 | Fault |
| Not Connect | 15 | 33 | Signal Ground |
| Logic Ground | 16 | 34 | Not Connected |
| Chassis Grd | 17 | 35 | +5 VDC |
| +5 V pullup | 18 | 36 | Select In |

Figure 9.15  IBM-compatible Parallel Port Connector

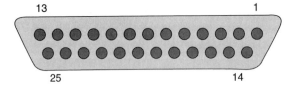

When IBM adopted the Centronics port, it decided to use a female DB25 connector on the back of the computer to attach the printer. The cable that makes the connection between the computer and the printer has a male DB25 on the end that plugs into the back of the computer and a male 36-pin Centronics connector that plugs into the printer. This cable is now simply referred to as a *parallel printer cable* and usually comes with a new printer or can be purchased from any retailer that sells computer supplies. Figure 9.15 illustrates the DB25S connector that is located on the back of the computer. Table 9.11 lists the pin assignments.

As shown in Table 9.11, the Centronics-compatible port consists of eight data lines, eight ground lines, and nine hardware flow control lines. Understanding the actual function of the hardware flow control signals is only necessary for an individual who designs peripherals around the port. Teaching de-

**Table 9.11**  Parallel Port Compatibility Mode (Centronics mode)

| Pin | Centronics | Nibble | Byte | EPP | ECP |
|---|---|---|---|---|---|
| 1 | Strobe | — | HostCLK | Write# | HostClk |
| 2–9 | D0-D8 | D0-D8 | D0-D8 | D0-D8 | D0-D8 |
| 10 | Ack | PrtClk | PrtClk | Intr | PeriphClk |
| 11 | Busy | PrtBusy | PrtBusy | Wait | PeriphAck |
| 12 | Paper Error | AckDataReq | ActDataReq | AckDataReq | AckReverse |
| 13 | Select | Xflag | Xflag | Xflag | Xflag |
| 14 | AutoFeed | HostBusy | HostBusy | Dstrb# | HostAck |
| 15 | Fault | DataAvail | DataAvail | DataAvail | PeriphRequest |
| 16 | Initialize | — | Int | RevRequest | — |
| 18–25 | Ground | Ground | Ground | Ground | Ground |

sign is not the intent of this book. For this reason, only the names of the hardware flow control signals are given in this table.

Even though the SPP uses hardware flow control lines, their state (1 or 0) is determined by software. The application or operating system using the port is responsible for controlling the state of the output flow control lines and reading the state of the input flow control lines. As you can imagine, this greatly affects the overall speed capabilities of the port.

The Compatibility mode, or Centronics mode as it is commonly called, uses the parallel port as the unidirectional printer port that was originally defined by Centronics. This mode can only send data from the port to the printer. All of the hardware flow control lines are controlled by software. The Compatibility mode can transfer data from the port to the printer at a rate up to 150 KB per second.

The Nibble and Byte modes are directly compatible with the Centronics mode but require a parallel port with bi-directional capabilities. Just like the Compatibility mode, these modes use software to control not only the data, but all of the hardware flow control signals as well. The data and hardware flow control names used by the Nibble and Byte modes are listed in Table 9.11, columns 3 and 4.

The Nibble mode allows bi-directional transfer of 4 data bits at a time. *Bi-directional* means that data can be transferred to and from the peripheral. This transmission can take place at a rate up to 15 KBps. No hardware modifications are required to incorporate this mode because it is accomplished through software, not hardware.

The Byte mode uses the eight data lines designated by the Centronics mode, but, similar to the Nibble mode, allows data to flow in both directions—from the peripheral to the port and vice versa. Just like the Nibble mode, no additional hardware is required because Byte mode is accomplished through software. The data transfer rate for Byte mode is the same as that for the Compatibility mode.

Unlike the COM ports, the parallel port(s) is not configured using Device Manager. The parallel port that is integrated into the motherboard is configured in the CMOS. A parallel port that is installed in an expansion slot is most often hardset on the card. The motherboard's integrated parallel port is configured using the CMOS Setup program, which was covered in chapter 8. The available options are listed in section 9.10.1.

### 9.4.3 Enhanced Parallel Port (EPP)

As stated previously, the Enhanced Parallel Port (EPP) uses the same DB25S connector that is used by the SPP. In fact, EPP is fully compatible with all of the SPP standards and can be used in all of the SPP modes. In the EPP mode, however, everything changes. First, the hardware flow control signals are controlled by hardware, not by software. Second, all of the hardware flow control pin-outs of the DB25S connector are reassigned to different signals. The names of EPP hardware flow control signals are given in column 5 of Table 9.11. Third, EPP mode allows addressing of up to 64 different devices that can be connected to the single DB25S connector by using a device called an *EPP hub*. EPP has a transfer rate of between 500 KB per second up to 2 MB per second.

The only peripherals that use all of the advantages offered by the EPP parallel port are EPP-compatible devices. The motherboard's integrated parallel port is configured for EPP mode using the CMOS Setup program, which was discussed in chapter 8. The available options, along with explanations, are given in section 9.10.

### 9.4.4 Enhanced Capabilities Port (ECP)

The Enhanced Capabilities Port (ECP) mode allows for bi-directional communication between the parallel port and peripherals. When the parallel port is configured for ECP mode, it is not downwardly compatible with the SPP or EPP modes. Similar to EPP, the hardware flow control signals are fully supported by hardware. The names of the ECP hardware flow control signals are given in column 6 of Table 9.11.

ECP offers some features that make it a desirable port to have. First, the ECP can be programmed to use one of the computer's DMA channels. Second, the ECP supports data compression with a ratio of up to 64 to 1. Third, like the EPP, ECP supports peripheral addressing, which means up to 128 ECP peripherals can be connected to the ECP. Fourth, the ECP uses a FIFO memory that allows the port itself (not the program) to complete data transfers.

The data transfer rate for ECP is the same as that for EPP when compression is not used. The only peripheral that will work on the parallel port that is programmed for ECP mode is an ECP-compatible device. The motherboard's integrated parallel port is configured for ECP mode using the CMOS Setup program. The available options are covered in section 9.10.

## 9.5 KEYBOARDS

One of the primary input devices on a compatible computer is the keyboard. There are three basic styles of keyboards used in compatible computers: the 87-key standard, the 101-key enhanced (extended), and the 104-key Windows 9X. A typical 104-key keyboard is shown in Figure 9.16. The 87-key will not be covered because it was the standard keyboard for the original IBM-PC that came out in 1981 and is no longer used with any of the new computers.

Of the three, the 101-key (enhanced or extended) keyboard is probably the most widely used. The Windows 9X 104-key keyboard is rapidly replacing it, however. These two keyboards are basically the same. The only real difference between them is three additional keys on the Windows 9X keyboard that are unique to Windows 9X. These three keys are located between the Alt and Ctrl keys. The two keys with the Microsoft symbol will open the Windows 9X Start menu. The additional key will open the Options menu for a highlighted (selected) item.

The layout of the letters and numbers on the enhanced keyboard is the same as that on a standard QWERTY typewriter keyboard, except the number keys 0 and 1 have been added. These number keys are not on a standard typewriter keyboard. The enhanced keyboard also has two Control keys (Ctrl), two Alternate keys (Alt), and one Escape key (Esc). The function of each of these keys is defined by the application that runs the computer. There are also two Shift keys that, when pressed in conjunction with another key, give the user access to shifted letters and the punctuation above the top-row number keys. Other keys on the keyboard include the two Enter keys (Enter) that tell the computer to accept the information previously typed on keyboard, and a single Tab key (Tab) that moves the cursor a predefined number of character positions, depending on the application program.

A small keypad that serves two functions is located on the right of the keyboard. One function is that of a standard desktop calculator keypad. The second function is that of an edit keypad that is used to position the cursor and insert and delete characters. One of the keys on this small keypad is labeled *NumLock*. This key is used to switch the small keypad between its calculator and edit functions. You

**Figure 9.16** 104-key Keyboard

**Figure 9.17** DIN5 and Mini-DIN6 Pin Locators (Note: The Mini-DIN is smaller than shown)

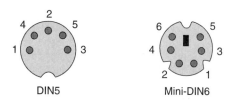

cannot use the keypad as an edit and calculator keypad at the same time. When the computer is initially turned on, the keypad usually defaults to the calculator function. Some CMOS setups allow the user to define the default states of this keypad when the computer is booted.

Between the main letter/number keys and the calculator/edit keypad is a separate edit keypad. This keypad is used to position the cursor and delete and insert characters in most applications that use text. The Delete key can also be used in Windows to delete the highlighted item.

The 12 horizontal keys across the top of keyboard that are labeled F1 through F12 are called **function keys.** The interpretation of these keys is **software defined,** which means the keys will do different things in different programs. Consult the user's reference manual for each software package for the functions of these keys.

The keyboard port is located on the compatible computer's motherboard. The keyboard attaches to the connector on the motherboard using either a male DIN5 or Mini-DIN6 connector. Adapters are available at most computer supply stores that will convert between the two. The DIN5 and Mini-DIN6, with pin locations, is shown in Figure 9.17 and the signal assigned to each pin is listed in Table 9.12.

Referring back to Table 9.1, we find the keyboard port is assigned I/O addresses 060H and 064H. Table 9.2 indicates that the interrupt assigned to the keyboard is IRQ1. This is simply nice-to-know information because the user cannot reassign the keyboard's address and interrupt. In fact, IRQ1 is not available on the expansion bus.

The keyboard can be configured in both CMOS and Windows 9X/2000. The CMOS setting was given in chapter 8, while the steps used to configure the keyboard in Windows 9X/2000 are given in Procedure 9.8.

**Table 9.12** Keyboard Pin Assignments for the DIN-5 Keyboard Connector

| Pin | DIN5 | Mini-DIN6 |
|---|---|---|
| 1 | Keyboard clock | Keyboard Data |
| 2 | Keyboard data | Not Connected |
| 3 | Keyboard reset | Ground |
| 4 | Ground | + 5 VDC |
| 5 | + 5 V | Keyboard Clock |
| 6 | No Pin | Not Connected |

---

**Procedure 9.8: Configuring the keyboard using Windows 9X/2000**

1. Click on the Start Button on the Task Bar.
2. Move the mouse pointer up to Settings and then over to Control Panel.
3. Click on Control Panel.
4. After the Control Panel window opens, locate and double-click on the Keyboard icon.
5. After making the desired settings, click on the OK button to accept the changes or the Cancel button to abort without making changes.

---

## 9.6 POINTING DEVICES (MICE AND TRACKBALLS)

The mouse and trackball are the primary input devices used by the shells of Windows 3.X, Windows 9X, and Windows 2000. Figure 9.18 shows a two-button mouse; Figure 9.19 shows a trackball. It should be obvious that the mouse gets its name from its physical appearance. The trackball gets its name from the fact that the pointer on the screen tracks as the large ball is moved with the hand. Even though the two pointing devices look somewhat different, both operate on the same principle.

The operator moves the large ball on the top of a trackball, which turns two small paddle wheels that have an optical sensor placed on each side. As the ball moves, the paddle wheels turn, breaking the light beam and creating digital pulses. The mouse works on exactly the same principle except the trackball is underneath the mouse. As the mouse is moved across a flat surface, the trackball underneath the mouse turns, and the rest is the same. Figure 9.20 shows an inside look at a typical mouse.

All mice fall into one of two categories: serial or PS/2. A serial mouse connects to one of the computer's programmable serial ports, while a PS/2 mouse uses a dedicated port. The advantage of using a PS/2 mouse port is that it does not use any of the resources assigned to the computer's programmable serial ports. Similar to the serial ports, the PS/2 mouse requires the use of a range of I/O addresses and an interrupt. The address and IRQ assignments for the COM and PS/2 mouse ports are listed in Tables 9.1 and 9.2, respectively.

A pointing device that is designed to connect to a serial port uses either a female DB9S or DB25S connector. A pointing device that is designed to connect to a PS/2 mouse port has a male Mini-DIN 6 connector. Like the keyboard, adapters are available at computer supply stores that can adapt one connector to the other. The pin-out for the COM port is listed in Table 9.5. The pin-out for the PS/2 mouse port is shown in Figure 9.21, along with its connector.

Because MS-DOS does not support a mouse directly, a TSR must be loaded in order for a DOS application program to use the mouse. This is also true if a DOS application runs in Windows 9X DOS mode. When a new mouse is purchased, the DOS mouse driver is usually supplied with the mouse along with installation instructions.

**Figure 9.18**  Two-button Mouse

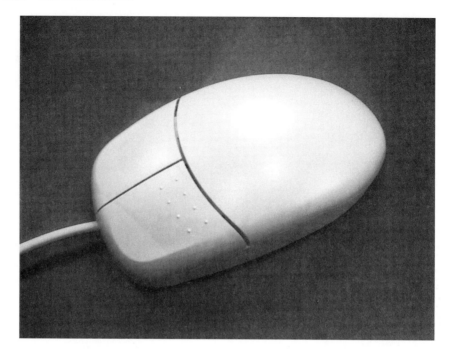

**Figure 9.19**  Trackball

Most pointing devices can be purchased with two or three buttons located at the top edge or sides of the device. The buttons are visible on the mouse and trackball shown in Figures 9.18 and 9.19. These buttons are easily pressed (clicked) with the fingertips to select different options. The functions of the left and right buttons on the pointing device when using Windows 3X/9X/2000 were covered in chapters 5 and 6. If a center button is available, its function varies depending on the

**Figure 9.20** Inside a Typical Mouse

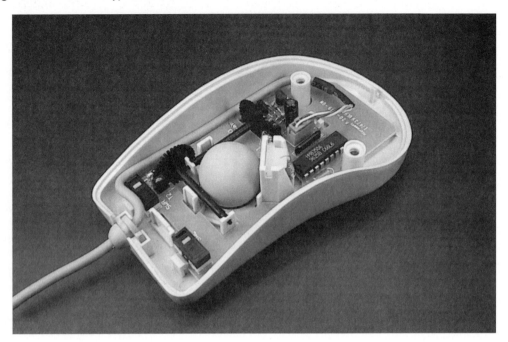

**Figure 9.21** PS/2 Mouse Port Connector and Pin Assignments

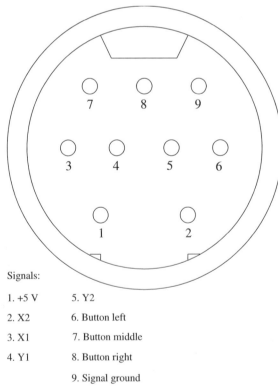

Signals:

1. +5 V
2. X2
3. X1
4. Y1
5. Y2
6. Button left
7. Button middle
8. Button right
9. Signal ground

*Source:* Dan L. Beeson, *Assembling and Repairing Personal Computers* (Upper Saddle River, NJ: Prentice Hall, 1997) p. 220.

Input/Output 313

---

> **Procedure 9.9: Setting the pointing device's user-defined options in Windows 9X/2000**
>
> 1. Click on the Start button on the Task Bar.
> 2. Move the mouse pointer up to Settings and then over to Control Panel.
> 3. Click on Control Panel.
> 4. Locate and then double-click on the Mouse icon.
> 5. Configure the pointing devices.
> 6. To accept the changes, click on the OK button. To abort without changing the current configuration, click on the X in the upper right-hand corner of the window or click on the Cancel button.

---

> **Procedure 9.10: Cleaning the pointing devices**
>
> 1. Shut down the computer using the correct procedures.
> 2. Remove the access retainer ring that surrounds the ball by rotating it in the direction indicated on the retainer ring or in the pointing device's documentation. Store the retainer ring in a safe location.
> 3. Once opened, position the pointing devices so that the ball falls into the hand.
> 4. Once removed, clean the ball using warm, soapy water. Once cleaned, dry the ball thoroughly with a lint-free towel. Store the ball in a safe location.
> 5. Use a can of pressurized air that is designated for computer use to blow out the cavity from which the ball was removed. Cans of pressurized air are available at most stores that sell computer supplies.
> 6. Using isopropyl alcohol and a cotton swab, clean the rollers inside the mouse cavity and allow to air-dry thoroughly.
> 7. Replace the ball and retainer ring.
> 8. Apply power to the computer and check the pointing devices for proper options.

---

application and the pointing device. For this reason, always consult the documentation for the application program or pointing device.

There are several features for the pointing device that can be configured by the user under Windows 9X/2000. The actual features offered depend on the pointing devices used. To configure the user-defined pointing device options, follow the steps outlined in Procedure 9.9.

### 9.6.1 Pointing Device Maintenance

Most pointing devices such as mice and trackballs are fairly inexpensive devices that are simply replaced when they fail. However, they do need to be cleaned occasionally because dirt and grime transferred from the ball build up on the motion-sensing device surfaces inside the mouse. This is usually indicated when the pointer being controlled by the mouse or trackball does not respond in one or both directions or responds with a jerky motion. The steps to follow to clean a typical pointing device are covered in Procedure 9.10.

## 9.7 SOUND

You have probably heard the term *multimedia* a few times over the last couple of years. More and more computers and applications are moving toward multimedia to present information. A **multimedia** application conveys information to the user using more than one media of delivery. The more forms of media by which a human can receive information, the more likely the information will be communicated. A few examples of multimedia are printed outputs, visual outputs, modems, and sound. Sound is becoming one of the most common forms of media used by applications.

> **Procedure 9.11: Determining the resources used by the sound card using Windows 9X/2000**
>
> 1. Click on the Start button and then move up to Settings and over to Control Panel.
> 2. Click on Control Panel.
> 3. After the Control Panel window opens, locate and then double-click on the System icon.
> 4. With Windows 2000, click on the Hardware tab.
> 5. Click on the Device Manager tab or button.
> 6. Click on the plus (+) sign beside the category labeled "Sound, video, and game controllers."
> 7. Locate and then double-click on the Gameport Joystick listing.
> 8. Click on the Resource tab for the resource listing.
> 9. Write down the address, DMA, and IRQ assignments.
> 10. Close all open windows by clicking on the Cancel button or on the X in the upper right-hand corner.

The small internal speaker that is standard equipment on all compatible computers does not have the ability to meet the high demand for quality sound. In addition, the hardware is such that the software would have to be very complex to create even the simplest of sound effects. For this reason, special adapter cards called *sound cards* meet the demand for quality sound.

Currently, the most widely used sound cards are 16 bit, although there are 8-bit cards available. Most current sound cards are designed around the PCI expansion bus because of its advanced support of PnP technology. No matter which expansion slot is used, the 16-bit sound card provides excellent audio reproduction in the form of sound effects, voice, and music. Most also have the ability to input and record sounds through a **musical instrument digital interface (MIDI,** pronounced "med-ie"). MIDI-compatible musical instruments that connect to the sound card's MIDI port can be purchased.

Both 8-bit and 16-bit sound cards use interrupts and DMA to transfer information with the application. Sound cards were not around when the AT came out, so they are not part of the original I/O address map. Not being part of the original I/O address map means that a sound card has to fight with other devices that were not part of the original I/O address map for available addresses, interrupts, and DMA channels.

To help alleviate the problem, most hardset 16-bit sound cards allow the ports (sound and MIDI) to be configured to several different addresses. For example, the Sound Blaster-16, manufactured by Creative Labs, Inc., can be configured to set the sound port to address 220H*, 240H, 260H, or 280H. The MIDI port can be configured to address 330H* or 300H. The sound port can also be configured to use IRQ2, IRQ5*, IRQ7, or IRQ10 interrupt signals. The 8-bit DMA channel can be configured for DMA0, DMA1, or DMA3. The 16-bit DMA can be configured for DMA5, DMA6, or DMA7.

The configuration of a hardset sound card is usually a hit-or-miss process that involves leaving the address at the defaults (indicated by the asterisk in the preceding paragraph) and trying it to see if it works. If it does not, try another address, interrupt, or DMA channel until you find a combination that works. Once you have the correct combination of addresses, interrupt, and DMA that seems to work, write it down because some applications that use the sound card will ask for this information.

PnP has alleviated a lot of stress in the configuration of the sound card's resources. The only major problem is that PnP may assign the card resources that some DOS applications may not be able to use. Also, because DOS applications will not check the PnP configuration information of Windows 9X/2000, the resources used by the sound card will have to be entered manually into the configuration settings of DOS applications. For this reason, the PnP resources assigned to the sound card should be documented. To find this information using Windows 9X/2000, follow the algorithm listed in Procedure 9.11.

**Figure 9.22** Game Port

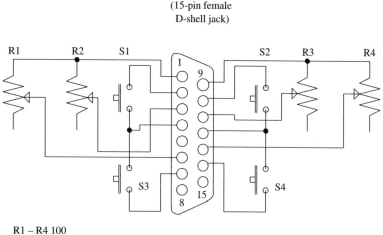

R1 – R4 100
S1 – S4 N.O. PB switches
Pins 1, 8, 9, and 15 = +5 V
Pins 4, 5, and 12 = GND
Joystick has male 15-pin D-shell plug

*Source:* Dan L. Beeson, *Assembling and Repairing Personal Computers* (Upper Saddle River, NJ: Prentice Hall, 1997) p. 201.

Just like other enhancements that have been added to the compatible since the IBM-AT, there are no set standards for the reproduction of sound effects, voice, or music. In this situation, a particular company's product usually becomes the standard that everyone else copies. Creative Labs and its line of *Sound Blaster* sound cards is the defacto standard when it comes to sound card adapters. Most applications that support sound usually have a driver for a Sound Blaster card. When you purchase a computer system with sound capabilities, or you purchase a sound card to install yourself, make sure the card is a Sound Blaster or is **sound blaster-compatible.**

## 9.8 JOYSTICKS AND GAME PADS

A *joystick* and *game pad* are input devices that connect to the computer's *game port* (if one is installed). The game port was once a part of the standard set of I/O ports located on the MI/O card or integrated on the motherboard. Now, the game port is most often found on the sound card adapter.

If your computer does not have a game port, one can be purchased for less than $10 at a computer parts store. Adding a game port does not require any address or interrupt configuration changes because the game port uses fixed resources. Simply plug the card into an open 8- or 16-bit slot in the expansion bus and it is ready to go.

Although a single joystick or game pad is usually used, the standard game port is configured to accept two. If the card does not have two inputs, a special Y-connector cable can be used to connect two standard, PC-compatible joysticks to one port.

The game port connector in Figure 9.22 shows how the joystick and its controls are electrically connected. The pin-out of the game port is listed in Table 9.13. The game port is the DB15 female connector on the back of the computer. (The Mini DB15S [small] female 15-pin is for video.) The procedure for configuring the controller under Windows 9X/2000 is given in Procedure 9.12. The game port is located at I/O addresses 200H through 20FH.

**Table 9.13** Game Port Pin-out

| Pin | Signal |
|---|---|
| 1 | +5 VDC |
| 2 | Button 1 |
| 3 | Joystick one X position control |
| 4 | Ground |
| 5 | Ground |
| 6 | Joystick one Y position control |
| 7 | Button 2 |
| 8 | + 5 VDC |
| 9 | + 5 VDC |
| 10 | Button 4 |
| 11 | Joystick two X position control |
| 12 | Ground |
| 13 | Joystick two Y position control |
| 14 | Button 3 |
| 15 | + 5 VDC |

**Procedure 9.12: Setting up the game controller with Windows 9X/2000**

1. Click on the Start Button on the Task Bar.
2. Move the mouse pointer up to Settings and then over to Control Panel.
3. Click on the Control Panel icon.
4. After the Control Panel window opens, locate and double-click on the Game Controller icon.
5. Once the desired settings have been made, click on the OK button to place the changes into effect or the Cancel button to abort without making changes.

## 9.9 CONFIGURING A HARDSET MI/O CARD

One of the most popular I/O cards used in 286-, 386-, and early 486-compatible computers is called an *MI/O card* (or simply an I/O card). The MI/O adapter card integrates an IDE hard drive interface (HDC), floppy drive controller (FDC), two-serial ports (PORTA and PORTB), one-parallel/printer port, and a game port on a single adapter card. With all of these functions, this card costs less than $20. A hypothetical MI/O adapter card is shown in Figure 9.23.

When the computer system is purchased, the installed I/O card will be preconfigured to operate in the computer. If no other adapters are installed later, there should be no need to change the default settings.

If other adapters such as internal modems, additional serial or parallel ports, or other IDE or floppy adapters are installed, the MI/O card will probably have to be reconfigured. Also, if any of these items are installed in the computer and the MI/O adapter card has to be replaced, the new MI/O card will usually have to be reconfigured to operate in the computer.

# Input/Output

**Figure 9.23** MI/O Card

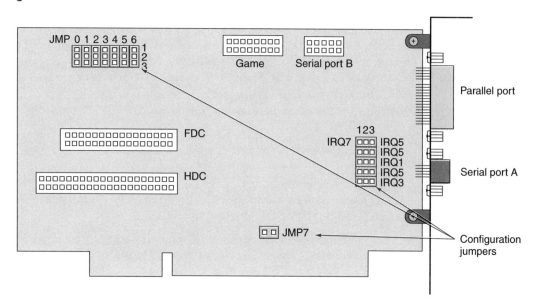

**Figure 9.24** MI/O Card Setup Sheet

**JUMPER SETTINGS**   Note **Denotes default setting

**FDC**
1-2 = Disable
2-3 = Enable**

**PRN (Printer)**

| JMP1 | JMP2 | |
|---|---|---|
| 2-3 | 1-2** | PRN1 |
| 2-3 | 2-3 | PRN3 |
| 1-2 | 1-2 | Disable |
| 1-2 | 2-3 | PRN2 |

**COM (Serial) Port A and PortB**

| JMP3 | JMP4 | JMP5 | Port A | Port B |
|---|---|---|---|---|
| 2-3 | 2-3 | 2-3 | COM2 | COM1 |
| 2-3 | 2-3 | 1-2** | COM1 | COM2 |
| 2-3 | 1-2 | 2-3 | COM2 | Disable |
| 2-3 | 1-2 | 1-2 | COM1 | Disable |
| 1-2 | 2-3 | 2-3 | Disable COM1 | |
| 1-2 | 2-3 | 1-2 | Disable COM2 | |
| 1-2 | 1-2 | 2-3 | COM3 | COM4 |
| 1-2 | 1-2 | 1-2 | Disable disable | |

**IDE JMP6**
1-2   Disable
2-3   Enable**

Port A's IRQ is set by shorting the 1-2 side of the IRQ jumper block
Port B's IRQ is set by shorting the 2-3 side of the IRQ jumper block

Game (JMP7)   Short to enable game**
              Open to disable game

The act of reconfiguring an MI/O card usually involves only installing or removing jumpers or changing the state of switches on a DIP-switch. The problem is not the act of reconfiguring, but reading the charts that tell you how. Figure 9.24 shows a typical chart that shows how to reconfigure a hypothetical MI/O card.

The card illustrated in Figure 9.23 uses a set of six 3-post jumpers (JMP0 through JMP6) to set the configuration of the hard drive, serial port, parallel port, and floppy disk controller. The card uses a set of five 3-post jumpers to set the IRQ for the parallel port and the serial ports. A single 2-post jumper (JMP7) is used to enable or disable the game port. Notice that the posts on the 3-post jumpers are numbered 1-2-3. In the charts that show how to position the jumper, if 1-2 is written under a jumper number, it indicates that a jumper should be placed between post numbers 1 and 2. On the other hand, if 2-3 is written under a jumper number, it indicates that a jumper should be placed between post numbers 2 and 3.

Let us work some problem examples on configuring the hypothetical MI/O card illustrated in Figure 9.23 and the data in Figure 9.24. Problems 9.3 through 9.6 refer to one or both of these figures.

**Problem 9.3:** In what position must the jumpers be placed on JMP1 and JMP2 to address the I/O card's printer port 2 (PRN2)?

**Solution:** A jumper should be placed between post 1 and post 2 on JMP1 and between post 2 and post 3 on JMP2.

**Problem 9.4:** In what position must the jumpers be placed on JMP3, JMP4, and JMP5 to address the I/O card's serial port A for COM1 and serial port B for COM2?

**Solution:** The jumper on JMP3 should be placed on posts 2 and 3. The jumper on JMP4 should be placed between posts 2 and 3. The jumper on JMP5 should be placed on posts 1 and 2.

**Problem 9.5:** In what position should the jumper be placed for the I/O card's serial port A to use IRQ4?

**Solution:** Positioning a jumper on the left side of the IRQ jumper block sets serial port A's IRQ. The left side of the jumper corresponds to posts 1 and 2. The desired interrupt is IRQ4, so a jumper will be placed between posts 1 and 2 next to IRQ4.

**Problem 9.6:** In what position should each jumper be placed to configure the AT-I/O card illustrated in Figure 9.24 to the following:

    Printer = PRN1–IRQ7
    Serial Port A = COM3–IRQ4
    Serial Port B = COM4–IRQ3
    IDE hard drive enabled
    Game port enabled

**Solution:** Using the charts in Figure 9.24, place the jumpers as follows: JMP1=2-3; JMP2=1-2; JMP3=1-2; JMP4=1-2; JMP5=2-3; JMP6=2-3; JMP7=Shorted (Jumped); PRN IRQ=1-2; PORTA IRQ4=1-2; PORTB IRQ3=2-3

## 9.10 CONFIGURING THE SOFTSET INTEGRATED PERIPHERAL PORTS

Motherboards currently have the standard I/O ports installed on the motherboard itself. Believe it or not, it is actually cheaper for manufacturers to do it this way and reduce the number of expansion slots and adapter cards. The term used to identify a motherboard with the standard ports installed is "With I/O."

The standard I/O ports that are integrated on most motherboards with I/O are two EIDE (hard drive), one floppy drive, one parallel port, and two programmable serial ports. Newer motherboards also have two USB ports, a PS/2 mouse port, and maybe an IEEE 1394 FireWire. All of the integrated ports are usually softset and configured using the computer's CMOS Setup program, which was discussed in chapter 8. The options and explanations from chapter 8 are simply repeated in the next section.

### 9.10.1 Typical Integrated Peripherals CMOS Options

This section repeats the information contained in the "Integrated Peripherals" section of chapter 8. Configuration options on the computer I am currently using are listed. The titles of these listings may vary from computer to computer, but they should be basically the same. For more detailed information, refer to chapter 8.

# Input/Output

**IDE HDD Block Mode:** Set to enable to automatically detect if the hard drives installed in the computer support block transfer; otherwise, disable.

**IDE Primary/Secondary Master/Slave PIO:** Use to select the PIO mode used by the BIOS to transfer information with the IDE device attached to the primary and secondary. This should be set according to the IDE devices' documentation. If the Auto option is selected, the BIOS will automatically determine the best mode. PIO is covered in chapter 11.

**IDE Primary/Secondary Master/Slave UDMA:** Select Auto only if the installed hard drive supports UDMA/33 and the OS has the required drivers for UDMA. UDMA is discussed in chapter 11.

**On-Chip Primary/Secondary PCI IDE:** Use to enable or disable the motherboard integrated primary and secondary IDE ports. Disable this feature only if an IDE adapter is used instead of the onboard IDE.

**USB Keyboard Support:** Enable this option if a USB keyboard is attached; otherwise, disable.

**USB Latency Timer (PCI CLK):** Sets the number of PCI clock cycles the USB controller maintains control of the PCI bus.

**Onboard FDC Controller:** Enables or disables the motherboard integrated floppy disk controller. Disable this option only if an FDC adapter is used or no floppy drives are installed in the computer. FDCs are discussed in chapter 10.

**Onboard UART 1 / 2:** Selects the address and interrupt used by the two integrated serial ports. This option can also be used to disable the motherboard's integrated serial ports.

**IR Duplex Mode:** Select either full duplex (two directions at the same time) or half duplex (two directions but not at the same time) used by an infrared device connected to one of the serial ports. Set this to Disabled if an IR device is not being used.

**Use IR Pins:** Selects the pins used to transmit and receive IR information. Set this according to the documentation supplied with the IR device.

**Onboard Parallel Port:** Selects the base port address and interrupt used by the motherboard's integrated parallel. This option can also be used to disable the integrated parallel port.

**Parallel Port Mode:** Selects either the EPP or ECP modes for the parallel port.

**ECP Mode Use DMA:** If the ECP mode is selected, this option selects the DMA channel to use.

## 9.11 FINDING OUT WHICH RESOURCES ARE IN USE

Before you install a new adapter card in your computer, you need to know the I/O resources already in use. If your OS is MS-DOS 6.0 or above or Windows 3.1 or above, Microsoft Diagnostics (MSD) can be used. If your OS is Windows 9X, the Device Manager is most often used. If the OS is Windows 2000, System Information is usually used. All of the methods will be discussed in the following sections.

### 9.11.1 Microsoft Diagnostics (MSD)

Microsoft Diagnostics is not really a diagnostic program, but rather an audit program. It displays basic information about the resources in your computer. Example 9.2 shows an MS-DOS command line to run MSD. Once the program is started, a menu similar to the one shown in Figure 9.25 will be displayed.

**Example 9.2:** MS-DOS command line example

```
C:\WINDOWS>MSD
```

To display the information represented by each item in the Main menu, press the key on the keyboard that corresponds to the highlighted letter in each menu listing. For example, to display the LPT Port information, press the letter *L* on the keyboard. This procedure will cause a window similar to the one shown in Figure 9.26 to appear. The displayed information will depend on the number of LPT ports installed in the computer. According to Figure 9.26, this computer has two parallel ports: LTP1 at address 3F8H (Hex) and 278H.

**Figure 9.25** MSD Main Menu

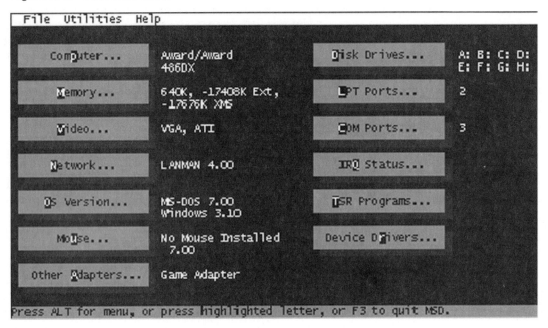

**Figure 9.26** MSD Showing LPT Resources

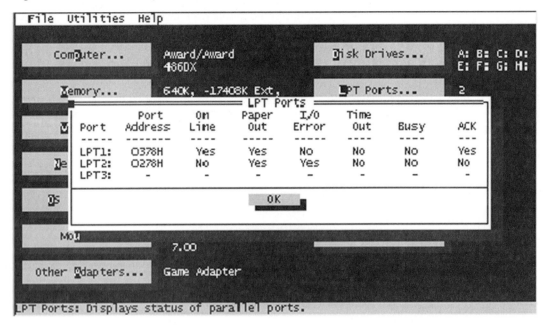

To display the information on the serial ports, press the C key on the keyboard when the Main menu is displayed. This will cause a window similar to the one shown in Figure 9.27 to appear. This computer has three serial ports installed at COM1 (03F8H), COM2 (02F8H), and COM3 (3E8). This option is also used to identify the UART used by each COM port. In Figure 9.27, COM1 and COM2 use 8250, while COM3 uses a 16550.

To get accurate information, MSD must be executed at a true MS-DOS prompt. If using MSD with Windows 9X, shut down to DOS mode using the Shut Down option in the Start menu. If running Windows 2000, the computer will have to be booted with a DOS diskette with MSD on the diskette. To my

Input/Output

**Figure 9.27** Using MSD to Display COM Port Resources

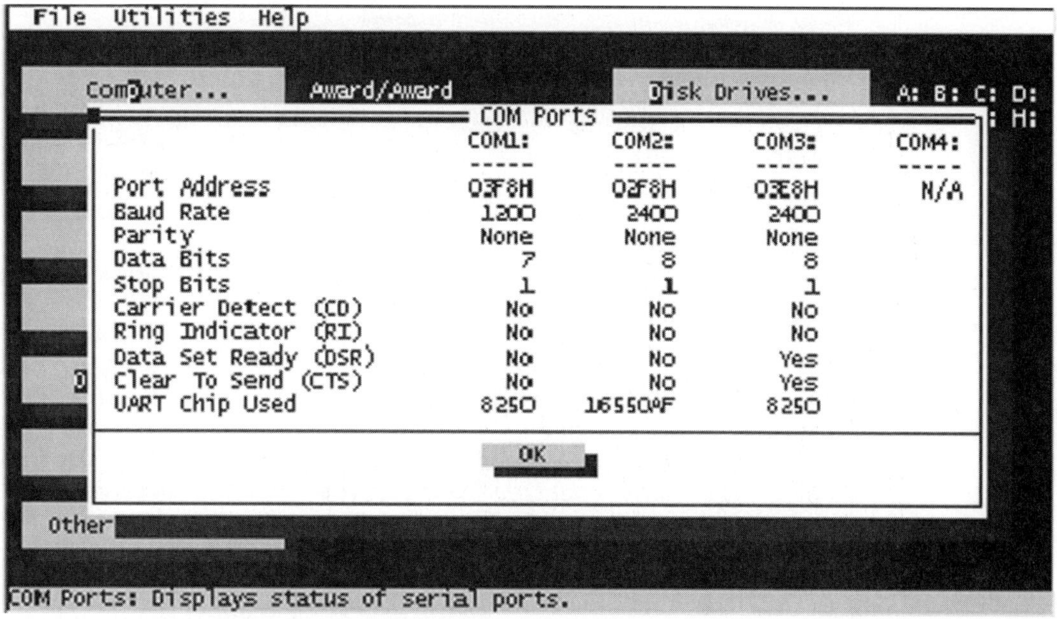

```
File  Utilities  Help

   Computer...        Award/Award           Disk Drives...      A: B: C: D:
   ╔══════════════════════ COM Ports ══════════════════════╗   H:
   ║                       COM1:    COM2:    COM3:    COM4: ║
   ║                       -----    -----    -----    ----- ║
   ║  Port Address         03F8H    02F8H    03E8H     N/A  ║
   ║  Baud Rate            1200     2400     2400           ║
   ║  Parity               None     None     None           ║
   ║  Data Bits            7        8        8              ║
   ║  Stop Bits            1        1        1              ║
   ║  Carrier Detect (CD)  No       No       No             ║
   ║  Ring Indicator (RI)  No       No       No             ║
   ║  Data Set Ready (DSR) No       No       Yes            ║
   ║  Clear To Send  (CTS) No       No       Yes            ║
   ║  UART Chip Used       8250     16550AF  8250           ║
   ║                                                        ║
   ║                           [  OK  ]                     ║
   ╚════════════════════════════════════════════════════════╝

 Other

COM Ports: Displays status of serial ports.
```

---

### Procedure 9.13: Opening Device Manager

1. Click on the Start button.
2. Move up to Settings and then over to Control Panel.
3. Click on Control Panel. A window similar to that shown in Figure 9.28 will display.
4. Locate and then click on the System icon.
5. On Windows 2000, click on the Hardware tab.
6. Click on the Device Manager tab on Windows 9X or on the button on Windows 2000. A window similar to that shown in Figure 9.29 will open.

---

knowledge, you cannot harm anything using MSD. Explore all of the Main menu options and see all of the information about your computer that you can discover.

### 9.11.2 Windows 95/98/2000's Device Manager

Windows Device Manager is probably the tool that is most often used by the computer technician. This should be evident by the number of procedures it has been used in up until this point. Following is a list of some of the tools available though Device Manager:

- Displays the resources used by each port;
- Changes or updates drivers;
- Indicates devices whose resources or driver are incompatible;
- Indicates devices that do not have a driver;
- Allows configuration settings on certain devices; and
- Provides basic diagnostic tests on certain devices.

Device Manager is available with Windows 95, 98, NT, and 2000. The procedure used to open Device Manager is basically the same for all the operating systems and is given in Procedure 9.13.

As shown in Figure 9.29, Device Manager displays all of the hardware categories that are installed in your computer. A click on the small plus (+) sign beside a category will display the ports or peripherals contained in the category. Two important categories are System Devices and Computer. The System

**Figure 9.28** Windows 98 Control Panel

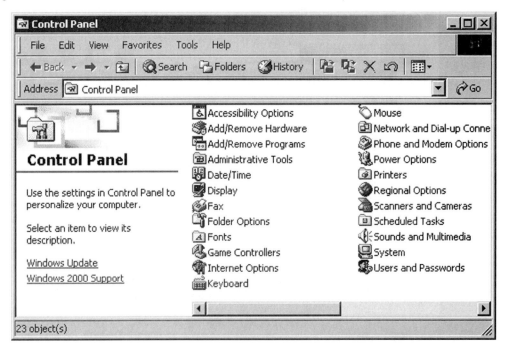

**Figure 9.29** Windows 98 Device Manager

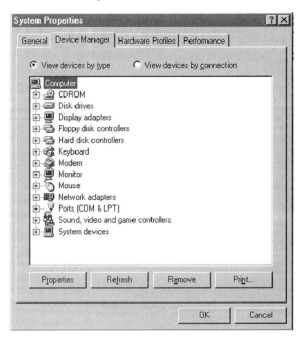

Devices category (usually located toward the bottom of the list) displays all of the system devices such as PIC, DMA, PIC, chipset, and internal speaker. An example of this category is shown in Figure 9.30.

Under Windows 9X, the Computer category will display a compilation of the resources used by all of the ports and the system RAM installed in the computer. Under Windows 2000, all of this information is moved to a utility called System Information.

If you are using Windows 9X, open up the Computer category by double-clicking on the Computer icon. The Computer Properties window, similar to the one shown in Figure 9.31, will display. From this

**Figure 9.30** System Devices in Windows 98

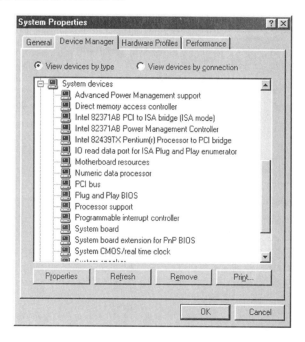

**Figure 9.31** Device Manager Computer Properties

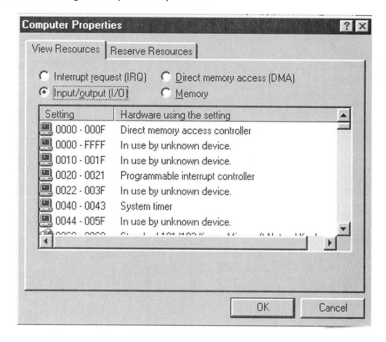

window you can display currently assigned IRQs, DMAs, I/O addresses and memory by simply clicking the left mouse button on the individual option buttons (small circles) that are beside the resource of interest. Figure 9.32 displays the IRQ assignments.

If you are using Windows 2000, the information displayed in the Computer category has been moved to System Information. To open this listing, follow the algorithm given in Procedure 9.14. Note that Windows 98 also contains the System Information option.

**Figure 9.32** IRQ Assignments

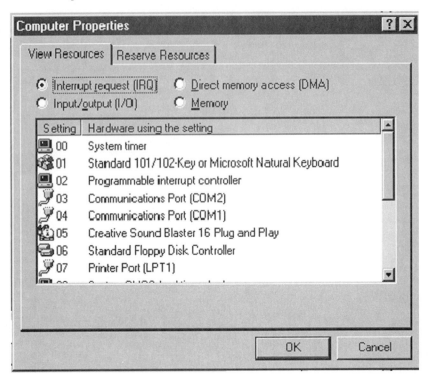

---

**Procedure 9.14:   Opening Windows 2000 System Information**

1. Click on the Start button.
2. Move up to Programs, over to Accessories, over to System Tool, and then click on System Information. A window similar to the one shown in Figure 9.33 should open.
3. Click on the plus (+) sign beside the Hardware Resources listing.

---

In addition to displaying the resources used by the ports installed in the computer, Device Manager can also be used to detect resource conflicts and faulty driver problems rapidly. If the exclamation point (!) or question mark (?) is displayed by a hardware category, it indicates the I/O device is not working at all or is not working correctly.

The exclamation point usually indicates that there is a resource conflict between two ports or that a port is using a defective driver. If the driver is the problem, the driver will have to be reinstalled by right-clicking on the problematic device listing, clicking on the Properties option, and then clicking on the "Driver" tab.

If the problem is a resource conflict, it can usually be corrected by changing the conflicting resource(s) used by one of the ports. If the port is hardset, the adapter will have to be removed from the expansion slot and reconfigured according to the documents provided with the adapter. If the port is one of the motherboard's softset ports, its resources will have to be configured using the CMOS Setup program. If the conflicting port is PnP, its resources can be reconfigured using Device Manager. This process is covered in the next section.

The question mark (?) identifies a hardware category called *Other Devices*. Windows will automatically create this category when it does not have a driver installed for PnP devices. This can occur during the PnP detection phase when Windows 9X/2000 is installed and it cannot find a driver for a device on the Windows installation CD. The category is also added if a new PnP device is installed in the computer and the user, for various reasons, cancels the installation of its driver. To correct this problem, follow the steps outlined in Procedure 9.15. Before you perform the procedure, you must have the correct

Input/Output    325

**Figure 9.33**   Displaying Port Resources Using Windows 2000

---

### Procedure 9.15:   Other Devices Category

1. Click the left mouse button on the small plus (+) sign beside the Other Devices category.
2. Click the left mouse button once on a device that appears within the category.
3. Click the left mouse button on the Remove button at the bottom of the Device Manager window.
4. Repeat steps 2 and 3 for each device in the Other Devices category.
5. Click on the Refresh button at the bottom of the Device Manager window.
6. If the removed PnP device is not detected when the Refresh button is clicked on, choose the Shut Down and Restart computer option from the Shut Down option on the Task Bar.
7. Follow the instructions on the screen as Windows 95 re-detects the PnP devices.
8. If the driver is installed correctly, Windows 9X/2000 should now place the PnP devices in the correct category.

---

driver(s) for the PnP adapter(s) either on a CD-ROM or a diskette. The Other Devices category is shown in Figure 9.34.

### 9.11.3   Using Windows 95's Device Manager to Softset PnP Ports

Window 95's Device Manager can be used to softset PnP cards. This sometimes becomes necessary if Windows 9X/2000 assigns resources to the PnP adapter that Windows 3.X and MS-DOS application cannot find or use. Device Manager can also be used to change the default resources that Windows 98/2000 assigns to the port if the resources have been changed. The steps to follow are outlined in Procedure 9.16.

There are a few very important things I should mention before Producedure 9.16 is used. First, do not attempt to change PnP or default port configurations if all devices are working correctly. Second, before you change resources, make sure the resources you are planning to assign are not already in use. Third, make sure the resources you are planning to assign to the port are compatible with the software that will use the port. Fourth, before you make any changes, document the port's original configuration. And fifth, make a backup of your Windows 9X/2000 Registry files just to be on the safe side.

**Figure 9.34** Other Devices

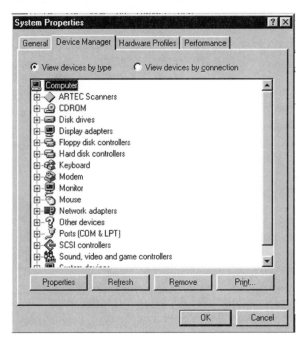

---

### Procedure 9.16: Softsetting PnP Ports

1. Open the Device Manager using the procedure steps outlined in section 9.11.2.
2. Click the left mouse button on the plus sign (+) beside the hardware category that contains the port whose resources you want to change.
3. Click the left mouse button once on the port whose resources you wish to change.
4. Click the left mouse button on the Properties button located at the bottom of the Device Manager window. This will open a Properties window for the port whose resources you wish to change.
5. Click the left mouse button on the Resources tab located toward the top of the port's resource window.
6. Remove the check from the Use automatic settings checkbox by clicking the left mouse button on the small box.
7. Choose from a set of default configurations by using the small scroll button to the right of the Settings based on text window. As you move through the selections, the resources in the Resource type and Settings text window will show the new values. To make the changes, click the left mouse button on the OK button at the bottom of the Properties window.
8. Some ports allow you to make changes to individual resources. If this is the case, double-click the left mouse button on the resource you wish to change. A scroll box will appear with a list of resources you can select.

---

## 9.12 ADAPTER CARD REMOVAL AND INSTALLATION

Removing and installing adapters, especially PnP, is a fairly easy procedure. Following is a generic procedure that can be followed to remove and install most adapters. However, if documents are provided with the adapter, they should supersede the procedure listed in this section.

## Procedure 9.17: Adapter card removal and installation

**Removal**

1. Turn the computer's power switch off and then unplug the power cord from the rear of the computer.
2. Remove the computer's cover and store the screws in a safe location.
3. If available, attach an anti-static mat to a bare metal surface on the computer chassis.
4. Wear an anti-static wristband attached to a bare metal surface on the computer chassis. If an anti-static wristband is not available, touch a bare metal surface on the chassis often to prevent static buildup.
5. Identify the adapter card that is to be removed.
6. You may wish to label the data cables with masking tape and a marker if you are not sure of their orientation should the adapter be reinstalled.
7. Remove all internal and external cables that are attached to the adapter.
8. Remove the screw that secures the adapter and store it in a safe location. This screw is located on the top of the metal plate that is attached to the back of the adapter.
9. Firmly grip the top rear of the adapter card at the metal mounting plate and the top front of the edge card.
10. Pull straight up on the card until it unseats from the expansion bus connector. The card may need to be gently rocked toward the front and rear.
11. Place the removed card on the anti-static mat or store in an anti-static bag.
12. If another card will not be mounted in the expansion slot, place a metal expansion slot cover over the hole in the back panel that was left open when the adapter was removed. These panels can be purchased at most computer supply stores.

**Installation**

1. If the installation procedure is included with the adapter card, it should be used instead of the following steps.
2. If the adapter is of the hardset design, configure the card's resources according to the documentation supplied with the adapter.
3. Follow steps 1 through 4 from the adapter card removal procedure.
4. Locate a free expansion slot that is the same type as that required by the adapter card to be installed, for example, ISA, PCI, or APG. If an upgrade is being performed, remove the adapter that is being upgraded.
5. Remove the metal plate that covers the expansion slot's opening on the computer's back panel. On some chassis, the metal plate is held in place by a mounting screw similar to those mounting the other adapter cards into the expansion bus. On new chassis, a metal knockout may cover the back panel opening. If the latter is the case, remove the knockout with a pair of pliers to avoid lacerations to the hands.
6. Once the metal plate is removed, insert the adapter card with the male edge card connector first and the mounting plate toward the rear of the computer. Refer to Figure 9.35.
7. Make sure the male edge card connector on the adapter card aligns with the channel in the expansion slot located on the motherboard. Also, make sure the key(s) on the adapter's male connector aligns with the key(s) in the female expansion slot.
8. Once aligned, push down firmly and evenly on the metal tab on the adapter's mounting plate and on the top-front edge of the card. Push until the male edge card connector on the adapter card is seated into the female edge card connector on the expansion slot. When seated correctly, the adapter card should not be at an angle in the expansion slot and the adapter card mounting plate tab will be aligned with the mounting screw hole on the chassis.
9. Secure the adapter with the correct sized screw, making sure the screw is not over-tightened.
10. Attach all of the required data cables, making sure that the connectors are oriented correctly with pin-1 on the cable and connector to pin-1 on the adapter connector.
11. Check the computer for proper operation before reinstalling the cover.

**Figure 9.35**

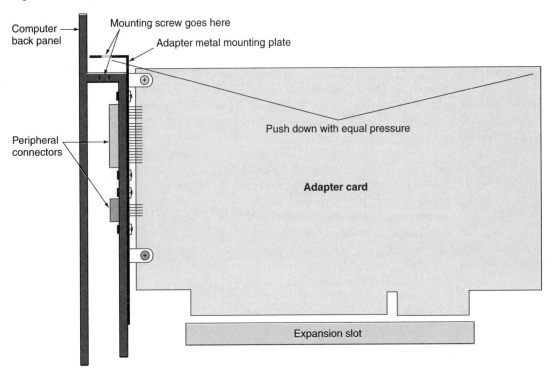

## 9.13 CONCLUSION

This chapter has introduced you to the different I/O devices that can be installed in a compatible computer and illustrated the procedures required to configure different I/O adapter cards. The chapter also introduced a lot of terminology used in relation to the I/O devices in the computer.

Even though all of the information in this chapter sounds complicated, if you are the kind of person who can follow written instructions and are not afraid to ask for help, you should be OK. Newly purchased adapter cards and peripherals usually provide a user's manual that will give you all of the needed procedures to install and use the adapter successfully. Most user's manuals also list a help number that you can call if things do not go according to plan. There are also hundreds of PCers who are dying to share their knowledge with you, just to show you how smart they are.

If you cannot, or will not, follow written instructions or if you are afraid to ask questions, do not get involved with computers. You will be in for a lot of headaches!

## Review Questions

1. Write the meaning of the following acronyms:
   a. MIDI
   b. IEEE
   c. PIC
   d. EIA
   e. DMA
   f. IRQ
   g. bps
   h. PnP

### Fill In the Blanks

1. An _____ _____ allows information to be brought from the outside world into the computer.
2. A _____ connects to the computer through a port.
3. The addresses assigned to the math coprocessor are _____ through _____ .
4. The device that controls the interrupts in the compatible computer is called a(n) _____ _____ _____ .

5. IRQ3 is assigned to _____ and _____.
6. The floppy drive(s) use DMA _____.
7. A port whose resources can be changed by the user may be _____, _____, or _____.
8. The computer resources available to I/O ports are _____, _____, and _____.
9. Switch banks that use the same pin spacing as DIP IC are called _____-_____.
10. Hardset cards are usually configured with _____ and _____.
11. Softset adapter cards are configured with _____.
12. PnP adapters cards are configured by the _____ _____.
13. In order to use PnP, the computer must have a PnP _____, the computer must be running a PnP _____ _____, and the adapter must be of _____ design.
14. Connectors that have pins are called _____ connectors.
15. Connectors that have the sockets that the pins plug into are called _____ connectors.
16. The keyboard connector can be either a _____ or a _____ connector.
17. The _____ serial port standard is used on the compatible computers.
18. The serial port on the compatible computer can use either a _____ or a _____ connector.
19. Transmission of information in two directions at the same time is called _____ _____.
20. _____ _____ flow control uses conductors in the cable.
21. _____ _____ refers to changing peripherals without turning off the computer.
22. Up to _____ devices can be connected to the USB port.
23. The _____ _____ _____ is a unidirectional port that does not use DMA.
24. The _____ mouse does not use one of the programmable serial ports.

## Multiple Choice

1. The addresses for the floppy disk controller are:
   a. 3F8H through 3FFH.
   b. 3F0H through 3F7H.
   c. 2F8H through 2FFH.
   d. 1F0H through 1F8H.
2. The addresses for serial port 3 (COM3) are:
   a. 3E8H through 3EFH.
   b. 2E8H through 2EFH.
   c. 2F8H through 2FFH.
   d. all of the above.
3. The DMA channel assigned to the floppy disk drive is:
   a. channel 0.
   b. channel 1.
   c. channel 2.
   d. channel 3.
4. Which two devices are usually used to set a hardset adapter card's configuration?
   a. the CONFIG.SYS and AUTOEXEC.BAT files
   b. the CONFIG.SYS file and the CMOS
   c. jumpers and switches
   d. the CMOS and the AUTOEXEC.BAT file
5. The purpose of the enhanced keyboard's edit keypad is to:
   a. position the cursor and insert and delete characters.
   b. change the configuration of the keyboard.
   c. change the configuration of an adapter card.
   d. edit the CONFIG.SYS file and the AUTOEXEC.BAT file.
6. The interrupt assigned to serial port 2 (COM2) is:
   a. IRQ0.
   b. IRQ1.
   c. IRQ2.
   d. IRQ3.
   e. IRQ4.
7. Which of the following is not a peripheral?
   a. printer
   b. input port
   c. display
   d. keyboard
8. EIA stands for:
   a. elementary interface adapter.
   b. Elastic Information Association.
   c. Electronic Industries Association.
   d. Electronic International Association.
9. The serial port can be identified on the back of the computer by a:
   a. 25-pin female connector.
   b. 9-pin male connector.
   c. 25-pin male connector.
   d. Both b and c are correct.
10. The parallel port(s) can be identified on the back of the computer by a:
    a. 25-pin female connector.
    b. 9-pin male connector.
    c. 15-pin female connector.
    d. 25-pin male connector.
11. The game port can be identified on the back of the computer by a:
    a. 25-pin female connector.
    b. 9-pin male connector.
    c. 15-pin female connector.
    d. 25-pin male connector.
12. If a single parallel port at I/O address 378H is installed in a compatible computer, it will be used as:
    a. printer 1 (LPT1).
    b. printer 2 (LPT2).
    c. printer 3 (LPT3).
    d. none of the above.
13. A game port is at I/O address:
    a. 1F0H through 1F8H.
    b. 200H through 20FH.
    c. 2E8H through 2EFH.
    d. 3F0H through 3FFH.

14. A game port uses IRQ number:
    a. 1.
    b. 2.
    c. 3.
    d. none of the above.
15. Plug-and-Play means:
    a. no hardware configuration is required by the installer to make the adapter work.
    b. the operating system automatically configures the adapter's resources.
    c. computer upgrades will be an easier task.
    d. all of the above answers are correct.
16. To use Plug-and-Play adapters, the computer must:
    a. be running an MS-DOS version 6.22 or better.
    b. use at least an 80286 microprocessor.
    c. be running a PnP operating system.
    d. have at least a 540-MB hard drive installed.

# CHAPTER 10

# Mass Storage Basics

## OBJECTIVES

- Define terms, acronyms, and expressions used in relation to a computer's mass storage.
- List the characteristics of 5¼" and 3½" floppy disks and drives.
- Describe why tape backup units are needed.
- Identify different tape formats used with the compatible computer.
- Explain the different floppy disk formats.
- Describe the proper care of floppy diskettes.
- Identify double-density and high-density 3½" diskettes.
- Describe the information contained on a disk's first cylinder.
- List the information contained within each directory entry.
- Describe the function of the FAT.
- Explain how a program is loaded into the RAM of the computer.
- Describe the proper cable connection for floppy drives.
- List the advantages of a CD-ROM drive.
- Describe the characteristics of CD-R and CD-RW drives.
- Explain the advantages of a DVD-ROM drive.
- Describe a computer's drive bays.

## KEY TERMS

mass storage
auxiliary storage
magnetic tape
read/write head
backup
restore
tape drive
disk drive
floppy diskette
hard disk
random access
track

sector
density
double-sided
cylinder
FORMAT
file allocation table (FAT)
directory
tracks per inch (TPI)
5¼" diskette
index hole
write-protect notch
5¼" floppy drive

3½" diskette
3½" floppy drive
volume boot-record
physical sector
logical sector
allocation unit
cluster
subdirectories
floppy disk controller (FDC)
compact disc-read-only memory (CD-ROM)

High Sierra format
compact disc-recorder (CD-R)
compact disc-read/write (CD-RW)
digital versatile disc read-only memory (DVD-ROM)
digital video disc
drive bay
internal drive bay
external drive bay

Computers operate on the same basic principle as do people. Information is brought into the human brain, processed, and an output occurs sooner or later as a response to the processed information. The input of the information can come through several input devices such as the eyes, ears, or nose, or by touch. No matter how hard we try, it is impossible to hold all of the input information the brain has to process even in a single day. Some of the information is very important, so it needs to be saved for further reference.

**332** Chapter 10

To help solve this problem, we are constantly coming up with methods to hold information in places other than our brain until we are ready to use it, for example, notebooks, tape recorders, books, and so on. In fact, the information you are reading now is a good example. You may not remember all of the material covered in this book, but you can always re-read the material if you need it. If a human brain cannot hold all of the information it needs to process, then there is no way a compatible computer can have enough memory to hold all of the information it will ever process.

Data must first be brought into the memory unit of a computer before the computer can execute a program or process the data. It is highly unlikely that a personal computer will have enough RAM to hold its main operating system plus all the application programs and data it will have to process. To alleviate this problem, a form of **mass storage,** or **auxiliary storage,** is used to hold programs and data on a nonvolatile media until the computer is ready to use them. The most widely used mass storage devices in microcomputers are tape drives, floppy disk drives, CD-ROMs, and hard drives. Hard drives will be discussed in detail in chapter 11.

Disk drives are currently the most popular form of mass storage, but that was not always the case. When computers were in their infancy, the magnetic tape dominated the computer industry. Even though magnetic tape has moved to the background, it still has its place in today's world of computing.

## 10.1 TAPE DRIVES

The **magnetic tape** that is used with computers works on the same basic principle as the VHS and cassette tapes that we use to record video and audio information. Magnetic tape is made from a strip of Mylar (a form of flexible plastic) that is coated with a thin layer of ferric (metal). When an electromagnetic field is applied to the surface of the magnetic tape by a **read/write head (R/W),** the particles of metal align themselves with the magnetic lines of force produced by the read/write head(s). This process is illustrated in Figure 10.1.

The magnetic tape is read by moving it over the same read/write heads that were used to record the information. As the magnetic particles on the surface of the tape switch directions, a small voltage is induced into the read/write heads that is amplified before it is sent to the computer. The magnetic tape is wound on a spool or placed in a small cassette, similar to a VHS tape.

**Figure 10.1** Magnetic Tape

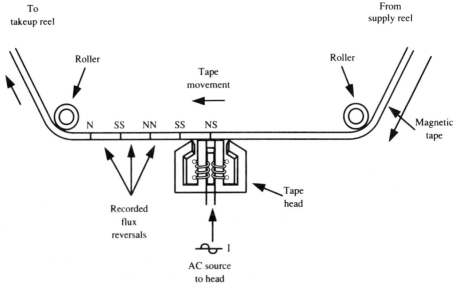

*Source:* Asser/Stiglian/Bahrenberg, *Microprocessor Theory and Servicing* (Upper Saddle River, NJ: Prentice Hall, 1997) p. 438.

# Mass Storage Basics

The nice thing about magnetic tape is that it can be used over and over again. The binary information used by the computer is stored on the tape magnetically. Once recorded, it can be played back into the computer's RAM for processing as many times as necessary. Because magnetic tape is in a continuous strip, its biggest disadvantage is that the information is stored on the tape sequentially (one bit after another). To get to the data at the end of the tape requires advancing the tape through all of the information that precedes the desired data.

As stated previously, magnetic tape was once the most widely used form of mass storage. Today it is most often used as a secondary backup system when the primary mass storage system fails. You might wonder why we need secondary storage. I am sorry to be the bearer of bad news, but sooner or later your primary mass storage device, usually the hard drive, is going to fail. When this happens, some or all of the information stored on the hard drive will be lost. Hopefully, you will have maintained an up-to-date copy of the information.

Copying all or some of the information on a hard drive onto a secondary storage device is called making a **backup.** The act of returning the backup information to the hard drive is called **restoring.** MS-DOS supplies a utility called *BACKUP* that can be used to back up and restore information from a hard drive to floppy diskettes. Because Windows 3.X is a parasite operating system, the DOS Backup utility is used to backup Windows 3.X files. Windows 95/98/2000 has a backup utility written by Seagate® that will back up to almost any mass storage device. This backup utility is located in the Accessories option of the Start menu.

Using the operating system to back up and restore the information stored on hard drive sounds great. However, a big problem arises because current hard drives exceed 10 gigabytes in storage capacity. It takes a lot of floppy diskettes and a lot of time to back up or restore all of the information on one of these huge hard drives.

A **tape drive** is a high-speed alternative to floppy diskettes when it comes to backing up the information stored on a hard drive. Tape drives are available in formats that will hold several gigabytes of information on a single tape. Also, tape drives can be purchased as either internal or external units.

An internal tape drive mounts inside the system unit in an external drive bay. Two popular styles of internal tape drives are shown in Figure 10.2. The actual installation procedure for an internal tape drive will vary depending on the drive. Internal tape drives use the floppy disk, EIDE, SCSI, or USB interface. The EIDE and SCSI interfaces are discussed in greater detail in chapter 11.

**Figure 10.2** Typical External Tape and Internal Tape Drives

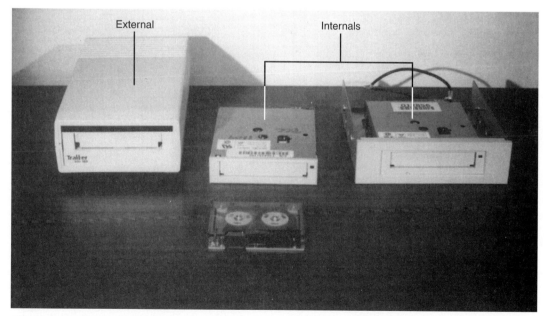

*Source:* Dan L. Beeson, *Assembling and Repairing Personal Computers* (Upper Saddle River, NJ: Prentice Hall, 1997) p. 229.

No matter which interface is used, documentation should be supplied with the drive that will provide installation instructions and the software that is required to make the drive operate. If the documentation is not provided, contact the manufacturer of the drive. It should be mentioned that an internal drive is considerably cheaper than an external drive with the same features because the internal tape drive does not require a case or power supply.

Figure 10.2 also shows a typical external tape drive. An external tape drive usually attaches to a parallel port, USB, or SCSI adapter. When a parallel tape drive is used, a special cable that allows a printer and tape drive to be attached to the same parallel port is usually supplied with the external tape drive. To install an external tape backup, simply plug the unit's power cord into a wall receptacle and connect the cable between the unit and the computer's parallel port. As with the internal tape drive, software is

**Table 10.1** QIC Mini Cartridge

| QIC-40 | 40 MB |
|---|---|
| QIC-80 | 125 MB, 200 MB, 400 MB, and 500 MB |
| QIC-3010 | 340 MB, 420 MB, 800 MB, and 1.1 GB |
| QIC-3020 | 680 MB, 850 MB, 1.6 GB, and 2.2 GB |
| QIC-3040 | 840 MB, 1 GB, and 2.5 GB |
| QIC-3050 | 1 GB, 1.3 GB, and 3.1 GB |
| QIC-3080 | 1.6 GB, 2 GB, and 5 GB |
| QIC-3095 | 2 GB, 4 GB, and 5 GB |
| QIC-3210 | 2.3 GB and 5.1 GB |
| QIC-3230 | 8 GB, 10 GB, and 20 GB |
| **QIC Data Cartridge** | |
| QIC-150 | 250 MB |
| QIC-525 | 525 MB |
| QIC-1000 | 1 GB |
| QIC-2GB | 2 GB |
| QIC-5GB | 5 GB |
| QIC-5010 | 13 GB |
| QIC-5210 | 15 GB |
| **Travan** | |
| TR-1 | 400 MB |
| TR-2 | 800 MB |
| TR-3 | 1.6 GB |
| TR-4 | 4 GB |
| TR-5 | 10 GB |

normally provided with the external tape drive to make it operate. The external tape backup is very easy to move from computer to computer.

The standards used by most of today's tape drives for the compatible PC are variations of the standards established by the *Quarter Inch Committee (QIC)*, which gets its name from the width of the tape. QIC-compatible drives are designed to handle either a 3½″ mini cartridge or a 5¼″ data cartridge. A popular variation of QIC is 3M's Travan standard, which uses a slightly wider tape width of 0.315″. The current storage capacities for the QIC and Travan standards are listed in Table 10.1. All standards are not downwardly compatible, so consult the label on the box or the documentation that is provided with the drive to determine which tape standards are supported.

I highly recommend the purchase of a tape drive if you store information on your hard drive that you cannot afford to lose. These high-speed drives are relatively easy to install and operate, and can spare you a lot of emotional agony.

## 10.2 DISK DRIVES

The most widely used method of mass storage on computers today is the **disk drive.** A disk drive stores information magnetically on a circular disk that is covered with a ferric surface, in much the same way as magnetic tape. The disks are made of Mylar or aluminum. The disks that are made of Mylar are called **floppy diskettes** because they are not rigid. The disks made of aluminum are called **hard disks** because they are rigid. We will save our discussion of hard drives for the next chapter.

### 10.2.1 Floppy diskettes

Floppy diskettes get their name from the fact that they are in the shape of a disk. Floppy diskettes are made of Mylar just like the magnetic tapes discussed previously. The circular shape allows the disk to rotate past the read/write heads. The head assembly moves in and out perpendicular to the rotating diskette to record and retrieve information. Unlike magnetic tape, the data on a floppy diskette is *not* placed on the surface sequentially and can be accessed randomly. **Random access** means that data may be accessed at any time—directly—without going through all of the information in sequence.

To facilitate data organization so it can be located rapidly, the surface of a diskette is divided into two units called **tracks** and **sectors.** A track is a circular ring that goes around the surface of the diskette. Sectors are multiple segments into which the track is divided. The tracks and sectors on the surface of a diskette are illustrated in Figure 10.3. The sectors actually hold the information stored on a diskette. No matter how many tracks are on a diskette's surface, each sector holds 512 bytes of user data.

The tracks and sectors shown in Figure 10.3 are for illustration purposes only. If you look at a floppy diskette, you will not be able to see the tracks and sectors because they are placed magnetically.

The floppy drives used in compatible computers use different numbers of tracks and sectors depending on their density. **Density** is a term that indicates the diskette's storage capacity. Information is usually recorded on both sides of diskettes, so they are sometimes referred to as **double-sided.** The track numbers on the top surface of the diskette are directly above the same track numbers on the bottom surface.

The term **cylinder** is used to describe the same track location on both sides of the diskette. In other words, track 0 on the bottom surface of the diskette is directly below track 0 on the top surface. Information is stored on both tracks before the read/write heads are moved to another track. For this reason, both track 0s are referred to as one unit called *cylinder 0.* The term *cylinder 5* refers to track 5 on the bottom surface and track 5 on the top surface.

In order for any operating system to access a diskette, the diskette must be in a track/sector format that the operating system understands. To magnetically place the tracks and sectors on the surface of the diskette of a compatible computer, a program called **FORMAT** is usually used.

Most of today's diskettes are preformatted prior to purchase using either an IBM-compatible or an Apple Macintosh® format. These two formats are not compatible with each other, so make sure you purchase the correct format. Because most diskettes will eventually have to be reformatted, the operating system also gives the user the ability to format the diskette. The actual procedure and format options will depend on the operating system and will be covered later in this chapter.

**Figure 10.3** Tracks and Sectors

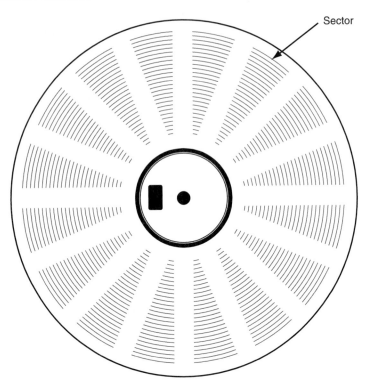

## 10.3 THE BASICS OF STORING DATA

When information is stored on the diskette, the operating system will assign it a track/sector address that is free. The OS finds the location of free and used space on the disk by using a table located on the first cylinder called the **file allocation table (FAT).** The function of the FAT is pretty important to understand, so we will discuss it in greater detail throughout section 10.5.

When information is stored on the diskette, the OS first checks the directory to see if the file already exists. If not, the FAT is accessed to determine where free space is available. After the information is saved on the diskette, the OS stores the file's name and starting location in a section of the first cylinder called the **directory.** Directories will be discussed in greater detail in section 10.5.

When a file is retrieved from the diskette, the operating system will check the directory for the file's name and starting track/sector. Once the file is located on the disk, the FAT tells the operating system how to put the file together as it is loaded into the computer's RAM.

This concludes our introduction into the processes of storing and retrieving information on a floppy diskette. A more in-depth discussion will follow in later sections. For now, let us discuss the floppy drive itself.

## 10.4 FLOPPY DRIVES

Floppy drives are available in two standard sizes: 5¼″ and 3½″. These numbers indicate the physical size of the diskettes that are used to store and retrieve information, not the actual drive size. Floppy drives are also rated by the number of tracks they can place within an inch on the surface of the diskette, or the **tracks per inch (TPI).** The 5¼″ disk drives are available in 48 TPI and 96 TPI, whereas all 3½″ disk drives are 135 TPI. This rating defines the spacing of the tracks, not the number of tracks actually used to store information. For example, a 135-TPI drive spaces the tracks $\frac{1}{135}$ of an inch apart, but the diskette is formatted with 80 tracks per surface.

## Mass Storage Basics

**Table 10.2** Floppy Diskette Formats

| Format | TPI | Trk/Side | Sides | Sec/Trk | Capacity |
|---|---|---|---|---|---|
| 5¼" Double-density | 48 | 40 | 2 | 9 | 360 KB |
| 5¼" High-density | 96 | 80 | 2 | 15 | 1.2 MB |
| 3½" Double-density | 135 | 80 | 2 | 9 | 720 KB |
| 3½" High-density | 135 | 80 | 2 | 18 | 1.44 MB |
| 3½" 2.88 MB | 135 | 80 | 3 | 36 | 2.88 MB |

**Figure 10.4** 5¼" Diskettes and 3½" Diskettes

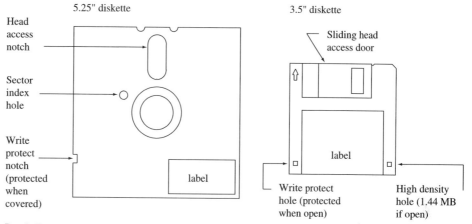

*Source:* Dan L. Beeson, *Assembling and Repairing Personal Computers* (Upper Saddle River, NJ: Prentice Hall, 1997) p. 168.

The floppy diskette formats used on compatible computers are listed in Table 10.2. Diskettes should be purchased that have the same storage capacity as the drive in which they are intended to operate. If necessary, the higher density floppy drives have the ability to format and read lower density diskettes, but this is not true the other way around. The procedure used to format a diskette using the operating systems covered in this book is given in section 10.6. The procedure for formatting a diskette should also be explained in the user's manual for the operating system utilized on your computer.

### 10.4.1  5¼" Diskettes

A 5¼" diskette is shown in the illustration on the left in Figure 10.4. The **5¼" diskette** is placed in a flexible plastic jacket, which does not provide a lot of protection. There are several physical features of the 5¼" diskette that stand out. The predominant feature is the large, circular hole in the center of the diskette. This hole is where the electrical motor that rotates the diskette within the jacket clamps onto the diskette when the drive door is closed.

Looking at the diskette from the top, just to the right of this large hole, is a smaller hole referred to as the **index hole.** The drive uses this reference to locate the first sector on each track when the diskette is originally formatted. It is also used to detect the rotational speed of the drive.

The **write-protect notch** is located on the right edge of the jacket. If this notch is covered with an opaque tab (no light will pass through), the disk can be read from but not written to. The large, oval

**Figure 10.5** Typical 5¼″ Drive and 3½″ Floppy Drives

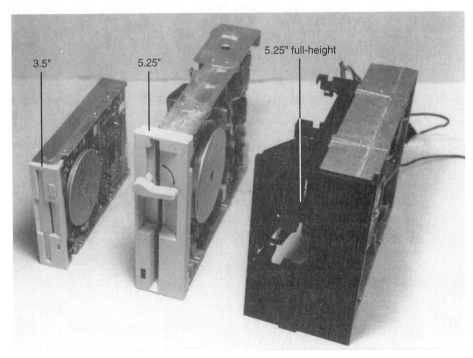

*Source:* Dan L. Beeson, *Assembling and Repairing Personal Computers* (Upper Saddle River, NJ: Prentice Hall, 1997) p. 166.

opening toward the front of the jacket is where the read/write heads access the diskette. The two notches to each side of the read/write head access hole are used for mechanical alignment.

### 10.4.2 The 5¼″ Disk Drive

Two of the three drives shown in Figure 10.5 are typical **5¼″ floppy drives.** The drive on the far right is called a *full-height 5¼″ drive* and the drive in the center is called a *half-height drive.* Full-height drives, which are no longer used, were installed in the original IBM-PC that was introduced in 1981. Even though the half-height drive is currently more popular than the full-height drive, it is also seldom used, but is nice to have around. Both the full-height and half-height drives have a width of 5⅞″ inches (150 mm). The full-height and half-height drives have a thickness of 3¼″ (83 mm) and 1⅝″ (42 mm), respectively.

The thin, horizontal opening in the center of the drive is the diskette load slot. This is the opening into which the 5¼″ diskette is inserted with the label side up. The drive latch has to be in the up position before the diskette is inserted. The drive latch mechanism shown in Figure 10.5 will not be the same on all 5¼″ floppy drives, but it will be similar to the one shown. The diskette can be inserted into the drive upside down without damage to the diskette or drive; however, data cannot be read from or written to the diskette. Once the diskette is inserted, the disk drive latch must be moved to the down position to engage the drive motor and read/write heads. The disk access indicator will illuminate any time the computer is accessing the drive. The disk drive latch should not be open when the access indicator is on.

### 10.4.3 The 3½″ Diskette

The illustration on the right in Figure 10.4 shows a 3½″ floppy diskette. The **3½″ diskettes** are placed in a hard plastic jacket that supplies good mechanical support. The jacket is about the same size as a shirt pocket, which makes it very easy to transport. A slide door that covers the read/write head access hole opens automatically when the diskette is inserted into the drive.

Note the write-protect window in the lower left-hand corner of the 3½″ floppy shown in Figure 10.4. A slide door can be positioned manually over the write-protect window to allow the diskette to be read from and written to. If the slide door is in the open position, the diskette can only be read from and not written to. The additional window in the lower right-hand corner of the figure indicates the diskette is

**Figure 10.6** Handling Floppy Diskettes

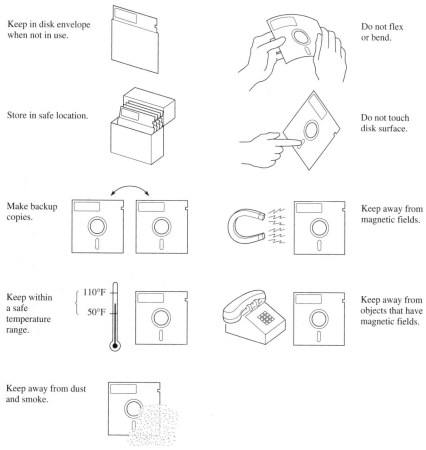

*Source:* Antarakos/Adamson, *Microcomputer Repair* (Upper Saddle River, NJ: Prentice Hall, 1996) p. 74.

a high-density (HD) diskette (1.44 MB) and should not be used on a DD drive (720 K). If this window is not present, it indicates the diskette is double-density.

### 10.4.4 The 3½″ Disk Drives

The image on the left in Figure 10.5 shows a typical **3½″ floppy drive.** The 3½″ floppy drive is much smaller and thinner than the 5¼″ floppy drive. The 3½″ floppy drive can be mounted in the computer either horizontally or vertically. The 3½″ diskette is inserted into a thin slot, similar to that for the 5¼″ drive. Also like the 5¼″ drive, a disk access light on the front of the drive illuminates when the computer is accessing the drive.

When the disk is inserted, it snaps into the drive, engaging the motor and read/write heads, so a door latch is not required. The diskette is inserted into the drive with the diskette's read/write access window first, and the metal ring turned toward the drive access indicator. The mechanics of the drive will not allow the diskette to be inserted upside down or backwards, so do not force it.

A button on the face of the drive ejects the diskette when pressed. The diskette eject button should not be pressed when the access indicator light is on.

### 10.4.5 Caring for a Floppy Diskette

People often handle floppy diskette carelessly and then wonder why they lose data. Floppy diskettes are made from plastic and information is stored magnetically. These two facts make the data very susceptible to being lost or the diskette itself being damaged if basic care is not taken. This basic care includes the following list along with Figure 10.6.

- Store floppy diskettes in a safe place, away from dust, moisture, magnetic fields (such as TV, speakers, and computer monitors), and extreme temperatures.

- Never expose diskettes to direct sunlight for long periods of time.
- Label each floppy diskette so that you can identify the information stored on it.
- Place the label on the front of the diskette, at the top, so that the label does not stick to any exposed surface of the diskette.
- Use a felt-tipped pen to write on a label that has already been placed on a 5¼″ diskette to prevent damaging the diskette inside the jacket.
- When handling a 5¼″ floppy diskette, never touch its exposed surfaces.
- Never open the 3½″ diskette's read/write head access door.

If you follow these simple steps, you can maintain the information on the diskette for extended periods of time. However, if the data you have on the diskette is very important, it is a good idea to keep multiple copies. This is especially true for diskettes that contain the drivers for your CD-ROM, video adapter, and sound card.

### 10.4.6 Which Size Floppy Drive Do I Need?

Most preassembled compatible computers have one floppy drive as standard equipment—a high-density (HD) 3½″ drive. If you are assembling your own computer and are going to purchase one floppy drive, the 3½″ HD is your best bet. This drive will allow you to read both 720-K and 1.44-MB 3½″ floppy disks. Another option is the 2.88-MB 3½″ drives that will read and write all 3½″ densities currently on the market. As far as I know, there are no commercial applications sold on 2.88-MB diskettes.

If you are going to install a second floppy drive, it will probably be best to install a HD 5¼″ drive. This will give the computer both a HD 5¼″ and a HD 3½″ drive. You will not have to worry about diskette size and density with this configuration.

## 10.5 THE FIRST CYLINDER

A large portion, if not all, of the first cylinder of a disk formatted by MS-DOS, Windows 3.X, Windows 9X, or Windows 2000 is reserved by the operating system and is not directly user accessible. This is true for all disk drives including both floppy drives and hard drives. This portion of the first cylinder contains valuable information that is placed on every floppy or hard drive disk when the disk is formatted. The operating system, not the user, automatically updates the reserved section of the disk. The information in this reserved section includes:

- A short machine language program called the **volume boot-record** (master partition boot-record on hard-drives);
- File allocation tables (FATs) that keep a record of how the storage space on the disk is allocated; and
- A directory that holds the names and other information about the files stored on the diskette.

We already know that when the compatible computer is booted, it starts running machine code programs that are located in the computer's BIOS. As discussed in chapter 5, the programs in the BIOS will:

1. Get the BIOS configuration information for the CMOS;
2. Initialize all of the programmable chips in the computer;
3. Run the POST; and
4. Try to load information from side 0, track 0, sector 0 from the boot drive that is selected by the user in the CMOS.

If the A: drive is selected as the boot drive and the diskette in this drive has been formatted by any of the operating systems covered in this book, a small machine code program (512 bytes) called the *volume boot-record* will be in this sector. If this program does not exist, it indicates the diskette has never been formatted. When a nonformatted diskette is detected, the BIOS will display an error message similar to, "Error reading drive A:."

If the volume boot-record does exist, the program is loaded into RAM and control of the computer is turned over to this program. The boot sequence used by hard drives is almost identical to the one used to boot from a floppy, and will be discussed in chapter 11.

Table 10.3  Format Information Obtained from the Volume Boot-record for the Indicated Drives*

| Drive Type | Bytes/ Sector | Sec/ AU | # of FATs | RD Entries | Total #Sectors | Sectors/ FAT | Sectors/ Track |
|---|---|---|---|---|---|---|---|
| DD 5¼″ | 512 | 2 | 2 | 112 | 720 | 2 | 9 |
| HD 5¼″ | 512 | 1 | 2 | 224 | 2,400 | 7 | 15 |
| DD 3½″ | 512 | 2 | 2 | 112 | 1,440 | 3 | 9 |
| HD 3½″ | 512 | 1 | 2 | 224 | 2,880 | 9 | 18 |
| 20 MB HD | 512 | 4 | 2 | 512 | 41,531 | 41 | 17 |

*AU = Allocation unit
\#  = Number
RD = Root directory

Once the volume boot-record program is loaded into the RAM and running, it sends the computer back to the default drive to load the additional files needed by the operating system. The names of the system files required to boot the operating system often vary among operating systems. The files required to boot MS-DOS and Windows were discussed in chapters 5 and 6. If any of the required boot files are not present, the volume boot-record program will display an error message similar to, "Non-System disk or disk error—Replace and press any key to continue." This error message does not mean that the diskette is defective; it is simply not a bootable disk.

The volume boot-record serves another function in addition to loading the main operating system from the disk. It also contains some very important information about how the disk itself is formatted. Some of the other information contained in the boot-record includes:

- Number of bytes/sector;
- Number of sectors/allocation unit;
- Number of FATs;
- Maximum number of root directory entries;
- Total number of sectors contained on the diskette;
- Number of sectors/FAT;
- Number of sectors/track; and
- Number of R/W heads.

Table 10.3 contains information that was obtained from the boot-record of the drive type listed using a DOS-based program called Debug. Debug is supplied with the operating system and has been around ever since DOS was first introduced. In fact, Debug is still available with Windows 9X.

*Debug* is a program that gives the user low-level control of the computer without using the operating system. One feature provided by Debug is the ability to load the information from the reserved section of a floppy disk or hard drive into the computer's RAM. After the data are loaded into the RAM, the user can use Debug to display the contents of the RAM on the computer's display. Debug gives the user low-level control of the computer, so it can be a valuable tool. It can also be very dangerous to use. Additional information about Debug is contained in Appendix D.

Learning how to load and control the Debug program is not really necessary, even for most computer technicians. For this reason, the procedure is not covered in this book. However, if you have a curious mind and a craving for knowledge, you might find several books written about Debug interesting. Most books that cover assembly language programming of the compatible computers also cover Debug.

The volume boot-record is located in logical sector 0 on every floppy disk formatted by the operating systems covered in this book, no matter what density. The location of the volume boot-record on hard drives varies depending on the location of the bootable partition, which is covered in chapter 11.

Following the volume boot-record on the first cylinder of the diskette are the FATs and then the root directory. The actual sector location and length of the FATs and root directory vary depending on the drive type and the options chosen when the diskette is formatted. The space and location used for the volume boot-record, FATs, and directory for several drives are listed in Table 10.4.

Table 10.4  Sector Locations of Volume Boot-record, FATs, and Root-Directory

| Drive Type | Master Boot-Record | | FATs | | Root Directory | |
|---|---|---|---|---|---|---|
| | Sector Number | Total Sectors | Sector Numbers | Total Sectors | Sector Numbers | Total Sectors |
| DD 5¼" | 0 | 1 | 1–4 | 4 | 5–B | 7 |
| HD 5¼" | 0 | 1 | 1–E | 14 | F–1C | 14 |
| DD 3½" | 0 | 1 | 1–6 | 6 | 7–D | 7 |
| HD 3½" | 0 | 1 | 1–12 | 18 | 13–20 | 14 |
| 20 MB HD | 0 | 1 | 1–52 | 82 | 53–72 | 32 |

### 10.5.1 Physical and Logical Sectors

When a diskette is formatted, the surface of the disk is divided into tracks and sectors. A diskette can be formatted in many ways, so let us just consider the 3½" HD diskette. The format parameters for this size drive are listed in Table 10.3.

When a 3½" HD diskette is formatted using the default parameters, data are recorded on both sides of the diskette, with 80 tracks on each surface. Both sides of the disk are used, so there is a total of 160 tracks on the diskette. The read/write (R/W) heads of the drive are connected together mechanically so both heads move in unison and are at the same track and sector position on both sides of the disk. Each step of the R/W heads moves both heads to the same track position on both sides. Because both the bottom and top tracks can be accessed at every position of the read/write heads, these tracks are treated as a single unit called a *cylinder*. A cylinder refers to both tracks that are under the R/W heads on both surfaces of the diskette at any one time. For example, cylinder 0 will equal track 0 on both the bottom and top of the diskette. This means there are a total of 80 cylinders on the 3½" HD diskette.

Using hexadecimal, these cylinders are numbered $0-4F_{16}$ on a 3½" HD floppy diskette. A cylinder consists of two tracks, so there are 36 sectors per cylinder. With 36 sectors per cylinder, and 80 cylinders, there are 2,880 sectors. Each sector can hold 512 bytes of information. Using a little math, a 3½" floppy diskette can hold $1,474,560_{10}$ bytes of data (512 × 2,880). Remember from chapter 2 that because this is storage capacity, this number is divided by 1,024, which equals 1,440 KB. This is why these diskettes and drives are often called *1.44-MB* drives.

A **physical sector** is the actual side, cylinder, and sector location on the disk. An example would be side 0, cylinder 20, sector 5. This indicates that the location of the sector is on track $20_{16}$ on the bottom side of the diskette (side 0) and the 6th sector from the index hole (starting from 0). This is a confusing way to locate sectors on the disk surface.

Instead, the sectors are usually numbered sequentially beginning with side 0, cylinder 0, sector 0, going through sector 2,879 ($0-B3F_{16}$). Using this format, the sector location is called the **logical sector** because this number does not directly give the side, cylinder, and sector locations. This is an important concept to understand because it is used by most of the current high-capacity hard drives. Logical sector numbers will be used during the remainder of this discussion.

### 10.5.2 What Is an Allocation Unit?

To speed up data transfer, the operating systems discussed in this book read and write data to and from the disk in a group of one or more sector(s) called an **allocation unit,** or **cluster.** Understanding allocation units is very important because they are not only used by floppies, but also by hard drives. An allocation unit will vary in size depending on the drive type being used and the way the diskette or hard drive is formatted. Column 3 of Table 10.3 lists the number of sectors per allocation unit for several drive types currently supported by the compatible computer's operating systems. A small 20-MB hard drive is also shown to demonstrate the hard drive use of allocation units. Larger capacity hard drives require larger allocation units. This subject is discussed further in chapter 11.

For a 3½″ HD diskette formatted on a compatible computer, each allocation unit is equal to a single sector and is assigned a consecutive number that begins with the number 2 and continues through $B410_{16}$ ($2-2881_{10}$). The reason why 2 is the first allocation unit will be explained later. The file allocation table (FAT) maps the allocation units of the disk. Basically, the FAT shows the operating system how to path a file through the allocation units. The function of the FAT will become clearer as our discussion continues.

Notice in Table 10.3 that floppy drives use allocation units of either one or two sectors, whereas the small 20-MB hard drive uses an allocation unit size of four sectors. Because the hard drive allocation unit is four sectors, information is stored and retrieved on the drive in blocks of 2,048 bytes (4 sectors × 512 bytes per sector). This means if a file is saved on the drive that is a single byte in length, it will occupy 2,048 bytes of disk space. If a file that is 2,051 bytes in length is stored on the 20-MB hard drive, it will require two allocation units. This means the file will occupy 4,096 bytes of storage space. As you can see, as the size of the allocation unit increases, the total storage space of the drive that is utilized becomes less efficient. This concept becomes very critical in the hard drives that are covered in chapter 11.

Let us leave the allocation unit sizes for now and discuss the section of the disk that keeps track of the information stored in each allocation unit. As stated previously, the section of the first cylinder is called the FAT. To help avoid confusion, the remainder of this discussion will deal only with the format used by a 1.44-MB 3½″ HD diskette and a drive formatted with 80 cylinders, 36 sectors/cylinder, and 512 bytes/sector.

### 10.5.3 The FAT

The FATs are created on the diskette when the disk is formatted. The FAT is so important that the compatible computer's operating system creates two copies on every diskette or hard drive during the format process. For a 3½″ HD diskette, each FAT requires nine sectors of disk space. This means that 18 total sectors are required to create the two copies of the FAT. These 18 sectors are placed consecutively immediately following the volume boot-record, so the FAT occupies logical sector numbers $1-12_{16}$. The operating system uses the information in the FAT to:

1. Find allocation units that are unused when the operating system stores or updates files;
2. Locate any clusters that were found defective when the diskette was formatted; and
3. Find what path to take to load each allocation unit of the file into RAM.

Each copy of the FAT is divided into fields that correspond directly to the addressable allocation units on the disk. For floppy diskettes and small hard drives, the fields are 12 bits long, with each field holding information about the allocation unit to which it is assigned. Because each field is 12 bits long, it is referred to as a *FAT12*. Using 12 bits means the FAT can supply information on 4,096 ($2^{12}$) allocation units. A drive that is formatted using the FAT12 system can use fewer than 4,096 allocation units, but never more than 4,096.

Both FATs are automatically updated by the operating system when new file information is recorded on the disk. If an error occurs when the operating system reads the first FAT, the OS will automatically read the information from the second. When this occurs, no error message will be displayed, so the operator will have no indication that one of the FATs is defective. If both FATs become defective, the data on the diskette will be unreadable. For this reason alone, the DOS command CHKDSK or SCANDISK should be used often to make sure the two copies of the FAT are the same. The Fix option of CHKDSK should be used if errors are detected.

As stated earlier, each FAT occupies nine consecutive sectors on a formatted 3½″ HD diskette. Each sector holds 512 bytes of data, so each FAT occupies 4,608 bytes of disk space (9 × 512).

Of these 4,608 bytes, the first 3 bytes in the FAT (0–2) are used to store information about how the diskette is formatted. After these 3 bytes, groups of 1½ bytes (12 bits) are used to store information about each allocation unit. When the operating system needs to read the information about a certain allocation unit from the FAT, it multiplies the allocation unit number by 1.5 (1½ bytes or 12 bits) to find the location of the least significant byte of the desired allocation unit information. This is why the number 2 is the first allocation unit number available—the operating system will skip over the 3-byte format information (2 × 1.5 = 3). For example, the information for allocation unit 6 will be found in byte 9 of the FAT (6 × 1.5 = 9).

When the operating system reads the FAT, it treats the entries as 16-bit numbers (2 bytes). However, FAT12 fields are only 1½ bytes (12 bits) in length. This means that when allocation unit information is loaded from the FAT, 4 additional bits from another FAT field are loaded. In addition, compatible computers place 16-bit numbers in memory with the least significant byte (LSB) first and the most significant byte (MSB) second.

When the FAT is loaded into memory for examination, the data are always displayed from low address to high address. This makes the numbers in the FAT appear backwards. With the numbers appearing backwards and the FAT displayed in bytes, and 4 additional bits from another FAT field, it is difficult for humans to interpret the actual data in the FAT without using a program to straighten it out.

To illustrate how confusing the FAT can get, Example 10.1 shows a hypothetical listing of what a FAT segment from a 1.44-MB diskette might look like if it was loaded into the RAM of the computer and displayed using Debug. Even though the FAT is a binary table, Debug displays the contents in hexadecimal. Notice that it is basically impossible to recognize the individual AT fields from this listing.

**Example 10.1:** Example of a partial listing of an actual FAT

|    | 0  | 1  | 2  | 3  | 4  | 5  | 6  | 7  | 8  | 9  | A  | B  | C  | D  | E  | F  |
|----|----|----|----|----|----|----|----|----|----|----|----|----|----|----|----|----|
| 00 | FD | FF | FF | 03 | 40 | 00 | 05 | 60–| 00 | 0A | 80 | 00 | 09 | F0 | FF | 0B |
| 10 | F0 | 00 | 0D | E0 | 00 | 12 | 00 | 01–| 11 | 0F | FF | 13 | 0F | FF | 15 | F0 |
| 20 | FF | F7 | FF | **00** | 00 | 00 | 00 | 00–| 00 | 00 | 00 | 00 | 00 | 00 | 00 | 00 |
| 30 | 00 | 00 | 00 | 00 | 00 | 00 | 00 | 00–| 00 | 00 | 00 | 00 | 00 | 00 | 00 | 00 |

Referring to Example 10.1, the numbers down the right and across the top are for reference only and are not part of the FAT display. These numbers will be used to aid in explaining the FAT and are not FAT field numbers. To find a byte number address, add the row number on the left to the column number across the top. For example, byte number $23_{16}$ is underlined and in bold type.

To show you how confusing this is, the location of entry $002_{16}$ in the FAT is at offset $03_{16}$. If you look at row 00 and column 3, you are looking at only part of the FAT entry for allocation unit $002_{16}$. The $03_{16}$ is only the least significant byte (8 bits) of the 12-bit number. The 0 in the number 40 contained at offset 04 is the most significant hexadecimal digit. Putting the two together shows that the FAT entry for allocation unit number $002_{16}$ is $003_{16}$.

As you can see, the information contained in Example 10.2 is basically useless, except to a few individuals who can translate the table into meaningful information. For this reason, there are several shareware programs available that will load the FAT into the computer's memory and straighten out the display so that it can be deciphered easily. Example 10.3 is a listing of the same FAT segment shown in Example 10.2 that has been translated into a more straightforward format. This straightened out format will be used during the remaining discussion on FAT.

**Example 10.2:** Straightened up FAT

| Allocation unit FAT field number | 000 | 001 | 002 | 003 | 004 | 005 | 006 | 007 |
|---|---|---|---|---|---|---|---|---|
| Contents | FFD | FFF | 003 | 004 | 006 | 00A | 009 | FFF |

| Allocation unit FAT field number | 008 | 009 | 00A | 00B | 00C | 00D | 00E | 00F |
|---|---|---|---|---|---|---|---|---|
| Contents | 00B | 00F | 00D | 00E | 012 | 010 | 011 | FFF |

| Allocation unit FAT field number | 010 | 011 | 012 | 013 | 014 | 015 | 016 | 017 |
|---|---|---|---|---|---|---|---|---|
| Contents | 013 | FFF | 015 | FFF | FF7 | FFF | 000 | 000 |

| Allocation unit FAT field number | 018 | 019 | 01A | 01B | 01C | 01D | 01E | 01F |
|---|---|---|---|---|---|---|---|---|
| Contents | 000 | 000 | 000 | 000 | 000 | 000 | 000 | 000 |

# Mass Storage Basics

> **Procedure 10.1: Procedure for loading a display of the FAT of a 1.44-MB diskette using Debug**
>
> 1. Place a formatted 1.44-MB diskette into the A: drive. This diskette can contain data but should be a scratch diskette. As the name implies, a scratch diskette is a diskette that does not contain valuable data.
> 2. From the DOS prompt, enter the following command. The placement of spaces is critical to the syntax of the command. For this reason, underscores are used to show the location of spaces in the command line. Replace all underscores with spaces.
>
>    C:\Windows>debug
>    -L_100_0_1_12
>    -D_100
>
> **Explanation:**
> The dash (-) is the Debug prompt. The line (L 100 0 1 12) loads into memory offset 100 the contents of drive 0 (A:) sector 1 and $12_{16}$ (18) sectors. Spaces are critical in this command. The line (D 100) dumps the contents of memory offset 100 and the next consecutive 128 bytes to the screen. To continue displaying the rest of the FAT, keep pressing the D key. Once you are finished, enter the following line. This will exit Debug and return the computer back to the DOS prompt.
>
> -Q

Table 10.5 Type of Allocation Unit Information Contained in Each FAT Field*

| | |
|---|---|
| 000 | Indicates that the allocation unit is unused and available |
| FF8–FFF | Indicates that this is the last allocation unit of a file |
| FF7 | Indicates that the allocation unit is defective |
| XXX | Indicates the location of the next cluster of a file |

*The individual fields in the FAT describe the usage of their corresponding allocation units. The contents of the FAT12 fields are interpreted as shown.

The DOS Debug program was used to load and display the FAT segment shown in Example 10.2. The steps used are outlined in Procedure 10.1. For more information on available Debug commands, consult Appendix D.

Note that because the contents of the FAT depend on the files stored on the diskette, the FAT entries displayed in Procedure 10.1 will be different for each diskette, except for the first 3 bytes. These bytes should be the same for all 1.44-MB floppies because they are a code that indicates the way the diskette was formatted. All of the other FAT fields will contain codes that depend on the type of information contained in the allocation unit represented by the FAT entry. These codes are listed in Table 10.5.

Let us discuss the Xs shown in Table 10.5 a little further. In Example 10.4, the FAT field for allocation unit number $00A_{16}$ contains the value $00D_{16}$. After the operating system loads the contents of allocation unit number $00A_{16}$ from the diskette, it needs to know where to go next. To find out, the operating system examines the FAT field for allocation unit number $00A_{16}$ and finds that this table entry contains the value $00D_{16}$. This tells the operating system to load the contents of allocation unit $00D_{16}$ after loading $00A_{16}$. If the operating system finds a number between FF8 and FFF in the FAT field for allocation unit number $00A_{16}$, it will stop loading data.

Before we continue our discussion of the FAT, let us discuss the root directory in greater depth and then tie everything together to see how data are stored and retrieved on disk.

## 10.5.4 The Directory

On a 3½" HD diskette, the root directory on the diskette is located in logical sectors $13_{16}$ ($19_{10}$) through $20_{16}$ ($32_{10}$), for a total of $14_{10}$ sectors. Each filename in the directory uses 32 rigidly defined bytes. Seven sectors will give a total of $7168_{10}$ bytes for the root directory. With 32 bytes/entry, up to 224 entries can be made in the root directory of a 1.44-MB diskette. **Subdirectories** are treated as files, so the number of entries in these directories is limited only by disk space.

If the directory (root or subdirectory) is loaded into the RAM of the computer using Debug, and examined, it will appear similar to the directory segment shown in Example 10.3. The characters represented by the ASCII code for the filename and extensions are displayed to the right. Note that this is only a small portion of the directory.

**Example 10.3:** Directory example

|    | 0  | 1  | 2  | 3  | 4  | 5  | 6  | 7  | 8  | 9  | A  | B  | C  | D  | E  | F  |          |     |
|----|----|----|----|----|----|----|----|----|----|----|----|----|----|----|----|----|----------|-----|
| 00 | 52 | 49 | 43 | 48 | 20 | 20 | 20 | 20-45 | 58 | 45 | 20 | 00 | 00 | 00 | 00 | RICH | EXE |
| 10 | 00 | 00 | 00 | 00 | 00 | 00 | 00 | 60-54 | 07 | **02** | **00** | 80 | 28 | 00 | 00 |     |     |
| 20 | 52 | 41 | 59 | 4D | 4F | 4E | FF | 20-45 | 58 | 45 | 20 | 00 | 00 | 00 | 00 | RAYMOND | EXE |
| 30 | 00 | 00 | 00 | 00 | 00 | 00 | 00 | 60-54 | 07 | 07 | 00 | 1F | 09 | 00 | 00 |     |     |
| 40 | 42 | 45 | 54 | 54 | 45 | 4D | 45 | 52-47 | 49 | 46 | 22 | 00 | 00 | 00 | 00 | BESSEMER | GIF |

Each 32-byte directory entry is divided into the following fields:

| 0–7   | Filename (8 bytes of ASCII) |
| 8–A   | Extension (3 bytes of ASCII) |
| B     | Attributes |
| 16–17 | Time (2 bytes) |
| 18–19 | Date (2 bytes) |
| 1A–1B | Starting allocation unit number (2 bytes) |
| 1C–1F | Total byte length of file (4 bytes) |

The first byte of the filename field may contain the following special information:

| 00 | Directory entry has never been used since format. |
| 05 | First character of filename in ASCII code is actually $E5_{16}$. |
| E5 | File has been erased. |

The bits in the attribute field are assigned as follows (1 in bit = true):

| 0 | Read-only |
| 1 | Hidden file—excluded from normal search |
| 2 | System file—excluded from normal search |
| 3 | Volume label—root directory only |
| 4 | Subdirectory |
| 5 | Archive bit—set on whenever file is modified |
| 6 | Not currently used |
| 7 | Not currently used |

The bits in the time field are assigned as follows:

| 0–4 | Binary number in 2-second increments |
| 5–A | Binary number of minutes |
| B–F | Binary number of hours |

The bits in the date field are assigned as follows:

| 0–4 | Day of the month |
| 5–8 | Month |
| 9–F | Year relative to 1980 |

As you can see, the actual directory on the diskette is hard to decipher. For this reason, all of the operating systems used on compatible computers display the contents of the directory in a more user-friendly format. The operating system command that displays the directory does not show the starting allocation unit number of the file. The operating system also does not display or allow the user to set some of the file's attribute bits. Knowledge of this information and the ability to control every aspect of the directory are critical to individuals who specialize in data recovery.

The same procedure used to load the FAT into the RAM using Debug can be used to load the root directory. The only difference is that the line that specifies the sector number and the number of sectors needs to be changed to read: (L 100 0 13 14).

### 10.5.5 Storing and Retrieving Information on a Disk

The starting allocation unit number of a file is not located in the FAT. It is stored in the directory with the filename. Once the operating system loads the data from the first allocation unit into RAM, it uses the number of this allocation unit as an entry point into the FAT. From the entry point on, each FAT field contains the number of the next allocation unit in the file, until the end-of-file mark (FF8–FFF) is encountered in the FAT. (Note: The operating system does not display the starting allocation unit number or attribute information when the directory is listed.)

When the operating system saves a file to the disk, it first checks the directory to see if the filename already exists. If the filename is present, the new data replace the old, using the same allocation units that were assigned to the original. If more allocation units are needed, the operating system will locate free allocation units by searching for the entry 000 in the FAT and will assign them to the file as needed. If the file occupies fewer allocation units than were previously needed, these will be marked unused (000) in the FAT.

If the filename is not present, the operating system starts reading the FAT to find the first free cluster (allocation unit). Once the operating system finds the first free cluster, it starts storing the file into that cluster. If more than one allocation unit is needed, the operating system finds the next free allocation unit in the FAT and then puts its number in the FAT field that is assigned to the allocation unit just written. This process continues until the last allocation unit of the file is written to the disk. The operating system then stores FF8–FFF (depending on the file type) in the FAT field to indicate that this is the last allocation unit of the file.

After the entire file is written to the diskette, the file's starting allocation unit number, attributes, size, creation date, and creation time are added to the directory listing. The directory listing is updated last in case an error is encountered or the creating procedure is terminated by the user.

### 10.5.6 Loading a Program

This section discusses the sequence followed by the compatible computer's operating system to load a program named RICH.EXE into the computer's RAM. The information given in Examples 10.2 and 10.3 will be used for the remainder of this discussion.

RICH.EXE is one of the programs in the hypothetical directory shown in Example 10.3. According to the chart of where information is located, the starting allocation unit number for this file is found at offset $1A_{16}$. To find the entry in the example, find 10 in the left column. Then follow this over to the column numbered 0A, and you should see the number 02 at the intersection. This number indicates the starting allocation unit number is $002_{16}$. The remainder of the file's allocation unit locations are found in the FAT shown in Example 10.2.

To start the load process, the operating system is told to launch the file using the procedure available with the operating system's shell. This procedure is outlined for DOS, Windows 3.X, Windows 9X, and Windows 2000 in chapters 5 and 6. We will launch the program from the DOS prompt. To launch RICH.EXE, a command line similar to the one shown in Example 10.4 can be used.

**Example 10.4:** Example of a command line used to launch the program RICH.EXE.

A>RICH

As with all DOS command lines, the Enter key must be pressed after typing the line to cause the operating system to execute the command. Once the Enter key is pressed, the operating system searches the directory for RICH.BAT, RICH.COM, or RICH.EXE. If a filename with one of these extensions does not exist, the operating system will display a "Bad command or file name" or a similar error message on the screen and return control of the computer back to the operating system's shell. In this example, the filename will be found under RICH.EXE. Once the file is found, the OS will get the starting allocation unit number from the directory entry RICH.EXE (002 for this example).

After the operating system gets the starting allocation unit number for the directory, the OS establishes some important information called the *program segment prefix (PSP)* in the RAM. The PSP is a block of data that the operating system places in the RAM addresses immediately preceding the program it is loading. This data helps the operating system regain control of the computer when the external program is terminated. It also contains information the program uses to determine the computer resources that are available to the program.

After the PSP is established, the operating system loads the contents of the allocation unit (AU) pointed to by the directory (002 in this example) into the computer's RAM. After the contents of this AU are loaded, the operating system goes to the location in the FAT that holds information pertaining to AU-002. In the FAT shown in Example 10.2, the FAT entry for 002 is 003. This number tells the OS to continue loading the program from AU number 003. After the OS loads the contents of AU-003, it then goes to the FAT entry for number 003, which tells the OS to load the contents of AU-004. The FAT entries in this example route the OS through the allocation units in the order of 003, 004, 006, 009, and 00F. Remember that AU-002 was specified by the file's directory entry.

As can be seen by this example, the load sequence starts with the AU number contained in the directory and continues until the OS detects the value FFF in a FAT field. This value indicates that this is the last AU used by the file. In this example, FFF is located in FAT for AU-00F. Once the end-of-file code (FFF) is reached in the FAT, control of the computer is turned over to the program RICH.EXE.

As you can see, the process used to load a file into the RAM of the computer is fairly straightforward. Notice also that the contents of a file do not have to be in consecutive allocation units. In fact, the more often a file is loaded into RAM, modified, and stored back on the disk, the more likely the file will spread over the disk surface in noncontiguous allocation units.

A file that is not stored in consecutive allocation units is called a *fragmented* file. Having fragmented files on the disk does not affect the computer's ability to load the program because the FAT tells the operating system exactly how to put the file back together as it is loaded into the computer memory. However, fragmented files do significantly increase the time required to load the file into the computer's RAM. For this reason, the OS supplies a utility program that will defragment the files on floppy diskette or hard drive. Defragmenting a hard drive is especially critical because the file sizes are so large. The defragmenting procedures for the operating systems covered in this book are covered in chapter 11.

If both FATs are lost, the information on the disk will be unreadable. This is because the operating system will not know how to put the files back together. For this reason, the FATs are a very popular target for virus programs, which are discussed in greater detail in chapter 15.

## 10.6 FORMATTING A DISK

Knowing how to format a disk with the computer's operating system is very important. Even though most of the floppy diskettes purchased today are preformatted, a time will arise sooner or later when you will need to reformat the diskette. For this reason, the procedures for formatting a diskette for DOS, Windows 3.X, and Windows 95/98 are covered in this section.

MS-DOS versions above 5.0 support two types of formats: an unconditional format and a quick format. The unconditional format will format the diskette completely—recording new tracks, sectors, directory, and FATs. The quick format will make a copy of the FATs and directory on the last few cylinders of the diskette and then place blank FATs and a directory in the correct location. This process is very quick but will not detect defective allocation units on the diskette. The quick format also allows the diskette to be unformatted by the operating system.

To format a disk using DOS, the FORMAT command is used. If you are using a DOS version above 4.0, several format options are available. The syntax for the FORMAT command line is shown in Procedure 10.2.

Referring to Procedure 10.2, the two most widely used switches are /s and /u. The /s switch formats the diskette and transfers to system files so the computer can be booted from the floppy. The /u switch

## Procedure 10.2: Formatting a diskette using MS-DOS

At the DOS prompt, type the following command line syntax. Underscores (_) show the location of spaces.

FORMAT_drive_[/switches]

## Procedure 10.3: Formatting a diskette with Windows 3.X

1. Double-click on the Main icon in the Program Manager window.
2. Double-click on the Winfile icon to open the File Manager.
3. On the Menu Bar of the File Manager, click on the Disk option.
4. From the items available in Disk Options menu, click on the Format listing.
5. From the pop-up Format window, select the disk to format and select the diskette's capacity.
6. To transfer the DOS system boot files to the floppy diskette, place a check in the appropriate box.
7. To perform a quick format, place a check in the appropriate box. I do not recommend using this option because the diskette is not actually formatted.
8. To format the selected diskette, click on the OK button. To exit without formatting, click on the Cancel button.

performs an unconditional format. A diskette formatted with the /u switch checks the integrity of the disk surface during the format procedure but does not allow the diskette to be unformatted, however. Example 10.5 illustrates the use of the FORMAT command to format a diskette in the A: drive using the unconditional format switch. The command is entered from the C prompt.

**Example 10.5:** Command line to perform an unconditional format of a diskette in the A: drive from the C: drive.

C:/>format a: /u

Formatting a diskette with Windows 3.X and Windows 9X/2000 is quite simple because of their GUI. Procedure 10.3 gives the steps and options to perform the format process with Windows 3.X. Procedure 10.4 is for Windows 9X/2000. Note that these procedures could also be used to perform a high-level format on a hard drive. For more information on this procedure, consult chapter 11.

## Procedure 10.4: Formatting floppies using Windows 9X/2000

1. Click on the Start button.
2. Move the pointer up to Programs and then over to Windows Explorer.
3. Click on Windows Explorer.
4. Right-click on the drive you wish to format in the left panel in Windows Explorer.
5. Choose the Format option displayed in the pop-down menu.
6. In the Format Drive window, use the scroll bar to select the capacity of the diskette being formatted.
7. Click on the Option button to choose Quick, Full, or Transfer System Files only. The Quick format does not perform a format of the diskette's surface and can be unformatted. The Full format formats the entire diskette and performs a surface scan for errors. Once a diskette has been formatted using the Full option, it cannot be unformatted. The Transfer System Files option will transfer the boot files to a previously formatted diskette.
8. If desired, enter a software label in the text box.
9. Remove or place a check in the check boxes beside No label, Display summary when finished, or Copy system files depending on the desired options.
10. Click on the Start button.
11. Once the format process in completed, click on the Close button.

**Figure 10.7** Floppy Drive Pin Connector

Knowing how to use the floppy drive through the computer's operating system is very important for individuals who are learning to build, repair, and upgrade computers. However, it is just as important that these individuals know the requirements of the interface between the floppy drive and the computer, which is our next topic.

## 10.7  THE FLOPPY CONNECTION

A complex IC called the **floppy disk controller (FDC)** is usually located on the motherboard as part of the chipset. The FDC is comprised of a set of ports that handle all of the operations of the floppy disk drive(s). The FDC and its associated ports are at I/O addresses 3F2H through 3F7H and use IRQ6 and DMA2. This is simply nice-to-know information because the resources used by the FDC are fixed and cannot be changed by the user.

The FDC in most compatible computers can actually control up to four drives, a little known fact because the operating systems covered in this book support only two floppy drives. If a third or fourth drive is installed, it will have to be supported by either a TSR or an application program. The drives controlled by the FDC do not necessarily have to be floppy drives. In fact, some internal tape drives are controlled by the FDC in the compatible computer. However, with MS-DOS and all versions of the Windows operating system, there can be only two floppy drives. Instead of actually having to access the two drives by address, all of the aforementioned operating systems allow the user to access the floppy drives under the names A: or B:.

A 34-pin male pin connector is located on the motherboard or MI/O adapter card. Its layout is similar to the one shown in Figure 10.7. A special cable designed for the floppy drive interface must be inserted into this connector and the connector on each floppy drive. The pin assignments for this cable are given in Table 10.6.

A 34-pin ribbon cable is used to make the connection between the motherboard or MI/O adapter card connector and the drive(s). There can be from three to five daisy-chained connectors on the cable. The pin connector on the cable that is spaced apart from the others is attached to the FDD connector on the motherboard or MI/O card. The pin-1 side of the cable is denoted by the cable's colored edge. The motherboard or MI/O card's FDD pin connector should have the number 1 written on the circuit card, identifying the pin-1 side of the connector. If this is not the case, the documentation that comes with the motherboard or I/O adapter card should have a drawing of the connector indicating the location of pin-1. When the cable is installed, the pin-1 side of the ribbon cable should mated to the pin-1 side of the motherboard or I/O adapter card connector. A typical floppy drive cable is shown in Figure 10.8. Notice that there is a twist in the cable between the outer set of connectors where the drives attach. This twist in the cable determines which drive is A: or B:. The A: drive is attached to one of the two connectors at the end of the cable, after the twist. If installed, the B: drive attaches to one of the two cable connectors before the twist. If only one drive is installed in the computer, it must be addressed as the A: drive and attached to one of the connectors at the end of the cable after the twist.

The 5¼" floppy drive attaches to the cable through a 34-pin male edge card connector, whereas the 3½" floppy drive uses a 34-pin male pin connector. The floppy cable will usually have both a female pin and an edge card connector on each side of the twist in the cable (see Figure 10.8). This allows either style drive to be assigned as either the A: or B: drive. If the cable has only two con-

Table 10.6  Pin Assignments for the Floppy Disk Port Connector

| Signal | Pin | Pin | Signal |
|---|---|---|---|
| Normal/High density | 2 | 1 | Ground |
| In use/head load | 4 | 3 | Ground |
| Drive select 3 | 6 | 5 | Ground |
| Index pulse | 8 | 7 | Ground |
| Drive select 0 | 10 | 9 | Ground |
| Drive select 1 | 12 | 11 | Ground |
| Drive select 2 | 14 | 13 | Ground |
| Motor on | 16 | 15 | Ground |
| Direction | 18 | 17 | Ground |
| Step | 20 | 19 | Ground |
| Write data | 22 | 21 | Ground |
| Write gate | 24 | 23 | Ground |
| Track 0 | 26 | 25 | Ground |
| Write-protect | 28 | 27 | Ground |
| Read data | 30 | 29 | Ground |
| Side select | 32 | 31 | Ground |
| Ready | 34 | 33 | Ground |

nectors on each side of the twist, both will be female pin connectors. If this is the case, an adapter can be purchased that will convert the pin to an edge card connector so that a 5¼" floppy drive can be attached.

The 5¼" floppy drive's edge card connector usually has a plastic key that allows it to be attached in only one direction. The 3½" floppy drive and FDD M/IO card pin connectors can be attached in either direction. Of course, only one direction is correct. If one of the connectors is backwards, the drive access light(s) illuminates immediately when power is turned on to the computer and stays illuminated. A backwards connector will not hurt the drive, motherboard, or MI/O adapter card but can possibly destroy information stored on a diskette. As a safety precaution, do not have a diskette in the drive when you apply power to a computer with a newly installed drive or if the disk drive cable has been removed and reattached.

To alleviate this problem, should it occur: (1) turn the computer off; (2) turn one connector around; and (3) reapply power to the computer. If the light still comes on and stays on as soon as the computer is turned on, repeat the procedure until it stops.

Even though the floppy drive has been the primary method of getting information into the computer's memory or hard drive for years, it appears to be in its waning years of usefulness. This is due mainly to its relatively small storage capacity. The current method most often used to deliver operating systems, applications, and drivers from the vendor to the consumer is by way of a media called CD-ROMs, which are covered later in the chapter. For now, let's examine the procedure used to remove and install a typical floppy drive.

**Figure 10.8** Typical Floppy Drive Cable

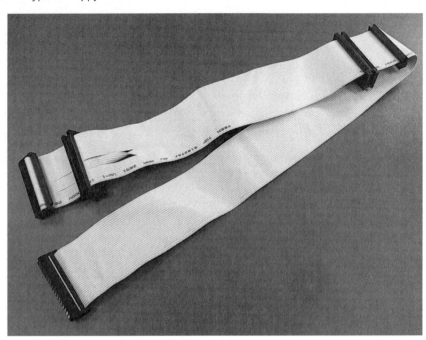

## 10.7.1 Floppy Drive Removal and Installation

The physical installation of a floppy drive is not a difficult task, mainly because no drivers are required. This is because all current compatible PC operating systems, including the BIOS, offer full support for these drives. For this reason, floppy drives are truly Plug-and-Play devices. The steps to follow to remove or install a floppy drive are given in Procedure 10.5.

### Procedure 10.5: Floppy drive removal and installation

**General**
1. Make sure the computer's power switch is in the off position and the power cord is unplugged from the rear of the computer.
2. Remove the computer chassis and place the screws in a safe location.
3. If available, attach an anti-static mat to a bare metal surface on the computer chassis.
4. If available, wear an anti-static wristband attached to the anti-static mat or to a bare metal surface on the computer chassis. If an anti-static wristband is not available, touch a bare metal surface on the chassis often to prevent static buildup.

**Removal**
1. Remove the FDD power and data cable from the rear of the drive.
2. Some desktop and tower computers mount the 3½″ floppy drive and hard drive in a removable drive bay that has to be removed to get to the screws that mount the drive. The removable bay is usually secured with two screws. One is located where the removable bay is attached to the computer's front panel. The other screw is usually located on the side of the bay that attaches to the 5¼″ bay. Once these screws are removed and stored in a safe location, remove the bay by sliding it toward the rear of the computer.
3. Remove the FDD mounting screws that secure the drive in the drive bay and store them in a safe location. The screws are located on both sides of the FDD. The drive may need to be supported from the back.
4. Remove the drive by gently pushing the rear of the FDD toward the front panel of the computer. This will move the drive out the front of the chassis where it can be removed entirely. Place the removed FDD on an anti-static mat or store the FDD in an anti-static bag.

**Installation**
1. Remove the drive from the anti-static bag and place it on the anti-static mat. If an anti-static mat is not available, touch the FDD's anti-static

bag to a bare metal surface on the computer's chassis.
2. Locate a drive bay that matches the drive's width and has an opening to the outside through the chassis. To gain access to the external drive bay opening, a plastic Bissell (faceplate) may have to be removed.
3. On a new chassis, a metal knockout may have to be removed from the front of the external drive bay in order for the drive to mount. Be careful removing the knockout.
4. Slide the FDD into the bay through the opening in the front of the chassis until the faceplate on the FDD is flush with the front panel of the computer. You may have to support the rear of the drive with your hand through the back of the drive bay.
5. While supporting the rear of the drive with your hand, install the mounting screws though the mounting slots in the side of the drive bay into the screw holes in the side of the FDD. Make sure the correct size screws are used. The correct screws are often supplied with a new drive.
6. Once mounted, connect the drive to the last connector on the cable (after the twist) if the FDD is going to be the A: drive. Use the connector before the twist if the drive is to be the B: drive.
7. Check all power connectors and data cables on the other internal peripherals.
8. Reinsert the power cord and turn the computer on.
9. Check the new drive for proper operation. If a problem exists, follow the troubleshooting procedures in this chapter.
10. If the drive seems to be operating properly, remove power from the computer, and then reinstall the computer and return the computer to service.

## 10.8 CD-ROM

**CD-ROM** is the acronym for **compact disc-read-only memory.** CD-ROMs are simply another form of mass storage and are not part of the computer's memory unit. The CD-ROM gets its name from the fact that it can only be read from and not written to. Software supplied on CD-ROM is purchased prerecorded from software vendors and cannot be written over.

You might ask, "Why buy them if they can only be read from?" The reason is that current CD-ROM discs can hold up to 650 MB of data. This density gives them the ability to hold hundreds of large graphics files, entire encyclopedia sets, very large games, and many other things. CD-ROMs and audio compact discs work on exactly the same principle. In fact, an audio CD can be played in a CD-ROM drive without problems if a sound card is installed in the computer and the operating system has the correct application available.

CD-ROM drives are the major reason for the boom in multimedia presentations. As previously stated, *multimedia* is information conveyed to the user by more than one form of media. It has been proven that the more forms by which information can be presented, the more information we will retain.

The most popular forms of multimedia are sound and motion video. These forms of media are not widely used in computers because of the high number of bytes that are required to store information in both of these formats. This is where the CD-ROM comes into play with its ability to store over 600 MB of information on a single disk.

The expression *multimedia computer* is used in reference to a computer that has a CD-ROM, sound card, and speakers as standard equipment. More often than not, the multimedia computer also has an internal modem installed.

If your computer does not have a CD-ROM drive, one can be purchased as either an internal or external unit at most places that sell computer hardware. A typical internal CD-ROM drive is shown in Figure 10.9. The internal units are cheaper because a case and power supply are not required. An internal CD-ROM mounts in a 5¼″ external drive bay (see section 10.10.1). Most external CD-ROMs connect to the computer with an interface called *SCSI*, but they are also available for parallel and USB ports. An internal CD-ROM can connect to the computer with a proprietary, SCSI, or EIDE interface. SCSI and EIDE interfaces are hard drive interfaces, which are covered in chapter 11. The proprietary interface is discussed in the next section.

CD-ROMs and audio CDs operate on the principle of light reflection. Pressed into the clear surface of the CD disk are valleys called *pits* and plateaus called *lands*. A very thin reflective coating, usually aluminum, is placed over the back of the CD to provide reflection. As the CD rotates, a laser

**Figure 10.9** CD-ROM

beam is focused though lenses onto the CD's surface. The beam is reflected off the pits and lands at a different phase, indicating 0s or 1s. The laser beam reflects off the surface back into a photo diode or transistor that detects the phase shifts as either 0s or 1s. This basic operation is illustrated in Figure 10.10.

Unlike information recorded on a floppy disk and hard drive, the information placed on a CD is in one continuous spiral that starts in the center of the disc and extends toward the outside. The actual data is placed in packets of 2 KB that are read as a contiguous unit. Even though information is not saved in the same format used by floppy disks and hard drives, a utility program called *MSCDEX* allows the user to access data on the CD in a manner similar to that used by floppies and hard drives.

### 10.8.1 CD-ROM Proprietary Ports

When CD-ROM drives were first introduced to the compatible computer community, there was no standard port for them to attach to. As a result, CD-ROM drive manufacturers developed their own proprietary ports for their CD-ROM drives. The CD-ROM manufacturers allowed sound card manufacturers to install their proprietary ports on their sound cards. The sound card, CD-ROM drive, and speakers were sold as part of a unit called a *multimedia kit*.

At the time, there were about four leading manufacturers that were making CD-ROM drives. Each drive manufacturer used its own proprietary interface. This meant that the only CD-ROM drive that could be connected to the sound card was the one for which it was designed. Some sound card manufacturers tried to sell cards that would handle all of the CD-ROM interfaces on the market. They did this by having multiple CD-ROM connectors on the sound card. As you can guess, this caused a lot of problems because if the sound card or CD-ROM became defective, users had to purchase a replacement for the defective component that matched the interface of the component that was not replaced.

Most of the CD-ROM drives manufactured today do not use proprietary interfaces. Instead, as stated previously, they are designed to connect to an EIDE, SCSI, parallel, or USB port.

Along with the interface standard, there are several other CD-ROM standards one needs to understand in order to fully comprehend the operation and terminology used with this technology. For this reason, the next few sections cover some of the most widely used standards.

**Figure 10.10** Basic Operation of a CD-ROM Drive

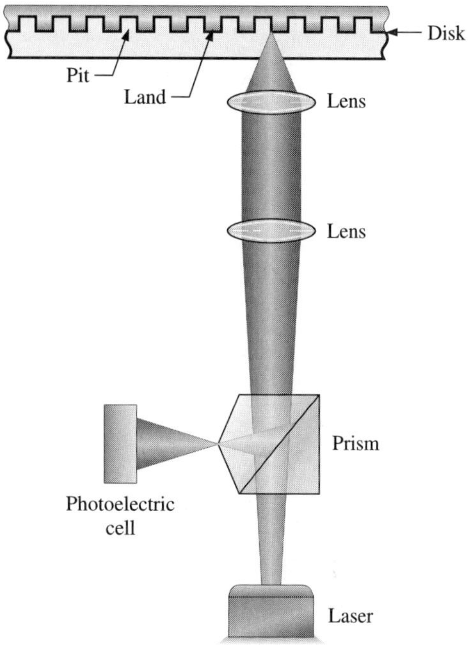

*Source:* Floyd, *Digital Fundamentals,* 7th ed. (Upper Saddle River, NJ: Prentice Hall, 2000) p. 659.

## 10.8.2 Other CD-ROM Standards

When CD-ROM drives were first introduced, there were no set standards as to the format that would be used to store information on the disk. If drive manufacturers used different formats, the CDs made for one manufacturer's drive could not be read by another manufacturer's CD-ROM drive.

To help resolve this major problem, leading CD-ROM manufacturers met in 1995 at the High Sierra Hotel and Casino in Lake Tahoe, California. The manufacturers agreed on a CD-ROM format that became known as the **High Sierra format.** Later, the International Standards Organization (ISO) brought out an international standard called *ISO 9660,* which used the High Sierra format as its backbone. Adoption of this standard has allowed CD-ROMs to be a cross-platform media that can be used with different operating systems and computers other than the compatibles.

Of all the specifications and standards used by the CD-ROM, the speed rating is probably one of the most confusing. For this reason, the speed rating is the next topic of discussion.

## 10.8.3 CD-ROM Speed

When CD-ROM drives first came out, they rotated the CD at the same speed as an audio CD player. This rotational speed allowed a data transfer rate of approximately 150 KB per second (KBps). Most CD-ROM drives are currently rated with a multiplier of this original transfer speed. For example, a 12× CD-ROM drive has a maximum transfer of 12× 150 KBps, or 1,800 KBps.

**Problem 10.1:** What is the maximum transfer speed of a 24× CD-ROM drive?

**Solution:** 24 × 150 KBps = 3,600 KBps

CD-ROM discs are rapidly becoming the preferred media for operating systems, applications, and drivers because of their large storage capacity and their relatively low cost. Every compatible PC currently manufactured comes with a CD-ROM drive as standard equipment. Through the years, the only limitation to the CD-ROM technology was that it had the ability to only read CD-ROM discs and not create them. However, this limitation is rapidly disappearing.

### 10.8.4 CD-R

One can imagine the need to have a transportable media that can not only be read from but also written to. Since the inception of the compatible PC in 1981, the most popular transportable media in the compatible PC has been the lowly floppy drive. During the infancy of the PC, the floppy diskette performed the tasks assigned to it quite nicely. It provided a means for vendors to sell their operating systems, applications, and drivers so the user could install them to a hard drive. It also provided a method to copy data that could be used by other computers. However, all of this occurred in the days before the graphical user interface (GUI), large applications, and sound and video requirements desired by today's computer users. The storage capacity requirements needed by current operating systems, applications, and data make the meager 1.44-MB storage capacity offered by the current 3½″ floppy drive standard equivalent to trying to fit a gallon of water into a thimble.

In the last few years, many contenders have tried to replace the floppy drive with the next standard for transferable read/write storage. These include drives with names such as Zip and Jaz, along with a few others. The companies that manufacture these drives have increased the storage capacities to over 2 GB, with larger capacities sure to follow. The only problem with these drives is that they are not included in the list of standard peripherals that are part of today's premanufactured compatible PCs. This limits the transfer of information only between computers that have the same proprietary drives installed. For example, I have a Zip drive on the computer at my home but none of the computers at my office have a Zip drive installed. On the other hand, all of the computers at my home and office have CD-ROMs installed.

Besides the floppy, the only other standard mass storage device included with today's compatible PC, which utilizes a transferable storage media, is the CD-ROM drive. The problem here is that the standard CD-ROM drive can only read from CDs that were pressed at the manufacturer or recorded on a newer device called a *CD-R* and *CD-RW*.

The **compact disc recorder (CD-R)** drive is basically a CD-ROM drive that has the ability to read and write information. When reading, the CD-R drive functions exactly like the standard CD-ROM drive found on almost all of the current compatible computers. When writing, the intensity of the CD-R drive's laser is increased in order to burn pits and lands onto the surface of a specially designed CD-R disc. Once recorded, information on the CD-R disc can be read by the CD-R drive itself or by most of today's standard CD-ROM drives.

CD-R drives can be used to copy the entire contents of other CDs or individual files and folders from other mass storage devices such as hard drives and floppies. What is really neat is the fact that when recording individual files and folders, all of the information intended to be recorded on the CD-R disc does not have to be recorded in one session. This feature allows other files and folders to be appended to previously recorded information until the CD-R disc is filled to capacity. The storage capacity of a current CD-R disk is the same as that of the pressed CD-ROM disc—650 MB.

Along with the ability to record data, most of the software that accompanies the CD-R drive allows the user to record up to 74 minutes of audio information using exactly the same format as that used by prerecorded audio CDs. This allows users to make copies of their favorite music CDs. The ability of the CD-R to record information in different sessions allows users to make compilations of their favorite songs from different CDs or from audio files stored on the computer's hard drive.

The only problem with CD-R technology is that it is a write-once-read-many (WORM) format. This means that the CD-R disc cannot be erased and used over and over again like a floppy diskette. This limitation of the CD-R is being resolved by the CD-RW.

### 10.8.5 CD-RW

Just like the CD-R, **compact disc read/write (CD-RW)** drives function exactly like a standard CD-ROM when they are reading a prerecorded or pressed CD. The major difference is that the CD-RW has the ability to erase and re-record information on a specially designed CD-RW disc. The secret to this trick is the CD-RW disc itself and the laser technology used in the CD-RW drive. The CD-RW disc responds to three different intensities of the laser beam: one for read, one for write, and the other for erase. The CD-RW drive adjusts the intensity of the beam depending on the mode selected by the program. The intensity used by the CD-RW to write is compatible with the surface of a CD-R disc. This allows

Mass Storage Basics

the CD-RW drive to record information on either media. However, the CD-RW disc can be erased and re-recorded.

If a new technology called *pack writing* is supported by the CD-RW drive and the supplied software, the CD-RW disc can be formatted to act like a standard floppy or hard drive. Once formatted, the computer operating system can read or write to the formatted CD-RW disc as if it was any other magnetic media.

### 10.8.6 CD-R and CD-RW Summary

To sum things up, CD-R and CD-RW drives, discs, and software are an excellent investment. If prices of the technology continue their current downward spiral, CD-Rs and CD-RWs will surely be the transportable read/write media of choice to replace the 3½" floppy disk. Even though the drives themselves are still relatively expensive, the CD-RW and CD-R discs are very reasonably priced. The current cost of a blank CD-R disc is around $1, while the cost of a CD-RW is around $2.

Installing the CD-ROM, CD-R, and CD-RW drives into a compatible computer is a fairly straightforward process, which is covered in section 10.8.9. At this time, let us discuss the installation of the driver that makes the CD-ROM drive work. Before we move on, however, it should be mentioned that duplicating copyrighted material without permission is against the law.

### 10.8.7 CD-ROM DOS Drivers

DOS, Windows 3.X, and Windows 9X DOS mode do not directly support a CD-ROM drive. As a result, a driver for the CD-ROM drive must be loaded before the drive will operate under the OS. The driver has to be loaded by a line in the CONFIG.SYS. This is true no matter which interface is used by the drive.

When a new drive is purchased, the DOS driver for the CD-ROM drive is usually supplied on a diskette that comes with the drive. If you do not have a diskette that contains the driver for the CD-ROM drive, you can usually obtain one from the CD-ROM drive manufacturer.

Unfortunately, the CD-ROM driver is often not supplied with preassembled computers. However, it is usually already installed in a directory on the computer's hard drive. If this is the case, you need to make a copy of the driver to a diskette no matter which operating system is running the computer.

You may wonder why you need a copy of the DOS CD-ROM driver on diskette if you are using an operating system that fully supports a CD-ROM drive such as Windows 9X. Sooner or later you will need to reinstall Windows 9X from scratch or you will need to run a program that will only run in DOS mode. When this time comes, you will have the DOS-based driver to install Windows 95 or use the DOS mode that is available with Windows 95 or 98.

To help identify the location of the DOS CD-ROM driver on the hard drive, you will need to display the contents of the CONFIG.SYS file. If you are running Windows 95 or 98, you can use the System Editor utility program to edit the contents of the DOS and Windows 3.X configuration files. The procedure for using System Editor was covered in chapter 6.

The line that loads the DOS CD-ROM driver should be the first or second line in the CONFIG.SYS file. It begins with the DEVICE= or DEVICEHIGH command. If there are no DEVICE commands in the CONFIG.SYS file, there is no DOS CD-ROM driver being loaded and the CD-ROM will not work in DOS mode. An example of a CONFIG.SYS command line to load the DOS CD-ROM drivers is given in Example 10.6.

**Example 10.6:** A typical CONFIG.SYS command to load the CD ROM driver for DOS

> Underscores (_) are used to show the location of spaces in the command line.
> DEVICE= C:\CDROM\DRIVER.SYS_/D:MSCD0001

The command shown in Example 10.6 consists of three or four parts. First, there is the DEVICE= command that is used to load drivers in the CONFIG.SYS file. Following this command is C:\CDROM\DRIVER.SYS, which is the path and name of the driver itself. Of course, this path and driver name can vary among drivers. Next in the command line is /D:MSCD0001, which identifies the number of the CD-ROM drive for which the driver is written. If another CD-ROM is installed in the computer with a different driver, it might use /D:MSCD0002 or something similar. The letters can vary from driver to driver, but the numbers will always identify the number of the drive. Make a note of this listing of the CONFIG.SYS because it will have to be matched exactly in the MSCDEX line in the AUTOEXEC.BAT file. Following the drive number, there may be switches that are indicated with the slash (/). Because switches vary among drivers, they will not be discussed here.

In addition to the CD-ROM driver loaded in the CONFIG.SYS file, another utility program is required before the CD-ROM can be accessed from MS-DOS or Windows 9X's DOS mode. The name of the file that contains this utility is *MSCDEX.EXE,* which should be provided with the software that comes with the drive. It can also be downloaded from Microsoft's Web site.

Unlike the CD-ROM driver, MSCDEX must be loaded through a command in the AUTOEXEC.BAT file. An example of the syntax and an explanation of the command options are given in Example 10.7. The function of MSCDEX is to add extensions to the DOS operating system that allow it to access the information on the CD-ROM. If this driver is not loaded, the CD-ROM will not operate even if the correct driver is loaded in the CONFIG.SYS file.

**Example 10.7:** MSCDEX syntax

d:\path\mscdex /switch

Example:

C:\windows\mscdex.exe /d:mscd0001 /v

| Switches | Explanation |
|---|---|
| /d:x | The device number. The characters and numbers must match those listed in the line in the CONFIG.SYS that loads the CD-ROM driver. |
| /L:letter | Designates a drive letter for the CD-ROM drive. This letter must not conflict with other drives already assigned and must be equal to or lower than the letter specified by the LASTDRIVE command in the CONFIG.SYS file. If a letter is not assigned, the CD-ROM will automatically take the next available letter after the last hard drive. |
| /v | Causes the computer to display information about the CD-ROM drive's assigned drive letter, memory allocation, and buffers. |
| /m:x | Allows the configuration of the RAM buffers used by the CD-ROM drive. |
| /k | Adds support for Kanji Japanese standard. |
| /s | Allows the CD-ROM drive to be shared on a network. |

### 10.8.8 CD-ROM Settings in Windows 9X

Although some people like the feature, one of the most annoying things to me about the CD-ROM in Windows 9X is the auto-insertion feature. This feature automatically launches CDs that have an AUTORUN.INF file in the CD's root directory when they are inserted into the CD-ROM drive. Holding the Shift key down on the computer when the disc drawer is withdrawn into the CD-ROM drive will defeat this option. To turn off this feature completely and to display other settings concerning the CD-ROM drive, follow the steps outlined in Procedure 10.6.

### 10.8.9 CD-ROM/R/RW Removal and Installation

Like floppy drives, all standard CD-ROMs are fairly easy to install. The only major difference is that the CD-ROM installs into an external 5¼″ drive bay and an additional cable is needed to play audio CDs through the sound card. Also, if the drive is to be used in a computer using MS-DOS, Windows 3.X, or Windows 9X DOS mode, a DOS-based driver will have to be installed in the CONFIG.SYS and AUTOEXEC.BAT files. In addition, an application program that supports the CD-R or CD-RW will have to be installed in order to obtain the write abilities of these drives. The removal and installation procedures are given in Procedure 10.7.

# Mass Storage Basics

## Procedure 10.6: Windows 9X CD-ROM settings

1. Click on the Start button, then move up to Settings, and then over to Control Panel.
2. After the Control Panel window opens, locate and then double-click on the System icon.
3. After the System Properties window opens, click on the Device Manager tab.
4. Under Device Manager, locate and then click on the plus (+) sign beside the CDROM listing.
5. Double-click on the CD-ROM drive for which the settings are to be changed.
6. Click on the Settings tab.
7. To disable Auto-Insert notification, remove the check beside this option.
8. The settings for Disconnect and Sync Data Transfer or for CD-ROM that use the SCSI interface should not be changed if the drive is working directly.
9. If the CD-ROM drive supports DMA, select this option.
10. Click on the OK button to accept changes or the Cancel button to exit without changes.
11. If changes were made, the computer will have to be restarted in order for the changes to take effect.

## Procedure 10.7: CD-ROM and DVD removal and installation

### General
1. Make sure the computer's power switch is in the off position and the power cord is unplugged from the rear of the computer.
2. Remove the computer's cover and place the screws in a safe location.
3. If available, attach an anti-static mat to a bare metal surface on the computer chassis.
4. Wear an anti-static wristband attached to the anti-static mat or to a bare metal surface on the computer chassis. If an anti-static wristband is not available, touch a bare metal surface on the chassis often to prevent static buildup.

### Removal
1. Remove the data cable, small audio cable, and power cable from the back of the CD-ROM drive.
2. Make a note of the position of the setting for the master/slave/cable-select jumper on the back of the drive.
3. Remove the mounting screws from both sides of the CD-ROM drive.
4. Gently push the drive from the back toward the front panel of the computer. Sometimes the drive will need a slight wiggle to free it from the front panel. The drive should come out of the computer from the front.
5. Place the drive in an anti-static bag for storage.

### Installation
1. Unpack the CD-ROM drive and place it on an anti-static mat connected to the chassis of the computer. If an anti-static mat is not available, touch the anti-static bag that contains the new drive to a bare metal surface on the computer chassis.
2. Move the configuration jumper on the back of the CD-ROM drive to either the master or slave position, depending on the computer in which the drive is going to be installed. For more information on the master/slave/cable-select setting, refer to chapter 11.
3. If installed, remove the plastic bezel from the computer's front panel to expose a free 5¼" external drive bay. On most new chassis, a metal knockout will also have to be removed in order to install the CD-ROM drive. Remove the knockout with a pair of pliers to avoid laceration to the hands.
4. Slide the drive into the chassis of the computer through the front panel until the faceplate of the drive is flush with the front panel of the computer.
5. While supporting the drive from the rear, align the mounting holes on the drive bay with the screw holes on the side of the drive. Once aligned, secure the drive with four mounting screws, two on each side. Make sure the correct screws are used and avoid over-tightening.
6. Install the data, power, and audio cables into the correct connectors on the back of the CD-ROM drive. Make sure the striped side of the data cable is toward the pin-1 side of the male pin connector on the drive and the motherboard connector.
7. If not installed, insert the other end of the data cable into the correct connector on the motherboard, sound card, or MIO board. Make sure the striped side of the cable is aligned with the pin-1 side of the connector.
8. Check all power and data cable connections on the other internal peripherals, insert the power

cord, and turn the computer's power switch to the on position.

9. If the CD-ROM drive is installed on Windows 9X or Windows 2000 computer, the drive should be recognized automatically. If the drive is installed in a computer running MS-DOS, Windows 3.X, or Windows 9X DOS mode, a DOS-based CD driver and MSCDEX must be installed and configured.

10. If the computer will not start the POST when power is applied, the data cable is probably inserted backwards into either the CD-ROM drive or the motherboard connector.

11. If the computer boots properly, attempt to read information from a known good disc.

12. If the drive seems to be working properly, shut down the operating system, turn the computer off, and replace the computer's cover.

## 10.9 DIGITAL VERSATILE DISC READ-ONLY MEMORY (DVD-ROM)

A relatively new mass storage technology referred to as **digital versatile disc read-only memory (DVD-ROM)** is on the horizon. Similar to the CD-ROM, the DVD-ROM standard used by computers is an adaptation of the popular digital-video-disks currently used to provide high-quality video and audio information in today's consumer market. I see DVD-ROM drives replacing the CD-ROM drives as the primary transportable storage media for the future because of two reasons.

First of all, a DVD-ROM drive can read a CD. This downward compatibility means that the adoption of the DVD technology should be practically painless. Second, current CD-ROMs have a capacity of approximately 650 MB. Although this seems like a large storage capacity, it is relatively small when compared to the storage capacities available with current DVD technology, which are from 4.38 to 15.9 GB. To achieve these massive storage capacities, DVD-ROMs use some very interesting techniques that include a more effective utilization of the disc surface, recording on both sides, and double-layered recording.

Before we continue, it should be noted that DVDs have the same physical dimensions as do CD-ROM drives. This allows the steps in Procedure 10.7 to also be used when removing or installing a DVD-ROM drive.

### 10.9.1 DVD-ROM Storage Capacity

DVD-ROM technology effectively utilizes the disc surface by using tighter tracks, a larger data area, smaller pit and land lengths, better error correction, and more efficient modulation techniques. All of these improvements translate to almost a 7× data density rate over that used by CD technology. Unlike CD-ROMs, DVD-ROMs allow information to be placed on both sides of the disc. And, last but not least, DVDs allow two layers of information to be placed on each side of the disc.

Layered recording is similar to looking at writing on a clear glass pane and also under the glass pane. Your eyes can focus on the writing on the glass and then focus on the writing under the glass. This is basically the way layered DVD works. The laser beam can be focused on the upper layer to read information and then focused to read the information on the lower layer. All of the techniques utilized by DVD yield the storage capacity variations listed in Table 10.7.

Table 10.7  Current DVD Storage Capacities

| | |
|---|---|
| Single-sided/Single-layer | 4.38 GB |
| Double-sided/Single-layer | 8.75 GB |
| Single-sided/Double-layer | 7.95 GB |
| Double-sided/Double-layer | 15.9 GB |

## 10.9.2 DVD-ROM Transfer Speed

DVD-ROM technology had a transfer rate of approximately 1.32 gigabytes per second (GBps) when it was first introduced. Because this was the speed first used by DVD-ROMs drives, it is referred to as the *1X* speed, similar to the speed for CD-ROM drives. This means the 2X DVD-ROM drive will have a transfer rate of 2.64 GBps (2 × 1.32 GBps) when reading DVDs. Even though 1X DVD-ROM drive's transfer rate is approximately nine times (9×) the transfer speed of a CD-ROM drive, DVD drives usually spin CDs faster than DVDs, which translates to a faster transfer rate for CDs than the 9× multiplier. To determine the actual transfer rates for CDs, consult the specifications that are provided with each DVD-ROM drive.

## 10.9.3 DVD-R and DVD-R/W

DVD recorders and DVD read/write drives are also available. These devices can record on both sides but do not have the ability to do double-layer recording. Even though this sounds great, the costs of these drives are still over $1,000 and the discs cost around $40. Current prices for a CD-RW drive are around $200 and the cost of a CD-RW disc is about $2. This price differential still makes the CD-RW media the current read/write favorite among PC users.

## 10.10 THE CHASSIS

All internal mass storage devices are mounted in the compatible computer's chassis into metal cages called *drive bays*. Like all of the other devices in the computer, there are terms and expressions that a technician needs to know, relative to drive bays, to repair and upgrade a compatible computer effectively. The following section covers those terms and expressions.

### 10.10.1 Drive Bays

The first mass storage device used in the compatible computer was the 5¼″ floppy drive. The original drive had a width of 5¾″ and a height of 3¼″. The drive was mounted in a rack, called a **drive bay,** inside the computer that had the same physical size as the 5¼″ drive. The drive bay had an opening to the outside of the chassis to expose the face of the drive. This feature allowed floppy diskettes to be inserted and removed. The original IBM-PC came equipped with two drive bays to facilitate the mounting of the same number of drives.

When hard drives were first introduced with the IBM-XT, the drives were manufactured with the same widths and heights as those used by the 5¼″ floppy drive. This allowed the hard drive to be installed into one of the two 5¼″ drive bays that were provided with the XT computer. However, this meant that only one 5¼″ drive could be installed.

During the years of the XT and compatibles, a 5¼″ drive was introduced that had the same width as the original drive, but had a height of only 1⅝″. This drive became known as the *half-height drive* and the original drive became known as the *full-height drive.* The size of the half-height drive allowed two drives to be mounted into a single full-height bay. This feature allowed the XT computer to once again support two floppy drives along with a hard drive.

As technology improved, the 3½″ drive was introduced. This drive has a width of 3¾″ and a height of 1″. The size of this drive is completely different than the size of either the full-height or half-height 5¼″ drives, so a new drive bay had to be developed to facilitate its mounting. Similar to the 5¼″ drive bay, this bay also had to have an opening to the outside of the chassis.

The physical size of these three drives became the drive size standards used by today's computers. Drive bays have to be designed specifically for a drive size, so they are referred to as 5¼″ full-height, 5¼″ half-height, and 3½″ drive bays. To facilitate ease of mounting and rapid adoption, all current internal mass storage devices are designed around one of these drive bay standards.

Even though a hard drive needs a bay in which to mount, it does not need an opening to the outside of the computer. In fact, the hard drive does not even have to be mounted in the front of the computer. This brought about the design of a bay that is contained within the computer without an opening to the outside of the chassis. The drive bay with this characteristic is called an **internal drive bay.** A bay that provides an opening to the outside of the computer is called an **external drive bay.**

**Figure 10.11** Drive Bays

The more drive bays a chassis contains, the larger its size. This is one of the major reasons why chassis come in a wide variety of sizes and shapes. A typical tower chassis' drive bays are shown in Figure 10.11.

One important specification of a computer case that is often overlooked is the number of available internal and external drive bays. How many drive bays are enough? I recommend three internal 3½″ drive bays, two external 3½″ drive bays, and three external 5¼″ drive bays. This configuration should satisfy any foreseeable need for mounting mass storage devices. Plastic covers called *bezels* can also be purchased to cover up the openings produced by any empty external drive bay.

Note that 3½″ mounting kits can be purchased that allow a 3½″ drive to be mounted into a half-height 5¼″ drive bay. A typical mounting kit is shown in Figure 10.12. The kit usually comes with a fake front that will cover up the portion of an external 5¼″ drive bay that is not occupied by the 3½″ drive.

## 10.11 TROUBLESHOOTING FLOPPY AND CD-ROM DRIVES

Troubleshooting floppy and CD-ROMs drives is fairly easy because they are so inexpensive. If these devices are determined to be defective, they are usually simply replaced. Unfortunately, there are a few other things that can make these drives appear defective. To help determine the problem, troubleshooting methods for both of these devices are given in the next two sections.

### 10.11.1 Troubleshooting Floppy Drives

The lowly floppy drive usually goes unnoticed until it is needed. When the computer's main operating system fails, the floppy disk is the device of last resort to boot the computer. For this reason, you need to check the floppy drive periodically to confirm that it is operational.

If you start getting read errors on all of your floppy diskettes or if a Drive Not Ready error message is displayed, these are good indications that the floppy drive or its interfaces are having problems. Diagnosing and repairing floppy drive problems is a fairly simple process. There are only a few things that can go wrong: a defective drive; dirty R/W heads; or problems with the data cable, power cable, FDD port, or driver programs. If the floppy drive is determined to be defective, it is simply replaced because of its low cost.

The first thing you should do when you get read errors from a diskette is to determine if the information on the diskette itself is corrupted. This is accomplished easily by trying to read a known good

**Figure 10.12**  3½" Mounting Kit

diskette in the drive. If the drive can read another diskette, odds are the information on the diskette is corrupted. If this is the case, hopefully you have additional copies of the information.

If the drive cannot read another diskette, try cleaning the drive R/W heads with a head-cleaning kit. Disk head-cleaning kits can be purchased for around $10 at most stores that sell computer supplies. If you do not have a head-cleaning kit, buy one and have it on hand. The heads on the floppy drive are very delicate, so do not try to clean the heads using any other method.

If head cleaning does not solve the problem, disconnect the power and remove the cover from the computer. Once the cover is removed, reseat the floppy cable connectors that are inserted on the motherboard or MI/O adapter card and also on the disk drive itself. If this does not fix the problem, try a new floppy drive cable, if one is available. Having spare cables for the floppy and hard drives is an expensive investment that will probably pay off sooner or later. If this doesn't work, the problem is either the port or the drive itself.

The only way to determine if the drive is defective is to replace it with another drive or try the drive in another computer. Trying the drive in another computer is the best option because it will definitely determine if the drive is defective. If this is not possible, most vendors will allow a drive to be returned within 30 days of purchase. If a local vendor has such a return policy, purchase a new drive and return it if it does not fix the problem. Also, if you are going to be in the business of repairing or upgrading compatible computers, it is a good idea to have a spare 3½" floppy drive on hand.

If the problem cannot be corrected with one of these procedures, the problem is probably in the port. In this case, purchase a new floppy disk control port that will insert into one of the computer's expansion slots. Remember to disable the integrated port on the motherboard or MI/O adapter card before the new adapter is installed in the computer.

### 10.11.2 Troubleshooting CD-ROM Drives

The technique used to troubleshoot CD-ROM drives is very similar to the technique used to troubleshoot floppy drives, but with one important difference. MS-DOS, Windows 3.X, and Windows 9X DOS mode requires that a CD-ROM driver along with MSCDEX be loaded before the CD-ROM will operate. If the CD-ROM will work in Windows 9X but will not work when the computer is in DOS mode, this is the problem.

If the drive is not working in Windows 9X, check the drive cables. If this does not fix the problem, try a known good cable, if one is available. Windows 9X sometimes gets confused on the position of the CD-ROM's door. To check for this, right-click on the letter for the CD-ROM in My Computer. On the displayed menu, click on the Eject option twice.

If the drive works in MS-DOS and DOS mode but does not work in Windows 9X, it usually indicates that Windows 9X driver is not compatible with the drive. In this case, you must load the correct driver with the CONFIG.SYS file. This is usually accomplished by simply removing the REM statement in front of the DEVICE= command that loads the driver. This can be done using Windows 9X's System Editor, which was discussed in chapter 6.

If the Error Reading message is displayed, first try cleaning the CD. Dust and fingerprints can cause this problem. If the drive can read another CD, it means the drive is not the source of the problem. The next step is to try cleaning the lens on the CD-ROM with a disc-cleaning kit. Like the head cleaner for the floppy drive, this kit is available at most retailers that sell computer supplies. If cleaning the lens does not fix the problem, remove the computer's cover and reseat the cable connectors that are associated with the CD-ROM drive. Make sure the power is turned off and the power cord is disconnected.

If an installation of a new CD-ROM causes the hard drive to stop working or the computer to not boot at all, the problem is usually that the cable is inserted backwards. Also, check the CD-ROM's ID to confirm that it does not conflict with any other IDE or SCSI device, depending on the interface. If the drive being installed is used, make sure the interface it uses is not designed around one of the proprietary interfaces used on sound cards.

### 10.11.3 Caring for CDs and DVDs

Even though CD and DVD discs are more rugged than floppies, they can be damaged easily. Prolonged exposure to heat and scratches are two major forms of damage. To prevent damaging those valuable CDs or DVDs, use the following precautions:

1. Always handle the discs by the outside edges to prevent fingerprints or smudges.
2. Never place the discs on a hard surface with the recorded surface facing down.
3. Never stack the discs directly on top of each other.
4. Always store the discs in a protective case or CD carrier.
5. Never leave the discs in hot, humid environments.
6. Never expose the discs to direct sunlight.
7. Write only on the printed area with felt pens or pens that are approved for labeling CDs. Never use a ballpoint pen or any other hard object.
8. Clean the discs with a soft, lint-free cloth, wiping from the center of the disc toward the outside edges. Do not use a circular motion.
9. Never use a chemical-based cleaner to clean the discs.

## 10.12 CONCLUSION

One important fact you should have gained from this chapter is that mass storage devices are not part of the computer's memory unit. Mass storage devices are simply a place to store data and programs until they are loaded into the computer's RAM for processing. The mass storage devices currently used on the compatible computers are tape drives, floppy disk drives, CD-ROMs, DVDs, and hard drives.

Although they are slow, tape drives are the preferred method of backup because they are relatively cheap and hold large amounts of information. The two most widely used tape drive standards are QIC and Travan. If an internal tape drive is installed in the computer, it usually uses the floppy, IDE, SCSI, or USB interface. If an external tape drive is installed on the computer, it usually attaches to the parallel port or uses the USB or SCSI interface.

The FDCs used in the compatibles can have up to four drives attached. However, MS-DOS and Windows support only two. The two floppy drives are called A: and B:. If a third or fourth drive is added, it must be accessed through a TSR or application program. Most newly purchased compatibles usually have only one floppy 3½″ floppy drive installed. This drive is the A: drive.

Floppy drives come in two physical sizes: 5¼″ and 3½″. The 5¼″ drives are available in two densities: 360 KB and 1.2 MB. The 3½″ drives are available in three densities: 720 KB, 1.44 MB, and 2.88 MB. The higher density drives can retrieve information stored on lower density diskettes.

A CD-ROM is not part of the computer's read-only memory. The CD-ROM is simply another form of mass storage. The CD-ROM's high density has made it a major factor in multimedia presentations. CDs are also currently the preferred method for delivering operating systems, applications, and drivers to the consumer.

CD-Rs and CD-RWs give the user the ability to create and copy CDs. The CD-Rs use WORM technology; the CD-RWs allow CDs to be erased and reused. Both of these technologies will read manufacturers' CDs but require specially designed discs to write.

DVDs provide at least seven times the storage capacity of CD-ROMs. In addition to reading high-capacity DVDs, these drives can also read CDs that are made by the manufacturers or created on CD-Rs or CD-RWs. The DVD's ability to hold massive amounts of data will probably make it the transportable media of choice on future-generation computers.

# Review Questions

1. Write the meanings of the following acronyms:
   a. QIC
   b. FAT
   c. TPI
   d. ISO
   e. CD-ROM

## Fill in the Blanks

1. _____ _____ devices hold information until it is loaded into the computer's RAM for processing.
2. The storage device most often used for backups is the _____ drive.
3. Drives record information in circles around the surface of the disk called _____.
4. To organize information on the surface of the disk, it is divided into _____ and _____.
5. Floppy drives that record information on both sides of a disk are called _____ _____.
6. A program called _____ places the tracks and sectors on the drive's surface.
7. A floppy drive specification that indicates the spacing between the tracks is _____ _____ _____.
8. A _____ is used to define the same track location on drives that record on multiple surfaces.
9. The HD 3½" disk drive can store up to _____ bytes of information.
10. The small machine language located on a formatted floppy disk that actually boots the computer is called the _____ _____ record.
11. Logical sectors are sequentially numbered from _____ to the _____ sector on the disk.
12. The _____ _____ _____ maps the allocation units on the disk.
13. The fundamental storage unit used by disk drives is called a(n) _____ _____.
14. The location of the first allocation unit used by a file is contained within the _____.
15. To check the FATs on the disk, the programs _____ or _____ should be used.
16. The DOS-based CD-ROM driver that must be loaded in the AUTOEXEC.BAT file is named _____.
17. A computer that presents information in several forms is usually called a _____-_____ computer.
18. The original standard that was used to store information on the CD is called the _____ _____ format.

## Multiple Choice

1. Alcohol and a cotton swab should be used to clean the heads on a floppy disk drive.
   a. True
   b. False
2. A computer tape drive works on basically the same principle as a VHS tape drive.
   a. True
   b. False.
3. To organize the data stored on a disk, the surface of the disk is divided into:
   a. tracks and clusters.
   b. top and bottom.
   c. tracks and sectors.
   d. file allocation tables and directories.
4. Before a disk can be used to store and retrieve data, it must first be:
   a. partitioned.
   b. cleaned.
   c. write-protected.
   d. formatted.
5. MS-DOS fully supports CD-ROM drives and will probably not need a driver.
   a. True
   b. False
6. MSCDEX is loaded in the AUTOEXEC.BAT file.
   a. True
   b. False
7. Tapes and disk drives store information in the form of:
   a. electrical energy.
   b. light waves.
   c. radio waves.
   d. magnetic fields.
8. An external tape backup unit:
   a. connects to the computer through a tape backup adapter card.

b. connects to the computer through a parallel, SCSI, or USB port.
c. is installed in a 5¼″ drive bay.
d. costs less than an internal tape backup with the same features.

9. Which of the following is not part of the information kept in the directory?
   a. filename
   b. starting allocation unit
   c. ending allocation unit
   d. file size

10. The term density indicates the:
    a. thickness of the storage media.
    b. spacing of the tracks.
    c. storage capacity of the drive.
    d. access speed of the drive.

11. The tracks on a HD 3½″ disk are spaced approximately:
    a. 0.1250″ apart.
    b. 0.0283″ apart.
    c. 0.1042″ apart.
    d. 0.0074″ apart.

12. The B: drive:
    a. is connected to the floppy disk cable before the twist.
    b. is connected to the floppy disk cable after the twist.
    c. does not require a cable to communicate with the floppy disk controller.
    d. is connected to a different floppy disk controller than the A: drive.

13. The FDC used in most compatible computers can have:
    a. one drive installed per controller.
    b. three drives installed per computer.
    c. from one to four floppy disks attached.
    d. only have two floppy disks attached.

14. The /L switch used with the MSCDEX drive:
    a. sets the length of the buffer drive.
    b. sets the loading format for the CD-ROM drive.
    c. is no longer used with CD-ROM drives.
    d. designates the drive letter used by the CD-ROM drive.

15. Which diskette has the ability to hold the most information?
    a. 5¼″ DD
    b. 3½″ DD
    c. 5¼″ HD
    d. 3½″ HD

16. An allocation unit is defined as:
    a. a group of one or more sectors.
    b. a group of one or more clusters.
    c. both sides of the diskettes.
    d. the same track on both surfaces of the disk.

17. An allocation unit (cluster) for a HD 3½″ floppy disk equals:
    a. one sector.
    b. two clusters.
    c. two sectors.
    d. one cluster.

18. MS-DOS and Windows can control:
    a. four floppy disk drives.
    b. two floppy disk drives.
    c. one floppy disk drive.
    d. a different number of drives depending on the user option selected in the computer's CMOS.

19. Which density 3½″ format can be read by a 2.88-MB drive?
    a. 720 KB
    b. 1.44 MB
    c. 2.88 MB
    d. All of the above are correct.

20. The index hole on the 5¼″ drive:
    a. indicates the location of the first sector on each track.
    b. indicates the location of the first track on the disk.
    c. prevents the drive from being written to if covered with an opaque tab.
    d. is used to align the diskette within the drive.

21. FAT in the acronym for:
    a. file access table.
    b. file attempt to transfer.
    c. file allocation tabulation.
    d. file allocation table.

22. Tape drives:
    a. store information sequentially.
    b. store information in tracks and sectors.
    c. allocate the surface of the disk.
    d. hold all of the starting files' cluster numbers.

23. The FAT is used to:
    a. tell MS-DOS the path to take to locate the clusters that belong to each file.
    b. indicate defective clusters.
    c. identify the location of free clusters on the disk.
    d. All of the above answers are correct.

24. A X32 CD-ROM drive will transfer information at a maximum speed of:
    a. 32 KBps.
    b. 4800 KBps.
    c. 150 KBps
    d. 4,800 KBps.

25. Files that are not stored in consecutive clusters:
    a. are fragmented.
    b. are not accessible.
    c. will cause an error message to be displayed when they are accessed.
    d. occupy more storage space than drives that are stored in consecutive clusters.

26. Which of the following is not contained in a directory entry?
    a. starting cluster number
    b. ending cluster number
    c. date of storage
    d. file attributes
    e. total length of the file

# CHAPTER 11

# Hard Disk Drives

## OBJECTIVES

- Define terms, acronyms, and expressions used in relation to hard drives.
- Name the two hard drive positioners and describe their characteristics.
- Describe the different hard drive controllers and list the characteristics of each.
- Describe the ATA and ATAPI standards.
- Explain current SCSI standards.
- Explain the procedures for configuring IDE and SCSI drives.
- State the reasons for performing low-level formatting, partitioning, and high-level formatting of hard drives.
- Explain the differences between FAT16 and FAT32 file systems.
- Describe the regular maintenance that should be performed on a hard drive.
- Explain drive compression.

## KEY TERMS

hard drive
Winchester
fixed drive
head crash
voice coil
closed loop
stepper motor
open loop
sector header
ST-506/412
daisy-chain
modified frequency modulation (MFM)
run length limited (RLL)
enhanced small device interface (ESDI)

advanced technology attachment-1 (ATA-1)
integrated drive electronics (IDE)
master
slave
enhanced integrated drive electronics (EIDE)
small computer system interface (SCSI)
primary channel
secondary channel
advanced technology attachment packet interface (ATAPI)
programmable input output (PIO)
bus-mastering DMA
cylinder-head-sector (CHS)

enhanced CHS (ECHS)
logical block addressing (LBA)
low-level format
interleave
FAT16
FAT32
partition
logical drive
master partition boot record (MPBR)
partition table
high-level format
fragmented
disk optimizer
compression
type number

Decreasing the access time of the disk drive, which increases the transfer rate, can be accomplished in two ways: first, by placing more sectors per track; and second, by increasing the rotational speed of the drive. Because diskettes are made of a flexible plastic, the magnetic coating cannot handle the data densities required to add more sectors. Floppy diskettes are made of Mylar (flexible plastic) and the heads are in contact with the diskette's surface, so increasing the speed is out of the question. This leaves one logical solution to increase the access time: make the diskettes out of something other than Mylar, which brought about the introduction of hard drives.

**Figure 11.1** A Typical Hard Drive

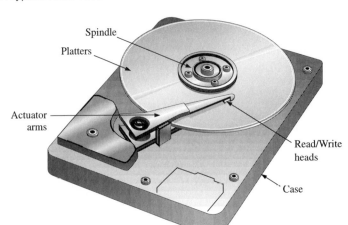

*Source:* Floyd, *Digital Fundamentals,* 7th ed. (Upper Saddle River, NJ: Prentice Hall, 2000) p. 654.

Currently, hard drives spin between 3,600 and 7,200 revolutions per minute (rpm), with faster speeds sure to follow. They also use rigid, circular platters that are usually made of aluminum. It is because of these rigid platters that these drives are called **hard drives.** Hard drives are also known as **Winchesters** and **fixed drives.**

There are currently two techniques used to coat the aluminum platters with a magnetic media. The first, and older method, is to coat the platters with metal oxide, similar to the coating on floppy diskettes. The second, and most current method, is electro-plating the surface with chrome.

As shown in Figure 11.1, hard drives can have one or more platters that are usually found in two diameters: 130 mm (5.1″) and 95 mm (3.7″). The drives with the 130-mm platters are called 5¼″ drives because they have the same width and height as a standard half-height 5¼″ floppy drive. The drives with the 95-mm platters are called 3½″ drives because they have the same width and height as the standard 3½″ floppy drive. Remember that 5¼″ and 3½″ floppy drives get their names from the size of the diskettes that they are designed to handle, not the physical size of the drive.

Hard drives have one R/W head for each surface of a platter. This means that a drive that has two platters has four read/write heads. Although some very expensive cartridge systems are available, most hard drives have the platters permanently mounted on a common spindle inside a sealed case. The sealed case should never be opened when servicing a hard drive in the field. The reason for this practice will become obvious as this discussion continues.

When power is not applied to the hard drive, the heads rest on the surface of the platters. When power is applied, the platters start to rotate up to the 3,600 rpm (or higher) speed, causing pressure to build underneath the heads, which lifts the heads off the surface of the platters. When the drive is spinning at full speed, the heads on the newer hard drives float from about 3 to 7 millionths of an inch above the platter's surface, as illustrated in Figure 11.2.

With such a small gap and high speed, if any particle of dirt or dust is contacted on the platter surface, it can cause the heads to oscillate and strike the surface of the platter. This is commonly called a **head crash** and could damage the platter's surface or the head assembly itself, making the drive unusable. Figure 11.2 illustrates the relationship between the distance the heads fly above the platter surface and some possible surface contaminants.

This is why the drive case must not be opened unless it is inside a class-100 or better clean room. In a class-100 clean room, 1 cubic foot of air cannot contain more than 100 0.5-micron particles. These rooms require special air filtration systems that evacuate and refresh the air continuously. Because these environments are excessively expensive, few companies are prepared to service hard drives. The cost

**Figure 11.2** Contamination of Surface

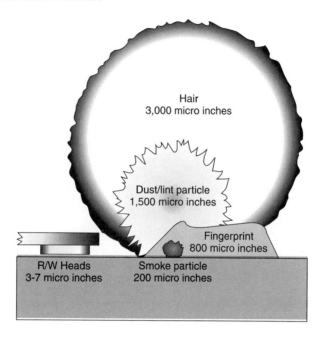

of hard drives is relatively cheap per million bytes of storage, so hard drives are simply replaced when they become defective.

Because the platters are rigid and enclosed in a sealed environment, the mechanics of the drive are more precise than those of floppy disk drives. This means that even though the coating method is basically the same, the density of hard drives can surpass that of floppy systems by several magnitudes. Currently, hard drive units can be purchased with densities of over 40 GB. As new electro-plating techniques and read/write technologies emerge, the densities will rise even further.

## 11.1 READ/WRITE HEAD POSITIONING

Currently there are two techniques used to move the read/write heads across the surface of the platters. One technique uses a voice coil that moves the R/W heads in much the same manner as the voice coil in a speaker system moves the speaker cones to reproduce audio. The second, and older, technique uses a stepper motor, which is the same method used by floppy disk drives.

### 11.1.1 Voice Coil

A **voice coil** uses a stationary, permanent magnet surrounded by coils of wire that are free to move across the top of the magnet (see Figure 11.3). When an electrical current is passed through the coil, it moves along the permanent magnet in a continuous, smooth motion. The movement will be in a direction that is determined by the direction of current through the coil. This is a smooth, continuous motion with no stops between the starting cylinder and destination cylinder. In a drive that uses this type of head positioning mechanism, the R/W heads are attached to the voice coil and the permanent magnet is stationary. The entire voice coil head positioning mechanism is located in the sealed chamber with the platters (see Figure 11.4).

Voice coil drives are **closed loop** systems, which means an electrical signal is sent back from the drive indicating the heads are positioned over the correct cylinder. To accomplish this, the voice coil drive dedicates one platter surface to head position information. The head position information is put on the platter when the drive is manufactured. The head that reads this position information has no

**Figure 11.3** Voice Coil Positioner

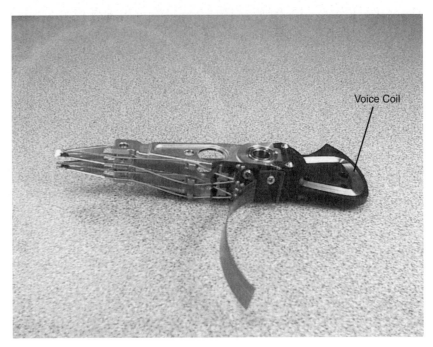

**Figure 11.4** Voice Coil Inside Enclosure

write capability, so the position information (called *index marks*) can never be over-written. When a command to seek a certain cylinder is received by the hard drive controller (HDC), the HDC sends the direction command to the hard drive. The drive's electronics send current to the voice coil to move the heads in the desired direction. As the heads move, the position feedback head sends a pulse to the HDC as each index mark is passed. Once the correct cylinder is reached, the heads are stopped directly above the desired cylinder. Compared to stepping motor drives, a drive that uses this type of positioner seeks very quickly, is immune to temperature variations, is not affected by mechanical wear, and can withstand a fairly hard shock while the computer is operating.

# Hard Disk Drives

**Figure 11.5** Stepper Motor

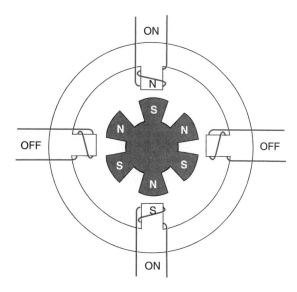

**Figure 11.6** Stepper Motor Mounting

## 11.1.2 Stepper Motor

A **stepper motor** is an electrical motor that has a permanent magnet rotor and a stator that usually has four sets of field coils. Figure 11.5 illustrates the basic makeup of a stepper motor. The *rotor* is the part of the motor that rotates. The *stator* is the part of the motor that is stationary. The permanent magnet rotor poles and the stator field coil magnetic poles are placed at precise angular displacements. When an electric current is passed through a set of the field coil windings, it produces a magnetic field that the permanent magnet rotor poles rotate to align with. As the field coils are switched, the rotor moves in precise angular steps to keep itself in alignment with the magnetic field produced by the field coils. The velocity, direction, and position of the stepper motor can be precisely controlled to a high degree of accuracy. Once the desired position is reached, the permanent magnet rotor locks on the metal of the field poles even when power is removed.

The stepper motor is mounted outside the sealed chamber. A typical stepper motor mounted on a hard drive is shown Figure 11.6. The motor's spindle penetrates the case through a sealed hole. The stepper motor spindle is linked to the heads mechanically by either a split-steel band coiled around the stepper motor spindle or with a gear mechanism.

The stepper motor positioner is an **open loop** system, which means there is no electrical feedback indicating if the heads are over the correct cylinder. A stepper motor positioner system depends on a track-zero sensor that the heads move to when a restore command is sent to the HDC. Once track zero is located, the HDC will keep up with the current head position. When the HDC receives a command to seek, the HDC simply moves the heads, in single steps, either in or out, with one step equal to one cylinder. These drives depend on software information recorded at the front of each sector, called a **sector header,** to determine if the heads are over the correct cylinder.

Stepper motor hard drives require more preventative maintenance to ensure their dependability. Because these drives are open loop, they can be affected by a variety of factors such as temperature and mechanical wear and shock while the computer is operating.

## 11.2 CONTROLLERS

In the old days, interfacing the hard drive to the compatible computer was usually accomplished by adding a hard disk adapter card in an open slot in the expansion bus. Today, the hard drive is usually attached to a connector located on the motherboard. The adapter card or motherboard connector is then cabled to a hard drive that is compatible with the controller. The compatible computer has used several types of interfaces, controllers, and associated hard drives over the years. They include modified frequency modulation (MFM), run length limited (RLL), enhanced small device interface (ESDI), small computer system interface (SCSI), integrated drive electronics (IDE), and enhanced IDE (EIDE).

Choosing a controller and hard drive will depend on several factors: (1) the amount of storage desired; (2) the access speed desired; (3) the type of computer the drive will run in; and, of course, (4) the price. This choice becomes academic if a computer already has a hard drive installed and a second one is to be added. In that case, a drive must be selected that is compatible with the controller and interface currently installed. Some adapters are only available in 16-bit cards, so they will run only on an AT-compatible. Some adapters require a computer VL bus or PCI interface. The main thing to remember is that the hard drive and the adapter card must match.

### 11.2.1 ST-506/412 Interface

When IBM designed the PC, it chose peripheral interfaces that were already on the market. This gave IBM the ability to use off-the-shelf peripherals that were available. The first interface chosen for the hard drive was the **ST-506/412** interface developed by Seagate Technology. Because IBM chose this established interface, all of the drive manufacturers that wanted to produce a hard drive for the IBM-PC and XT made their drives ST-506/412-compatible. This is probably the reason why ST-506/412 interfaces were used in over 90% of the compatible computers with hard drives under 40 MBs.

The ST-506/412 uses two cables to interface the adapter card to the drive. One cable is a 34-pin control cable and the other is a 20-pin data cable. The ST-506/412 interface has the ability to control two drives. If two drives are connected, the control cable is **daisy-chained** to two drives, but each drive requires a separate data cable. The last drive connected to the control cable daisy-chain must have a *terminating resistor* installed. The ST-506/412 interface is illustrated in Figure 11.7.

When using the ST-506/412 interface, the data encoding/decoding and separation take place on the adapter card by a special IC called a hard disk controller (HDC). Data communication with the drive takes place serially over the data cable so the data throughput (the speed at which the data go through the controller) is limited. The term *serial* means that all of the data are sent one bit after another. The ST-506/412 can control a drive only up to 140 MB.

Encoding is a method of synchronizing the transfer of data between the drive and controller. There are currently two encoding methods used: **modified frequency modulation (MFM)** and **run length limited (RLL).** Understanding MFM and RLL encoding is not really necessary to understand the differences among hard drives. RLL encoding places approximately 50% more data in the same space that is used by MFM. However, because RLL was not around when hard drives were originally installed on the compatible computer, MFM was the first encoding method used.

**Figure 11.7** ST-506/412 Interface

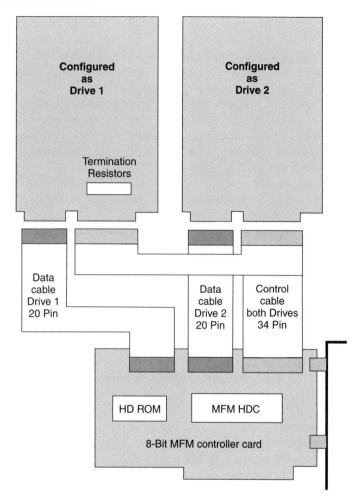

## 11.2.2 MFM Controller

The MFM adapter cards use the ST-506/412 interface and the MFM encoding technique. The MFM adapter cards have the HDC located on the adapter card and install into one of the computer's expansion slots. Hard drives using a MFM controller are formatted with 17 sectors per track and 512 bytes per sector. With this density, the transfer rate of MFM hard drives is around 522,240 bytes per second if the drive rotates at 3,600 rpm.

## 11.2.3 RLL Controller

RLL adapters use the ST-506/412 interface, which formats each track with 26 sectors and 512 bytes per sector. Like the MFM adapter card, the RLL HDC is located on the adapter card. With the increased density of RLL, the transfer rate of RLL hard drives is approximately 798,720 bytes per second at 3,600 rpm—a transfer rate slightly over 50% greater than that of MFM encoding. There are also more sectors per track, so the R/W heads do not have to seek as often, which decreases the time it takes to transfer data.

As stated earlier, MFM and RLL adapter cards both use the ST-506/412 interface. This makes it theoretically possible to increase the capacity of a hard drive by changing from an MFM adapter card to an RLL using the same drive. However, with the increased density, an RLL-capable drive must be used with an RLL controller.

### 11.2.4 ESDI

In the mid-1980s, Maxtor Corporation introduced the **enhanced small device interface (ESDI,** pronounced "es-dee") adapter card and hard drive. The HDC is still on the adapter card, but the encoding, decoding, and data separation circuits are built into the drive itself, so the HDC sends data to the drive in parallel. Parallel data transfer greatly increases the throughput (transfer rate) of the drive and reduces the amount of hardware needed for control.

The adapter card for the ESDI looks similar to the ST-506/412 in that it has a single 34-pin cable connection and two 20-pin connectors. In fact, an ST-506/412 drive could be connected to an ESDI adapter card, but it would not work. On the ESDI adapter card, the 34-pin connector is used for drive control and can be daisy-chained to two ESDI drives. The 20-pin connector is the data cable and each drive requires an individual cable.

The ESDI controllers usually format 34 sectors/track and 512 bytes/sector. At a speed of 3,600 rpm and with 34 sectors/track, the data transfer rate is 1,044,480 bytes per second. With all of the enhancements offered by ESDI, the transfer rates jumped from 10 MBps to 15 MBps.

At one time, ESDI was the HDC and hard drive combo of choice in the high-end (expensive) IBM-ATs and PS/2s running the 286 or 386 processor. However, ESDI was short-lived because of the introduction of the ATA standard, which offered cheaper drives and interfaces at approximately the same performance as ESDI.

### 11.2.5 ATA-1 (IDE)

The **Advanced Technology Attachment-1 (ATA-1)** standard specifies the attachment of a port to one or two hard drive(s) using a parallel interface. To accomplish this feat, the HDC, data encoding/decoding, and data separation circuitry are moved from the port to the hard drive. This innovation helps boost the data transfer rate by several magnitudes and also makes the drive easier to install and set up. Because most of the circuitry that controls the hard drive is on the hard drive itself, the drives and standards are more commonly called **integrated drive electronics (IDE).** The original IDE controllers and hard drives approach the transfer rates of the ESDI interface, but at a reduced price, which killed the ESDI hard drive technology. In fact, today's IDE drives far exceed the transfer rates that are offered by ESDI. Also, because the IDE port requires fewer components to support the IDE drive, the IDE port is most often integrated on the motherboard.

The IDE drive uses different track layouts than those used by an MFM controller and usually uses the RLL encoding format. This seemed to make them incompatible with the drives supported by the BIOS in the ATs and the compatibles. Smaller IDE drives emulate (act like) the MFM and RLL hard drive controllers already supported by the AT's BIOS. This feature means that no special software drivers are required to interface an IDE drive to an AT computer and no special ROMs are required by the IDE adapter.

IDE drives are formatted to 34 or more sectors per track and are low-level formatted at the factory. Their price and speed make the IDE and today's EIDE hard drives the hard drive of choice for the IBM-AT and above compatible computers.

The ATA-1 standard calls for a single IDE port, usually called a *channel,* that can handle up to two IDE drives. On most AT-compatibles using the 286 microprocessor, the IDE port is on an adapter card called a *multifunction I/O (MI/O)* card, which was discussed in chapter 9. On most compatibles using a 386 and above microprocessor, the IDE port is integrated on the motherboard and is part of the chipset.

ATA is a standard for internal mounted devices and is not designed to connect to peripherals outside the computer chassis. This means that all IDE devices have to be mounted inside the computer's case. A 40-pin ribbon cable is used to connect the port to the drive(s), as illustrated in Figure 11.8. If a single drive is used, the ATA-1 standard calls for the drive to be configured as a *stand-alone* using jumpers on the drive's circuit card. If two drives are connected to the channel, one is configured as the **master** and the other is configured as the **slave.** If two drives are used, a cable with three daisy-chained connectors is required. The setup and connection of IDE drives is discussed in section 11.6.

The ATA-1 standard also specifies the methods used to transfer data between the hard drive and the computer's RAM. These methods are referred to as *modes.* The modes specified call for two programmable input/output modes (PIO modes 0 and 1) and two DMA modes (0 and 1). These modes are discussed in greater detail in section 11.3.3.

**Figure 11.8** IDE/EIDE Cable

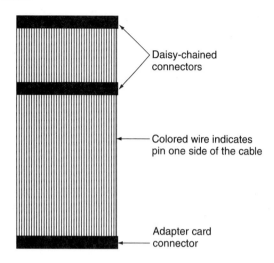

## 11.3 EIDE AND SCSI

There are currently two hard drive and port combos used on today's compatible computers. One is called **enhanced integrated drive electronics (EIDE)** and the other is the **small computer system interface (SCSI)**.

### 11.3.1 ATA-2 (Enhanced IDE)

Like the ATA-1, the ATA-2 (EIDE) standard is for internally mounted devices only and does not support devices mounted outside of the computer's chassis. Unlike the ATA-1, this standard specifies two IDE ports (channels) instead of one. One of the ports is called the **primary channel** and the other is called the **secondary channel**. Having two ports allows the connection of a maximum of four IDE or EIDE devices—two per port.

EIDE is downwardly compatible with IDE. This feature allows older IDE devices designed around the ATA-1 standard to be attached to either EIDE channel. Of course, these ATA-1 devices have to be set for stand-alone, master, or slave, depending on the configuration.

If a single EIDE device is connected to either of the EIDE channels, it must be configured as the master device, not as a stand-alone. If two devices are connected to either the primary or secondary port, one must be configured as the master device while the other is configured as the slave. If multiple IDE/EIDE devices are used, the computer will attempt to boot from the master device on the primary channel. In addition to supporting up to four IDE devices, ATA-2 also added the following features:

- ATAPI support for CD-ROM drives and tape drives;
- PIO modes 3 and 4;
- DMA modes 3 and 4;
- Bus-mastering DMA with a PCI expansion bus;
- Logical block addressing; and
- Auto-identify command.

All of these features are discussed in the following sections.

### 11.3.2 ATAPI

CD-ROMs and tape drives were originally SCSI devices that interfaced to the computer through an SCSI port. Unfortunately, the ATA-1 standard only addressed the connection of IDE hard drives. CD-ROM and tape drive manufacturers needed a method of adopting their SCSI devices over to the IDE interface. SCSI and its advantages are covered in section 11.7.

When the ATA-2 standard was introduced, it contained another standard called the **advanced technology attachment packet interface (ATAPI).** The ATAPI provides a means by which the SCSI command set is adapted to the command set used by EIDE. With this additional standard, all the manufacturers had to do to move to EIDE was to convert the SCSI interfaces over to the EIDE interface.

ATAPI CD-ROM drives and tape drives are interfaced to an EIDE channel in the exact same way as an IDE drive. However, two IDE devices cannot transfer data on the same channel at the same time. Thus, if a hard drive and a CD-ROM are connected to the same EIDE channel, the hard drive will have to wait for the CD-ROM or tape drive to finish transferring data before it has its turn. For this reason, it is recommended that the CD-ROM or tape drive be connected to the secondary EIDE channel and the hard drive(s) be connected to the primary channel.

Unfortunately, most preassembled computers connect the CD-ROM drive and hard drive to the primary channel to save the cost of an extra cable. If your computer is configured this way, it is highly recommended that you purchase another IDE cable and move the CD-ROM to the secondary channel. The configuration and attachment of IDE devices is covered in section 11.6.

### 11.3.3 IDE Transfer Modes

Two categories of modes are currently used to transfer information with the IDE and EIDE hard drive: **programmable input output (PIO)** and *direct memory access (DMA)*. To support any of the PIO or DMA modes mentioned in the following sections, the mode must be supported by the computer's BIOS, chipset, and the hard drive.

### 11.3.4 PIO

Originally, for MFM and RLL drives, data were transferred between the hard drive(s) and the computer's RAM using DMA. When DMA is used, data are transferred directly from the hard drive to the computer's RAM without going through the microprocessor. Even though this seems to be the route to go, there is a major problem in using this method on older machines.

The DMA controllers (DMACs) used on older compatible computers can only transfer 8 bits of data at a time. Because all IDE hard drives currently transfer data in widths of 16 bits, the 8-bit limitation of these DMACs is a major problem. To help alleviate this problem, most IDE devices support another mode of transfer called *PIO*.

PIO transfer is controlled totally by software. This means that all of the information transferred between the hard drive(s) and the memory must pass through the microprocessor. Even though PIO seems to be less efficient than DMA, with the data bus width and fast speeds of the microprocessor, PIO is several magnitudes faster than using the older 8-bit DMACs. Table 11.1 lists the four PIO modes that are available for transferring information with the IDE devices.

As stated previously, in order to use any of the PIO modes listed in Table 11.1, the mode must be supported by both the hard drive and system BIOS. Most hard drives, however, will work fine if a lower

Table 11.1  PIO Modes

| IDE Mode | Theoretical Transfer Rate |
| --- | --- |
| PIO mode 0 | 3.3 MBps |
| PIO mode 1 | 5.2 MBps |
| PIO mode 2 | 8.2 MBps |
| **EIDE Modes** | |
| PIO mode 3 | 11.1 MBps |
| PIO mode 4 | 16.6 MBps |

PIO mode is selected. For example, if a PIO mode 4 hard drive is installed in a computer that supports only PIO mode 3, then mode 3 can be used.

### 11.3.5 DMA

Computers that have a PCI expansion bus have greatly improved the DMACs used by the computer. This has brought back a resurgence in the use of DMA to transfer data between the computer's RAM and the hard drives.

Table 11.2 lists the nine DMA modes that are currently available. Note that most drives that use DMA will also work on a computer that supports only PIO. When this is the case, try the fastest PIO mode available. If you get any read errors when the computer is accessing the hard drive, try the next lower PIO mode until reliable transfer is achieved.

### 11.3.6 Bus-mastering DMA (ATA-4) and ATA/66

The PCI expansion bus and the ATA-4 standard add the ability to perform **bus-mastering DMA.** When bus-mastering DMA is used, the hard drive itself handles the data transfer between the drive and the computer's memory without going through the microprocessor. This technique greatly increases the data transfer speed of the drive. Bus-mastering DMA must be supported by the hard drive, motherboard, and operating system. The extra components needed on the drive and motherboard add considerably to the overall cost of a computer system.

Another alternative is UDMA/33 or UDMA/66 (ATA/33 or ATA/66). These methods of transfer bring the theoretical burst transfer rate of the EIDE interface up to 33.3 MBps and 66.6 MBps, respectively. Both UDMA/33 and ATA/66 require support by the hard drive, chipset, and the operating system.

Although Windows 95 and Windows 98 both support UDMA, it uses a single driver (ESDI_506.PDR) for all IDE hard drives. This driver is very inefficient when handling UDMA devices. For this reason, I highly recommend using the UDMA driver that is usually provided with the motherboard or UDMA hard

**Table 11.2** DMA Modes

| IDE DMA Modes—Single Word | Theoretical Transfer Rate |
|---|---|
| Mode 0 | 2.1 MBps |
| Mode 1 | 4.2 MBps |
| Mode 2 | 8.3 MBps |
| **DMA modes—Multiword** | |
| Mode 0 | 4.2 MBps |
| Mode 1 | 13.3 MBps |
| Mode 2 | 16.6 MBps |
| **Ultra DMA (DMA/33)** | |
| Mode 0 | 16 MBps |
| Mode 1 | 24 MBps |
| Mode 3 | 33.3 MBps |
| Ultra DMA/66 (ATA/66) | |
| Mode 0 | 66.6 MBps |

> **Procedure 11.1: Setting the BIOS hard drive transfer mode.**
>
> This procedure may vary depending on the BIOS running the computer.
>
> 1. If running Windows 3.X, 9X, or 2000, close all applications and choose the Restart option in the Shut Down menu.
> 2. During the restart sequence, watch for "Press Delete to run Setup" or a similar message to be displayed, and then press the appropriate key. If this message is not displayed, consult the documentation that came with the computer and follow the procedure to enter the CMOS Setup routine.
> 3. When the main CMOS Setup screen appears, use the cursor control and Enter keys to select the Integrated Peripheral option.
> 4. Use the keys that are displayed on the CMOS screen to select the desired PIO modes and also enable or disable UDMA for the Master/Slave Primary and Secondary IDE channels according to each hard drive's documentation.
> 5. Use the displayed keystrokes on the CMOS Setup screen to exit and save. If you do not wish to make changes, use the displayed keystrokes to exit without saving.

drive. If you buy a preassembled computer that uses drivers specified as UDMA/33, ATA/33, ATA/66, or UDMA/66, you need to make sure you have a copy of the UDMA driver in case you have to reinstall the operating system.

Just another note on ATA/66: A special 80-conductor cable is required to attain the 66.6-MBps speed. The 40 additional conductors are all connected to ground to help prevent cross-talk between the data conductors, which allows for faster data transfer speeds. Due to the 40 additional conductors, this cable costs considerably more than standard IDE cables. The cable is still terminated with the standard 40-pin female pin connector used by all IDE devices.

### 11.3.7 Setting the IDE or EIDE Transfer Mode

The original transfer mode used by the hard drive is set in the CMOS. This allows the computer to access the hard drives so the main operating system can be loaded. Once the operating system loads, the transfer mode can be changed by the computer's main operating system. Procedure 11.1 gives the steps to set the BIOS transfer mode on computers running Award's BIOS. The setup for other BIOSs should be similar, but the documentation that is provided with the motherboard should be referenced.

## 11.4 CHS (NORMAL), ECHS (LARGE), AND LBA

The hard drive is told where to locate information on its surface by being given a cylinder number, a head number, and a sector number (CHS). The CHS can be sent to the hard drive directly using the CHS method or developed in other forms called *ECHS* and *LBA* and then translated to a CHS by the hard drive. All of these methods are discussed in the following sections.

### 11.4.1 CHS

When **cylinder-head-sector (CHS)** addressing is used, the BIOS sends a cylinder number, head number, and sector number to the hard drive. In some BIOSs, CHS addressing is also called *normal*. CHS addressing was the original method used to address information on the hard drive and is still the method used to address information on floppies.

The standard BIOS that came in the AT and compatible computers supported hard drives that could have up to 1,024 cylinders, 16 heads, and 63 sectors per track. Each sector holds 512 bytes of user data, so using a little math indicates that the BIOS that uses CHS will only support a drive up to $528,482,304_{10}$ bytes ($512 \times 1,024 \times 16 \times 63$). This means that a BIOS that supports only CHS addressing cannot access a drive that has a capacity of over 521 MB.

Note that just because the BIOS supports drives up to 521 MB does not mean the operating system will access a drive with this capacity. Another major factor that determines the limitation of the drives

# Hard Disk Drives

---

### Procedure 11.2: Displaying the IDE hard drive resources

1. Click on the Start button on the Task Bar.
2. Move the mouse button up to Settings and then over to Control Panel.
3. Click on Control Panel.
4. When the Control Panel window opens, locate and double-click on the System icon.
5. In Windows 2000, click on the Hardware tab.
6. Click on the Device Manager tab or button.
7. Click on the plus (+) sign beside the Hard Drive Controllers category.
8. Double-click on the Primary IDE Controller or Secondary IDE Controller option.
9. Click on the Resources tab.
10. When finished, click on the Cancel button on each open window.

---

is the FAT used by the operating system. FAT systems used by hard drives are discussed in section 11.9.5. The file allocation table (FAT) was first introduced in chapter 10.

## 11.4.2 Enhanced CHS

To overcome the 521-MB barrier imposed by CHS addressing, another method called **enhanced CHS (ECHS)**, or *large* was introduced with the ATA standard. Using hard drives that support translation and enhanced BIOS that supports ECHS, a hard drive can have up to 1,024 cylinders, 256 heads, and 63 sectors. When these numbers are multiplied by 512 bytes per sector, we find that ECHS supports hard drives up to $8,455,716,864_{10}$ bytes.

To support ECHS, the computer must be running an *enhanced BIOS* that supports the additional head selection bits. An IDE drive that supports translation is also required. Translation is required because there will probably never be a hard drive that has 256 R/W heads. This many R/W heads would require the hard drive to have 128 platters. Hard drives that support ECHS usually have more then 1,024 cylinders, which seems to exceed the cylinder limitation imposed by the BIOS. Looks can be deceiving, however. IDE drives take an ECHS address from the BIOS and translate the address, on the fly, to a different CHS number for the actual location of the data. For example, an IDE hard drive may have 4,096 cylinders and 6 R/W heads and the BIOS thinks it has 1,024 cylinders and 24 R/W heads. An example of translation is if the BIOS sends a command to the drive to select cylinder 20 and head 12, the drive may translate this to cylinder 100 and head 3.

## 11.4.3 LBA

The current method of addressing information stored on a hard drive is called **logical block addressing (LBA)**. When LBA is used, the sector number is specified instead of a CHS. Using LBA, the sectors are numbered starting with LBA0, which coincides with cylinder 0, head 0, and sector 0. After LBA0, the sectors are numbered sequentially until the last sector on the hard drive is reached. A 28-bit binary number specifies the LBA sector number. This means that LBA will support sector numbers up to $268,435,456_{10}$. Once again, because each sector holds 512 bytes of data, LBA supports hard drives up to $137,438,953,472_{10}$ bytes (128 GB). The LBA request still has to be translated into a CHS for the operating system. This means that a BIOS that supports only ECHS will still have a drive limitation of 8 GB. However, a BIOS that supports LBA can theoretically have a capacity up to 128 GB.

The choice of CHS, ECHS, or LBA to address information on the drive is set as part of the CMOS Setup program, which was discussed in chapter 8. Most current BIOSs support the Auto ID command made available by ATA-2. If this is the case, the technique used to address the sectors on the drive is automatically detected during the boot sequence.

## 11.5 THE RESOURCES REQUIRED BY EIDE

Each EIDE channel requires a group of port addresses and an interrupt. EIDE consists of a primary and a secondary channel, so two sets of port addresses and two interrupts are required. Procedure 11.2 lists the steps to follow to display the resources used by hard drives that use Windows 9X/2000 Device Manager.

## 11.6 THE IDE CONNECTION

As stated previously, each IDE/EIDE channel can have one or two IDE devices attached. The connection is made from the IDE/EIDE connector on the motherboard or adapter card to the drive(s) by way of a 40-pin, flat ribbon cable. The pin-out for the IDE/EIDE connector is listed in Table 11.3.

A typical IDE/EIDE cable is shown in Figure 11.8. The cable may have two or three 40-pin female connectors mounted along the cable. Multiple connectors on a single cable are referred to as a *daisy-chain* cable. A three-connector daisy-chain cable will operate in a one- or two-drive system whereas a two-connector cable will only operate a single IDE drive. If your computer has a two-connector daisy-chain cable installed and you wish to add a second drive, you will have to purchase a three-connector cable.

To ensure signal integrity, the length of the cable should not be over 18″. One edge of the cable has a colored wire that indicates the pin-1 side of the cable and connectors. The cable can be installed on the drive two ways, but only one way will work. The pin-1 side to the pin connector on the rear of the IDE drive is usually on the same side as the power connector. However, the documentation that is provided with the drive should be consulted when installing the cable. The label on the top of the drive itself may also contain information on the cable orientation.

Table 11.3  Pin-out of IDE/EIDE Connector

| Signal | Pin | Pin | Signal |
|---|---|---|---|
| Reset | 1 | 2 | Ground |
| D7 | 3 | 4 | D8 |
| D6 | 5 | 6 | D9 |
| D5 | 7 | 8 | D10 |
| D4 | 9 | 10 | D11 |
| D3 | 11 | 12 | D12 |
| D2 | 13 | 14 | D13 |
| D1 | 15 | 16 | D14 |
| D0 | 17 | 18 | D15 |
| Ground | 19 | 20 | Key |
| DMARQ | 21 | 22 | Ground |
| I/O Write | 23 | 24 | Ground |
| I/O Read | 25 | 26 | Ground |
| I/O Chn. Rdy | 27 | 28 | Unused |
| DMACK | 29 | 30 | Ground |
| IRQ | 31 | 32 | Host 16-bt I/O |
| DA1 | 33 | 34 | Passed Diagnostics |
| DA0 | 35 | 36 | DA2 |
| HChip Sel00 | 37 | 38 | HChip Sel01 |
| Drive Active | 39 | 40 | Ground |

# Hard Disk Drives

As discussed previously, the IDE standard consists of only one channel whereas the EIDE consists of two channels: a primary and a secondary. When only one IDE drive is installed on a computer that uses IDE, it must be configured as a single drive. A single EIDE drive must be installed on the primary channel and configured as the master drive. When two drives are installed on the same channel of either an IDE or EIDE channel, one drive is configured as the *master drive* and the other is configured as the *slave drive*. The master drive on the single IDE or primary EIDE channel automatically takes the C: drive designator. The slave drive, however, takes a letter designator greater than C:, depending on how the master drive is partitioned, which is discussed in section 11.9.7. A typical EIDE setup with both channels filled is shown in Figure 11.9.

The drive's configuration is usually defined by a group of jumper blocks located near the connector where the ribbon cable connects the drive to the IDE channel. For example, Figure 11.10 shows the location of the configuration jumpers for most EIDE hard drives along with the configuration settings. In fact, the jumper configurations shown in Figure 11.10 are the standard settings for most current EIDE devices, including CD-ROMs. Most of the better drives have additional jumpers that allow the drive to run on an older IDE system. Consult the drive's documentation to determine if it supports this feature.

Referring to Figure 11.10, note the additional setting referred to as *cable select*. This setting allows the computer to configure the drive as either the master or the slave. I have never seen a compatible computer use this setting, but you should consult your computer's documentation to determine if this setting is required.

Although the EIDE device setting has become fairly standard, ATA-1 (IDE) hard drives use a different number of jumpers and different jumper settings to configure their drives as either single, master, slaved, or cable select. In fact, different models produced by the same company often use different configuration settings. When an IDE drive is purchased, the drive's configuration information is usually supplied on a small leaflet that is in the box with the new drive. It very important to keep this leaflet

**Figure 11.9** Typical EIDE Setup

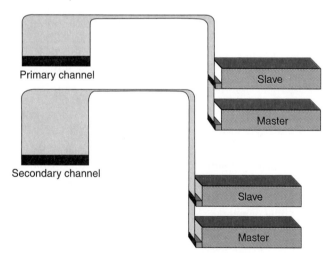

**Figure 11.10** EIDE Drive Configuration Jumpers

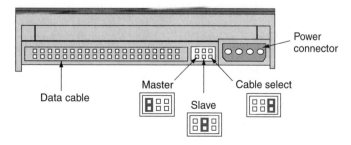

Table 11.4  Configuration of Hypothetical IDE (ATA1) Drive*

| Configuration | J4 | J5 |
|---|---|---|
| Single drive | 0 | 0 |
| Master drive | 0 | 1 |
| Slave drive | 1 | 0 |
| Cable select | 1 | 1 |

*1 = jumper installed
0 = jumper not installed

in a safe location for further reference. An example of the configuration information that might be provided with an IDE drive is given in Table 11.4.

Most of the newer EIDE drives have the jumper configuration information listed on the label attached to the top or bottom of the drive. If this information is not available, you can obtain the drive's configuration information from a third-party publication or by contacting the manufacturer of the drive.

**Problem 11.1:** How should the jumpers be installed to configure an IDE as the slave on a two-drive system using the setup information given in Table 11.4?

**Solution:** According to Table 11.4, a single jumper needs to be placed on J4.

When a new computer system is purchased, it usually has one EIDE hard drive and one EIDE CD-ROM drive installed. Both of these devices are usually connected to the primary IDE channel with the hard drive configured as the master and the CD-ROM drive configured as the slave. I recommend purchasing an additional cable and moving the CD-ROM to the secondary channel. On this channel, the CD-ROM can be configured as either the master or the slave. If you purchase an additional hard drive, it should be configured as the slave drive and placed on the primary channel. If you install a CD-ROM, CD-R, CD-RW, or DVD drive, it should be placed on the secondary channel.

## 11.7  SCSI

The small computer system interface (SCSI, pronounced "*scuzzy*") was developed from an interface originally created by Shurgart called the *Shurgart Associates System Interface (SASI, pronounced "sassy").* SCSI was first installed on the Apple Mac Plus and has been installed on every Mac since then. IBM adopted the SCSI bus as standard equipment in the PS/2 models 80 and 60 SX. However, on most compatible computers, SCSI is not standard equipment, so an SCSI adapter must be purchased and installed in one of the computer's expansion slots in order to use this interface standard.

SCSI is not really a controller, but a host adapter that has its own well-defined bus structure. An SCSI adapter card can control multiple SCSI devices that are connected to one another in a daisy-chained arrangement, as illustrated in Figure 11.11. The number of SCSI devices that can be controlled is dependent of the SCSI standard being used.

Unlike the ATA standards, SCSI is both an internal and an external standard. Internal SCSI devices are mounted within the computer's chassis and use the power provided by the computer's power supply. External SCSI devices are placed outside the computer and have their own case and power supply. Unfortunately, the ability to support both internal and external SCSI devices adds considerably to the cost on the SCSI interface.

SCSI is usually implemented with a single adapter that plugs into one of the computer's expansion slots or is embedded in the motherboard. However, multiple SCSI adapters can be installed as long as the assigned resources do not conflict with each other or other ports in the computer. Any controller that plugs into the expansion bus can theoretically be manufactured to be SCSI-compatible and used as a

# Hard Disk Drives

**Figure 11.11** Internal and External SCSI Devices that are Daisy-chained Together

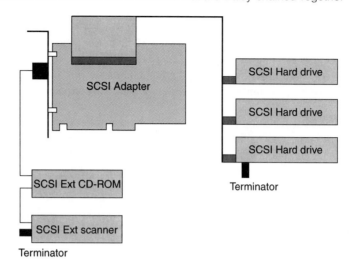

**Table 11.5** SCSI Transfer Rates

| SCSI Standard | 8-bit Transfer, 50-pin Cable | 16-bit Transfer, 68-pin Cable | SCSI Devices |
|---|---|---|---|
| SCSI 1 | 5 MBps | | 7 |
| Fast SCSI (SCSI 2) | 10 MBps | | 7 |
| Ultra SCSI (Fast-20) | 20 MBps | | 7 |
| Ultra-2 SCSI (Fast-40) | 40 MBps | | 7 |
| Fast Wide SCSI | | 20 MBps | 15 |
| Ultra Wide SCSI | | 40 MBps | 15 |
| Ultra-2 Wide SCSI | | 80 MBps | 15 |

device connected to the SCSI interface. Some SCSI devices that are currently available include hard drives, CD-ROMs, CD-Rs (recordable), CD-R/Ws (read/write), tape backups, and scanners.

There are several SCSI standards currently available. These include SCSI-1, Fast SCSI (SCSI-2), Fast-20 (Ultra SCSI), and Fast-40 (Ultra-2 SCSI). Most of these SCSI standards can be implemented using a 50-pin cable with 8-bit data transfer, which allows the connection of up to seven SCSI devices, or a 68-pin cable that allows 16-bit transfer and the connection of up to 15 SCSI devices. The latter is referred to as *Wide SCSI*. The pin-out for the 50-pin SCSI is listed in section 11.7.2.

The numbers in the SCSI standards indicate the maximum transfer speed when 8-bit data transfer is used. For example, Fast-20 (Ultra SCSI) indicates a data transfer rate of up to 20 MB per second (MBps). If 16 bit (2-byte) transfer is used, the transfer rate will be twice the indicated speed. If Wide Fast-20 is used (also called Ultra Wide SCSI), up to a 40-MBps transfer rate can be achieved. Table 11.5 lists the transfer speeds for the different SCSI standards currently available.

## 11.7.1 Why SCSI?

PCs are becoming more and more complex. They are usually multimedia computers that can have one or more hard drives, a tape backup unit, a network card, a scanner, a laser printer, a CD-ROM, etc. If SCSI devices are not used, each device requires its own adapter card and software driver. This does not count the adapter cards required for the video, floppies, printers, serial, etc. that also may be installed. There are not enough expansion slots to handle all of these devices, and getting these devices to cooperate can often be a nightmare because the hardware interface or software drivers can conflict with one another. If a major portion of these were SCSI devices, they could all be connected to a single SCSI adapter card that occupies only one expansion bus slot.

The SCSI controller uses a standard command set, with each target controller generating the actual signals needed to communicate with the connected peripheral devices. This eliminates any software conflicts because only one driver is required to access all of the peripherals connected to the SCSI bus. Also, because the SCSI controller is the only controller directly accessed by the computer, there is no hardware conflict. One other interesting fact is that the cable connection between the peripheral devices and the target controller uses standardized cable. This makes it possible to use SCSI-compatible devices on any computer that uses the SCSI bus.

The main reason why SCSI has found limited use on the compatible computer is the cost of SCSI devices. All of this may seem complicated when the HDC adapter card can be plugged directly into a spot in the expansion bus. This may be why the SCSI interface has found limited used in IBM-compatible computers. What seems interesting is the fact that seven SCSI hard drives could be connected to one SCSI adapter card. If these hard drives were all 13-GB drives, over 91 GB of storage space would be available.

With all of the flexibility added with the SCSI adapter bus, it is the peripheral interface (including hard disk) of choice for high-end workstations and network file servers.

## 11.7.2 The SCSI Connection

Two current connection standards are used to interface SCSI devices: the *single-ended* and the *differential*. Single-ended is the most widely used in the compatible computers because it is the cheapest to implement. Like most other parallel port standards, single-ended SCSI uses one ground wire per signal wire. This limits the overall cable length used to connect the SCSI devices together to 6 meters (20 feet) for SCSI and 3 meters (10 feet) for FAST SCSI. This limitation should not be a problem in most personal computer applications.

Differential SCSI uses a plus (+) and minus (−) wire for each data line instead of using a ground. This feature allows the cable that connects the SCSI devices together to be up to 25 meters (82.5 feet) in length. As you probably guessed, this feature adds considerably to the overall cost of the devices.

Note that devices that use single-ended SCSI and devices that use differential SCSI should never be connected together on the same cable. Doing so will cause the computer to shut down or not boot, and usually causes damage and smoke. For this reason, the international standard symbols, which are shown in Figure 11.12, are placed next to the port and device cable connectors to identify the interface for which they are designed.

Both 8-bit single-ended and differential internal SCSI devices interface by way of a 50-pin male connector on both the SCSI port and SCSI device. The cable for the standard SCSI interface is often called an *A-cable*. Figure 11.13 shows a 50-pin connector used on an A-cable. The pin-outs for both the single-ended SCSI and differential SCSI are listed in Table 11.6.

The 16-bit versions of SCSI use a D-shaped 68-pin connector for both internal and external SCSI devices. The cable, along with the connectors, is called a *P-cable*. A typical 68-pin P-cable connector

**Figure 11.12** Single-ended and Differential Symbols

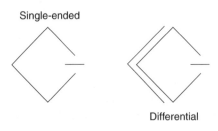

# Hard Disk Drives

**Figure 11.13** 50-pin SCSI Male Pin Connector

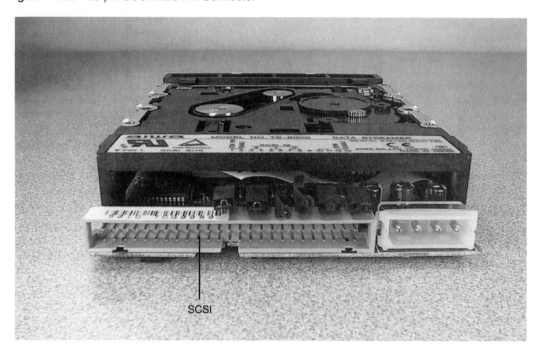

is shown in Figure 11.14. The pin-out for the single-ended P-cable is listed in Table 11.7. Note that adapters are available that will convert a P-cable connector to an A-cable connector. This allows a device designed around an A-cable to run on a computer that is designed around a P-cable interface.

Another SCSI interface is not discussed here because it has limited use in the compatible PC because of its price tag. This is a 32-bit interface that uses a cable called a *Q-cable*. A Q-cable simply consists of two P-cables that are stacked back to back.

## 11.7.3 SCSI Termination

To maintain the integrity of the data down the entire length of an internal or external SCSI daisy-chain requires termination. To accomplish this, a device called a *terminator* is used. The terminator is attached to the last device on both the internal and external SCSI chains. There are two types of terminators used in the SCSI standard: passive and active.

Passive terminators are designed for the SCSI-1 standard and are nothing more than individual resistors connected between each data line and +5 VDC. The device that generates the signal only has to bring the line to ground to represent a low and the resistor pulls the line back up to a high. An example of an external passive terminator is shown in Figure 11.15.

Although passive devices are relatively inexpensive, they can produce signal reflection when operating at the high speeds required by the SCSI-2 and Ultra SCSI. This is why SCSI-2 requires active termination. Active terminators use electronic devices that provide a constant current source over varying loads. This feature allows the chain to maintain almost constant high voltage levels and offers very little impedance to the signal. The lowered impedance practically eliminates signal reflection, which translates to faster speeds.

External SCSI devices that are designed to be daisy-chained have two 50-pin female Centronics connectors—one labeled *IN* and the other *OUT*. Individual cables are routed from the OUT connector to the IN connector on the next device down the chain. The last device on the chain must have a terminator similar to the one shown in Figure 11.15.

Like the IDE interface, internal SCSI devices use a single ribbon cable with multiple connectors. When passive termination is used, the last device on the chain will have to have a small terminator resistor pack installed in a socket on the device. If the last device provides active termination, a configuration jumper or switch is used to enable or disable the active termination.

**Table 11.6** Pin-outs for the Single-ended and Differential SCSI Interfaces

| Single-ended | | | | Differential | | | |
|---|---|---|---|---|---|---|---|
| *Signal* | *Pin* | *Pin* | *Signal* | *Signal* | *Pin* | *Pin* | *Signal* |
| Ground | 1 | 2 | D0 | Ground | 1 | 2 | Ground |
| Ground | 3 | 4 | D1 | +D0 | 3 | 4 | -D0 |
| Ground | 5 | 6 | D2 | +D1 | 5 | 6 | -D1 |
| Ground | 7 | 8 | D3 | +D2 | 7 | 8 | -D2 |
| Ground | 9 | 10 | D4 | +D3 | 9 | 10 | -D3 |
| Ground | 11 | 12 | D5 | +D4 | 11 | 12 | -D4 |
| Ground | 13 | 14 | D6 | +D5 | 13 | 14 | -D5 |
| Ground | 15 | 16 | D7 | +D6 | 15 | 16 | -D6 |
| Ground | 17 | 18 | Parity | +D7 | 17 | 18 | -D7 |
| Ground | 19 | 20 | Ground | +Parity | 19 | 20 | -Parity |
| Ground | 21 | 22 | Ground | DIFFSENS | 21 | 22 | Ground |
| Reserved | 23 | 24 | Ground | Reserved | 23 | 24 | Reserved |
| Open | 25 | 26 | TERMPWR | TERMPWR | 25 | 26 | Terminator Power |
| Reserved | 27 | 28 | Ground | Reserved | 27 | 28 | Reserved |
| Ground | 29 | 30 | Ground | +ATN | 29 | 30 | -Attention |
| Ground | 31 | 32 | ATN | Ground | 31 | 32 | Ground |
| Ground | 33 | 34 | Ground | +Busy | 33 | 34 | -Busy |
| Ground | 35 | 36 | Busy | +Ack | 35 | 36 | -Acknowledge |
| Ground | 37 | 38 | Ack | +Reset | 37 | 38 | -Reset |
| Ground | 39 | 40 | Reset | +Message | 39 | 40 | -Message |
| Ground | 41 | 42 | Message | +Select | 41 | 42 | -Select |
| Ground | 43 | 44 | Select | +C/D | 43 | 44 | -Control/Data |
| Ground | 45 | 46 | C/D | +REQ | 45 | 46 | -Request |
| Ground | 47 | 48 | REQ | +I/O | 47 | 48 | -I/O |
| Ground | 49 | 50 | I/O | Ground | 49 | 50 | Ground |

**Figure 11.14** P-cable 68-pin Connector

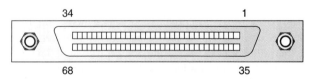

**Table 11.7** Pin-out of 68-pin PChannel

| Signal | Pin | Pin | Signal |
|---|---|---|---|
| Ground | 1 | 35 | D12 |
| Ground | 2 | 36 | D13 |
| Ground | 3 | 37 | D14 |
| Ground | 4 | 38 | D15 |
| Ground | 5 | 39 | Parity1 |
| Ground | 6 | 40 | D0 |
| Ground | 7 | 41 | D1 |
| Ground | 8 | 42 | D2 |
| Ground | 9 | 43 | D3 |
| Ground | 10 | 44 | D4 |
| Ground | 11 | 45 | D5 |
| Ground | 12 | 46 | D6 |
| Ground | 13 | 47 | D7 |
| Ground | 14 | 48 | Parity0 |
| Ground | 15 | 49 | Ground |
| Ground | 16 | 50 | Ground |
| TERMPWR | 17 | 51 | TERMPWR (Termination Power) |
| TERMPWR | 18 | 52 | TERMPWR |
| Reserved | 19 | 53 | Reserved |
| Ground | 20 | 54 | Ground |
| Ground | 21 | 55 | Attention |
| Ground | 22 | 56 | Ground |
| Ground | 23 | 57 | Busy |
| Ground | 24 | 58 | Acknowledge |
| Ground | 25 | 59 | Reset |
| Ground | 26 | 60 | Message |
| Ground | 27 | 61 | Select |
| Ground | 28 | 62 | Control/Data |
| Ground | 29 | 63 | Request |
| Ground | 30 | 64 | Input/Output |
| Ground | 31 | 65 | D8 |
| Ground | 32 | 66 | D9 |
| Ground | 33 | 67 | D10 |
| Ground | 34 | 68 | D11 |

**Figure 11.15** External passive SCSI Terminator

### 11.7.4 SCSI-ID

Because so many devices can be connected to the SCSI expansion bus, there has to be a way of identifying each device, which is the function of the SCSI-ID. All SCSI devices have a group of switches or jumpers that allow the device to be assigned a unique ID number from 0 to 7 for A-cable devices and from 0 to 15 for P-cable devices.

An A-cable device has three switches or jumpers while the P-cable has four. This number of switches is required because the ID is set in binary. Unfortunately, the switch arrangements can be reversed among different devices, so consult the device's documentation to determine the correct configuration. An example of an A-cable device's configuration chart is shown in Example 11.1.

**Example 11.1:** Typical SCSI device configuration chart

| SCSI-ID | SW2 | SW1 | SW0 |
|---|---|---|---|
| 0 | OFF | OFF | OFF |
| 1 | OFF | OFF | ON |
| 2 | OFF | ON | OFF |
| 3 | OFF | ON | ON |
| 4 | ON | OFF | OFF |
| 5 | ON | OFF | ON |
| 6 | ON | ON | OFF |
| 7* | ON | ON | ON |

*SCSI-ID 7 is reserved for the SCSI controller.
SW3 ON = Termination    OFF = No Termination

**Problem 11.2:** Referring to Example 11.1, what position must SW0, 1, 2, and 3 be placed in to set the device's ID to 5 with active termination?

**Solution:** The switches must be set as follows:

SW0 = ON
SW1 = OFF
SW2 = ON
SW3 = ON

## 11.8 INSTALLING THE DRIVE

Mechanical installation of a hard drive is fairly straightforward for anyone with the slightest mechanical inclination. The installation procedure for internal SCSI and IDE drives is basically the same, so only one procedure will be given in Procedure 11.3.

## 11.9 THE SOFTWARE SIDE OF A HARD DRIVE'S INSTALLATION

Performing the hardware installation of a hard drive is a fairly straightforward process. However, once the hard drive is configured and installed, the installation process is only half over. After the hard drive has been hardware-installed, it is time to software-install the drive. The process of

---

### Procedure 11.3: Mechanical installation of a hard drive

1. If a hard drive upgrade is being performed, back up all of the important information on all of the hard drives in the computer. If the boot hard drive is being replaced, make a copy of the DOS CD-ROM, video, and sound card drivers.
2. Use anti-static procedures whenever work is being performed inside the computer.
3. Turn the computer off and unplug the computer from the power receptacle.
4. Remove the computer chassis cover and place the screws in a safe place.
5. Locate an empty drive bay that matches the form factor of the hard drive being installed.
6. If a spare power connector is not available, a power splitter may be required.
7. Configure the drive as either master or slave for EIDE, depending on the application, or set a unique ID if the drive is SCSI.
8. If the drive being installed is the last device on the SCSI daisy-chain, it must be terminated. The termination can be selected through a jumper on the device or through a terminating resistor pack, so consult the device's documentation.
9. Insert the drive into the bay from the back with the electronic components down and the interface connector toward the rear of the computer.
10. Securely mount the drive in the bay by using at least three screws—two screws on one side and one on the other. Do not over-tighten the screws because this may damage the drive.
11. Insert the data interface cable into the drive making sure the pin-1 side of the cable is inserted into the pin-1 side of the connector, as illustrated in Figure 11.16.
12. Make sure the other end of the data interface connector is inserted into the correct port on the motherboard or adapter.
13. Insert the drive's power connector as illustrated in Figure 11.16.
14. Check all cable connections to the motherboard and the internal peripherals.
15. Plug in the computer and turn it on.
16. If the computer does not boot, the drive does not spin, or multiple beep codes are sounded, make sure that the cables are not reversed. Also make sure that the cables are seated correctly in the connectors.
17. If the computer supports the ATA-2 standard, monitor the boot sequence to see if the newly installed drive is detected.
18. If the installed drive is SCSI, the device ID should be displayed when the SCSI adapter is detected during the boot sequence.
19. If the drive seems to function correctly, replace the computer cover.
20. Before the drive can be used, it must be software-installed.

**Figure 11.16** Installing the Interface and Power Cable

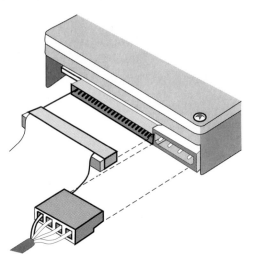

software installation includes a low-level format (when required), establishing an interleave factor (when required), partitioning, and a high-level format. All of these processes will be discussed in the following sections.

## 11.9.1 The Low-Level Format

During the **low-level format** process, tracks and sectors are created on the surface of the disk, defective sectors are detected, and the drive's interleave is set. Interleaving, which is the way the sectors are numbered on the disk, is discussed in section 11.9.4.

When a low-format is performed on a hard drive, it does not create the master Boot Record, FATs, or root directory. This means the operating system's FORMAT utility cannot be used. To accomplish a low-level format requires the use of a special utility program that is provided by hard drive adapter ROMs, the BIOS, or the drive manufacturers. The following list contains a few important topics to consider before a low-level format is attempted:

1. *Do not* attempt to low-level format any ATA or SCSI drive before reading sections 11.9.2 and 11.9.3.
2. When a hard drive is low-level formatted, you will lose all of the data on the drive. For this reason, perform a backup on the drive, if possible. Of course, if the reason for the low-level format is because the drive is not accessible or the drive is new, a backup is not required.
3. Allow the drive to run for approximately 30 minutes before attempting the low-level format to allow the surface of the platters to rise to operating temperature.
4. Mount the drive in the position in which it is going to operate.

Most MFM and RLL controllers contain the low-level formatter on ROMs installed on the hard drive adapter card. They are accessed using a DOS utility program called Debug. Debug is a program that gives the user binary level control of the computer. Using Debug is very dangerous for the untrained individual, but learning its features can be of great benefit to individuals who are technically and software inclined.

The adapter ROM low-level formatter for the MFM and RLL drives is usually located at address $C8005_{16}$. However, consult the documentation that comes with the adapter for its actual location. Procedure 11.4 gives the steps required to start the low-level formatter on most MFM or RLL drives.

Some BIOSs contain low-level formatters for MFM and RLL drives as part of the CMOS Setup program. If this is the case, you can select this option to low-level format the MFM or RLL drive. However, *never* use this option to low-level format an IDE, EIDE, or SCSI drive. Doing so may damage the drive. For information relative to formatting these drives, read the following sections.

# Hard Disk Drives

> **Procedure 11.4: Low-level formatting an MFM or RLL drive**
>
> 1. At the DOS prompt type debug and then press the Enter key.
> 2. A dash (-) will appear on the screen. This is the Debug prompt.
> 3. Type g=c800:5 and then press Enter.
> 4. The low-level formatter should start. If it does not, it indicates the formatter is not available or the drive controller is not an MFM or RLL.
> 5. Follow the instructions displayed by the low-level formatter.

### 11.9.2 Low-level Formatting an IDE or EIDE Drive

IDE and EIDE drives are low-level formatted by the manufacturer. Although it is rare, the low-level format on an IDE/EIDE drive may be lost. When this occurs, all of the information on the hard drive will be inaccessible, so keep backups. Never try to low-level format an IDE or EIDE hard drive using the low-level format program available with some BIOSs.

The corruption of an IDE/EIDE drive's low-level format is usually indicated when the drive loses information. Another indication is when a large number of defective allocation units are detected when drive maintenance utilities such as CHKDSK and SCANDISK are executed. Another indication of a corrupted low-level format is when the drive will no longer partition or high-level format. If any of these errors are detected, low-level formatting the hard drive often corrects the problem.

As stated, the low-level format program and setup for the drive can usually be obtained from the drive's manufacturer at no charge except for shipping. Most hard drive manufacturers have a Web site, so the shipping charge can be avoided if the low-level format utility is downloaded from the Internet. Before you call the drive's manufacturer or visit the manufacturers' Web site, make sure you have the model number of the drive available. This information is usually located on a label attached to the top of the drive. Unfortunately, the drive will usually have to be removed from the computer to see the label.

If there is not a low-level format utility available from the drive's manufacturer, there are some third-party low-level IDE/EIDE formatting utilities available from various vendors. The best known of all the third-party low-level formatters available on the market is probably Ontrack's Disk Manager. If a low-level formatter is not available from the manufacturer, Disk Manager is probably the next-best option.

### 11.9.3 Low-level Formatting an SCSI Hard Drive

Like (E)IDE, the low-level format utility provided with some BIOSs should never be used to low-level format an SCSI drive. An SCSI drive is low-level formatted through the SCSI adapter controller. Also like (E)IDE, the low-level format utility is usually unique among different SCSI adapter manufacturers.

When an SCSI adapter is purchased, the low-level format utility is usually provided on a diskette or is located in the BIOS ROMs installed on the adapter itself. To access the low-level format routine in the adapter's BIOS, the DOS utility Debug is usually used. The procedure to use the ROM-based low-level format utility should be provided in the documentation that comes with the SCSI adapter. If this procedure or diskette is not provided, consult the SCSI adapter card manufacturer.

### 11.9.4 Interleave

The drive's interleave is set as part of the low-level format process. Interleave is a process of numbering the sectors on the hard drive to improve data transfer rates. If the interleave is set wrong, the transfer rate of the drive can be greatly degraded.

As stated earlier, most MFM and RLL hard drives rotate at 3,600 rpm or higher and are available with 17, 26, 32, 34, or 36 or more sectors per track. The drives are all soft-sectored (identified magnetically), with each sector containing 512 bytes of user data. Just like the floppies mentioned in chapter 10, the operating system does not access the drive by sectors, but in a group of consecutive sectors called a *cluster,* or *allocation unit.* Hard drives use four or more sectors for each allocation unit. An allocation unit is located in consecutive *sector numbers* on each platter. When the drive is accessed for data transfer, the contents of an allocation unit are always transferred. For some of the older hard disk controllers (HDCs), this multiple sector access created a problem.

To understand the reasoning around the need to interleave a hard drive, we must first have a basic knowledge of the process used to store and retrieve information from the drive surface. A typical read/write sequence for a hard drive is listed in Example 11.2.

**Example 11.2:** Typical read/write sequence used with hard drives

1. The operating system's hard disk driver program, which controls the hard drive, sends a command sequence to the hard drive controller (HDC) telling it to seek to a certain cylinder, select a head, and either read or write a sector of data.
2. The R/W heads move to the cylinder; the HDC selects the desired head and then starts reading the sector information to determine when the correct sector is under the R/W heads.
3. Once the correct sector is located, the entire sector is read or written. After the sector of data is transferred from the drive to the HDC, and then to memory or vice versa, the operating system must again command the HDC to seek a cylinder, select a head, and either read or write a sector of data. It must do this at least the number of times required to transfer the number of sectors that belong to an allocation unit.

While the operating system is sending the HDC the required command sequence, the platters are still rotating. The IBM/PC/XT and some AT computers, using a standard MFM or RLL controller, cannot transfer the data and send the required command sequence to the HDC before the next consecutive sector header passes under the R/W heads. This means the platter will have to make a complete revolution to access the next consecutive *physical sector*. The same problem can arise if the data throughput of the controller is not fast enough to transfer the data before the next physical sector passes underneath the R/W heads.

To prevent this, a technique called **interleaving** is used. Interleaving is illustrated in Figure 11.17. Because hard drives are soft-sector, in the same manner as floppies, the sector numbers are not defined by hardware but are established when the drive is low-level formatted. The operating system reads sector information recorded at the front of each sector to determine the sector under the R/W head. The sectors do not have to be consecutive, but can be interleaved (staggered) on the track.

The interleaved sectors are called *logical sectors* because they are no longer physically sequential. The separation of the sector is called the *interleave factor*. If the physical sector numbers and logical sector numbers are the same, the interleave is 1:1. The left-hand drawing in Figure 11.17 is an example of an interleave of 1:1. The right-hand drawing is an example of an interleave of 1:3 because consecutive logical sectors occur every three physical sectors. If the correct interleave is chosen, the HDC can transfer the sector data to or from memory and receive the next DOS command just before the next consecutive logical sector number passes under the R/W heads.

If an interleave is chosen that is too small, the drive will still have to make a complete revolution to read the next logical sector. On the other hand, if the interleave is too large, the R/W heads will have to

**Figure 11.17** Interleaving

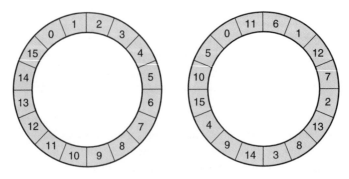

# Hard Disk Drives

wait for the correct sector to pass under them. Choosing the correct interleave for a computer's hard drive-HDC combination is determined by the speed of the computer, the transfer rate of the HDC, and the access time of the hard drive itself. Programs that select the correct interleave factor for the computer/drive combination are available from many software vendors.

If the computer uses an IDE or EIDE drive, the drive interleave is usually set by the manufacturer and should not be changed by the user. Most current IDE and EIDE hard drives use an interleave of 1:1.

## 11.9.5 FAT16 and FAT32

Just like floppy drives, hard drives store and retrieve information in allocation units. Also, just like floppy drives, file allocation tables (FATs) that are located at the beginning of each drive partition control the allocation units. Current hard drives use either the FAT16 or FAT32 system. These FAT systems are similar to the FAT12 system discussed in chapter 10. The only major difference is the number of table entries available and the number of sectors per allocation unit.

The name **FAT16** indicates that 16 binary bits are used to determine the maximum number of entries in the FAT. The 16 binary bits yield a maximum of $65,536_{10}$ table entries ($2^{16}$). Because we are discussing storage capacity, 65,536 can also be expressed as 64 K. Remember, no matter the media, each sector holds 512 bytes (0.5 KB) of data. This means if an allocation unit equals one sector, the drive can store only 32 MB of data (64 K $\times$ 0.5 K) per partition. It should be obvious that the allocation units used by a hard drive have to be greater than one sector. Table 11.8 lists the different allocation unit sizes and hard drive partition capacities used by FAT16.

As shown in Table 11.8, the overhead for large drives using FAT16 is huge. Remember that information is stored and retrieved to and from a disk in allocation units, despite the size of the file being saved. This means if a 5-byte file is saved on a 1.6-GB drive using FAT16, 32 KB of storage space will be used.

**Problem 11.3:** How much storage space will an 86-KB file occupy on a 1.6-GB hard drive using the FAT16 system?

**Solution:** A 1.6-GB drive uses 32-KB allocation units. The drive stores in multiple allocation units, so three allocation units, or 96 KB, of storage space will be required to store an 86-KB file.

Windows 95b, Windows 98, and Windows 2000 are currently the only Microsoft operating systems that support FAT32. Windows 2000 also supports a file system called the *New Technology File System (NTFS)* that is also available with Windows NT Workstation and Server. NTFS is discussed in section 11.9.6. The only non-Microsoft operating system that supports FAT32 is a free operating system called *Linux*.

Table 11.8 Allocation Unit Size Related to Partition Capacity When Using FAT16*

| Allocation Unit | | Partition |
|---|---|---|
| Sector | Size | Capacity |
| 2 | 1 KB | 64 MB |
| 4 | 2 KB | 128 MB |
| 8 | 4 KB | 256 MB |
| 16 | 8 KB | 512 MB |
| 32 | 16 KB | 1,024 MB or 1 GB |
| 64 | 32 KB | 2,048 MB or 2 GB |

*Drive sizes that fall between drive capacities must use the larger allocation unit size. For example, an 800-MB drive uses 16-KB allocation units.

Table 11.9  Allocation Unit Sizes for FAT32

| Allocation Unit | | Partition |
| --- | --- | --- |
| Sector | Size | Capacity |
| 8 | 4 KB | 8 GB |
| 16 | 8 KB | 16 GB |
| 32 | 16 KB | 32 GB |
| 64 | 32 KB | 2 TB |

As the name implies, **FAT32** uses 32 bits to define the number of FAT entries. Of the 32 bits available with FAT32, the current version of Windows 9X uses only 21 of the bits. This allows up to $2^{21}$ allocation units, or $2,097,152_{10}$ (2,048 G). The smallest allocation unit size used by FAT32 is 4,096 bytes (4 KB). Using a little math, it is easy to see that this allocation unit size allows single partitions of up to 8,192 KB, or 8 GB (2,048 G $\times$ 4 KB).

The major advantage of FAT32 is its small allocation unit size. A 2-GB drive using FAT16 uses 32-KB partitions, whereas the same hard drive formatted with FAT32 uses 4-KB partitions. Allocating space on the surface of the disk in 4-KB increments is a lot more efficient than allocating space on the hard drive in 32-KB increments. The allocation unit sizes currently available with FAT32 are listed in Table 11.9.

**Problem 11.4:** How much storage space will be required by a 33-KB file stored on a 2-GB hard drive using FAT16? How much space will be required on the same capacity hard drive using FAT32?

**Solution:** A 33-KB file will require 64 KB (two 32-KB allocation units) of space on the drive using FAT16, whereas 36 KB (nine 4-KB allocation units) of storage space will be required on a drive using FAT32.

Note that DOS, Windows 3.X, and Windows 95 OSR1 are not able to access the hard drive directly once it has been formatted using FAT32. This includes any Startup diskette that was made under Windows 95 OSR1. Any application program that accesses the hard drive directly will have problems if the application does not support FAT32. This includes third-party disk optimizers or backup programs.

### 11.9.6 New Technology File System (NTFS)

To offer some downward compatibility, Windows NT supports FAT16, while Windows 2000 supports both FAT16 and FAT32. Although the FAT system has served the PC community for years, it has some major drawbacks when it comes to the security required by the business community. In fact, the FAT system offers little or no security whatsoever. To access a drive using the FAT system, even with password protection, simply requires booting the computer from a floppy. An individual can then delete, rename, and copy any file. Even the password and boot protection offered by the CMOS can be easily defeated by removing the battery or by using the clear CMOS jumper. Once the CMOS has been cleared, all password protection is removed.

To overcome this and other shortcomings of the FAT system, Microsoft gave Windows NT another file system called the *New Technology File System (NTFS)* and passed it along to Windows 2000. Some of the improvements offered by NTFS over FAT include the following:

- Increased security;
- Unicode support;
- Individual file or folder compression;
- Cluster addressing with a 64-bit binary number;
- High degree of fault tolerance; and
- Support of extremely large drives and partitions.

**Table 11.10** Default Cluster Sizes Used by NTFS

| Drive Size (GB) | Cluster Size (KB) | Number of Sectors |
|---|---|---|
| 0–4 | 4 | 8 |
| 4–8 | 8 | 16 |
| 8–16 | 16 | 32 |
| 16–32 | 32 | 64 |
| 32+ | 64 | 128 |

One major security issue resolved by NTFS is being able to access the drive using a DOS boot disk. NTFS is not compatible with FAT, so if the computer is booted from a DOS diskette, all NTFS partitions are invisible. Also, Windows NT/2000 requires that an administrator be assigned during the installation process. The administrator can set up user accounts and control which account can access what drives and what folders. Once ownership is established, the drive or folder cannot be accessed by anyone other than the owner. This ownership stays in place even if the drive is removed and reinstalled on another computer running Windows NT/2000.

Unicode is an internationally adopted code that is meant to replace all of the current binary codes used to represent numbers, characters, and punctuation, similar to ASCII. Most countries have adopted Unicode to allow text information to be transmitted among computers. While ASCII uses only 8-bit binary, Unicode uses 64-bit, which provides 18,446,744,073,709,551,616 character codes. This provides plenty of combinations for unique codes for every character and punctuation symbol for every language in the world. Because of this feature, all major software writers and Internet Web browsers are adopting Unicode. Windows NT/2000 support Unicode, so files can be given names in any language.

**Compression** is the act of making files occupy less storage space. Once compressed, the file must be uncompressed before it can be used. File compression has been around for years. In fact, Microsoft first supported compression with MS-DOS 6.0. The biggest drawback of compression is that it takes time to compress and uncompress the files as they are being stored or retrieved from the hard drive. When an FAT system is used, the entire drive has to be compressed, which slows down access to the drive. NTFS, however, allows individual files or folders to be compressed. Single file and folder compression allows files that are not used often to take up less disk space but allows files that are used often to be left in their uncompressed form so they can be accessed faster.

FAT16 only allows up to 64-K ($2^{16}$) allocation units to be in a single partition. Although better, FAT32 allows up to 4-G ($2^{32}$) allocation units. Even though FAT32 seems to provide a large number of allocation units, it is small compared to the 64-bit that NTFS provides to keep up with cluster in an NTFS partition. This many bits allows 16-T clusters ($2^{64}$) to be addressed in a single NTFS partition, which should support any size drive during the next decade.

Fault tolerance is the ability to recover from a failure. FAT systems have a very low degree of fault tolerance. This is especially true if the computer locks up or a problem occurs during a disk write operation. It is imperative to maintain a backup of all of your important information saved in a FAT partition. NTFS keeps a record of the write process that can be used to recover the file in the event of a system crash.

Because Windows NT offers an almost unlimited number of clusters, it would seem that an individual sector (512 bytes) should be used to efficiently utilize the storage capacity of the drive. However, the tradeoff to using small cluster sizes is that it slows down the access speed considerably because there are so many clusters the OS has to go though. To speed up access, the size of the cluster needs to be increased. Even though the user can define the cluster size used by NTFS, there are default sizes that are used to maximize the speed-versus-efficiency of the drive. The size of the cluster is chosen automatically according to the drive capacities listed in Table 11.10.

There are a few other advantages of the NTFS system that I am not going to mention because they are very technical in nature and are beyond the scope of this text. Hopefully I have provided enough information to give you insight into the advantages of using an NTFS file system over an FAT. Unfortunately, Windows 95 and 98 do not support NTFS.

**Figure 11.18** Two Partitioned Physical Drives

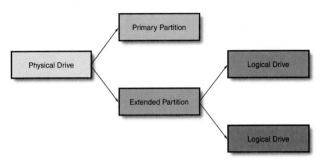

## 11.9.7 Partitions

When MS-DOS was first written, the only hard drive available for the IBM-PC was a 5-MB unit. Looking into the future, Microsoft wrote MS-DOS to access up to a 32-MB hard drive directly. With the hard drive technology available at that time, this seemed like plenty of headroom. However, as the demands placed on personal computers kept increasing and hard drive technology kept improving, the 32-MB limit was quickly exceeded. This gave the impression that MS-DOS could not run a hard drive that was over 32 MB, but this is not the case.

IBM designed the PC to be capable of being controlled by several different operating systems. This developed the need to have several different operating systems available on the hard drive at the same time. These operating systems need to be kept separate. Instead of buying individual hard drives for each operating system, another method, called *partitions,* is used separate the operating systems on a single drive.

Each hard drive can be split into one or more **partitions** (divisions). When a hard drive is partitioned, its cylinders are divided into **logical drives.** After the number of partitions has been selected, a high-level format must be performed on each logical drive. The high-level format assigns each sector to the logical drive, adds the FATs and directory, and places the volume boot record at the beginning of each partition.

When using MS-DOS versions below 3.3, a special TSR that is loaded with an AUTOEXEC.BAT or CONFIG.SYS file is required to gain access to the individual partitions. MS-DOS 3.3 and above will access the partitions automatically. Once low-level formatted, partitioned, and high-level formatted, a 64-MB drive could logically be divided into two 32-MB drives, or any combination with partitions less than 32 MB. MS-DOS will access the logical drives as the C: drive and D: drive, thinking that they are separate drives, as illustrated in Figure 11.18. Dividing a single drive into multiple partitions gives the ability to break the partition size limit of any of Microsoft's operating systems.

MS-DOS 3.3 and above allow the drive to be split into two categories or partitions called *primary* and *extended.* The boot files for an operating system must be contained within a primary partition. Up to four primary partitions can be on a single hard drive. The primary partition from which the computer boots is called the *active partition.* The active partition is selected through a utility program called FDISK, which will be discussed later.

All drives must have at least one primary partition whether they are bootable or not. In other words, if two drives are installed, the second drive must also have a primary partition even though it may never boot the computer. If used, there can only be one optional extended partition and it can never be made active. Even though the extended partition cannot be bootable, it can be divided into multiple logical drives.

Each primary partition, along with each logical drive set up in the extended partition, is assigned a unique drive letter designator (C through Z). For example, if a hard drive is divided into primary and extended partitions, and the extended partition is split into two logical drives, the drive will use three drive letters—one for the primary and two for the extended. If it is the first hard drive, it will use letters C, D, and E. If a second drive is installed, its first primary partition will take the next consecutive letter following the last letter used by the first drive. In the example given, the first partition on the second drive will take the letter F.

# Hard Disk Drives

**Problem 11.5:** What drive letter will be used by the primary partition of the primary slave drive if the primary master drive is divided into four partitions?

**Solution:** The primary master is the first drive in an IDE system, so it will take the letters C, D, E, and F. This means the first partition on the primary slave will take the letter G.

Microsoft operating systems that use FAT16, including Windows 95a, set the size of an individual partition to a maximum of 2 GB when using 32-KB allocation units. If the drives on a computer running Windows NT are formatted using FAT instead of NTFS, the partition size is also limited to 2 GB. The partition sizes offered by operating systems that support FAT32 have practically eliminated the need to split a drive into multiple partitions because they offer a partition size of up to 2 TB when using 32-KB allocation units.

All versions of MS-DOS above 3.2 and all versions of Windows 95, 98, NT, and 2000 use a partition utility called FDISK, which is covered in section 11.9.8, to set up an FAT partition. A GUI utility, called *Disk Administrator* in Windows NT and *Disk Management* in Windows 2000, is used to manage both FAT and NTFS partitions. Disk Management is discussed in section 11.9.9.

The partitions on a hard drive, even if there is only one, are controlled by a special sector on the hard drive called the **master partition boot record (MPBR).** Similar to the volume boot record (VBR) used by floppy diskettes, the MPBR is located at logical sector 0 of every hard drive that is low-level formatted and partitioned. The MPBR is the machine language program that is loaded, by the BIOS, into the computer's RAM each time the computer boots from the hard drive.

Along with starting the boot sequence from the hard drive, the MPBR also contains the **partition table.** This table holds the starting and ending cylinder numbers of each partition on the disk. The partition table also specifies which partition will gain control of the computer during boot-up. This partition is called the *bootable partition* and must contain the operating system boot files. When the partitions are configured using FDISK, the bootable partition is referred to as the *active partition.*

Partitioning the drive only modifies the information in the hard drive's partition table and does not actually divide the physical drive into the logical drives. The process of assigning the sectors to a logical drive, creating the FATs and multiple root directories, and placing the VBR at the beginning of each partition is a feat performed by a high-level format. The high-level format must be performed on each partition before it can be used.

## 11.9.8 Partitioning an FAT Drive with FDISK

The DOS utility FDISK is used to set partitions on a hard drive that uses the FAT system. There are two versions of FDISK. One version supports only FAT16 and the other supports both FAT16 and FAT32. The latter is available only with Windows 95b, Windows 98, and Windows 2000. Because these two versions are basically the same from a user's point of view, only the FAT32 version of FDISK is covered in this section.

To launch FDISK, simply type FDISK at the DOS prompt. If the version of FDISK being used supports FAT32, a message will be displayed informing you that your computer has drives larger than 512 MB. If this message is not displayed when you launch FDISK, you have an FAT16 version of FDISK and FAT32 is not available. If supported, FAT32 is selected by answering yes to the question "Do you wish to enable large disk support?" If you answer no to this question, FAT16 will be used. No matter which version you use, the primary FDISK display contains a menu with the numbered options shown in Example 11.3.

**Example 11.3:** FDISK options

> **Note:** Listing 5 will only appear on a computer that has multiple hard drives.
>
> 1. Create DOS partition or Logical DOS Drive
> 2. Set active partition.
> 3. Delete partition or Logical DOS Drive.
> 4. Display partition information.
> 5. Change current fixed disk drive.

As stated in Example 11.3, listing #5 will only appear on a computer that has multiple hard drives installed. If visible, it is the first option that needs to be considered. This is the option used to select the hard drive to be partitioned. Because hard drive letters are logically assigned, a number is used to select the drive instead of a letter. On an (E)IDE system, the primary master is selected by the number 1, the primary slave by the number 2, and so on.

In Example 11.3, option #1 is used to create a DOS partition or logical drive. The partition created can be either a primary or an extended partition. A hard drive has to have a primary DOS partition in order to be accessed. On the other hand, if logical drives are to be created on the drive, an extended partition will also have to be created. When option #1 is selected, another menu will be displayed with the numbered options shown in Example 11.4.

**Example 11.4:** FDISK menu to create a partition or logical drive menu

1. Create Primary DOS Partition
2. Create Extended DOS Partition
3. Create Logical DOS Drive(s) in the Extended DOS Partition

If option #1 is selected in Example 11.4, a question will appear asking if you wish to use the entire disk storage capacity as the primary DOS partition. If the answer is yes, an extended partition and logical drive(s) cannot be created and the primary partition will automatically become the active (boot) partition. If the answer is no, a screen will appear with information similar to that shown in Example 11.5. From this screen, you enter the size desired for the primary DOS partition. In this example, the primary DOS partition is being set to use 50% of the total disk space.

**Example 11.5:** FDISK primary partition configuration screen

Create Primary DOS Partition
Current Fixed disk drive: 1

Total disk space is 3086 Mbytes
(1 Mbyte = 1048576 bytes).

Maximum space available for partition is 3087 Mbytes (100%).
Enter the partition size in Mbytes or percent of disk space (%) to create a Primary DOS Partition. . . . . . (50%)

Press ECS to return to Fdisk Options

If the primary DOS partition does not use the entire disk surface, an extended DOS partition needs to be created. To begin the procedure, select option #1 from the FDISK options screen shown in Example 11.3. From the next screen select option #2, "Create Extended DOS Partition," from the Options menu shown in Example 11.4. Once this option is selected, the computer will display a text screen similar to the one shown in Example 11.6.

**Example 11.6:** Creating an extended DOS partition.

Create Extended DOS Partition
Current fixed disk drive: 1

| Partition Status | Type Size in Mbytes | Percentage of Disk Used |
|---|---|---|
| C: 1    PRI DO  S21 | 50 | |

Total disk space is 3087 Mbytes (1 Mbyte = 1048576).
Maximum space available for partition is 1543 Mbytes (50%)
Enter partition size in Mbytes or percent of disk space to create an Extended Partition. . . . . . (1543)

Press ESC to return to Fdisk Options.

# Hard Disk Drives

From the screen shown in Example 11.6, the user can see the size and letter assigned to the primary DOS partition and also enter the desired size of the extended DOS partition. In Example 11.6, the size of the extended DOS partition is set to use the remainder of the drive's storage space. Here the size is entered in megabytes to illustrate that the size can be entered using either a size or percentage format.

Once the ESC key is pressed to exit the screen that is used to create an extended DOS partition, most current versions of FDISK will then display a message indicating there are no logical drives in the extended partition and asking if you want to create one. You should answer yes to this question. A screen similar to the one shown in Example 11.7 will then display.

**Example 11.7:** Creating logical drives

---
Create Logical DOS Drive(s) in the Extended DOS partition.

Total partition space is 1543 Mbytes (1 Mbyte = 1048576 bytes). Maximum space available for logical drive is 1543 Mbytes (100 %)

Enter the logical drive size . . . . . . . (100%)

Press ESC to return to Fdisk Options
---

From the screen shown in Example 11.7, you can assign all of the space of the extended DOS partition to one logical drive or you can divide the extended DOS partition into two or more logical drives. In Example 11.7, the entire extended DOS partition is assigned to one logical drive. As each logical drive is created, it is assigned a drive letter that follows the letter assigned to the previous drive. After all of the space in the extended DOS partition has been assigned to logical drives, press the ESC key to return to FDISK options.

Once the primary and extended partitions have been created along with the logical drives, then the next step is to select an active partition. The active partition is where the computer will attempt to load the operating system if the drive is selected as the boot drive in the CMOS setup. In almost all applications, the active partition will be the primary DOS partition. To select the active partition, select option #3 from the FDISK screen shown in Example 11.3. Once selected, a screen similar to the one shown in Example 11.8 will be displayed. In this example, the primary DOS partition is set as the active partition.

**Example 11.8:** Setting the active partition

---
Set Active Partition

Current fixed disk drive: 1

| Partition | Status | Type | Size in MB | Percentage Of Disk Space |
|---|---|---|---|---|
| C: 1 | A | PRI DOS | 1543 | 50 |
| 2 | | EXT DOS | 1543 | 50 |

Total disk space is 3086 megabytes

Enter the number of the partition you want to make active..[1]

Press ESC to return to Fdisk Options.
---

Once the active partition has been selected, press the ESC key to exit all screens. A message will display informing you that the computer must be reset in order for the partitions to take effect. Reboot the computer from the hard drive by performing a warm-boot using the Ctrl-Alt-Del key sequence.

After the computer reboots, the software installation of the drive is only half complete. The next step that has to be performed before the drive is accessible is that each partition must be given a high-level format. Before discussing the high-level format, let us discuss partitioning a drive using Windows NT or Windows 2000.

### Procedure 11.5: Opening Windows 2000 Drive Management

1. To use Disk Management, you must be logged on as the administrator or as a member of the administrative group.
2. Click on the Start button located on the Task Bar.
3. Move up to Settings, and then over to Control Panel.
4. Click on Control Panel.
5. After the Control Panel window opens, locate and then click on the Administrative Tools icon.
6. After the Administrative Tools window opens, locate and then click on the Computer Management icon.
7. After the Computer Management window opens, locate and then click on the plus (+) sign beside the Storage icon (this may already be open).
8. Click on the Disk Management icon.
9. A window similar to the one shown in Figure 11.19 will display.
10. For information on the use of Disk Management, click on Action on the left side of the Menu Bar, and then click on Help.

**Figure 11.19** Windows 2000 Disk Management

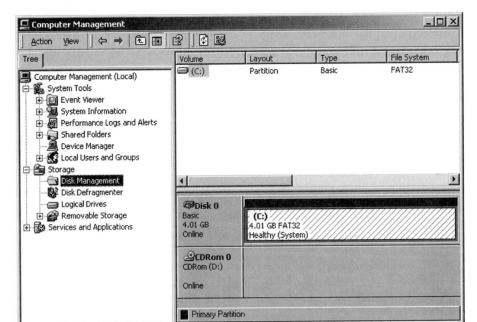

## 11.9.9 Partitioning with Windows NT and 2000

During the installation process of Windows NT/2000, you will be asked if you wish to format the drives using either FAT or NTFS. If the drives installed have already been formatted and partitioned using FAT, a message will display asking if you wish to convert each partition to NTFS. Because Windows NT and 2000 can access information using either FAT or NTFS, some of the partitions can be formatted using FAT and others can be formatted using NTFS.

If the drive has not been partitioned previously, the entire capacity of the drive is set as the primary partition up to the maximum partition size. For FAT16 systems using Windows NT or Windows 95a, this will be up to 2 GB. If the boot drive needs to be partitioned, it should be done using FDISK before Windows 9X, NT, or 2000 is installed. If multiple hard drives are installed, they should be formatted and partitioned using Windows NT Drive Administrator or Windows 3000 Disk Management. To open the Windows 2000 Disk Management Utility, follow the steps outlined in Procedure 11.5.

### 11.9.10 High-level Format

On any DOS version above 3.3, including Windows 95/98/NT/2000, the FORMAT command can be used to perform the high-level format on each hard drive partition. With DOS versions below 3.3, a utility diskette must be used to perform the high-level format. The **high-level format** actually sets up the Boot Record and FATs on the drive in each partition. It also scans each sector header to find those that were marked defective during the low-level format and enters this information in the respective FAT. The high-level format must be performed even if the drive is partitioned to one logical drive. When a high-level format is performed, all data will be lost in the formatted partition.

### 11.9.11 When to Low-level Format, Partition, and High-level Format

Any time a new MFM or RLL controller and hard drive are installed in a computer, the low-level format (if not already done), partition, and high-level format process must be completed before the hard drive will be accessible by operating system. The ESDI and IDE drives should never require a low-level format, but still must be partitioned and high-level formatted. If a stepper motor drive is installed, this process will have to be performed on a regular basis to ensure data dependability.

As stated earlier, the stepper motor drives are open loop. With continued use, the head alignment of these drives will change due to mechanical wear. To prevent total loss of data, these drives should be low-level formatted, partitioned, and high-level formatted at least every 2 years or the first time a read error message is displayed. This procedure will place the cylinders at the new alignment of the R/W heads. With a little added preventative maintenance, the life span of the stepper motor drive is comparable to that of a voice coil drive.

When a low-level format, partition, or high-level format is performed on a drive that contains data, *all of the data* on the hard drive will be lost. Keep a copy of the data stored on the hard drive so it can be recopied.

## 11.10 PREVENTATIVE MAINTENANCE

With all of the important information stored on the hard drive, it is imperative that the hard drive be maintained. Eventually the hard drive will either fail or will have to be replaced, so it is very important to make regular backups of all of the important information stored on the computer's hard drives.

In addition to regular backups, two additional preventative maintenance items should be performed on drives that use the FAT partition to ensure that the drives remain reliable over an extended period of time. These maintenance items include (1) checking the FATs and surface of the disks for errors, and (2) defragmenting the information on the drive.

### 11.10.1 Checking the Drive for Errors

Part of the preventative maintenance on a hard drive is to check the condition of the FATs, directory, and file structure. When MS-DOS or Windows 3.X is used, a utility called CHKDSK is run that will check and repair the FATs. CHKDSK does not check the integrity of the directory and files structure, however.

When Windows 9X is used, a utility called SCANDISK is used to check the hard drives. SCANDISK will not only check for errors between the two FATs, it will also check the integrity of the hard drive surface. SCANDISK should be run at least once a week as part of the preventative maintenance schedule for hard drives. Procedure 11.6 lists the steps to follow to run SCANDISK under Windows 9X.

### 11.10.2 Optimizing an FAT Drive

Hard drives can contain hundreds of files. A large portion of these files will be data files that belong to a word processor, database, or spreadsheet application program. When the files are loaded, altered, and then stored back on the disk, they become fragmented. The term **fragmented** is used to indicate that the files are no longer in consecutive allocation units. The operating system has no problem accessing the files because the FAT tells the operating system exactly how to connect the files together. However, the more fragmented the file, the longer it will take to retrieve and save the file on the disk.

### Procedure 11.6: Running Windows 9X's SCANDISK

1. Click on the Start button.
2. Move up to Programs.
3. Move over to Accessories.
4. Move over to System Tools.
5. Click on the ScanDisk option.
6. Click on the drive you wish to scan for errors.
7. Click on either "Thorough" or "Standard".
8. Click on "Automatically fix errors".
9. Click on the Start button.

**Note:** The drive should not be used while SCANDISK is running.

### Procedure 11.7: Defragmenting a hard drive with Windows 9X

1. Before using Disk Defragmenter, disable the computer screen-saver program (if used).
2. Click on the Start button on the Task Bar.
3. Move up to Programs.
4. Move over to Accessories.
5. Move over to System Tools.
6. Click on Disk Defragmenter.
7. Use the scroll button to select the drive to be defragmented.
8. Click on the OK button to begin the defragmenting process.

**Note:** Do not use the hard drive while defragmenting is in process.

To help speed up disk access, a **disk optimizer,** such as Norton's Speed Disk, MS-DOS 6.0's DEFRAG, or Windows 9X Disk Defragmenter should be used weekly to place all of the files on the disk in consecutive clusters. This does not increase the disk space but will speed up the access. The steps to follow to defragment a hard drive using Windows 9X are listed in Procedure 11.7.

### 11.10.3 Backing Up Information

As stated previously, it is a fact of life that sooner or later the hard drive in your computer will have to be replaced. One major reason for replacing the hard drive is to go to a bigger size. Another reason is that at some time, the hard drive will fail. All kinds of things can go wrong on a hard drive to cause data to become corrupted or accidentally erased. Hopefully, when your time comes, you will have a backup copy of the information stored on your hard drive.

MS-DOS and Windows 3.X supply a utility called BACKUP that will back up the contents of the hard drive to floppy diskette. Of course, with the size of today's drives, this process will take a long time and requires hundreds of floppy diskettes. If you use a tape drive instead of floppies, a secondparty backup program for the particular drive will have to be used with these two operating systems.

Windows 98 provides a program called Microsoft Backup that supports most of the current backup device standards. This backup utility is powerful and easy to use. Not only does it allow the user to back up information on tape drives, it also allows backups to be performed to other hard drives on the computer or even network hard drives. The option listing to run Microsoft Backup is located in the System Tools option of the Start menu.

## 11.11 COMPRESSION

The longer you are involved with computers, the more you will realize that there is no such thing as too much hard disk storage. Today's operating systems, application programs, and data files most likely require several megabytes of hard disk storage space. In fact, purchasing a computer with less than 4 GB of hard disk storage is not a wise move. Even 100 GB of hard disk storage will eventually become cramped.

Buying a larger hard drive would seem to be the logical answer to the space problem. However, even with the declining prices of hard drives, this can be an expensive solution. A cheaper solution to increase the amount of information stored on the hard drive is to compress the stored information.

# Hard Disk Drives

Third-party (other than IBM and Microsoft) compression programs have been around for several years. One of the most popular is called STACKER, which is written by Stac, Inc. Microsoft finally started supporting data compression technology with MS-DOS 6.0's utility called DoubleSpace. Both STACKER and DoubleSpace load a TSR that provides on-the-fly data compression that is transparent to the user. Both will more than double the amount of free storage space available on the hard drive. The only hassle of installing either of these utilities is the time required to originally compress all of the hard drive's information. Once compressed, all data that is written or read from the hard drive will be compressed and uncompressed by the compression software.

This book is not going to attempt to explain the process used to compress the information stored on the drive. Data compression has been around for several years and it is a proven technology that has been tested over and over again. Many users still remain horrified about on-the-fly data compression. Their fears have been fueled by the many horror stories about users who have lost megabytes of information in attempting to compress their hard drives. However, most of the problems with data loss are caused by something other than the data compression software. But the association exists, so the compression software usually gets the blame. The most common cause of lost data is that the user failed to maintain his or her hard drive by keeping it defragmented and its surface free of errors.

Hard drive compression is a cheap, safe, and easy way to more than double the amount of available space on a hard drive. Because all of the information going to and from the drive has to pass through the compression TSR, drive transfer is slowed slightly but usually goes unnoticed by the user.

Compression software completely changes the information stored on the hard drive, so it is a good idea to back up your data before installing any compression software. A backup utility such as Windows 98's Microsoft Backup can do this. Even after the drive is compressed, it is very important to get into the habit of optimizing the drive and checking the integrity of the disk surface on a weekly basis. MS-DOS 6.2 and Windows 9X provide one utility program called SCANDISK for checking the drive's integrity, and one called DEFRAG for optimizing the drive. If you use compression software other than DoubleSpace, use the integrity checker and optimizer provided with that software.

With the introduction of MS-DOS 6.0, Microsoft started adding a compression utility as part of the operating system. The original compression utility was called DoubleSpace because the storage capacity of the drive practically doubles after compression. The biggest problem with DoubleSpace is its inability to compress drives greater than 2 GB because it only uses 16-K allocation units. Windows 9X provides a 32-bit compression utility called Drive Space III that allows compression drives up to 2 GB because it uses 32-K allocation units. Currently, hard drives that are formatted using FAT32 are not supported by Drive Space III, however.

When a hard drive is compressed, the drive is automatically divided into two partitions. One partition will not be compressed and will hold information such as the system boot files and Windows swap file. The other partition is compressed and is called the *compressed volume file (CVF)*. The CVF contains all of the compressed data in the form of one large file. Both the uncompressed and compressed drives are given drive letters similar to those of standard partitions.

## 11.12 TYPE NUMBERS

If you are still using a PC- or XT-compatible computer, you will have to run either an MFM or RLL controller with BIOS ROMs installed on the hard drive adapter card. This is because the main BIOSs of these computers do not fully support hard drives.

The main BIOS in compatibles above the AT have full support for a limited number of drives. The basic characteristics of several once-popular drives are placed in a table in the computer's BIOS. Each entry in the table is assigned a **type number.** The following hard drive characteristics are included in the table: a

- Number of read/write heads;
- Number of cylinders;
- Location of where to start write precompensation (if used); and
- The landing zone (if used).

The entries for the number of heads, cylinders, and sectors have been discussed in previous sections of this chapter and will not be discussed here. *Write precompensation* is a method used to adjust for the difference in data density between the inside and outside cylinders when information is written to the drive. Precompensation is rarely used in current drives because today's hard drives use a technique called *zone density recording*. This technique splits the cylinders into 16 recording zones. The sectors per cylinder vary from zone to zone, with the outside cylinders containing the most sectors and the inside cylinders containing the least. This technique is possible because the hard drive controller selects the actual sector on the platter's surface after it translates the sector selection sent by the operating system.

The *landing zone* is the cylinder number to which the heads are sent when power is removed from the drive. The heads of most IDE and EIDE drives are mechanically parked and locked down when power is removed, so the landing zone entry is not used in currently manufactured hard drives.

The type number for the drive installed in the computer is selected with the CMOS Setup utility and is saved as part of the configuration settings in the CMOS. The characteristics for each of the type numbers may be different among ROM manufacturers, but generally the first 14 type numbers are the same as those originally dictated by IBM. Unfortunately, there is little uniformity among various BIOS manufacturers after that point. Table 11.11 lists the 14 common type numbers and the characteristics of each.

The most popular compatible BIOS writers, American Megatrends, Inc. (AMI) and Phoenix/Award, have 46 type numbers supported by their BIOS and provide the drive parameters. To supported drives that are not listed in Table 11.11, a Type 47 is added to both the AMI and Phoenix/Award BIOS ROMs that allows the user to enter the cylinder, head, sector, precompensation, and landing zone parameters of their hard drive, if known. This information is usually contained in the documentation supplied with new drives or is on the drive label. If you do not know the information for a drive already installed in your compatible computer, contact the drive manufacturer for this information. Currently, type 47 is the most widely used method of defining hard drive parameters.

Older IDE hard drives use a universal translator to translate parameters for one of the listed type numbers into those required by the IDE drive. To set up the CMOS for one of these IDE drives, choose

Table 11.11  Typical Type Number Information Contained in the BIOS

| Type | Cylinders | Heads | Write Precompensation |
|---|---|---|---|
| 1 | 306 | 4 | 128 |
| 2 | 615 | 4 | 300 |
| 3 | 615 | 6 | 300 |
| 4 | 940 | 8 | 512 |
| 5 | 940 | 6 | 512 |
| 6 | 615 | 4 | 0 |
| 7 | 462 | 8 | 256 |
| 8 | 722 | 5 | 0 |
| 9 | 900 | 15 | 0 |
| 10 | 820 | 3 | 0 |
| 11 | 855 | 5 | 0 |
| 12 | 855 | 7 | 0 |
| 13 | 306 | 8 | 128 |
| 14 | 733 | 7 | 0 |

Hard Disk Drives 405

a type number that is close to the rated capacity of the drive. If the drive will partition and high-level format, it should work.

The parameters used by IDE drives above 100 MB usually have to be defined by entering the required parameters in the CMOS type 47 entry. In fact, all hard drives currently manufactured for the compatible PC are type-47 drives. One major problem exists with using type-47 hard drives. Pre-assembled computers very rarely provide the parameters of their installed hard drives. This means that when the contents of the CMOS disappear, the drive will not be able to be accessed until its parameters are obtained. If you are using a Type-47 drive, it is imperative that you document the drive parameters and store the information in a safe place. Fortunately, the type-47 dilemma was practically eliminated with the introduction of the ATA-2 (EIDE) standard.

As discussed previously, the EIDE (ATA-2) standard added a drive ID command to the IDE command set. With this command, a program can literally ask the hard drive for its parameters. This command was rapidly added to the BIOS program so the computer could auto-detect the hard drive parameters during the boot process. The auto-detection or type entry of the hard drive parameters is selected in the Standard setting of the CMOS Setup program.

## 11.13 TROUBLESHOOTING HARD DRIVES

Because the hard drive is the primary source for almost all information, including the operating system and most applications, it is definitely considered a critical resource. As with all other devices, the problem with a hard drive can either be hardware or software. Following are some common hard drive problems, along with their solutions.

*The hard drive is dead.*

An active LED on the front panel that does not illuminate, and a drive's spindle motor that does not rotate when power is first applied to the computer are both signs of a dead hard drive. If this is the case, perform the following steps. Between each step, apply power to the computer and check for proper operation. If the step does not resolve the problem, remove power from the computer and then perform the next step.

1. Remove power from the computer and then remove the computer's cover.
2. Make sure the power plug is seated snugly in the hard drive connector. If a "Y" connector is being used to supply driver power, remove the connector and then use a connector directly from the computer's power supply.
3. Make sure the interface data cable is seated snugly in the hard driver connector and is not reversed on any device on the daisy-chain or hard driver port.
4. If the drive is SCSI, check to make sure the daisy-chain is terminated correctly.
5. Remove the data cable from the drive but leave the power connected. Apply power to the computer. If the drive now spins, disconnect the cable from any other devices on the channel to determine the defective device.
6. Enter the computer's CMOS Setup program. Make sure the integrated IDE port has not been disabled.
7. Try the Auto-Detect option on the main CMOS Setup screen. If the drive is now detected, restart the computer.
8. If available, consult the drive's documentation and visit the drive manufacturer's Web site to obtain additional troubleshooting steps.
9. If all else fails, replace the drive.

*The drive spins but the computer cannot access the drive.*

1. If the computer and drive support the ATA-2 standard, reset the computer. Then determine if the drive is being detected during the boot sequence. When the drive is detected, a message is usually displayed.
2. If the drive or computer does not support auto-detect, enter the CMOS Setup program and confirm that the correct drive parameters are entered in the Standard CMOS settings.
3. Remove power from the computer and then remove the computer's cover.

4. Reseat all of the data and power connectors on all of the devices located on the daisy-chain.
5. If the drive is SCSI, make sure the daisy-chain is terminated correctly.
6. If available, consult the drive's documentation and visit the drive manufacturer's Web site to obtain additional troubleshooting steps.
7. If all of these steps fail, replace the drive.

*The hard drive's performance is degraded.*

1. Run an up-to-date virus scan on the drive and allow the virus program to repair all detected viruses.
2. Defragment the drive.
3. If the drive is (E)IDE, check to see if the drive's port is not running in the DOS Compatibility mode by going to the Properties listing for the drive using Device Manager. DOS compatibility mode is discussed in section 14.13.1.
4. If available, consult the drive's documentation and visit the drive manufacturer's Web site to obtain additional troubleshooting steps.

*The operating system is reporting read errors, or too many bad sectors are detected by ScanDisk.*

**Note:** After each step, check the drive for proper operation before performing the next step.

1. If possible, run an up-to-date virus scan on the drive and allow the virus program to repair all infected files.
2. If possible, back up all of the important information on all hard drives installed in the computer.
3. Remove power from the computer and then remove the computer's cover.
4. Reseat all of the data and power connectors on all of the devices on the hard drive's daisy-chain.
5. Perform a high-level format. Maintain backups because this step will destroy all of the information on the drive.
6. Perform a low-level format using the formatter provided by the drive's manufacturer. Once performed, repartition the driver using FDISK and then perform a high-level format on each partition. This step will destroy all of the information on the drive.
7. If available, consult the drive's documentation and visit the drive manufacturer's Web site to obtain additional troubleshooting steps.
8. If these steps do not resolve the read error, replace the drive.

### 11.13.1 Troubleshooting DOS Compatibility Mode and Hard-Disk Driver

One problem I have seen with Windows 9X is that when something happens to the 32-bit hard disk driver supplied by Windows 9X, the operating system automatically shifts to the 16-bit real-mode drivers. When this happens, the performance of the hard drive is degraded immensely. Windows 9X will start using the real-mode driver for one of the following reasons:

- There is a resource conflict with other devices.
- Windows 9X did not detect the hard drive controller, or the hard drive controller listing is accidentally removed from Device Manager.
- The default 32-bit driver becomes corrupted.
- An incompatible driver is installed.
- The hard driver is configured incorrectly.

The default 32-bit IDE driver used by Windows 9X for both the primary and secondary channels is ESDI_506.PDR. The default 32-bit driver for SCSI is SCSIPORT.PDR. One or both of these drivers will be installed automatically during the installation process of Windows 9X. To display or update the Windows 98 hard drive drivers being used by a computer, follow the steps listed in Procedure 11.8.

The use of the DOS compatibility mode is usually indicated by an exclamation mark beside the hard drive category under Device Manager. If the exclamation mark is present, open the category by clicking on the plus (+) sign, then double-click on each listed item, and then read the information in the device status panel. If it indicates that the DOS compatibility mode is being used, perform the following steps:

1. Check the settings in the BIOS to determine if the correct drive parameters are being used. Also make sure the right PIO or DMA mode is being selected in the CMOS. If selected, turn off the Block Transfer option.
2. For (E)IDE drives, check Device Manager to determine if ESDI_506.PDR is the driver being used.

## Hard Disk Drives

3. Run an up-to-date virus scanner on the hard drive(s) and allow it to repair detected viruses.
3. Check to make sure the "Disable all 32-bit drivers" option is not selected. This can be determined by following Procedure 11.9.
4. Determine if a hard disk driver is being loaded though the CONFIG.SYS or SYSTEM.INI files. If so, place a REM before the suspected line in the CONFIG.SYS file or a semi-colon (;) before the suspected line in the SYSTEM.INI files, and then reboot the computer. Windows 9X's System Editor can be used to display the contents of both of these files.
5. This step will destroy all of the information on the hard drive. Maintain a current backup.
   - Boot the computer from the Windows 9X Setup disk.
   - Reformat all of the partitions on the drive that contains the Windows 9X operating system.
   - Using FDISK, remove all of the partitions on the hard drive and then reset the computer.
   - Using FDISK, repartition the drive to the desired parameters.
   - Reinstall the DOS mode CD-ROM driver.
   - Reinstall Windows 9X.
   - Reinstall all applications.
   - Restore the application's data from the most current backup. Do not restore a previous version of the Windows 9X operating system.
6. If all of this fails, the problem is most likely with the hard drive controller itself.

---

### Procedure 11.8: Displaying or updating the hard disk drivers used by Windows 98

1. Click on the Start button and then move up to Settings and over to Control Panel.
2. Click on Control Panel.
3. Locate and then double-click on the System icon.
4. Click on the Device Manager tab.
5. Click on the plus (+) sign beside the "Hard disk controller" category.
6. Double-click on either the "Primary IDE controller (dual fifo)" or "Secondary IDE controller (dual fifo)" listings.
7. Click on the Drivers tab.

**To view driver information:**
1. Click on the Driver Files Details button.
2. Write down the name of the driver along with the path. Store this information in a safe place. The true hard-disk driver should be the first one listed. The driver "IOS.VXD" is a virtual driver used by Windows 9X.
3. Close all open windows by using the Cancel, OK, or X buttons, depending upon which applies.

**To update or change the Windows 98 hard disk drivers:**
1. Click on the Update Drivers button.
2. When the Update Wizard starts, click on the Next button.
3. Click on "Search for better drivers than the one your device is using" to allow Windows 98 to automatically look for a new driver, or, "Display a list of all drivers in a specific location, so you can select the driver you want" to specify the location of the driver manually.
4. Follow the displayed instructions, depending on the option chosen.
5. Restart Windows for the changes to take effect.

---

### Procedure 11.9: Determining if 32-bit drivers are disabled

1. Click on the Start button, then move up to Settings, and then over to Control Panel.
2. Click on Control Panel.
3. After the Control Panel opens, locate and then double-click on the System icon.
4. After the System Properties window opens, click on the Performance tab.
5. Click on the File System button.
6. Click on the Troubleshooting tab.
7. If checked, remove the check from the "Disable all 32-bit protected mode disk drivers option."
8. To abort without making changes, click on the Cancel or X button on each open window.
9. To accept changes, click on the OK button on each open window.
10. Restart the computer for the changes to take effect.

## 11.13.2 Hard Drive Troubleshooting Conclusion

I realize this troubleshooting section does not contain all of the problems that can arise with hard drives, but I can only give information on the problems that I have encountered. If the problem you encounter is not listed or the correction steps do not solve the problem, consult other technical documents.

An ideal place to do research is at your local library or bookstore. Also, always read the documents that are supplied with the device and follow the listed installation procedures for the hardware and software. Consult the device's documents for troubleshooting tips and the telephone number for the manufacturer's help line. If documents are not available, another source for information is the FAQ and troubleshooting sections of the drive manufacturer's Web site. There are also hundreds of user groups on the Internet that are valuable resources for troubleshooting.

## 11.14 CONCLUSION

Hard drives are very fast and store lots of data when compared to the current versions of floppy drives, CD-ROMs, and other alternative storage devices. They are relatively inexpensive when the amount of data they store is considered. The density, speed of data transfer, and the availability of the desired data are all major advantages of the hard drive. Most of the new operating systems and applications programs, such as word processors and database systems, are written to be installed on a hard drive.

Hard drives are electro-mechanical devices that can fail for either electronic or mechanical reasons. Even with the best of care, all hard drives will eventually fail. Because a clean-room environment is required to repair them, they are *not* considered a field-serviceable item. To prevent a total loss of all data on the hard drive when failure occurs, keep a copy (floppy or tape backup) of all of your important information.

There are currently two popular standards used to interface to the compatible computer. The most popular, and cheapest, standard is the EIDE. It allows up to four IDE devices to be connected through two channels called primary and secondary. The second popular standard is called SCSI. There are several versions of this standard that support different numbers of devices, word sizes, and transfer speeds. SCSI is very popular on network servers.

To keep the data on the hard drive reliable and the transfer speed as fast as possible, the MS-DOS command CHKDSK or SCANDSK should be used to check the drive's FATs. SCANDISK is the preferred method because it also checks the integrity of the drive's surface. The files on the drive should be optimized with MS-DOS 6.0's utility Defrag or Windows 9X Defragmenter at least once per week.

## Review Questions

1. Write the meaning of the following acronyms
    a. HDC
    b. MFM
    c. RLL
    d. ESDI
    e. SCSI
    f. IDE
    g. EIDE
    h. ATA
    i. ATAPI
    j. CD-ROM
    k. DMA
    l. PIO
    m. CHS
    n. ECHS
    o. LBA

### Fill in the Blanks

1. Hard drives are also known as _____ and _____ drives.
2. The _____ _____ positioner dedicates a platter surface for head positioning information.
3. The SCSI transfers data in either _____ bits or _____ bits.
4. ATA-1 is also known as _____ _____ _____.
5. The hard drive is accessed in units of more than one sector called _____ _____.
6. ATA-2 is also known as _____ _____ _____ _____.
7. ATA-2 added support for the SCSI command set with the _____ _____ _____ _____ standard.

8. FAT16 will allow a total of _____ allocation units on a hard drive.
9. The original hard drive interface for the compatible computer was the _____ .
10. The ports for the EIDE interface are called _____ _____ and _____ _____ .
11. _____ _____ _____ is totally under the control of software and all the data going to and from memory goes through the microprocessor.
12. The technique of transferring information directly between a port and memory is called _____ _____ _____ .
13. PIO mode 3 supports data transfer of up to _____ MBps.
14. Multiword DMA mode 2 can handle data transfers up to _____ MBps.
15. In an EIDE system that is configured to boot from the hard drive, the computer will attempt to boot from the _____ _____ unless changed in the CMOS.
16. When the Windows 9X operating system uses the real mode drivers provided to support MS-DOS, this is referred to as _____ _____ mode.

## Multiple Choice

1. The two types of head positioner are:
   a. hard and soft.
   b. fast and slow.
   c. voice coil and stepper motor.
   d. in and out.
2. The type of controller used on most AT-compatible computers is the:
   a. IDE.
   b. MFM.
   c. ESDI.
   d. SCSI.
3. The disk controller for an IDE drive is located:
   a. on the adapter card plugged into the computer's expansion bus.
   b. on the motherboard.
   c. inside the microprocessor.
   d. on the drive itself.
4. Which of the following is not actually a disk controller?
   a. IDE
   b. MFM
   c. ESDI
   d. SCSI
5. Interleave is the:
   a. number of sectors on a track.
   b. spacing of logical sectors on a track.
   c. number assigned to the physical sectors on a track.
   d. type of controller being used.
6. What does an interleave of 5 indicate?
   a. The sector numbers are spaced five physical sectors apart.
   b. The drive is running at five times its normal speed.
   c. The computer reads one sector every five revolutions.
   d. There are five sectors per cluster (allocation unit).
   e. Answers a and c are both correct.
7. Logical drives are:
   a. the number of hard drives that DOS thinks is available.
   b. the number of hard drives installed in the computer.
   c. a hard drive that is capable of performing logic.
   d. the complement of a physical drive.
8. Which of the following must be done to select the number of logical drives a hard disk will be divided into, assign the number of cylinders each logical drive will have, and place the partition table at cylinder 0, sector 0?
   a. perform a low-level format
   b. perform a high-level format
   c. use a partition utility (FDISK or PART)
   d. use a defragmenting utility
9. Which of the following must be done to create sectors on a new hard drive, check the integrity of the surface, set the interleave factor, and mark defective sectors?
   a. perform a low-level format
   b. perform a high-level format
   c. partition the disk
   d. remove power
10. Which of the following actually divides the surface of the physical hard drive into the number of logical drives (partitions) selected and places the boot-record, FATs, and directory in the correct locations?
    a. low-level format
    b. high-level format
    c. partition
    d. sharp knife
11. A low-level format on a hard drive will do all of the following except:
    a. check the integrity of the disk surface.
    b. establish sector headers.
    c. establish the FATs.
    d. establish the interleave factor.
12. The MS-DOS utility FDISK is used to partition the disk. When the disk is partitioned:
    a. the number of partitions are selected.
    b. it places the partition table (record) on side 0, track 0, sector 0.
    c. the starting and stopping points of each partition are defined.
    d. the partition that the computer will boot off is identified.
    e. all of these answers are correct.
13. When the hard drive is high-level formatted:
    a. the FATs are placed on the hard drive.
    b. bad sectors that were identified during the low-level format are entered into the FATs.
    c. the integrity of each platter's surface is checked.
    d. all of the above answers are correct.
    e. only answers a and b are correct.
14. When a hard drive is optimized:
    a. the files are placed in consecutive allocation units (clusters).
    b. the files are placed in nonconsecutive allocation units.
    c. the storage space on the drive approximately doubles.
    d. a special TSR is required.
15. Compressing a hard drive means:
    a. the hard drive is pressed to reduce its physical size.
    b. the information of the drive is altered to make it occupy less space.

c. the speed of data transfer will increase.
d. the drive needs to be optimized.
16. Regular maintenance on a hard drive should include:
   a. optimizing the information and scanning the surface for errors.
   b. changing the oil and checking the air pressure.
   c. cleaning the surface of the disk.
   d. replacing the filter.

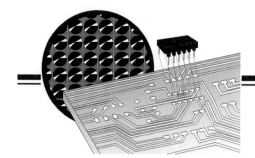

# CHAPTER 12

# Video Adapters and Monitors

## OBJECTIVES

- Define terms, expressions, and acronyms used relative to computer monitors.
- Describe the basic components of a monitor.
- Describe the basic components of a video display adapter.
- Perform calculations relative to video systems.
- Describe the characteristics of the monitors and adapter cards used with the compatible computers.
- Explain video text and graphics modes.
- Describe the differences among the different compatible monitors and adapter cards.
- List the expansion buses used by a video adapter.
- Explain troubleshooting procedures for common video problems.

## KEY TERMS

cathode ray tube
cathode
phosphor
monochrome
deflection yoke
scan line
frame
horizontal sync
vertical sync
video display controller (VDC)
text mode
graphics mode
pixel

resolution
Video Electronics Standards
  Association (VESA)
monochrome display adapter
  (MDA)
Hercules graphics adapter (HGA)
monochrome graphics adapter
  (MGA)
triad
color graphics adapter (CGA)
enhanced graphics adapter (EGA)
analog
RAMDAC

video graphics array (VGA)
super video graphics array (SVGA)
multisync
Windows accelerated adapters
dot-pitch
field
interlaced
noninterlaced
refresh rate
WYSIWYG
advanced graphics port (AGP)
degaussing coil

Most of the information received by humans comes through their eyes. This is why most computers transfer information to humans using a device called a video monitor or display. Computer video monitors have gone from row upon row of blinking lights, through teletypes, to the CRT monitors used on today's computers.

## 12.1 VIDEO BASICS

CRT is an abbreviation for **cathode ray tube.** The CRT is the main component of a computer monitor. The purpose of the monitor is to convert the binary codes of the computer into a visual form that means something to humans. The CRTs used in computer monitors work on the same principle as the CRTs used in the television sets we watch every day. To understand the terminology used when dealing with a computer's monitor, it is necessary to have a basic understanding of the operation of CRTs and broadcast TV.

At the very end of the CRT's neck is one or more *electron guns,* or **cathodes,** as illustrated in Figure 12.1. The electron gun produces a cloud of electrons that is negatively charged, while the front of the CRT is positively charged. The attraction between the negative and positive charges causes the electron cloud to move in the form of a beam toward the face of the CRT.

During its travel toward the face, the beam is focused, its intensity is controlled, and the electronic circuits that control the CRT's overall operation position the beam. After going through this maze of controls, the beam finally strikes the front inside surface of the CRT. The electron beam itself is invisible. To create a visible light, the inside face of the CRT is coated with **phosphor,** which emits light when the electron beam strikes it. The number of electrons that strike the phosphor and the electrons' speed control the intensity of light emitted by the phosphor. The color of the emitted light depends on the phosphor's chemical composition.

**Figure 12.1** Basic CRT

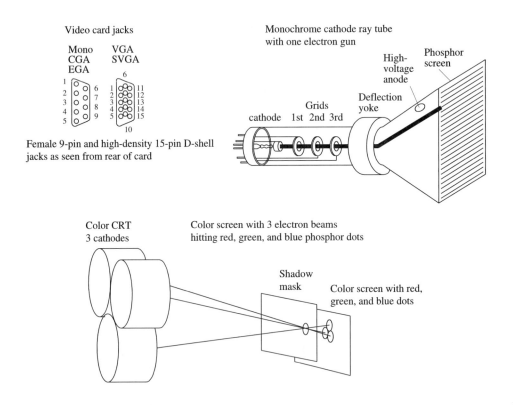

*Source:* Dan L. Beeson, *Assembling and Repairing Personal Computers* (Upper Saddle River, NJ: Prentice Hall, NJ, 1997) p. 140.

A computer CRT that displays in one color has a single gun and is called a **monochrome** monitor (one color + black). A color CRT usually has three guns and the inside surface of the CRT is coated with thousands of red, green, and blue phosphor dots. Color CRTs are called RGB(I), TTL, VGA, SVGA, multisync, or analog.

The beam is positioned on the screen magnetically by a device mounted on the neck of the CRT called a **deflection yoke.** The deflection yoke's location is shown in Figure 12.1. The electronic circuits that control the deflection coils first position the beam in the top left-hand corner of the screen. The deflection coils then move the beam across the screen from left to right to produce a single, very thin line on the face of the CRT called a **scan line.** Once the beam reaches the right-hand side of the screen, it is turned off (blanked), then returned to the left, and moved down one scan line. The beam is turned back on and moved across the face of the CRT again to produce a second scan line. This process continues until the beam reaches the bottom-right corner of the screen, producing hundreds of scan lines. At the bottom right-hand corner, the beam is once again turned off and returned to the upper left-hand corner. This process is illustrated in Figure 12.2. This entire process is repeated continuously as long as power is applied to the CRT monitor and its control circuits.

As the beam moves across each scan line, its intensity is varied to produce different shades that represent the video (picture) for that single scan line. All of the scan lines, with their video, added together produce a single still picture that is called a **frame.** Continuous frames are sent by the broadcast station to the electronic circuits inside the CRT along with synchronizing signals. This continuous update (refresh) makes the images seem stationary if each frame is the same, or in a smooth motion if the image changes from frame to frame. At the present time, broadcast TVs use 30 frames every second with 525 scan lines per frame.

Motion pictures, TVs, and computer monitors all use a series of frames projected rapidly on a screen. The human eye does not see the individual still pictures, but detects them as a smooth, continuous motion. For reasons that will be explained later in this chapter, most people are able to detect the 30-frame per second frame rate of broadcast TVs and low-end computer monitors. To alleviate this problem, a technique called *interlaced video* is used. Interlaced video will be discussed later in this chapter.

To keep new images coming to the display, broadcast TV's video camera scans at 30 frames per second, just like the TV set. At the end of each scan line and at the bottom of the screen, timing pulses that are generated by the camera keep the monitor in step (synchronized). The timing pulse at the end of each scan line is called the **horizontal sync.** The timing pulse that occurs at the bottom of the screen is called the **vertical sync.**

A computer's monitor does not receive its video and synchronizing information from a video camera but from a special integrated circuit called a **video display controller (VDC).** The VDC is usually

**Figure 12.2** Noninterlaced Video Scan Lines

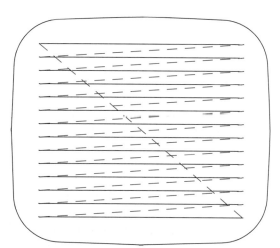

mounted on an adapter card installed in one of the computer's expansion slots, but it can also be integrated into the motherboard.

Like everything else in the computer, the VDC is controlled by a program. The program sends the data to be displayed for each frame to the circuitry on the video display adapter. Once the information is sent to the video display adapter, the program continues on with some other task. If the desired image was not somehow captured on the video display adapter, it would be gone at the end of the frame. This would cause the display to go blank so quickly that we would not even detect something had ever been displayed. This means that the VDC performs other functions besides simply sending frames to the monitor. These functions include holding the information to be displayed, sending the individual frames to the CRT monitor, and generating the required horizontal and vertical sync pulses.

A special RAM called *video RAM* holds the information the program wishes to display. Some motherboard chipsets include the video circuitry and video RAM while other others utilize a portion of the computer's system RAM as video RAM. However, most compatible computers use a video display adapter card that plugs into an expansion slot on the motherboard. Using this setup, the video RAM is part of the video display adapter and none of the system RAM is used.

The application program sends the information to be displayed to the video RAM and not directly to the monitor. The video display controller (VDC) constantly reads the contents of the video RAM to send the individual frames to the monitor and generates the sync pulses to make the whole thing work. As long as the operating system or application does not change the contents of the video RAM, the VDC sends the same frame over and over so that the image appears stationary. As soon as the operating system or application changes the contents of the video RAM, the new frame is sent to the CRT the next time that the VDC scans the video RAM. If the frames constantly change, it makes the image appear in motion.

### 12.1.1 Video BIOS

Just like everything else in the computer, video has evolved tremendously since the PC was introduced in 1981. New video standards and video display adapters seem to be introduced every year. Because there are so many video standards and manufacturers, it is impossible for the system BIOS to support all of them. In fact, a simple monochrome adapter and a color graphics adapter are the only video adapters supported in current compatible computer systems' BIOS. Both of these video adapters will be discussed in later sections. You might be wondering, "If these are the only two adapters fully supported in the BIOS, how do all of the other video adapters work?"

As discussed in chapter 7, memory addresses C0000H through EFFFFH of the upper memory area (UMA) are reserved for adapter ROMs. Each time the computer boots, the main system's BIOS scans these memory pages looking for ROMs. If the system BIOS finds ROMs in these pages, the programs contained in these ROMs supersede the programs contained in the system BIOS. This is the feature currently used by all video adapters above the color graphics adapter (CGA). The BIOS on most video cards uses the address range C0000H through C7FFFH. Note that today's advanced video adapters use address ranges other than those just mentioned.

Most of the video ROMs are designed to be shadowed. This feature is utilized only if it is enabled in the computer's CMOS. Shadowing the video will greatly increase the speed execution for this very important driver program, so it should definitely be enabled using the BIOS Setup program that was discussed in chapter 7.

### 12.1.2 Modes of Operation

Most VDCs used on the IBM-compatible computers can be operated in several modes. The two primary modes are **text** and **graphics.** The actual number of each of these modes depends on the VDC and monitor chosen.

All VDCs currently sold for the compatible computer support one or more text modes. If the VDC is operated in one of the text modes, it will only display those characters that are sent to the video RAM in ASCII. Sending only ASCII to the VDC's RAM makes the video access fast because less information is transferred between the application program and the VDC.

Text modes are given in the form of characters per row by the number of rows. An example of a text mode is 80×25 (pronounced 80 by 25). This example indicates that there are 80 characters displayed

Table 12.1  VGA Video Adapter Video Modes*

| Mode Number | Type | Text | Pixel | Colors |
|---|---|---|---|---|
| 0,1 | Text | 40×25 | | 16 |
| 2,3 | Text | 80×25 | | 16 |
| 4,5 | Graphics | 40×25 | 320 × 200 | 4 |
| 6 | Graphics | 80×25 | 600 × 200 | 4 |
| 7 | Text | 80×25 | | 1 |
| D | Graphics | 40×25 | 320 × 200 | 16 |
| E | Graphics | 80×25 | 640 × 200 | 16 |
| F | Graphics | 80×25 | 640 × 350 | 1 |
| 10 | Graphics | 80×25 | 640 × 350 | 16 |
| 11 | Graphics | 80×30 | 640 × 480 | 2 |
| 12 | Graphics | 80×30 | 640 × 480 | 16 |
| 13 | Graphics | 40×25 | 320 × 200 | 256 |

*Hexadecimal is used to represent the 7-bit binary display mode number.

on a single row and 25 rows. To calculate the total number of characters that can be displayed, simply multiply the two numbers together. Therefore, it can be easily determined that an 80×25 text mode will allow a total of 2,000 (80 × 25) characters to be displayed on the computer's CRT.

**Problem 12.1:** How many characters can be displayed on the screen if a 40 × 25 text mode is selected?

**Solution:**  40 × 25 = 1,000 characters

If the VDC is switched to one of the graphic modes, the image displayed on the screen is produced with thousands of small squares. The smallest square that the VDC can control, in the selected graphics mode, is a picture element called a **pixel.** The smaller the pixel, the higher the picture's **resolution,** which is the sharpness of the picture. Like text modes, pixel resolution is given in the form of the number of pixels per row by the number of rows. An example of this syntax is 320×200 (320 by 200), which indicates there are 320 pixels per row and 200 rows. To find the total number of pixels displayed in a chosen graphics mode, multiply the two numbers together. For example, for a pixel resolution of 320 × 200, there are 64,000 pixels on the screen.

**Problem 12.2:** How many pixels are displayed by the VDC if a pixel resolution of 640 × 480 is used?

**Solution:**  640 × 480 = 307,200 pixels

When the VDC is running in one of the graphics modes, the program has to update all of the pixels on the screen to change the displayed image. This requires a lot of the computer's time when compared to the text mode. It also requires the VDC to have more video RAM installed to handle all of the pixel information.

To switch the VDC to the different video modes, a unique binary code for each mode is sent to one of the VDC's control registers. This binary code consists of a 7-bit binary number that provides a total of 128 total video modes. These codes are used to select both text and graphics modes. All current compatible PC graphics adapters (that I know of) support the video modes that were established when IBM introduced the video graphics array (VGA) video adapter. These standard VGA video modes are listed in Table 12.1.

Table 12.2 Current VBE Video Modes*

| Mode Number | Type | Text | Pixels | Colors |
|---|---|---|---|---|
| 100 | Graphics | — | 640 × 400 | 256 |
| 101 | Graphics | — | 640 × 480 | 256 |
| 102 | Graphics | — | 800 × 600 | 16 |
| 103 | Graphics | — | 800 × 600 | 256 |
| 104 | Graphics | — | 1,024 × 768 | 16 |
| 105 | Graphics | — | 1,024 × 768 | 256 |
| 106 | Graphics | — | 1,280 × 1,024 | 15 |
| 107 | Graphics | — | 1,280 × 1,024 | 256 |
| 108 | Text | 80×60 | | |
| 109 | Text | 132×25 | | |
| 10A | Text | 132×43 | | |
| 10B | Text | 132×50 | | |
| 10C | Text | 132×60 | | |
| 10D | Graphics | — | 320 × 200 | 32,768 |
| 10E | Graphics | — | 320 × 200 | 65,636 |
| 10F | Graphics | — | 320 × 200 | 16,777,216 |
| 110 | Graphics | — | 640 × 480 | 32,768 |
| 111 | Graphics | — | 640 × 480 | 65,536 |
| 112 | Graphics | — | 640 × 480 | 16,777,216 |
| 113 | Graphics | — | 800 × 600 | 32,768 |
| 114 | Graphics | — | 800 × 600 | 65,536 |
| 115 | Graphics | — | 800 × 600 | 16,777,219 |
| 116 | Graphics | — | 1,024 × 768 | 32,768 |
| 117 | Graphics | — | 1,024 × 768 | 65,536 |
| 118 | Graphics | — | 1,024 × 768 | 16,777,216 |
| 119 | Graphics | — | 1,280 × 1,024 | 32,768 |
| 11A | Graphics | — | 1,280 × 1,024 | 65,536 |
| 11B | Graphics | — | 1,280 × 1,024 | 16,777,216 |

*All mode numbers are in hexadecimal.

After the VGA mode numbers listed in Table 12.1, video manufacturers use different codes for additional modes above VGA. In fact, video cards manufactured by the same company often have different mode codes. Because of this diversity, it is impossible for the BIOS and operating system to have the necessary drivers to make every video card on the market operate. For this reason, new video adapters provide a diskette or CD-ROM that contains the necessary drivers.

The **Video Electronics Standards Association (VESA)** is attempting to standardize video modes for graphics adapters above VGA. Having one standard set of video modes will allow a common driver for all video adapters, no matter who the manufacturer may be. To support these additional codes, the video adapter's BIOS must support *video BIOS extensions (VBEs)*.

The VBE adds an additional 8 bits to specify the video modes, although currently only 1 bit is used. The other 7 bits are reserved for future expansion. This additional bit means that all VBE video modes will be above 100 H. Table 12.2 lists the VBE video modes.

Most currently manufactured video adapters support VBE, plus additional graphics modes under the original mode selection format. For this reason, video drivers are still supplied with each video card and usually must be installed into the operating system in order for all of the modes supported by the graphics adapter to be available to the operating system.

## 12.1.3 Video RAM

When the operating system or application program uses one of the VDC text modes, the application sends information to the video memory in Enhanced ASCII. Remember, when Enhanced ASCII is used, each character requires 1 byte of memory.

**Problem 12.3:** How many characters can be displayed using the 80×25 text mode?

**Solution:** With 80 columns and 25 rows, 2,000 characters can be displayed.

Each character that can be displayed in a given text mode also requires an additional byte of memory for its attributes. An attribute is the appearance of the character being displayed. The attributes that are available for text mode are foreground color, background color, high intensity, underline, and blinking. This means that all text modes require 2 bytes of video RAM per character—1 for the ASCII character and the other for the character's attribute. Table 12.3 lists the bits used to control each attribute byte.

**Problem 12.4:** How much video RAM is required to display 80×25 text mode, including attributes?

**Solution:** 4 KB of video RAM is required: 2 KB for the ASCII characters and 2 KB for each character's attributes.

Table 12.3  Text Mode Attribute Bits

| Bit Number | Attribute |
|---|---|
| D7 | 0 = Nonblinking; 1 = Blinking |
| D6–D4 | Background color |
| D3 | 0 = Normal intensity; 1 = High intensity |
| D2–D0 | Foreground color |

Table 12.4  Number of Bits of Video RAM Required for the Listed Number of Colors

| Colors | Bits | Bytes | Multiplier |
|---|---|---|---|
| 2 | 1 | 8 pixels per byte | 0.125 |
| 4 | 2 | 4 pixels per byte | 0.25 |
| 16 | 4 | 2 pixels per byte | 0.5 |
| 256 | 8 | 1 byte per pixel | 1.0 |
| 65,536 | 16 | 2 bytes per pixel | 2.0 |
| 16,670,216 | 24 | 3 bytes per pixel | 3.0 |

When the operating system or application switches the VDC to one of its graphics modes, the binary patterns stored in the memory on the video adapter determine if the pixel is on or off. When on, the binary patterns also determine the color. The more bits of video memory used for each pixel, the more colors are available. The amount of video RAM on the VDC is given in bytes. Remember from chapter 1 that 1 byte equals a group of 8 bits. Table 12.4 lists the number of bits or bytes of video RAM required for the different color resolutions available on current compatibles.

To calculate the minimum amount of video RAM required to display a certain pixel resolution in a given number of colors, the total number of pixels is multiplied by the amount of video RAM required to control the pixel. Some color resolutions require less or more than 1 byte of memory. Table 12.4 provides a multiplier to make the calculation easier.

The actual memory on the video adapter is usually greater than the calculated amount because current video adapters can have 4 KB, 16 KB, 32 KB, 128 KB, 256 KB, 1 MB, 2 MB, 4 MB, or 8 MB of video RAM or more. If the calculated number equals 1.2 MB, an adapter card with 2 MB must be used. In addition to the memory required by each pixel, current graphics cards that have 3D capabilities require additional video RAM to support 3D calculations.

**Problem 12.5:**  What is the minimum amount of video RAM required to display a pixel resolution of 320 × 200 at 64 K different colors?

**Solution:**  320 × 200 equals 64,000 pixels. When 64 K colors are used, 2 bytes of video memory are required to control each pixel, which means that 64,000 × 2 = 128,000 bytes of video RAM are required.

**Problem 12.6:**  What is the minimum amount of video RAM required to display a pixel resolution of 800 × 600 at 16 M colors?

**Solution:**  800 × 600 = 480,000 pixels. To display 64 M colors requires 3 bytes of VRAM per pixel. 480,000 × 3 = 1,440,000 bytes of video RAM.

## 12.2  VIDEO IN THE COMPATIBLE COMPUTER

Installing a monitor (CRT) on a compatible computer requires two things: a monitor and a video adapter card. The monitor can be either color or monochrome, but the VDC must match the style of monitor chosen, or vice versa. As stated previously, the VDC can be a part of the motherboard, but most compatibles require the video adapter to be installed in one of the computer's expansion slots.

The monitor is a separate unit that interfaces, by way of a round cable, to a connector provided by the computer. The computer's video connector is usually located at the rear of the computer and is either a female DB9 or a Mini-DB15S. The choice of connector is based on the video standard chosen. These standards fall into either the monochrome or color category.

**Figure 12.3** DB9S Connector used by MDA, CGA, and EGA

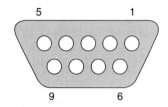

**Table 12.5** MDA/MGA/HGA Connector Pin Assignments

| Signal | Pin |
|---|---|
| Ground | 1 |
| Ground | 2 |
| Not connected | 3 |
| Not connected | 4 |
| Not connected | 5 |
| Intensity | 6 |
| Video signal | 7 |
| Horizontal sync | 8 |
| Vertical sync | 9 |

### 12.2.1 Monochrome Monitors

The monochrome monitor and monochrome VDC combo is the cheapest that can be purchased for the compatible computers. The monochrome monitor's inside surface is coated with one-color phosphor and has only one cathode. The most popular monochrome video display adapter is the **monochrome display adapter (MDA).** The MDA card usually installs in one of the computer's expansion slots. The monochrome monitor connects to the output of the MDA through a DB9 female connector that is located on the back of the computer. The MDA card is a text-only card that displays only those characters sent to it in ASCII. The MDA card does not have the ability to display graphic images on the screen and will not work with programs that generate a graphics output.

The **Hercules graphics adapter (HGA)** and the **monochrome graphics adapter (MGA)** are also available for the monochrome monitor. The MGA and HGA adapters display graphics on a monochrome monitor in shades of gray, similar to a black-and-white TV. The application program running the computer must support the MGA graphics command set to display graphics with the MGA or the HGA. These cards are totally compatible with the MDA and even use the same pin-out (same pin assignments). The connector used by the MDA/MGA/HGA is a female DB9S connector that is located on the back of the computer (Figure 12.3). The pin-outs for the MDA/MGA/HGA connectors are given in Table 12.5.

### 12.2.2 Color Monitors

As stated earlier, the inside surface of a color CRT is coated with thousands of red, green, and blue phosphor dots, as illustrated in Figure 12.4. A group of red, green, and blue phosphor dots is usually arranged in a triangular pattern called a **triad.** Relative to color monitors, each triad is usually considered to be

**Figure 12.4** Color CRT Triad

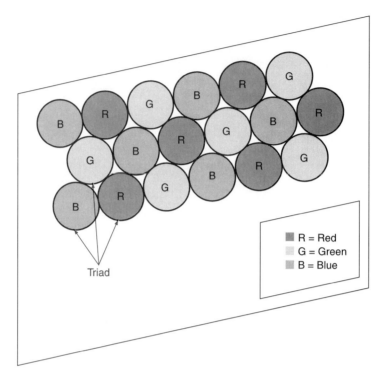

a single phosphor dot. Groups of one or more triads are used to create pixels. It is important to understand that a pixel and phosphor dots are not the same. A single triad is the smallest that a pixel can be; however, it can definitely be larger than one triad. The program and color VDC card chosen control the size of each pixel.

Color CRTs use three electron guns—one for each color of phosphor. The guns are turned on or off or varied in intensity to generate the different hues. The color VDCs that will be discussed in this chapter are the color graphics adapter (CGA), enhanced graphics adapter (EGA), video graphics array (VGA), and super video graphics array (SVGA).

### 12.2.3 Color Graphics Adapter (CGA)

The **color graphics adapter (CGA)** card and monitor were the first color combo sold for the IBM-compatible computers. CGA uses the same interlaced 525 scan lines with 30 frames per second used by broadcast TV. CGA can display only 16 colors and has five modes of operation: two text modes and three graphics modes. The two text modes are 40×25 and 80×25. In the text mode, CGA automatically displays the characters on the screen when the ASCII characters that represent the characters and their attributes are stored in the adapter's video RAM. The text can be displayed in up to 16 foreground colors on a background of 8 colors.

The three graphics modes available with CGA are 160×200, 320×200, and 640×200. The pixels in the 160×200 mode can be displayed in 16 different colors. The 320×200 mode allows 14 colors and the 640×200 can be displayed in only 2 colors.

The monitor used with the CGA card goes by several names: a CGA monitor, an RGBI (red, green, blue, and intensity) monitor, or a TTL (transistor-transistor-logic) color monitor. The acronym *TTL,* when referring to monitors, means that the video image sent from the adapter card to the monitor is simply composed of binary 1s and 0s.

In any of the graphics modes, the binary patterns stored in the video RAM define the color of each pixel. The more colors in which each pixel can be displayed, the more video RAM is required. The CGA has 16 KB of video RAM installed on the card. This is enough memory to display 160×200 in 16 colors, 320×200 in 4 colors, or 640×200 in 2 colors.

Table 12.6  CGA Adapter Pin Assignments

| Signal | Pin |
| --- | --- |
| Ground | 1 |
| Ground | 2 |
| Red video signal | 3 |
| Green video signal | 4 |
| Blue video signal | 5 |
| Intensity | 6 |
| Not connected | 7 |
| Horizontal sync | 8 |
| Vertical sync | 9 |

MDA, CGA, and EGA (discussed next) all use the same female DB-9 connector to connect the monitor to the VDC, which is shown in Figure 12.3. The signals assigned to each pin are different for each of these VDCs. Table 12.6 lists the pin assignments for CGA.

### 12.2.4 Enhanced Graphics Adapter (EGA)

The **enhanced graphics adapter (EGA)** card and the EGA monitor were the first attempt at high-resolution video for the compatible computers. To increase the resolution, more scan lines were added to the CRT and more memory was added to the adapter card. More scan lines make the triads smaller, which means smaller pixels, which translates to availability of higher resolutions for EGA over CGA. More memory means more colors are available with EGA than with CGA.

EGA supports both text and graphics modes and is totally compatible with CGA. The term *CGA-compatible* indicates that all programs written for CGA will work on EGA without modification. In fact, all current video adapters (that I know of) have maintained compatibility with CGA.

EGA has a maximum pixel resolution of 640 × 200 at 16 colors with 64 K of video RAM installed. EGA adds secondary signals to control each of the CRT's guns. These secondary signals give EGA the ability to display $2^6$ (64) colors. However, with just 64 K of video RAM installed, 64 colors can only be displayed at a maximum pixel resolution of 320 × 200. The EGA uses the same connector that is shown in Figure 12.3, but with different pin assignments. Table 12.7 lists pin assignments used by the EGA connector.

The EGA standard was only on the market for approximately 1 year before the next video standard, VGA, was introduced. Because of the major improvements offered by the VGA standard, EGA disappeared from the PC scene before it had a chance to be adopted in many computers.

**Problem 12.7:**  How many bytes of video RAM are required to display 640 × 200 pixels in 16 colors?

**Solution:**  Referring to Table 12.4, we find the color of 2 pixels is determined by each byte of video RAM. 640 × 200 equals 128,000 pixels × 0.5 = 64,000 bytes of video RAM.

### 12.2.5 Analog Monitors

MDA, MGA, HGA, CGA, and EGA all use a digital format to send video information to the monitor. A digital format allows each of the monitor's guns to be turned off or on. When turned on, each gun can be varied in only two or three steps of intensity: two steps for MDA and CGA and three steps for EGA. Simply sending binary 1s and 0s to the monitor controls these steps. Higher numbers of conductors used to send the information translates into more colors.

Table 12.7  EGA Connector Pin Assignments

| Signal | Pin |
|---|---|
| Ground | 1 |
| Secondary red video | 2 |
| Red video | 3 |
| Green video | 4 |
| Blue video | 5 |
| Secondary green video | 6 |
| Secondary blue video | 7 |
| Horizontal sync | 8 |
| Vertical sync | 9 |

Using digital formats makes control of the video system easy, but limits the number of colors that can be displayed. Monochrome can display in three colors (or hues): black, low intensity, and high intensity. CGA can display up to 16 colors (in text mode), while EGA can display up to 64 colors. To display more colors, another format called *analog* had to be adopted by both the video adapters and the monitors they controlled.

The term **analog** basically means that a signal can be varied in an infinite number of steps between two points. A dimmer switch used on lighting systems is a good example of true analog. However, in computers, the term *analog* means that the signal can be varied in many discrete (individual) steps. A comparison of the way that a computer represents analog is a TV's turner, which can select the many channels available on today's televisions in a variety of ways.

To increase the number of colors above EGA's 64, most modern color monitors and adapters use analog signals instead of digital. In fact, the analog monitors used on today's computers are capable of displaying a near-infinite range of colors. The VDC, not the monitor, usually limits the number of colors that can be displayed. The amount of memory installed on the analog adapter card is the main factor that determines the number of colors.

Information stored in the adapter's video RAM is still binary or digital. The video adapters that are designed to run an analog monitor contain a special circuit commonly called a **RAMDAC.** The DAC stands for digital-to-analog converted. The function of the RAMDAC is to take the digital information supplied by the video RAM and convert it into the analog information being sent to the monitor.

### 12.2.6 Video Graphics Array (VGA)

**Video graphics array (VGA)** was the first analog standard established for the compatible computers. A VGA card has 256 KB of video RAM installed and can display up to 256 colors, depending on the graphics mode chosen. VGA displays a maximum $640 \times 480$ pixel resolution in 16 colors; 256 colors can be displayed at a maximum resolution of $640 \times 400$. VGA fully supports all CGA and EGA pixel resolutions (graphics modes). If a VGA card is used, a VGA analog monitor is required. VGA uses a female Mini-DB15S connector to interface (connect) the adapter card to the monitor, which is shown in Figure 12.5. The pin-out for the VGA adapter is listed in Table 12.8. Note that all video standards above the VGA use this same connector and pin-out, so it will not be given again.

**Problem 12.8:**  What is the minimum amount of video RAM required to display a pixel resolution of $320 \times 200$ in 256 colors?

**Solution:**   $640 \times 400 \times 1$ byte/pixel $= 256{,}000$ bytes

**Figure 12.5** VGA Mini-DB15S Connector

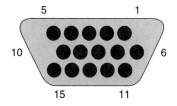

**Table 12.8** VGA Connector Pin-out

| Signal | Pin |
|---|---|
| Red video | 1 |
| Green video | 2 |
| Blue video | 3 |
| ID bit 2 | 4 |
| Ground | 5 |
| Red return | 6 |
| Green return | 7 |
| Blue return | 8 |
| Key (blank) | 9 |
| Sync return | 10 |
| ID bit 0 | 11 |
| ID bit 1 | 12 |
| Horizontal sync | 13 |
| Vertical sync | 14 |
| Not connected | 15 |

## 12.2.7 Super VGA (SVGA)

The **super video graphics array (SVGA)** standard is where most of the confusion starts. At the present time, any video card that has the ability to display a pixel resolution over 640 × 480 at 16 colors is considered SVGA. SVGA cards can be purchased with as low as 256 K of video RAM installed or can go over 8 MB.

The low-end SVGA cards have 256 KB of video RAM installed. This amount of memory allows a maximum pixel resolution of 800 × 600 at 16 colors. This resolution is well above VGA's maximum pixel resolution of 640 × 480. When 256 colors are displayed, however, VGA's 640 × 400 is still the highest resolution available.

The next higher SVGA adapter has 512 KB (0.5 MB) of video RAM installed. This amount of memory allows it to display up to a 1,024 × 768 at 16-color pixel resolution. The 512-KB VGA card's maximum color resolution is 256 colors at 800 × 600.

> **Procedure 12.1: Displaying the video adapter resources**
>
> 1. Click on the Start button on the Task Bar.
> 2. Move the mouse pointer up to Settings and then over to Control Panel.
> 3. Click on Control Panel.
> 4. After the Control Panel window opens, locate and double-click on the System icon.
> 5. On computers running Windows 2000, click on the Hardware tab after the Systems Properties window opens.
> 6. In Windows 9X, click on the Device Manager tab. In Windows 2000, click on the Device Manager button.
> 7. Click on the small plus (+) sign beside the Display Adapter category icon.
> 8. Click on the name of the video adapter listed in this category and then click on the Properties button.
> 9. Click on the Resources tab to display the adapter's resources.
> 10. Click on the small X in the upper right-hand corner of each window or click on the Cancel button to close all open windows.

Further up the ladder are the 1-MB, 2-MB, 4-MB, and 8-MB SVGA cards. SVGA cards with 1 MB of video RAM can display a maximum pixel resolution of 1,024 × 768 at 256 colors. The 2-MB SVGA cards are sometimes called 24-bit cards because they can use up to 24 bits to select each pixel's color, which gives the ability to display up to 16,777,216 color combinations. These 2-MB cards have a maximum pixel resolution of 1,280 × 1,024.

The resolutions offered by SVGA require the monitor to synchronize with scan lines starting with CGA's 565 to over 1,000. In order for a monitor to support a different number of scan lines, it must be able to adjust automatically to different horizontal and vertical sync pulses. Most of today's monitors have this ability and are often called **multisync** monitors.

**Problem 12.9:** How much video RAM is required to display a pixel resolution of 1,024 × 768 at 16 million colors?

**Solution:** 1,024 × 768 = 786,432 pixels. Referring to Table 12.4, 16 MB colors require 3 bytes of memory per pixel. 3 × 786,432 equals 2.25 MB of video RAM.

## 12.3 VIDEO ADAPTER RESOURCES

As discussed previously in chapters 7 and 9, the video adapter uses addresses in both the memory and I/O address maps. In the memory address map, the video adapter uses addresses C0000 through C7FFF for its BIOS. It also uses address range B0000 through B7000 for a monochrome text memory and B8000 through BFFFFH for a color text memory. A graphics video adapter uses A0000 through AFFFF for graphics memory. Along with these standard memory ranges, most of today's video adapters also use memory address ranges above the upper address limit of the system RAM.

The I/O addresses are 2B0 through 2BF, 3B0 through 3BB, and 3D through 3DF. All of these ranges are required for programming registers to specify the graphics and text modes and for status register. Along with memory and I/O addresses, most current video adapters also use one of the system's IRQs. Procedure 12.1 can be used to display the actual resources used by the video adapter on a computer running Windows 9X/2000.

## 12.4 WINDOWS ACCELERATED CARDS

In the early days of computer display monitors, everything was done in text, which consisted of merely transferring ASCII characters along with their attribute codes to the video RAM. Using the 80×25 text mode required the operating system or application to update only 4 KB of video RAM to change the entire display. When Windows 3.X came along, the GUI was introduced to the compatible computer, which requires the VDC to run in a graphics mode. The default graphics mode for Windows 3X is 640×480 at 16 colors. This required the application to update 614,400 bytes of memory to change the image on the screen. As you can see, the graphics mode takes over 150 times longer to update the video RAM than does the text mode.

As discussed previously in chapter 5, Windows gets its name from the fact that all applications run inside a rectangular area on the screen called a *window*. When Windows 3.X was first introduced, the pixels that defined each window had to be drawn on the CRT by the operating system. Needless to say, this slowed down the CRT updates considerably. The GUI used by Windows 3.X turned the video into one of the major bottlenecks of the computer.

To help reduce the video bottleneck, video adapter manufacturers used a feature for the VDCs that allowed the VDC to draw windows without operating system intervention. The operating system sends a series of commands to tell the VDC to draw a window, including the window's size and location. Once the command set is sent by the operating system, the VDC actually draws the window on the display. The video graphics adapters that support this feature have the ability to draw windows several magnitudes faster than video graphics adapters that do not. This feature is known as **Windows accelerated adapters**. As far as I know, all current graphics adapters for the compatible PC support Windows acceleration.

### 12.4.1 3D Graphics Cards

Displayed images on current monitors can have height and width, but not depth. This means that the computer's monitor can display only two-dimensional (2D) images. Over the last few years, game programmers have given a three-dimensional (3D) feel to games by adding very complex variations of shading and texture to 2D images. The application program itself performs the process of displaying images to appear 3D. As you can imagine, this eats up a lot of processing power and a lot of program code. The larger the program, the slower the video will be updated.

Video adapter manufacturers are rapidly moving toward 3D accelerated video adapters that are similar to the Windows accelerators. This involves adding extra processing power to the VDC and additional video RAM on the video adapter to handle image texturing. Adding these additional features means the application can send 2D images to the video adapter along with a set of commands that tell the VDC how to add texturing. Like Windows accelerators, the actual process of displaying the image in 3D is performed by the VDC. As you can imagine, turning the processing of the 3D image over to the VDC speeds up 3D animation by several magnitudes.

In order for the application to take advantage of the 3D accelerators, the application must use the command set supported by the 3D adapter. Because 3D adapters are still in their infancy, there is no current standard command set for 3D acceleration. This means that 3D applications written for one 3D accelerator may not run correctly on another. Hopefully, this drawback will be resolved in the near future.

Note that most current 3D accelerators also support 2D acceleration. This feature means that 2D applications and operating systems should run on 3D video cards without problems. This includes games and applications that are written to generate 3D images through software.

## 12.5 MONITOR SPECIFICATIONS

As with most computer peripherals, the monitor has many important specifications. These include size, viewable area, dot-pitch, sync rates, interlaced, noninterlaced, and power-saving features. The problem is that most individuals do not understand the specifications and must rely on the integrity of a salesperson to choose the correct monitor for them. Hopefully, the next few sections will resolve this situation.

### 12.5.1 Display Size

The size of the display is measured in inches. The measurement is made diagonally from any corner to the opposite corner. For example, a 14″ monitor measures 14″ from the tip of the upper right-hand corner to the tip of the lower left-hand corner, or vice versa.

A portion of the monitor's face is hidden behind the cover, and it is extremely hard to focus the beam on the very outside edges of the CRT, so most monitors use a rating referred to as *viewable area*. As the name implies, the viewable area is the portion of the face of the CRT, measured diagonally from corner to opposite corner, that the user can actually see. As you can guess, the viewable area is the most important size rating.

### 12.5.2 Dot-Pitch

The triads on many 14" SVGA monitors (measured diagonally) may be too large for a pixel resolution of 1,024 × 768. To use a SVGA card that can display 1,024 × 768 pixel resolution requires a monitor with very small phosphor dots. This resolution's introduction brought about the specification used in monitors called the monitor's **dot-pitch.** A monitor's dot-pitch is measured in millimeters (.001 meter = 1 mm) and refers to the spacing of the phosphor dots that make up the individual triads.

To display at 1,024 × 768 pixel resolution on a standard 14" monitor requires a dot-pitch of no more than 0.28. If a 14" monitor is used with a larger dot-pitch, the displayed image will smear when the 1,024 × 768 pixel resolution is selected. A 17" monitor can have a 0.31 dot-pitch and can display 1,024 × 768 with no problems. However, a 17" SVGA monitor with a 0.31 dot-pitch will cost a lot more than a 14" SVGA monitor with a 0.28 dot-pitch. A dot-pitch of 0.31 can be used on a 14" SVGA monitor if 800 × 600 is the maximum pixel resolution used.

### 12.5.3 Calculating Size Relative to Resolution

To determine if a certain monitor will display a given pixel resolution, a derivative of a formula called the Pythagorean Theorem is used. Pythagorean Theorem states that the hypotenuse of a right triangle squared equals the sum of the squares of the other two sides ($c^2 = a^2 + b^2$). Understanding the Pythagorean Theorem is really not necessary as long as you can use the formula given in Example 12.1.

**Example 12.1:** Formula for determining if a monitor will display a given pixel resolution

Formula: $D = $ (Square root $[(640 \times 0.31)^2 + (480 \times 0.31)^2] \times 0.039$

$$D = [\sqrt{(H \cdot P)^2 + (V \cdot P)^2}] \cdot 0.039$$

where:

$D = $ The display size in inches
$H = $ The number of horizontal pixels
$V = $ The number of vertical pixels
$P = $ The dot pitch in millimeters

If the answer produced by the formula shown in Example 12.1 is greater than the size of the monitor, the monitor will not display the given resolution. Note that the 0.039 multiplier is used to convert the dot-pitch from millimeters to inches, which is the measurement used to specify the monitor's size. To help understand this formula, let us work a couple of problems.

**Problem 12.10:** Can a 14" monitor with a 0.31 mm dot-pitch display a 640 × 480 pixel resolution?

$$D = \text{(Square root } [(640 \times 0.31)^2 + (480 \times 0.31)^2] \times 0.039$$
$$D = 9.672"$$

**Solution:** Yes, because 9.672" is less than 14".

**Problem 12.11:** Can a 14" monitor with a 0.31 dot-pitch display a 1,024 × 768 pixel resolution?

$$D = \text{(Square root } [(1,024 \times 0.31)^2 + (768 \times 0.31)^2] \times 0.39$$
$$D = 15.48"$$

**Solution:** No, because 15.48" is larger than 14".

**Figure 12.6** Interlace Video

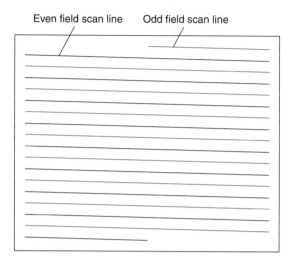

## 12.5.4 Interlaced or Noninterlaced Video

Understanding the reasons why broadcast TV has to use 30 frames per second is beyond the scope of this book and is not really necessary to know in the first place. However, using this speed did require engineers to overcome one major obstacle. Sending images to the screen at 30 frames per second takes exactly 1/30 second to complete all of the frame's 525 scan lines. At this rate, the phosphor at the top of the screen starts to fade before the scan lines at the bottom are completed. This gives the appearance of a dark wave moving from the top to the bottom on the screen.

To solve this problem, the engineers came up with a method by which the beam is brought down the screen twice for each frame. During the first trip down the screen, the odd-numbered scan lines are displayed. During the second trip, the even-numbered scan lines are displayed. Each trip down the screen is called a **field.** Using this method, it takes two fields to create each frame. The frames are still displayed at a rate of 30 frames per second, but the flicker is eliminated because the screen is refreshed at 60 fields per second. This method is referred to as **interlaced** video. When CRT monitors were being developed, interlaced video was chosen because this technology was readily available. An illustration of interlaced video is shown in Figure 12.6.

To increase the resolution of the displayed image, the number of scan lines must be increased. This can be accomplished by bringing the electron beam across (horizontal sweep) the display faster or down (vertical sweep) the display slower. The problem with bringing it down the screen slower is that it will decrease the field rate, approaching the point of a display flicker. For this reason, the method used to increase the scan line's number is to increase the horizontal sweep.

There is also an advantage to increasing the vertical sweep rate. Increasing the vertical sweep increases the field rate. You might wonder, "Why increase the field rate?" As more scan lines are added, the monitor's resolution increases and the individual scan lines become thinner. With more scan lines, the 30-frame-per-second frame rate would make the display flicker slightly, even with interlaced video. To compensate for the flicker caused by the higher resolution, the frame rate is increased. Current frame rates vary from 30 frames per second to over 85 frames per second on the high end.

Increasing the frame rate using interlaced video has one major drawback. Interlaced video requires the vertical sweep to be twice as fast as the frame rate. When 30 frames per second are used, the frequency going to the deflection coils that control the vertical position would be 60 sweeps per second. With a frame rate of 72 per second, a vertical sweep rate of 144 sweeps per second would be required.

Again, for reasons that are beyond the scope of this book, as the frequency goes up, the strength of the magnetic field produced by a coil goes down. It would be impossible to use interlaced video to handle frame rates from 30 to 72 frames per second because the strength of the magnetic field controlling the vertical position would vary considerably between the lower and upper ends. For this reason, the maximum frame rate for interlaced video is currently about 42.5 frames per second.

**Noninterlaced** video monitors start the beam in the upper left-hand corner and bring it down the screen one scan line at a time. Because the frames are completed in one pass, the frame and field rates are the same.

Today's monitors have the ability to display either interlaced or noninterlaced video. At low frame rates, monitors use interlaced video. However, at high frame rates the monitor will switch to noninterlaced, which allows a high-resolution, flicker-free display.

### 12.5.5 Refresh Rate

One confusing setting used by today's video adapters is the refresh rate. To understand this setting, one must understand the terms frame rate and field rate. **Frame rate** is the speed that the video adapter updates the displayed image whereas **field rate** is the speed that the video adapter sends the image to the monitor. The field rate determines the number of times the electron beam makes a trip down the surface of the screen to keep the image from flickering. For this reason, the field rate is referred to as a video adapter's **refresh rate.**

Since the refresh rate is being controlled by repetitive signals applied to the vertical deflection yoke, the rate is usually expressed in Hertz. As an example, the refresh rate for television is 60 Hz while the frame rate is 30 frames per second. Remember from the previous section that the value of the refresh rate (field rate) is twice the frame rate when interlaced video is used. When noninterlaced video is used, the values of the frame rate and refresh rate are the same.

The video adapter determines the number of refresh rates available. Setting the refresh rate used by the adapter is done automatically by the selection of a graphics mode, through a DOS command line, or through Display Properties in Windows 9X/NT/2000. The procedure for setting the refresh rate through a DOS command line or Display Properties will vary between adapters. For this reason, the procedure and syntax will be provided in the documentation supplied with the video adapter. The procedure used to open Display Properties is covered later in this chapter.

Note that the attached monitor must support the refresh rate selected for the video adapter. If a refresh rate is selected for the adapter that is not supported by the monitor, the displayed image will usually become unreadable and, in rare instances, the monitor may be damaged. If a wrong refresh rate is selected when running Windows 98/NT/2000, start the operating system in Safe Mode (Chapter 6) and change the refresh rate using Display Properties.

## 12.6 EXPANSION BUS

In the old days, we were happy to get information processed at the speed of a computer in any form. Early computers were designed to process only numbers and text information, so engineers developed a code that used 1s and 0s to represent numbers and text. This is the Enhanced ASCII that was discussed in chapter 1.

The first VDCs were designed to use nothing but ASCII, so the first operating systems were written to use only ASCII to send information to the computer's display. All current versions of MS-DOS operate the VDC in the standard 80×25 text mode. This mode requires MS-DOS to update only 4 KB of video RAM to change the entire display. Displaying only numbers and letters is fine in some applications, but the demands placed on today's machines have definitely limited the use of ASCII.

With ASCII and text mode, there is no way to display graphics information like you see throughout this text. There are also times when the appearance of the text needs to change to emphasize certain points. To alleviate the limitations of text mode and ASCII, most popular operating systems, such as Windows 9X and most current applications, operate the VDC in one of its graphics modes.

In graphics mode, all information, whether pictures or text, is actually displayed in graphics form. Displaying the information graphically allows applications to show text and graphics information on the computer's screen exactly how it will appear when it is printed. Displaying information this way has become known as a **WYSIWYG** (wiz-ze-wig) display. WYSIWYG is the acronym for *what you see is what you get*.

The problem with displaying everything in graphics mode is that the application has to update hundreds of thousands of bytes of video RAM instead of the 4 KB required by MS-DOS's 80×25 text mode. With all of the video RAM that has to be updated by these GUIs (graphics users interfaces), video has turned into one of the computer's major bottlenecks. One of the factors that determines the size of the bottleneck is the expansion bus used by the video adapter.

The speed of the expansion bus and the number of data bits are directly proportional to the speed at which the operating system or application can update the displayed image. On the other hand, high-speed video may not be necessary in some computer applications. The expansion slots available to the video adapter include ISA, VL bus, PCI, and AGP.

The ISA expansion bus operates at around 8 MHz and transfers data in 16-bit groups, or 2 bytes. Using a little math, ISA has the ability to transfer data at a rate of 16 MBps (8M × 2). This relatively slow transfer rate limits current ISA video adapters to MDA, CGA, EGA, MGA, and VGA. The first Windows 3.X accelerators were designed around the ISA expansion bus.

The VL bus operates at around 40 MHz and transfers data in 32-bit groups, or 4 bytes. At this rate, VL bus can transfer data at rates of 160 MBps. With the loading effects of the VL bus and additional I/O wait states, VL bus will never really achieve this speed, but it is still a lot faster than ISA. VL bus is the predominant expansion bus used by video adapters in the 486 personal computers.

The current version of PCI operates at 33 MHz and transfers data in 32-bit groups, or 4 bytes. This speed and group size give PCI the ability to transfer data at rates of 132 MBps (33 M × 4). Using PCI allows data transfer rates of over eight times that of ISA. Because PCI practically eliminates wait states and adds other features such as bus mastering and stacked interrupts, PCI is currently the predominant expansion bus used by video adapters, and probably will be for a few years to come.

### 12.6.1 Advanced Graphics Port (AGP)

All major applications for the compatible PC are currently written using a GUI. There is also a lot of work in progress involving 3D imaging, and video game writers are always in search of faster and faster video frame rates. What all of this boils down to is that the transfer rates offered by PCI are not sufficient for the video requirements of some applications being developed for today's personal computers. To alleviate this problem, Intel has established a new video standard and proprietary slot dedicated strictly to video. *Proprietary* means that only video adapters designed to the new Intel specification are able to operate in this slot. This slot and standard are referred to as the **advanced graphics port (AGP),** which is shown in Figure 12.7.

As the name implies, AGP is really not an expansion bus because it is used strictly as a port for an AGP video card. AGP offers a much higher transfer rate than PCI because it operates at the computer's local bus speed. A peak transfer rate on a 66-MHz local bus is approximately 266 MBps (66,000,000 · 4 bytes) at what is called the *X1 mode.*

The AGP standard also defines X2 and X4 modes that allow the AGP slot to run at twice and four times the local bus. In order to use the X2 and X4 modes, these modes have to be supported by the

**Figure 12.7** AGP Slot

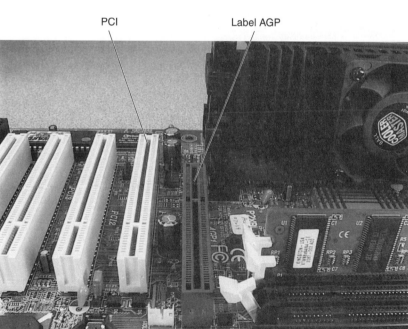

**Table 12.9** Pin-out of the AGP Slot

| Signal | B | A | Signal | Signal | B | A | Signal |
|---|---|---|---|---|---|---|---|
| Spare | 1 | 1 | +12 V | Vddg3.3 | 34 | 34 | Vddg3.3 |
| +5 V | 2 | 2 | Spare | AD21 | 35 | 35 | AD22 |
| +5 V | 3 | 3 | Reserved | AD19 | 36 | 36 | AD20 |
| +USB | 4 | 4 | -USB | Ground | 37 | 37 | Ground |
| Ground | 5 | 5 | Ground | AD17 | 38 | 38 | AD18 |
| INTB | 6 | 6 | INTA | C/BE2 | 39 | 39 | AD16 |
| Clock | 7 | 7 | Reset | Vddg3.3 | 40 | 40 | Vddg3.3 |
| Request | 8 | 8 | Grant | IRDY | 41 | 41 | FRAME |
| +3.3 V | 9 | 9 | +3.3 V | 3.3 Vaux | 42 | 42 | Reserved |
| ST0 | 10 | 10 | ST1 | Ground | 43 | 43 | Ground |
| ST2 | 11 | 11 | Reserved | Reserved | 44 | 44 | Reserved |
| RBF | 12 | 12 | Pipe | +3.3 V | 45 | 45 | +3.3 V |
| Ground | 13 | 13 | Ground | DEVSEL | 46 | 46 | TRDY |
| Reserved | 14 | 14 | Reserved | Vdd93.3 | 47 | 47 | STOP |
| SBA0 | 15 | 15 | SBA1 | PERR | 48 | 48 | Spare |
| +3.3 V | 16 | 16 | +3.3 V | Ground | 49 | 49 | Ground |
| SBA2 | 17 | 17 | SBA3 | SERR | 50 | 50 | PAR |
| SB_STB | 18 | 18 | Reserved | C/BE1 | 51 | 51 | AD15 |
| Ground | 19 | 19 | Ground | Vdd93.3 | 52 | 52 | Vdd93.3 |
| SBA4 | 20 | 20 | SBA5 | AD14 | 53 | 53 | AD13 |
| SBA6 | 21 | 21 | SBA7 | AD12 | 54 | 54 | AD11 |
| Key | 22 | 22 | Key | Ground | 55 | 55 | Ground |
| Key | 23 | 23 | Key | AD10 | 56 | 56 | AD9 |
| Key | 24 | 24 | Key | AD8 | 57 | 57 | C/BE0 |
| Key | 25 | 25 | Key | Vdd93.3 | 58 | 58 | Vdd93.3 |
| AD31 | 26 | 26 | AD30 | AD_STB0 | 59 | 59 | Reserved |
| AD26 | 27 | 27 | AD28 | AD7 | 60 | 60 | AD6 |
| +3.3 V | 28 | 28 | +3.3 V | Ground | 61 | 61 | Ground |
| AD27 | 29 | 29 | AD26 | AD5 | 62 | 62 | AD4 |
| AD25 | 30 | 30 | AD24 | AD3 | 63 | 63 | AD2 |
| Ground | 31 | 31 | Ground | Vdd93.3 | 64 | 64 | Vdd93.3 |
| AD_STB1 | 32 | 32 | Reserved | AD1 | 65 | 65 | AD0 |
| AD23 | 33 | 33 | C/BE3 | Reserved | 66 | 66 | Reserved |

AGP graphics card. If available, the X2 mode will mean a peak transfer rate of 532 MBps and the X4 mode translates to a speed of 1,064 MBps, both on a 66-MHz local bus. Along with these features, the AGP standard also allows the video adapter to access the system memory directly to hold 3D-texture information. This feature allows the AGP card to hold frames and perform amazing 3D graphics calculations.

The AGP slot looks similar to a PCI expansion slot, but the AGP connector is offset farther from the back edge of the motherboard and the key is in a different location. Figure 12.7 shows an AGP slot and its relationship to the PCI expansion bus. The pin-out of the AGP slot is given in Table 12.9.

## 12.7 THE VIDEO DRIVER

All currently manufactured compatible PC graphics adapters (that I know of) support the 80×25 text mode and the 640×480 at 16-color graphics mode. Beyond these two standard modes, all other video modes are pretty much up in the air for different graphics adapters. Because there are so many different graphics and text modes used by different video adapters, it is impossible for the operating system to support them all. For this reason, operating systems such as Windows 9X/2000 default to the 640×480 at 16 color graphics modes when they are installed. Also, the 640×480 at 16 colors is the only mode available when Windows 9X is operating in the Safe mode.

To give the operating system the ability to use the additional text and graphics modes available with a given graphics video adapter, a driver that is unique to the video adapter must be added to the operating system. For this reason, most video adapters purchased for the compatible PC are supplied with a diskette or CD-ROM that contains the correct driver.

If a newly installed video adapter is PnP, Windows 9X/2000 will detect it the first time it is booted after the installation. The Windows 9X/2000 Setup Wizard will automatically launch and will eventually inform the user to insert the media with the driver into the correct drive. If the video adapter is not PnP, the user will have to install the video driver manually. The user will also have to manually install the video driver after the initial installation of Windows 9X/NT/2000. This procedure will be covered in the next section.

The video adapter may be used with other operating systems, so the diskette or CD-ROM that is supplied with the adapter may have multiple drivers for several operating systems. As you probably guessed, it is imperative that the driver match the operating system running the computer. For this reason, it is very important to consult documentation that is supplied with the adapter for the correct installation procedure for the operating system.

Video adapter manufacturers update drivers for their products constantly. In fact, when you purchase your new video adapter, an updated driver is probably already available by the time you open the box. To help support their products, all major video card manufacturers maintain a Web site on the Internet. Along with troubleshooting aids and installation procedures, these Web sites also provide free, updated drivers that are available for download. After the new video adapter and supplied driver are installed, you should visit the manufacturer's Web site as soon as possible to check for updated video drivers. In fact, you should visit the site monthly to check for updated drivers.

Before you visit one of these sites, make note of the model number of the video adapter. In addition to the model number, some video drivers are specified by the chipset on the video adapter. Document any numbers on the large chips on the video adapter, if present, before the adapter is installed, otherwise you will have to remove the adapter from the computer to gain access to these numbers. Web sites sometimes provide a program that will identify an installed video adapter's chipset without removing it from the computer.

### 12.7.1 Setting up the Video in Windows 95/98/2000

Most of the user-defined settings for the video are established through the Properties option of the Display Settings menu. This menu can be accessed using several methods, but the most often used is to simply right-click on an empty spot on the desktop. Once the Display Settings menu appears, click on the Properties option to open the Display Properties window.

**Figure 12.8** Typical Windows 98 Display Setting Window

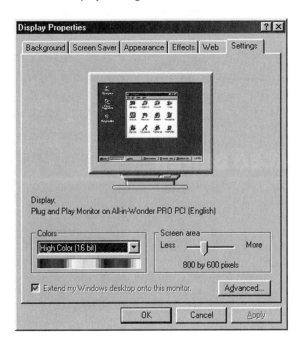

The options available in the Display Properties window may vary from computer to computer because they are open to modification by the display driver when it is installed. Explanations of the options added by the driver are usually discussed in the documentation supplied with the video adapter. Information on the other options on the Display Properties window can be obtained through the Windows 9X/2000 Help menu by typing "display" in the Index option. The only option we will discuss is the Display Settings option.

The Display Settings option is used to change the color and pixel resolution. Display Settings is also used to change the type of video monitor and to manually install or upgrade the display driver. Both of these settings are fundamental in determining the number of color and pixel resolutions that are available. Figure 12.8 shows a typical Display Settings window using Windows 98.

To install a new video driver or to change the monitor type using Windows 98, follow the steps outlined in Procedure 12.2.

---

### Procedure 12.2: Changing video driver or monitor type in Windows 98

**General procedures:**
1. Back up any important information.
2. Right-click on an open spot on the desktop. A pop-up menu should appear.
3. Click on the Properties option.
4. After the Windows Properties window opens, click on the Settings tab.
5. Click on the Advanced button.

**Changing the video driver:**
1. Click on the Adapter tab or the Monitor tab, depending on the driver(s) you wish to update.
2. Make a note of the current driver listed to the left of the Change button. If the new driver causes the video to fail, reboot the computer in the Safe mode and change back to the driver that is working.
3. Click on the Change button.
4. After clicking on the Change button, the Update Device Driver Wizard window will appear. Click on the Next button.
5. When the next window of the Update Driver Wizard displays, click on the desired option to either allow Windows 98 to automatically search for an update driver or to install a driver manually by choosing from a list of drivers.

Video Adapters and Monitors

**Automatic search:**
1. To allow Update Driver Wizard to automatically search for an update driver, click on, "Search for a better driver than the one you are using," and then click on the Next button.
2. Place a checkmark beside the storage devices that you wish Update Driver Wizard to search for by clicking on the box. Once the appropriate boxes have been checked, click on the Next button. Update Driver will search the selected devices for an updated driver.

**Manual installation:**
1. In the Update Driver Wizard window, click on "Display a list of drivers in a specified location, so you can select the driver you want," and then click on the Next button.
2. When the next window displays, a text box will display the driver(s) currently in use. To display a list of all manufactured drivers currently installed or available on the Windows 98 CD-ROM, click on "Show all drivers."
3. To load an updated driver from a diskette or CD-ROM, click on the Have Disk button and then select the location of the driver.
4. Once the updated driver has been located, click on the Next button and follow any additional instruction displayed by the Update Drivers Wizard.
5. Windows will have to be restarted before the changes will take effect.

## 12.8 VIDEO TROUBLESHOOTING

Just like everything else we have discussed so far, problems with video are categorized as either hardware or software. The problem is being able to distinguish between the two. Hopefully the following section will help in this determination. Before we get started in the troubleshooting procedures, we need to talk about the basics:

1. Before you make any changes in hardware settings or drivers, always back up important information on the computer's hard drive. Even though the installation should go without problems, always be prepared for the worst.
2. Keep an updated driver for the video adapter installed in the computer. Updated drivers can usually be easily obtained from the adapter manufacturer's Web site. In fact, you should make sure free drivers are available through the Internet before you purchase a video adapter.
3. Before a new driver is installed, document the name of the driver currently installed. This allows the old driver to be reselected if the new driver fails to function. To reinstall the old driver, restart Windows 9X in the Safe mode and follow the steps outlined in Procedure 12.2.
4. Most of today's monitors have a power LED that provides a valuable troubleshooting tool that can be used to isolate problems with the monitor itself. The power LED is usually located on the front of the monitor in the lower right-hand corner. When the power is turned on and the monitor is receiving video information from the computer, this LED should be green. If the LED is amber, it usually indicates the monitor is turned on but is not receiving video information from the computer. If the LED is flashing amber, it usually indicates the monitor is in standby mode. Standby mode is used to turn off the high voltage on the monitor when the computer has not been used for a predefined period of time. This feature helps conserve power.

    To determine if an operational monitor supports the green and amber feature, simply disconnect the interface cable from the back of the computer and turn the monitor on. If the power LED turns from green to amber, this feature is supported. If the power LED remains green, the monitor does not support this feature and you will have to follow all of the procedures given under each troubleshooting category.
5. Consider purchasing a spare video adapter, an invaluable aid in troubleshooting video problems. An inexpensive 2D card is all that you need and should cost less than $30.
6. If video failures occur, check the manuals that are provided with the monitor and video adapter for troubleshooting procedures. If none are available, the following generic procedures may be followed. These procedures are for the most common failures.

*No Video Is Displayed*

1. Make sure the monitor is plugged in and turned on. This should be indicated by either a green or an amber LED located somewhere on the monitor's front panel.

2. Determine if the monitor is receiving video information. This is usually indicated by the monitor's power LED. If the LED is green, the monitor is receiving video information and the problem is with the monitor. If the LED is amber, the monitor is turned on but not receiving video information, so the problem is probably in the interface cable, computer, or video adapter.
3. If the monitor's power LED is green, disconnect the monitor's cable from the connector on the back of the computer. If the LED now turns amber, make sure the monitor's brightness and contrast settings are set correctly. If so, the monitor is probably defective.
4. If the monitor's power LED is amber, check the monitor's interface cables. Make sure they are seated snugly in both of the connectors on the back of the computer and the back of the monitor, if a connector is used.
5. Determine if the computer is turned on and plugged in. The power LED on the front of the computer should be illuminated and the power supply fan should be running. If not, follow the procedure to troubleshoot a dead computer that was covered in chapter 15.
6. If power is getting to the computer, determine if the computer is booting by pressing the Reset button. Listen carefully for beep codes. If the computer beeps more than once, consult the beep codes listed in chapter 15.
7. If the computer does not beep at all, or shows no sign of booting, then the problem is with the computer and not the monitor. Troubleshooting procedures for a dead computer can be found in chapter 15.
8. If the beep code indicates a defective video adapter, check the CMOS setting for the correct video type.
9. If the beep code still indicates a defective video adapter, remove the computer's cover and reseat the video card. If this does not correct the problem, try swapping the video card with a spare, if available.

*Video Rolls Vertically or Horizontally*

1. Restart the computer. If the monitor's display rolls during the DOS portion of the boot sequence, the monitor is probably defective.
2. Reseat the video cable connectors.
3. If the monitor's display does not roll during the DOS boot sequence, boot Windows 9X into the Safe mode by pressing the F8 key when the POST beep is heard. If the computer monitor does not roll, the wrong graphics mode or graphics driver is selected.
4. If the monitor still rolls, replace the monitor.

*Monitor Displays the Wrong Color*

1. Reset the video cable connectors.
2. Restart the computer in Safe mode.
3. Check the video mode setting and driver setting in Windows 9X Display Manager.
4. If available, swap the video card with a spare.
5. If available, swap the monitor with a known good monitor.

### 12.8.1 Degaussing

The entire process of displaying information on the front of the CRT operates on the principle of the repulsion and attraction of magnetic fields. Due to the constant application of these controlling magnetic fields, the front of the CRT will become magnetized if not compensated for. A magnetized CRT is apparent when color splotches appear around the CRT's edges. To help prevent this from happening, a coil called a **degaussing coil** is wrapped around the outside edge of the CRT. A typical degaussing coil is shown in Figure 12.9.

When power is first applied to the CRT, alternating current is automatically supplied to the degaussing coil for a brief period of time to demagnetize the CRT. However, the duration of the alternating magnetic field produced by the internal degaussing circuitry is sometimes not enough to completely demagnetize the CRT. For this reason, most of the better monitors supply a control that allows the user to activate the degaussing coil while the CRT is operational. This causes a distorted image to be displayed, but will not damage the CRT. However, it should be noted that repeatedly activating the degaussing function might damage the circuitry.

**Figure 12.9** Degaussing Coil

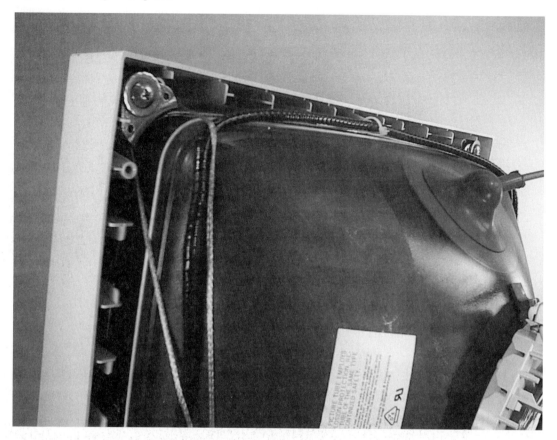

**Figure 12.10** External Degaussing Coil

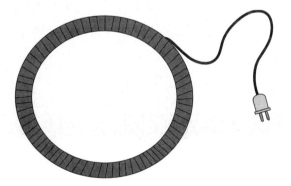

Color splotches may also appear if the CRT is reoriented to a new position after a long period of time. Under this condition, the automatic and manual degaussing features are often not sufficient to correct the problem. If this situation ever occurs, an external degaussing coil is usually required. An illustration of an external degaussing coil is shown in Figure 12.10.

## 12.9 CONCLUSION

Choosing a video adapter and monitor depends on the application of the computer. Most business computers do not need high-speed video, high pixel resolution, high color resolution, or 3D acceleration. However, computers that are used for games or graphics applications may need all of these advanced features.

Choosing the monitor and VDC for your computer is a very important decision. Nearly all of today's applications are written with the expectation of running on a color monitor. Low-end color monitors (CGAs) have poor resolution and their refresh rate is so slow that they can be more of a headache than a cheap monochrome monitor that displays at a higher resolution.

SVGA is currently the most widely used video standard on new computers. SVGA covers such a wide range of pixel resolutions, so purchase an SVGA system that meets your specific needs. To get the most out of your video system from the application programs currently being written, an SVGA with 4 MB of video RAM should be the minimum choice.

When purchasing an SVGA monitor system, try to avoid SVGA monitors with a dot-pitch greater than 0.31 for a 14″ display. (Large monitors can have a larger dot-pitch but will cost more.) When you buy a system that is already put together, make sure you know the resolution of the video system installed. Just because the system is advertised as having SVGA does not mean much. Also remember that the VDC card and monitor must match—you cannot run an EGA monitor with a VGA card.

The newest video graphics adapter is AGP. This standard offers speeds that far surpass those offered by PCI, but at a higher cost. One of today's trends that is disturbing to me is that manufacturers are moving the AGP circuitry from the video adapter into the chipset. A connector is mounted on the motherboard to provide the connection to the monitor. This feature reduces the overall cost of the computer because the AGP slot is not installed on the motherboard and a separate AGP graphics card does not have to be installed. The major problem with this feature is that it limits the ability to upgrade the computer to newer video standards. Also, if the video circuits in the chipset become defective, the only option is to revert to a PCI video adapter. For these reasons, I highly recommend purchasing a motherboard or computer that has the AGP slot and not the integrated port.

# Review Questions

1. Write the meaning of the following acronyms:
   a. CRT
   b. VDC
   c. VESA
   d. VBE
   e. HGA
   f. MDA
   g. MGA
   h. CGA
   i. EGA
   j. VGA
   k. SVGA
   l. DAC
   m. WYSIWYG
   n. AGP

## Fill in the Blanks

1. A _____ monitor displays information in one color plus black.
2. The _____ coating on the inside surface of the CRT emits light when struck by an electron beam.
3. The CRT's _____ _____ magnetically positions the electron beam.
4. The individual pictures displayed by televisions and CRTs are called _____.
5. The information displayed by a television and CRT consists of hundreds of _____ lines.
6. A _____ is the name for each time the beam makes a single trip down the scan.
7. Interlace video uses two _____ per _____.
8. A _____ _____ pulse keeps the vertical information displayed on the CRT synchronized with the information coming from the video adapter.
9. The _____ _____ on the video adapter holds information that is being displayed.
10. The video BIOS is usually located between addresses _____ and _____.
11. The two categories of modes used by video adapters are _____ and _____.
12. A picture element is known as a _____.
13. VGA mode 10 is a _____ mode that displays a resolution of _____ × _____ pixels at _____ colors.
14. VBE mode 11B is a _____ mode that displays a resolution of _____ × _____ pixels at _____ colors.
15. A video mode of 16,777,216 colors requires _____ bytes of video memory per _____.
16. The _____ _____ _____ displays only text.
17. A(n) _____ monitor can display an infinite number of colors.
18. A monitor's _____ _____ specification determines the monitor's resolution.

## Multiple Choice

1. The image on the front of the CRT consists of:
   a. several hundred individual lines that run horizontally.
   b. several hundred individual lines that run vertically.
   c. several hundred lines that run both horizontally and vertically.
   d. All of the above answers are correct.
   e. None of the above answers are correct.
2. The unit in the CRT that emits the electron cloud is called the:
   a. electron emitter.
   b. anode.

## Video Adapters and Monitors

    c. cathode.
    d. deflection yoke.
3. An electron beam is visible.
    a. True
    b. False
4. _____ video paints a frame in a single pass down the screen.
    a. Interlaced
    b. Noninterlaced
    c. Single-pass
    d. One-pass
5. The electronic circuit that is actually responsible for displaying images on the CRT is the:
    a. microprocessor.
    b. video RAM.
    c. video display controller.
    d. RAMDAC.
6. To hold the information between each frame:
    a. the video adapter card always has RAM installed.
    b. the video adapter card usually has RAM installed.
    c. some of the computer's conventional RAM is used to hold the video information.
    d. some of the computer's ROM is used to hold the video information.
7. When interlaced video is used:
    a. the monitor's horizontal sync is twice the frame rate.
    b. the monitor's vertical sync is twice the frame rate.
    c. the number of scan lines per frame is doubled.
    d. the frame rate is doubled.
8. A pixel is:
    a. the smallest dot the adapter card can control in the mode it is operating in.
    b. a single phosphor dot on the front of the CRT.
    c. a group of three phosphor dots called a triad.
    d. a term used with monochrome monitors.
9. The smaller the pixel:
    a. the lower the resolution.
    b. the higher the resolution.
    c. the slower the monitor.
    d. the more the flicker.
10. To display a pixel in 16,670,216 colors, each pixel requires:
    a. 1 bit of memory.
    b. 1 byte of memory.
    c. 2 bytes of memory.
    d. 3 bytes of memory.
    e. 2 bits of memory.
11. The color video standard that only displays information in 64 colors is:
    a. VGA.
    b. EGA.
    c. CGA.
    d. XGA.
12. A pixel and a triad are the same.
    a. True
    b. False
13. The current video standards used by the compatible computer support the video modes that were introduced with CGA.
    a. True
    b. False
14. The monitors designed around this standard can display information in an infinite number of colors:
    a. VGA
    b. CGA
    c. EGA
    d. HGA
    e. MDA
15. A pixel resolution of $1{,}024 \times 768$ indicates:
    a. 1,024 pixels are available.
    b. 768 pixels are available.
    c. 786,432 pixels are available.
    d. 786,432 colors are available.
16. How much video RAM is required to display information in the $80 \times 25$ text mode?
    a. 2K
    b. 4K
    c. The amount of video RAM depends on the number of colors.
    d. The amount of video RAM depends on the number of attributes.
17. How much video RAM is required to display $1{,}024 \times 768$ pixel resolution at 256 colors?
    a. 196,608 bytes
    b. 700 Kb
    c. 786,432 bytes
    d. None of the above answers are correct.
18. With noninterlaced video:
    a. the vertical sync rate is twice the frame rate.
    b. the horizontal sync rate is twice the frame rate.
    c. the vertical sync rate and the frame rate are the same.
    d. the horizontal sync rate and the frame rate are the same.
19. What does the acronym WYSIWYG refer to?
    a. Information is displayed as it will appear when it is printed.
    b. This word means nothing because it is misspelled.
    c. The information being displayed is in graphics mode.
    d. The information being displayed is in text mode.
20. What is the preferred video card for a Pentium computer?
    a. EISA
    b. VL bus
    c. ISA
    d. PCI
21. Which graphics standard is a port?
    a. VGA
    b. AGP
    c. CGA
    d. VL bus
22. A pixel:
    a. is a single phosphor dot.
    b. is an $8 \times 8$ matrix of phosphor dots.
    c. is the same size in all graphics modes.
    d. varies in size depending on the graphics mode.
23. A graphics mode of $640 \times 350$ is a total of:
    a. $640 \times 350$ pixels.
    b. 350,640 pixels.
    c. 640,350 pixels.
    d. 307,200 pixels.

24. If the 1,024 × 768 graphics mode displaying in 65,536 colors is selected:
    a. 1 MB of video RAM is required.
    b. 2 MB of video RAM is required.
    c. 4 MB of video RAM is required.
    d. 8 MB of video RAM is required.
25. A card that supports commands to draw windows is called:
    a. Windows accelerated.
    b. Windows drawer.
    c. Accelerated Windows drawer.
    d. Windows compatible.
26. A 15″ monitor that has a dot pitch of 0.28 mm can display a resolution of 1,152 × 864.
    a. True
    b. False
27. To reduce flicker:
    a. Use a lower refresh rate.
    b. Use a higher resolution.
    c. Use higher refresh rate.
    d. Use a lower number of colors.
28. The monitor mode that requires two trips down the screen per frame is called:
    a. VGA.
    b. AGP.
    c. interlaced.
    d. noninterlaced.
29. The term *resolution* indicates the:
    a. sharpness of the picture.
    b. size of the picture.
    c. color of the picture.
    d. number of attributes available.
30. An analog monitor:
    a. has the ability to produce more colors than does a CGA monitor.
    b. was originally called VGA monitor.
    c. can control each gun in many steps.
    d. All of the above answers are correct.
    e. Only answers b and c are correct.
31. Which of the following standards has the ability to produce the highest resolution image?
    a. SVGA
    b. VGA
    c. CGA
    d. HDTV
32. The timing pulses that keep the monitor in sync with the output of the adapter are:
    a. the PCI clock and system clock.
    b. vertical sync and horizontal sync.
    c. VGA and AGP.
    d. monitor clock and adapter clock.
33. A monitor's dot-pitch refers to:
    a. the spacing of the phosphor dots on a color monitor.
    b. the spacing of the pixels on a color monitor.
    c. the size of the characters displayed on a color monitor.
    d. the color of the character being displayed.

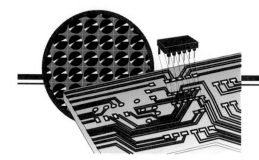

# CHAPTER 13

# Printers

## OBJECTIVES

- Define terms and expression relative to printers.
- Explain the difference between fixed fonts and true-type fonts.
- Describe the characteristics of impact printers.
- Describe the characteristics of nonimpact printers

## KEY TERMS

hard copy
printer
impact
nonimpact
font
pitch
characters per inch (CPI)
point size
fixed font
pica
elite
true type font (TTF)

fully formed
dot matrix
tractor feed
friction feed
sheet feeder
continuous form (fanfold)
draft
near-letter-quality (NLQ)
letter-quality (LQ)
impact dot matrix
print head

stepping motor
bi-directional
daisy-wheel
thermal printer
light amplification by stimulated emission of radiation (LASER)
image drum
toner
dots per inch (DPI)
inkjet
emulating

There is a great need to generate a permanent record of the data generated by a computer. This permanent record may be the listing of a program, a document produced by a word processor, inventory output, checks, and so on. Monochrome or color monitors that are used as the primary output device on most computers cannot produce a permanent record. The computer needs to have the ability to get information onto paper (referred to as a **hard copy**) to create a permanent record of the computer's output. This hardcopy output is accomplished by a device call a **printer.**

## 13.1  TERMINOLOGY

There are two overall categories of printers: **impact** and **nonimpact.** Impact printers rely on the same technique used on old-fashioned typewriters. A print head of some kind hammers against a ribbon, the ribbon is forced onto a sheet of paper, and the force of the impact deposits some of the ink from the ribbon onto the paper. Nonimpact printers, on the other hand, produce characters on the paper using techniques other than mechanical force. These printers still use print heads, but the print head does not strike a ribbon to produce print.

**Figure 13.1** 9 × 7 Dot Matrix

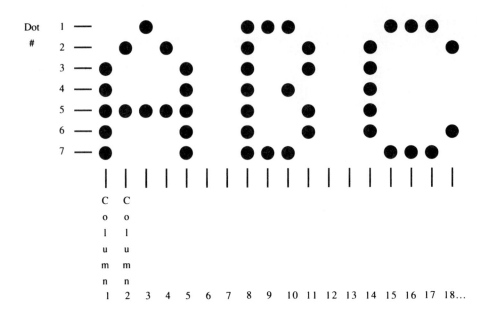

*Source:* Asser/Stigliano/Bahrenburg, *Microcomputer Theory and Servicing* (Upper Saddle River, NJ: Prentice Hall, 2001) p. 551.

The type style (appearance) of the printed character is called the character's **font.** Fonts include such names as Times Roman, Courier, Old English, Koffee, and many others. There are literally hundreds of fonts with some of the more popular word-processing programs and operating systems.

The size and spacing of printed characters is called the character's **pitch.** A character's pitch is usually given in one of two units of measurement: **characters per inch (CPI)** or **point size.** Fonts that use CPI are sometimes called **fixed fonts** because the fonts and pitches available are stored in a ROM installed inside the printer. The software running the computer sends a series of control codes to the printer to tell the printer which font and pitch to use. Once the application program has selected the font and pitch, the application program sends ASCII to the printer to identify the characters to be printed. The most popular CPI pitches are those used in standard bar typewriters—10 CPI, called **pica,** and 12 CPI, called **elite.** The larger the number of CPIs, the smaller the character pitch. The printer, not the program, determines the number of fixed fonts available to the application program.

Point size is a unit of measurement normally used with **true-type fonts (TTFs)**. When TTFs are used, characters are sent to the printer in graphics form instead of ASCII. A printer with graphics capabilities is needed to use TTFs. A TTF is a scaleable font whose font and pitch are changed in a graphics file created in the computer's memory or disk drive by the application program running the computer. This graphics file is transferred to the graphics printer, which prints the images as if it were a picture, not a text document.

As previously stated, a TTF's pitch is given in point size, not in CPI. Point size is usually based on 1/72 of an inch per point. A point size of 12 means the character's size is approximately 0.1666" (12 × 1/72). The larger the point size, the larger the character's pitch. When TTFs are used, the program itself, not the printer, determines the number of fonts and pitches available to the application program. There are hundreds of TTFs available for all versions of the Windows operating system.

Printers produce characters that are either **fully formed** or in a **dot matrix.** The fully-formed printers are impact printers that have the individual characters embossed on some type of surface. A mechanical force causes the embossed image to strike a ribbon, which is impregnated with ink. The ribbon is driven into a sheet of paper that reproduces the character. The types of fully-formed printers covered in this chapter are daisy-wheel and line printers.

Printers that produce characters in a dot matrix can be either impact or nonimpact. The characters printed using the dot matrix technique are formed with a group of dots arranged in columns and rows

**Figure 13.2** Typical Friction Feed

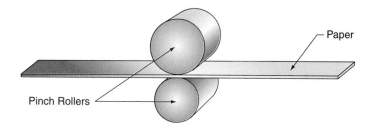

**Figure 13–3** Typical Tractor Feed Mechanism

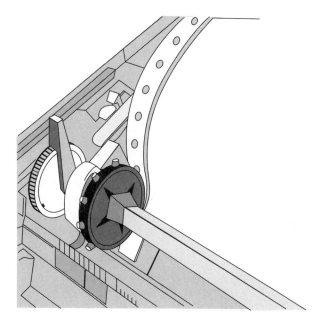

*Source:* Courtesy IBM.

(matrix). Figure 13.1 shows a typical 9 × 7 dot matrix pattern. The printers covered in this chapter are dot matrix, thermal, inkjet, and LASER.

Paper is moved through the printer by either **tractor feed** or **friction feed.** Friction feed pinches individual sheets of paper between two rollers, similar to a standard typewriter. As the rollers move, the paper goes up or down depending on the rollers' direction of rotation. Inkjet and LASER printers currently use this type of paper feed. Friction feed provides the ability to use single sheets of paper. Most printers that use this method are equipped with a **sheet feeder** that automatically feeds the individual sheets of paper into the printer. Figure 13.2 shows a friction feed.

Tractor feed uses **continuous form (fanfold)** paper that has equally spaced holes along the outer edge. Once printed, perforations allow the paper to be separated into individual sheets and the tractor holes removed. Once removed, the paper appears as standard typewriter paper. The holes in the paper are positioned into the printer's tractor assembly. When the tractor assembly moves, it pulls or pushes the paper through the printer. Once the paper is loaded, the tractor feed automatically moves the sheets through the printer with no operator intervention. A tractor feed mechanism is shown in Figure 13.3.

Several terms are used when dealing with the quality of the print produced by printers. **Draft** print is the lowest quality of print, and is usually used for the proof versions of documents. If a printer has the ability to generate draft-quality print, it will not cause as much wear on the ribbon or will use less ink or toner. Draft print usually prints a lot faster than other print qualities. **Near-letter-quality (NLQ)** print closely resembles the quality that would be produced if the documents were created on a standard typewriter. **Letter-quality (LQ)** printers produce documents that cannot be distinguished from documents generated on a typewriter. In fact, the letter quality produced by today's printers exceeds the quality of the print produced by mechanical typewriters.

## 13.2 IMPACT PRINTERS

As stated previously, impact printers form their printed output by mechanically striking a ribbon into the paper. The first impact printers used on microcomputers were simply modified electric typewriters. The technique used by impact printers to produce print is very similar to that of their ancestors. The most popular impact printers are the dot matrix and the daisy-wheel.

### 13.2.1 Impact Dot Matrix

Even though the technology is considered outdated, the **impact dot matrix** printer (simply called *dot matrix*), is still widely used in microcomputers. As the name implies, the dot matrix printer forms its characters with a group of dots. The dots are formed by electrically forcing a wire pin into a ribbon that strikes the paper. The pins are physically arranged in a vertical line inside an assembly called a **print head.** The print head is stepped across the paper by a special motor called a **stepping motor.**

As the print head moves from column to column, some or none of the pins are fired into the ribbon to form one column of dots that will eventually produce the character. The print head is then advanced to the next column and the pins needed to produce the dots in that column are fired. The print head moves across the line of character columns until the entire line is printed. The paper is then advanced to the next line of text. A typical dot matrix printer and print head are shown in Figure 13.4.

The number of pins in the print head and the spacing between the pins determine the quality of the print produced. Dot matrix printers are currently available with 9-, 16-, and 24-pin print heads. Most 9-pin dot matrix printers can produce print in draft and NLQ forms. The 16- and 24-pin print heads can produce print in draft and LQ forms. The typical arrangements of the pins on 9- and 16-pin dot matrix print heads are shown in Figure 13.5.

**Figure 13.4** Typical Dot Matrix Printer and Print Head

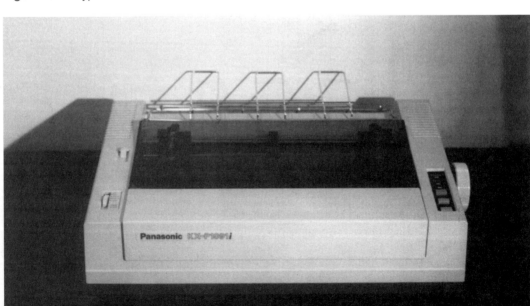

(Continued)

**Figure 13.4** (Continued)

*Source:* Dan L. Beeson, *Assembling and Repairing Personal Computers* (Upper Saddle River, NJ: Prentice Hall, 1997) p. 212.

**Figure 13.5** Typical Layout of 9- and 16-pin Print Heads

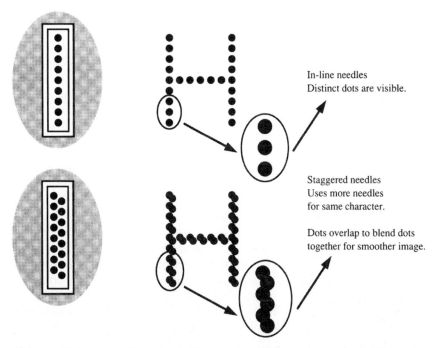

*Source:* Asser/Stigliano/Bahrenburg, *Microcomputer Theory and Servicing* (Upper Saddle River, NJ: Prentice Hall, 2001) p. 551.

Today's dot matrix printers are **bi-directional,** which means the print head will print from left to right or from right to left. Most new dot matrix printers can position the paper by either friction or tractor feed, depending on the preference of the operator.

All of today's dot matrix printers can operate in either text mode or graphics mode. In text mode, the application program sends ASCII characters to the printer and the printer's electronics take care of firing the pins in the print head to produce the characters. The only fonts and pitches available are those in the printer's ROM.

**Figure 13.6** Daisy Wheel Print Wheel

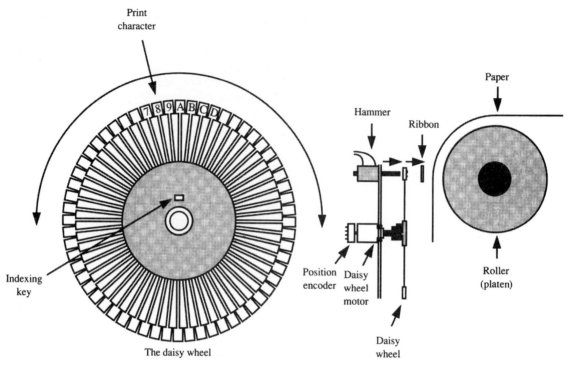

*Source:* Asser/Stigliano/Bahrenburg, *Microcomputer Theory and Servicing* (Upper Saddle River, NJ: Prentice Hall, 2001) p. 553.

When a program selects the graphics mode, the program that controls the printer controls the firing of the individual pins. In this mode, the printer works similar to a TV set or computer monitor. The images are formed by an array of individually controlled dots. The printer's graphics mode allows pictures to be formed along with the use of TTFs.

### 13.2.2 Daisy-Wheel

The impact **daisy-wheel** printer uses fully-formed characters embossed at the end of an arm. The arms are arranged in a circle on a print wheel that looks similar to the petals of a daisy. A typical daisy print wheel, along with the basic mechanics of the printer, is shown in Figure 13.6. Daisy-wheel printers are letter-quality printers. Because fully-formed characters are used, the only way the pitch or font can be changed is by changing the print wheel. To print a particular character, the wheel is rotated until the proper petal is aligned in front of a hammer mechanism. The hammer is fired against the petal and the petal hits the ribbon, which is forced into the paper, creating an image of the character. The characters are fully formed, so the graphics mode is not available with these printers.

Daisy-wheel printers are noisy and slow. The only redeeming virtue of this technology is its letter-quality printing and its ability to print on multi-part paper. Among its disadvantages is the fact that it cannot print graphics. Current nonimpact printers produce text quality that surpasses that of the daisy-wheel. In fact, today's standards for print quality rival typesetting.

## 13.3 NONIMPACT PRINTERS

There are several types of nonimpact printers used with microcomputers. As previously stated, nonimpact printers produce characters by some method other than mechanical force. All nonimpact printers we will discuss produce characters in a dot matrix, similar to the impact dot matrix printer discussed previously. However, because the impact dot matrix printer was the first to utilize the technique of producing characters and images with dots, it carries the name. The nonimpact printers we will discuss in the following section are the thermal, LASER, and inkjet.

### 13.3.1 Thermal

A **thermal printer** is a character-at-a-time printer that produces its characters in a dot matrix. The thermal printer's print head consists of a vertical row of very tiny heating elements that heat and cool in a fraction of a second. The process that produces the characters on paper is similar to that of a dot matrix printer, but with no mechanical force. Instead, the paper used by a thermal printer is coated with a chemical that changes to a dark color when heat is applied. Instead of firing a pin into a ribbon, print is produced when the chemical changes color. These printers are cheap, tiny, and very quiet. This makes them ideal for the very small laptop and notebook computers.

### 13.3.2 LASER

**Light amplification by stimulated emission of radiation (LASER)** printers work on the same basic principle as the photocopiers found in just about every office in the United States.

The LASER printer's ability to produce print is based on the fact that some materials react strangely to light. Selenium, along with a few other complex organic compounds (that we do not need to know), modify their electrical conductivity when they are exposed to light.

Copiers and LASER printers profit from these characteristics by focusing an optical image on a selenium-coated, photo-conductive drum called an **image drum.** The image drum is given a static electrical charge before being exposed to the image.

A dry powder, called **toner,** is then spread across the drum, but it only sticks to the areas that are charged. The toner that sticks to the drum assumes the same static charge as the drum. The paper that will eventually hold the image is passed between the image drum and a wire that is charged to a potential opposite that of the toner. This literally pulls the toner off the drum onto the paper. The paper is then squeezed between two hot rollers, which causes the toner to "fuse" into the paper, creating permanent print.

The copy machines found in most offices use a bright light that reflects off the information to be copied to expose the image drum. A LASER printer uses a laser beam as the light source. The laser beam is usually scanned across the image drum with a rotating mirror. The laser beam is rapidly switched on and off by digital information coming from the computer to form a single, minuscule dot at a time on the image drum. This process is repeated thousands of time as the image drum rotates. LASER printers use a LCD-shutter (liquid crystal device) or an array of LEDs (light-emitting diodes) to modulate the light striking the image drum. The basic process and internal works of a LASER printer are illustrated in Figure 13.7.

Most LASER printers produce images in a 300 **dots per inch (DPI)** array or higher. This tiny dot size allows these printers to produce some of the highest quality print available on microcomputers. The biggest advantage of the LASER printer is its speed. In fact, the speed of LASER printers is rated in pages per minute, whereas other printers are rated in characters per second.

The biggest disadvantage of LASER printers is that they are considered a high-maintenance item. They are not only expensive to purchase, but are also expensive to maintain. The toner cartridge has to be replaced every 2,000 or so hard copies. The image drum will also have to be replaced sooner or later because the coating will eventually wear off.

Some LASER printers use a lower quality image drum that is replaced with the toner as a single unit at about 2,000 hard copies. This replacement unit runs between $50 and $100. Other LASER printers use a high-quality image drum that is replaced as a separate unit at approximately 25,000 hard copies, at a cost of about $250. The toner is replaced as a cartridge after approximately 2,000 hard copies and costs about $25. If you purchase either style of LASER printer, you will spend a considerable amount of money to get the quality and speed offered by these printers.

### 13.3.3 Inkjet

**Inkjets** are currently the preferred printer for the compatible computer. Inkjet printers produce images by spraying droplets of ink onto paper. Even though it sounds unlikely, inkjet printers work well enough to rival, and even surpass, LASER printers in quality. Inkjet printers produce their characters in much the same way as do the dot matrix printers. Instead of driving a ribbon into the paper

**Figure 13.7** Typical Layout of a LASER Printer

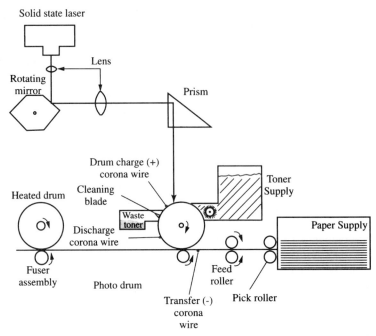

*Source:* Asser/Stigliano/Bahrenburg, *Microcomputer Theory and Servicing* (Upper Saddle River, NJ: Prentice Hall, 2001) p. 555.

**Figure 13.8** Typical Operation of an Inkjet Printer

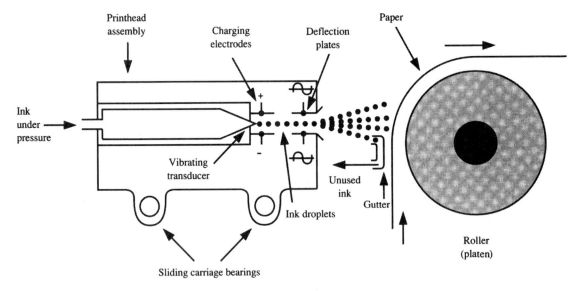

*Source:* Asser/Stigliano/Bahrenburg, *Microcomputer Theory and Servicing* (Upper Saddle River, NJ: Prentice Hall, 2001) p. 553.

with a pin, the inkjet squirts ink into place via tiny nozzles, each one corresponding to a print wire of the impact dot matrix printer. An illustration of the printing process used by the inkjet is shown in Figure 13.8.

Without a ribbon to blur the image, inkjet printers can produce top-quality print. Moreover, the "laser-less" technology of the inkjet printers translates to a lower price tag and lower overall cost of operation. However, inkjets are several magnitudes slower than LASERs because they rely on a traveling print head to scan across the sheet, rather than lightning-fast optics.

Because inkjets are nonimpact printers, they are much quieter than their impact dot matrix printer counterparts. For best results, inkjets require special paper with controlled absorbency. This results in a higher cost per page. If you try to get by "on the cheap" with paper that is too porous, the inks will spread into a blur.

The biggest advantage of inkjet printers is their ability to produce high-quality color print at a relatively low cost. In order to perform this amazing feat, the print head has three ink nozzles with shades of the colors red, green, and blue. As the three colors of ink are squirted toward the same dot position on the paper, the inks blend to form different colors. Precise control of the amount of ink coming from each nozzle can produce an almost unlimited number of hues.

Due to its high-quality printer and color capabilities at a reasonably low price, the inkjet printer is currently the predominant printer for the compatible PC. This is especially true for home use. The only disadvantages of the inkjet printer are the speed of print and the cost of the ink cartridges.

## 13.4 EMULATION

Printers all do basically the same thing, so it would seem that any printer that is compatible with the compatible computer printer's interface should work with any software. While this may have been true in the days of ASCII and fully-formed printers—when fonts and pitch were changed by changing hardware, and when graphics were not available—times have changed.

Nowadays, printers have multiple typefaces and special formatting features that require the computer to send special codes called *escape sequences* (an ASCII escape character followed by one or more characters of command, control, and other codes). No program can possibly know all of the codes for all available printers. However, most programs know the codes used by several of the most popular ones. If your printer can make use of the same codes by **emulating** (acting like) a more common printer, your software should be able to get the most out of your printer.

Printing graphics images is far more complex than printing text. While text coding is reasonably straightforward, because nearly all applications can output ASCII characters for text, there is no similar standard for graphics images. To print any kind of bit image, a printer must be told where to place each dot on the page. The instructions might take the form of a bitmap, or they may be a code in a page-description language.

As already stated, in graphics mode the printer is told explicitly where each dot goes. Unfortunately, a number of coding systems are used to assign dot locations. At one time, every manufacturer had its own system, but today, most printers emulate one of the industry leaders. These include printers such as the Epson dot matrix line, the IBM Graphics or Proprinters, or the Hewlett-Packard line of LASERS and inkjet printers.

To ensure that your printer is compatible with the most popular application programs being written for the compatible computers, choose a printer that has the ability to emulate one of the mentioned industry standards.

## 13.5 SETTING UP A PRINTER USING WINDOWS 9X/2000

When using DOS and Windows, the individual applications support the printer's graphics modes. This means that a printer driver that is unique to each application must be installed. The documentation that is supplied with the application program contains the procedure to install the driver.

When running Windows 9X/2000, the printer's graphics modes are supported by the operating system. This feature means that only one driver per printer needs to be installed in the operating system. Once this is accomplished, all Windows 9X/2000 applications can use this driver.

The process of installing a new printer and its drivers, or upgrading an existing driver, is made easy by using Windows 9X/2000's Wizards. Procedure 13.1 lists a generic process to install a new printer and driver. If available, the driver installation procedure that is contained in the printer's documentation should supersede these procedures.

### Procedure 13.1: Installing a new printer and driver

1. Back up any important information.
2. If a document that contains the installation procedure is supplied with the printer, follow that procedure.
3. Shut down Windows 9X/2000 using the correct procedure and then remove power from the computer.
4. Insert the end of the printer cable that has the male DB25 connector into the parallel port's female DB25 pin connector on the back panel of the computer. Screw the finger screws that are located on each side of the cable connector into the screw holes on each side on the connector on the back of the computer. These screws should only be finger-tight.
5. Insert the male Centronics connector into the female Centronics connector on the back of the printer. Lift the metal snap rings that are located on each side of the connector on the back of the printer until they snap into the slots in the side of the connector on the printer cable.
6. If required, insert the power cord into the power receptacle that is usually located on the back panel of the printer.
7. Plug the printer's power cord or power module into a proper power receptacle.
8. Apply power to the computer and then turn on both the computer and the printer.
9. If the printer is PnP, the Windows 9X/2000 operating system should recognize the printer during the initial boot sequence and automatically start the "Add Printer Wizard." If not, the Wizard will have to be started manually. To start the Wizard: (a) click on the Start Button; (b) move up to Settings; (c) move to Printers; and (d) click on Printers.
10. After the Printers window opens, double-click on the Add Printer icon. This will start the Add Printer Wizard.
11. After the Add Printer Wizard window displays, click on the Next button.
12. Select whether the printer is a local printer (connected directly to the computer) or a network printer that will be accessed over a LAN, and then click on the Next button.
13. If a driver diskette is not supplied with the printer, locate the printer manufacturer in the left-hand windowpane and the model of the printer in the right-hand windowpane. Once the correct printer is selected, click on the Next button.
14. If a driver diskette is provided with the printer, insert the supplied diskette or CD into the correct drive and then click on the Have Disk button. Use the browse functions to locate the correct printer driver for Windows 9X/2000. Once located, click on the OK button.
15. Select the port to which the printer is attached from a list of the available ports displayed by the Add Printer Wizard, and then click on the Next button.
16. When asked if you want to print a test page, answer yes to determine if the printer and interface are working correctly.

## 13.6 TROUBLESHOOTING PRINTER PROBLEMS

Like most everything else in the computer, printer problems boil down to hardware or software. The hardware that may cause problem includes the printer, cable, and port. The software includes the operating system, print driver, and the application itself. The following sections list possible printer problems and feasible solutions. Before you follow any of these procedures, consult the troubleshooting section of the printer's documentation to determine if there is a procedure for the situation.

*The Printer Does Not Print.*

1. Make sure that the printer is turned on and plugged into a working power receptacle.
2. Make sure the power cord is plugged snugly into the back of the printer.
3. Make sure the printer is not off-line or is not paused.
4. Check to see if the printer's data cable is inserted snugly into the connectors in the back of the computer and printer.
5. Try to run the printer self-test. The procedure should be in the printer documentation. If the self-test does not run, try removing the data cable from the printer.
6. Try another printer cable, if available.
7. If the printer has a fuse holder or circuit breaker that is accessible without removing the printer's cover, make sure the fuse is not blown or the circuit breaker is not tripped.

*The Printer Prints Incorrect or Garbled Characters.*

1. Check the printer cable to make sure it is seated snugly in the connectors on both the printer and computer.
2. Turn the printer off and then back on.
3. Restart the computer.
4. Check to make sure the printer cable is not routed over electrically noisy appliances such as fans or stereo speakers.
5. Try a new printer cable, if available.
6. Check for the correct printer driver.
7. Remove and reinstall the printer driver.

*The Paper Constantly Jams.*

1. Check the paper path through the printer to determine if it is clear and free of debris.
2. Check the condition of the platens and pressure rollers.
3. Check the printer manuals for paper feed cleaning instructions.

*The Printed Characters are Light or Faded.*

If the printer is inkjet:

1. Check for an empty or low ink cartridge.
2. Check to see if there is an ink-conserving option turned on.
3. Check the control that adjusts the distance of the print head from the paper to determine if it is set correctly (if available).

If the printer is LASER:

1. Make sure the draft option is not selected by the application using the printer.
2. Check the level of the toner.
3. Check the condition of the image drum.
4. Make sure toner-saving options are not selected (if supported).

If the printer is dot matrix or daisy-wheel:

1. Make sure the draft option is not selected by the application using the printer.
2. Check the condition of the ribbon.

## 13.7 CONCLUSION

When buying a printer, match the printer to your essential applications, decide how much waiting you can tolerate, and set a budget. The strengths and weaknesses of each printer will help guide you. If you do not need graphics but want high-quality text, an inexpensive daisy-wheel printer will do. If you have to be at your desk while the printer does its dirty work, you will favor a quiet, nonimpact model. If you need to print on multi-part paper, you will need an impact printer. Most current inkjet printers will give you letter-quality print at an affordable price. If you need typeset quality and speed, the LASER is the answer, but you will pay the price.

## Review Questions

1. Write the meaning of the following acronyms.
    a. CPI
    b. TFF
    c. LED
    d. LCD
    e. LASER

### Fill in the Blanks

1. The transfer of computer to paper is often called a _____ _____.
2. Pitch is a term used to indicate the _____ and _____ of the printed characters.

3. A _____ printer produces characters by striking an ink-impregnated ribbon onto the paper.
4. The _____ _____ printer produces characters using pins.
5. The _____ printer works on the same principle as most copy machines.
6. Printed information that appears the same or better than that produced by a standard typewriter is called _____ _____ .
7. The unit of the printer that actually produces the printed output is called the _____ _____ .
8. Most current printers provide _____ and _____ modes.

## Multiple Choice

1. A character's font is the:
   a. size of the characters.
   b. color of the characters.
   c. style of characters.
   d. spacing of the characters.
   e. Answers a and d are both correct.
2. Which printer produces characters by spraying droplets of ink onto paper?
   a. inkjet
   b. LASER
   c. daisy-wheel
   d. dot matrix
   e. line printer
3. Which printer works on the same basic principle as a copy machine?
   a. inkjet
   b. LASER
   c. daisy-wheel
   d. dot matrix
   e. line printer
4. Which font type specifies pitch in point size?
   a. fixed fonts
   b. variable fonts
   c. true-type fonts
   d. false-type fonts
5. The two basic categories of printers are:
   a. dot matrix and fully-formed.
   b. impact and nonimpact.
   c. fast and slow.
   d. tractor feed and friction feed.
6. Which of the following is a character-at-a-time printer?
   a. dot matrix printer
   b. thermal printer
   c. daisy-wheel printer
   d. All of these answers are correct.
7. Thermal printers:
   a. need special paper that changes color when heated.
   b. burn the characters into the paper.
   c. make a lot of noise.
   d. are the fastest of all printers.
8. Character pitch is the:
   a. size and spacing of printed characters.
   b. appearance of characters displayed on a monitor.
   c. attributes of characters displayed on a monitor.
   d. shape of printed characters.
9. A printer that uses a tractor feed:
   a. always uses standard typewriter paper.
   b. is faster than all other printers.
   c. uses continuous form paper.
   d. is slower than all other printers.
10. Emulation means:
    a. one device is a copy of the other.
    b. devices made by one company act like devices made by another.
    c. the same company makes two devices that act the same.
    d. Answers b and c are both correct.
11. A backup of important information does not need to be made before installing a new printer driver.
    a. True
    b. False
12. The inkjet printer is the fastest printer available for the compatible computer.
    a. True
    b. False
13. Which printer can create multiple copies using multi-part (carbonless) paper?
    a. dot matrix
    b. ink jet
    c. thermal
    d. LASER
    e. All of the above are correct.
14. The term *point size:*
    a. refers to the number of characters that will be printed within an inch.
    b. refers to the number of dots that determine the size of the printed character.
    c. refers to the style of print.
    d. is only used by postscript fonts.
15. A daisy-wheel printer is an example of a:
    a. dot matrix printer.
    b. fully-formed printer.
    c. line printer.
    d. LASER printer.
16. CPI stands for:
    a. characters per inch.
    b. cycles per interval.
    c. computer printer interface.
    d. character printer interface.

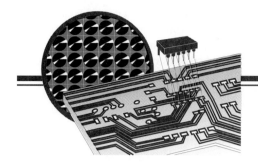

# CHAPTER 14

# Computer Network Basics

## OBJECTIVES

- Define terms, expressions, and acronyms used relative to networks.
- Define terms used relative to network media.
- Name and describe three network topologies.
- Explain the Ethernet protocol.
- Recognize the ISO network model.
- Identify network hardware.
- List the characteristics of ThickNet, ThinNet, 10BaseT, 100BaseT, FDDI, and ATM.
- Describe the characteristics of the NetBEUI, IPX/SPI, and TCP/IP protocol suites.
- Explain the IP portion of the TCP/IP suite.
- Define terms, expressions, and acronyms used relative to the Internet.
- Explain modem protocols and standards.

## KEY TERMS

keypunch operator
Holorith card
card reader
dumb terminal
network
local area network (LAN)
wide area network (WAN)
bandwidth
topology
workstation
nodes
server
bus topology
terminator
star topology
hub
critical resource
ring topology
token ring
attenuation
repeater
bridge
router
subnet

gateway
International Standards Organization (ISO)
Open System Interconnection (OSI)
physical layer
data link layer
network layer
transport layer
session layer
presentation layer
application layer
Ethernet
carrier sense multiple access with collision detection (CSMA/CD)
preamble
destination address
source address
type field
data field
frame check field
transceiver
ThickNet (10Base5)

Ethernet adapter card
ThinNet (10Base2)
10BaseT
Fiber Distributed Data Interface (FDDI)
Asynchronous Transfer Mode (ATM)
Network Interface Card (NIC)
data link control (DLC)
medium access control (MAC)
organizational unique identifier (OUI)
peer-to-peer
client/server
100BaseT
network server
domain
client
Network BIOS Extended User Interface (NetBEUI)
workgroup

Internetwork packet exchange/sequence packet exchange (IPX/SPX)
transmission control protocol/Internet protocol (TCP/IP)
IP address
Internet Assigned Number Authority (IANA)
Internet Corporation of Assigning Names and Numbers (ICANN)
dynamic host configuration protocol (DHCP)
Internet domain name (IDN)
domain naming service (DNS)
binding
sharing
mapping
modem

451

Early computers were usually big, expensive, and very complex. They required a generous amount of space, consumed large amounts of power, and usually had only one terminal for program input. These computers required a highly trained operator to actually control the computer's day-to-day operation. Programmers submitted jobs to **keypunch operators** who punched **Holorith cards** to represent the programs and data. From the keypunch operator, the Holorith cards were submitted to the computer's operator. The operator fed the programs into the computer through a **card reader.** The programmers usually did not have access to the computer itself. It was common for programmers to never actually see the computer for which they were writing the programs.

Computer technology has improved immensely through the years. It is now possible for a mainframe computer to handle multiple input and output devices at the same time and run several application programs simultaneously. A mainframe computer allows many remote terminals to be attached through which individual programmers can access the computer directly. The terminals used with mainframes are often referred to as **dumb terminals** because they are nothing more than addressable keyboards and monitors with no actual processing power. The information entered at the keyboard is fed through a cable to the mainframe computer. The actual processing of the information is performed by the very expensive and fast mainframe.

The introduction of the microprocessor chip by Intel in 1971 greatly changed the world of computers and networking. When microcomputers are connected together instead of being dumb terminals, each computer has the ability to not only input and output information, but also to process, share, store, and retrieve information as well. In most cases, connecting several microcomputers together eliminates the need for a very expensive mainframe. The connection of computers and peripherals together so they can communicate with each other and share resources is referred to as a **network.** Computers that are connected together in a network to share software and hardware are called either a **Local Area Network (LAN)** or a **Wide Area Network (WAN),** depending on the proximity of the computers.

LANs are networks in which all of the attached computers are in the same locale. This does not mean that the computers have to be in the same room. In fact, LANs can have computers located on all of the floors of an office complex or even in several buildings on a college campus. WANs are networks in which the computers are connected remotely in different geographical locations, such as in different cities, states, or countries. Whether the network is a LAN or a WAN, all of the computers in the network have to be connected together somehow to share information and resources.

## 14.1 MEDIUM (MEDIA)

Medium (plural, *media*) is one of those words that has a different definition depending on its application. In many fields, including networking, a *medium* is a substance through which information is passed from one point to another.

Digital information that is transferred over a computer network is usually in the form of electrical current or light pulses. Copper wire most often carries the electrical current pulses, whereas glass or plastic most often carries the light pulses. All media exhibit attenuation to the signal. The term *attenuation* refers to signal loss over distance. Usually, the less attenuation a medium has, the more expensive it will be.

Along with attenuation, the speed at which a medium can transfer data is also an important determining factor in its overall cost. Another factor that must be considered when selecting a medium for a network is its susceptibility to electro magnetic interference (EMI). All of these factors are important and are discussed in the following sections.

Currently, the media most often used in networking are coaxial, twisted-pair, and fiber optic cables. One very important characteristic that determines which one of these media is selected is a rating called bandwidth.

### 14.1.1 Bandwidth

The speed at which networks move information among computers is rated in megabits per second (Mbps). Most current networks move data at 10 Mbps or more. The medium that connects the computers together on the network must be able to reliably handle the speed of the network. To determine a medium's maximum reliable speed capabilities, a specification called **bandwidth** is used.

The term *bandwidth,* relative to networking, has a slightly different meaning than its meaning relative to other forms of electronic communication. Because a network always transfers data at the same

**Figure 14.1** Drawing of Coaxial Cable

speed, there is no need to specify the low side of the medium's bandwidth. Also, because a network transfers binary information in Mbps, network media are rated in maximum Mbps.

Networks move binary bits from one place to the next in millions of bits per second, so it is imperative that the selected medium has the ability to operate at the network's speed. For this reason, the bandwidth of a medium must meet or exceed the operating speed of the network.

### 14.1.2 Coaxial Cable

Coaxial cable is constructed using a center copper conductor that is encapsulated in an insulating jacket. Sometimes the copper conductor is plated with tin to inhibit oxidation. The insulating jacket that covers the center conductors is itself covered with a braided metal conductor that is incorrectly called a *shield*. This is incorrect because a true shield is never used as a conductor. The braided conductor is covered with yet another insulating jacket that forms the outside cover of the coax. Figure 14.1 shows a section of typical coax cable.

One of the characteristics of coax cable is that it has fixed impedance (resistance) to a signal over distance. This characteristic is used to accomplish what is referred to as *impedance matching,* which means that if all of the impedances of all of the devices used in the communication channel match, the channel will have maximum power transfer.

The current coaxial cables used in compatible computer networking are called *RG-8* and *RG-58*. Both of these coaxial cables have an impedance of approximately 50 ohms and a bandwidth of 10 Mbps. The difference is that RG-58 has an outside diameter of approximately 0.19″ (5 mm) while RG-8's outside diameter is approximately 0.4″ (10 mm). Because RG-58 has a smaller diameter, it has a higher attenuation but is considerably cheaper than RG-8.

### 14.1.3 Twisted Pair

Another widely used metallic medium used in compatible computer networking is twisted-pair cable. Twisted-pair cable is the same type of cable used by telephone systems. Twisted pair gets its name due to the fact that electrical conductors are grouped in pairs (twos) and each pair of conductors is twisted together. One of the electrical conductors in each pair carries the data and the other electrical conductor of the pair returns the signal. Twisting the signal-carrying conductor with the signal's return conductor dramatically reduces susceptibility to electro magnet interference (EMI), the electrical noise generated by all electrical/electronic devices.

There are currently two categories of twisted-pair cables used in compatible computer LANs: category 3 (CAT-3) and category 5 (CAT-5). CAT-3 cable has a bandwidth of 10 Mbps, whereas CAT-5 cable has a bandwidth of 100 Mbps. The only difference between the two categories is that CAT-5 has more twist per inch, which allows it to operate at a higher frequency but also adds considerably to the cost.

Along with the bandwidth, twisted-pair cable is also categorized as either unshielded twisted pair (UTP) or shielded twisted pair (STP). As the name implies, the conductors in the STP cable are enclosed in a metal foil shield that makes it less susceptible to EMI but more expensive than the UTP. Currently, UTP is far more popular than STP. Typical STP and UTP 6-pair cables are shown in Figure 14.2.

### 14.1.4 Twisted-pair T-carriers

To provide long-distance, high-speed transfer to private and business communities, the telephone system developed a set of multi-channel, high-speed dedicated twisted-pair lines called *T-, DS-,* or *E-carriers*. T-carriers are primarily used in the United States and Japan, so we will discuss this type.

T-carriers are numbered T1 through T4 depending on the transfer speed, with the slowest being the T1. A T1-carrier consists of 24 multiplexed (time-shared) channels. Each channel transfers information

**Figure 14.2** Illustration of UTP and STP Cable

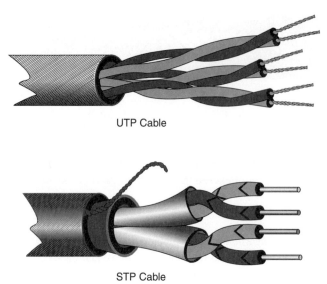

UTP Cable

STP Cable

**Table 14.1** Currently Available T-carriers

| Number | T1 Used | Channels | Speed |
|--------|---------|----------|-------|
| T1 | 1 | 24 | 1.544 Mbps |
| T2 | 4 | 96 | 6.312 Mbps |
| T3 | 28 | 672 | 44.736 Mbps |
| T4 | 168 | 4,032 | 274.176 Mbps |

at a rate slightly over 64 Kbps, including timing bits. This rate gives the T1 the ability to transfer information at a speed of 1.544 Mbps if all 24 channels are used. Due to the expense of a T1, it is often split into single or multiple channels called *fractional T1s.*

The other carriers, T2 through T4, simply consist of multiple T1-carriers. Table 14.1 lists the currently available T-carriers showing the number of T1s, channels, and data transfer rates.

## 14.1.5 Fiber Optics

Fiber optics uses a small glass or plastic tube to transmit information in the form of light. The main difference between glass and plastic is that plastic has a higher attenuation than glass but is far less expensive.

The basic construction of a fiber optic conductor consists of a glass or plastic center tube, called the *core,* that carries the information. A translucent plastic coating called *cladding* surrounds and is in direct contact with the core. The function of the cladding is to increase the reflection of the outside circumference of the core.

The core and cladding are further enclosed in a protective tube called a *jacket,* or *buffer.* A jacket that is in direct contact with the cladding is referred to as a *tight jacket* (or buffer), whereas a jacket that is not in direct contact with the cladding is referred to as a *loose jacket* (or buffer). A typical tight jacket fiber optic conductor is shown in Figure 14.3.

Fiber optic conductors are further categorized into multi-mode and single-mode. *Multi-mode* is the cheapest because it allows multiple light wavelengths to pass, which translates into higher attenuation. Multi-mode fiber optic cable also has a larger diameter, which makes it easier to manufacture. The

**Figure 14.3** Construction of a Typical Fiber Optic Conductor

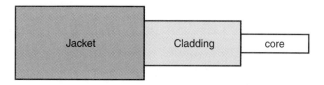

**Table 14.2** SONET OC Speed Rankings

| OC | Speed in Mbps |
|---|---|
| OC-1 | 51.84 |
| OC-3 | 155.52 |
| OC-6 | 311.04 |
| OC-9 | 466.56 |
| OC-12 | 622.08 |
| OC-18 | 933.12 |
| OC-24 | 1244.16 |
| OC-36 | 1866.24 |
| OC-96 | 4976.00 |
| OC-192 | 9952.00 |

multi-mode fiber optic used most often in computer networking uses a 62.5-micron core surrounded by 125-micron-thick cladding. A standard infrared light emitting diode (LED) is usually used to generate the light transmitted over a multi-mode conductor.

A *single-mode* fiber optic conductor will only allow light moving at one wavelength to pass. Because single-mode fiber is designed around one specific light wavelength, it exhibits very low attenuation to the light that it is designed to carry. Single-mode cable is very thin and its light source is usually a LASER-LED. Both of these features make single-mode fiber optics cable considerably more expensive than multi-mode. The single-mode cable used most often in networking has an 8.3-micron core with 125-micron-thick cladding.

The low attenuation feature of fiber optics allows information to travel greater distances than would be traveled using conductors made from metal. Another major advantage of fiber optic cable is its immunity to EMI. This feature makes fiber optic cable immune to lightning strikes or any electrical noise. And, last but not least, another major advantage of fiber optic cable is that it has a much higher bandwidth than do metallic conductors.

Even though fiber optic cable seems to have all of the advantages, there are two disadvantages that limit its use. The first disadvantage is that fiber optics cable is considerably more expensive than its metallic counterpart. Secondly, metallic conductors are much easier to terminate with connectors and splice together.

### 14.1.6 Synchronous Optical Network (SONET)

SONET is an international standard for high-speed digital communication over fiber optic cable. SONET speeds are ranked using a specification called Optical Carrier (OC), beginning with OC-1. Table 14.2 lists current OC rankings for SONET.

**Figure 14.4** Basic Data Packet

### 14.1.7 Serial Transfer

All networks (that I know of) transfer data in serial form. Remember from chapter 9 that serial information is transferred over a single conductor 1 bit at a time. When copper is used, two conductors are required—one to carry the data and the other as the return conductor. If fiber optics is used, only a single conductor is needed. Even though parallel transfer is several magnitudes faster than serial, serial is most often used in networking because of one major advantage: cost. At the 2-year technical college where I teach, miles of network cable connect the entire campus. If this was parallel cable, the cost of this cabling would be astronomical.

### 14.1.8 Packets and Frames

On most LANs, only two computers can transfer information on the medium at any one time. You might wonder, "How can two computers transfer millions of bytes of information without tying up the network?" The answer to this question is that the information to be transferred is broken down into small chunks called *packets,* or *frames.*

If a computer needs to transfer information with another computer, it sends a single packet and then releases the network. If no other computer needs the network, another packet is transferred. If a computer is waiting to send data, it will transfer a packet when the other computers release the network. Using this technique, each computer usually gets its turn no matter how much data are transferred by other computers on the network.

To transfer data between two computers, the packets must contain information in addition to the data to be transferred. This other information is usually contained with a packet header and a trailer that encapsulates the actual data being transmitted. Among other things, the packet header contains the network address of the destination computer and originating computers. The trailer contains error-detection information that the receiving computer can use to determine if the data it received are the same as the data that were sent. The basic makeup of a packet is shown in Figure 14.4. In order for two computers to transfer data, the information contained in the packet must be assembled in the same way and the length of the packet must be a fixed length. The packet's size and information are determined by network protocols.

## 14.2 NETWORK PROTOCOLS

As stated previously, a *protocol* is an agreed-upon way of doing something. In order for computers to communicate with each other on a network, they must all use the same set of protocols. The protocols used in networking fall into one of three categories: software, hardware-software, and hardware.

*Software* protocols are the programs that control the overall operation of the network. The major software protocols that we will discuss are the network operating system (NOS) and client software.

*Hardware-software* protocols control how packets are assembled. These protocols also determine how the computers on the network are logically addressed. Because these protocols determine more than one aspect of the network, protocols that fall within this category are called *suites,* or *stacks.* The protocol stacks that we will discuss are NetBEUI, IPX/SPX, and TCP/IP. Of course, these are acronyms whose meaning will be given in the section on each suite.

*Hardware* protocols actually get the information on and off the network. This information includes the electrical characteristics, the way the network is physically connected together, and the method used to detect when another computer is using the network. Hardware protocols are further divided into two categories: the network topology and the network interface card (NIC). The topologies we will discuss are the bus, star, and ring. The NICs that we will discuss are Ethernet, token ring, and FDDI.

**Figure 14.5** Typical Bus Topology

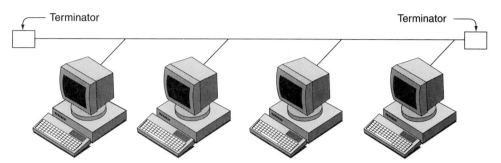

## 14.3 NETWORK TOPOLOGY

A network's **topology** is the way that computers and devices are physically connected together on the network. Three basic network topologies are discussed in this chapter: bus, star, and ring. Each of these topologies has unique characteristics, such as the amount of cabling, expansion potential, message size, traffic volume, cost, bandwidth, reliability, and simplicity. A small network will probably use a single topology, whereas an advanced LAN or WAN will probably use combinations of these topologies to provide the best overall performance.

Each device that is capable of communicating with other devices on the network is called a **node.** Some examples of network nodes are workstations, servers, hubs, bridges, routers, and gateways. A **workstation** is a PC that is connected to the network. Individuals use the workstations to access and use the resources and services offered by the network. A *server* is a dedicated computer that provides services and resources to the workstations. All of the other aforementioned nodes will be discussed in greater detail throughout this chapter.

To select the different nodes on the network, all nodes (expect hubs and repeaters) are addressable units. This is a network address that should not be confused with the address provided by the microprocessor's address bus. The methods used to determine the network address vary among networks.

### 14.3.1 The Bus

The **bus topology** utilizes a coaxial cable that runs the entire length of the network to which all of the devices are attached. Figure 14.5 illustrates the bus topology. Control of the coaxial cable does not have to be centralized to a hub, as is the case with a star topology. The coaxial cable itself is passive, so the network adapter in each computer has to provide enough drive for the signal to travel the entire length of the cable. Once the signal hits the end of the cable, its power has to be dissipated by a resistor called a **terminator** to prevent the signal from reflecting back into the cable and corrupting the data.

The biggest advantage of the bus topology is that it is relatively easy to set up and it usually costs less than other topologies. The biggest disadvantage is that the cable, terminator, and any network interface card can cause the network to fail. When the failure of any portion of the network causes the entire network to fail, it is referred to as a **critical resource.** Another disadvantage with the bus topology is that it is the most difficult topology to troubleshoot when it fails.

### 14.3.2 The Star

The predominant feature of the **star topology** is that each device is connected to a *central device,* similar to the points of a star, as illustrated in Figure 14.6. The central device, called a **hub,** controls the communication within the LAN. All communication among the outside computers must pass through the hub.

The only disadvantage of the star topology is that it is more expensive due to the cost of the hub and the additional cabling required. Although more expensive than a bus topology, the star topology is rapidly becoming the preferred topology among PC networks. The main reason for its popularity is its ease of troubleshooting. The only critical resource is the hub, and every device is connected to a central location. By simply disconnecting nodes from the hub, most network hardware problems can be isolated easily.

**Figure 14.6** Star Topology

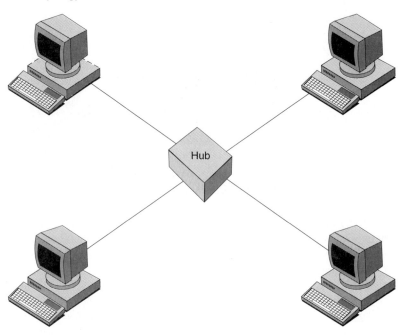

**Figure 14.7** Ring Topology

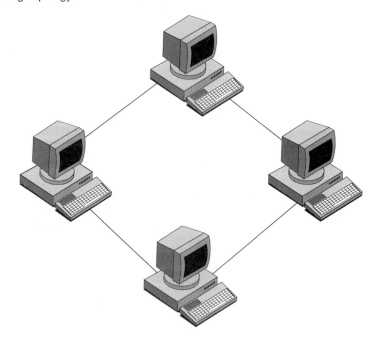

### 14.3.3 The Ring

The **ring topology,** as its name implies, interconnects nodes in a closed-loop configuration, as shown in Figure 14.7. A message is transmitted in one direction, going from computer to computer around the ring. Eventually the node that originally transmitted the message will receive the message.

Each computer acts as a repeater, retransmitting the message to the next computer in the loop. Each computer also shares the responsibility of identifying if the circulating message is addressed to itself or to another computer. In either case, the message is eventually received by the destination computer and

**Figure 14.8** Example of Network Using All Three Topologies

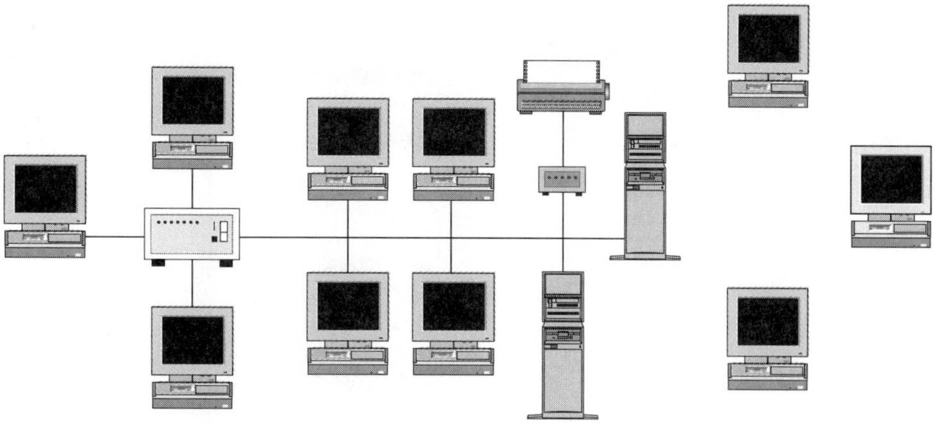

returned to the original source computer. The source computer then verifies that the message circulated around the ring is identical to the message that was originally transmitted.

The destination computer typically sets acknowledgment bits within the message header so that the source computer has a way of verifying the message was in fact received. The critical resource in a ring topology is the interconnecting cable links and each networked computer. A networked computer that malfunctions is typically replaced or bypassed.

In the ring topology, the technique most often used to prevent more than one computer from using the network is called *token passing,* or **token ring.** A *token* is a special bit pattern or packet that circulates around the ring from computer to computer when the network is not being used. Any computer that wants to use the network takes possession of the token and holds it when the token comes up. The computer then transmits its packet onto the ring. The data packet circulates to its destination computer on the ring and returns to the computer that originally transmitted the data. The data are then verified to determine if they are the same as originally transmitted and that they were received by the destination computer. The token is then passed forward to the next node. This technique basically eliminates two nodes trying to use the network simultaneously.

### 14.3.4 Combining Topologies

LANs do not have to use one topology. In fact, large LANs usually consist of one or more topologies. Figure 14.8 shows a LAN that uses all three topologies discussed in this chapter. To connect different topologies together on the same network usually requires special devices called *bridges, routers,* and *gateways.*

### 14.3.5 Repeaters, Bridges, Routers, and Gateways

All of the mediums used in networking exhibit attenuation. **Attenuation** means the strength of the signal that is carrying the data decreases over distance. To increase the distance of a network past the limitation of the medium, a device called a **repeater** is used. A repeater is simply a device that receives, amplifies, and then retransmits the information on the network. There are two types of repeaters available. One simply amplifies and then retransmits all of the information it receives, including the noise. The other, called an *intelligent repeater,* removes the electrical noise before the signal is retransmitted. Of course, the intelligent repeater costs more.

We now know that a LAN does not have to use one topology. In fact, most large LANs that have developed over the years usually consist of more than one topology. The general philosophy is why tear something apart that is still working just to go to another topology. On the other hand, why use an older, more expensive topology when it comes time to add more computers to the network? A device called a **bridge** is used to connect a LAN that uses different topologies but similar data packets.

As the size of the LAN becomes larger and larger and more computers are connected to the network, another problem arises. More computers on the network means that more network traffic is generated.

**Figure 14.9** OSI Networking Model

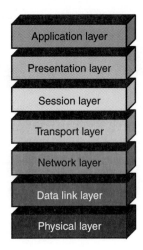

The more network traffic, the slower the network becomes. To help solve this problem, large networks are usually divided into individual addressable segments with a device called a **router.**

A router allows multiple LANs to be connected together in manageable segments called *subnets*. All of the computers on each subnet must share transfer time with other computers on the same subnet, but do share the same media with other computers on other subnets. One subnet's information is not transferrable to any of the other subnets on the LAN unless information is specifically addressed to another computer on another subnet. This technique of splitting a network into more manageable subnets speeds up the large LANs considerably because all of the computers on the network are not sharing the same media.

Last, but not least, there needs to be a device that translates hardware and software protocols among completely different networks that are attached together on a WAN, such as the Internet. This is a function of a device called a **gateway.** Some people also refer to the computer that makes the connection to the Internet as a gateway.

## 14.4 NETWORK STANDARDS

When microcomputer networks started to take hold, there were several hardware and software manufacturers doing their own thing. Unfortunately, this led to incompatibility among different vendors' network hardware and software. To provide some form of standardization among network vendors, the **International Standards Organization (ISO)** developed the **Open System Interconnection (OSI)** standard.

The OSI standard provides specifications that network hardware and software manufacturers can use to make their hardware or software compatible with that of other vendors. If the network's hardware and software are developed using the ISO-OSI standard, the network designer can use different vendors' hardware and software to develop the network. For this reason, the ISO-OSI standard is the most widely used in the microcomputer networking community. Figure 14.9 shows the ISO-OSI network model used in microcomputer networks.

The first four layers from the bottom of the OSI model are hardware levels that define the actual circuits that are used to get data on and off the network. The upper three layers are software layers that define the programs needed to run the network. A brief description of the responsibility of each layer follows.

>   **Physical layer:** The physical layer is the bottom layer of the ISO model. This layer defines the mechanical and electrical rules that determine how the data are transmitted and received. This layer defines the minimum and maximum voltage levels, current levels, and impedance levels of the network. The physical layer also contains the circuitry that detects network traffic and determines the topologies available.

**Data link layer:** This layer defines the mechanism by which data packets are transmitted among stations to achieve error-free transfer. This is accomplished by adding error-checking information to the packets sent to it by the network layer and checking for errors in packets coming from the physical layer.

**Network layer:** This network layer determines logical addressing and routing by matching a logical network address to a physical network address. If used, network routers operate within this layer of the OSI model. The network layer also reassembles packets from the transport layer into a smaller size if they are too large for the data link layer to handle.

**Transport layer:** This layer ensures the reliability and efficiency of the end-to-end movement of data within a network. Layers above the transport layer are no longer concerned with the hardware aspect of the network and deal only with the software aspect. The tasks performed by the transport layer are error detection and recovery, multiplexing (time-sharing) of information onto and off of the network, and assembling and unassembling data packets.

**Session layer:** This layer actually starts and ends the session with another computer on the network. It does this by verifying user passwords, indicating or detecting the first packet that belongs to the block of information to be transferred, determining when the last packet is sent or received, and handling the recovery from transfer interruptions or crashes.

**Presentation layer:** This layer takes care of network security, file transfer, and formatting functions. It also handles protocol conversion among different computers using different data transfer formats such as ASCII or EBCDIC, data encryption, data compression, and so on. This layer also determines if the required resource is a network or local resource. If the application layer requests a local resource, the information is not passed to the session layer.

**Application layer.** This layer is where the specific application program that performs the end-user task is defined, including database management, programs, word processing, spreadsheets, electronic mail (e-mail), and so forth. This layer initiates the request to use either a network or a local computer resource.

## 14.4.1 IEEE

Using the OSI standard as a foundation, several committees of the Institute of Electrical/Electronic Engineers (IEEE) developed a set of standards for LAN topologies and access methods. The most popular standard used in microcomputers was developed from Xerox's Ethernet LAN. IEEE used Xerox's Ethernet to establish the IEEE 802.3 standard, which is based on the first four layers of the ISO-OSI network model. The Ethernet standard is the predominant LAN used with microcomputers today.

## 14.4.2 Ethernet

The network specification that has almost become synonymous with the microcomputer LAN scene over the years is **Ethernet.** Xerox Corporation originally developed Ethernet in the late 1970s. It was upgraded in 1980 by the combined efforts of Xerox Corporation, Digital Equipment Corporation (DEC), and Intel Corporation. Ethernet's popularity stems from its simplicity. It is simple to configure, is changeable, and is readily expandable. The Ethernet standard defines the type of cable used, the maximum length of a signal cable without a repeater, the format used for the data packet, and how data collisions are avoided.

## 14.4.3 CSMA/CD

To prevent data collisions, which would exist if more than one adapter tries to place information on the network at the same time, Ethernet uses a method called **carrier sense multiple access with collision detection (CSMA/CD).** This simply means that before an Ethernet adapter transmits data, it listens for current activity on the network. If no other Ethernet adapter is transmitting data, the listening adapter begins transmitting to another Ethernet adapter connected to the network. All Ethernet adapters that are connected to the network receive that transmission, typically a data packet. However, each Ethernet adapter is assigned a unique *network address* and the transmitted data packet contains the network address of the Ethernet adapter that is to receive the data. All of the Ethernet adapters that have a different network address than that specified by the data packet will simply discard the data.

**Figure 14.10** Ethernet Data Packet

**Figure 14.11** Illustration of the NIC Connection to the Vampire Tap

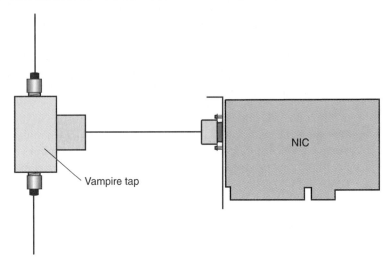

Due to the length of a network, two adapters still might try to place data on the network at the same time, which will cause a collision. To detect collisions, each Ethernet adapter listens to the information it is trying to place on the network. If the data become corrupted and a collision is detected, each of the cards trying to place information on the network will automatically stop transmitting. Each card waits a random length of time before attempting to send its information again. This technique prevents the two cards from trying to retransmit at the same time.

### 14.4.4 Ethernet Data Packet

Figure 14.10 illustrates a typical Ethernet data packet. The data packet starts with a **preamble** of 8 bytes that is used for packet synchronization. The **destination address** is the address of the workstation that the information is being sent to. The **source address** is the address of the workstation where the data packet originated. The **type field** is a code used to indicate the format of the data (ASCII, EBCDIC, and so on). The **data field** can hold from 46 bytes to 1,500 bytes of data. The **frame check field** ensures that the data in the other fields arrived correctly.

### 14.4.5 ThickNet (10Base5)

Ethernet was first implemented in a bus topology using 50-ohm, RG-8 coaxial cable. As previously stated, RG-8 coax cable has a thickness about the size of a human thumb. At points that run in multiples of 2.5 meters, metal collars, called *vampire taps,* are attached to the cable to make the electrical connection to the coax cable's electrical conductors. Inside the vampire tap is a unit called a **transceiver** that contains the electronic circuits responsible for getting electrical information on and off the network.

Attached to the transceiver is a cable, referred to as an *AUI (attachment unit interface),* which supplies the signal to the Ethernet network interface card (NIC) installed in one of the computer's expansion slots. The cable is attached to the Ethernet NIC and the vampire tap using a set of DB15 connectors. Figure 14.11 illustrates this connection between the NIC and the vampire tap. The DB15S connector on the metal plate of a typical NIC card is shown in Figure 14.12.

This arrangement is commonly known as **ThickNet,** or **10Base5.** The name *ThickNet* is used because of the large diameter of the coaxial cable. The name *10Base5* is used because ThickNet operates

**Figure 14.12** Ethernet Adapter Showing 10Base5 DB15 Connector

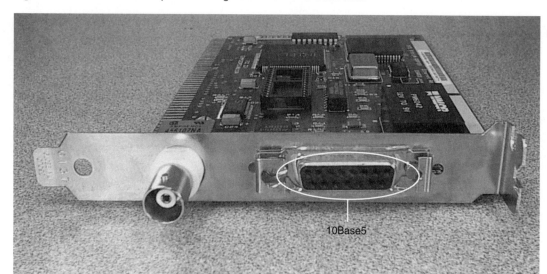

**Table 14.3** 10Base5 Specifications

| | |
|---|---|
| Maximum length of a segment without a repeater | 500 meters (1,650 feet) |
| Maximum total length of network | 2,500 meters (8,250 feet) |
| Maximum number of segments | 5 |
| Number of repeaters | 4 |
| Maximum number of segments with nodes | 3 |
| Maximum nodes per segment | 30 |
| Maximum length of AUI cable | 50 meters (165 feet) |

at 10 Mbps and can be run in segments of up to 500 meters (1,650 feet) in length. To prevent signal reflection, 50-ohm terminators are required on each end of the cable. To increase the length of a 10Base5 network beyond the 500-meter limit, devices called *repeaters* can be used. The agreed-upon specifications used by ThickNet are listed in Table 14.3.

The major problem with ThickNet is that the coax cable used to connect workstations on the network is expensive and fairly large in diameter. The size of the cable makes it difficult to route through wiring conduits in existing buildings.

Other Ethernet standards have been developed that reduce the wire size to make cable routing easier and less expensive. Three other popular standards are thin Ethernet (ThinNet), 10BaseT, and 100BaseT.

### 14.4.6 ThinNet (10Base2)

**ThinNet 10Base2** is implemented in a bus topology but uses a smaller 50-ohm coax cable referred to as *RG-58*. Unlike ThickNet, the transceiver and cable tap are part of the ThinNet NIC card and are not a separate unit. The RG-58 cable is about half the diameter of the RG-8 cable used with ThickNet and looks similar to the coax used in cable television connections.

As stated, ThinNet is connected in a bus topology. Unlike ThickNet, a single coax conductor does not run the entire length of the network segment. The cable is cut in minimum lengths of 2.5 meters to

**Figure 14.13** ThinNet NIC BNC Connector

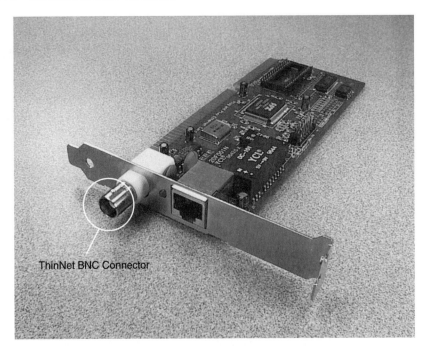

**Figure 14.14** BNC "T" and Terminator

help avoid signal degradation, which could cause the network to fail. A special connector called a *BNC connector,* shown in Figure 14.13, is attached to each end of the short cable that connects the computers together on the network. Another special connector called a *BNC-T,* shown in Figure 14.14, is used to connect to the adapter card. Two cables are connected to each side of the BNC-T. Each cable connected to the BNC-T will connect to a BNC-T on the next computer on the network, which connects to the next, which connects to the next, and so on. A device called a *terminator,* shown in Figure 14.14, must be installed on the very last computer's BNC-T on each end of the network. The connection of a workstation using ThinNet with the bus topology is illustrated in Figure 14.15.

ThinNet is also called *10Base2*. It gets this name because it operates at 10 Mbsp. 10Base2 can be run in segments of up to 185 meters (610.5 feet) without a repeater. Table 14.4 lists other specifications for 10Base2.

## 14.4.7 10BaseT

Another popular method of implementing Ethernet is called **10BaseT**. The *10* in this name indicates a 10-Mbps transmission rate and the *T* indicates twisted-pair cable is used instead of coax. Even

**Figure 14.15** Back-plane Showing ThinNet Connection

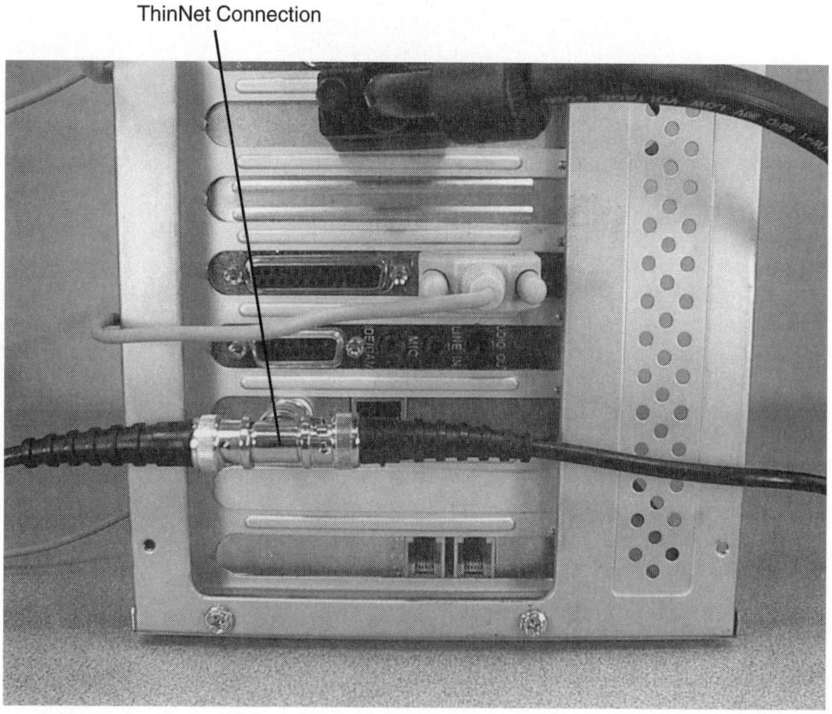

**Table 14.4** 10Base2 Specifications

| Maximum length of a segment without a repeater | 185 meters (610.5 feet) |
| --- | --- |
| Maximum length of network | 925 meters (3053 feet) |
| Maximum number of segments | 5 |
| Number of repeaters | 4 |
| Number of segments with nodes | 3 |
| Maximum nodes per segment | 30 |

though CAT-3 cable is designed for 10BaseT, CAT-5 cable can also be used. 10BaseT can also use UTP or STP cable.

10BaseT is implemented in a star topology with each length of twisted-pair cable terminated with a male RJ-45 connector. The twisted-pair cable used with 10BaseT has to be at least category 3 (CAT-3) cable that is rated at 10 Mbps. The twisted-pair cable can be either unshielded twisted pair (UTP) or shielded twisted pair (STP). The STP cable has better EMI characteristics but is more expensive. Figure 14.16 shows one end of a 10BaseT CAT 3 UTP cable showing the RJ-45 connector.

The operation of the 10BaseT Ethernet network is basically the same as that of the standard Ethernet discussed previously. 10BaseT used CSMA/CD to control network traffic. 10BaseT can be implemented in either a star or star-bus topology. The star topology utilizes a single hub, whereas the star-bus connects multiple hubs together. A typical star topology hub with several RJ-45 (10BaseT) cables attached is shown in Figure 14.17.

The hubs used by 10BaseT can be either active or passive. The passive hubs are cheap and greatly limit the overall length of the network. The active hub also serves as a repeater, which helps extend the overall length of the network. Even though an active hub extends the length, there are still standards

**Figure 14.16** RJ-45 Connector

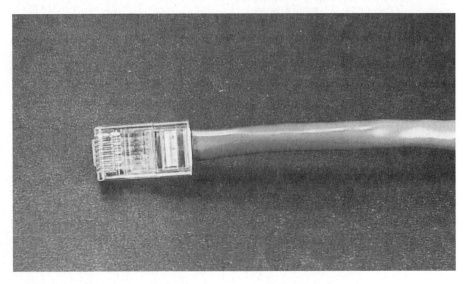

**Figure 14.17** Typical 10BaseT Hub

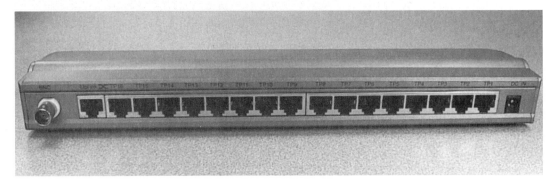

**Table 14.5** 10BaseT Maximum Specifications

| | |
|---|---|
| Maximum number of segments | 1,024 |
| Length of a segment | 100 meters (330 feet) |
| Number of hubs connected together | 4 |

that have to be followed to ensure that a 10BaseT network operates reliably over its entire operating range. Table 14.5 lists the accepted segment, hub, and distance specifications.

### 14.4.8 10BaseFL

10BaseFL is an implementation of the Ethernet standard using fiber optic cable. Although the transmission speed is the same as that of 10BaseT, 10BaseFL is often used when the network needs to run in an environment that is electrically noisy, such as outdoors where the cable may be struck by lightning. Another advantage of 10BaseFL is the fiber's low attenuation, which makes it possible to traverse cables over long distances without a repeater. Table 14.6 lists the maximum specifications for 10BaseFL.

Table 14.6  10BaseFL Specifications

| Number of segments | 1,024 |
|---|---|
| Maximum length per segment | 2,000 meters (6600 feet) |
| Maximum number of hubs connected together | 4 |

### 14.4.9 Fast Ethernet (100BaseT)

The latest Ethernet standard that is rapidly become the network standard of choice for new installations is **100BaseT**. As the name implies, 100BaseT operates at a speed of 100 Mbps and is implemented in a star or star-bus topology using multiple hubs connected together. To set up a network using 100BaseT, all of the network adapters and hubs must be rated at 100 Mbps. The twisted-pair cable must be category 5 (CAT5) cable that is rated at 100 Mbps. The distance and number of segment specifications for 100BaseT are the same as those for 10BaseT.

## 14.5 OTHER NETWORK HARDWARE PROTOCOLS

Along with Ethernet, there are many other network hardware protocols used in LANs and WANs. Of these, the most popular are token ring, FDDI, and ATM. These will be discussed briefly in the following sections.

### 14.5.1 Token Ring

Token ring is an IEEE 802.5 network hardware protocol standard that is supported mainly by IBM. This topology uses a data packet called a *token* that passes around the ring when no data are being transferred. When a computer wishes to place a packet of information on the ring, it waits for the token. When the token is received, the computer holds the token so that no other computer can place information on the network.

The packets used by token ring contain the logical network address of the destination computer along with the logical network address of the originating computer, which is holding the token. The computer holding the token places the packet it wishes to transmit on the ring. The next computer on the network, called the *nearest downstream neighbor* (NDN), receives the packet. The NDN checks the addressing information in the packet to determine if it is the destination computer. If not, the computer passes the information to its NDN. This process continues until the computer to which the information was addressed receives the packet. Once this happens, the computer sets bits in the packet header, indicating that the packet has been received. The destination computer then passes the packet to its NDN. The data will eventually circulate back around to the originating computer. Once received, the computer checks the packet to see if it has been received and also checks to see if the packet contains the same data that were transmitted. Token ring provides a very high degree of data integrity and totally eliminates data collisions.

Contrary to its name, token ring is actually physically wired as a star topology but is electrically wired as a ring. This technique is used because the star topology is the easiest to troubleshoot when the network goes down. Token ring uses hubs that are called *Multistation Access Units (MSAUs)*. Because each token ring adapter is a repeater, the MSAU itself is a passive device that simply acts as a central connection point to connect the network together. Even though the topology is star, token ring is electrically connected as a ring through the MSAU, as shown in Figure 14.18.

### 14.5.2 Fiber Distributed Data Interface (FDDI)

**Fiber Distributed Data Interface (FDDI)** is a network hardware protocol that uses a dual ring topology and fiber optics cable as its medium. The two rings are referred to as the *primary* and *secondary* rings. The primary ring is used to transfer data, whereas the secondary ring takes over the data transfer if the primary ring fails. This dual-ring feature adds a high degree of fault tolerance to FDDI.

**Figure 14.18** Connecting Token Ring in a Physical Star and Electrical Ring with a MSAU

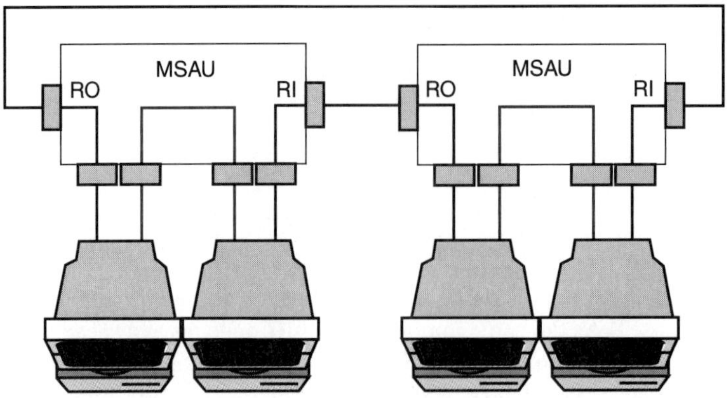

**Table 14.7** ATM Speeds and Media

| 25 Mbps | CAT-5 Twisted Pair |
|---|---|
| 1.544 Mbps | T1 Carrier |
| 44.736 Mbps | T3 Carrier |
| 51.84 Mbps | OC1 SONET |
| 155.52 Mbps | OC2 SONET |
| 622.08 Mbps | OC12 SONET |
| 2.4 Gbps | OC48 SONET |

FDDI uses a token passing scheme similar to that of token ring, however, there is one big difference. FDDI does not wait for the data packet to circulate back around to the transmitting computer before the computer releases the token to its NDN. This feature allows multiple packets to circulate around the ring at any one time, enhancing the network's overall transfer speed by several magnitudes.

### 14.5.3 Asynchronous Transfer Mode (ATM)

The fastest and most expensive network hardware protocol is **Asynchronous Transfer Mode (ATM).** ATM works similar to a telephone system. The hubs used in an ATM network operate like the electronic switches that form a virtual circuit between two telephones. The header of an ATM packet does not contain the logical network address of the destination computer. Instead, it contains a binary number that tells the ATM routers how to create a path to the destination computer. This is similar to dialing someone's number on the telephone. The virtual circuit allows the two computers to exchange information without sharing the media with any other computer. This feature greatly enhances the speed of transmission. Current ATM speeds and media are listed in Table 14.7.

## 14.6 THE NETWORK INTERFACE CARD (NIC)

The **Network Interface Card (NIC)** installs in one of the computer's expansion slots on most compatible computers. On computers that are sold as workstations or servers, the NIC may be integrated into the motherboard. The NIC occupies the physical and data link layers of the OSI model and performs most of the work of getting data on and off the network in the correct form. The adapter card assembles and

**Figure 14.19** Typical Ethernet NIC

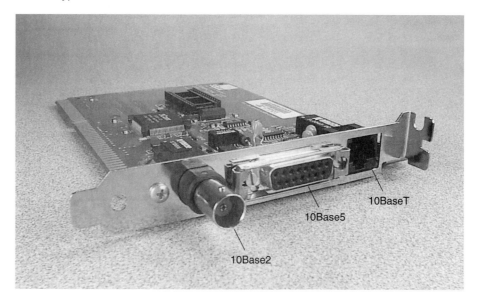

---

### Procedure 14.1: Displaying the resources of the NIC using Windows 9X

1. Click on the Start button.
2. Move up to Settings and then over to Control Panel.
3. Click on Control Panel.
4. Locate and then double-click on the System icon in the Control Panel window.
5. Click on the Device Manager tab.
6. Click on the plus (+) sign beside Network Adapter.
7. Double-click on the network adapter.
8. Click on the Resources tab.
9. When finished, close all open windows.

---

disassembles data packets, sends out source and destination network addresses, detects transmission errors, and prevents transmission while other units on the network are transmitting. The typical Ethernet NIC shown in Figure 14.19 has connectors for ThinNet, ThickNet, and 10BaseT cables.

### 14.6.1 Adapter Card Resources

Each NIC installed in the computer—there can be more than one—requires the use of a range of I/O addresses and a single interrupt. Because the NIC is not part of the original AT-I/O addressing map, the actual I/O address assignments and IRQ number vary from machine to machine.

The NIC's resources can be hardset, softset, or PnP. A 10-Mbps NIC can be purchased in either an ISA or PCI design. The advantages of using a 10-Mbps PCI NIC include the PnP features of PCI, the ability to stack IRQs, and the availability of more I/O addresses. The 100-Mbps Ethernet NICs are available only in PCI.

When you use hardset or softset NIC, make sure the resources you assign the card are not already in use by another device. The manufacturer's default settings for NIC cards usually use the same IRQ as the default settings of most sound cards. This means one of these cards must have the IRQ reassigned if both cards are installed in the same computer. A PnP card's resources are configured by the Windows 9X/2000 operating system, so you should not have to worry about resource conflicts. To check a network adapter's resources on a computer running Windows 9X, follow the steps outlined in Procedure 14.1.

> **Procedure 14.2: Using WINIPCFG to display a network adapter's MAC address***
>
> 1. Click on the Start button.
> 2. Click on Run.
> 3. In the Run window text box, type WINIPCFG.
> 4. The DLC address is displayed in hexadecimal to the right of the heading "Adapter Address."
> 5. Close all open windows by clicking on the X in each window's upper right-hand corner.
>
> *Microsoft's TCP/IP protocol suite must be installed on the computer.

### 14.6.2 The Network Address

Just as all of the devices in a computer are assigned unique binary addresses so the microprocessor can select them individually with its address bus, all NICs attached to the network must have a unique binary network address so they can be identified on the network. This unique network address consists of two parts: a physical and a logical address.

The logical network address is a unique identifier in the form of a name or number that is assigned to each node by the user. This is usually done during the software configuration of the network and will be discussed later.

The physical network address is assigned to each NIC when it is manufactured and usually cannot be changed by the end user. On networks that conform to the OSI model, the physical network address is called the **data link control (DLC)** address. On Ethernet networks, the DLC address is also called the **medium access control (MAC)** address. No matter which name is used, every NIC on the network must have a unique physical network address. To avoid confusion, I will use the DLC address for the remainder of this discussion.

The DLC address is a 48-bit (6 byte) binary number that provides a total of $281,474,976,710,656_{10}$ DLC address combinations ($2^{48}$). Of the 48-bit binary DLC address, the most significant 24 bits (3 bytes) are called the **organizational unique identifier (OUI).** IEEE assigns an OUI to each NIC manufacturer that makes OSI-compliant NICs. Once the OUI is assigned, the manufacturer uses the remaining 24 bits to identify each NIC it produces. The combination of the 24-bit OUI and 24-bit NIC ID gives the manufacturer the ability to produce $16,777,216_{10}$ NICs with unique DLC addresses within each assigned OUI. Once 90% of the addresses are used, the manufacturer can request an additional OUI from IEEE.

Using the combination of the OUI and NIC-ID means that all of the NIC cards produced in the world that follow the OSI standard have a unique DLC or MAC address. This fact allows NIC cards to coexist with one another on the worldwide network of computers called the Internet.

Some network adapters provide a utility on the diskette that displays the adapter's DLC address. If the TCP/IP protocol stack is installed in Windows 9X, a utility called *WINIPCFG* is provided that will display the DLC address, along with some other important information that will be discussed in later sections. The procedure to install TCP/IP is also discussed in section 14.8.3. Procedure 14.2 lists the steps to display the DLC address using WINIPCFG.

## 14.7 THE NETWORK OPERATING SYSTEM

Having all of the computer hardware in the world is useless unless there is software to make it operate. This holds true for networks as well. There are two main categories of network operating systems (NOS): *peer-to-peer* and *client/server.*

### 14.7.1 Peer-to-Peer

A **peer-to-peer** network operating system, such as Windows 9X, is one in which all of the computers on the network are clients and all have the ability to perform the functions of a server. In Windows 9X, the client software and most network components are installed automatically with the NIC's driver. However, they can be added, removed, and configured manually by the user at anytime. These procedures will be discussed in later sections. Once the network components have been added, *File and Print Sharing* must be installed to enable the server capabilities of Windows 9X/2000.

Without one centralized computer to validate users and control the access to network resources, peer networks most often use *shared-level access*. When this technique of access control is used, the individual users of each computer control the shareable resources of that computer through the use of passwords. A unique password is usually assigned for read-only access and another unique password is assigned for full access. To use the shared resource at one of the mentioned access levels, a network user must know the correct password. The procedures for setting up, sharing, and assigning passwords on a peer network using Windows 9X are covered in section 14.9.

One computer on a peer-to-peer network may be chosen to hold a majority of the network's data files. This computer usually has a larger hard drive storage capacity than the other computers on the network and is called a *file server*. The file server on a peer-to-peer network can also be used as a network workstation.

The advantages of a peer-to-peer network operating system are that it is easy to set up, all servers can also be used as a workstation, an expensive computer does not have to be dedicated to handle the server's responsibilities, and each computer user controls access to his or her computer's resources.

The disadvantages are that the number of passwords can become unmanageable and the overall security of the network is low. These disadvantages make networks that use a peer-to-peer network operating system practical only on networks with 10 or less computers.

## 14.7.2 The Client/Server NOS

A **client/server** network operating system (NOS) employs dedicated computers called **network servers** that control the overall operation of the network. The network servers provide centralized access security; centralized file storage of applications and data; shared resources such as printers, scanners, and fax/modems; and messaging services such as e-mail, video conferencing, and chat. Unlike the peer-to-peer system, these specialized computers are not usually used as workstations.

A single server can provide all of the services available on the network or can spread them out to individual dedicated servers, depending on the size of the network. Dedicated servers are usually named according to the network service they provide, such as file server, print server, e-mail server, and application server.

Depending on the network services provided, servers usually have the most sophisticated CPU(s), the most memory, the largest disk storage, and more. The features that are offered by the network server usually make it the most expensive computer on the network. Each computer that is used as a network server runs the network operating system. This computer can be an upper-end microcomputer, a minicomputer, or even a mainframe.

All of the other computers connected to the network are referred to as **clients** or *workstations*. These computers will run client software that will allow them to access the resources offered by the servers as if they were their own. To control access, users are assigned to a domain. A **domain** is the grouping of users that share the same network login security. When users log into the domain, they are given permissions that control their level of access to the network server's disk drives, printer(s), fax/modems, scanners, plotters, and other peripherals that are shared over the network.

The actual level of access is determined by the network server's operating system and an individual who is usually called the *network administrator*. The workstations themselves usually cost much less than the server because they do not need all of the expensive peripherals attached. The computer that is used as a network's server is usually not, and in most instances cannot, be used as a workstation.

Two client/server network operating systems currently dominate the compatible computer LAN community: Novell's Netware and Microsoft's Windows NT Server.

Novell's *Netware* has become one of the best-known server-based network operating systems in the microcomputer networking community. Novell offers several network starter kits that can be configured to run a star, ring, bus, or combination topology network. Several hardware protocols can be used to send information over the network, but the most widely used is Ethernet. Netware currently uses Novell's proprietary IPX/SPX protocol suite, but may adopt TCP/IP as the default suite for Netware 5.0.

The Netware network operating system (NOS) runs on a dedicated server. This means that the computer that is assigned the server responsibilities cannot be used as a workstation. Netware also uses a text shell similar to MS-DOS's COMMAND.COM. These two features make the hardware required by the server less expensive than that required by other NOSs such as Windows NT Server and Windows 2000 Server.

Microsoft's Windows NT/2000 Server NOS is rapidly gaining wide acceptance in the networking community, mainly because it is fairly easy to install and set up. The main feature of Windows NT/2000 Server is that it uses a GUI shell identical to the one used by Windows 9X. Windows NT/2000 Server NOS should not be confused with Windows 9X because NT/2000 is a completely different operating system. Windows NT/2000 is a full-blown, 32-bit operating system that uses preemptive multitasking. Unlike a Novell Netware server, a computer running Windows NT/2000 Server can also be used as a workstation, similar to a peer-to-peer server. However, this is not recommended because it will degrade the server performance.

Because Windows NT/2000 Server NOS uses a GUI and has the ability to run applications, it requires lots of memory, a color monitor, and a graphics adapter. Unlike most operating systems, Windows NT/2000 Server can be installed on computers that run the DEC-Alpha RISC microprocessor. Another big advantage of Windows NT/2000 Server over Netware is its ability to run up to four microprocessors in a symmetric multiprocessing (SMP) arrangement.

### 14.7.3 Combining Network Operating Systems

Most large networks combine both the peer-to-peer and client/server network operating systems to get the advantages offered by both. Most large networks may even use multiple servers that run different client/server operating systems. At the technical college where I teach, the workstations run either Windows 9X or Windows NT Workstation, which provides the advantages of a peer-to-peer network. All of the workstations at the college log on to either a Windows NT server or a NetWare server, or both. These servers validate user names and provide the other advantages offered by a client/server network. To use multiple network operating systems on the same network requires the installation and correct configuration of client software on each workstation.

### 14.7.4 Client Software

A client is a computer that uses the services provided by a server. Unlike a server, the client computer's operating system does not make the computer a client. A computer becomes a client when client software that is compatible with the operating system already running on the computer and the NOS running on the servers is installed. Once installed, the client software captures calls for the operating system to use either local or network resources. The client software decides if the requested resource is a local resource or a network resource and redirects the call to the correct location.

The two main client software packages supported by Windows 9X are Microsoft client and Netware client. Microsoft's technical documents often refer to its client software as the *redirector;* Novell often calls its client the *requester.* Both clients are automatically installed on the computer when an NIC is installed. Client software can be removed and added, or the properties can be set using the steps listed in Procedure 14.3.

### 14.7.5 Network Security

The major problem with the peer-to-peer network operating system is trying to maintain network security. Each computer on a peer-to-peer operated network shares the responsibility of the server, so security levels must be established at each workstation. Because many individuals may use the same workstations, network security becomes a nightmare for a large LAN environment. A peer-to-peer network operating system is ideal for small LAN applications, however.

On a client/server-based network, access to the network and its shared resources is controlled by a server. Before the user can even attempt to access the network, he or she must have an account on the login server that contains his or her username and password. The login server on a Windows NT/2000 network is called the *domain server* and the login server on a Netware network is called the *preferred server.* To log on to a network, the name of the login server must be entered into the Properties settings for the correct client software.

When users attempt to log on to the network, the login server validates their username and password. Once the user has been validated, information in the user account determines which resource he or she

Computer Network Basics 473

---

### Procedure 14.3: Removing, adding, or changing the properties of the client software

1. Click on the Start button.
2. Move up to Settings and then over to Control Panel.
3. Click on Control Panel.
4. Locate and then double-click on the Network icon.
5. The Network options window will open.

**Removing a client:**
1. Click on the client to remove.
2. Click on the Remove button.
3. Follow the displayed instructions.

**Adding a client:**
1. Click on the Add button.
2. Click on the Client option and then click on the Add button.
3. After the Select Client options window opens, click on Microsoft in the left windowpane.
4. In the right windowpane, select either Client for Microsoft networks or Client for Netware networks by clicking on the desired option.
5. Click on the OK button.
6. Follow the displayed instructions.

**Changing a client's properties:**
1. Click on the name of the client in the top windowpane of the Network window and then click on the Properties button.
2. The only option for Microsoft client is to make the computer log on to a Windows NT Server. When this option is selected, enter the name of the domain that contains the user accounts in the appropriate window. Obtain this information from the network administrator.
3. The options available with Network client vary a great deal from version to version. Because this is the case, consult the network administrator for the correct setting in the Netware Properties window.
4. Once completed, click on the OK button for each network window to accept changes. Click on the Cancel button to abort without changes.
5. Restart Windows 9X for the changes to take effect.

---

can access and to what level. The network administrator usually maintains the login server and the user accounts. All of the security is maintained on the network server and not at the individual workstations. This degree of network security makes the network server operating system the most widely used in large LANs.

## 14.8 PROTOCOL STACKS

In order for computers to communicate on the network, they must all use the same protocols. As discussed previously, the OSI model defines several layers that are used to establish communication. To communicate, the protocols that control the different layers of the OSI model must be the same. For this reason, network protocols are grouped together in *stacks,* or *suites.* These protocol stacks are used to control different layers of the OSI model. Three protocol suites are most often used in the networking of compatible computers: NetBEUI, IPX/SPX, and TCP/IP.

### 14.8.1 NetBEUI

**Network BIOS Extended User Interface (NetBEUI)** is the default protocol suite used by a Microsoft peer-to-peer network. NetBEUI identifies computers on the network by a computer name and a MAC address. As each computer logs onto the network, it broadcasts its presence on the network by sending its computer name and MAC. The other computers respond to the broadcast by supplying their computer name and MAC address. Each computer builds a database of computer names and MACs so it will know which computers are available to transfer data.

The biggest advantage of NetBEUI is that it is fast and easy to set up because only two things must be supplied by the user during the network software installation process: a computer name and a workgroup name. A **workgroup** is two or more computers grouped under the same logical workgroup name, which makes it easier to share resources with other computers in the workgroup.

The biggest disadvantage of NetBEUI is that there is no way to identify a subnet address, so NetBEUI is a nonrouteable protocol suite. This means a network using only the NetBEUI protocol suite cannot be divided into segments to decrease network traffic.

### 14.8.2 IPX/SPX

**Internetwork packet exchange/sequence packet exchange (IPX/SPX)** is a proprietary protocol suite used by Novell's Netware. Because IPX/SPX is a proprietary protocol suite, Microsoft developed a totally compatible suite called *NWLink*.

IPX/SPX is a routable protocol suite that consists of two protocols. IPX controls network addressing and routing, whereas SPX makes sure the data arrive at their destination correctly. The IPX address consists of two main parts that determine a computer location on the network. One part is the network address and the other is the MAC. IPX/SPX routers keep up with network addresses and MAC addresses using the *router information protocol (RIP)*. The RIP determines the path that information takes through the network routers to communicate with another computer on the network. The routers also run the *service advertising protocol (SAP)* to keep up with the network address and the MAC address of the different servers, along with the services they offer.

IPX/SPX is a very powerful suite that offers very reliable data transfer, along with being routable. Due to the small packet size, IPX/SPX is also a very fast protocol suite. The only major disadvantage of IPX/SPX is that it is not compatible with the Internet.

### 14.8.3 TCP/IP

**Transmission control protocol/Internet protocol (TCP/IP)** is by far the most used protocol suite because of its use by the global Internet. This is due mainly to the fact that it is routable over very large WANs. Because of its dominance in the LAN community, we will discuss TCP/IP in greater detail.

The TCP portion of the suite is used to control the transport layer of the OSI model. TCP is a *connection-oriented* protocol because it confirms the connection between the sending and receiving computers before any packets are transferred. A *connectionless-oriented* protocol such as NetBEUI and IPX/SPX sends the packets without confirming the connection. TCP also uses the *reliable* method of transfer because the receiving computer acknowledges the error-free receipt of each packet to the sender.

The IP portion of the TCP/IP protocol suite controls the network layer of the OSI model. It defines the logical network address of each computer on the network. Unlike the DLC address that is programmed into the NIC by the manufacturer, the IP is user-definable. In other words, the IP is a logical address that is controlled by software, whereas the DLC address is a physical address that is built into the NIC.

The **IP address** is currently a 32-bit binary number that is used to specify a specific network segment and node on that segment. To make the IP more manageable, the 32-bit number is split into 4 bytes (octets) that are separated by a period. Due to its appearance, this arrangement is often called a *dotted quad,* or *IPv4.*

Because binary is usually too complicated for the average person to comprehend, computers process the binary information to allow the IP address to be entered and displayed into the TCP/IP configuration setup in decimal form. Decimal is used to represent each 8-bit binary octet, so each octet of the dotted quad can have 256 combinations—0 through 255. An example of a typical dotted quad IP address is "175.157.93.158."

To allow routing, the IP address consists of two parts. From the most significant digit moving toward the least is the *network ID,* which defines an individual network or subnet. From the least significant digit moving toward the most is the *node ID,* which defines a unique node on the network or subnet that is defined by the network ID. A binary number, called the *subnet mask,* defines where the network ID and node ID meet. Like the IP address, the subnet mask is entered into the setup for the TCP/IP protocol suite in decimal form.

To help understand the subnet mask, both the IP address and subnet mask must be in binary form. To identify the bits in the IP that define the network ID, 1s are used in the subnet mask. On the other hand, to identify the bits of the IP that define the node ID, 0s are used in the subnet mask. Example 14.1 illustrates the IP address "175.157.93.158" and the subnet mask "255.255.255.0."

Computer Network Basics

**Example 14.1:** How the subnet mask works

Converting the example IP into binary:

175.157.93.158—10101111 . 10011101 . 01011101 . 10011110

Converting the subnet mask into binary:

255.255.255.0—11111111 . 11111111 . 11111111.00000000

Using the subnet mask to define the network ID and node ID:

Network ID = 11001111 . 10011101 . 01011101. 00000000 or 175.157.93.0
Node ID = 00000000 . 00000000 . 00000000 . 10011110 or 0.0.0.158

No matter how large the network, every computer on the network must have a unique IP address. If you are setting up the TCP/IP protocol suite on a private network that will never be attached to the Internet, the assignment of IP addresses can be established locally according to the network administrator's desired scheme. If a computer is going to be attached to the Internet, an IP must be used that does not conflict with any other IP address on any other computer connected to the Internet in the entire world.

This is a good place to mention that the IP addresses provided by IPv4 are rapidly being consumed on the Internet by the large demand of the international community. Due to this fact, there is a move under way to implement a 6 octet (6-byte) IP address scheme called the *IPv6*. This is currently being tested on an experimental basis on a section of the Internet called *6Bone*. For more information of the 6Bone experimental network and IPv6, visit www.6bone.net.

## 14.8.4 Obtaining an Internet IP Address

The United States has been the custodian of the Internet since its conception by the U.S. Department of Defense. The National Science Foundation currently funds the **Internet Assigned Number Authority (IANA),** which is responsible for assigning Internet IP addresses. Since its conception, the Internet has grown from a connection of U.S. educational institutions and government agencies into a conglomerate of organizations, companies, and individuals located throughout the world. To be truly international, the organizations and individuals that rely on the Internet should fund it. For this reason, the United States will stop funding the Internet in the near future and move it toward becoming self-sufficient. At that time it will be managed by an international not-for-profit organization called **Internet Corporation of Assigning Names and Numbers (ICANN)** (www.icann.net).

Even when control and funding of the Internet are transferred to ICANN, the assignment of Internet IP addresses will still be handled by IANA. IP addresses are actually assigned by three regional registries that are located in the Asia-Pacific region (www.apnic.net), the United States (www.arin.net), and Europe (www.ripe.net). These registries assign a block of IP addresses to *Internet service providers (ISPs)* for an annual fee that depends on the block size. Individuals and organizations receive IP addresses from the various ISPs. The block sizes available for ISPs are called class A, class B, and class C.

A *class A* block uses the most significant octet to identify the network ID and the remaining 3 octets to specify the node ID. The use of 3 octets gives a class A block the ability to specify 16,777, 216 node IDs. A class A license is identified by most significant octet values between 0 and 127.

A *class B* block uses the 2 most significant octets to identify the network ID and the remaining 2 octets to specify the node ID. Using 2 octets provides a class B with $2^{16}$ node ID combinations, or 65,536. The most significant numbers of a class B IP address block begin with an octet value of 128 and go though 191.

The *class C* block of IP addresses uses the 3 most significant octets to identify the network ID and the remaining least significant octet to define the node ID. Because an octet can have only 256 combinations, this is the size of class C block. A class C block has a most significant octet value between 192 and 223.

**Problem 14.1:** What is the IP block class for the following IP addresses? 207.157.93.158

**Solution:** The MS octet is between 192 and 223, so this is a class C block.

**Table 14.8** Reserved IP Address Ranges for Private Networks Using the TCP/IP Protocol Suite

| Class A | 10.0.0.0 through 10.255.255.255 |
|---------|---------------------------------|
| Class B | 172.16.0.0 through 172.31.255.255 |
| Class C | 192.168.0.0 through 192.168.255.255 |

---

### Procedure 14.4: Configuring the IP address in Windows 9X manually

1. Determine if the network uses static or dynamic IP addresses. If static IP are used, obtain an ID and subnet mask from the network administrator.
2. Click on the Start button, move up to Settings, and then over to Control Panel.
3. Click on Control Panel.
4. After the Control Panel window opens, locate and then double-click on the Network icon.
5. Click on the TCP/IP protocol that you wish to configure (there may be more than one), then click on the Properties button.
6. Select the "Obtain an IP address automatically" for dynamic or "Specify an IP address" for static IP address assignment.
7. If static IP is selected, enter the computer's assigned IP address (in dotted quad form) in the space provided. Also enter the assigned subnet mask (in dotted quad form) in the space provided.
8. Click on the OK button to accept the changes or Cancel (X) to abort.
9. Restart the computer for the changes to take effect.

---

Within each of these classes of IP addresses is a range of reserved IP addresses. IANA requests that this range be used for private networks that do not intend to connect to the Internet. Using IP addresses within these reserved ranges will prevent IP conflicts with other licensed ISPs if a decision is made later to connect the network to the Internet. This will prevent a conflict with one or more node IDs that mistakenly do not get changed to an IP address within the assigned range. The reserved private network ranges for each class are listed in Table 14.8.

Whether the node's IP address is assigned locally for a private LAN or through an ISP for a node attached to the Internet, it is entered into the node's TCP/IP configuration settings as either static or dynamic. A static IP address is assigned and entered into the TCP/IP setup manually. A static IP remains the same until it is changed manually. The advantage of using a static IP is that it is easier to monitor the activities of nodes on the network. Servers on the Internet must use a static IP address.

If the node running TCP/IP is configured for dynamic IP, a server running the **dynamic host configuration protocol (DHCP)** service automatically assigns the node a dynamic IP address each time the node logs onto the network. Because all of the computers that belong to a LAN are not usually logged on at the same time, dynamic assignments support more computers with fewer IP addresses. On a private LAN, there is no great advantage to using dynamic IP addresses. However, this is the preferred and recommended method that an ISP uses to assign IP addresses for workstations logged onto the Internet.

The default TCP/IP setting for Windows 9X uses dynamic IP addresses. If there is a server running DHCP services on the network, the computer should find the server automatically and obtain an IP address after the computer is restarted during the initial installation of the suite. If a server running DHCP is not present, a message will be displayed indicating that a server running DHCP is not available. In this case, the IP will have to be entered manually. Procedure 14.4 lists the steps to enter an IP address in Windows 9X manually.

The other options available under Windows 9X TCP/IP properties are for networks that are attached to the Internet. These options are not usually required for computers that are on a private network or attached to the Internet through a modem. Entering the IP addresses for DNS servers, Internet gateway, and WINS servers are among these options. DNS servers will be discussed in the next section.

As discussed previously, gateways convert protocols. To connect to the Internet, a gateway is usually required to convert the hardware protocol used by the network into the protocol used by the medium

**Table 14.9** Internet Domain Name Top-level Codes*

| | |
|---|---|
| Commercial | com |
| Educational and research | edu |
| Government | gov |
| Military agency | mil |
| Nonprofit organization | org |
| International government | int |
| Internet Service Provider | net |

*A country code may follow the top-level code. An example of an Internet domain name that uses the country code is "www.aaa.com.uk," where "uk" is the country code for United Kingdom. For a complete list of country codes, visit www.iana.org/cctld.html.

that makes the connection to the Internet, for example, converting from Ethernet to one of the T-carriers. Because a gateway will be a node on the LAN, it will be assigned an IP address, just like any workstation or server. If not provided by the DHCP server, the IP address for the gateway must be entered manually into the configuration settings for the TCP/IP suite.

*Windows Internet Naming Service* (WINS) allows the use of NetBEUI computer names to identify computers on a network using the TCP/IP protocol suite instead of IP addresses. When a computer asks for another computer by name, it references a server running WINS that matches the requested computer name to its IP address. To use WINS, the computer must be logged on to a Windows NT/2000 server that is running WINS and the server's IP address must be entered in the proper configuration settings for the TCP/IP suite.

## 14.8.5 Internet Domain Names and URLs

To help simplify the process of identifying computers on a large network and on the Internet, a device called an **Internet domain name (IDN)** is usually used instead of an IP address. An IDN consists of a hostname, a domain, and a top-level code with each separated by a period. An example of an IDN is "home.microsoft.com." The hostname portion of the IDN is a name assigned to each computer on the TCP/IP network. Hostnames can be duplicated as long as the computers belong to different domains. One common host name is "WWW" that is used by most Web servers on the Internet. The domain portion of the IDN gives the name of the domain to which the computer belongs. The top-level code is a standardized designator that defines the category of the host. Table 14.9 lists the seven top-level codes currently used on the Internet.

*Universal resource locator (URL),* which is often incorrectly called an *Internet address,* is a method used to address servers on the Internet instead of using an IP address. A URL is given in the format "protocol://IDN/directory/filename.extension." An example of a URL is "http://www.microsoft.com/windows98/updates/default.asp." The protocol and Internet domain name portions of the URL are required, whereas the directory and filename and extensions are optional and are only needed to specify a file if the server does not automatically provide it. Note that most browsers automatically use the http protocol if another is not provided. The commonly used protocol designators are listed in Table 14.10.

The URL does not actually replace the IP address on the Internet or private network. When a computer attempts to address another computer on the network using a URL instead of an IP address, it is routed to a special server on the network that provides a **domain naming service (DNS).** DNS servers keep a database that is used to match the IDN supplied by the URL to the actual IP address of the computer being selected. Before a computer's IDN is available on a private network or the Internet, it must first be added to the database on the DNS server. The network administrator will enter the IDNs on a private network. On the Internet, this is accomplished by registering the IDN, for a small fee, at www.networksolutions.com.

Table 14.10  Internet Transfer Protocol Designators

| | |
|---|---|
| http | Hypertext transfer protocol. HTTP is the transfer protocol used by the World Wide Web. If a user does not define a protocol, HTTP is usually used by default. |
| gopher | Gopher document or menu. Gopher Internet sites contain menus to aid in locating information on the Internet. |
| ftp | File transfer protocol. FTP is used to transfer files over the Internet. |
| news | Newsgroup |
| telnet | A telnet site allows the user to remotely control another computer. |
| WAIS | Wide Area Information Search. Allows the indexing and searching of databases over the Internet |
| file | Used to locate a file on a local hard drive |

### Procedure 14.5: Configuring DNS

1. Click on the Start button, then move up to Settings, and then over to Control Panel.
2. Click on Control Panel.
3. After the Control Panel window opens, locate and then double-click on the Network icon.
4. Click on the adapter whose TCP/IP protocol is being configured and then click on Properties.
5. Click on the DNS Configuration tab.
6. Click on the Enable DNS option (only select this option if DNS is being used on the network).
7. Enter the computer's assigned host name in the appropriate text box.
8. Enter the computer's assigned domain name in the appropriate text box.
9. Enter each DNS server's IP address in the text box under the "DNS Server Search Order." Click on the Add button between each entry.
10. Once completed, click on the OK button.
11. Click on the OK button on the Network window.
12. Choose "Restart Windows" when the option box displays.
13. After Windows restarts, close all open windows.

If the computer is connected to the Internet through a modem, DNS is provided by the ISP. On computers permanently attached to the Internet or running on a large network using DNS, the DNS options on the workstation are either automatically configured by the DHCP server or configured manually. To configure DNS manually, follow the steps outlined in Procedure 14.5.

## 14.9  BINDING, SHARING, AND MAPPING WITH WINDOWS 9X

After the NIC, client, and protocol suites have been installed and configured, the time comes to set up binding, sharing, and mapping. This can be done automatically with the use of script files, but it is most often done manually on computers running Windows 9X.

### 14.9.1  Binding

**Binding** is used to indicate which installed network hardware and software components go together. Each network adapter is bound to one or more protocol suites and each protocol suite is bound to one or more clients. Procedure 14.6 lists the steps to follow to configure binding in Windows 9X.

### 14.9.2  Sharing

**Sharing** is the act of making a computer's resources available on the network. Before any of the computer's resources are available for sharing on a peer network with Windows 9X, file and print sharing must first be enabled. This is accomplished by following the steps listed in Procedure 14.7.

## Procedure 14.6: Binding with Windows 9X

1. Click on the Start button, then move up to Settings, and then over to Control Panel.
2. Click on Control Panel.
3. After the Control Panel window opens, locate and then double-click on the Network icon.
4. On the Network window, click on the adapter or protocol suite on which you wish to configure binding, and then click on the Properties button.
5. Click on the Binding tab.
6. Click on the check boxes beside the protocols or clients you wish to bind.
7. Click on the OK button on all open windows to accept the changes. Click on the Cancel button to abort without changing.
8. Restart the computer for the changes to take effect.

## Procedure 14.7: Enabling file and print sharing

**Caution:** You must have a Windows 9X CD available before starting this procedure.

1. Click on the Start button, then move up to Settings, and then over to Control Panel.
2. Click on Control Panel.
3. After the Control Panel window opens, locate and then double-click on the Network icon.
4. Once the Network window opens, click on the File and Print Sharing button located in the General tab.
5. If file and print sharing has not been previously installed, follow the installation instructions displayed by Windows 9X.
6. Once installed, the File and Print Sharing window will display. Click on the check boxes to enable the desired sharing options.
7. To accept the changes, click on the OK button on all open windows. To abort without changing, click on the Cancel button.
8. Restart the computer for the changes to take effect.

## Procedure 14.8: Selection of user-level or shared-level access

**Caution:** You must have a Windows 9X CD available before starting this procedure.

1. Click on the Start button, then move up to Settings, and then over to Control Panel.
2. Click on Control Panel.
3. After the Control Panel window opens, locate and then click on the Network icon.
4. Once the Network window opens, click on the Access Control tab.
5. Select the desired access control by clicking on the appropriate option dot.
6. If user-level access is selected, enter the name of the Windows NT domain server, which will provide the list of users, in the text box.
7. To accept changes, click on the OK button on all open windows. To abort without changing, click on the Cancel button.
8. Restart Windows for the changes to take effect.

Once file and print sharing have been installed or enabled, two types of sharing are available with Windows 9X: *shared-level access* and *user-level access.* Shared-level access is the only sharing method available with peer-to-peer networks using Windows 9X, whereas user-level or shared-level is available with Windows 9X computers that are logged onto a server.

When shared-level access is used, access to a computer's resources is controlled by assigning one or two passwords to the resource. One password is used if read-only or full-access privileges are allowed for the shared resource. Two passwords are required if read-only and full-access privileges are allowed. Shared-level access is the default method used by Windows 9X when the network components are originally installed on the workstation.

User-level access requires the computer to log on to a Windows NT/2000 server. If user-level access is selected, the workstation receives a list of users that have a user's account on the server each time that a resource is selected for sharing. The workstation operator then selects the network users he or she wants to give permission to access the workstation's shared resource, and to what level. To select user-level or shared-level access in Windows 9X, follow the steps listed in Procedure 14.8.

---

### Procedure 14.9: Printer and fax/modem sharing

**Caution:** You must have a Windows 9X CD before starting this procedure. Also, fax applications, such as Microsoft fax, must have been installed previously on the computer.

1. Click on the Start button, then move up to Programs, and then over to Windows Explorer.
2. Click on Windows Explorer.
3. After the Windows Explorer window opens, locate and then click on the Printer folder in the left windowpane.
4. In the right windowpane, locate and then right-click on the printer or fax/modem to be shared.
5. Form the pop-up option menu that appears, click on the Sharing option. If this option is not available, print sharing is not enabled.
6. Enter the correct password or select the users to determine the level of access.
7. Click on the OK button to accept changes. Click on the Cancel button to abort without changing.

---

### Procedure 14.10: Sharing drives and folders

1. Click on the Start button, then move up to Programs, and then over to Windows Explorer.
2. Click on Windows Explorer.
3. If sharing a drive, right-click on the drive letter in the left windowpane.
4. If sharing a file folder, click on the drive that contains the folder in the left windowpane, and then right-click on the folder in the right windowpane.
5. From the pop-up options menu, select the Sharing option.
6. After the drive or folder properties window opens, click on the Share As option dot to enable sharing.
7. Enter the appropriate passwords or select the users depending on the access control selected and the level of access.
8. When finished, click on the OK button to accept the changes or the Cancel button to abort without changing.

---

Once the access control method has been selected for the workstation, it is time to actually select the computer resources that will be shared on the network. With Windows 9X, sharing can be configured through either My Computer or Windows Explorer. I will use Windows Explorer for the procedures in this section because it gives more control. With both methods, there are four resources that can be shared on a computer running Windows 9X: printers, fax/modems, disk drives, and file folders.

Print sharing allows the workstation to share the computer's printer or fax/modem. Currently, Window 9X cannot share a computer's modem on the network, but there are third-party utilities that offer this feature. The process of sharing a printer with Windows 9X is fairly simple to configure and is covered in Procedure 14.9.

Sharing a computer's files can range from the level of sharing an entire disk or CD-ROM drive down to sharing a single folder. Individual files within a folder cannot be shared, however. Procedure 14.10 lists the process used to share drives and folders.

### 14.9.3 Mapping

**Mapping** allows a workstation to use disk and CD-ROM drives that have been shared by other computers on the network as if they belong to the workstation. To map drives using Windows 9X, follow the steps outlined in Procedure 14.11.

## 14.10 THE INTERNET

Probably the number one reason why an individual purchases a computer is to have the ability to access a worldwide network of computers called the Internet, or Net. The Internet is a seemingly endless source of information from education institutions, religious organizations, governments, private agencies, companies, and individuals all over the world.

> **Procedure 14.11: Mapping network drives**
>
> 1. Double-click on the Network Neighborhood icon located on the desktop.
> 2. Double-click on the name of the computer within the workgroup that is sharing the resource you wish to map. If the computer's name is not displayed, it may be a member of another workgroup or the computer may not have file and print sharing enabled.
> 3. If the computer is a member of another workgroup: double-click on Entire Network; double-click on the name of the workgroup; double-click on the name of the computer.
> 4. If the selected computer is sharing drivers, folders, or printers, the name of the resource will appear under the computer name.
> 5. Right-click on the drive or folder you wish to map, and then select the Map Network Drive option. A pop-up window will appear with the workstation's next available drive letter. If a drive letter above this letter is desired, use the scroll button to advance through the available drive letters.
> 6. Click on the check box beside "Reconnect at login" to automatically map the drive each time the computer logs onto the network.
> 7. Click on the OK button to accept the changes or the Cancel button to abort without changing.

For most individuals, access to the Internet is through a modem, which will be discussed in the next section. A user accesses the Internet though an ISP that has an Internet on-ramp (connection). America-Online, AT&T, Compuserve, Microsoft Network, and Prodigy are national ISPs that provide access to the Net. There are also hundreds of smaller companies that provide access by modem through the larger ISPs.

Once connected, the user can surf the Net through gophers, FTPs, usernets, telnets, or the World Wide Web sites. A search of the Net for information can be made using Archies, Veronicas, WHOIS, WAIS, or Web search engines.

Individuals can communicate on the Net through e-mail, FAQs, chat, form submittals, user groups, and video conferencing. If you are the average person, these last few sentences probably sound like a foreign language.

Just like most professional occupations, the Internet has its own terminology that one has to learn in order to understand the technology. If you are going to work in an occupation or use the technology, you need to learn to speak the language. Table 14.11 lists a small sample of the most often-used Internet terms. A complete list would be impossible because new terms and meanings are added every day.

## 14.11 MODEMS

As stated previously, serial is the method most often used to transmit information over a network. It would not make sense to run serial cables all over the country when there are already serial lines that run all over the world. You may ask "Where are these serial lines?" The answer is, "The telephone system, of course."

There is only one major problem to overcome to allow the phone system to be used with network computers. The standard telephone system is designed to handle analog signals in the form of audio (sounds), not digital signals in the form 1s and 0s. To use the analog phone system, there has to be a device that can take the digital serial data coming from one of the computer's COM ports and convert it to audio to send over the phone system. Because the digital information is changed to make it compatible with the phone system, it has to be *modulated*. The same device also needs the ability to take the modulated signal coming from another computer through the phone lines and convert it to digital so it can be received by the computer's serial port. This technique is called *demodulation*. The device that performs these tasks is a **modem** (*m*odulator/*dem*odulator).

A modem allows a computer to connect to a network remotely using the standard phone system instead of connecting to the network's topology directly. This gives the remote computer the ability to have all of the advantages of a LAN without being physically connected to it all the time.

Unfortunately, choosing a modem is not a simple task. In addition to manufacturers and models, you must consider standards for transfer rates, error correction, data compression, and expansion bus type. Where once you had only two standards to choose between—Bell 103 and 212A (300 bps and 1,200

**Table 14.11** Common Internet Terms

| | |
|---|---|
| Archie: | An Internet tool used to search FTP sites for filenames |
| Download: | Copying a file from a host computer to your computer |
| FTP: | File transfer protocol; a network of computers on the Internet that allows only the uploading and downloading of files |
| Gopher: | A network of servers on the Internet that provides menus that help find information |
| Hypertext: | Text that provides automatic links between key elements and Internet sites. Hypertext allows the user to move through information nonsequentially. Hypertext is usually underlined and displayed in a different color than the rest of the text on the Web page. |
| Surf: | Moving from site to site on the Internet in search of information |
| Telnet: | Servers on the Internet that allow a user, with the correct client software, to logon and use the remote computer |
| Upload: | Copying files from the workstation computer to a host computer |
| USENET: | A worldwide network of news groups that post user messages and bulletins |
| Veronica: | A gopher search program that allows users to search gopher sites for keywords |
| WHOIS: | A program that allows users to search databases of specific URLs |
| WAIS: | World Area Information Servers; a system that allows the user to search large databases using key words |
| Web Browser: | A client program, such as Microsoft's Internet Explorer and Netscape's Navigator, that lets the user access WWW servers |
| World Wide Web: | WWW; a network of computers on the Internet that allows the user to use hyperlinks, graphics, and sound to present information. To access a WWW server, the workstation must be running Web browser applications. |

bps)—you must now consider MNP-1 through MNP-10, and V.22 through V.42bis standards. Sounds like Greek, doesn't it?

Distinguishing among standards involves a host of acronyms, a hodgepodge of speeds, and more features than a new automobile's list of available options. Today, you can find modems that push data at rates higher than 56 kilobits per second (Kbps). Some modems also combine a FAX while others also serve as an answering machine. As usual, before I go any further with our discussion on modems, we need to talk a little about the medium over which modems must transmit their information—the telephone system.

## 14.11.1 A Few Facts about the Phone System

The standard telephone system is not designed around digital information. It is designed around the spoken human voice. The human voice produces sound waves that move at different frequencies according to their pitch. The higher the pitch of the sound waves, the higher the frequency. Telephones convert the frequencies produced by the voice into electrical frequencies so they can be transferred over electrical conductors, amplified, and routed by the phone system's electronic circuits.

When electronic circuits and conductors are designed, they are designed to transfer frequencies around a certain bandwidth. Unlike the bandwidth discussed for network media, most other communication media bandwidths are given as a low-end and high-end frequency of operation. The wider the bandwidth (separation between low and high frequency) the circuitry is designed to carry, the more it will cost. Also, the higher the high-end frequency the circuitry is designed to carry, the higher the cost.

To keep the cost of the phone system low, and because it is designed around the spoken voice, the agreed-upon bandwidth is approximately 30–3,000 Hz. As you can probably guess, this bandwidth greatly limits the speed of the tones used to represent digital serial communication.

To allow modems to exceed the bandwidth of the telephone system, the modems take the asynchronous information coming from one of the computer's serial ports and modulate it into synchronous data that are actually transmitted over the phone system. The actual techniques used to perform this amazing feat are not really important to understand, even for computer technicians. Basically, the modulation techniques currently used by modems are forms of *amplitude shift keying (ASK), frequency shift keying (FSK), phase shift keying (PSK),* and *quadrature amplitude modulation (QAM)*. ASK changes the amplitude of the carrier, FSK changes the frequency of the carrier, PSK shifts the phase of the carrier, and QAM is a combination of PSK and ASK.

Currently, the most widely used form of modulation is variations of QAM. Modems that use this technique generate what is referred to as a *carrier* that is below the high-end bandwidth of the phone system. The binary bits that are transmitted are added to the carrier using forms of ASK and PSK, allowing multiple bits to ride on a single cycle of the carrier. If four bits ride on each cycle of the carrier, then a 2,400 Hz carrier will transfer data at a rate of 9,600 bit-per-second (bps) (4 × 2,400). To differentiate the two speeds, modem manufacturers use the term *baud* to specify the frequency of the carrier and *bits per second (bps)* to indicate the speed of the data. This is confusing under MS-DOS and Windows 3.X because the COM port speeds are set as baud because the rate of the carrier and the rate of the data are the same. To avoid this confusion, Windows 9X sets the speed of the COM ports in bps.

### 14.11.2 Hayes-Compatible

The Hayes Corporation, which is no longer in business, pioneered the first modems that were controlled by an onboard microprocessor. This feature made the modem a microcomputer that had the ability to run programs. For this reason, Hayes called its modems with this feature the *Hayes Smart Modem*. The Hayes Smart Modem gave the user the ability to send the modem commands to make it perform all kinds of amazing feats that will be discussed in the remaining sections.

Most of today's modem manufacturers make their modems Hayes Smart Modem-compatible or, simply, Hayes-compatible. The term *Hayes compatible* means that modems manufactured by companies other than Hayes Corporation recognize the same basic command set used by the Hayes Smart Modem. In addition to the Hayes command set, these modems usually require additional commands for maximum modem performance. However, because the Hayes command set is supported by most current modems, communication applications usually support Hayes-compatible modems. If you plan on buying a new modem, make sure it is Hayes-compatible. We next discuss a few other standards that one must understand about modems.

### 14.11.3 Modem Standards

Standards are your best assurance that a given modem can successfully connect with other modems anywhere in the world. The standards you choose will determine how fast (and reliably) your modem can transfer data. Bell originally started setting standards for modems in the United States to make sure the modems were compatible with its phone system. The Bell 103 standard was the first modem communication standard that was widely adopted. Although seldom used today, it still remains the standard of last resort in the United States when all else fails. Bell 103 set the standard for receiving and sending data at 300 bps. Bell 212A is the second widely adopted standard and is still in wide use today. This standard allows data transmission at 1,200 bps.

Current modem standards are set by an international standard organization called the *Comité Consultatif International Téléphonique et Télégraphique,* which roughly translates to the *Consultative Committee for International Telephone and Telegraph,* or *CCITT*. CCITT has recently changed its name to the *International Telecommunications Union, Telecommunications group,* or *ITU-T*. Due to the name change, current international modem standards can be listed under either CCITT or ITU.

All of the high-speed, nonproprietary standards used on today's modems are sanctioned by CCITT/ITU-T. The CCITT/ITU-T standards all begin with the letter *V*, followed by a period, and then

**Table 14.12** Modulation Protocols and Speeds in Bits per Second (bps)

| Protocol | Speeds |
|---|---|
| CCITT V.21 | 300/150/110 |
| CCITT V.22 | 1,200/600 |
| CCITT V.22 bis | 2,400 |
| CCITT V.32 | 9,600/4,800/2,400 |
| CCITT V.32 bis | 14,400/12,000/9,600/7,200 |
| ITU-T V.34 | 33,600/28,800/26,400/24,000/21,600/19,200/16,800/14,400 |
| ITU-T V.90 | 56,000 download/33,600 upload |

a 2-digit number. If the letters *bis* follow the number, it indicates that the standard has been revised once. The CCITT/ITU-T standards currently used by modems include V.21, V.22, V.22bis, V.32, V.32bis, V.34, V.34+ (V.34 amended), V.42, V42.bis, and V.90.

Along with CCITT/ITU-T, a company by the name of Microcom, Inc. also sets many of the standards used by today's modems. The modem standards set by Microcom are called *Microcom Network Protocol (MNP)* standards and are numbered 1 through 5. All of the standards set by either CCITT/ITU-T or MNP fall into one of three modem protocols: modulation, error correction, and compression.

### 14.11.4 Modulation Protocol

As discussed previously, the process of converting the digital information from one of the computer's serial ports into the tones used by the telephone system is modulation. The process of converting the tones back into digital information is demodulation. The same protocol that a modem uses to modulate the data has to be reversed by the modem on the other end of the phone line to demodulate the data if the modems are going to be able to communicate with each other.

Current international modulation protocols standards include CCITT/ITU-T's V.21, V.22, V.22bis, V.32, V.32bis, V.34, V.34+, and V.90. Modulation protocols not only establish the actual way the information is converted from digital information into tones, and vice versa, but they also determine the available speeds for sending and receiving data. As stated previously, the actual process used to modulate the data is not really important to understand. What is important to know is that a modem that sends data using the V.22 modulation protocol can only be received by a modem that also uses the V.22 modulation protocol. Table 14.12 lists the current CCITT/ITU-T modulation protocols and the speeds that are available for each.

### 14.11.5 Error Correction

Error correction for digital information transfer over phone lines has been around for years. Error-correction protocols are used to upload (send) and download (receive) files with a computer running an application program called a *Bulletin Board Service (BBS)*. On each computer, a program that is running in the computer's memory handles the error correction. Some of the names given to the error-correction protocols used by BBSs are YMODEM, GMODEM, XMODEM, and ZMODEM. In order for the computer to transfer data using error correction, the transmitting and receiving computers have to be running the same error-correcting protocol program.

Modems were once very dumb electronic devices, but this is not the case today. In fact, the modem is really a microcomputer that consists of a microprocessor, memory, input, and output. Placing a microprocessor in the modem gives it the ability to run programs. Having the ability to run programs means that the modem can perform the error correction instead of having it done by a program running inside the computer. This is by far the most efficient method.

Table 14.13  Modem Fallback

| V.90 > V.34 > V.32bis > V.32 > V.22bis > V.22 > V.21 |
|---|

Microcom, Inc. was the first company to introduce error-correction protocols that were incorporated into modems. The most popular Microcom error-correction protocols are MNP 1 through 4. CCITT/ITU-T also established an error-correcting protocol called V.42.

The use of an error-correction protocol is not required for two modems to communicate with each other. In fact, all that is required is that both modems use the same modulation protocol. Although not a requirement, error-correction protocols are very important to ensure that the high-speed serial data transmitted by today's modems arrive error-free.

If an error-correction protocol is used, it allows modems to communicate at a speed that is 20% faster than the modem's fastest connect speed. To accomplish this feat, however, the computer's serial port bps rate should be set at least 20% higher than the fastest connect speed of the modem. For example, if the modem is a V.32bis (14.4 Kbs) and the modem supports V.42 error correcting, the serial port to which the modem is connected needs to be set to at least 17,280 bps, if that speed is available. If not, set the computer serial port to the next higher speed.

### 14.11.6  Data Compression

Data compression has been used in computer communication and data storage for a long time. Like error correction, data compression was, and still is, performed by a program running in the computer's RAM. Data compression basically allows binary information to be represented with fewer binary bits. The data are compressed for transmission or storage and then uncompressed when received or retrieved.

Because today's modems are microcomputers that have the ability to run programs, compression has been added to their repertoire of features. Modem compression is performed by the modems and not by an application program running in the computer's RAM. The compression standards in current use are Microcom's MNP-5 and the ITU-T V.42bis. These compression protocols offer compression rates of up to 400% ($\times 4$).

Just because compression is available to your modem does not mean it is going to send data 400% faster than the bit speed. First of all, both sending and receiving modems have to support error compression, and data that have already been compressed cannot be compressed further. However, before error correction has a chance of being used by your modem, you *must* set the bps rate of the computer's serial port at four times the modem's maximum bps speed.

### 14.11.7  Modem Negotiation and Fallback

Like the compatible computers, modems are downwardly compatible, which means that a V.90 modem supports all of the modulation protocols of the V.34, V.32bis, V.22bis, V.22, and V.21. Most V.90 modems also support the MNP and ITU-T error-correcting and data-compression protocols. When two modems attempt to connect, they automatically negotiate which modulation, error-correction, and compression protocol they both have in common. This is what is going on when you hear all of those strange sounds coming from an internal modem when it first connects to another.

Before the connection can be made, the first thing the modems must negotiate is a modulation protocol. When the remote modem answers the phone, it sends back a connect tone to the originating modem. When the origination modem hears this tone, it sends a tone to the remote modem indicating the selected modulation protocol. The originating modem will normally try to connect at its modulation protocol's highest speed. However, the user has the ability to set the modulation protocol and speed using modem commands. If the remote modem supports the originating modem's requested modulation protocol, the connection is made immediately. Otherwise, the originating modem will "fallback" to the lower modulation protocols until it gets a connect signal from the modem it is dialing. Table 14.13 shows the fallback path for a V.90 modem.

Once the modulation protocol is selected by the modems, V.32 modems (and up) also negotiate the speed within the selected modulation protocol. The modems start at the highest available speed

**Figure 14.20** External Modem

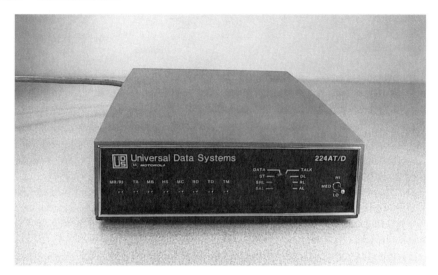

and choose the one that allows for reliable communication over existing telephone line conditions. These modems can also change to a lower speed within the selected modulation protocol in the middle of connection. This speed change usually occurs because of an increase in telephone line noise. This feature is called *retaining* because it helps maintain the connection over dynamically variable line conditions.

Once the modems agree upon modulation protocol and speed, they then negotiate an error-correction and data-compression protocol. These two protocols are not required for two modems to communicate with each other.

### 14.11.8 Internal and External Modems

An external modem connects to one of the serial ports (COM) on the back of the computer. A typical external modem is shown in Figure 14.20. The external modem is entirely self-contained and can be moved easily from one computer to another. Because the external modem is a self-contained unit, it usually costs more than an internal modem with the same features. Installing an external modem involves the simple process of plugging the modem into any 120-vac wall plug and connecting a standard serial cable between the modem and the serial port of the computer. Once the modem is installed, the communication program is configured as to which serial port to use.

An internal modem has the modem circuits and serial port circuits integrated on a single card and installs in one of the computer's expansion slots. The internal modem is *not* easily transported from one computer to the next. Because the internal modem card contains the serial port, it must be configured so that it will not conflict with another serial port already in the computer. Figure 14.21 shows a typical internal modem.

The relatively slow speed of the modem makes it an ideal adapter for the ISA expansion bus. Most modems purchased for ISA are either hardset or PnP. The PnP limitations of the ISA bus are forcing most current PnP modems to be manufactured for the PCI expansion bus, however. To display the resources used by an internal modem's serial port on a computer running Windows 9X, use the steps outlined in Procedure 14.12.

### 14.11.9 The Modem AT Command Set

As stated previously, Hayes compatible modems recognize the same command set. The user can send commands directly to the modem by running a terminal mode program. When in terminal mode, information typed from the keyboard is sent directly to one of the computer's serial ports that has a modem

**Figure 14.21** Internal Modem

---

**Procedure 14.12: Displaying an internal modem's serial port resources**

1. Click on the Start button, then move up to Settings, and then over to Control Panel.
2. Click on Control Panel.
3. After the Control Panel window opens, locate and then double-click on the System icon.
4. Once the System Properties window opens, click on the Device Manager tab.
5. In Device Manager, click on the plus (+) sign beside the Modem icon.
6. Click on the name of the modem whose resources you wish to display.
7. Click on the Resources tab.
8. Close all open windows.

---

attached. A communication application program such as Windows 3.X's Terminal or Windows 95's Hyper Terminal initiates the terminal mode.

With a terminal mode program running, the user types commands to the modem from the computer's keyboard. All modem commands must be preceded by the attention command *AT*. You can use the *AT* command by itself to check the communication link between the computer and the modem by typing *AT* and pressing the Enter key. The modem will respond with an OK to the screen if the computer and modem are talking to each other. If an OK does not appear, it usually indicates that the COM port's address or interrupt is set incorrectly. It could also mean that the modem was not initialized correctly.

The command set for today's modems is complex and lengthy. For example, the modem in my computer has 138 commands. This large number of commands prohibits a detailed explanation here, but an exact listing of each command and its function is usually supplied with a modem when it is purchased. A set of commands that are recognized by most modems is given in Table 14.14. The commands recognized by the modem installed in your computer are probably more extensive.

**Problem 14.3:** What terminal mode command would be used to tell the modem to dial the number 555-1234 using tone dial?

**Solution:** Starting with the required prefix AT, then referring to the commands listed in Table 14.14, the command will be: ATDT555-1234.

**Table 14.14** Common AT Modem Commands

| | |
|---|---|
| A | Answer incoming call |
| A/ | Repeat last command (Don't preface with AT. Enter usually aborts.) |
| D | Dial the following number and then handshake in originate mode.<br>*Dial Modifiers* (These are common but most modems will have more.)<br>  **P**  Pulse dial<br>  **T**  Touch tone dial<br>  **W**  Wait for second dial tone<br>  **,**   Pause for time specified in register S8 (usually 2 seconds)<br>  **;**   Remain in command mode after dialing<br>  **!**   Flash switch-hook (Hang up for a half second as in transfering a call) |
| E | Will not echo commands to the computer (also E0) |
| E1 | Will echo commands to the computer (so one can see what one types) |
| H | On hook (hang up, also **H0**) |
| H1 | Off hook (phone picked up) |
| I | Inquiry, Information, or Interrogation<br>(This command is very model specific. The I0 Command usually returns a number or code, while higher numbers often provide much more useful information.) |
| L | Speaker loudness (**L0** off or low volume) |
| L1 | Low volume |
| L2 | Medium volume (usual default) |
| L3 | Loud or high volume |
| M | Speaker off (**M0**) (**M3** is also common, but different on many brands) |
| M1 | Speaker on until remote carrier detected (until the other modem is heard) |
| M2 | Speaker is always on (data sounds are heard after CONNECT) |
| O | Return online (**O0** see also **X1** as dial tone detection may be active) |
| O1 | Return online after an equalizer retrain sequence |
| Q | Quiet mode **Q0** displays result codes, user sees command responses (e.g. OK) |
| Q1 | Quiet mode, result codes are suppressed, user does not see responses |
| Sn? | Query the contents of S-register *n* |
| Sn=r | Store the value *r* in S-register *n* |
| V | Nonverbal (Numeric result codes **V0**) |
| V1 | Verbal English result codes (e.g., CONNECT, BUSY, NO CARRIER, etc.) |
| X | Hayes Smartmodem 300 compatible result codes (**X0**) (Many have more than 4) |
| X1 | Usually adds connection speed to basic result codes (e.g., CONNECT 1200) |
| X2 | Usually adds dial tone detection (preventing blind dial and sometimes **ATO**) |

(Continued)

Table 14.14 (Continued)

| | |
|---|---|
| X3 | Usually adds busy signal detection |
| X4 | Usually adds both busy signal and dial tone detection |
| Z | Reset modem to stored configuration (Z0, Z1, etc. for multiple profiles) Same as &F (factory default) on modems without NVRAM (nonvolatile memory) |
| &C0 | Carrier detect (CD) signal always on |
| &C1 | Carrier detect indicates remote carrier (usual prefered default) |
| &D0 | Data terminal ready (DTR) signal ignored (*See your manual* on this one!) |
| &D1 | If DTR goes from On to Off the modem goes into command mode (*some* modems) |
| &D2 | Some modems hang up on DTR On to Off transition (usual prefered default) |
| &F | Factory defaults (most modems have several defaults **&F1, &F2,** etc.) |
| &P | (**&P0**) U.S./Canada pulse dialing 39% make/61% break ratio |
| &P1 | U.K./Hong Kong pulse dialing 33% make/67% break ratio |
| &T | Model-specific self-tests on some modems |
| &V | View active (and often stored) configuration profile settings (or **ATI4**) |
| &W | Store profile in NVRAM (&W0 &W1, etc. for multiple profiles) |
| &Zn=x | Store number *x* in location *n* for AT DS on some modems |

### S Registers

| Register | Range | Default | Function |
|---|---|---|---|
| S0 | 0–255 rings | 1–2 | Answer on ring number (*don't answer if 0*) |
| S1 | 0–255 rings | 0 | If S0>0, this register counts incoming rings |
| S2 | 0–127 ASCII | 43 + | Escape to command mode character, S2>127 no ESC |
| S3 | 0–127 ASCII | 13 CR | Carriage return character |
| S4 | 0–127 ASCII | 10 LF | Line feed character |
| S5 | 0–32,127 ASCII | 8 BS | Backspace character |
| S6 | 2–255 seconds | 2 | Dial tone wait time (blind dialing, see **X**n) |
| S7 | 1–255 seconds | 30–60 | Wait time for remote carrier |
| S8 | 0–255 seconds | 2 | Comma pause time used in dialing |
| S9 | 1–255 1/10 sec. | 6 | Carrier detect time required for recognition |
| S10 | 1–255 1/10 sec. | 7–14 | Time between loss of carrier and hang up |
| S11 | 50–255 millisec. | 70–95 | Duration and spacing of tones when tone dialing |
| S12 | 0–255 1/50 sec. | 50 | Guard time for pause around +++ command sequence |

## 14.11.10 The Modem Initialization String

When a communication program is loaded, it sends a set of AT commands to the modem, which gets the modem ready to use. This set of AT commands is called the *modem's initialization string*. A modem's initialization string has to be correct or the modem may not work at all. If it does work, it will not work efficiently. Example 14.2 contains the initialization string used by my communication software to initialize the modem.

**Example 14.2:** Modem initialization string

```
ATX4&C1&D2W1M1L1S0=0S7=47
```

As you can see in Example 14.2, there is a lot to an initialization string. In the old days, modem users had to come up with the correct string by going through the modem's user manual and trying the old hit-and-miss routine. Most modems purchased today come with their own communication software. This software will support several modems, including the one purchased. To come up with the correct initialization string usually requires users to select their modems from a list of modem model numbers.

If your modem is not listed, try one of the Hayes modems that matches the options of the modem installed on your computer. You should eventually come up with the right combination.

To connect to the national service providers like America Online, Compuserve, Microsoft Network, or Prodigy, a communication package that is written specifically to connect to these board services is required. These packages can be purchased from most software retailers. However, they can usually be obtained free by calling a toll-free telephone number or from other board service members. These communication packages are very user-friendly and take care of all of the modem commands for you.

## 14.11.11 Setting Up the Modem in Windows 98

The modem initialization string and configuration are usually set by the communication application that uses the modem. The documentation for the application should give the procedure steps to follow. Other Windows 98 applications use the initialization string and configurations established though the Windows 98 operating system itself. Procedure 14.13 outlines the steps to follow to set up a modem with Windows 98.

---

### Procedure 14.13: Modem setup with Windows 98

1. Click on the Start button, then move up to Settings, and then over to Control Panel.
2. Click on Control Panel.
3. After the Control Panel window opens, locate and then double-click on the Modem icon.
4. After the Modem Properties window opens, the Add, Remove, Properties, and Dialing Properties option buttons are displayed, along with a list of the modems installed in the computer.
5. Click on Add the Software Components for New Modems Installed in the Computer.
6. Click on the desired modem in the list and then click on the Remove buttons to remove a modem's software components.
7. Click on the Dialing Properties button to modify how each call is dialed. The settings available with this button are used to:
   - Establish configurations for multiple locations for portable computers;
   - Enter digits required to get an outside line if calling through a company's PBX;
   - Enter the characters required to disable call waiting during the session (usually *70);
   - Select the use of touch dial or tone dial (default); and
   - Enter calling card numbers for long-distance calls.

Computer Network Basics

8. Click on the Properties button. Once the modem Properties window opens, click on the following tabs:
   - General tab to change the speaker volume and serial ports maximum connect speed;
   - Distinctive Ring tab for telephones that support this service; and
   - Forwarding tab to enter a number for call forwarding;
   - Click on the Connection tab to set the default frame size for the modem's serial port and caller preferences.

9. Click on the Port Settings button to set the UARTs first-in-first-out (FIFO) buffer setting for read and write operations.
10. Click on the Advanced button to change the modem's initialization string and to enable error correction and compression, and to set the type flow control.
11. Click on the View Log button to view a log of the last work session.
12. If any changes were made to any of the indicated options, click on the OK button on all open windows to accept the changes or the Cancel button to abort without changing.

## 14.12 CONCLUSION

A network is the connection of computers that share resources. The topology is the way a network is connected together. The most predominant topologies are bus, star, and ring. Computers that are connected together in the same locale are called local area networks (LANs). LANs that are remotely connected together over great distances are called wide area networks (WANs).

Ethernet, the most widely used protocol in networks, uses a CSMA/CD to control network traffic and can be connected using ThickNet, ThinNet, 10BaseT, or 100BaseT formats. Ethernet can be connected in a bus, star, or ring, or a combination of topologies.

A network's operating system is the software that makes the network operate. There are basically two categories of network operating systems: network server and peer-to-peer. A network server operating system has a single computer that controls the entire operation of the network. The network server is a critical resource; if it fails, the whole network fails. A network that uses a peer-to-peer operating system allows each computer of the network to share the responsibilities of the server. On a peer-to-peer network, if any computer fails, the network should continue to function at a reduced capacity.

The network server is the predominant network operating system in large LANs because of a high degree of network security. The peer-to-peer operating system is usually used in small LANs because it is easy to use and setup.

The TCP/IP protocol suite is rapidly becoming the predominant suite used in networking because TCP/IP is the only suite used by the Internet. The TCP/IP currently in use is called the dotted quad, or IPv4. A new version called the IPv6 is currently under development. In networking, it is very important to understand the operation and configuration of the TCP/IP stack.

Modems are devices that modulate information to send over telephone lines and demodulate information coming from telephone lines. Most current modems are called Hayes-compatible because they recognize the same AT command set developed for this modem. In order for two modems to communicate with each other, both must use the same modulation protocol. Error-correction and compression protocols are two options used by modems. Modem protocol standards are currently set by CCITT/ITU-T.

## Review Questions

1. Write the meaning of the following acronyms.
   a. LAN
   b. WAN
   c. ISO
   d. OSI
   e. IEEE
   f. EMI
   g. SONET
   h. NetBEUI
   i. IPX/SPX
   j. TCP/IP
   k. MSAU
   l. CSMA/CD
   m. FDDI

n. ATM
o. NIC
p. MAC
q. DLC
r. OUI
s. NOS
t. GUI
u. IANA
v. ICANN
w. ISP
x. DHCP
y. DNS
z. WINS
aa. URL
bb. FTP
cc. IDN
dd. WAIS
ee. ASK
ff. PSK
gg. FSK
hh. QAM
ii. CCITT
jj. ITU-T
kk. MNP

## Fill in the Blanks

1. An addressable display that depends on the processing power of a mainframe computer is called a _____ _____ .
2. A network that is set up over multiple cities is called a _____ _____ network.
3. The speed rating of a network medium is called _____ .
4. _____ is the name used for signal loss over distance.
5. The RG designator is used to identify _____ cables.
6. A T3 carrier consists of _____ T1s and has the ability to transfer information at _____ Mbps.
7. _____-_____ fiber optic cable allows light of different wavelengths to pass.
8. A _____ _____ is used to send light over a single-mode fiber optic conductor.
9. Sending information 1 bit at a time over a single conductor is called _____ .
10. A typical network packet consists of a _____ , _____ , and _____ .
11. A _____ is an agreed-upon way of doing something.
12. The three predominant network topologies are _____ , _____ , and _____ .
13. 10BaseT segments may be up to _____ feet ( _____ meters) in length.
14. 10Base2 uses a smaller cable size than 10Base5 and usually is called _____ .
15. Before a computer on a token ring network can transmit data, it must first obtain the _____ .
16. Fast Ethernet is also called _____ .
17. 10Base2 is connected in a _____ topology.
18. When a star topology is connected using multiple hubs, it is called a _____-_____ topology.
19. ATM creates a _____ circuit between the two computers.
20. _____ uses two rings called _____ and _____ .
21. The DLC address is _____ bits long, which allows for _____ combinations.
22. The resources required by an NIC is a set of _____ and a(n) _____ .
23. The part of a DLC that is assigned to a manufacturer is called the _____ _____ _____ .
24. A _____ server holds user accounts on a Windows NT network.
25. The _____ server hold user accounts on a Netware network.
26. The access control that requires each resource to have a password is called _____ _____ _____ .
27. The individual who controls user accounts on a client/server network is called the _____ _____ .
28. In order for a computer to access a server, it must first be running the correct _____ software for the operating system installed on the computer.
29. NetBEUI uses a _____ _____ and _____ to identify computers on the network.
30. The agency that assigns IPs on the Internet is called the _____ _____ _____ _____ _____ .
31. The _____ _____ _____ _____ _____ service dynamically assigns IP addresses if the computer is configured correctly.
32. Making a computer's resources available on a network is called _____ .
33. The protocols used by modems are _____ , _____ , and _____ .
34. A _____-compatible modem is supported by the most applications and operating systems.

## Multiple Choice

1. A topology is:
   a. the network operating system.
   b. the way a network is connected together.
   c. the protocol used.
   d. the appearance of the network from the top.
2. Which of the following uses basically the type of cable used by the telephone company?
   a. ThickNet
   b. ThinNet
   c. 10BaseT
   d. FlatNet
3. Which of the following is required at each end of bus topology using ThinNet?
   a. a server
   b. a terminator
   c. a token
   d. a knot
4. A file server is:
   a. used by Novell Netware.
   b. used on some peer-to-peer operated networks.

c. only used on a token ring network.
d. only used on a bus network.

5. If a network critical resource fails:
   a. the entire network will fail.
   b. just the critical resource will fail.
   c. half the workstations will cease to function.
   d. all computers on the network will fail to operate.

6. The star topology:
   a. uses a central unit called hub.
   b. does not have a critical resource.
   c. is the fastest of all topologies.
   d. is the cheapest of all topologies.

7. Microsoft refers to its client software as a:
   a. repeater.
   b. requester.
   c. redirector
   d. router.

8. A device that allows a network to be divided into addresses subnets is called a:
   a. router
   b. gateway.
   c. repeater
   d. requester.

9. The layer of the OSI model that is responsible for starting and ending a network session is called the:
   a. data link layer.
   b. network layer.
   c. transport layer.
   d. session layer.
   e. presentation layer.

10. The layer of the OSI model that establishes network logical addresses is the:
    a. data link layer.
    b. network layer.
    c. transport layer.
    d. session layer.
    e. presentation layer.

11. Which layer performs error detection and recovery?
    a. physical layer
    b. data link layer
    c. network layer
    d. transport layer
    e. application layer

12. The layer of the OSI model that initiates the request for a network resource is the:
    a. application layer.
    b. transport layer.
    c. network layer.
    d. data link layer.
    e. physical layer.

13. ThinNet:
    a. uses 50-ohm coax cable.
    b. is cheaper than ThickNet.
    c. has the transceiver and cable trap on the adapter card.
    d. All of the above are correct.

14. The physical layer of the ISO model:
    a. ensures the reliability and efficiency of the end-to-end movement of the data.
    b. is the layer by which the user communicates directly with the network.
    c. takes care of network security.
    d. defines the mechanical and electrical characteristics of the network.

15. Bus topology:
    a. uses a single cable run to connect all of the workstations.
    b. uses a central node or hub.
    c. connects the computer together in a circle.
    d. All of the above answers are correct.

16. 10BaseT is implemented in:
    a. a bus topology.
    b. a star topology.
    c. a ring topology.
    d. all three network topologies.

17. Ethernet:
    a. is a network hardware protocol.
    b. is a network operating system.
    c. uses a network server.
    d. is a peer-to-peer operating system.

18. Novell:
    a. is a peer-to-peer network operating system
    b. is a network server operating system.
    c. is a protocol.
    d. is a topology.

19. ThickNet makes the connection to the coax cable through a:
    a. DB25S connector.
    b. DB9P connector.
    c. vampire trap.
    d. data trap.

20. The TCP/IP suite confirms the connection between the sending and receiving computers. This is called:
    a. a reliable connection.
    b. connectionless oriented.
    c. an IP connection.
    d. connection oriented.

21. The experimental network established on the Internet to test IPv6 is called:
    a. EIP.
    b. IPv6+.
    c. Top Dog.
    d. 6Bone.

22. A subnet mask:
    a. is another name for the IP.
    b. uses names instead of IPs.
    c. identifies the IP's network ID and node ID.
    d. allows NetBEUI to be used over IPX/SPX.

23. A static IP:
    a. will automatically change each time the computer logs onto the network.
    b. has to be entered into the computer manually.
    c. is automatically set when the NIC's driver is installed, but if it needs changing it will have to be done manually.
    d. is assigned by a server running DHCP.

24. The core used on single mode cable is:
    a. 62.5 microns.
    b. 82.5 microns.
    c. 125 microns.
    d. 8.3 microns.
    e. 8.3″
    f. 62.9″

25. The URL does away with the IP.
    a. True
    b. False
26. Assigning one computer's shared hard drive to be used as a resource of another is called.
    a. sharing
    b. mapping
    c. using
    d. redirecting
27. What ITU-T modulation standard sets the 56 Kbps speed?
    a. V.34+
    b. V.80
    c. V.36
    d. V.32bis
    e. None of these is correct.
28. The CCITT V.34 standard provides a maximum uncompressed bit speed of:
    a. 1,200 bps.
    b. 14,400 bps.
    c. 28,800 bps.
    d. 57,600 bps.
    e. 33,600 bps.
29. The Hayes-compatible command *ATH* will tell the modem to:
    a. dial a number.
    b. hang up the phone.
    c. place the current caller on hold.
    d. hold on for dear life.

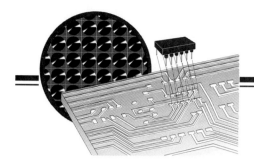

# CHAPTER 15

# Diagnostics

## OBJECTIVES

- Define terms and expressions used relative to diagnostics.
- List the characteristics of ROM-based diagnostics.
- Recognize beep codes.
- Describe POST codes.
- Identify IBM numerical error codes.
- State the characteristics of disk-based diagnostics.
- Describe a computer virus.
- List the basic procedure used to troubleshoot a dead computer.

## KEY TERMS

ROM-based diagnostics
beep code
error code
POST code

POST card
disk-based diagnostics
system properties
Device Manager

Dr. Watson
virus
virus checker

I recently received a call from a former student who purchased a new inkjet printer for his computer. After he installed the new printer and its drivers, the printer would only print crazy characters. What was the problem? Well, it could have been his brand new printer, the printer cable, the printer driver, printer port, or the application that was trying to use the printer. How can one determine the source of the problem with so many different variables that can cause the same symptoms?

To understand computers, you have to understand the concept that hardware does not work without software, and software is meaningless without hardware. The two go together and are inseparable. The scenario just mentioned occurs quite often in the field of computers. A new application is installed and the computer's printer stops working. A change is made to the operating system's configuration file, and the serial ports die. Was the problem caused by the change that was made or was it simply time for the hardware to give up the ghost? How can you tell what is causing the problem—hardware or software? Diagnostic programs are usually the answer to this question.

Diagnostic programs are written specifically to check the hardware in a computer. These are programs that will operate the hardware in a certain way every time they are executed. If you can eliminate the hardware as the source of the problem with diagnostics, the problem will usually be in the software. If the computer fails a diagnostic test, the problem is in the computer's hardware. Diagnostic programs not only indicate the device that failed, but they will usually give the corrective action necessary to help remedy the problem. Diagnostics for the compatible computers are available in two forms: ROM-based and disk-based.

## 15.1 ROM-BASED DIAGNOSTICS

**ROM-based diagnostics** are contained in the computer's BIOS ROM ICs. The compatible computers have a small self-test that automatically runs each time the computer is booted. This small, ROM-based diagnostic program is called the *power-on self-test, (POST)*. There are full and abbreviated versions of the POST. The full version is executed during a hard-boot sequence and the abbreviated version is performed during the soft-boot sequence, both of which were covered in chapter 2.

Most of the tests performed by the POST are not user-definable. In other words, in most cases you cannot select what is tested and what is not. Most current BIOSs do allow limited control by permitting the user to decide whether the floppies or extended memory are tested. The options to make these changes are available though listings in the CMOS Setup program. It may be desirable to skip over these tests to decrease the computer's boot time. The settings that control the POST were discussed in chapter 8.

You can tell when the full version of the POST is running because the RAM being tested is counted off in the upper left-hand corner of the computer's monitor. If the POST completes without errors, the computer will sound one short beep over the internal speaker. If the computer does not beep, or beeps more than once, something is usually wrong. The sequence of beeps is referred to as a **beep code.** The duration and number of beeps indicate what major component in the computer failed. If the computer does not beep and there is no display on the computer's monitor, the problem is usually caused by incorrect CMOS settings, CPU, RAM, ROM, or the power supply. Troubleshooting this condition is covered at the end of this chapter.

The problem with the beep codes is that, unfortunately, they are not standardized among the many brands of BIOSs written for the compatible PC. Table 15.1 contains a list of the beep codes for the BIOSs written by AMI, Award, IBM, and Phoenix, which are currently the most popular. Your computer may use different beep codes, so check your computer owner's manual or visit the BIOS writer's Web site.

Table 15.1  Beep Codes for BIOS ROMs Written by AMI, Award, IBM, and Phoenix*

| AMI BIOS | | |
|---|---|---|
| *Beep Code* | *Location of Error* | *Recommended Action* |
| 1 beep | Refresh failure | 1. Reseat the memory modules. 2. Replace the memory modules. 3. Replace the motherboard. |
| 2 beeps | Memory parity error | 1. Reseat the memory modules. 2. Replace the memory modules. |
| 3 beeps | Base 64 K failure | Reseat or replace the memory modules. |
| 4 beeps | Timer not operational | Replace the motherboard. |
| 5 beeps | Processor error | Replace the motherboard. |
| 6 beeps | Gate A20 failure | 1. Reseat the keyboard connector. 2. Replace the keyboard. |
| 7 beeps | Processor exception interrupt error | Replace the motherboard. |
| 8 beeps | Display memory read/write failure | Replace the video adapter. |
| 9 beeps | ROM checksum error | 1. Reseat the BIOS. 2. Contact the motherboard manufacturer for replacement BIOS ROMs. |
| 10 beeps | CMOS shutdown register r/w error | Replace the motherboard. |
| 11 beeps | Cache memory bad | Replace the motherboard. |

(Continued)

**Table 15.1** (Continued)

| AWARD BIOS ||| 
|---|---|---|
| *Beep Code* | *Location of Error* | *Recommended Action* |
| 1 long and 2 short beeps | Video controller | 1. Reseat the video adapter. 2. Replace the video adapter. |
| Any other | Memory failure | 1. Reseat the memory modules. 2. Replace the memory modules. |

| IBM (Generic) BIOS |||
|---|---|---|
| *Beep Code* | *Location of Error* | *Recommended Action* |
| No beep and no display | Power supply | See Introduction, PS troubleshooting. |
| Continuous beeping | Power supply | See Introduction, PS troubleshooting. |
| Repeated short beeps | Power supply | See Introduction, PS troubleshooting. |
| 1 long and 2 short beeps | Display adapter | See Chapter 12, troubleshooting. |
| 1 long and 1 short beeps | Motherboard | 1. Check the power connections. 2. Replace the motherboard. |
| 2 short beeps | Keyboard | 1. Check keyboard connection. 2. Replace the keyboard. |
| Normal beep, no display | Display adapter or monitor | 1. Reseat the monitor cable. 2. Try another monitor. 3. Replace the video adapter. |
| Normal beep, drive is not accessed | Disk adapter or drive | 1. Check the disk drive cable and power connections. 2. Replace the disk drive. |

| PHOENIX BIOS (Version 4.0) |||
|---|---|---|
| *Beep Code* | *Location of Error* | *Recommended Action* |
| 1-2-2-3 | BIOS ROM checksum | 1. Reseat the BIOS. 2. Contact the motherboard manufacturer for a replacement BIOS. |
| 1-3-1-1 | Test DRAM refresh | 1. Reseat the RAM modules. 2. Replace the RAM modules. |
| 1-3-1-3 | Keyboard controller | 1. Reseat the keyboard connector. 2. Replace the keyboard. 3. Replace the motherboard. |
| 1-3-4-1 | RAM failure, address line | 1. Reseat the memory modules. 2. Replace the memory modules. |
| 1-3-4-3 | RAM failure, low byte | 1. Reseat the memory modules. 2. Replace the memory modules. |
| 1-4-1-1 | RAM failure, high byte | 1. Reseat the memory modules. 2. Replace the memory modules. |
| 2-1-2-3 | ROM copyright notice | Contact the motherboard or BIOS manufacturer for replacement BIOS chips. |
| 2-2-3-1 | Unexpected interrupt | 1. Reseat the expansion cards. 2. Check IRQ configurations in the CMOS and operating system. |
| 1-2 | ROM checksum failure | 1. Reseat the BIOS chips. 2. Contact the motherboard or BIOS manufacturer for replacement BIOS chips. |
| 1 | No errors, boot OS | |

*The beep codes listed in this table may vary from computer to computer depending on the manufacturer and version of BIOS used. For this reason, it is highly recommended that you visit the Web site for the BIOS in your computer and download the correct beep codes.

**Table 15.2** IBM Numerical Error Codes*

| Error Code | Location of Error |
|---|---|
| 02x | Power supply |
| 1xx | Motherboard |
| 20x | RAM |
| 30x | Keyboard |
| 4xx | Monochrome adapter |
| 5xx | Color adapter |
| 6xx | Floppy disk adapter or unit |
| 7xx | Math coprocessor |
| 9xx | Primary printer adapter |
| 11xx | Primary serial adapter |
| 12xx | Secondary serial adapter |
| 13xx | Game adapter |
| 14xx | Alternate printer adapter |
| 17xx | Fixed disk adapter or unit |
| xxxxx | ROM |

*Note: The $x$ is replaced by a number in the actual code.

**Problem 15.1:** A computer running an AMI BIOS sounds five beeps during the POST. What is the problem?

**Solution:** According to Table 15.1, this beep code indicates a processor error.

Along with the beep codes, most of the current compatible's BIOS ROMs will display an error message on the screen indicating the source of the POST failure. Most of the error messages displayed by AMI, Award, and Phoenix BIOSs are self-explanatory. IBM's BIOS does not display an error message but displays a 3- to 5-digit number called on **error code.** These error codes have to be interpreted to determine the faulty hardware. Table 15.2 lists the numerical error codes displayed by IBM's BIOS and the indicated hardware failure.

**Problem 15.2:** What failure is indicated if a 1701 error message is displayed on an IBM computer?

**Solution:** Referring to Table 15.2, the fixed disk is the indicated failure.

### 15.1.1 Post Codes

When a test is started during the POST, the BIOS sends an 8-bit binary code, called a **POST code,** to port address 80H that indicates which test is in progress. If a special adapter card, called a **POST card,** is installed in one of the computer's ISA expansion slots, it will capture and display the POST code as two hexadecimal digits. If the computer fails during one of the tests of the POST, the POST card will still display the number of the test that was running when the failure occurred. When this happens, the displayed POST code can then be interpreted to determine which unit caused the failure.

Table 15.3  POST Code Errors*

| AMI |
|---|
| www.ami.com/amibios/support |
| **Award and Phoenix** |
| http://www.phoenix.com/pcuser/ |
| **MrBIOS** |
| www.mrbios.com/literat.html |

*Note: Most of the documents on these Web sites require Adobe's Acrobat Reader, which can be downloaded at no charge from www.adobe.com/prodindex/.

Unfortunately, the failures indicated by the POST codes that are displayed by the POST card are different among BIOS writers. Because 8 bits are available for each POST code, the different BIOS POST programs can generate up to 256 different combinations. This large number makes it very difficult to list all of the POST codes for all of the different BIOS writers. The URL for the location of the POST codes used by the BIOSs written by AMI, Award, Phoenix, and MrBIOS are listed in Table 15.3.

## 15.2  DISK-BASED DIAGNOSTICS

**Disk-based diagnostics** are loaded from a disk into the RAM of the machine. Before disk-based diagnostics can load, the computer must pass the POST. Once loaded, most disk-based diagnostics will provide most of the options contained in the following list. One thing is certain—if your computer can pass all of the options available with these diagnostics, the hardware is probably not the source of the computer's problem.

- RAM;
- ROM;
- Serial port(s);
- Parallel port(s);
- Math processor;
- Monitor;
- Video adapter;
- Modem;
- Keyboard;
- Mouse;
- Game port;
- Computer's performance;
- Display the IRQ assignments;
- Display the DMA assignments; and
- Loop-back plugs to completely check the serial and parallel ports.

Several vendors sell disk-based diagnostics. All that I have tried seem to do the job they were written to do, so the decision on which to buy basically boils down to user preference. Some vendors give away shareware or "lite" versions that can be downloaded from their respective Web sites. This is a nice feature because it gives individuals the ability to "try before they buy." To help determine which diagnostic package to purchase, Table 15.4 contains a list of the URLs for the Web sites of some of the more popular companies that write disk-based diagnostics programs.

Unfortunately, I cannot recommend a certain vendor's disk-based diagnostics program over another because all of those listed do the job for which they were written. Personally, I prefer diagnostics that run under MS-DOS or boot into their own operating system from a floppy diskette. This makes the

**Table 15.4** Diagnostic Program Vendors

| Advanced Personal Systems | www.syschk.com |
|---|---|
| American Megatrends, Inc. (AMI) | www.ami.com |
| CyberMedia | www.cybermedia.com |
| Data Depot | www.datadepo.com |
| DiagSoft, Inc. | www.diagsoft.com |
| McAfee.com Corporation | www.mcafee.com |
| Mijenix Corporation | www.mijenix.com |
| Quarterdeck Corporation | www.quarterdeck.com |
| Scope Diagnostics LTD | www.scopediag.co.uk |
| Symantec Corporation | www.symantex.com |
| TouchStone Software Corporation | www.touchstonesoftware.com |
| Ultra-X, Inc. | www.uxd.com |
| Watergate Software | www.ws.com |
| Windsor Technologies | www.windsortech.com |

diagnostics program easily transportable and gives it the greatest chance of booting on a computer that is having a problem. I also look for diagnostics programs that provide loop-back plugs. These are special connectors that are installed on the computer's serial and parallel ports that allow the diagnostic program to fully check these ports for proper operation. Before you make your decision on which diagnostics program to purchase, visit the Web sites that are listed in Table 15.4, or other Web sites you may know of, and download evaluation copies and decide which you prefer.

There is one diagnostic program that we are going to discuss that has limited abilities. The reason is that this disk-based diagnostics program has been provided by Microsoft operating systems since MS-DOS 6.0. The name of this program is Microsoft Diagnostics.

### 15.2.1 Microsoft Diagnostics (MSD)

Microsoft provides a small, disk-based diagnostic program called *Microsoft Diagnostics* that uses the filename MSD.EXE. The program installs automatically with MS-DOS 6.X and Windows 3.X during the operating system's installation processes. The MSD program is not installed as part of Windows 9X but is available on CD-ROM. From the CD, the program can be executed or copied onto the computer's hard drive. Unfortunately, MSD is not provided with Windows NT or Windows 2000.

On computers that run only MS-DOS, the program is located in the DOS directory. On computers that run Windows 3.X, the program is located in the Windows subdirectory. On computers that run Windows 9X, the program is usually located on the CD at \tools\oldmsdos\msd.exe. On computers running Windows 9X, it is highly recommended that the computer be in DOS mode before MSD is executed. Running it from the Window 9X DOS prompt will cause the program to return incorrect information.

MSD is not truly a diagnostics program because it does not give the user the ability to run in-depth checks on any of the computer's hardware. MSD does provide the auditing of the basic hardware and name and address location of certain drivers installed in the computer. The result of this auditing allows the user to see information pertaining to the following:

- Microprocessor;
- Chipset;

# Diagnostics

**Figure 15.1** Microsoft Diagnostics Main Menu

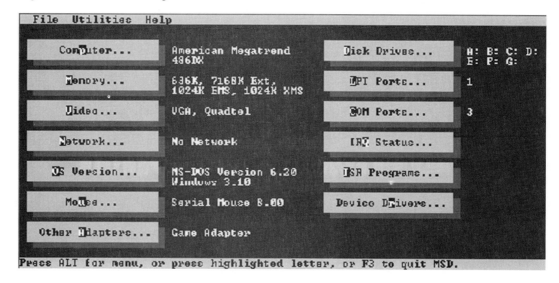

**Figure 15.2** Microsoft Diagnostics COM (Serial)

| COM PORTS | | | | |
|---|---|---|---|---|
| | COM 1: | COM 2: | COM 3: | COM 4: |
| Port address | 03F8H | 02F8H | O3E8H | N/A |
| Baud rate | 1200 | 1200 | 2400 | |
| Parity | None | None | None | |
| Data bits | 7 | 7 | 8 | |
| Stop bits | 1 | 1 | 1 | |
| Carrier detect (CD) | No | No | No | |
| Ring Indicator (RI) | No | No | No | |
| Data set ready (DSR) | Yes | Yes | No | |
| Clear to send (CTS) | Yes | Yes | No | |
| UART Chip used | 16550AF | 8250 | 8250 | |

- Memory map;
- Video adapter;
- Network information (where applicable);
- Mouse information;
- Game port information;
- Disk drive type and letter assignments;
- Parallel port (LPT) address assignments and flow control status;
- Programmable serial port (COM) address assignments, flow control status, and word size configurations;
- Basic information about the Windows operating system (Note: The Windows version number will probably be wrong when running Windows 9X);
- In-depth IRQ information including assignments, service routine addresses, status, and the handler (BIOS or OS);
- TSR names and address locations; and
- Device driver names and locations.

Figure 15.1 shows the MSD Main menu and the available options. To access the information shown on each item, simply press the highlighted letter in each menu item. For example, to list the COM ports (serial), press the C key. The COM port listing will appear, similar to the screen shown in Figure 15.2.

**Problem 15.3:** If another serial port with the configuration shown in Figure 15.2 is installed in the computer, what port should it be assigned to?

**Solution:** The information shown in Figure 15.2 indicates that COM1, COM2, and COM3 are already installed in the computer, so the new serial port will have to be COM4.

MSD is totally safe. You cannot hurt anything in your computer by running it. You should select each option and try to interpret the information MSD is giving you about the selected item. You will be surprised at what you can learn about the computer from this simple program.

## 15.3 WINDOWS 9X/2000 SYSTEM PROPERTIES

Throughout this textbook we have been using a very powerful tool in the troubleshooting arsenal provided by Windows 9X/2000 called **System Properties.** System Properties gives the user the ability to manage the computer resources through Device Manager and also allows the user to customize some of the computer's performance settings. To open the System Properties window, follow the steps outlined in Procedure 15.1.

### 15.3.1 Device Manager

On computers that run Windows 9X/2000, the computer's memory, peripherals, ports, and drivers are controlled by the user through **Device Manager.** Along with controlling the resources and drivers, Device Manager also indicates whether each port or peripheral is functioning properly. To open Device

---

**Procedure 15.1: Opening the System Properties window**

1. Click on the Start button, then move up to Settings, and then over to Control Panel.
2. Click on Control Panel.
3. After the Control Panel window opens, locate and then double-click on the System icon.

---

Figure 15.3  Windows 9X Device Manager

Manager on a Windows 9X computer, first open the System Properties window and then click on the Device Manager tab. If Windows 2000 is being used, open System Properties, click on the Hardware tab, and then click on the Device Manager button. Once Device Manager is open, the Windows 9X System Properties window, similar to the one shown in Figure 15.3, will appear.

When Device Manager is open, it will display a list of all of the Windows 9X/2000 controlled hardware categories that are currently installed in the computer. Device Manager does not display ports or peripherals that are controlled strictly by an application and not by Windows 9X, however.

Device Manager gives an instant indication of categories that contain a port or peripheral that is having one of three problems that Windows 9X/2000 can detect. These three problems are resource conflicts, incorrect or defective drivers, and no driver is loaded to support the device. A resource conflict or defective driver is indicated by an exclamation mark beside the category, whereas the absence of a driver is indicated by a question mark. Note that Device Manager does not detect problems other than defective or missing drivers and resource conflicts.

From the main Device Manager window, each category can be perused easily to display information relative to its ports and peripherals. To display the properties of an individual port or peripheral, use the steps outlined in Procedure 15.2.

### Procedure 15.2: Displaying category information using Device Manager

1. Open System Properties using Procedure 15.1.
2. In Windows 2000, click on the Hardware tab.
3. Once the System Properties window is displayed, click on the Device Manager tab in Windows 9X or the Device Manager button in Windows 2000.
4. Locate and then click on the plus (+) sign beside the desired category in the main Device Manager display.
5. A listing of all the ports and peripherals installed under that category will display. For example, a typical listing of the ports and peripherals under the "Hard disk controllers" category is shown in Figure 15.4.
6. Once the ports and peripherals under the category are listed, double-click on the name of the port or peripheral to display its properties, or click on the name and then click on the Properties button. A typical display for the "Primary IDE controller (dual fifo)" listing is shown in Figure 15.5.

Figure 15.4  Listings Under the "Hard Drive Controllers" Category

**Figure 15.5** Typical "Primary IDE Controller" Properties Display

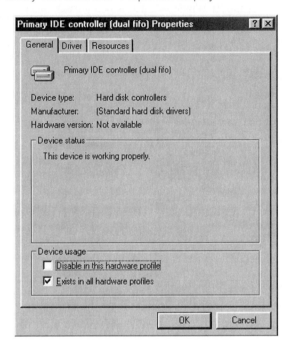

---

### Procedure 15.3: Changing a port's resources using Device Manager

**Note:** Always document the port's currently used resources before making any changes.

1. From the port's Properties window, click on the Resource tab.
2. Click on the check box labeled "Use automatic setting" to remove the checkmark.
3. Once the check is removed, click on the scroll button beside the text box labeled "Setting based on" to choose from a list of basic configurations, or double-click on the actual resource to change its setting individually.
4. Once the resource has been changed, click on the Change Settings button or on the Cancel button to exit without making changes.

---

When the selected device's Property window displays, it will default to the General information tab. This General information tab will display the name, manufacturer, and version number of the selected port or peripheral. Under the General information tab there is also a frame called "Device status" that indicates if the device is working properly or if there is a problem. If a problem is detected, the probable cause will be displayed within this frame. Remember that the only two problems that Device Manager can detect are a resource conflict and missing or defective drivers.

The other tabs under the port or peripheral's Properties window will vary depending on the category and device chosen. In Figure 15.5, the tabs available for the "Primary IDE controller" are General, Driver, and Resources. Along with these, tabs may be included for Settings and Diagnostics.

The Resource tab, available under a port's Properties window, is used to display any conflicts and is also used to change or set the resources used by PnP and softset ports. If the user can change the resources used by a port, a small check box will be displayed with the label "Use automatic settings." The procedure to follow to change a port's resources is listed in Procedure 15.3. An example of a COM port that can be softset through the Resource tab is shown in Figure 15.6.

**Figure 15.6** Resource Tab for Softset COM Ports

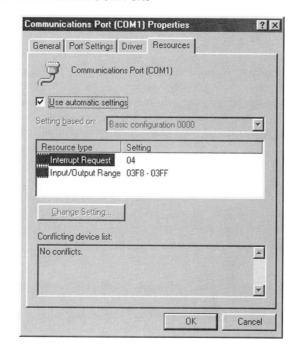

### 15.3.2 System Performance

The Performance tab that is available under System Properties is used to trim the settings of some of the computer's devices to optimize them relative to the computer's application. This tab includes settings for the file system and video performance along with the virtual memory settings.

Under the File System tab, there are settings to optimize the performance of the hard drive(s), floppy disk(s), CD-ROM(s), removable disk(s) (if installed), and basic file system troubleshooting aids. Most of the settings under the Troubleshooting tab will greatly affect the system performance, so it is highly recommended that only technicians and advanced users who have knowledge of these settings make any modifications. As an aid to understanding the settings offered under the Troubleshooting tab, Table 15.5 lists the options along with a basic description. This information was obtained directly from the Microsoft Windows 95 Resource Kit.

**Table 15.5** Disk Performance-troubleshooting Options*

| File System Option | Description |
| --- | --- |
| **Disable New File Sharing And Locking Semantics** | This option alters the internal rules for file sharing and locking on hard disks, governing whether certain processes can have access to open files in certain share modes that guarantee a file will not be modified. This option should be checked only in the rare case that an MS-DOS-based application has problems with sharing under Windows 9X. This sets SoftCompatMode=0 in the registry. |
| **Disable Long Name Preservation for Old Programs** | This option turns off the tunneling feature, which preserves long filenames when files are opened and saved by applications that do not recognize long filenames. This option should be checked in the rare case that an important legacy application is not compatible with long filenames. This sets PreserveLongNames=0 in the registry. |

(Continued)

Table 15.5 (Continued)

| File System Option | Description |
|---|---|
| **Disable Protected Mode Hard Disk Interrupt Handling** | This option prevents Windows 9X from terminating interrupts from the hard disk controller and bypassing the ROM routine that handles these interrupts. Some hard disk drives might require this option to be checked in order for interrupts to be processed correctly. If this option is checked, the ROM routine handles the interrupts, which slows system performance. This sets VirtualHDIRQ=1 in the registry. (This setting is off by default for all computers in Windows 95, which is the reverse of Windows 3.x.) |
| **Disable All 32-bit, Protected-Mode Disk Drivers** | This option ensures that no 32-bit disk drivers are loaded in the system, except the floppy driver. Typically, you would check this option if the computer does not start due to disk peripheral I/O problems. If this option is enabled, all I/Os will go through real mode drivers or the BIOS. Notice that in this case, all disk drives that are visible only in protected mode will no longer be visible. This sets ForceRMIO=1 in the registry. |
| **Disable Write-Behind Caching For All Drives** | This option ensures that all data are flushed continually to the hard disk, removing any performance benefits gained from disk caching. This option should be checked only in the rare cases when you are performing risky operations and must ensure prevention of data loss. For example, a software developer who is debugging data at Ring 0 while creating a virtual device driver would check this option. This sets DriveWriteBehind=0 in the registry. |

*Copied from the Microsoft Windows 95 resource kit documentation

---

**Procedure 15.4: Opening Windows 98 and 2000's System Information**

1. Click on the Start button on the Task Bar.
2. Move up to Programs.
3. Move over to Accessories.
4. Move over to System Tools.
5. Click on System Information.
6. Click on the plus (+) sign in the left pane beside each category to see additional listings.
7. Click on a listing in a category to display information pertaining to it in the right pane.

## 15.4 WINDOWS 98/2000's SYSTEM INFORMATION

System Information is a very useful utility that is provided by Windows 98/2000 to give the user one place to go to glean configuration information about all of the computer's resources. System Information is a compilation of data obtained in Device Manager and the registry. To open System Information, follow the steps outlined in Procedure 15.4.

System Information will display a window similar to the one shown in Figure 15.7. The options available in the left pane are System Information, Hardware Resources, Components, and Software Environment. The System Information is the default display when the window originally opens. As can be seen in Figure 15.7, this window is basically an overall view of the computer's environment. The Hardware Resources category is the place to go to list conflicts, DMA and IRQ assignments, and the computer's I/O and memory addressing maps. The Components category contains in-depth information on all of the categories listed in Device Manager.

System Information is a valuable tool in determining the software and hardware environments for a given computer. It is also a good place to go to provide information to technical support personnel when the need arises. In fact, the information can be printed or saved in a file in text form so the information can be faxed directly to technical support. This is accomplished by using either the Save or Print options available on System Information window's Tool Bar. The Software Environment category con-

Figure 15.7 Windows 98 and 2000's System Information

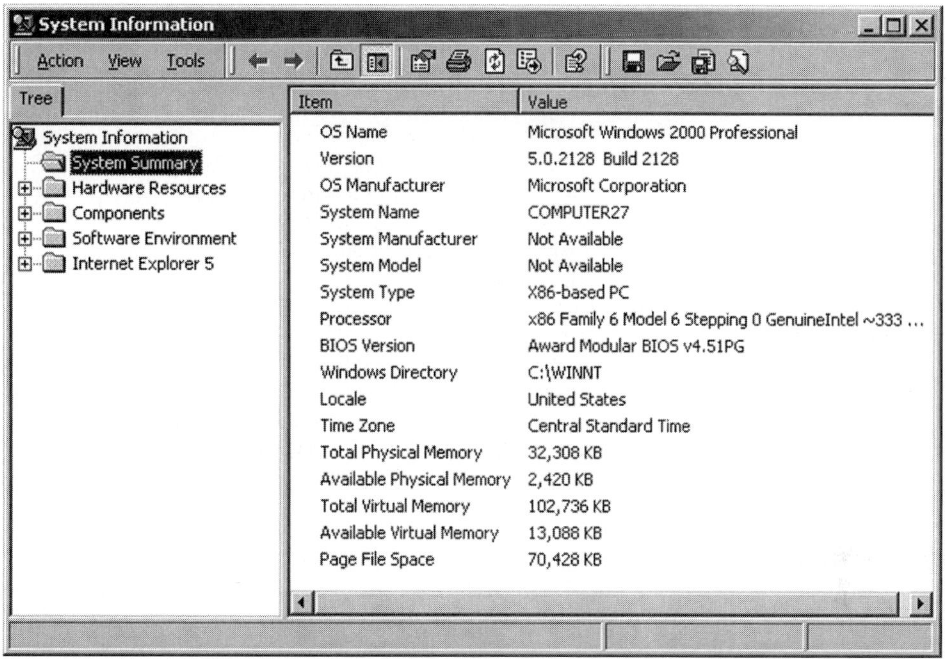
System Information
Action View Tools
Tree
System Information
System Summary
Hardware Resources
Components
Software Environment
Internet Explorer 5
Item
Value
OS Name
Microsoft Windows 2000 Professional
Version
5.0.2128 Build 2128
OS Manufacturer
Microsoft Corporation
System Name
COMPUTER27
System Manufacturer
Not Available
System Model
Not Available
System Type
X86-based PC
Processor
x86 Family 6 Model 6 Stepping 0 GenuineIntel ~333 ...
BIOS Version
Award Modular BIOS v4.51PG
Windows Directory
C:\WINNT
Locale
United States
Time Zone
Central Standard Time
Total Physical Memory
32,308 KB
Available Physical Memory
2,420 KB
Total Virtual Memory
102,736 KB
Available Virtual Memory
13,088 KB
Page File Space
70,428 KB

### Procedure 15.5: Adding Dr. Watson to the Task Bar

**Note:** Dr. Watson is only available on computers running Windows NT or Windows 98.

1. Click on the Start button and then click on Run.
2. In the Run text box, type "c:\windows\drwatson" (without the quotes).
3. Click on the OK button to add Dr. Watson to the System Tray or the Cancel button to abort.

tains listings for the 16- and 32-bit drivers, the task currently running, and programs that launch automatically at startup. For information on the options available, type "system information" in the Index tab of the Help menu.

## 15.5 DR. WATSON

Windows NT/2000 and Windows 98 both offer another tool called **Dr. Watson** to help in diagnosing software problems. This diagnostic aid automatically takes a snapshot of the computer's software environment that is in place when an application crash occurs. It stores this information in a log file that can be viewed to help determine what was going on at the time of the crash. However, the created log is of little use to the average individual, but is a valuable troubleshooting tool when consulting Microsoft's technical support.

To allow Dr. Watson to automatically take snapshots of the system when a crash occurs, it must be added to the System Tray located in the right-hand corner of the Task Bar. This is a simple process that can be performed using Procedure 15.5.

Once added to the System Tray, Dr. Watson can also take a manual snapshot of the system. Right-click on the Dr. Watson icon located to the right on the Task Bar, and then select Dr. Watson from the pop-up options menu. The "Open Log File" can be selected on the same pop-up option menu.

**Figure 15.8** Keyboard's General Tab Showing the Troubleshooter Button

## 15.6 WINDOWS 2000 TROUBLESHOOTER

Microsoft has provided a useful troubleshooting tool with Windows 2000 called *Troubleshooter*. This option is available on all devices listed under each category of Device Manager. Troubleshooter gives the user a step-by-step procedure to follow in the event a device is not functioning correctly. An example of the General tab for the computer's keyboard, showing the Troubleshooter button, is shown in Figure 15.8.

## 15.7 VIRUS SCANNERS

A computer **virus** is an unauthorized program, that when run, attaches itself to or overwrites an existing program. Once a computer is infected, the virus will eventually divulge its presence by deliberately causing the computer to malfunction. Sometimes the virus will cause a minor malfunction that is humorous; other times the virus will deliberately destroy software or hardware.

Even though it is not true, a lot of people believe they can get a virus from simply reading e-mail or downloading a graphics file. No matter what you have heard, this is not possible. Programs spread viruses, not data files such as e-mail and graphics files, which are simply data files used by programs. Unfortunately, most e-mail providers now allow program files to be attached to e-mail. Most individuals will unknowingly download and execute an attached program file that contains a virus and mistakenly think the infection was caused by the e-mail.

As stated, computers are infected with a virus through the execution of a program that contains the virus. The programs that spread the virus can come from many sources, such as co-workers, downloads, and even friends. The individual who gives you the disk or the program file that contains the virus usually does it unknowingly.

Virus programs are created by hackers who write these programs (if you want to call them that) and attach the virus to some seemingly harmless program such as a game. Once the program is run on a computer, the virus will then attach itself to some very important program on the hard drive such as the Volume Boot Record, COMMAND.COM, WIN.COM, or the system files MSDOS.SYS and IO.SYS. Once a computer is infected, the virus can replicate by attaching itself to legitimate programs. One pop-

Diagnostics 509

ular method of passing viruses is through the Boot Record of a floppy diskette. If the computer attempts to boot from the infected diskette, the virus will be spread to the computer.

Most viruses do not do their ugly work immediately. They lay dormant until some triggering mechanism, such as a certain date, sets them off to do their dirty work. Before this happens, you have probably passed the infected program on to your friends, they have passed it to theirs, and so on. This sounds very similar to the spreading of a human virus, doesn't it?

Fear of a virus infecting your machine is probably the best reason to buy off-the-shelf licensed software instead of getting software from other sources. Any program that comes from second sources can be infected. As a word of advice, if you download important files from the Internet such as hardware drivers, only download first-source files from the actual manufacturer's Web site and always check them for viruses before they are installed.

To help with the fight against the spread of viruses, software is available from several vendors that will scan for the presence of a virus. If a virus is detected, these programs will try to remove (clean) the virus without damaging the original program.

MS-DOS 6.0 provides a **virus checker** called *Microsoft Anti-virus (MSAV),* which can be installed as an MS-DOS-based program or a Windows-based program. There is only one disadvantage to using this software: it cannot be automatically updated to keep up with the ever-expanding list of viruses.

Windows 9X and Windows NT do not provide a virus-scanning program with the operating system, so one will have to be purchased from a second source. McAfee.com Corporation (www.mcafee.com) and Symantec Corporation (www.symatec.com) offer two of the best virus scanners I have seen. Both companies offer free trial versions of their virus-checking software that can be downloaded from their Web sites. These companies also provide free updates of their virus signature files from their respective Web sites. This is a necessary feature because new viruses are introduced almost daily.

If you are going to deal with the Internet or the transfer of files or programs by diskette, I highly recommend that you purchase and regularly use a virus-scanning program. In fact, running a virus scan of your hard drives should be part of your preventative maintenance routine. Some of the other things you can do to help avoid infecting your computer with a virus are provided in the following list:

- Never attempt to boot the computer from the floppy drive unless you are certain that the diskette's boot record is virus free.
- Never run programs that were given to you by anyone. If you must, make sure they are first checked with an up-to-date virus scanner.
- When using hyperlinks on a Web site, make sure the link is to a file with an htm (html) extension.
- Never download files attached to e-mail with an "exe" or "com" extension. If you must, check the file with an up-to-date virus scanner before it is run.
- Never use a screen-saver that is given to you by anyone.
- Because MSWord documents allow embedded macrocodes, do not read any files with a "doc" extension until they are checked with an up-to-date virus scanner.

I am not trying to scare you away from enjoying the entertainment and information available on the Internet or from the on-line community. In fact, I highly recommend that you get connected and start your on-line experience as soon as possible. But I do recommend that you purchase a good virus-scanning program and use it regularly.

## 15.8 TROUBLESHOOTING A DEAD COMPUTER

*Troubleshooting* is basically the art of locating and repairing problems. Troubleshooting procedures are covered throughout this book, especially in the sections for the port and peripheral, so the intent of this section is to simply tie up loose ends by covering the basic troubleshooting procedure used for a totally dead computer. A dead computer is indicated by a blank display, and no beep codes are sounded when the computer is turned on.

Believe it or not, troubleshooting a dead computer is usually easier than troubleshooting one with intermittent problems. This is because there are only a few things that can prevent the computer from attempting to boot. A basic algorithm to aid in troubleshooting a dead computer is provided in Procedure 15.6.

### Procedure 15.6: Troubleshooting a dead computer

1. Make sure the computer's power cord is plugged into an operational power receptacle and into the back of the computer.
2. Apply power to the computer and check the power LED on the front of the chassis and listen for the power supply fans. If either is not present, follow the power supply troubleshooting procedure outlined in the Introduction.
3. Use an anti-static wristband and static mat when working inside the computer.
4. Remove power from the computer and then remove the chassis cover. Inspect the motherboard for loose connections. Push in on all connectors going to the motherboard, adapters, and peripherals. Make sure the power supply connector is seated snugly in the motherboard connector. Apply power and try booting the computer.
5. Remove power from the computer and then look for missing or loose ICs and memory modules. Sometimes chips can actually work their way out of sockets. This is especially true in tower chassis. Also, sometimes people do things out of malice by deliberately loosing or removing ICs. I have also seen the memory stolen from computers on more than one occasion. If any missing ICs or memory modules are found, replace them, apply power, and then try to boot the computer. Note: Some memory modules and IC sockets are deliberately left empty depending on the upgrade options available with the computer.
6. Remove power from the computer and then check for the correct position of motherboard jumpers and switches. For some reason, people have a tendency to move these around when a computer is not working and try to fix it themselves. If any switches are found to be in the wrong position, correct the problem, apply power, and then try booting the computer.
7. If a POST card is available, remove power from the computer and then install the POST card into an empty slot in the expansion bus. Apply power to the computer while watching the display on the POST card. Once the POST failure occurs, document the code being displayed and refer to the POST code chart provided from the BIOS writer's Web site. A POST card is a valuable aid when troubleshooting a dead computer. Once the problem indicated by the POST card has been corrected, apply power and then try to boot the computer.
8. Remove power from the computer and then reseat the BIOS ROM, memory modules, and CPU. Once completed, apply power to the computer and attempt to boot the computer.
9. Remove power from the computer, and then remove all expansion cards. Reapply power. If the computer attempts to boot by sounding a beep code, remove power from the computer and then reinsert the cards one at a time, beginning with the video card. After installing each card in the appropriate expansion slot, apply power to the computer. If the computer attempts to boot, remove power and reinstall another card. Repeat the process until the defective card is located.
10. Remove power from the computer and then remove all peripheral data cables going to the motherboard. However, do not remove the motherboard's power connector(s) or the connector going to the internal speaker. If the computer now attempts to boot by sounding a beep code, reinstall each peripheral connector, one at a time, until the defective peripheral is located.
11. Clear the CMOS with the clear CMOS jumper located on the motherboard. If there is not a jumper available, remove the battery on the motherboard and reinstall it after approximately 20 seconds. If a CMOS failure error message is displayed during the next boot sequence, allow the computer to run the CMOS Setup program and choose the "Load the Default Setting" or a similar option. If the computer now boots, restore the CMOS configuration from your documented hardcopy.
12. If all of this fails, replace the motherboard.

## 15.9 CONCLUSION

All IBM-compatible computers run a power-on-self-test (POST) each time the computer is turned on or when the Reset button is pressed. These diagnostics are not in-depth and only determine if the computer is capable of loading the operating system. If errors are detected during the POST, the computer will usually display an error message on the computer's display indicating the source of the malfunction. Also, most compatible computers will sound a series of beep codes by way of the computer's internal speaker. Last, but not least, the computer will also send a POST code indicating the test that was in progress to I/O address 80H.

To determine if the computer's hardware is functioning at top efficiency, a full-blown version of disk-based diagnostics should be used. Disk-based diagnostics will check most of the major functions of the computer's hardware and will usually check the computer's performance compared to other computers.

System Properties is a valuable tool available to Windows 9X users. Through System Properties, the user can locate nonfunctioning peripherals and ports, find resource conflicts, change resources used by PnP and softset devices, and optimize the computer's performance.

A virus is a program written to cause a computer to malfunction. If any outside disks are used on a computer, or programs are downloaded from the Internet or a BBS, the program or disk should be checked for a virus before any attempt is made to access the information. A program that detects and corrects viruses is called a virus scanner.

## Review Questions

1. What are the meanings of the following acronyms?
   a. POST
   b. CMOS
   c. CPU
   d. RAM
   e. ROM
   f. BIOS
   g. MSD

### Fill in the Blanks

1. A _____ program is used to check the computer's hardware for correct operation.
2. The small diagnostic program contained within the BIOS that automatically runs each time a hard-boot is performed is called the _____ _____ _____ _____ .
3. To have the best chance of locating defective hardware, _____-_____ diagnostics should be used.
4. If a computer is running AMI BIOS and 11 beeps are sounded during the POST, a _____ _____ error is indicated.
5. If an IBM computer displays a 1301 error code during the POST, a _____ _____ failure is indicated.
6. The BIOS will send a binary code to port address _____ indicating the test that is currently in progress.
7. A device that allows disk-based diagnostics to fully check out serial and parallel ports is called a _____ _____ plug.
8. The small auditing program provided by Microsoft is called _____ _____ .
9. _____ _____ allows the user to configure the ports and peripherals controlled by Windows 9X.
10. A program that is provided by Windows 98 and NT that will automatically take snapshots of the computer's memory when an application crashes is called _____ _____ .

### Multiple Choice

1. If a computer running AMI BIOS sounds two beeps each time it is booted:
   a. the display adapter is defective.
   b. the power supply is defective.
   c. the keyboard is defective.
   d. a memory parity error has occurred.
2. If a computer running Phoenix BIOS beeps once, then three times, then once, and finally three times, the failure is probably the computer's:
   a. RAM.
   b. keyboard.
   c. ROM.
   d. hard drive.
3. What is the first thing one should check when a computer is dead?
   a. the video monitor
   b. the power source
   c. the floppy drive(s)
   d. the position of the Reset button
4. POST codes are displayed on the computer's screen.
   a. True
   b. False
5. If the number 502 appears on the display of an IBM computer when power is supplied:
   a. the display adapter is defective.
   b. the network adapter card is defective.
   c. the floppy adapter is defective.
   d. the power supply is defective.
6. Disk-based diagnostics:
   a. are contained in the computer's ROM and are sometimes called the POST.
   b. can never check the operation of the computer's serial ports.
   c. are loaded into the computer's ROM from a disk.
   d. are loaded into the computer's RAM from a disk.
7. A virus can infect a computer if the user simply reads an e-mail.
   a. True
   b. False
8. Microsoft Diagnostics (MSD):
   a. will check the computer's RAM for proper operation.
   b. will check the computer's serial port(s) for proper operation.
   c. will check the computer's parallel port(s) for proper operation.
   d. None of the above answers is correct.
9. A computer virus:
   a. cannot be detected before its damage is done.

b. is the name given to programs written to deliberately cause the computer to malfunction.
   c. is usually spread from computer to computer by sharing software.
   d. Both answers b and c are correct.
   e. Both answers a and b are correct.

10. If the RAM passes the test in the POST, it has to be error free.
    a. True
    b. False

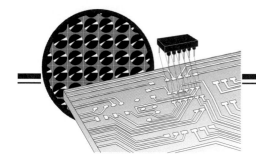

# CHAPTER 16

# Interrupt and DMA Basics

## OBJECTIVES

- Define terms and expression used relative to interrupts.
- Explain the difference between 8-bit and 16-bit interrupts.
- List the basic algorithm for an interrupt request.
- Describe the software polling method of accessing I/O devices.
- State the interrupt method of accessing I/O devices.
- Define terms and expressions used relative to DMA.
- Describe the differences between 8-bit and 16-bit DMA.
- Explain the basic steps the PC follows for a DMA sequence.
- Describe the difference between a transfer using DMA and a bus-mastering hard drive.

## KEY TERMS

software polling
interrupts
interrupt service routine
interrupt vector table
type number
internal interrupt
software interrupt

hardware interrupt
RESET
nonmaskable (NMI)
interrupt request (IRQ)
programmable interrupt controller (PIC)
interrupt acknowledge (INTA)

8-bit interrupt
16-bit interrupt
legacy IRQ
direct memory access controller (DMAC)
bus-master
hold acknowledge (HLDA)

Chapter 9 introduced you to the basic concept of interrupts and DMA. That introduction was just enough to help understand I/O devices. To fully understand I/O, a more in-depth discussion of interrupts and DMA is necessary. The more you understand how things work and tie together, the more capable you will be in configuring and diagnosing hardware. We will begin our discussion with interrupts and finish with DMA. Before continuing, you may want to return to chapter 8 to review the basic information on interrupts and DMA that was introduced there.

513

## 16.1 SOFTWARE OR HARDWARE INTERRUPTS

It has already been established that all microcomputers consist of four basic units: the microprocessor, memory, input, and output. Accessing the memory in the computer is accomplished easily by the microprocessor because the memory units are inside the microcomputer and are connected directly to the computer's buses. I/O transfer is another matter, however. I/O devices are usually quite slow, are usually controlled by external forces, and are not under direct control of the microprocessor like the memory in the machine. Due to these facts, the I/O devices do not connect directly to the computer buses. They communicate with the buses through ports. The computer communicates directly with the port, not with the device itself. The two techniques used to determine when the communication takes place with I/O ports are called *software polling* and **hardware interrupts.**

### 16.1.1 Software Polling

Using the **software polling** method, all I/O transfers are controlled by the program alone. The program constantly reads input devices, whether they have changed or not, and constantly updates output devices, whether they need it or not. This method works quite well in small computers in which the microprocessor does not have a lot to do. But in larger computers with more sophisticated I/O arrangements, the time it takes to read each input device and update each output becomes cumbersome and, in some cases, is prohibitive.

A good analogy of I/O transfer using the polling method is a classroom lecture. When polling is used, students cannot ask questions during a lecture unless the instructor asks an individual student, by name, if he or she has a question. Only the student whose name is called can ask a question at that time. The instructor stops the lecture at given time intervals and polls each student for questions. Each student is polled, whether he or she has a question or not.

As can be seen from this analogy, software polling takes a lot of computer time but requires very little computer hardware to accomplish. A faster but more costly method used by I/O devices, to get the microprocessor's attention, is a method called interrupts.

### 16.1.2 Interrupt Basics

A more effective way of handling I/O transfer involves a method called **interrupts.** An interrupt is a technique by which an I/O device or program can signal the microprocessor, through pins on the control bus or special instructions, that something needs to be taken care of. The microprocessor can also interrupt itself when some fault conditions are detected. When an interrupt occurs, the microprocessor usually suspends its current operation and takes care of the interrupt before proceeding with the operation it was performing before the interrupt occurred.

An analogy of the interrupt method of I/O transfer is the way formal classroom lectures might work. The instructor lectures at a constant rate, never asking for questions from the students. When a student has a question, he or she raises a hand so the instructor can see it. When the instructor sees the raised hand, he or she:

- Suspends the lecture (making a note of the current location);
- Lets the student ask the question;
- Answers the question; and
- Continues with the lecture at the same spot at which it was suspended.

As you can probably tell, the use of interrupts is a more efficient method than polling but requires more computer hardware to implement. Added hardware translates to a higher cost for the computer.

### 16.1.3 Software Polling or Interrupts

The option of using the software polling or the interrupt method to accomplish I/O transfer is up to the engineers who design the computer. Some computers use only software polling; some computers use only interrupts; others use both. The compatible computers use both methods to handle I/O transfer.

Software polling is the primary method used to update the video adapter's memory, while interrupts are used for the computer's time of day, floppy drive(s), hard drive(s), keyboard, serial port(s), parallel port(s), memory parity failure, and the math coprocessor.

## 16.2 INTERRUPTS IN DETAIL

The information in this section is not really necessary to have a basic understanding of the compatible computer's interrupts. However, to fully understand the concept of the operation of the compatible PC and the reasons why things are the way they are, it is fairly important for you to understand the concepts introduced in this section.

The Intel-compatible microprocessors, used in the compatible computers, use three main categories of interrupts: internal, software, and hardware. Regardless of which category of interrupt is used, the microprocessor uses the following algorithm when it receives an interrupt:

1. Suspend running the current program.
2. Save the current program's location and status in a section of memory called the *stack*. The location of the stack is kept in a register within the microprocessor called the *stack pointer*.
3. Jump to a program that is called an **interrupt service routine.** Simply stated, an interrupt service routine is a program that is written specifically to take care of the interrupt. The location of the service routine is obtained through the use of a type number and an interrupt vector table, which are discussed later.
4. Once the interrupt service routine is completed, the information that was saved on the stack is retrieved and control of the computer is returned back to the program that was originally running before the interrupt occurred.

The microprocessors used in the compatible computers find the memory address of the interrupt service routine by using a table located in the first 1,024 bytes of the computer's RAM called the **interrupt vector table** or, simply, *vector table*. The first 1,024 bytes of the computer's RAM will place the vector table between hexadecimal addresses 0 and 3FF, which is within the address range of the computer's volatile RAM. The vector table is located in volatile RAM, so it will be lost each time the computer is turned off. This is no problem, however, because a copy of the table is kept in the computer's nonvolatile ROM and is copied to the correct addresses each time the computer is booted. Placing the vector table in the computer's RAM makes it possible for the operating system, TSRs, and applications to modify the contents of the table, if needed.

Every entry in the vector table is used as a pointer to the memory address of a unique interrupt service routine. To hold the address, each table entry requires four consecutive memory addresses. Four addresses per entry and 1,024 addresses means that the vector table can hold a maximum of 256 entries (1,024/4). When an interrupt occurs, the actual location that the microprocessor goes to in the vector table is determined by a number from 0 through 255 (0 through FF hexadecimal) that is called the **type number.** The microprocessor takes the supplied type number and multiplies it by 4, then uses the results as a pointer to the first of four addresses in the vector table, which contains the location of the interrupt service routine. An example of finding the vector table address, given the type number, is shown in Problem 16.1.

---

**Problem 16.1:** What four addresses will be used in the vector table for a Type A interrupt?

**Solution:** $A_{16} \times 4 = 28_{16}$ (remember this is base 16 arithmetic and not base 10). The four consecutive addresses that will be used for a Type A interrupt are $28_{16}$, $29_{16}$, $2A_{16}$, and $2B_{16}$.

---

The four addresses obtained from multiplying the type number are not the actual address of the interrupt service routine. The contents of these four consecutive addresses contain the address of the interrupt service routine. A few of the type number assignments, along with the interrupt category used by the compatible computers, are listed in Table 16.1.

Table 16.1  Partial List of Interrupt Type Number Assignments and Categories

| Hex Type Number | Interrupt | Category |
|---|---|---|
| 0 | Divide by 0 | Internal |
| 1 | Machine code single step | Internal |
| 2 | Nonmaskable interrupt | Hardware |
| 4 | Signed number overflow | Internal |
| 5 | Screen print | Software |
| 8 | IRQ0—Programmable Interval Timer (PIT) | Hardware |
| 9 | IRQ1—Keyboard | Hardware |
| A | IRQ2—Cascade from second PIC | Hardware |
| B | IRQ3—ISA COM2 and COM4 | Hardware |
| C | IRQ4—ISA COM1 and COM3 | Hardware |
| D | IRQ5—Second parallel port (sound card) | Hardware |
| E | IRQ6—Floppy disk drive | Hardware |
| F | IRQ7—First parallel port | Hardware |
| 10 | Set video mode | Software |
| 12 | Get memory size | Software |
| 13 | Use to access the hard drive | Software |
| 14 | Initialize serial port | Software |
| 16 | Read keyboard buffer | Software |
| 17 | Output a character to the printer | Software |
| IA | Get time of day | Software |
| 1F | Set video graphics mode | Software |
| 20 | Used to terminate DOS programs | Software |
| 21 | MS-DOS functions call | Software |
| 27 | Terminates a TSR | Software |
| 70 | IRQ8—CMOS real-time clock | Hardware |
| 71 | IRQ9—Software redirect of IRQ2 | Hardware |
| 72 | IRQ10—Not assigned | Hardware |
| 73 | IRQ11- Not assigned | Hardware |
| 74 | IRQ12—Not assigned | Hardware |
| 75 | IRQ13—Math coprocessor | Hardware |
| 76 | IRQ14—Primary EIDE | Hardware |
| 77 | IRQ15—Secondary EIDE | Hardware |

## Interrupt and DMA Basics

**Problem 16.2:** What is the type number used by the first parallel port, and what four addresses in the vector table will contain the address of the interrupt service routine?

**Solution:** According to Table 16.1, the first parallel port is a Type F interrupt. $F_{16} \times 4 = 3C_{16}$, which yields an address combination of $3C_{16}$, $3D_{16}$, $3E_{16}$, and $3F_{16}$.

### 16.2.1 Internal and Software Interrupts

**Internal interrupts** are generated by the microprocessor. Some internal interrupts are used for single-stepping a machine code program for debugging purposes (type 1). Other internal interrupts detect if the programmer decides to do something stupid such as divide by 0 (type 0) or exceed the range of the microprocessor's signed numbers (type 4).

**Software interrupts** are generated by an interrupt instruction that is part of a machine code program. Microsoft operating systems uses the type $21_{16}$ software interrupt extensively to access real-mode hardware drivers. The type $20_{16}$ software interrupt is used to terminate real-mode transient programs. Type $27_{16}$ is used to designate a real-mode program as a TSR. Type 5 causes the current image displayed on the screen to be printed to the printer. Type $13_{16}$ is the main interrupt used to access information on each of the computer's hard drives.

These are just a few of the software interrupts available to programmers. If you are going to really get into machine language, assembly language, or C language programming on the compatible, get a good book on how to use these interrupts. This will make your life a lot simpler.

### 16.2.2 Hardware Interrupts

**Hardware interrupts** are generated by I/O devices through the control bus. The I/O device can use pins on the microprocessor to signal that it needs service, similar to raising your hand to ask a question. The microprocessors used in the compatible computers have three hardware interrupt pins. One, called **RESET,** has been discussed in several sections in this textbook. The other two are called **non-maskable (NMI)** and **interrupt request (INTR).** The pins used by these interrupts are shown for the Pentium® compatible microprocessor in Figure 16.1. Referring to this figure, we find the RESET pin

**Figure 16.1** P54C Pentium Pin-out

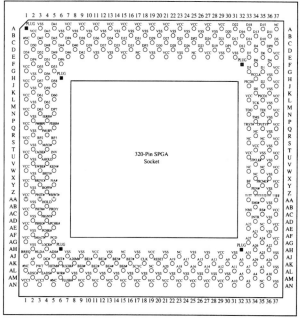

*Source:* Asser/Stigliano/Bahrenburg, *Microcomputer Theory and Servicing* (Upper Saddle River, NJ: Prentice Hall, 2001) p. 653.

at the intersection of column 20 and row AK. The NMI is located at column 33 and row AC and the INTR is located at column 34 and row AD. Even though the Pentium-compatible microprocessor is the only one shown, all Intel-compatible microprocessors have pins that are used for these three interrupts. The pin-outs of other Intel-compatible microprocessors are shown in chapter 3.

The RESET interrupt has the highest priority of all of the interrupts. When the Reset pin is asserted, it forces the microprocessor to stop what it is doing and return to the Reset program in the BIOS. This pin is automatically asserted when the computer is first turned on by a signal from the power supply called *power good*. It is also asserted by pressing and releasing the RESET button located on the front panel of most compatible PCs above an AT. This is the only interrupt that does not use the vector table. When this pin is pulsed (brought to a 1 and then back to a 0), the microprocessor will always go the address memory $FFFF0_{16}$ looking for an instruction's operational code. The address $FFFF0_{16}$ will always be located in the computer's ROM BIOS.

The nonmaskable interrupt is the next highest in priority. The microprocessor cannot ignore this interrupt. NMI is used in the compatible computer by the circuits that detect RAM parity or ECC failures and by the math coprocessor. Nonmaskable is always a type 2 interrupt. This means the location for the interrupt service routine will always be contained in the vector table at addresses $8_{16}$, $9_{16}$, $A_{16}$, and $B_{16}$.

Last, but not least, is the interrupt request pin (INTR). This interrupt pin is the workhorse of the interrupts and is covered in the next section.

## 16.3 INTERRUPT REQUEST

The most used microprocessor interrupt pin in the compatible computers is the interrupt request pin (INTR). This pin is used to interrupt the microprocessor for the timer, keyboard, floppy disk(s), hard disk(s), serial port(s), parallel port(s), sound cards, some video adapters, network cards, and whatever else may need to interrupt the microprocessor in the future. How the microprocessor handles all of these interrupts with a single INTR pin is an interesting feat that will be discussed later in this section, but first some basics.

The reason why the INTR pin is so popular is that it is *maskable*, which means that a programmer can tell the microprocessor to either ignore or honor this pin by using special instructions within a program. This is a very powerful feature because there are times when an interrupt will cause problems.

To handle all of the interrupts that need to use the maskable INTR pin, one or two special ICs called **programmable interrupt controllers (PICs)** are used. On older motherboards, the PIC(s) can be located on the motherboard easily under the part number 8259. On today's motherboards, the PICs are integrated into the chipset, however.

Each PIC can support up to eight interrupts. The ports that need to interrupt the microprocessor connect to the PIC, not to the microprocessor. The PIC then connects to the microprocessor's INTR pin. The PIC contains internal memory, called *registers*, which are programmed with the type number for each of the eight interrupts it supports.

When one of the ports generates an interrupt request to the PIC, it, in turn, signals the microprocessor though the INTR pin. Once the microprocessor senses a signal on the INTR pin, it issues two cycles called **interrupt acknowledge (INTA)** cycles. This only occurs if the program has enabled the INTR pin. When the PIC detects the second INTA cycle, it places the type number that was programmed into it for the interrupt on the data bus. The microprocessor then reads the data bus and uses the type number as a pointer into the vector table using the method discussed previously. As you can see, this is very complicated, but it is important to have a basic understanding of the process.

### 16.3.1 Original 8-bit Interrupts

There is only one PIC installed on the PC/XT-compatible computer's motherboard, which gives these computers the ability to handle only eight interrupts. The eight interrupts are connected to the single PIC through pins called IRQ0, IRQ1, IRQ2, IRQ3, IRQ4, IRQ5, IRQ6, and IRQ7. The PIC then provides a single pin called INTR that connects to the INTR pin of the microprocessor.

IRQ0 was originally connected to an 8254 *programmable interval timer (PIT)*. This interrupt was used to time memory refresh but now is mainly used to generate beep codes from the computer's in-

Table 16.2  Original 8-bit Interrupt Assignments

| IRQ | Assignment |
|---|---|
| IRQ0 | Time of day |
| IRQ1 | Keyboard |
| IRQ2 | Not assigned |
| IRQ3 | Serial port #2 (COM2) |
| IRQ4 | Serial port #1 (COM1) |
| IRQ5 | Hard drive(s) |
| IRQ6 | Floppy drive(s) |
| IRQ7 | Parallel port (printer port) |

ternal speaker. IRQ1 was and still is connected to the keyboard controller port. This interrupt indicates each time that a key on the computer's keyboard is pressed. Both the PIT and keyboard ports were originally individual ICs on the motherboard but now they are both incorporated into the motherboard chipset. Because both of the ports are always located on the compatible PC's motherboard, there is no reason for them to be available on pins of the expansion bus.

The remaining PIC's interrupts, IRQ2 through IRQ7, are available to 8-bit adapters installed in one of the computer's ISA expansion slots. Because they were originally developed for the 8-bit expansion bus, they are assigned pins on the large 64-pin connector on all ISA slots and are referred to as **8-bit interrupts**. The pin assignments are listed in chapter 4, Table 4.2.

Having IRQ2 through IRQ7 as part of the expansion bus seems to indicate that any 8-bit or 16-bit adapter card that plugs into the ISA expansion bus will have access to these interrupts. Unfortunately, this is not the case because all but one of these IRQs of the PIC are assigned to a specific port. The original assignments for these eight interrupts are listed in Table 16.2. In the introduction of the AT-compatible computer, some of these assignments were changed, however.

As you can see in Table 16.2, IRQ2 is the only 8-bit interrupt that remains unassigned. This created a problem for any additional I/O card that needed to use interrupts because there was only one available on the expansion bus. This posed a major problem for hardware engineers who design expansion cards for the compatible computers.

## 16.3.2 Current 8-bit and 16-bit Interrupts

When the AT came out, IBM decided to add a second PIC on the motherboard that provided additional interrupts. This gives the compatible computers above an XT the ability to handle a total of 16 interrupts. Even though the pins are named IRQ0 through IRQ7 on the PIC, IBM chose to label the signals on the computer IRQ8 through IRQ15. The second PIC cannot attach directly to the INTR pin because the first PIC is already connected to it. To solve this problem, IBM decided to attach the second PIC to the IRQ2 input of the first PIC, as illustrated in Figure 16.2. Because these additional eight interrupts are added to the small connector on the ISA expansion slot, which is only available to 16-bit adapter cards, the interrupts added by the second PIC are called **16-bit interrupts.** Along with adding additional IRQs, the assignments on some of the original 8-bit interrupts are reassigned. The devices assigned to both 8-bit and 16-bit interrupts are listed in Table 16.3.

Even though IRQ2 was lost, it seems like the addition of these interrupts should have been the solution to the interrupt problem, but this is not the case. All legacy 8-bit devices such as parallel ports (printer), additional serial ports, floppy drives, and so on still fight over the 8-bit IRQ lines originally assigned to the PC. This really becomes a problem when adding additional COM and parallel ports or when adding 8-bit devices that were not part of the original interrupt assignments.

As discussed in chapter 9, the interrupt setting on a hardset adapter is set by either switches or jumpers. When set manually, it is imperative that the interrupt be configured so it will not conflict with

**Figure 16.2** AT PIC Block Diagram

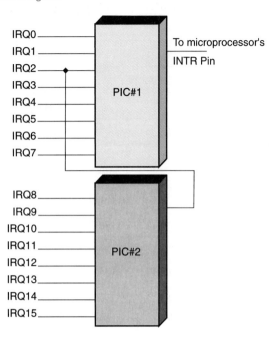

**Table 16.3** AT and Above Computer Interrupt Request (IRQ) Assignments

| | 8-bit Interrupts |
|---|---|
| IRQ0 | Timer |
| IRQ1 | Keyboard |
| IRQ2 | Interrupt controller |
| IRQ3 | COM2 and COM4 (Serial ports 2 and 4) |
| IRQ4 | COM1 and COM3 (Serial ports 1 and 3) |
| IRQ5 | LPT2 (parallel/printer port 2) (Sound card)* |
| IRQ6 | Floppy disk |
| IRQ7 | LPT1 (parallel/printer port 1) |
| | 16-bit Interrupts |
| IRQ8 | Real-time clock |
| IRQ9 | System (PCI = available) |
| IRQ10 | Available (network card)* |
| IRQ11 | Available (Sound card)* |
| IRQ12 | Available (P/S2 Mouse)* |
| IRQ13 | Math coprocessor |
| IRQ14 | Hard drive (primary EIDE) |
| IRQ15 | Available (secondary EIDE) |

*Not part of the original I/O address map

## Procedure 16.1: Determining the IRQ pin(s) used by the PCI interrupts

1. Click on the Start button, then move up to Settings, and then over to Control Panel.
2. Click on Control Panel.
3. After the Control Panel window opens, locate and then double-click on the System icon.
4. After the System Properties window opens, click on the Device Manager tab.
5. Under Device Manager, double-click on the Computer category. This is the top listing under Device Manager.
6. Click on the Interrupt Request (IRQ) option button.
7. Using the scroll buttons on the right-hand side of the windowpane under the "Hardware using the setting" column, look for IRQ listing "IRQ Holder for PCI Steering." These are the IRQs being used by the PCI expansion bus.

any other port already installed in the computer. Also, a hardset adapter's interrupt must be set so that the application program can find it.

On the other hand, PnP devices are automatically configured by Window 9X/2000 so the adapter's address and interrupt does not conflict with other ports installed in the computer. This is only true for ISA PnP devices if there are resources available. Remember from chapter 4 that the addressing range of the ISA expansion bus is limited to addresses below 400H, and the only interrupts available are those listed in Table 16.3. The PCI expansion bus offers an I/O addressing range up to FFFFH and also makes a few changes with the legacy interrupts that make PnP more feasible.

### 16.3.3 Interrupts on the PCI Bus

Among many others, the PCI bus has overcome two of the most important limitations of the ISA bus: ISA's 3FFH maximum address range and the shortage of interrupts.

There are four interrupt pins available on the PCI expansion bus: interrupts A, B, C, and D. The actual pin assignments on the PCI connector for these interrupt pins are listed in chapter 4, Table 4.6. These interrupt pins are routed, through PCI control logic, to one or more of the standard IRQ pins, usually IRQ11 and IRQ12. The actual IRQ pins used vary from machine to machine. To determine the IRQ(s) used, follow the steps outlined in Procedure 16.1.

A typical Device Manager window showing the IRQ(s) assignment for the PCI expansion bus interrupts is shown in Figure 16.3. Unlike the legacy IRQs that are usually controlled by the system BIOS, the IRQs assigned to the PCI bus are controlled by the operating system. This allows devices on the PCI expansion bus to share interrupts without creating conflicts. This can be seen in Figure 16.3 by the multiple listing for IRQ09.

### 16.3.4 Legacy or PnP

The term **legacy IRQ** refers to the original AT computer IRQ assignments. Most fixed ports that were designed before the introduction of the PCI expansion bus required the use of one of these interrupts. If an interrupt is going to be used for a legacy device, it cannot be assigned to a PnP device, and vice versa.

The assignment of the IRQs, as either PnP or legacy ISA, is accomplished through the CMOS Setup program, which was discussed in detail in chapter 8. The configuration settings are found under the main setup screen listed under a title similar to "PnP/PCI Configuration." If an IRQ in this setup option is set to legacy ISA, it will not be available for PnP. On the other hand, if the IRQ is set for PCI/ISA PnP, the IRQ will be open to assignment to PnP devices.

### 16.3.5 Which Interrupt to Use?

If the adapter is a PCI-PnP-compatible device, the assignment of the IRQ will be done automatically by the Windows 9X operating system. If the adapter is not PnP or plugs into the ISA bus, the IRQ assignment becomes more complex.

**Figure 16.3** Device Manager Window Showing IRQ Assigned to PCI

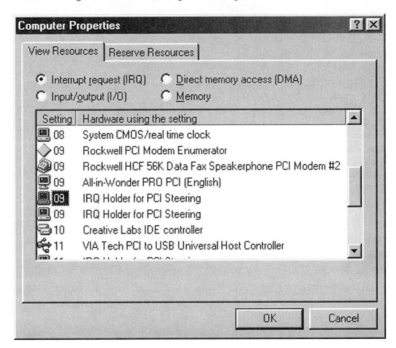

Basically, for these adapters, any unused interrupt between 2 and 15 on a 16-bit adapter card, or 2 and 7 on an 8-bit adapter card, can be used. The problem is finding interrupts that are free. Utilities such as Windows 95/98 Device Manager, Microsoft Diagnostics (MSD), Norton's Syinfo®, or Quarterdeck's Manifest® will identify them. Also, disk-based diagnostics such as Microsoft Diagnostics, QAplus®, Checkit®, Window Sleuth®, and Microscope® will also identify the interrupts that are available.

What do you do if one of the mentioned utilities is not available? Ports, processors, and disks are usually assigned to the standard interrupts already listed in Table 16.3. Check off the devices you have installed in your computer from the appropriated IRQ list (PC/XT or AT and above). This should help you determine which interrupts might be free. If you have to change an interrupt from its original setting, check the documentation that comes with the new card to see if any modifications are required during software installation to indicate the IRQ used.

Before you change any jumpers or switches on any card, make sure the original settings are documented. This will allow you to return the card to its original configuration if the need arises. Good luck, and may the force be with you.

## 16.4 DIRECT MEMORY ACCESS

Knowledgeable compatible PC technicians have used the term **bus-master** for years. In recent years, it has found its way into the repertoire of words used by PC laypeople. Basically, a bus-master is any device that can control the computer's buses. Of course, the microprocessor is definitely a bus-master. The second bus-master has been on the compatible computer's motherboard since its inception in 1981. This device is called a **direct memory access controller (DMAC).** With the introduction of the MCA, EISA, and PCI bus, the ability for a peripheral to be a bus-master was added.

As previously discussed in chapters 9 and 11, DMA allows the transfer of data directly between the peripheral and the computer's RAM without going though the microprocessor. To accomplish this feat, the microprocessor must be stopped and isolated, and another bus-master will take over control of the system bus to perform the data transfer.

The process of stopping and isolating the microprocessor is the function of a microprocessor control bus pin called *HOLD*. On the Pentium-compatible microprocessor that is shown in Figure 16.1, the HOLD pin is located at column 4, row AB. When a device needs to isolate the microprocessor, it brings

Table 16.4  DMA Assignments

| 8-bit DMAs | |
|---|---|
| DMA0 | Available |
| DMA1 | Available (sound card) |
| DMA2 | Floppy disks |
| DMA3 | Available |
| **16-bit DMAs** | |
| DMA4 | First DMAC cascade |
| DMA5 | Available (sound card) |
| DMA6 | Available |
| DMA7 | Available |

the HOLD pin to a logic 1. Once the microprocessor senses this logic level, it will suspend the current operation and simply switch off all of its output pins. Of course, the device that needs to control the computer's buses needs to be informed when the microprocessor has isolated itself. To provide this indication, the microprocessor uses another pin on the control bus called **hold acknowledge (HLDA).** On the Pentium-compatible microprocessor, the HLDA pin is located at the intersection of column 3, row AJ.

Once the microprocessor is switched off, the DMAC or bus-mastering peripheral must provide the control line information to the port and the address to the memory. It must also provide signals, called *device acknowledge (DACK)*, to the port that will enable it for data transfer without an address. The reason for this is that the computer address bus cannot provide an address that enables a memory storage location and port at the same time.

The address, control, and transfer information is programmed into the DMAC before it is used. To do this, the DMAC must have a set of I/O addresses. The addresses used by the DMACs in the compatible computer are listed in chapter 9, Table 9.1.

The DMAC(s) used in the original compatible computer were 8237s. The 8237 can handle data transfers in 8-bit groups. The 8237 can control DMA requests from four ports or peripherals. The PC/XT-compatible computer uses only one 8237 while PCs above the XT use two. The two DMACs allow the computer to control eight DMA requests. Actually, the number is seven because first DMAC connects to one of the inputs of the second DMAC. The current devices assigned to these DMA are listed in Table 16.4.

As discussed in chapter 9, the original DMAs are connected to the 8-bit portion of the ISA bus, so they are called 8-bit DMAs. The DMAs added by the second DMAC are connected to the small ISA connector, so they are called 16-bit DMAs even though the data transfer was originally only 8 bits.

There are a couple of major differences between the 8237 used in the PC/XT/AT-compatible computers and those used today. First, the DMACs are now integrated into the motherboard's chipset. Second, the DMACs can now be programmed to transfer data in 8-, 16-, or 32-bit groups. The second feature has brought about a resurgence in the use of DMAC to transfer information with peripherals.

A typical DMA sequence begins with the requesting device generating a device request (DRQ) signal. The DRQs are found on the ISA bus on pins DRQ0 though DRQ7, which are listed in chapter 4, Table 4.2. Note that an ISA pin for DRQ4 is omitted because it is used as a redirect for the first DMAC.

The DRQ signals are connected directly to the inputs of the DMAC. Once the DMAC senses the DRQ signal, it is asked to use the computer's buses by generating a signal on its hold request (HRQ) pin. This pin is connected to the microprocessor HOLD pin. Once the microprocessor senses the signal on the HOLD pin and switches off its output pin, it generates the HLDA signal, which is connected to both DMACs.

Once the DMAC requesting the DMA cycle detects the HLDA signal, it generates a signal called *device acknowledge* (DACK0-4 on the ISA bus) that is used to enable the I/O device that is requesting the DMAC cycle. The DMAC also provides a preprogrammed address on the address bus to select a memory location along with the appropriate control bus signals. Once the DMA transfer is completed, the DMAC removes the hold signal and the microprocessor switches its output pins back on and once again takes over control of the computer's buses.

This is the same basic procedure used by bus-mastering devices such as hard drives that are designed around the ATA-4 standard. The only difference is that the DMAC on the hard drives provides the address and control information instead of the DMACs on the motherboard.

Like the IRQs, the DMA can be assigned to the legacy ISA devices or to PCI/ISA PnP devices through the CMOS Setup program. This process for launching the Setup program and the locations of the DMA configuration settings are covered in chapter 8.

## 16.5 CONCLUSION

Most I/O devices used on the compatible computers signal the microprocessor to which they need to transfer information by use of interrupts. There are three categories of interrupts used by the Intel-compatible microprocessors: hardware, software, and internal.

The hardware interrupts used by the Intel-compatible microprocessors are further divided into three additional categories called reset, interrupts request (INTR), and nonmaskable (NMI).

Reset forces the microprocessor into programs in the computer's ROM each time power is turned on to the computer or when the RESET button is pressed. The main use of the NMI interrupt is to detect a memory parity or ECC failure. All other I/O devices use the microprocessor's IRQ pin. To sort out which device generates the interrupt request, a special IC called a programmable interrupt controller (PIC) is used.

The PC- and XT-compatible computers use one PIC, so they can handle eight interrupts. Compatible computers above the XT have two PICs, so they can handle up to 16 interrupts. On today's computers, the PICs are integrated within the chipset. The AT-compatibles split the interrupts into two groups: 8-bit interrupts and 16-bit interrupts. These interrupts can be assigned to either legacy ISA or PCI/ISA PnP using the CMOS Setup program.

If a new adapter card is installed in the computer that uses the same interrupt assignment as a device already installed, an interrupt conflict will exist and both devices may not operate. For this reason, most adapter cards give the user the ability to change interrupt assignments to avoid a conflict with previously installed cards. PnP adapters allow the Windows 9X operating system to assign the IRQ so it will not conflict with other devices.

The microprocessor, DMACs, and bus-mastering peripherals are all considered bus-masters. A bus-master is any device that can control the computer's buses. There are currently two DMACs located within the motherboard chipset. The current DMAC can be programmed to transfer data in 8-, 16-, or 32-bit groups. Like the IRQs, the DMA channels can be configured as legacy ISA or PCI/ISA PnP.

## Review Questions

1. Write the meaning of the following acronyms.
    a. IRQ
    b. INTR
    c. DMA
    d. INTA
    e. DACK
    f. NMI
    g. PIC
    h. DMAC
    i. PnP

### Fill in the Blanks

1. The interrupts originally assigned to the ISA bus are referred to as _____ ISA.
2. The PICs and DMAs are now currently part of the _____.
3. A device that needs to interrupt the microprocessor in the compatible PC must be connected to one of the _____ _____ _____.

4. When the compatible microprocessor senses a signal on the INTR pin, it responds by generating two _____ _____ cycles.
5. When the _____ _____ method is used to transfer information with a port, it is controlled totally by software.
6. To get the microprocessor's attention, _____ are often used.
7. The original program's memory location and status are saved on the _____ when an interrupt occurs.
8. The _____ _____ _____ holds the address of the _____ _____ routine.
9. A _____ number is used as a pointer into the _____ _____ table.
10. The interrupt vector table can hold _____ entries.
11. A type 13 interrupt belongs to the _____ category and is used to access the _____ _____.
12. The INTR pin is used most often because it is _____.
13. The microprocessor responds to the HOLD signal with a signal called _____ _____.
14. Any device that can control the computer's buses is called a _____ _____.
15. The floppy drives use DMA _____.
16. The DMACs used in current compatibles provide _____ DMA channels.

## Multiple Choice

1. When software polling is used to access I/O devices:
   a. the application program determines when information is transferred with the I/O device.
   b. each I/O device uses a pin on the microprocessor to signal when it needs to transfer information.
   c. it requires more hardware to implement, therefore it costs less.
   d. it saves time and memory space.
2. When hardware interrupts are used:
   a. the application program determines when information is transferred with the I/O device.
   b. each I/O device uses the control bus to signal the microprocessor when it needs to transfer information.
   c. it requires less hardware to implement than software polling.
   d. it will slow the transfer down but saves considerable money.
3. A program written to take care of an interrupt is called a(n):
   a. interrupt program.
   b. programmable interrupt controller.
   c. interrupt service routine.
   d. interrupt vector table.
4. The number that is used as a pointer into the vector table is called the:
   a. interrupt number.
   b. type number.
   c. IRQ number.
   d. vector table number.
5. An interrupt that is generated by the microprocessor itself is called:
   a. software interrupt.
   b. hardware interrupt.
   c. interrupt request.
   d. shareware interrupt.
6. A peripheral that can transfer information directly with the computer's memory is called a(n):
   a. hard drive.
   b. bus-mastering peripheral.
   c. multi-word DMA.
   d. interrupt request.
7. The IRQ assigned to the secondary IDE channel is:
   a. IRQ5.
   b. IRQ14.
   c. IRQ15.
   d. IRQ8.
8. The PCI expansion bus connects directly to IRQ11 and IRQ12.
   a. True
   b. False
9. The microprocessor and DMAC can share the computer's buses at the same time.
   a. True
   b. False
10. If an IRQ is used by a hardset ISA card, it must be:
    a. configured as PCI/ISA PnP in the CMOS settings.
    b. configured as a hardset ISA in the CMOS settings.
    c. configured as a softset ISA in the CMOS settings.
    d. configured as legacy ISA in the CMOS settings.
11. The three categories of interrupts used by the Intel-compatible microprocessors are:
    a. fast, faster, and fastest.
    b. IRQ, DMA, and INTR.
    c. software, hardware, and internal.
    d. ROM, CMOS, and RAM.
12. A floppy disk drive uses:
    a. IRQ5.
    b. IRQ6.
    c. IRQ4.
    d. an interrupt determined by the user.
13. On current compatibles, the primary hard drive uses:
    a. IRQ14.
    b. IRQ15.
    c. IRQ7.
    d. an interrupt determined by the user.
14. The second serial port (COM2) uses type number:
    a. 1.
    b. 3.
    c. 4.
    d. 11 ($B_{16}$).
15. The pin number used to reset the Pentium-compatible is:
    a. 1.
    b. column 4, row Z.
    c. column 20, row AK.
    d. column AK, row 20.
16. In all compatible computers, the first 1,024 RAM addresses are reserved for:
    a. the interrupt vector table.
    b. the application program.
    c. MS-DOS.
    d. Windows.

17. Each entry in the interrupt table requires:
    a. 32 consecutive memory addresses.
    b. 16 consecutive memory addresses.
    c. 8 consecutive memory addresses.
    d. 4 consecutive memory addresses.
18. The software interrupt type number used to terminate a program is:
    a. 20.
    b. 21.
    c. 22.
    d. 23.
19. The interrupt used by a PnP adapter card is usually changed:
    a. by the application program running the computer.
    b. by the operating system.
    c. by the user.
    d. None of the above is correct.
20. The interrupt used by a hardset adapter card is usually changed:
    a. by the application program running the computer.
    b. by the operating system.
    c. by the user.
    d. None of the above is correct.

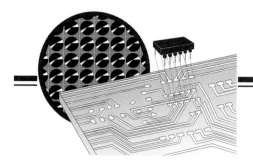

# APPENDIX A
# Powers of 2 and 16

| Power | 2 | 16 |
|---|---|---|
| 0 | 1 | 1 |
| 1 | 2 | 16 |
| 2 | 4 | 256 |
| 3 | 8 | 4,096 |
| 4 | 16 | 65,536 |
| 5 | 32 | 1,048,576 |
| 6 | 64 | 16,777,216 |
| 7 | 128 | 268,435,456 |
| 8 | 256 | 4,294,967,296 |
| 9 | 512 | 68,719,476,736 |
| 10 | 1,024 | 1,099,511,627,776 |
| 11 | 2,048 | 17,592,186,044,416 |
| 12 | 4,096 | 281,474,976,710,656 |
| 13 | 8,192 | 4,503,599,627,370,496 |
| 14 | 16,384 | 72,057,594,037,927,936 |
| 15 | 32,768 | 1,152,921,504,606,846,976 |
| 16 | 65,536 | 18,446,744,073,709,551,616 |
| 17 | 131,072 | 295,147,905,179,352,825,856 |
| 18 | 262,144 | 4,722,366,482,869,645,213,696 |
| 19 | 524,288 | 75,557,863,725,914,323,419,136 |
| 20 | 1,048,576 | 1,208,925,819,614,629,174,706,176 |
| 21 | 2,097,152 | 19,342,813,113,834,066,795,298,816 |
| 22 | 4,194,304 | 309,485,009,821,345,068,724,781,056 |
| 23 | 8,388,608 | 4,951,760,157,141,521,099,596,496,896 |
| 24 | 16,777,216 | 79,228,162,514,264,337,593,543,950,336 |
| 25 | 33,554,432 | 126,765,060,022,822,940,149,670,320,5376 |
| 26 | 67,108,864 | 20,282,409,603,651,670,423,947,251,286,016 |
| 27 | 134,217,728 | 324,518,553,658,426,726,783,156,020,576,256 |
| 28 | 268,435,456 | |
| 29 | 536,870,912 | |
| 30 | 1,073,741,824 | |
| 31 | 2,147,483,648 | |
| 32 | 4,294,967,296 | |
| 33 | 8,589,934,592 | |
| 34 | 17,179,869,184 | |
| 35 | 34,359,738,368 | |
| 36 | 68,719,476,736 | |
| 37 | 137,438,953,472 | |
| 38 | 274,877,906,944 | |
| 39 | 549,755,813,888 | |
| 40 | 1,099,511,627,776 | |

| | |
|---|---|
| 41 | 2,199,023,255,552 |
| 42 | 4,398,046,511,104 |
| 43 | 8,796,093,022,208 |
| 44 | 17,592,186,044,416 |
| 45 | 35,184,372,088,832 |
| 46 | 70,368,744,177,664 |
| 47 | 140,737,488,355,328 |
| 48 | 281,474,976,710,656 |
| 49 | 562,949,953,421,312 |
| 50 | 1,125,899,906,842,624 |
| 51 | 2,251,799,813,685,248 |
| 52 | 4,503,599,627,370,496 |
| 53 | 9,007,199,254,740,992 |
| 54 | 18,014,398,509,481,984 |
| 55 | 36,028,797,018,963,968 |
| 56 | 72,057,594,037,927,936 |
| 57 | 144,115,188,075,855,872 |
| 58 | 288,230,376,151,711,744 |
| 59 | 576,460,752,303,423,488 |
| 60 | 1,152,921,504,606,846,976 |
| 61 | 2,305,843,009,213,693,952 |
| 62 | 4,611,686,018,427,387,904 |
| 63 | 9,223,372,036,854,775,808 |
| 64 | 18,446,744,073,709,551,616 |
| 65 | 36,893,488,147,419,103,232 |
| 66 | 73,786,976,294,838,206,464 |
| 67 | 147,573,952,589,676,412,928 |
| 68 | 295,147,905,179,352,825,856 |
| 69 | 590,295,810,358,705,651,712 |
| 70 | 1,180,591,620,717,411,303,424 |

# APPENDIX B

# Microprocessor Comparison Charts

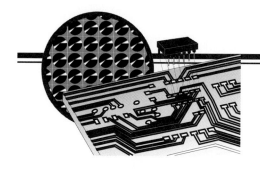

Microprocessor comparison

## PRE PENTIUM® INTEL PROCESSORS

| MPU | 8086 | | 8088 | | 80286 | | 386 SX | 386 DX | 486 SX | 486 DX | 486 SX2 | | | 484 DX2 | | | | 486 DX4 | | | |
|---|---|---|---|---|---|---|---|---|---|---|---|---|---|---|---|---|---|---|---|---|---|
| Approx. number of transistors | 30,000 | | 30,000 | | 134,000 | | 275,000 | 275,000 | 1.2 M | 1.2 M | 1.2 M | | | 1.2 M | | | | 1.6 M | | | |
| Approx. transistor size | 3 | | 3 | | 1.5 | | 1 | 1 & 0.8 | 1 & 0.8 | | | | | | | | | | | | |
| Address Bus Size in Bits | 20 | | 20 | | 24 | | 24 | 32 | 32 | 32 | 32 | | | 32 | | | | 32 | | | |
| Physical Memory | 1 MB | | 1 MB | | 16 MB | | 16 MB | 4 GB | 4 GB | 4 GB | 4 GB | | | 4 GB | | | | 4 GB | | | |
| Virtual Memory | 0 | | 0 | | 0 | | 64 TB | 64 TB | 64 TB | 64 TB | 64 TB | | | 64 TB | | | | 64 TB | | | |
| Data bus | 16 | | 8 | | 16 | | 16 | 32 | 32 | 32 | 32 | | | 32 | | | | 32 | | | |
| Internal data bus | 16 | | 16 | | 16 | | 32 | 32 | 32 | 32 | 32 | | | 32 | | | | 32 | | | |
| Clock (speeds in MHz) C=MPU core, M=Multiplier, B=System bus | C: 5,8,10 | B: 5,8,10 | C: 5,8 | B: 5,8 | C: 6,8,10,12.5 | B: 6,8,10,12/5 | C: 16,25,30,33 / 16,25,33,40 | C: 16,25,33,40 | C: 16,20,25,33 / B: 16,20,25,33 | C: 25,33,40,50 / B: 25,33,40,50 | C: 50 | M: 2 | B: 25 | C: 50,66,80 | M: 2,2,2 | B: 25,33,40 | | C: 75,100,120,150 | M: 3,3,3,3 | B: 25,33,40,50 | |
| Avg clock cycles/bus cycle | 4 | | 4 | | 3 | | 2 | 2 | 1 | 1 | 1 | | | 1 | | | | 1 | | | |
| Word size | 8,16 | | 8,16 | | 8,16 | | 8,16,32 | 8,16,32 | 8,16,32 | 8,16,32 | 8,16,32 | | | 8,16,32 | | | | 8,16,32 | | | |
| Number of pipelines | 1 | | 1 | | 1 | | 1 | 1 | 1 | 1 | 1 | | | 1 | | | | 1 | | | |
| Math Coprocessor | EXT | | EXT | | EXT | | EXT | EXT | NA | INT | NA | | | INT | | | | INT | | | |
| L1 | 0 | | 0 | | 0 | | 0 | 0 | 8 KB | 8 KB | 8 KB | | | 8 KB | | | | 16 KB | | | |

Microprocessor comparison

| MPU | Org P5 | | P54C | | P55C | | P6 | | Pentium® II | | Intel @ Celeron™ | | Intel @ Xeon™ | |
|---|---|---|---|---|---|---|---|---|---|---|---|---|---|---|
| | | | | | | | Intel | | | | | | | |
| Approx. number of transistors | 3.1 M | | 3.2 M | | 3.2 | | 5.5 M w/o L2 | | 7.5 M w/o L2 | | 7.5 M w/o L2 | | 7.5 M w/o L2 | |
| Address Bus Size in Bits | 32 | | 32 | | 32 | | 36 | | 32 | | 32 | | 32 | |
| Physical Memory | 4 GB | | 4 GB | | 4 GB | | 64 GB | | 64 GB | | 64 GB | | 64 GB | |
| Virtual Memory | 64 TB | | 64 TB | | 64 TB | | 64 TB | | 64 TB | | 64 TB | | 64 TB | |
| Data bus in bits | 64 | | 64 | | 64 | | 64 | | 64 | | 64 | | 64 | |
| Internal data bus in bits | 32 | | 32 | | 32 | | 32 | | 32 | | 32 | | 32 | |
| Clock | C | B | C | M | B | C | M | B | C | M | B | C | M | B | C | M | B |
| All speeds are in MHz<br>C = MPU core<br>M = Multiplier<br>B = System bus | 60<br>66 | 60<br>66 | 75<br>90<br>100<br>120<br>133<br>150<br>166<br>200 | 1.5<br>1.5<br>2.0<br>2.0<br>2.0<br>2.5<br>2.5<br>3.0 | 50<br>60<br>50<br>60<br>66<br>60<br>66<br>66 | 166<br>200<br>233 | 2.5<br>3.0<br>3.5 | 66<br>66<br>66 | 150<br>166<br>180<br>200 | 2.5<br>2.5<br>3.0<br>3.0 | 60<br>66<br>60<br>66 | 233<br>266<br>300<br>333<br>350<br>400<br>450 | 3.5<br>4.0<br>4.5<br>5.0<br>3.5<br>4.0<br>4.5 | 66<br>66<br>66<br>66<br>100<br>100<br>100 | 266<br>300<br>300A<br>333<br>366<br>400 | 4.0<br>4.5<br>4.5<br>5.0<br>5.5<br>6.0 | 66<br>66<br>66<br>66<br>66<br>66 | 400<br>450 | 4.0<br>4.0 | 100<br>100 |
| Avg clock cycles/bus cycle | 0.5 | | 0.5 | | 0.5 | | 0.3 | | 0.3 | | 0.3 | | 0.3 | |
| Word size | 8,16,32 | | 8,16,32 | | 8,16,32 | | 8,16,32 | | 8,16,32 | | 8,16,32 | | 8,16,32 | |
| Number of pipelines | 2 | | 2 | | 2 | | 3 | | 3 | | 3 | | 3 | |
| Math Coprocessor | INT | | INT | | INT | | INT | | INT | | INT | | INT | |
| L1 | 8K / 8K | | 8K / 8K | | 16KB / 16KB | | 8 KB / 8KB | | 16KB / 16KB | | 16KB / 16KB | | 16KB / 16KB | |
| On Board L2 | 0 | | 0 | | 0 | | 256 / 1 MB | | 256 KB | | 0 / 128 KB* | | 512KB/1MB/2MB | |
| L2 to core speed | | | | | | | - | | - | | 1:1 | | 1:1 | |
| Dynamic Execution | NO | | NO | | NO | | YES | | YES | | YES | | YES | |
| MMX | NO | | NO | | YES | | NO | | YES | | YES | | YES | |
| 3DNow! | NO | | NO | | NO | | NO | | NO | | NO | | NO | |
| Multiprocessing | 2 | | 2 | | 2 | | 4 | | 2 | | 4 | | 4 | |

Microprocessor comparison

| Manufacturer | \multicolumn{13}{c}{Advanced Micro Devices} |
|---|---|---|---|---|---|---|---|---|---|---|---|---|---|
| **MPU** | \multicolumn{4}{c}{**K5**} | \multicolumn{3}{c}{**K6®**} | \multicolumn{3}{c}{**K6®-II**} | \multicolumn{3}{c}{**K6®-III**} |
| Pin Compatible with | \multicolumn{4}{c}{P54C} | \multicolumn{3}{c}{P55C} | \multicolumn{3}{c}{P55C} | \multicolumn{3}{c}{P55C} |
| Number of Pins | \multicolumn{4}{c}{269} | \multicolumn{3}{c}{321} | \multicolumn{3}{c}{321} | \multicolumn{3}{c}{321} |
| Connector | \multicolumn{4}{c}{Socket 5 / Socket 7} | \multicolumn{3}{c}{Socket 7} | \multicolumn{3}{c}{Socket 7 / Super Socket 7} | \multicolumn{3}{c}{Super Socket 7} |
| Approx. number of transistors | \multicolumn{4}{c}{4.3 M} | | | | | | | | | |
| Address Bus Size in Bits | \multicolumn{4}{c}{32} | \multicolumn{3}{c}{32} | \multicolumn{3}{c}{32} | \multicolumn{3}{c}{32} |
| Physical Memory | \multicolumn{4}{c}{4 GB} | \multicolumn{3}{c}{4 GB} | \multicolumn{3}{c}{4 GB} | \multicolumn{3}{c}{4GB} |
| Virtual Memory | | | | | | | | | | | | | |
| Data bus in bits | \multicolumn{4}{c}{64} | \multicolumn{3}{c}{64} | \multicolumn{3}{c}{64} | \multicolumn{3}{c}{64} |
| Internal data bus in bits | \multicolumn{4}{c}{32} | \multicolumn{3}{c}{32} | \multicolumn{3}{c}{32} | \multicolumn{3}{c}{32} |
| Clock | PR | C | M | B | C | M | B | C | M | B | C | M | B |
| All speeds are in MHz<br>PR = Pentium Rating<br>C = MPU core<br>M = Multiplier<br>B = System bus | 75<br>90<br>100<br>120<br>133<br>166 | 75<br>90<br>100<br>90<br>100<br>116.7 | 1.5<br>1.5<br>1.5<br>1.5<br>1.5<br>1.75 | 50<br>60<br>66<br>60<br>66<br>66 | 166<br>200<br>233<br>266<br>300 | 2.5<br>3.0<br>3.5<br>4.0<br>4.5 | 66<br>66<br>66<br>66<br>66 | 266<br>300<br>300<br>333<br>333<br>350<br>366<br>380<br>400 | 3.5<br>4.0<br>3.0<br>4.0<br>3.5<br>3.5<br>4.0<br>4.0<br>4.0 | 66<br>66<br>100<br>66<br>95<br>100<br>66<br>95<br>100 | 400<br>450 | 4.0<br>4.5 | 100<br>100 |
| Avg clock cycles/bus cycle | \multicolumn{4}{c}{NA} | \multicolumn{3}{c}{NA} | \multicolumn{3}{c}{NA} | \multicolumn{3}{c}{NA} |
| Word size | \multicolumn{4}{c}{8,16,32} | \multicolumn{3}{c}{8,16,32} | \multicolumn{3}{c}{8,16,32} | \multicolumn{3}{c}{8,16,32} |
| Number of pipelines | \multicolumn{4}{c}{6} | \multicolumn{3}{c}{7} | \multicolumn{3}{c}{10} | \multicolumn{3}{c}{10} |
| Math Coprocessor | \multicolumn{4}{c}{INT} | \multicolumn{3}{c}{INT} | \multicolumn{3}{c}{INT} | \multicolumn{3}{c}{INT} |
| L1 – Instruction / data | \multicolumn{4}{c}{16KB / 8KB} | \multicolumn{3}{c}{32KB / 32KB} | \multicolumn{3}{c}{32KB / 32KB} | \multicolumn{3}{c}{32K/32K} |
| On Board L2 | \multicolumn{4}{c}{0} | \multicolumn{3}{c}{0} | \multicolumn{3}{c}{0} | \multicolumn{3}{c}{256K} |
| L2 to core speed | \multicolumn{4}{c}{NA} | \multicolumn{3}{c}{NA} | \multicolumn{3}{c}{NA} | \multicolumn{3}{c}{1:1} |
| Dynamic Execution | \multicolumn{4}{c}{YES} | \multicolumn{3}{c}{YES} | \multicolumn{3}{c}{YES} | \multicolumn{3}{c}{YES} |
| MMX | \multicolumn{4}{c}{NO} | \multicolumn{3}{c}{YES} | \multicolumn{3}{c}{YES} | \multicolumn{3}{c}{YES} |
| 3DNow! | \multicolumn{4}{c}{NO} | \multicolumn{3}{c}{NO} | \multicolumn{3}{c}{YES} | \multicolumn{3}{c}{YES} |
| Multiprocessing | \multicolumn{4}{c}{N0} | \multicolumn{3}{c}{NO} | \multicolumn{3}{c}{NO} | \multicolumn{3}{c}{NO} |

# Microprocessor Comparison Charts

| Manufacturer | VIA Cyrix | | | | | | | | | | | | |
|---|---|---|---|---|---|---|---|---|---|---|---|---|---|
| **MPU** | **6x86 (M1)** | | | | **6x86MX** | | | | **MII** | | | | |
| Pin Compatible with | P54C | | | | | | | | P55C | | | | |
| Number of Pins | | | | | | | | | 296 | | | | |
| Connector | Socket 5 / Socket 7 | | | | Socket 7 | | | | Socket 7 | | | | |
| Approx. number of transistors | | | | | | | | | | | | | |
| Address Bus Size in Bits | 32 | | | | 32 | | | | 32 | | | | |
| Physical Memory | 4GB | | | | 4 GB | | | | 4 GB | | | | |
| Virtual Memory | | | | | | | | | | | | | |
| Data bus in bits | 64 | | | | 64 | | | | 64 | | | | |
| Internal data bus in bits | 32 | | | | 32 | | | | 32 | | | | |
| Clock | PR | C | M | B | PR | C | M | B | PR | C | M | B | |
| All speeds are in MHz<br>PR = Pentium Rating<br>C = MPU core<br>M = Multiplier<br>B = System bus<br><br>100 | 120<br>133<br>150<br>166<br>200 | 100<br>110<br>120<br>133<br>150 | 2.0<br>2.0<br>2.0<br>2.0<br>2.0 | 50<br>55<br>60<br>66<br>75 | 166<br>166<br>200<br>233<br>266 | 133<br>150<br>155<br>188<br>208 | 2.0<br>2.5<br>2.5<br>2.5<br>2.5 | 66<br>60<br>66<br>75<br>83 | 300<br>300<br>333<br>366<br>400<br>433 | 225<br>233<br>262<br>250<br>285<br>3.0 | 3.0<br>3.5<br>3.5<br>2.5<br>3.0<br>3.0 | 75<br>66<br>75<br>100<br>95 | |
| Avg clock cycles/bus cycle | | | | | | | | | | | | | |
| Word size | 8,16,32 | | | | 8,16,32 | | | | 8,16,32 | | | | |
| Number of pipelines | | | | | | | | | 2 | | | | |
| Math Coprocessor | INT | | | | INT | | | | INT | | | | |
| L1 – Instruction / data | 16 KB Unified | | | | 64 KB Unified | | | | 64 KB Unified | | | | |
| On Board L2 | 0 | | | | 0 | | | | 0 | | | | |
| L2 to core speed | NA | | | | NA | | | | NA | | | | |
| Dynamic Execution | YES | | | | YES | | | | YES | | | | |
| MMX | NO | | | | YES | | | | YES | | | | |
| 3DNow! | NO | | | | NO | | | | NO | | | | |
| Multiprocessing | NO | | | | NO | | | | NO | | | | |

# Appendix B

| Manufacturer | Integrated Device Technology | | | | | |
|---|---|---|---|---|---|---|
| **MPU** | *WinChipC6* | | | *WinChip2* | | |
| Pin Compatible with | P54C & P55C | | | P55C | | |
| Number of Pins | 296 | | | 296 | | |
| Connector | Socket 7<br>Socket 5 | | | Socket 7 | | |
| Approx. number of transistors | | | | | | |
| Address Bus Size in Bits | 32 | | | 32 | | |
| Physical Memory | 4 GB | | | 4 GB | | |
| Virtual Memory | | | | | | |
| Data bus in bits | 64 | | | 64 | | |
| Internal data bus in bits | 32 | | | 32 | | |
| Clock | C | M | | B | C | M |
| All speeds are in MHz<br>PR = Pentium Rating<br>C = MPU core<br>M = Multiplier<br>B = System bus | 180<br>200<br>255<br>240 | 3.0<br>3.0<br>3.5<br>4.0 | 60<br>66<br>75<br>60 | 255<br>240<br>250<br>266<br>300 | 3.5<br>4.0<br>3.0<br>4.0<br>3.0 | 75<br>60<br>83<br>66<br>100 |
| Avg clock cycles/bus cycle | | | | | | |
| Word size | 8,16,32 | | | 8,16,32 | | |
| Number of pipelines | 1 | | | 1 | | |
| Math Coprocessor | INT | | | INT | | |
| L1 – Instruction / data | 32K/32K | | | 32K/32K | | |
| On Board L2 | NO | | | NO | | |
| L2 to core speed | NA | | | NA | | |
| Dynamic Execution | NO | | | NO | | |
| MMX | YES | | | Yes | | |
| 3DNow! | No | | | Yes | | |
| Multiprocessing | No | | | NO | | |

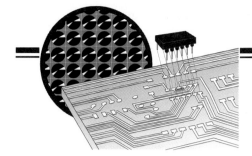

# APPENDIX C

# DOS Quick Reference

## Entry Form Notation

Entries that are given in capital letters must be spelled exactly as shown. The case used when entering the information from the computer's keyboard does not matter, however.

The user defines entries that are shown as lower case. Once again, the actual case used to enter the information into the computer does not matter.

## Additional notations

| | |
|---|---|
| [ ] | Brackets indicate the entry is optional |
| ... | More options may be added |
| d: | Drive name or destination drive name |
| s: | Source drive |
| t: | Target drive |
| /x | Switch |
| /x:n | Switch with a number |

## Command Entry Syntax

The underline indicates that no spaces are allowed between characters.

command [d:][filespec] [/switches...]
command [s:][sfilespec] [t:][tfilespec] [/switches]

| | |
|---|---|
| **command** | Command name. This can be any internal or external MS-DOS command or any external program name with the extensions COM, EXE, or BAT. |
| **filespec** | File specifications, including the path. |
| **sfilespec** | The specifications for the source file(s), including path. |
| **tfilespec** | The specifications of the target file(s), including path. |
| **switches** | Optional switches. |
| **pathname** | Path to the directory containing the external file if the file is not located in the default directory. |

## Filename Form

Filenames can be up to eight characters in length, not including an optional three-character extension. The filename can be entered in either lower- or uppercase letters. MS-DOS automatically converts filenames to uppercase letters.

### Valid characters
a–z   A–Z   0–9   ( )   &   #   %   ,   _   @   { }   ~   '   !

535

## Appendix C

**Invalid characters—DO NOT use the following as part of a filename.**
? . , ; : = * / \ + " < > or a space

## File Specifications (filespec or pathname)

| | |
|---|---|
| filename.ext | If located on the default drive in the default directory |
| d:filename.ext | If not located on the default drive. |
| \directory\...\filename.ext | If not located in the default directory. |
| d:\directory\....\filename.ext | If not located on the default drive or the default directory. |

### Syntax for the preceding file specifications:

**d:** is the drive designator.
**\** tells MS-DOS to begin following the path from the root directory toward the specified file.
**directory** is any subdirectory on the disk.
**filename.ext** is any valid MS-DOS filename including the extension.

Information can be routed to different I/Os using the COPY command. The following names are given to the I/O devices:

## Valid Device Names:

**CON** = console or keyboard; **PRN** = parallel printer #1; **LPT1** = parallel printer #1; **LPT2** = parallel printer #2; **LPT3** = parallel printer #3; **AUX** = serial port #1; **COM1** = serial port #1; **COM2** = serial port #2

**Example:**

| | |
|---|---|
| A>COPY CON RICH.TXT | Copies information entered from the keyboard to the file RICH.TXT that is located on the default drive in the default directory. |

## MS-DOS Commands

This is not a complete list of all of the commands available with MS-DOS. If you need additional information concerning other commands, consult the DOS reference manual.

### CHDIR or CD (Internal)

Displays or changes current directory.

**Syntax:**
CHDIR [d:][\][pathname]
CD [d:][\][pathname]

**Example:**
CD \PLAY\TIME\

### CHKDSK (External)

Checks integrity of FATs, provides status of disk contents and unused conventional RAM.

**Syntax:**
CHKDSK [d:] [/x]

The /x represents one of the following switches:

/F     Fix errors
/V     Display messages

DOS Quick Reference

**Example:**
  CHKDSK C: /F

### CLS (Internal)
Clears the screen and returns the MS-DOS prompt to the upper left-hand corner.

**Syntax:**
  CLS

**Example:**
  CLS

### COMP (External)
Compares file(s)

**Syntax:**
  COMP [s:][filespec1] [t:][filespec2]

**Example:**
  COMP C:\PLAY\TIME.EXE A:TIME.EXE

### COPY (Internal)
Copies file(s) specified. Device names can be substituted for filespecs.

**Syntax:**
  COPY [s:]sfilespec [t:]
  [tfilespec]

**Examples:**
  COPY A:\TIME.EXE C:\PLAY\*.*
  COPY A:*.EXE C:*.*

### DATE (Internal)
Displays and sets the date.

**Syntax:**
  DATE
  DATE mm-dd-yy

**Examples:**
  DATE
  DATE 01-21-99

### DELETE or DEL or ERASE (Internal)
Deletes file(s) specified.

**Syntax:**
  DEL [d:]filespec
  ERASE [d:]filespec

**Example:**
  DEL C:\PLAY\TIME.EXE

## DIR (Internal)

Lists requested directory entries.

**Syntax:**

DIR [d:][filespec] [/x...]

DIR [d:][pathname] [/x...]

The /x represents one or both of the following switches:

/P    Page mode
/W    Wide Display mode

**Example:**

DIR C:\WINDOWS /P

## DISKCOMP (External)

Compares the contents of two disks of the same density.

**Syntax:**

DISKCOMP [s: [t:]] [/x]

The /x represents one or more of the following switches:

/1    Compare side 1 only
/8    Compare the first 8 sectors per track only
/R    Sound bell when a user response is required

**Example:**

DISKCOMP A: B:

## DISKCOPY (External)

Copies disks of same density. When the target and source drive letters are the same, DOS will prompt when to swap diskettes. DISKCOPY will also format target diskette if unformatted.

**Syntax:**

DISKCOPY s: t: [/x]

The /x represents one or more of the following switches:

/V    Verify the copy
/F    Force a format of the destination disk
/1    Compare side 1 only
/R    Sound bell when a user response is required

**Example:**

DISKCOPY A: A: /V

## ERASE

See DELETE command.

## FC

Compares two or more files for differences.

**Syntax:**

FC [/x} s:filespec t:filespec

The x represents one or more of the following switches:

/A    Display the first and last lines in a group of lines that differ
/B    Compare binary

| | | |
|---|---|---|
| /C | Ignore case | |
| /L | Compare ASCII | |
| /LP n | Set internal buffer to *n* lines | |
| /N | Display the number of lines that differ in two ASCII files. | |
| /T | Do not treat tabs as spaces | |
| /W | Compress tabs and spaces | |

**Example:**

FC /L C:\PLAY\TEST.EXE A:TEXT.EXE

## FDISK

A DOS utility program that is used to define hard drive partitions.
Note: This use of the FDISK utility is covered in detail in chapter 11 of the text.

**Syntax:**

FDISK

**Example:**

FDISK

## FIND

Finds a sequence of characters in an ASCII file.

**Syntax:**

FIND [/x] "string of charters" [filespec]

The /x represents one or more of the following switches:

| | |
|---|---|
| /C | Display the number of times the string occurs |
| /N | Precede each line that contains the string with a line number |
| /V | Display each line that does not contain the string |

**Example:**

FIND "Now is the time" PLAY.TXT

## FORMAT (External)

Formats a disk to receive MS-DOS files.

**Syntax:**

FORMAT [d:][/x...]

The /x represents one or more of the following switches:

| | |
|---|---|
| /v | Specify the volume label (syntax: /v:*label*) |
| /q | Delete the FAT and root directory but do not actually format the surface of the diskette and perform a surface scan for errors. |
| /u | Unconditionally format the entire diskette and perform a surface scan for defective sectors. Diskettes formatted with this switch cannot be unformatted. |
| /n | Suppress the on-screen prompts (useful when FORMAT is used in a batch file) |
| /b | Reserve space for the system files so they can be transferred later. |
| /s | Transfer system files (DOS.COM, IO.COM, and COMMAND.COM) to the formatted diskette |
| /1 | Format a single side of a floppy diskette |
| /4 | Format a 360-K diskette in a 1.2-MB drive |
| /8 | Format a 5 1/4" floppy diskette to 8 sectors per track |
| /F:n | Format a lower density diskette in a 1.44-MB drive. N is the number of kilobytes to format, i.e., 720 indicates formatting the diskette to 720 K. |

**Example:**
   FORMAT A: /U

### LABEL (External)
Displays, creates, changes, or deletes disk volume label.

**Syntax:**
   LABEL [d:][label]

**Example:**
   LABEL
   LABEL C:OSR2

### LOADHIGH
Used in the AUTOEXEC.BAT file to attempt to load TSR into the upper memory area. LOADHIGH must be used in conjunction with the memory managers HIMEM or EMM386, or similar.

**Syntax:**
   LOADHIGH=TSR

**Example:**
   LOADHIGH MOUSE.COM

### MEM
Displays the computer memory configuration.

**Syntax:**
   MEM [/x]

The x is replaced by one of the following switches:

| | |
|---|---|
| /C | Classify programs by memory usage |
| /D | Display the status of all programs, internal drivers, and other information. |
| /M | Display a detailed listing of a module's memory use. The module's name must follow the /M separated by a colon (/M:module). |
| /P | Display memory information one page at a time |

**Example:**
   MEM \P \C

### MKDIR or MD (Internal)
Creates a new subdirectory.

**Syntax:**
   MKDIR [d:]pathname

**Example:**
   MD C:\EXAMPLE

### MODE (External)
Allows the user to change the default settings used for the computer's serial, parallel, display, and keyboard ports.

**Syntax:**
   Printer Port:
   MODE LPTn[:] [COLS=c] [LINES=1] [RETRY=r]

**Example:**
    MODE LPT1: COLS=80 LINES=25 RETRY=3

**Serial Port:**
    MODE COMm[:] [BAUD=b] [PARITY=p] [DATA=d] [STOP=s] [RETRY=r]

**Example:**
    MODE COM1: BAUD=9600 PARITY=E DATA=8 STOP=2 RETRY=3

**Device status:**
    MODE [device] [/STATUS]

**Example:**
    MODE LPT1 /STATUS

**Redirect Printing:**
    MODE LPTn[:]=COMm[:]

**Example:**
    MODE LPT1=COM2

**Prepare Code Page:**
    MODE device CP PREPARE=((yyy[...]) [drive:][path]filename)

**Select Code Page:**
    MODE device CP SELECT=yyy

**Refresh Code Page:**
    MODE device CP REFRESH

**Code Page Status:**
    MODE device CP [/STATUS]
    Display mode:MODE [display-adapter][,n]

**Example:**
    MODE:80

**Keyboard:**
    MODE CON[:] [RATE=r DELAY=d]

**Example:**
    MODE CON: RATE=20 DELAY=10

## MOVE (External)

Moves a file or files from a source location to a target location. The MOVE command can also be used to move the contents of one directory into another. After the move, the source directory is deleted.

**To move a file or files:**
    MOVE [/Y | /-Y] [s:]sfilespec [d:]tfilespec

**To move a directory:**
    MOVE [/Y | -Y}] [s:]sdirectory [d:]tdirectory

**Switches:**
- /Y      Do not prompt when creating a directory or overwriting the destination
- /-Y     Display a confirmation prompt when creating a directory or overwriting a destination

**Example:**
MOVE C:\PLAY.EXE C:\TWO\*.*

## MSCDEX

Loads the MS-DOS extensions used to access a CD-ROM drive. This command is usually located in the AUTOEXEC.BAT file. The driver for the CD-ROM drive must be loaded in the CONFIG.SYS file using the DEVICE command.

**Syntax:**
MSCDEX [/x] [/D:device] [/L:letter] [/M:buffers]

**Switches:**
- /D:device     The label used by the CD driver that was loaded in the CONFIG.SYS file. A typical label is MSCD0001.
- /L:letter      Allows the user to change the default drive letter assigned to the CD-ROM drive.
- /M:buffers    Allows the user to change the number of 2-KB buffers allocated to the CD-ROM drive.

**Example:**
MSCDEX /D:MSD000

## NLSFUNC (External)

Loads country-specific information from the indicated file.

**Syntax:**
NLSFUNC sfilespec

**Example:**
NLSFUNC COUNTRY.SYS

## PATH (Internal)

Sets a command search path for files not located in the default directory.

**Syntax:**
PATH[=][d:[path][;[d:][path]...]

**Example:**
PATH=C:\DOS;C:\WINDOWS

## PAUSE

Suspends the processing of a batch file and displays the message "Press any key to continue."

**Syntax:**
PAUSE

**Example:**
PAUSE

## PRINT (External)

Prints hard copy of ASCII files.

**Syntax:**

PRINT [d:]filespec [/x]

The /x represents one or more of the following switches:

| | |
|---|---|
| /A | Abort print cancels the specified file and all following file specifications until an /S switch is encountered. |
| /C:n | Copy (causes PRINT to produce *n* copies). |
| /F | Form feed (causes PRINT to issue a form feed at the end of each copy). |

**Example:**

PRINT C:\WINDOWS\README.TXT

## PROMPT (Internal)

Changes the MS-DOS system prompt.

**Syntax:**

PROMPT [text] [options]

Options include the following characters to create special prompts. Each selected option *must* be preceded by a dollar sign ($).

| | |
|---|---|
| $ | The $ character |
| t | Display current time |
| d | Display current date |
| p | Display working directory and default drive |
| v | Display DOS version number |
| n | Display default drive |
| g | Display > character |
| l | Display < character |
| b | Display | character |
| _ | CARRIAGE RETURN-LINEFEED |
| s | Display a space (leading only) |
| e | ASCII code X'1B" (escape) |
| q | Display = character |
| h | A backspace |

**Example:**

PROMPT War Eagle $p$g

Displays the text **War Eagle** followed by the default drive and directory followed by the > character.

## REM (Internal)

Turns an entry in a batch file or the CONFIG.SYS file into a remark. Remarks are skipped over when the commands in these files are being processed. The REM command can also be used to add comments in the indicated files.

**Syntax:**

REM statement

REM [comment]

**Example:**

REM

This is an example of a remark statement

## RENAME or REN (Internal)

**Syntax:**

RENAME [d:]filespec filename

REN [d:]filespec filename

**Example:**
   REN C:\WINDOWS\SYSTEM.INI SYSTEM.BAK

## RMDIR or RD (Internal)
Removes an empty directory. If the directory contains files other than (.) and (..), this command will cause an error message.

**Syntax:**
   RMDIR [d:]pathname
   RD [d:]pathname

**Example:**
   RD C:\PLAY

## SETVER (Internal)
Allows MS-DOS to return the DOS version number requested by an application. The SETVER command can also be used to display the current listing of version numbers returned and add and delete version numbers to the listing.

**Display current version listing:**

**Syntax:**
   SETVER

**Add an entry to the listing:**
   SETVER filename n.nn

**Example:**
   SETVER PLAY.EXE 6.22

**Delete an entry in the listing:**
   SETVER filname /D

**Example:**
   SETVER PLAY.EXE /D

## SYS (External)
Transfers hidden system files IO.SYS and DOS.SYS to the specified drive. The diskette in the specified drive must be formatted. COMMAND.COM must be copied to make the diskette bootable.

**Syntax:**
   SYS d:

**Example:**
   SYS C:

## TIME (Internal)
Displays and sets the system time.

**Syntax:**
   TIME
   TIME hh[:mm[:ss[.cc]]]

**Syntax:**
   TIME hh:mm:ss

**Example:**
TIME 12:32:51

## TREE (External)
Displays subdirectory paths on a disk.

**Syntax:**
TREE
TREE [d:][/F]

The /F switch causes TREE to list the files contained in each subdirectory.

**Example:**
TREE C:

## TYPE (Internal)
Displays the contents of the ASCII file specified.

**Syntax:**
TYPE [d:]filespec

**Example:**
TYPE C:\README.TXT

## VER (Internal)
Displays MS-DOS version number currently running the computer.

**Syntax:**
VER

**Example:**
VER

## VERIFY (Internal)
Verifies that data is correctly written to disk when turned ON. Default = OFF.

**Syntax:**
VERIFY [ON]
VERIFY [OFF]

**Example:**
VERIFY ON

## VOL (Internal)
Displays the disk volume label.

**Syntax:**
VOL [d:]

**Example:**
VOL C:

## XCOPY (External)
Copies files, including those in subdirectories.

**Syntax:**
XCOPY [s:]sfilespec [t:] [/x]

The /x represents one or more of the following switches:

| | |
|---|---|
| /A | Copy only files with the archive bit set |
| /D:mm-dd-yy | Copy files modified on or after the specified date |
| /E | Copy empty directories |
| /P | Prompts if new file has to be created |
| /S | Copy only subdirectories that contain files |
| /V | Verify each file after it is copied |
| /W | Waits for the user to press a key before each file is copied |

**Example:**

XCOPY C:*.EXE A:

## Hard Disk Commands

### BACKUP

Creates backup files from the hard disk to floppy diskettes.

BACKUP s:[pathname] t: [/x...]

| | |
|---|---|
| /s | Back up contents of all subdirectories |
| /m | Back up only files that have changed since the last backup, and turn off the archive attribute of the original files. |
| /a | Add backup files to an existing backup disk without deleting existing files. (The /a switch is ignored if the existing backup disk contains backup files that were created by using the BACKUP command from MS-DOS 3.2 or earlier.) |
| /f[:size] | Format the backup disk to the size you specify. With this switch, you direct backup to format the floppy diskettes that do not match the default size of the drive. |
| /d:date | Back up only the files modified on or after the specified date. |
| /t:time | Back up only the files modified at or after the specified time. |
| /l[:[d:][path]logfile] | Create a log file and add an entry to that file to record the backup operation. |

# APPENDIX D

# Debug

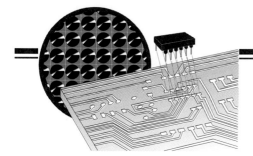

Debug is a utility debugging program provided by MS-DOS. It provides a controlled testing environment for binary and executable object files. Debug commands may be aborted at any time by pressing Ctrl+C.

## Commands

**debug** [filename]

| Debug Commands | Function |
|---|---|
| A[address] | Assemble |
| D[range] | Dump |
| E*address*[list] | Enter |
| F*range list* | Fill |
| G=address | Go |
| L | Load |
| N*filename* | Name |
| Q | Quit |
| R[register-name] | Register |
| T[=address][value] | Trace |
| U[range] | Unassemble |
| W | Write |

Following is the syntax and some examples of the Debug commands. This is only a partial list of the available commands. The following notations are used in the syntax:

| Bold letter | Must be entered as shown |
| Bracket | Optional entry |
| Italics | User-defined |

### A(ssemble)

Assembles 8086/88 mnemonics into memory.

This command creates machine code from assembly language statements. All numeric values are in hexadecimal format.

**Syntax:**

**a** [[*segment:*][*address*]]

**Parameter:**

*segment:address*

Specifies the location where the assembly language program will be assembled. The segment may be entered as hexadecimal or one of the segment registers may be used. If the segment is not specified, the current data segment will be used. The address is the logical address within the segment. If a logical address is not specified, assembly will start where it last stopped.

**Examples:**

| | |
|---|---|
| a 1000:1234 | Will start assembling in segment 1,000 at logical address 1234. |
| a cs:1234 | Will start assembly in the current code segment at logical address 1234. |

## D(ump)

Displays the contents of a range of memory addresses.

**Syntax:**

d [*range*]

**Parameter:**

*range*

Specifies the starting and ending addresses of the memory area whose contents you wish to display. If a range is not specified, Debug will display the contents of 128 bytes, starting at the end of the last range displayed by the DUMP command. If the segment is not specified, the current data segment is used.

**Examples:**

| | |
|---|---|
| d cs:1000 1100 | Displays 256 bytes of the current code segment starting at logical address 1000. |
| d 1235:0100 0110 | Displays 16 bytes of segment 1235 beginning at logical address 0100. |

## E(nter)

Enters data into memory at the address specified.

**Syntax:**

e *address*[*list*]

**Parameters:**

*address*

Specifies the first memory location where you want to enter data.

*list*

Specifies the data you want to enter into successive bytes of memory.

Notes: If an address is specified without a list parameter, Debug displays the address and its contents followed by a period (.) and waits for your input. At this point, you can perform one of the following actions:

The contents of the address can be replaced by typing a new value after the period.
Advance to the next address by pressing the Space bar.
Return to the preceding byte by pressing the Hyphen key.
Stop the enter command by pressing the Enter key.

**Examples:**

e cs:100     Enter from the current code segment starting at logical address 100.

Debug will display the first byte in the following format:

0458:0100 AC.__

To change the contents to 4C, type 4C at the cursor, as follows:

0458:0100 AC.4C__

Consecutive addresses may be changed or examined by pressing the Space bar between addresses instead of Enter. If the Space bar is pressed without entering new data, the original contents will not be affected.

The contents of multiple address may be changed using a single Enter command. Each entry in the list must be separated by a space.

   e 1000:100 23 CB 9A 45 98 9B 00 12 8C

If text is entered, it must be enclosed in quotation marks.

   e ds:100 "This will be entered as text"

## F(ill)

Fills addresses in the specified memory range with specified values.

**Syntax:**

f *range list*

**Parameters:**

*range*

   Specifies the starting and ending addresses of the memory area to be filled.

*"list"*

   Specifies the data to be entered. *List* can be either hexadecimal numbers or a string enclosed in quotation marks.

**Examples:**

   f 04ca:0200 0300 45 cc dd ee ff 10 54 1a 33

In response, Debug fills memory locations 04ca:0200 through 04ca:0300 with the values specified. Debug repeats the values until all of the 100-H bytes are filled.

## G(o)

Runs the machine code program currently in memory.

**Syntax:**

g [=*address*] [*breakpoints*]

**Parameters:**

=*address*

   Specifies the address in the program currently in memory at which you want execution to begin. If you do not specify *address,* Debug begins program execution at the current address in the CS:IP registers.

*breakpoint*

   Specifies 1 to 10 temporary breakpoints that you can set as part of the GO command.

**Examples:**

   g=cs:0100     Debug runs the program in memory in the current code segment.

## L(oad) Loads a file from diskette into memory.

**Syntax:**

l

Notes:  An absolute sector load can be performed with this command but it will not be used or covered in this class. To use the LOAD command with the indicated syntax, the Debug N(AME) command *must* first be used to indicate the name of the file to load.

## Examples:
l          Loads the currently named file into memory.

## N(ame)
Specifies the name of the file when saving or loading files to or from disk. Used in conjunction with the L(OAD) and the W(RITE) commands.

**Syntax:**

**n** [*d:*][*path*]*filename*

**Parameters:**

[*d:*][*path*]*filename*

Specifies the location and name of the file to be saved (w) or loaded (l).

**Examples:**

n shelia.com

Specifies the file RICH.COM on the default drive in the default directory. Will be created with the (W)RITE command if it does not exist.

n b:\two\start.exe

Specifies the file START.EXE on the B: drive in the subdirectory TWO.

## Q(uit)
Stops the Debug session without saving the information currently in memory.

**Syntax:**

**q**

**Example:**

q

Note: The MS-DOS prompt should be displayed.
To save information before you quit Debug, see the W(RITE) command.

## R(egister)
Displays or alters the contents of one or more μP registers.

**Syntax:**

**r** [*register-name*]

**Parameter:**

*register-name*

Specifies the name of the register whose contents are to be displayed or altered.

Notes: If you specify a register name, Debug displays the 16-bit value of that register in hexadecimal and displays a colon as the prompt. If the register is to be changed, type the new value and press Enter; otherwise, just press Enter to return to the Debug prompt with the register unaltered.

**Valid register names:**

**ax, bx, cx, dx, sp, bp, si, di, ds, es, ss, cs, ip, f**

If you type the **f** character, Debug displays the current setting of the status register using a two-letter code. To change the setting of a flag, type the appropriate two-letter code from the following table:

| Flag Name | Set | Clear |
|---|---|---|
| Overflow | OV | NV |
| Direction | DN | UP |
| Interrupt | EI(enable) | DI |
| Sign | NG(negative) | PL |
| Zero | ZR | NZ |
| Auxiliary Carry | AC | NA |
| Parity | PE(even) | PO |
| Carry/Borrow | CY | NC |

To view the contents of all registers and the machine code instruction pointed to by the current CS:IP location, simply type **r**.

Examples:  **r**  Displays the contents of all of the µP registers.
 **rcx**  Displays the contents of the count register.

## T(race)

Executes one or more machine code instructions, displays the contents of all registers, and decodes the next instruction pointed to by the CS:IP.

**Syntax:**

**t** [=*address*][*number*]

**Parameters:**

=*address*

Specifies the address at which Debug is to start single-stepping instructions. If =*address* is omitted, the program will begin single-stepping from the current CS:IP.

# APPENDIX E

# Answers to Odd-Numbered Chapter Review Questions

## Introduction

### Acronyms
a. APM = advanced power management
c. KWH = kilowatt hour

### Fill in the Blanks
1. volt
3. electrons
5. copper, aluminum
7. cable
9. ohm, Ω
11. fuse
13. direct current
15. frequency
17. DC power supply
19. capacitor
21. circuit common
23. red
25. uninterruptable

### Wire colors:
List the color wire
1. red  3. yellow  5. orange

### Multiple Choice
1. a
3. d
5. d
7. b
9. d
11. d
13. b
15. c
17. b
19. b
21. c
23. a
25. b

## Chapter 1

### Problems
1. a. 101011101 = 349
   c. 1100011100101010 = 50986
3. a. 1 1001 1101 0100 = 19D4
   c. 1001 0010 1011 = 92B
5. a. 2.23 μ (micro)
   c. 89 M (meg)
   e. 9.5 G (gig)
7. a. 1
   c. 1

### Multiple Choice
1. c
3. c
5. a
7. c
9. b
11. c
13. a
15. a
17. a
19. a

553

## Chapter 2
### Acronyms
a. ALU = arithmetic logic unit
c. IC = integrated circuit
e. RAM = random-access memory
g. SRAM = static random-access memory
i. POST = power-on self-test
k. COBOL = common business-oriented language

### Fill in the Blanks
1. arithmetic logic unit
3. address
5. fetch, execute
7. read-only memory
9. firmware
11. booted
13. RESET
15. machine
17. assembler, interpreter, compiler

### Multiple Choice
1. d
3. b
5. d
7. b
9. a
11. e
13. a
15. c
17. c
19. b

## Chapter 3
### Acronyms
a. ZIF = zero insertion force
c. CISC = complex instruction set computing
e. SECC = single-edge contact cartridge
g. FSB = frontside bus
i. PR = Pentium rating

### Fill in the Blanks
1. local
3. downwardly
5. pipeline
7. virtual
9. frontside
11. heat sink
13. symmetric multiprocessing
15. ATX

### Multiple Choice
1. b
3. c
5. b
7. b
9. c
11. b
13. d
15. a
17. b
19. a
21. c
23. c
25. c
27. a
29. d
31. b
33. a
35. d

## Chapter 4
### Acronyms
a. IBM = International Business Machines
c. BIOS = basic input/output system
e. AT = advanced technology
g. MCA = Micro-Channel architecture
i. VESA = Video Electronics Standards Association
k. PCI = peripheral component interface

### Fill in the Blanks
1. PC bus
3. PCI
5. open technology
7. turbo
9. 0000, FFFF
11. VL bus, open
13. Plug, Play
15. one-half

### Multiple Choice
1. b
3. b
5. a
7. e
9. b
11. c
13. b
15. c
17. b
19. d
21. a
23. c
25. b
27. c

## Chapter 5
### Acronyms
a. API = application program interface
c. EULA = End User License Agreement
e. TSR = terminate and stay resident

### Fill in the Blanks
1. 8, 3
3. End User License Agreement

# Answers to Odd-Numbered Chapter Review Questions

5. version
7. attributes
9. filename
11. internal
13. Program Information
15. WIN.INI
17. CONFIG.SYS
19. AUTOEXEC.BAT

## Multiple Choice
1. b
3. a
5. d
7. a
9. e
11. c
13. d
15. a
17. a
19. d
21. c
23. b

# Chapter 6

1. Windows Explorer
3. desktop, icons
5. Help
7. file association
9. System Editor
11. registry
13. USER

## Multiple Choice
1. b
3. a
5. b
7. b
9. b
11. b
13. c
15. b
17. c
19. b
21. c
23. a
25. b
27. b
29. b
31. a
33. e
35. d

# Chapter 7

## Acronyms
a. RAM = random-access memory
c. UMA = upper memory area
e. HMA = high memory area
g. EMM = expanded memory manager
i. DRAM = dynamic random-access memory
k. FPM = fast page mode
m. SDRAM = synchronous dynamic random-acess memory
o. DRDRAM = direct RamBus dynamic random-access memory
q. BASIC = Beginner's All-purpose Symbolic Instruction Code
s. DIMM = dual inline memory module
u. DIP = dual inline package

## Fill in the Blanks
1. conventional memory
3. cache
5. memory map
7. C
9. 64
11. EMM386
13. upper memory area
15. error-checking code
17. virtual

## Multiple Choice
1. c
3. d
5. b
7. b
9. a
11. c
13. a
15. b
17. c
19. b
21. b
23. d
25. b
27. c
29. a
31. b
33. d
35. a

# Chapter 8

## Acronyms
a. BIOS = basic input/output system
c. ESCD = extended system configuration data
e. ROM = read-only memory
g. EEPROM = electrically erasable programmable read-only memory
i. RTC = real time clock

## Fill in the Blanks
1. Setup
3. 70, 71
5. Del, boot
7. Clear CMOS
9. electrically erasable programmable read-only memory
11. master partition boot record

## Multiple Choice
1. b
3. b
5. b
7. b
9. a
11. b

## Chapter 9

### Acronyms
a. MIDI = musical instrument digital interface
c. PIC = programmable interrupt controller
e. DMA = dynamic memory access controller
g. bps = bits per second

### Fill in the Blanks
1. output port
3. 0F8H, 0FFH
5. COM2, COM4
7. hardset, softset, PnP
9. DIP switches
11. software
13. BIOS, operating system, PnP
15. female
17. RS-232C
19. full duplex
21. Hot-swapping
23. enhanced parallel port

### Multiple Choice
1. b
3. c
5. a
7. b
9. d
11. c
13. b
15. d

## Chapter 10

### Acronyms
a. QIC = Quarter Inch Committee
c. TPI = tracks per inch
e. CD-ROM = compact disk read-only memory

### Fill in the Blanks
1. Mass storage
3. tracks
5. double-sided
7. tracks per inch
9. 1.44 MB
11. first, last
13. allocation unit
15. CHKDIS, SCANDISK
17. multimedia

## Multiple Choice
1. b
3. c
5. b
7. d
9. c
11. d
13. c
15. d
17. a
19. d
21. d
23. d
25. a

## Chapter 11

### Acronyms
a. HDC = hard disk controller
c. RLL = run length limits
e. SCSI = small computer system interface
g. EIDE = enhanced integrated drive electronics
i. ATAPI = advanced technology attachment packet interface
k. DMA = direct memory access
m. CHS = cylinder head sector
o. LBA = logical block addressing

### Fill in the Blanks
1. fixed drives, Winchester
3. 8, 16
5. allocation units
7. advanced technology attachment packet interface
9. ST 506/412
11. programmable input/output
13. 11.1
15. primary master

### Multiple Choice
1. c
3. d
5. b
7. a
9. a
11. c
13. e
15. b

## Chapter 12

### Acronyms
a. CRT = cathode ray tube
c. VESA = Video Electronics Standards Association
e. HGA = Hercules graphics adapter
g. MGA = monochrome graphics adapter
i. EGA = enhanced graphics adapter
k. SVGA = super video graphics array
m. WYSIWYG = what you see is what you get

# Answers to Odd-Numbered Chapter Review Questions

## Fill in the Blanks
1. monochrome
3. deflection coils
5. scan
7. fields, frame
9. video ram
11. text, graphics
13. graphics, 640 × 350, 16
15. 3, pixel
17. analog

## Multiple Choice
1. a
3. b
5. c
7. b
9. b
11. b
13. a
15. c
17. c
19. a
21. b
23. d
25. a
27. c
29. a
31. a
33. a

# Chapter 13

## Acronyms
a. CPI = characters per inch
c. LED = light-emitting diode
e. LASER = light amplification by stimulated emission of radiation

## Fill in the Blanks
1. hard copy
3. impact
5. laser
7. print head

## Multiple Choice
1. c
3. b
5. b
7. a
9. c
11. b
13. a
15. b

# Chapter 14

## Acronyms
a. LAN = local area network
c. ISO = International Standards Organization
e. IEEE = Institute of Electrical and Electronic Engineers
g. SONET = synchronous optical network
i. IPX/SPX = internetwork packet exchange / sequenced packet exchange
k. MSAU = multi-station access unit
m. FDDI = fiber distributed data interface
o. NIC = network interface card
q. DLC = data link control
s. NOS = network operating system
u. IANA = Internet assigned numbers authority
w. ISP = Internet service provider
y. DNS = domain naming service
aa. URL = universal resource locator
cc. IDN = Internet domain name
ee. ASK = analog shift keying
gg. FSK = frequency shift keying
ii. CCITT = Comité Consultatif International Téléphonique et Télégraphique
kk. MNP = MicroCom Network Protocol

## Fill in the Blanks
1. dumb terminal
3. bandwidth
5. coax
7. multi-mode
9. serial
11. protocol
13. 330, 100
15. token
17. bus
19. virtual
21. 64, 18,446,744,073,709,551,616
23. organizational unique identifier
25. preferred
27. network administrator
29. computer name, mac
31. dynamic host configuration protocol
33. modulation, compression, error correction

## Multiple Choice
1. b
3. b
5. a
7. c
9. d
11. d
13. c
15. a
17. a
19. c
21. d
23. b
25. b
27. e
29. b

## Chapter 15
### Acronyms
a. POST = power-on self-test
c. CPU = central processing unit
e. ROM = read-only memory
g. MSD = Microsoft Diagnostics

### Fill in the Blanks
1. diagnostics
3. disk-based
5. game adapter
7. loop back
9. Device Manager

### Multiple Choice
1. d
3. b
5. a
7. b
9. d

## Chapter 16
### Acronyms
a. IRQ = interrupt request
c. DMA = direct memory access
e. DACK = device acknowledge
g. PIC = programmable interrupt controller
i. PnP = Plug-and-Play

### Fill in the Blanks
1. legacy
3. programmable interrupt controller
5. software polling
7. stack
9. type, vector table
11. software, hard drive
13. hold acknowledge
15. 2

### Multiple Choice
1. a
3. c
5. d
7. b
9. b
11. c
13. b
15. d
17. d
19. b

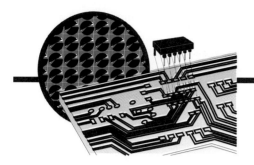

# ACRONYMS

Definitions for most of the following acronyms are located in the end-of-book Glossary.

## A

**A20**—address line 20
**ABIOS**—advanced BIOS
**ac**—alternating current
**ACPI**—advanced configuration and power interface
**AGP**—advanced (accelerated) graphics port
**AI**—artificial intelligence
**AIFF**—audio interchange file format
**AIU**—attachment unit interface
**ALU**—arithmetic logic unit
**ANSI**—American National Standards Institute
**APA**—all-point addressing
**API**—application program interface
**APM**—advanced power management
**ASCII**—American Standard Code for Information Interchange
**ASIC**—application-specific integrated circuit
**ASK**—amplitude shift keying
**ASME**—American Society of Mechanical Engineers
**ASPI**—advanced SCSI programming interface
**AT**—advanced technology
**ATA**—advanced technology attachment
**ATAPI**—advanced technology attachment packet interface
**ATM**—asynchronous transfer mode
**AUI**—attachment unit interface
**AVI**—audio video interleaved
**AWG**—American wire gauge

## B

**b**—bit
**B**—byte
**BASIC**—beginner's all-purpose symbolic instruction code
**BBS**—bulletin board service
**BCD**—binary coded decimal
**BIOS**—basic input/output system
**BLOB**—binary large objects
**BNC**—British naval connector/Bayonet-Neill-Concelman/Baby N Connector/Bayonet nut coupler
**BPS**—bits per second

## C

**CAD**—computer-aided design
**CADD**—computer-aided design and drafting
**CAE**—computer-aided engineering
**CAM**—computer-aided manufacturing
**CAMM**—cache/memory management unit
**CAS**—column address strobe
**CATV**—community antenna TV (cable TV)
**CCD**—charge coupled device
**CCITT**—Consultative Committee International Telegraph and Telephone
**CD**—compact disc
**CD-R**—compact disk-recordable.
**CD-ROM**—compact disc read-only memory
**CD-RW**—compact disk-read/write
**CGA**—color graphics adapter
**CHS**—cylinder-head-sector
**CIS**—card information structure
**CISC**—complex instruction set computer
**CMOS**—complementary metal oxide semiconductor
**CPGA**—ceramic pin grid array
**CP/M**—control program for microprocessors
**CPS**—characters per second
**CPU**—central processing unit
**CRC**—cyclic redundancy check
**CRT**—cathode ray tube

**CS**—chip select
**CSMA/CD**—carrier sense multiple access with collision detection
**CUA**—common user access

## D

**DAC**—digital to analog converter
**DAT**—digital audiotape
**dB**—decibel
**DBMS**—database management system
**DC**—direct current
**DCE**—data communication equipment
**DDE**—dynamic data exchange
**DDRSDRAM**—double data rate synchronous dynamic random-access memory
**DHCP**—dynamic host configuration protocol
**DIB**—dual independent bus
**DIMM**—dual inline memory module
**DIN**—Deutch industrial normal
**DIP**—dual inline package
**DLL**—dynamic link library
**DMA**—direct memory access
**DMAC**—direct memory access controller
**DNS**—domain naming service
**DOS**—disk operating system
**DPI**—dots per inch
**DPMI**—DOS protected mode interface
**DPMS**—DOS protected mode system
**DPSK**—differential phase shift keying
**DRAM**—dynamic random-access memory
**DRDRAM**—direct rambus dynamic random-access memory
**DSP**—digital signal processor
**DTA**—drive transfer address
**DTE**—data terminal equipment
**DTMF**—dual-tone modulated frequency
**DTR**—data terminal ready
**DUT**—device under test
**DVD**—digital versatile disk; digital video disk
**DWORD**—double-word (4 bytes)

## E

**EBCDIC**—extended binary coded decimal interchange code
**ECC**—1.) error checking and correction; 2.) error correcting code
**ECHS**—enhanced cylinder-head-sector
**ECP**—enhanced capabilities port
**EDO DRAM**—extended data output dynamic random-access memory
**EEPROM**—electrically erasable programmable read-only memory
**EGA**—enhanced graphics adapter
**EIA**—Electronics Industries Association
**EIDE**—enhanced integrated drive electronics
**EISA**—extended industry standard architecture
**EMF**—electromotive force
**EMM**—expanded memory manager
**EMS**—expanded memory specification
**EOS**—error correcting code on SIMM
**EPP**—enhanced parallel port
**EPROM**—erasable programmable read-only memory
**ESCD**—extended system configuration data
**ESD**—electrostatic discharge
**ESDI**—enhanced small device interface
**EULA**—End User License Agreement

## F

**FAQ**—frequently asked question
**FCB**—file control block
**FAT**—file allocation table
**FDDI**—fiber distributed data interface
**FIFO**—first-in/first-out
**FLOPS**—floating-point operations per second
**FM**—frequency modulation
**FPM**—fast page mode
**FPU**—floating point unit
**FSK**—frequency shift keying
**FSB**—frontside bus
**FTP**—file transfer protocol

## G

**GAL**—generic array logic
**GHz**—gigahertz
**GIF**—graphics image format
**GOP**—gigaoperations per second
**GPI**—graphics program interface
**GPIB**—general purpose interface bus
**GUI**—graphics user interface

## H

**H**—hexadecimal
**HCL**—hardware compatibility list
**HDTV**—high-definition television
**HGA**—Hercules graphics adapter
**HDD**—high-density diskette

**HMA**—high-memory area
**HTML**—hypertext mark-up language
**HTTP**—hypertext transfer protocol
**Hz**—hertz

## I

**IANA**—Internet Assigned Numbers Authority
**IBM**—International Business Machines
**IC**—integrated circuit
**ICANN**—International Corporation of Assigned Names and Numbers
**IDE**—integrated drive electronics
**IDN**—internet domain name
**IEEE**—Institute of Electrical and Electronic Engineers
**I/O**—input/output
**IP**—internet protocol
**IPS**—instructions per second
**IPX/SPX**—Internetwork packet exchange/sequenced packet exchange
**IrDA**—Infrared Data Association
**IRQ**—interrupt request
**ISA**—industry-standard architecture
**ISDN**—integrated service digital network
**ISO**—International Standards Organization
**ISP**—internet service provider
**ITU**—International Telecommunications Union

## J

**JEDEC**—Joint Electronics Devices Engineering Council
**JPEG**—Joint Photographic Expert Group

## K

**Kb**—kilobits
**KB**—kilobytes
**KHz**—kilohertz

## L

**LAN**—local area network
**LB**—local bus
**LBA**—logical block addressing
**LCC**—leadless chip carrier
**LCD**—liquid crystal diode
**LED**—light-emitting diode
**LIFO**—last-in/first-out
**LIM EMS**—Lotus-Intel-Microsoft Expanded Memory Specification
**LLC**—logical link control
**LPT**—line printer
**LSB**—least significant byte (byte)
**LSb**—least significant bit
**LSD**—least significant digit
**LZ**—landing zone

## M

**MAC**—medium access control
**MASM**—micro assembler
**MB**—megabyte
**MCA**—microchannel architecture
**MCB**—memory control block
**MCGA**—multi-color graphics adapter
**MCI**—media control interface
**MDA**—monochrome display adapter
**MDI**—multiple document interface
**MFLOPS**—million floating-point operations per second
**MFM**—modified frequency modulation
**MGA**—monochrome graphics adapter
**MHz**—megahertz
**MIDI**—musical instrument digital interface
**MIMD**—multiple instruction multiple data
**MIPS**—million instructions per second
**MMU**—memory management unit
**MMX**—multimedia extensions
**MNP**—microcom networking protocol
**MOS**—metal oxide semiconductor
**MOSFET**—metal oxide semiconductor field effect transistor
**MPC**—multimedia personal computer
**MPEG**—motion picture expert group
**MPU**—microprocessor unit
**MSAU**—multi-station access unit
**MSB**—most significant byte (byte)
**MSb**—most significant bit
**MSI**—medium scale integration
**MTBF**—mean time between failure

## N

**NetBEUI**—net BIOS extended user interface
**NetBIOS**—network basic input/output
**NIC**—network interface card
**NMI**—nonmaskable interrupt
**NTFS**—new technology file system
**NTSC**—National Television Standards Committee
**NVRAM**—nonvolatile RAM

## O

**OCR**—optical character recognition
**OEM**—original equipment manufacturer

OLE—object linking and embedding
OS—operating system

## P

PAL—programmable array logic
PBX—private branch exchange
PC—personal computer
PCB—printed circuit board
PCI—peripheral component interface
PCMCIA—Personal Computer Memory Card International Association
PDA—personal digital assistant
PGA—programmable grid array
PIC—programmable interrupt controller
PIF—program information file
PIO—programmable input/output
PIT—programmable interval timer
PLCC—plastic leaded chip carrier
PLL—phase lock loop
PMMU—page memory management unit
PnP—plug-and-play
POST—power-on self-test
PPGA—plastic pin grid array
PPI—programmable peripheral interface
PPP—point-to-point protocol
PROM—programmable read-only memory
PSK—phase shift keying
PSP—program segment prefix

## Q

QAM—quadrature amplitude modulation
QIC—Quarter Inch Committee

## R

RAID—redundant array of inexpensive drives
RAM—random-access R/W memory
RAS—row address strobe
RDRAM—Rambus dynamic random-access memory
RFI—radio frequency interference
RGB—red/green-blue
RIMM—Rambus in-line memory module
RISC—reduced instruction set computer
RLL—run length limited
ROM—read-only memory
RS—receive send
RTC—real-time clock

## S

SCSI—small computer system interface
SDRAM—synchronous dynamic random-access memory
SECC—single-edge connector cartridge
SEPP—single-edge processor package
SIMD—single instruction multiple data
SIMM—single inline memory module
SIP—single inline package
SIPP—single inline pin package
SLIP—serial line internet protocol
SMART—self-monitoring analysis and report technology
SMP—symmetric multi-processing
SMTP—simple mail transfer protocol
SONET—synchronous optical network
SQL—structured query language
SRAM—static random-access R/W memory
SSE—streaming SIMD extensions
SSI—small-scale integration
STP—shielded twisted pair
SVGA—super video graphic array

## T

TCP/IP—transmission control protocol/internet protocol
TIFF—tagged image file format
TIGA—Texas Instruments graphics architecture
TPI—tracks per inch
TSR—terminate and stay resident
TTL—transistor/transistor logic

## U

UART—universal asynchronous receiver transmitter
UDMA—ultra direct memory access
UMA—upper memory area
UMB—upper memory block
UPC—universal product code
UPS—uninterruptible power supply
URL—universal resource locator
USART—universal synchronous/asynchronous receiver transmitter
USB—universal serial bus
UTP—unshielded twisted pair

## V

VBE—VESA BIOS extension
VCPI—virtual control program interface
VDT—video display terminal

# Acronyms

**VESA**—Video Electronics Standards Association
**VFS**—virtual file system
**VGA**—video graphics array
**VLB**—VESA local bus
**VLIW**—very long instruction word
**VLSI**—very large scale integrated
**VMM**—virtual memory manager
**VRAM**—video random-access R/W memory
**VxD**—virtual device driver

## W

**WAN**—wide area network
**WINS**—Windows Internet Naming Service
**WORM**—write once read many
**WWW**—World Wide Web
**WYSIWYG**—what you see is what you get

## X

**XGA**—extended graphics array
**XMM**—extended memory manager
**XMS**—extended memory specification
**XT**—extended technology
**XVGA**—extended video graphics array

## Z

**ZIF**—zero insertion force

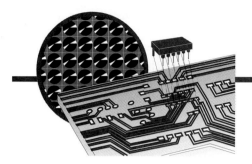

# GLOSSARY

**802.3**—The IEEE standard that defines Ethernet and the CSMA/CD protocol.

**10Base2 (ThinNet)**—An Ethernet LAN standard (IEEE 802.3) operating at 10 Mbps implemented in a bus topology using RG58 coax cable and running in segments of up to 185 meters.

**10Base5 (ThickNet)**—An Ethernet LAN standard (IEEE 802.3) operating at 10 Mbps implemented in a bus topology using RG8 coax cable and running in segments of up to 500 meters.

**10BaseFL**—The IEEE 802.3 Ethernet LAN standard operating at 10 Mbps that is implemented in a star topology using fiber optic cable.

**10BaseT**—An Ethernet LAN standard (IEEE 802.3) operating at 10 Mbps that is implemented in a star topology using unshielded twisted-pair wiring (twisted pair is the same as telephone cable). See *LAN* and *Ethernet*.

**100BaseT (Fast Ethernet)**—An Ethernet LAN standard (IEEE 802.3) operating at 100 Mbps that is implemented in a star topology using UTP or STP cable.

**3D-Now!**—The name given to the SIMD instructions added to the AMD K6-2 microprocessor's instruction set to support multimedia. These instructions have also been added to IDT's WinChip2 microprocessor.

**506/412**—The interface used to connect an MFM hard drive to a compatible computer.

**1394 (FireWire)**—An IEEE serial port standard that supports up to 67 devices. The standard also supports hot-swapping and isochronous operation.

## A

**ACPI (Advanced Configuration Power Interface)**—A Microsoft standard that allows the operating system to manage the power of the computer and motherboard devices.

**adapter**—A printed circuit card that inserts into an expansion slot on the motherboard that contains the circuitry necessary to form the I/O ports that interface with peripherals.

**address**—(1) A unique binary combination assigned by engineers to each memory location and port in a computer. (2) A unique binary identifier assigned to each node on a network.

**address bus**—A group of electrical conductors provided by the microprocessor that is used to select a device for communication. The communication of data takes place on the microprocessor's data bus.

**addressing mode**—The method by which a microprocessor addresses data to be processed.

**AGP (accelerated graphics port)**—An open-standard, dedicated graphics expansion slot developed by Intel. The AGP is not compatible with ISA or PCI. The AGP runs at the speed of the computer's local bus. See *ISA* and *PCI*.

**algorithm**—A step-by-step process used to solve a problem.

**allocation unit**—The fundamental storage unit used by disk drives. Information is stored and retrieved from disk drives in allocation units. The size of an allocation unit varies depending on the density of the disk drive. Current sizes range from 512 bytes used by some floppy drives to 32 KBs used by some hard drives.

**ALU (arithmetic logic unit)**—The unit inside the CPU where the math, logic, and bit manipulation is performed.

**ampere**—The unit of measurement of electrical current flow. One ampere equals $6.25 \times 10^{18}$ electrons flowing past a point in a circuit in 1 second.

**analog**—A signal whose level can change continuously between its limits, as opposed to digital, which is a signal that has a fixed number of stepped settings between the limits. See *digital*.

**(ANSI) American National Standards Institute**—An institute responsible for setting many of the standards used in computers.

**APA (all-point addressing)**—Another name for the graphics mode used with video adapters. Used to address the individual pixels on the video screen.

**API (application program interface)**—Another name for the computer's operating system.

**APM (advanced power management)**—A computer specification that allows the operating system to place a

565

peripheral into a low-power standby mode if it has been idle for a predefined period of time.

**application program**—Software written toward a specific application around an operating system such as MS-DOS and Windows. Examples include word processing, database management, and spreadsheets.

**Archie**—A software program that searches anonymous (open) FTP sites for certain filenames. The computer must be connected to the Internet to use this program.

**ARCNET (attached resource computer network)**—One of the first LANs invented, it uses inexpensive cards with a star-type wiring scheme. It is fairly slow at 2.5 Mbits/sec and is not an IEEE standard. See *LAN, star,* and *IEEE*.

**array**—An arrangement of elements in one or more dimensions.

**ASCII (American Standard Code for Information Interchange)**—A binary code used to represent characters, numbers, and control codes for communication between a computer and peripherals.

**ASIC (application-specific integrated circuits)**—Integrated circuits that are designed for a specific application.

**assembly language**—A very low-level language that is one step above machine language. It is written using instruction mnemonics instead of binary codes.

**asynchronous communications**—A method of transferring information without the source and destination being synchronized with one another. Data take the form of digital words with no time relationship to one another. Each word is identified by one or more start bits (indicating the beginning of a word) and one or more stop bits (indicating its termination).

**AT (advanced technology)**—The name given to IBM and compatible computers that run the 80286 microprocessor. The AT computer set most of the standards used in current compatible PCs.

**ATA-1 (advanced technology attachment) (IDE)**—The first standard used to define the IDE disk interface and command set. See *IDE*.

**ATA-2 (EIDE)**—An enhanced IDE standard. The ATA-2 defines faster data transfer modes and the use of logical block addressing to access large capacity drives.

**ATAPI (advanced technology attachment packet interface)**—An extension to the ATA specification that defines the protocols needed for devices such as CD-ROMs and tape drives to use the IDE interface.

**ATM (asynchronous transfer mode)**—A high-speed data transfer mode used by some LANs that operates similar to a telephone system, creating virtual circuits between computers. See *LAN*.

**attenuation**—The degradation of a signal over distance. The amount of attenuation is determined by the medium through which the information is transferred.

**attribute**—Information stored in the directory entry of a filename that defines whether the filename is read-only, system, hidden, a label, or a subdirectory.

**AUTOEXEC.BAT**—A batch file that executes a set of MS-DOS commands each time the computer is booted.

## B

**bandwidth**—Frequency or bit-carrying capacity of a medium that is usually measured in hertz or bits per second (bps).

**base memory**—The portion of memory that resides within the first megabyte address range (also called conventional memory). To remain compatible with the PC/XT/AT architecture, only the first 640 KB of addresses are accessible as user memory. A minimum of 64 K of base memory is required to load MS-DOS.

**BASIC (Beginner's All-Purpose Symbolic Instruction Code)**—A high-level programming language that can be used to write user programs. Uses terms and expressions used in English language.

**batch file**—A text file containing one or more MS-DOS commands, identified by the extension .BAT. MS-DOS, searches the root directory for the batch file AUTOEXEC.BAT every time the computer is reset. This means the DOS commands in this file will be executed automatically each time the computer is reset.

**Baud rate**—(1) The rate at which data is transferred between two pieces of equipment in a serial format. Basically, baud rate means bits per second in asynchronous communication. (2) The rate at which the carrier in a digital communication system changes its state to convey information. Each state change can encode any number of digital bits (from less than 1 to about 128 in today's systems), allowing the actual information throughput in bits per second to be faster than the baud rate.

**BCD (binary coded decimal)**—A coding system that represents the decimal numbers from 0 to 9 by using 4-bit groups of binary to represent each digit.

**BEDO DRAM (burst extended data output DRAM)**—DRAM that depends on the burst reads of a cache controller to reduce the memory access time to a 5-1-1-1 read timing sequence.

**beep code**—See *POST*.

**benchmark**—A point of reference from which measurements can be made.

**binding**—Assigning a network protocol suite to a network adapter. See *protocol suite*.

**BIOS (basic input/output system)**—The basic machine code program that is usually located in the computer's ROM. The computer will run these programs after a reset.

**bit**—A single binary digit.

**BNC (British national connector)**—A connector used in some LANs. See *LAN*.

# Glossary

**boot**—To load the resident (MS-DOS) instruction program into memory by setting the system unit power switch to the on position, pressing the RESET button (when applicable), or pressing Ctrl-Alt-Del. Derived from the phrase "pull up by one's own bootstraps," a reference to the need for a computer to load its own operating system.

**bridge**—A device used to divide a single LAN into individual addressable segments.

**browser**—Client software that allows the user to access certain servers on the Internet or Intranet. A browser was originally used to access servers on the Internet's World Wide Web. However, in recent years browsers have been adapted to access FTP sites as well as gopher sites. The computer must be networked to the Internet or local Intranet to use this software.

**buffer**—A section of RAM that is on the peripheral or is part of the user RAM that is a temporary storage location for different peripherals such as hard drives.

**bus**—A data path shared by many devices (e.g., multipoint line) with one or more conductors for transmitting signals, data, or power.

**bus-master**—A device, such as a hard-drive and DMAC, that has the ability to control the computer's address, data, and control buses to transfer information directly with the computer's RAM.

**byte**—(1) A group of 8 consecutive binary bits. (2) The fundamental storage unit used by compatible computers.

## C

**cache**—Memory used to improve the performance of hard disk and high-speed microcomputers. The most used information is stored in SRAM memory, which takes less time to access than the disk. Also, high-speed 386 (and above) computers use an SRAM cache to hold prefetched data from the slower DRAM. See *SRAM* and *DRAM*.

**CAD (computer-aided design)**—Programs that assist in architectural and engineering design.

**CAM (computer-aided manufacturing)**—A concept that allows designs to go directly from a CAD program to a programmable machine that produces a finished product.

**CCITT (Consultative Committee International Telegraph and Telephone)**—An international association that sets worldwide communications standards. CCITT recently changed its name to the International Telephone Union.

**CD-ROM (compact disk - read-only memory)**—A storage media that works on the same basic principle as CD audio players. CD-ROM is a read-only media that has the ability to hold over 650 million bytes of information.

**CDR-ROM (CD-ROM recorder)**—A CD-ROM that has the ability to write information onto a recordable CD.

**CGA (color graphics adapter)**—The standard low-resolution color display card used in IBM/PC-compatible computers. In the medium resolution graphics mode, it will display $320 \times 200$ pixels in 4 colors from a palette of 16. In the high-resolution graphics mode, it will display $640 \times 200$ pixels in black and white.

**CHS (cylinder-head-sector)**—One of the methods used to address the information stored on a disk drive. Most hard drives currently use logical block addressing instead of CHS. See *LBA*.

**CIS (card information structure)**—A specification that describes how a computer automatically configures PCMCIA cards. See *PCMCIA*.

**CISC (complex instruction set computer)**—A microprocessor that has one instruction for each operation. CISC processors support the instruction set of previous-generation processors.

**cladding**—A plastic coating around the glass core of a fiber optics conductor that adds in reflection and protection.

**client**—Software that allows a computer's operating system to access a network server.

**cluster**—The minimum number of sectors accessed by the disk operating system when data are transferred with a disk drive. See *allocation unit* and *sector*.

**CMOS (complementary metal oxide semiconductor)**—A logic family that uses both P-channel and N-channel MOS transistors to form logic gates. This type of logic consumes very little power. CMOS SRAM is used in most ATs to store setup information about the computer's configuration.

**coaxial cable (coax)**—A cable consisting of a single conductor enclosed in an insulator that is surrounded by a metallic shield and wrapped in an insulating external jacket. Ethernet uses coaxial cable to implement 10Base2 and 10Base5, which are both rated at 10 Mbps. See *Ethernet, 10Base2,* and *10Base5*.

**cold-boot**—Forcing a complete reinitialization of the microprocessor or computer by removing and then restoring power, or by pressing a hard RESET switch. Also called hard-boot.

**communication protocol**—The rules governing the exchange of information between devices on a data link.

**compiler**—A specialized machine code program that converts programs written in one computer language to another. For example, a compiler will convert a high-level computer language such as Pascal into the machine language understood by the microprocessor.

**composite**—Video signals that contain both the image and sync components of the picture.

**computer language**—The rules and regulations associated with a specific way of communication for computers.

**COM(X)**—The name used by most operating systems to access the compatible computer's programmable serial ports. The X is a number between 1 and 4 depending on the selected port, for example: COM2.

**configuration**—The devices that make up the system, subsystem, or network.

**configure**—To set up a computer for the correct input/output devices and operating system to make sure they interact with each other efficiently.

**control bus**—A group of electrical conductors used in a computer to control the communication of data.

**conventional memory**—See *base memory*.

**cookie**—Information that may be stored on a user hard drive by Web sites visited by the user. This information allows the Web site to save information that is unique to the individual.

**CP/M (control program for microprocessors)**—A very popular disk operating system in the late 1970s and early 1980s originally written for the 8080 and Z80 MPUs. MS-DOS is a spin-off of this operating system.

**CPS (characters per second)**—A speed rating used with serial communication. In serial communication, most characters consist of 10 bits.

**CPU (central processing unit)**—The unit in the computer that processes the information. In personal computers, the CPU is also called the microprocessor.

**CRC (cyclic redundancy check)**—An error-detection scheme in which the block check character is the remainder after dividing all of the serialized bits in a transmission block by a predetermined binary number.

**crosstalk**—The unwanted transmission of a signal on a channel that interfaces with another adjacent channel.

**CRT (cathode ray tube)**—The picture tube used by most televisions and computer displays.

**CSMA/CD (carrier sense-multiple access with collision detection)**—The IEEE 802.3 standard utilized by Ethernet that senses communication and collisions on a network.

**current**—The flow of electrons that performs electrical work created by voltage. The unit of measurement is the ampere.

**cursor**—The active point on the screen, usually a blinking underline, where the next character typed will appear.

**cylinder**—All hard disk or diskette tracks that can be read or written without moving the drive's read/write heads.

# D

**daisy-chain**—A set of similar devices connected along a single cable. Both hard and floppy disk drives are usually daisy-chained on a single interface cable, as are SCSI devices.

**database**—A collection of data that can be immediately accessed and operated upon by a data processing system for a specific purpose.

**data bus**—A group of conductors used by a computer to transfer data between a selected device and the CPU. The device can be either memory or I/O and is selected by the computer's address bus.

**data compression**—A means of packing information into fewer bytes by eliminating redundant data.

**DBMS (database management system)**—Application program to manage large amounts of data.

**debug**—To change or format any data, as in troubleshooting.

**default**—The original or automatically selected condition.

**delete**—The process of removing files from one or more of a computer's mass storage devices.

**DHCP (dynamic host configuration protocol)**—A service that provides automatic IP assignments to computers logging onto a network. Most Internet service providers (ISPs) use this service to better utilize their allotment of Internet IP addresses.

**dhrystone**—A standard benchmark for measuring the processing speed of a computer. Pronounced "dry stone."

**diagnostic**—A program designed to detect faults in hardware.

**DIB (dual independent bus)**—An independent bus used by some microprocessor such as the Intel® Pentium® Pro and the Pentium® II microprocessors to access the L2 cache.

**die**—An integrated circuit.

**digital**—A signal whose level can only vary between a few (usually two) discrete levels. Most signals in a computer system are digital.

**DIMM (dual-inline memory module)**—A 168-pin memory module standard that stores data in groups of up to 64 bits.

**DIN (Deutsche industry normal)**—A round connector used to connect the keyboard to the compatible computer's keyboard port. This can be either a standard DIN or mini-DIN. The mini-DIN is also used to connect a PS/2 mouse to a PS/2 mouse port.

**DIP (dual-inline package)**—A rectangular-shaped integrated circuit with two parallel rows of pins extruding from the longest sides. They come in sizes ranging from 4-pin to 64-pin configurations.

**DMA (direct memory access)**—A means by which a peripheral device, such as a disk drive, can transfer data directly to or from the computer's memory without going through the microprocessor.

**DMAC (direct memory access controller)**—The actual chip that handles the DMA. See *DMA*.

**DMM (digital multi-meter)**—A piece of electronic test equipment used to measure voltage, current, or resistance.

**DNS (domain naming service)**—Database servers that allow computers to use an Internet domain name (IDN) to select a computer instead of an IP address. See *IDN*.

**domain**—A group of users that share the same network login security.

**DOS (disk operating system)**—System that usually controls the overall operation of the computer above the BIOS level.

**dot matrix printer**—A printer that transfers many series of dots by striking a ribbon to the printer paper. These dots form letters, numbers, and symbols.

**dot-pitch**—A color display monitor specification that gives the spacing, in millimeters, of the triads that are used to display images. Simply stated, the smaller the dot-pitch, the better the monitor.

**dotted-quad**—Another name for the current IP address standard used by the TCP/IP stack. A dotted quad uses 4 bytes of binary to address computers on the network. This allows 4,294,967,296 logical address combinations. See *IP address* and *TCP/IP*.

**download**—Transferring information from a computer to another device or computer. See *upload*.

**DPI (dots per inch)**—The standard used to measure the spacing between the dots used by printers. The greater the DPI, the better the resolution.

**DRAM (dynamic random-access R/W memory)**—Volatile memory used as the main memory in most microcomputers with over 16 KB of RAM. This memory uses multiplexed address pins and requires constant access (refresh) or the data will be lost.

**driver**—A small program used to control a hardware device. It can be part of the BIOS or loaded in and added to DOS.

**DVD (digital video disk or digital versatile disk)**—A high-density video or data standard that is similar to CD-ROM but allows recording of information on both sides of the disk and also allows information to be recorded in layers.

## E

**EBCDIC (extended binary coded decimal interchange code)**—A code developed by IBM to transfer information between computers and peripherals. This code is not compatible to ASCII.

**ECC (error checking and correction)**—System that checks data for errors as it is being transferred. When necessary, ECC performs "on-the-fly" correction or single bit errors.

**ECC RAM (error checking and correction random-access memory)**—RAM that performs ECC without additional hardware. See *ECC*.

**ECP (enhanced capabilities port)**—A parallel port standard that allows the use of DMA, device addressing, and high-speed transfer.

**edge card connector**—A connector that consists of exposed conductors that run along the edge of the printed circuit board. The edge card connector inserts into another connector on another printed circuit board to connect the two together electrically.

**EDO DRAM (enhanced data output dynamic random-access memory)**—Memory that allows data from one read cycle to be extended into the addresses portion of the next read cycle, effectively reducing the number of clock cycles required for each read cycle by one.

**EEPROM (electrically erasable programmable read-only memory)**—See *Flash ROM*.

**EGA (enhanced graphics adapter)**—Adapter that provides graphics of 640 × 350 pixels. Provides 16 colors from a palette of 64. Usually supports both CGA and HGA graphics.

**EIA (Electronic Industries Association)**—Association that is usually responsible for standardization of communication ports.

**EIDE (enhanced integrated drive electronics) (ATA-2)**—See *ATA-2*

**EISA (enhanced industry-standard architecture)**—A 32-bit microcomputer bus developed by a consortium of compatible manufacturers to run the 80386 and 80486 microprocessors. These manufacturers developed EISA because IBM charges excessive royalties to any competitors who use the MCA slot designed by IBM. See *MCA* and *ISA*.

**EMI (electron magnetic interference)**—Interference caused by magnetic fields generated by electrical current flow that produces electrical noise in other devices. This electrical noise can cause devices to malfunction.

**EMM (expanded memory manager)**—A DOS TSR loaded with the CONFIG.SYS file that is used to manage expanded memory if used.

**EMS**—See *LIM EMS*.

**EPP (enhanced parallel port)**—A parallel port standard that specifies bi-directionality between the port and peripheral. EPP is totally compatible with SPP. See *SPP*.

**EPROM**—A special type of ROM that can be removed from the equipment and erased using high-intensity, short wavelength ultraviolet light. Once erased, EPROM can be programmed using special programming equipment.

**ESD (electrostatic discharge)**—The discharging of static electricity. If ESD occurs through most electronics devices, they will be damaged.

**ESDI (enhanced small device interface)**—A high-speed interface standard used between a computer and peripherals. Used to interface most extremely high-density hard disk drives.

**Ethernet**—A LAN architecture able to connect up to 1,024 personal computers and workstations to a single file server. Developed by Xerox, Digital, and Intel, it connects terminals via coaxial or twisted-pair cable using a bus or star topology. It has its own collision detection that allows several peripherals to run off the same cable. Ethernet is currently the most popular adapter.

**expanded memory**—A memory paging method that allows access to memory greater than the 1-MB memory limit of the 8088 and 8086 MPUs by using bank switching. This memory requires a special interface card and special driver program to access. See *LIM EMS*.

**expansion slot**—A series of connectors mounted on the motherboard, or back-plane, into which controller cards can be inserted to enhance the computer's capabilities. There are three types: (1) 8-bit for IBM-PC and XT-compatible cards, (2) 16-bit for AT-compatible, (3) 32-bit for the EISA, MCA, or PCI.

**extended memory**—The portion of memory in an AT system that resides above the 1-MB address range of the 8088 microprocessor. Programs running the processor in the protected mode can only access extended memory. This mode is only available to 80286 processors and above. Extended memory is not accessible to current versions of MS-DOS (version 4 or below) but it can be accessed by operating systems such as Windows 3.X, Windows 95, Windows 9X, Windows 2000, Windows Me, OS/2, UNIX, and XENIX.

**external commands**—DOS commands that are not loaded into the computer's memory when the computer is booted with DOS. To execute these commands, you must: (1) have the DOS diskette with the desired command inserted into the default disk drive, (2) be in the directory in which DOS is located on a hard drive, or (3) use the PATH command to set the path to where the external DOS commands are located on the hard drive.

## F

**FAQs**—Acronym for frequently asked questions.

**FAT (file allocation table)**—A record used by MS-DOS to assign disk storage space. See *allocation unit*.

**FAT12**—Indicates that 12 binary bits are used to select allocation units on a disk. When a disk is formatted using FAT12, 4,096 allocation units can be selected.

**FAT16**—Indicates 16 binary bits are used to identify allocation units on the disk. When a disk is formatted using FAT16, there can be up to 65,536 allocation units on the disk.

**FAT32**—Seems to indicate that 32 binary bits are used to identify allocation units on the disk. However, when FAT32 is used, 4 bits are reserved, which leaves 28 bits to identify allocation units on the disk. When a disk is formatted using FAT32, up to 268,435,456 allocation units can be selected.

**fiber optic**—A glass or plastic conductor that uses light to transfer information.

**FIFO (first-in first-out)**—A memory buffer that stores information and retrieves it in the order in which it is stored.

**file server**—A central computer that stores files and software shared by LAN users. When it holds database operations as well, it is called a database server. A computer dedicated to LAN printing is called a print server.

**firmware**—Another name for programs stored in ROMs.

**Flash ROM**—A form of electrically erasable programmable read-only memory (EEPROM) that is used as the BIOS of the computer or as the ROMs in peripherals such as modems. Flash ROMs can be erased and reprogrammed in the equipment in which they operate.

**FLOPS (floating-point operations per second)**—Measure that indicates the actual math speed of microprocessors/microcomputers.

**FM (frequency modulation)**—A method of transmitting data in which the frequency of the signal is changed to represent the data.

**FPU (floating point unit)**—Circuitry inside the microprocessor that is separate from the arithmetic logic unit, which uses special instructions to handle complex math operations. A math coprocessor is the name given to the FPU when it is a separate IC.

**frame**—The grouping of binary bits used to transfer serial information. The frame usually consists of a header, data, and a trailer.

**frequency**—The rate at which a signal repeats itself. Frequency is usually measured in cycles per second or Hertz.

**Frontside Bus (FSB)**—The bus used to connect the microprocessor to the other units in the computer. This is called the local bus on microprocessors that use Socket4, Socket5, and Socket7. The frontside bus actually consists of the microprocessor's address, data, and control buses. See LB (*local bus*).

**FSK (frequency shift keying)**—A frequency modulation technique in which one frequency represents a logic 1 and a second, a logic 0.

**FTP (file transfer protocol)**—The protocol used to transfer files over the Internet.

**FDX (full duplex)**—Transmission in either direction, at the same time.

## G

**gateway**—A device used on a network that translates hardware and software network protocols. A gateway is used to connect together networks that use different topologies or different packets.

**gender**—The characteristic of a connector. The connector that has the pins is referred to as a male connector, whereas the connector that has the receptacles into which the pins plug is called a female connector.

**GIF (graphics interchange format)**—A graphics file format developed by CompuServe that uses 256 colors (8-bit).

**G (gig)**—The metric prefix for $10^9$ (1,000,000,000) or $2^{30}$ (1,073,741,824) when binary is used.

**gigabyte**—A storage capacity of 1,073,741,824 bytes.

**GUI (graphical user interface)**—A method by which the user controls the options of an operating system or application program through the use of icons. See *icon*.

**gun**—A slang expression for the electronic guns in a CRT that discharge a beam of electrons that strikes phosphor on the inside surface of the screen, emitting light.

## H

**handshaking**—Exchange of predetermined signals between two devices establishing a connection.

**hardware**—The physical components of a computer system; e.g., ICs, keyboard, video monitor, and printer.

**HCL (hardware compatibility list)**—A list of devices directly supported by an operating system.

**HDD (high-density diskette)**—High-density 5.25″ and 3.5″ disks are able to store 1.2 and 1.44 megabytes of information, respectively, and can only be read by high-density drives.

**HDX (half duplex)**—Transmission in either direction, but not at the same time.

**hexadecimal**—A numbering system used in computers to represent binary numbers and codes. Each hexadecimal digit has the ability to represent 16 numerical combinations, 0 through F.

**HGA (Hercules graphics adapter)**—MDA card that allows graphics to be displayed on a TTL monochrome monitor. Programs must be written to support this card.

**high-level language**—A programming language that allows a more human interaction with the computer. Programs written in high-level language must be interpreted or compiled to machine language.

**HMA (high-memory area)**—The name given to the first page of extended memory (page $10^{16}$) when an 80286 or above microprocessors operate in the real mode.

**hot-swapping**—The act of removing and replacing peripherals without removing power. Hot-swapping should only be performed with peripherals and ports designed with this feature.

**HTML (hypertext mark-up language)**—A protocol used on the World Wide Web section of the Internet to transfer Web pages in text form. Once received, a program called a browser constructs and displays the Web page from information contained in an HTML document.

**hub**—The center of a star topology where nodes are connected.

**hyperlink**—Text or images on a Web page that link to another Web page, image, or other text. When the user clicks on the hyperlink, the browser automatically goes to the indicated link.

**HZ (hertz)**—The unit used to measure the rate of change of a signal. Hertz stands for cycles per second.

## I

**IBM®**—The registered trademark of International Business Machines.

**IC (integrated circuit)**—An entire circuit that is constructed on a single piece of semiconductor material.

**icon**—A picture used to represent a program, operating system command, or application program option.

**IDE (integrated drive electronics)**—Also known as ATA-1. An interface between µC's and hard disk drives that moves the hard disk controller from an adapter card to the drive. Provides nearly the performance of the ESDI at a reduced price.

**IDN (Internet domain name)**—The primary method used to select computers on the Internet. The IDN is in the form of "hostname.domain.class."

**IEEE (Institute of Electrical/Electronic Engineers)**—Group that usually standardizes electronic components such as the resistor color code.

**INF**—A text configuration file used by Windows 95 and 98 applications.

**INI**—A text configuration file used by the Windows 3.X operating system and Windows 3.X applications.

**interleave**—To arrange parts of one sequence of things or events so that they alternate with parts of one or more other sequences of the same nature so that each sequence retains its identity.

**Internet**—A global network of computers that all use the TCP/IP protocol stack. Individuals install client software on their computer to access different categories of servers attached to the Internet. These servers include, but are not limited to, WWW, FTP, e-mail, chat, gopher, and telnet servers.

**internet**—A network of networks.

**interpreter**—A machine language program that translates high-level languages into machine language one statement at a time. Interpreters never create a true machine language program.

**interrupt**—See *IRQ*.

**Intranet**—A LAN that provides features that are similar to those offered by the Internet.

**I/O (input/output)**—Unit used to connect the processed information inside the computer to the outside world.

**IP address**—A unique number assigned to each computer on a network using the TCP/IP protocol. The number is split into four sections that are separated by a period. For this reason, it technically is called a dotted-quad. An example of an IP address is 206.145.211.75.

**IPS (instructions per second)**—Unit that measures the throughput of microprocessors.

**IPv4**—See *dotted quad*.

**IPv6**—A new form of IP address standard currently under development that will use 6 bytes (48 bits) of binary to logically address computers using the TCP/IP suite. This will allow 2,814,749,76,710,656 address combinations. See *TCP/IP* and *IP address*.

**IrDA (infrared data association)**—A protocol that uses one of the computer's programmable serial ports to interface infrared devices.

**IRQ (interrupt request)**—Control signals used by ports to signal the microprocessor when they need service.

**ISA (industry-standard architecture)**—The standard 64-pin, 8-bit data bus used on all IBM PC/XT hardware-compatible computers. Also expanded to a 16-bit data bus for the IBM PC/XT-compatible computers.

**isochronous**—Digital information that is sent at a constant, predetermined rate.

**ISP (Internet service provider)**—A company that provides access to the Internet, usually for a monthly fee.

## J

**JPEG (joint photographic expert group)**—A standard by which pictures are compressed and decompressed to occupy less space on a disk drive or take less time to send over a network link. Files that are stored using this standard usually have a .JPG extension.

## K

**kilobyte**—A term that refers to a computer's storage capacity. A kilobyte (KB) is defined as 1,024 bytes.

## L

**L1 cache (level 1 cache)**—Cache memory that is mounted on the same semiconductor wafer as the microprocessor circuitry. Usually consists of a separate data cache and instruction cache.

**L2 cache (level 1 cache)**—Cache memory that is located outside the microprocessor wafer and is usually accessed at a slower speed than the L1 cache.

**LAN (local area network)**—Connecting computers together so they can share peripherals and information.

**LB (local bus)**—The microprocessor's address, data, and control buses.

**LBA (logical block addressing)**—A method of addressing information on SCSI and all current EIDE hard drives. Each sector is assigned a number starting with 0 until the last sector on the hard drive is reached. The information is addressed by sector number instead of using CHS. LBA can be used to access hard drives up to 128 GB.

**LCC (leadless chip carrier)**—A standardized package used by ICs.

**LIM EMS (Lotus, Intel, and Microsoft expanded memory specification)**—The specifications developed by Lotus, Intel, and Microsoft for the expanded memory card and software driver program required to access the memory above the 1-MB limit in computers with 8088 or 8086 microprocessors. See *expanded memory*.

**loose jacket**—A type of fiber optics cable in which the jacket is not attached to the cladding of the fiber. There is usually a silicon gel between the jacket and the fiber's cladding.

**LZ (landing zone)**—A designated area to which the hard drive's read/write heads are sent when power is removed.

## M

**machine language**—The binary language that a microprocessor understands. All other languages are translated to machine code.

**main storage**—Program-addressable storage from which instructions and other data can be loaded directly into registers for subsequent execution of processing.

**Manchester encoding**—A method used by Ethernet to encode data transferred on the network. Manchester encoding uses voltage transitions to represent 1s and 0s.

**mass storage**—Auxiliary storage in a computer system, as differentiated from RAM. Mass storage usually refers to floppy and fixed disks and magnetic tape.

**math coprocessor**—See *FPU*.

**MCA (Micro-Channel architecture)**—The microcomputer bus developed by IBM for the PS/2 series of computers. IBM has a patent on this bus.

**MCGA (multi-color graphics adapter)**—An IBM video standard that supports text and graphic display modes with a resolution of 300 × 200, with 16 colors from a palette of 262,144. Usually supported in non-IBM computers by a VGA adapter card. See *VGA*.

**MDA (monochrome display adapter)**—The card inside a computer that holds all the circuitry to drive a TTL monochrome monitor. This card displays only text characters.

**medium (pl., media)**—The substance through which information is transferred.

**MFLOPS**—Million FLOPS. See *FLOPS*.

**MFM (modified frequency modulation)**—A format used to record information on disk drives that uses timing bits only between consecutive 0s.

**MGA (monochrome graphics adapter)**—A graphics video card that controls a monochrome monitor.

**MHz (megahertz)**—One megahertz equals 1,000,000 cycles in 1 second.

**micron**—One millionth of a meter.

**MIDI (musical instrument digital interface)**—Standard used to interface digital musical instruments to the microcomputer.

**MIPS**—1 million IPS. See *IPS*.

**MMU (memory management unit)**—An IC used to manage the extend memory-addressing capabilities that allows sophisticated memory-paging and program-switching capabilities. Supported by the OS/2 operating system but not by current versions of MS-DOS (4.0 and below).

**MMX (multimedia extension)**—57 additional SIMD instructions added to the Intel Pentium-compatible microprocessors and above that allow faster video and sound processing.

**MNP (Microcom networking protocol)**—A set of standards for various aspects of the operation of modems including error control, data compression, line optimization, and data-rate negotiation.

**modem**—Short for *mo*dulator-*dem*odulator, this device converts data from one of the computer serial ports into tones that can be sent over phone lines. It also receives tones from another modem and converts them back to serial data for the computer's serial port.

**modulation**—The varying of a carrier wave to encode information. For example, amplitude modulation varies the strength or amplitude of the carrier waves in response to the changes in the data being transmitted.

**monochrome**—A monitor that displays one color plus black.

**MOS (metal oxide semiconductor)**—A semiconductor technology that consumes small amounts of electrical power but operates at relatively low speeds.

**mouse**—A hand-held device that controls the movement of a pointer on the computer's screen. As the device is moved on a desktop, the corresponding movement occurs on the screen. In order for the mouse to operate in MS-DOS, a software driver program must be installed.

**MPBR (master partition boot record)**—A small machine language program placed on the hard drive disk when it is partitioned. This program is loaded by the BIOS each time the computer is booted. The MPBR is the program that loads the operating system files into the computer's memory.

**MTBF (mean time between failures)**—A specification indicating the expected failure rate of a given product.

**Multimedia**—Using more than one medium (media) to transfer information.

**multi-mode cable**—Fiber optic cable that allows light of multiple wavelengths (colors) to pass. Multi-mode cable is cheaper and has a higher signal loss than single-mode fiber optic cable.

**multiplexing**—A method in which multiple signals share the same conductor or medium.

**multiprocessing**—A computer that contains multiple CPUs that allow different program modules to be executed on different CPUs. In order to use multiprocessing, it must be supported by the microprocessor, motherboard, and operating system.

**multiscan (Multisync®)**—A type of video monitor that can display the output from display adapters that have a variety of vertical and horizontal scanning (sync) frequencies.

**multitasking**—Running multiple programs on a single microprocessor using a time-share basis. The more programs that are multitasked, the slower all the programs will run. In order to use multitasking, it must be supported by the microprocessor and operating system.

**multithread application**—Programs that are written in modules that can be operated on by a CPU independent of other program modules. Multithread programs can be executed by a single CPU or multiple CPUs. See *multiprocessing*.

## N

**NetBEUI**—A protocol stack used by peer-to-peer Microsoft networks. NetBEUI identifies computers on the network by name and MAC address. NetBEUI does not have the ability to identify a network segment ID, so NetBEUI is a nonroutable protocol stack.

**network**—See LAN.

**nibble**—(1) A group of four consecutive binary bits. (2) The number of binary bits represented by each hexadecimal digit.

**NIC (network interface card)**—An adapter card that connects a computer to a LAN.

**node**—Any device—PC or terminal—on a LAN. See *LAN*.

## O

**object code**—Another name for a machine language (code) program. Also, the name of a machine code program before it has been linked to an operating system.

**OEM (original equipment manufacturer)**—A manufacturer that sells its products to a retailer instead of directly to the public.

**ohm ($\Omega$)**—The unit used to measure resistance. See *resistance*.

**oscilloscope**—A piece of electronic test equipment used to visually display voltage levels relative to time.

## P

**packet**—See *frame*.

**passive**—A media or device that moves information from one point to the next without modification or amplification. Most passive media attenuate the data with distance.

**PC (personal computer)**—The name given to the original microcomputer developed by IBM in 1981. Also the generic name given to all computers used for personal use.

**PC-card**—See *PCMCIA*.

**PCI (peripheral component interface)**—A local bus developed by Intel that allows data to be transferred in groups of 32 bits. The PCI bus runs at one-half the speed of the local bus up to 33 MHz. The PCI bus can run asynchronous with the microprocessor's local bus.

**PCMCIA (Personal Computer Memory Card International Association)**—A Plug-and-Play expansion interface currently used with laptop and notebook computers. The name has recently been changed to PC-card. See *PnP (Plug-and-Play).*

**peer-to-peer**—A network in which any node can access any other and has equal control over data. Workstations share administrative duties, and no workstation is subject to permission from a central computer.

**PGA (programmable grid array)**—An integrated circuit that has the pins extruding from the bottom instead of the sides. The pins are arranged in rows and columns.

**pin-out**—The electrical signal assignments given to the pins of an IC or a connector.

**PIO (programmable input/output)**—A method used to transfer information between a hard drive and the computer's RAM that is a combination of IRQ and software. When PIO is used, all information passes through the microprocessor.

**pipeline**—The path an instruction and data take through the microprocessor during processing.

**pipeline burst cache**—Cache memory that is accessed in bursts (more than one read or write).

**pixel**—A term used when the video display adapter is operating in one of its graphics modes. It is the smallest addressable unit on the video display whose size depends on the graphic mode selected.

**PLCC (plastic-leaded chip carrier)**—A package used to hold the integrated circuits of a semiconductor wafer.

**PMMU (page memory management unit)**—Special circuit designed to handle expanded memory. See *MMU*.

**PnP (Plug-and-Play)**—A technology in which the operating system sets the resources used by a port.

**point-to-point (link)**—A connection between two, and only two, pieces of equipment.

**POST (power-on self-test)**—A basic hardware and firmware test automatically initiated by the BIOS in a compatible computer after a hard- or soft-boot. If no errors are detected during the POST the computer's internal speaker should beep once. If more than one beep is heard, it indicates that one of the tests in the POST, has failed. The number and duration of beeps are an indication of which test failed and are called a beep code.

**power (electrical)**—The rate of doing work. The amount of power is equal to the number of volts times the number of amperes ($P = V \times I$).

**PPI (programmable peripheral interface)**—A programmable integrated circuit used to interface I/O devices to a microprocessor.

**PPP (point-to-point protocol)**—A protocol that establishes how a serial line, usually a telephone, connects to the Internet.

**protected mode**—The native mode of Intel's 80286 and above processors that enables the processor to perform as designed. This mode provides the ability to use the full addressing range and instruction set of these powerful MPUs. MS-DOS versions below 5.0 use the real mode and do not support this mode of operation. See *real mode*.

**protocol**—A formal set of conventions that governs the formatting and timing of message exchange between two communicating systems.

**protocol stack**—A set of protocols used to control the transport and network layers of the OSI networking model. The most popular protocol stacks used in the compatible PC are currently TCP/IP, IPX/SPX, and NetBEUI.

**protocol suite**—See *protocol stack*.

**PSP (program segment prefix)**—The information automatically stored by MS-DOS in addresses immediately preceding a user-defined program. This information is used to return control of the computer back to DOS when the program is terminated.

## Q

**QAM (quadrature amplitude modulation)**—A modulation technique used by modems in which both the phase and amplitude of the transmitted signal are modulated.

**queue**—(1) A waiting line. (2) Temporary storage inside the microprocessor where pre-fetched instructions are placed.

## R

**RAID (redundant array of inexpensive drives)**—A hard drive data storage technique that is used to increase data transfer rates and fault tolerances by spreading a file over multiple drives.

**RAM (random-access read/write memory)**—The memory inside the computer where the DOS operating system and application programs reside while they are being executed.

**real mode**—Mode of operation for 80286 and above processors that emulates an 8086/8088 microprocessor.

**redirector**—Another name for Microsoft's network client software.

**repeater**—A device set at intervals along a circuit to regenerate or boost the signal, preventing its decay over distance. Repeaters regenerate digital signals without change; they simply boost analog signals.

**requester**—Another name for Novell's Netware client software.

**reset**—A signal on the control bus used to initialize the microprocessor. This signal will force the 80x86-compatible microprocessors to go to address FFFF:0000 in the BIOS looking for an operational code. This signal is automatically asserted when the computer is turned on or when the RESET button is pressed. Also referred to as hardware reset, hard-boot, and cold-boot.

**resistance**—Opposition to electrical current flow, measured in ohms.

**RFI (radio frequency interference)**—Unwanted radio wave.

**RIMM (Rambus in-line memory module)**—The memory modules used for direct Rambus DRAM.

**ring**—The LAN topology that links workstations to one another in a circular fashion, with network control distributed throughout the loop. The more workstations there are in the ring, the slower the network. Should any workstation fail, the LAN may crash. Once the failed PC is shut off, the network resumes the loop and acts as if the PC is not there.

**RISC (reduced instruction set computer)**—A new group of microprocessors that have recently been introduced into the marketplace. These μPs have a smaller instruction set and faster throughput than an equivalent CISC. See *CISC*.

**RLL (run length limited)**—A data encoding/compression format used to store data on some high-density, hard disk drives. Holds 50% more data than MFM encoding. Only RLL-certified drives should be used with RLL encoding.

**ROM (read-only memory)**—The nonvolatile memory in the computer that contains the BIOS. See *BIOS*.

**router**—A device that is used to connect multiple LANs together into individual addressable subnets.

**RS232C**—A serial interface standardized by the Electronics Industries Association (EIA). The compatible computer's programmable serial ports use the RS-232C standard.

**RTC (real-time clock)**—The integrated circuit that keeps time in the computer even when power is turned off.

## S

**S-100 bus**—Slot with 100 pins; the first standardized computer expansion bus used with microcomputers.

**SCSI (small computer system interface)**—An interface adapter used in microprocessors usually for interfacing hard disk drives and other peripheral devices. A single SCSI adapter can handle up to seven SCSI peripherals.

**SDRAM (synchronous dynamic random-access memory)**—A high-speed DRAM that approaches the data transfer rate of SRAM.

**SECC (single-edge connector cartridge)**—The package used by Intel's Pentium® II microprocessors. It consists of the microprocessor mounted on a circuit card with a two-tier edge card connector similar to the one used by the EISA expansion bus.

**sector**—Part of a track on a floppy or hard disk drive surface. Each sector holds 512 bytes of data. Floppy disks may have 8, 9, 15, 18, or 36 sectors per track, while hard drives have more than 16.

**serial transmission**—The most common transmission mode in which information bits are sent sequentially on a single data channel.

**server**—A computer or device that provides services to client computers on a network.

**shadow RAM**—On an AT-type system, the memory between 640 K and 1,024 K that is not generally accessible to most MS-DOS programs. May be used to shadow video ROM and the computer's BIOS to improve the overall performance of the computer.

**short-haul modem**—A signal converter that conditions a digital signal to ensure reliable transmission over private telephone lines without interfering with adjacent pairs in the same cable.

**SIMD (single instruction, multiple data)**—Microprocessor instructions that operate on multiple pieces of data simultaneously rather than each data byte individually. These instructions most often apply to multimedia data.

**SIMM (single in-line memory module)**—A group of DRAMs mounted on a circuit card used to upgrade the memory in a computer that uses an edge card connector for the RAM memory.

**simplex transmission**—Data transmission in one direction only.

**single mode**—Fiber optic cable that is designed to transfer light at a single wavelength (color). The single mode cable used in networking typically has a diameter of 8.3 microns.

**SIP (single in-line package)**—A packaging method used for both passive and active electronic devices. This package is similar to a DIP but mounts vertically on the printed circuit card with pins extruding only from one side. See *DIP*.

**SIPP (single in-line pin package)**—Memory modules that are similar to SIMMs but have pins on one side instead of an edge card connector. See *SIMM*.

**SLIP (serial line Internet protocol)**—An older protocol that establishes how a serial line, usually a phone line, connects to the Internet. The newer protocol is called PPP.

**slot 1**—A female edge card connector mounted on the motherboard designed to accept either a Pentium® II, Pentium® III, or Celeron™ microprocessor. Slot 1 is also called a 242-contact connector.

**slot 2**—A female edge card connector mounted on the motherboard designed to accept a Pentium® II Xeon™ microprocessor. Slot 2 is also referred to as a 330 contact connector.

**software**—Computer programs that can be easily changed without changing hardware.

**SONET**—An internal standard for transferring information over optical fiber. SONET standards are given an optical carrier (OC) rating number according to speed.

**source code**—A computer program written in any language other than machine code. Source code programs have to be assembled, compiled, or interpreted into machine language before they can be run by the microprocessor.

**SPGA (staggered pin grid array)**—An IC package in which the pins coming out of the bottom are in rows that are offset from each other at an angle.

**SPP (Standard Parallel Port)**—The unidirectional parallel port standard originally used on the compatible PC.

**SRAM (static random-access read/write memory)**—RAM memory that is volatile but does not have to be refreshed. Usually used in computers with less than 16 KB of RAM and as high-speed cache memory.

**SSE (streaming SIMD extension)**—The name given to the 70 multimedia instructions developed by Intel for the Pentium III microprocessor. These are in addition to the MMX instruction.

**ST-506/412**—A hard drive interface standard originally developed by Seagate Technologies; the first hard drive interface used by the IBM-XT and early AT-compatible computers.

**star**—A network topology in which all workstations are wired directly to a central service unit or workstation that oversees the connections among workstations. Problem nodes are easily identifiable, but the network cannot operate if the server fails.

**STP (shielded twisted-pair)**—Network cable that is enclosed in a metal foil called a shield; used to reduce RFI.

**subnet mask**—A 32-bit binary number that identifies the bits of the IP address that are the network ID and the node ID. The subnet mask is entered into the setup of the TCP/IP suite in decimal form. See *TCP/IP*.

**superscaler microprocessor**—A microprocessor that uses multiple pipelines to process instructions and data. See *pipeline*.

**SVGA (super video graphics array)**—A general description for video display standards that exceed the resolution and palette of VGA. Common SVGA standards are 640 × 480 resolution with 256 colors, 800 × 600 resolution with 16 or 256 colors, 1,024 × 768 with 16 colors, and 1,024 × 768 with 256 colors. Emulates VGA, EGA, CGA, MCGA, MDA, and HGA.

**synchronous**—A method of transferring information requiring the source and destination devices to operate in lock-step (synchronized in time).

**syntax**—The rules that govern the way something is written; includes correct spelling and punctuation.

**syntax error**—An error that occurs when something is not written correctly.

## T

**T1**—T carrier consisting of 24 channels and a transfer rate of 1.544 Mbps. See *T carrier*.

**T2**—T carrier consisting of 96 channels and a transfer rate of 6.312 Mbps. See *T carrier*.

**T3**—T carrier consisting of 672 channels and a transfer rate of 44.763 Mbps. See *T carrier*.

**T4**—T carrier consisting of 4,032 channels and a transfer rate of 274.176 Mbs. See *T carrier*.

**T carrier**—A high-speed leased telephone line that is used for both voice and data transfer. T carriers are rated as a T1, T2, T3, and T4 depending on the number of channels and data transfer rate. The faster the transfer rate, the more expensive the lease.

**TCP (transmission control protocol)**—A protocol used to transport layers of the ISO model to establish error-free transfer of data on a network using the TCP/IP stack. See *TCP/IP* and *IP* address.

**TCP/IP (transmission control protocol/Internet protocol)**—The protocol suite that is used by the Internet. See *TCP* and *IP* address.

**termination**—(1) Connecting the ends of cables to resistors to provide cable loading and reduce signal reflection. (2) Placing a connector on the end of a cable.

**Thick Ethernet cable**—Also called ThickNet. Thick Ethernet cable is a 0.4″-diameter coaxial cable that allows the longest runs among workstations of any cable type. It is useful for linking several buildings in a single LAN.

**Thin Ethernet cable**—Also called ThinNet. Thin Ethernet cable is coaxial cable with a 0.25″-diameter. Thin Ethernet is currently the most common Ethernet cable available.

**throughput**—A measure of the total amount of useful processing carried out by a computer system in a given time period.

**TIFF (tagged image format)**—A graphic standard used by graphics files. Files that use this format usually have the .TIF extension.

**tight jacket**—A term used to describe fiber optics cable in which the jacket is physically in contact with the fiber's cladding.

**token ring**—A LAN in ring form, based on either star or ring topology, that requires a workstation to receive a group of bits called a token before transmitting. The token is passed in a circular fashion from workstation to workstation.

**topology**—The configuration among nodes, including the wiring scheme (ring, bus, or star).

**TPI (tracks per inch)**—A measurement that gives an indication of the track spacing for disk drives.

**TSR (terminate-and-stay resident)**—The name given to hardware driver programs that are loaded into the computer to control hardware that is not supported by the BIOS or MS-DOS. These programs stay resident in memory and only run when needed. An example is the driver program that runs a mouse.

**TWAIN**—A data transfer standard supported by most scanners.

**twisted pair**—Cable with a pair of insulated wires twisted around each other and sheathed. The twisting cancels interference for one wire to another (standard phone wire).

## U

**UART (universal asynchronous receiver transmitter)**—The integrated circuit that forms the programmable serial ports used in the compatible computers.

**UDMA (ultra direct memory access)**—A method used to transfer information between a hard drive and memory directly without going through the microprocessor.

**UMA (upper-memory area)**—The name used for memory address between 640 K and 1,024 K.

**upload**—Transferring information from a device to a computer.

**(UPS) uninterruptible power supply**—Source that provides power to a device when regular power fails. The UPS runs off rechargeable batteries that are charged when regular power is on.

**URL (universal resource locator)**—A name used instead of an IP to address computers on the Internet. The URL is in the form of host.domain.type. An example of a URL is www.microsoft.com.

**USB (universal serial bus)**—A PnP serial bus standard that allows up to 127 devices to be connected. The transfer rate of the USB is up to 12 Mbps. USB also supports hot-swapping.

**user-friendly**—Easy to use.

**UTP (unshielded twisted pair)**—Cable used in networking; usually used to connect a network in a star topology.

**utility**—A program that helps the user run, enhance, create, or analyze other programs, operating systems, or equipment.

## V

**vampire tap**—The device that connects nodes to an Ethernet 10Base5 network cable.

**VCPI (virtual control program interface)**—A standard for control programs.

**Veronica**—A software program that searches Gopher sites on the Internet for requested information. Similar to Archie for FTP sites.

**VESA (Video Electronics Standards Association)**—A consortium of video adapter manufacturers that establish video standards.

**VGA (video graphics array)**—IBM PC/XT/AT-compatible computer high-resolution display card. Provides a variety of text and graphics display modes up to 640 × 480 resolution, and 16 colors from a palette of 262,000. Usually supports EGA, CGA, MCGA, MDA, and HGA graphics. See *SVGA*.

**virtual mode**—An operating mode used by the 80386 and above processor that gives the processor the ability to do multitasking (running more than one DOS program at the same time). Windows 386 currently supports this mode of operation.

**virus**—An unauthorized program that attaches itself to or overwrites an existing program. Once a computer is infected, the virus will eventually divulge its presence by deliberately causing the computer to malfunction.

**VLB (VESA local bus)**—An interface bus used on 486 computers that provides 32 data lines running at the speed of the microprocessor's local bus.

**VRAM (video random-access memory)**—Memory specifically designed for video adapters.

## W

**wafer**—A circular piece of silicon material on which individual integrated circuits, called die, are constructed.

**WAN (wide area network)**—Connecting remote sites—and/or LANs—into a single system across one or more networks. See *LAN*.

**warm-boot**—A partial reset of the computer, controlled by software, that reinitializes the system and the RAM but skips a large portion of the POST. Holding down the Ctrl and Alt keys and simultaneously pressing the Del key performs a warm-boot when using MS-DOS. The process has to be repeated twice when using Windows 3.X or Windows 9X.

**WAV file (Windows audio volume)**—A file format developed by Microsoft that allows the storing and reading of audio information.

**whetstone**—A standard benchmark for measuring how fast a computer can calculate floating-point numbers. Pronounced "wet stone."

**word**—Relative to compatible computers, a group of 16 consecutive binary bits. A word requires 2 bytes of memory to store.

**workgroup**—A group of computers on a peer network that are arranged under a logical workgroup name, which makes it easy to share resources.

**workstation**—A PC or terminal that serves a single user in a network, as contrasted with the file server, which serves all users.

**worm**—A self-sustaining program that usually transmits vital information stored on the hard drive to unauthorized individuals. This information can be anything from passwords to database information. Unlike a virus, a worm usually removes itself once its mission has been completed.

**WORM drive (write once/read many)**—An optical mass-storage device similar to a CD-R that can be written to but not erased. These drives usually have the ability to store gigabytes of information.

**WWW (World Wide Web)**—A group of computers on the Internet that support the hypertext transfer protocol (HTTP). To access the service provided by the WWW, a computer must use a Web browser. When a computer accesses a server on the WWW, the server sends a text document written in a formatting language called hypertext markup language (HTML). The browser uses the HTML document to display a Web page.

**WYSIWYG (what you see is what you get)**—A document that is printed as it appears on the computer's display.

## X

**X86**—A generic term that represents all of the Intel and compatible microprocessors used in the compatible PC. Even though the current Intel processors use names instead of numbers, they are still considered part of this generic term.

**XGA (extended graphics array)**—An IBM video standard that supports resolutions up to 1,024 × 768 at 256 colors.

**XMM (extended memory manager)**—A driver that provides access to the computer's extended memory. This driver has to be loaded with the CONFIG.SYS file if Windows 3.X is used, whereas it is loaded automatically by Windows 9X.

**XMS (extended memory specification)**—A Microsoft specification for the extended memory manager to access extended memory in a computer running an 80286 and up processor.

**XT (extended technology)**—The designator given by IBM to its second release of the popular IBM computer. The biggest feature added by the XT was support for hard drives.

## Z

**ZIF (zero insertion force)**—A socket for integrated circuits that provides a handle that can be raised to allow the IC to be inserted and removed by exerting very little pressure. When the handle is lowered, the IC is physically latched into the socket and cannot be removed.

**zip drive**—A disk drive manufactured by Iomega that supports a removable 100-MB diskette that is similar to a 3 1/2" floppy diskette. The two are not interchangeable, however.

**zone recording**—A data recording method used by current hard drives in which the outside cylinders contain more sectors than the inside cylinders. Zone recording assigns groups of consecutive cylinders into units called zones, with each successive zone having fewer sectors than the previous zone.

# INDEX

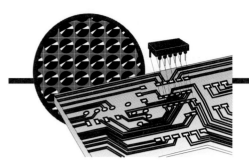

100baseT, 467
10base2. *See* thinnet
10base5. *See* thicknet
10baseFL, 466
10baseT, 464–66
242-contact slot connector (SC242). *See* slot-1
3.5" diskette, 338
3.5" floppy drive, 339
330 contact slot connector (SC330). *See* Slot 2
5.25" diskette, 337
5.25" drive, 338
586. *See* pentium
8.3 format, 158
80286, 78–79
80386, 80–83
80486, 84–85
80486DX2/DX4, 85–87
8086, 75–77
8088, 77
accelerated graphics port (AGP), 429
adapter BIOS, 234
address bus, 59, 60
addressing mode, 67
advanced graphics port. *See* accelerated graphics port (AGP)
Advanced Micro Devices (AMD)
 athlon, 106
 K5, 103
 K6. *See* AMD K6
 K6-II, 104
 K6-III, 105
advanced technology attachment packet interface (ATAPI), 375–76
advanced technology attachment-1. *See* integrated drive electronics (IDE)
allocation unit, 342, 391
alternating current (ac), 10–12
 cycle, 10
 frequency, 10
 hertz, 10
 transformer, 11
AMD athlon, 106
AMD K5, 103
AMD K6, 104
AMD K6-2, 104
AMD K6-III, 105
american standard code for information interchange (ASCII), 46–47

 device control codes, 47
 enhanced ASCII, 48–49
 parity, 48
ampere, 3
apparent power, 28
application program interface (API), 154
arithmetic logic unit (ALU), 56
assembly language, 66–67
 assembler, 67
 mnemonic, 66
AT motherboard form factor, 118
AT power supply, 19
ATA-2. *See* enhanced integrated drive electronics (EIDE)
athlon. *See* AMD athlon
ATM, 468
ATX motherboard form factor, 119
ATX power supply, 19
autoexec.bat, 185
auxiliary storage. *See* mass storage
background program, 177
backup, 333
bandwidth, 452–53, 482
base, 38
base address, 296
base-10. *See* decimal
base-2. *See* binary
base-memory. *See* conventional memory
basic input output system (BIOS), 63, 121, 132, 181
 adapter, 234
 configuration, 183, 270–272
 shadowing, 243–44
binary, 37, 39–40
 base-2 weights, 39
 binary to hexadecimal conversion, 44
 bit, 39, 45
 byte, 45
 byte-word, 45
 decimal to binary conversion, 42
 decimal weights, 39
 double word, 45
 hexadecimal to binary conversion, 45
 nibble, 45
 quad-word, 45
 reason for binary, 40
 word, 45
binding, 478
BIOS configuration, 183
bit, 39, 45

boot, 181
boot sequence, 181
booting, 64–65
booting Windows 9X, 214
bottle neck, 140
bridge, 459
burst extended data output (BEDO) memory, 248
bus cycle, 246
bus master, 377–78, 522
bus network topology, 457
bus voltage, 92
byte, 45, 232
byte-word, 45
cable, 293
cache, 262–63
 cache controller, 263
 L1, 262
 L2, 262
 TAG RAM, 263
cache controller, 263
cache memory, 83
capacitor, 15, 244
carrier sense multiple access with collision detection (CSMA/CD), 461
cathode ray tube (CRT), 412–14
 cathode/gun, 412
 deflection yoke, 413
 frame, 413
 horizontal sync, 413
 phosphor, 412
 scan line, 413
 vertical sync, 413
CD-R, 356
CD-ROM, 353–55
 high sierra format, 355
 ISO-9660, 355
CD-RW, 357
celeron. *See* PII celeron
centeral processing unit (CPU), 56
CGA, 420–21
checksum, 49
chipset, 109–115
 identifying, 117
 P6, PII, and PIII, 115
 socket5 & socket7, 110
circuit, 3
 load, 3
circuit breaker, 9
circuit common, 18–19

579

CISC. *See* complex instuction set computing
click, 174
client/server network, 471–72
   client, 472
closed loop, 369
cluster. *See* allocation unit
CMOS
   chipset setup, 276
   clear, 279
   configuration, 270–272
   ECSD, 279
   features setup, 275
   hdd smart capability, 275
   HDD smart capability, 275
   integrated peripherals, 278
   OS/2 compatible mode, 275
   PnP, 277
   power managment, 276–277
   setup, 272–273
   8/16 bit I/O recovery time, 276
   assign IRQ to VGA, 275
   auto configuration (memory), 276
   boot sequence, 274
   boot up num lock, 275
   boot up system speed, 275
   CPUfan off in suspend, 277
   date, 273
   daylight savings. *See*
   delay for HDD, 275
   DMA n assigned to, 277
   dose mode, 277
   DRAM enhanced paging, 276
   DRAM leadoff timing, 276
   DRAM page idle time, 276
   DRAM read burst, 276
   DRAM timing, 276
   DRAM write burst timing, 276
   ECP mode use DMA, 278
   enable external cache, 274
   enable internal cache, 274
   fast EDO leadoff, 276
   fast RAS to CAS delay, 276
   floppy, 274. *See*
   floppy disk seek, 275
   gate A20 option, 275
   halt on, 274
   hard-drive, 273
   HDD power down, 277
   IDE HDD block mode, 278
   IDE primary/secondary master/slave PIO, 278
   IDE primary/secondary master/slave UDMA, 278
   IR duplex mode, 278
   IRQ n assigned to, 277
   IRQ8 clock event, 277
   main menu, 273
   memory, 274
   memory hole 15MB to 15MB, 276
   memory test above 1 MB, 275
   onboard FDC controller, 278
   onboard parallel port, 278
   onboard UART 1/2, 278
   on-chip primary/secondary PCI IDE, 278
   OS select for DRAM > 64MB, 275
   parallel port mode, 278
   parity/ECC check, 275
   PCI IDE IRQ map to, 277
   PCI IRQ activated by, 277
   PCI/VGA palette snoop, 275
   PM control by APM, 277
   PnP OS installed, 277
   power management, 277
   primary/secondary IDE INT, 277
   PS/2 mouse fuction control, 275
   quick boot, 274
   refresh RAS assertion, 276
   reset configuration data, 277
   resources controlled by, 277
   SDRAM RAS to CAS delay, 276
   SDRAM speculative read, 276
   security option or password check, 275
   shadowing, 275
   SRAM (CAS lat/RAS to CAS), 276
   standby mode, 277
   suspend mode, 277
   swap floppy drives, 275
   system BIOS cacheable, 276
   system BIOS shadow cacheable, 275
   throttle duty cycle, 277
   time, 273
   typematic rate, 275
   USB keyboard support, 278
   USB latency timer, 278
   use IR pins, 278
   used mem base addr, 277
   VGA active monitor, 277
   video, 274
   video BIOS cacheable, 276
   video off after, 277
   video off method, 277
   virus warning, 274
   ZZ active in suspend, 277
   standard setup menu, 273–274
coaxial cable, 453
color graphics adapter. *See* CGA
column address strobe (CAS), 245
COM. *See* command file
com port. *See* serial port
command file (COM), 163
command prompt. *See* DOS prompt
command.com, 166
common. *See* circuit common
compatible computer, 45, 73
compiler, 68
complementary metal oxide semiconductor. *See* CMOS
complex instruction set computing (CISC), 94
computer, 55
   analog, 55
   arithmetic logic unit (ALU), 56
   control/timing unit, 56
   digital, 55
   input port, 56
   output port, 56
   peripheral, 56
computer protection
   surge suppresson, 27
   UPS, 26
conductor, 3–4, 6
   American Wire Gauge (AWG), 4
   cable, 5
   flat cable or ribbon cable, 5
   wire, 4
config.sys, 185–87
configuration, 183
   autoexec.bat (DOS), 185
   BIOS, 183
   config.sys (DOS), 185–87
   contol/progman/winfile, 192
   DOS, 184–87
   regedit. *See* regedit
   registry files, 218
   system.dat, 218
   system.ini (Win3.X), 188–89
   TSR, 187–88
   user.dat, 218
   win.ini, 192
connector, 293–94
   female, 293
   male, 293
continuous form, 441
control bus, 60
control panel, 204
control.ini, 192
control/timing unit, 56
conventional memory, 233
cooperative multitasking, 177
core voltage, 92
critical resource, 457
current, 3
   ampere, 3
cycle, 10
cyclic redundancy check (CRC), 50
   CRC-16, 50
   CRC-32, 50
cylinder, 335
cylinder head sector (CHS), 378
daisy wheel printer, 444
data bus, 60
data link control (DLC), 470
data link layer, 461
DC power supply, 12, 17
   + 12 VDC, 19
   + 5 VDC, 19
   + 5 VDC standby, 19
   +3.3 VDC, 19
   -12 VDC, 19
   -5 VDC, 19
   AT, 19
   ATX, 19
   compatible PC, 21
   connectors, 24
   inverter, 17
   linear, 17
   physical size, 21
   regulator, 17
   removal and installation, 31, 33
   switching, 17
   troubleshooting, 29
decimal, 37, 38–39
dedicated data bus (DDB), 263
default, 166
deflection yoke, 413
desktop, 172
device manager, 503
devicehigh, 239

# Index

diagnostics, 495
  device manager, 503
  disk based, 499–500
  Dr. Watson, 507
  POST, 496–99
  ROM based, 496–98
diagnostics XE "system properties"
  system properties, 502
die, 56
digital, 37
digital clock, 12
digital signal, 12
digital versatile disc (DVD), 360–61
diode, 15
DIP switch, 290
direct current (DC), 12
direct memory access, 288. *See* dma. *See* DMA
direct memory access controller (DMAC), 522
direct rambus DRAM (DRDRAM), 248
directory, 160–63, 336, 345–47
  path, 163
  root, 160
  sub-directory, 160–63, 346
disk based diagnostics, 499–500
disk operating system (DOS), 63, 132, 154
  application, 154
  command.com, 166
  external command, 166
  internal command, 166
  making a bootable disk, 170
  reason for learning, 169
  shutting down, 171
DLL. *See* dynamic link library
DMA, 288, 522–24. *See* direct memory access
domain server, 472
DOS. *See* disk operating system
DOS application, 80
DOS mode, 208
DOS prompt, 208
DOS shell, 169
dot matrix, 440
dot matrix printer, 442–44
dot pitch, 426
double data rate SDRAM (DDR SDRAM), 249
double word, 45
double-click, 174
downwardly compatible, 74
Dr. Watson, 507
dragging, 174
drive bay, 362
  bezel, 362
  external, 361
  internal, 361
driver, 187
dual independent bus (DIB), 98
dual inline memory module (DIMM), 254
dual inline package (DIP), 57, 76, 249
dumb terminal, 452
dynamic link library (DLL), 164
dynamic random access memory (DRAM), 244–45
  BEDO, 248

DDRSDRAM, 249
DRDRAM, 248
EDO, 248
FPM, 247
SDRAM, 248
ECC. *See* error correcting code
ECSD, 279
ehanced cylinder head sector (ECHS), 379
ehanced small device interface (ESDI), 374
electrical current. *See* current
electrical erasable programmable read only memory (EEPROM), 280
electrical ground. *See* circuit common
electrical safety, 29
electrical shock, 9–10
electro static discharge (ESD), 259
electromotive force (EMF), 3
  volt, 3
electron gun/cathode, 412
electronic devices
  capacitor, 15
  diode, 15
  inductor, 14
  resistor, 13
  semiconductor, 16
  switch, 13
  transistor, 16
EMM386, 239
emulation, 447
end users license agreement (EULA), 154
energy star, 26
engineering notation, 51
enhanced ASCII, 49
enhanced graphics adapter (EGA), 421
enhanced industry standard architecture (EISA), 138–40
enhanced integrated drive electronics (EIDE), 375
enhanced mode, 172
EPROM, 280
erasable programmable read only memory. *See* EPROM
error correcting code (ECC), 246–47
error correction on SIMM (EOS), 247
ESD safety precautions, 259
ethernet, 461–62
  100baseT, 467
  10baseFL, 466
  10baseT, 464–66
  CSMA/CD, 461
  thicknet, 462–63
  thinnet, 463–64
executable file (EXE), 163
execute phase, 59, 74
expanded memory, 235–36
expanded memory manager (EMM), 235
expansion bus, 121
  EISA, 138–40
  ISA, 135
  MCA, 135–36
  PC-bus, 130
  PCI, 144–45
  PCMCIA, 148
  VL-bus, 144
explorer. *See* windows explorer

extended configuration setup data. *See* ECSD
extended data output (EDO) memory, 248
extended memory, 236–37
extended memory manager (XMM), 236
extended memory specification (XMS), 236
external commands, 166
fast page mode memory (FPM), 247
FAT16, 393
FAT32, 394
FDDI, 467–68
fdisk, 397–99
fetch phase, 59, 74
fiber optics, 454–55
  multi-mode, 454
  single mode, 455
  SONET, 455
field, 427
file, 157
file allocation table (FAT), 340
file allocation unit (FAT), 343–45
file association, 177
file attribute, 158
file manager, 175
filename, 157
filename extension, 157
filp flop, 40
firewire, 303
firmware, 63, 155
fixed drive. *See* hard drive
fixed port, 289–90
flash ROM, 280–281
floating point unit (FPU), 84
floppy drive, 336–40
  3.5" diskette, 338
  3.5" drive, 339
  5.25 drive, 338
  5.25" diskette, 337
  controller, 350
  index hole, 337
  TPI, 336
  write protect notch, 337
flow-control, 299
  hardware, 299
  software, 299
folder, 202
font, 440
foreground program, 177
form factor (motherboard), 120
formatting, 160
Formula
  solve for I given P and V, 7
  solve for I given V and R, 6
  solve for KHW, 8
  solve for P given V and I, 7
  solve for R given V and I, 6
  solve for V given I and R, 6
  solve for V given V and I, 7
  solve for VA given W and PF, 28
fragmented file, 348, 401
frequency, 10
friction feed, 441
front side bus (FSB), 98
FSB. *See* front side bus
fundamental storage unit, 232
fuse, 8
game port, 315

# 582  Index

gateway, 460
gig (base10), 51
gig (base2), 50
graphical user interface (GUI), 172
graphics mode, 415–17
handshake. *See* flow-control
hard copy, 439
hard drive, 368
   CHS, 378
   compression, 402–3
   controller, 372
   disk optimizer, 402
   ECHS, 379
   FAT16, 393
   FAT32, 394
   high level format, 401
   interleave, 391–93
   interleave factor, 392
   LBA, 379
   low level format, 390–91
   partition, 396–97
   stepper motor, 372
   type number, 403–5
   voice coil, 370
hard drive controller, 372
   ESDI, 374
   MFM, 373
   RLL, 373
   ST506/412, 372
hardset adapter, 290–91
hardware, 65
hardware interrupt, 517–22
   interrupt request, 518–21
   nonmaskable, 518
   reset, 518
heat sink, 88
hercules graphics adapter (HGA), 419
hertz, 10, 51
hexadecimal, 42–45
   binary to hexadecimal conversion, 44
   converting hexadecimal to decimal, 44
   hexadecimal to binary conversion, 45
high level format, 401
high memory area (HMA), 237–38
high-level language, 68
himem.sys, 237, 239
hkeys, 218
horizontal sync, 413
hot swap, 300
hub, 457
I/O
   address map, 285
   integrated peripherals, 278
   pnp CMOS setup, 277
I/O address map, 284
IA-64. *See* intel architecture 64
IBM personal computer (PC), 129
IBM XT, 132–33
IBM-PC, 130
icon, 172
IEEE 1394. *See* firewire
IEEE 802.3. *See* ethernet
image drum, 445
impact printer, 439
impedance matching, 453
inductor, 14

industry standard architecture (ISA), 135
   addressing range, 135
information file (INI), 178, 188
INI. *See* information files (INI)
inkjet printer, 445–47
input port, 56, 284
input/output (I/O) port, 284
insulator, 4
integraded ports, 121
integrated drive electronics (IDE), 374
   cable select, 381
   DMA, 377
   master, 374, 381
   PIO, 376–77
   slave, 374, 381
   stand alone, 374
   UDMA, 377
integrated peripheral, 278, 293
Intel architeture 64, 103
Intel compatible, 74
interlaced, 427
internal commands, 166
internal interrupt, 517
Internet, 480–81
Internet Assigned Number Authority (IANA), 475
Internet Corporation of Assigning Names and Numbers (ICANN), 475
interpreter, 68
interrupt, 60, 286, 514
interrupt acknowledge (INTA), 518
interrupt request, 518–21
interrupt request (IRQ), 287, 286–88
interrupt service routine, 515
interrupt vector table, 515
inverter, 17
IP address, 474–76
   class A, 475
   class B, 475
   class C, 475
   DHCP, 476
   IANA, 475
   subnet mask, 474
IPv4, 474
IPv6, 475
IPX/SPX, 474
ISA. *See* industry standard architecture
jumper, 290
K5. *See* AMD K5
K6. *See* AMD K6
K6-2. *See* AMD K6-II
K6-III. *See* AMD K6-III
kernal, 233
key, 133
keyboard, 309
kilo (base10), 51
kilo (base2), 50
kilowatt-hour (KWH), 7
LAN, 452
laser printer, 445
   image drum, 445
   toner, 445
least significant digit (LSD), 38
legacy IRQ, 521
linear DC power supply, 17
load, 3

loadhigh, 239
local area network. *See* LAN
local bus, 85
logical block addressing (LBA), 379
logical sector, 342
long file names, 204–205
lotus, intel, microsoft (LIM), 235
low level format, 390–91
LPT. *See* parallel port
LPX motherboard form factor, 119
M1. *See* VIA/Cyrix M1
MAC. *See* DLC
machine language, 65–66
   operational code, 65
magnetic tape, 332–33
manufactured ROM, 280
mapping, 480
mass storage, 332
master partition boot record, 397
math coprocessor, 77–78
maximize, 173
MCA. *See* microchannel
medium, 452
medium access control. *See* DLC
meg (base10), 51
meg (base2), 50
memory, 244–45
   access time, 245–46
   actual, 241
   bank, 254–57
   byte, 232
   cache, 83, 262–63
      L1, 262
      L2, 262
      TAG RAM, 263
   conventional, 233
   DIMM, 254
   DIP, 249
   DRAM, 244–45
   ECC, 246–47
   EMM386, 238
   expanded, 235–36
   extended memory, 236–37
   HMA, 237–38
   installation
      72&30-pin SiMM, 259–61
      choosing modules, 257–59
      DIMM, 261–62
   manager (DOS), 238–40
   map, 232
   mem command, 241–42
   non-parity, 247
   nonvolatile, 62
   page, 232
   parity, 246
   random access memory (RAM), 64
   refresh, 245
   reserved, 243
   ROM, 62–63
   shadow, 243–44
   stored program concept, 62
   troubleshooting, 265–66
   UMA, 234
   UMB, 235
   unused (UMA), 234–35
   virtual

# Index

win9x, 265
virtual memory
    swap file, 264
    win3.x, 264
volatile, 62
memory access time, 245–46
memory manager, 239
memory map, 232
menemonic, 66
menu bar, 173, 200
mezzanine, 144
micro, 52
microchannel architecture (MCA), 135–36
microcomputer (MCU), 58
microprocessor
    80286, 78–79
microprocessor, 57, 84–85
    80386, 80–83
    80486DX2/DX4, 85–87
    8086, 75–77
    8088, 77
    address bus, 59, 60
    addressing mode, 67
    AMD athlon. *See* AMD athlon
    AMD K5, 103
    AMD K6, 104
    AMD K6-II, 104
    AMD K6-III, 105
    control buss, 60
    cyrix M1, 106
    cyrix M2, 106
    data bus, 60
    execute, 59
    execute phase, 74
    fetch, 59
    fetch phase, 74
    floating point unit (FPU), 84
    high-level language, 68
    local bus, 85
    machine language, 65–66
    multitasking, 78
    P55C pentium, 92
    pentium, 89–93
    pentium II, 97–99
    pentium pro, 94–97
    PII celeron, 99–100
    PII xeon, 100–101
    PIII, 101–102
    pipeline, 75
    program, 59
    protected mode, 79
    queue, 74
    read cycle, 61
    real mode, 78
    register, 57
    superscaler, 89
    winchip C6, 108
    winchip2, 108
    write cycle, 61
milli, 52
minimize, 173
MMX. *See* multi media extensions
modem, 481–90
    AT command, 487
    CCITT, 483
    compression, 485
    error correction, 484
    initialization string, 490
    ITU-T, 483
    modulation techniques, 483
modified frequency modulation (MFM), 373
monochrome, 413
monochrome display adapter (MDA), 419
monochrome graphics adapter (MGA), 419
most significant digit, 38
motherboard, 108–109
    form factor, 120
    removal and installation, 121
motherboard form factor
    AT, 118
    ATX, 119
    LPX, 119
    NLX, 119
mouse, 310
    click, 174
    double-click, 174
    dragging, 174
    right click, 200
MROM. *See* manufactured ROM
MS-DOS boot sequence, 182
multi media extensions (MMX), 93
multi-media, 313
multi-mode, 454
multiplexing, 76
multiprocessing. *See* symmetric multi-processing
multisync, 424
multitasking, 78, 81, 238
    cooperative, 177
    preemptive, 210
musical instrument digital interface (MIDI), 314
nano, 52
netBEUI, 473–74
network, 452, 468
    binding, 478
    bridge, 459
    client, 472
    client/server, 471–72
    connection oriented, 474
    connectionless, 474
    critical resource, 457
    data link layer, 461
    FDDI, 467–68
    frame/packet, 456
    gateway, 460
    hardware protocol, 456
    hardware software protocol, 456
    hub, 457
    mapping, 480
    network layer, 461
    node, 457
    OSI, 460
    peer-to-peer, 470–71
    physical layer, 460
    reliable, 474
    repeater, 459
    router, 460
    server, 457
    software protocol, 456
    subnet, 460
    token ring, 467
    topology, 457
    transport layer, 461
network interface card (NIC), 468–70
network layer, 461
new technology file system. *See* ntfs
nibble, 45
NLX mother board form factor, 119
non-impact printer, 439
non-interlaced, 428
nonmaskable interrupt, 518
NTFS, 390–91
object code compatible, 75
ohm, 6
omega, 6
opcode. *See* operational code
open loop, 372
open technology, 132
operating system, 63
operational codes (opcode), 65
Organizational Unique Indentifier (OUI), 470
output port, 56, 284
over-clocking, 93
overlay, 233
P5. *See* pentium
P54C, 90
P55C. *See* pentium
P6. *See* pentium pro
parallel port, 304–8
    ECP, 308
    EPP, 308
    SPP, 305
parasite operating system, 172
parity, 48
partition, 396–97
    extended, 396
    fdisk, 397–99
    logical drive, 396
    master partition boot record, 397
    primary, 396
    table, 397
path, 163
PC-bus, 130
peer-to-peer network, 470–71
Pentium, 89–93
    P54C, 90
    P55C, 92
    socket5, 91
    socket7, 92
Pentium II, 97–99
Pentium III, 101–102
Pentium pro, 94–97
Pentium rating (PR), 103
peripheral, 56, 130, 283
peripheral components interface (PCI), 144–45
    addressing range, 145
personal computer memory card international association (PCMCIA), 147–48
Personal System/2. *See* microchannel architecture (MCA)
phosphor, 412
physical layer, 460
physical sector, 342
PIF. *See* program information file
PII. *See* pentium II
PII celeron, 99–100

PII Celeron
  single edge processor package (SEPP), 99
PII xeon, 100–101
  SC330, 101
PII Xeon
  slot-2, 101
PIII. *See* pentium III
pin-out, 76
pins, 56
pipeline, 75
pitch, 440
pixel, 415, 420
PnP
  CMOS setup, 277
PnP adapter, 291–92
port
  address, 284
  connectors, 293–94
  DMA, 288
  fixed, 289–90
  hardset, 290–91
  input, 284
  input/output (I/O), 284
  interrupt, 286–88
  output, 284
  PnP, 291–92
  softset, 291
  standard, 292–93
power, 7–8
  apparent, 28
  kilowatt-hour (KWH), 7
  volt ampere, 7
  watt, 7
power management, 26
  ACPI, 26
  energy star, 26
  standby mode, 26
power management
  APM, 26
power on self test (POST), 496–99
  beep code, 496
  POST card, 498
  POST code, 498–99
power supply. *See* DC power supply
power supply troubleshooting, 29
  erratic behavior, 30
  no power, 31
  safety, 29
  technical, 31
power-on-self-test (POST), 63, 132
preemptive multitasking, 210
preferred server, 472
primary IDE, 293
printer, 439
  bi-directional, 443
  daisy wheel, 444
  dot matrix, 440, 442–44
  dots per inch, 445
  draft, 441
  emulation, 447
  friction feed, 441
  fully formed, 440
  impact, 439
  inkjet, 445–47
  laser, 445
  letter quality, 442

near letter quality, 441
non-impact, 439
print head, 442
stepper motor, 442
thermal, 445
tractor feed, 441
Procedure
  adapter card removal and installation, 327
  binding, 479
  calculating UPS size, 29
  CD & DVD removal and installation, 359
  CD-ROM settings, 359
  change monitor type, 432
  changing port resources, 504
  client software, 473
  configuring DNS, 478
  configuring IP, 476
  configuring keyboard, 310
  configuring Windows 9X APM, 26
  converting hexadecimal to decimal, 43
  copying needed file to boot disk, 171
  creating Win9X startup disk, 224
  determining sound card resources, 314
  determining to location of the ECSD, 279
  determining whether power supply or loads are defective, 32
  device manager, 503
  DIMM installation, 262
  displaying COM port interrupts, 297
  displaying DMA assignments, 289
  displaying I/O map with Win9X, 287
  displaying IRQ used by PCI, 521
  displaying IRQs with Win9X, 288
  displaying memory map (Win9X), 244
  displaying USB resources, 303
  DOS configuration file backup, 193
  downloading tweak UI, 221
  Dr. Watson, 507
  entnering control panel, 204
  FD removal and installation, 352
  file and print sharing, 479
  format using DOS, 349
  format using Win3.X, 349
  format using Win9x/2K, 349
  game controller setup, 316
  help menu (Win9X/2000), 201
  identifying chipset, 117
  installing system policy editor, 220
  installing windows 9X, 226
  launching system editor, 209
  loading and displaying fat, 345
  loading DOS PIF file (Win3.X), 178
  mapping drives, 481
  modem resources, 487
  motherboard removal and installation, 122
  opening win2K system information, 324
  opening windows explorer, 206
  other devices, 325
  parallel port resources, 305
  pointing device cleaning, 313
  pointing device setup, 313
  power supply removal and installation, 32
  printer driver, 448
  printer/fax sharing, 480
  registry backup and restore, 222
  registry backup using DOS, 223

removing items from startup menu, 224
removing items from systray, 225
resource of NIC, 469
restoring Win98 registry backups, 222
setting up unique DOS configurations, 209
sharing devices and folders, 480
shutting down Windows 3.X, 180
shuttng down DOS, 171
SIMM installation, 261
softset pnp ports, 326
start up folder, 203
starting PIF editor (Win9X), 209
system information, 506
system properties, 502
troubleshooting dead computer, 510
user/shared level access, 479
using winipcfg, 470
video adapter resources, 424
viewing memory map (Win2K), 244
Win3.X configuration file backup, 193
Win3.X virutal memory configuration, 264
Win9X modem setu, 490
Win9X/2K com port configuration, 301
Window 3.X file associations, 177
progman.ini, 192
program, 59, 65
program group, 174
program information file (PIF), 178, 208
program manager, 174
program segment prefix (PSP), 348
programmable interrupt controller (PIC), 287, 518
programmer, 59, 65
protected mode, 79
protection devices, 8–9
  circuit breaker, 9
  fuse, 8
protocol stack, 473, 474–78
  IPX/SPX, 474
  netBEUI, 473–74
  TCP/IP, 474–78
protocol suite. *See* protocol stack
quad-word, 45
queue, 74
quick start menu, 200
rambus in-line memory module (RIMM), 249
RAMDAC, 422
random access memory (RAM), 63–64
  fundamental storage unit, 232
read cycle, 61
read only memory (ROM), 62–63
  EEPROM, 280
  EPROM, 280
  flash ROM, 281
  motherboard, 121
read/write heads, 332
real mode, 78
reduced instruction set computing (RISC), 94
refresh, 245
refresh rate, 428
regedit, 218
register, 57
registry editor. *See* regedit
registry files, 218
regulator, 17
release number. *See* version number

# Index

repeater, 459
reserved characters (Win9X/2000), 205
reserved memory, 243
reserved punctuation symbols (MS-DOS), 157
reset, 64, 518
resistance, 6
  ohm, 6
resistor, 13
resolution, 415
reverse engineer, 132
ribbon cables. *See* cable
ring network topology, 459
RISC. *See* reduce instruction computing
ROM based diag, 496–98
root-directory, 160
router, 460
row address strobe (RAS), 245
run length limits (RLL), 373
safe mode, 216–217
SC242. *See* slot-1
scan line, 413
secondary IDE, 293
sector, 335
sector header, 372
semiconductor, 16
serial, 295
serial port, 295
  character, 299
  RS232C standard, 297
  UART, 299
server, 457
service pack, 212
shadow RAM, 243–44
shadowing (CMOS), 275
sharing, 478–80
shell, 165
shock. *See* electrical shock
short circuit, 4
shortcut, 202
SIMD. *See* single instruction multiple data
single edge processor package (SEPP), 99
single inline memory module (SIMM), 249–51
single instruction multiple data (SIMD), 102
single mode, 455
single-edge-connector (SEC), 97
slot-1, 98
slot-2, 101
slot-A, 106
small computer system interface (SCSI), 382–88
  a-cable, 384
  differential, 384
  ID, 388
  p-cable, 384
  q-cable, 385
  single ended, 384
  terminator, 385
SMP. *See* symmetric multi-processing
socket7. *See* pentium
softset adapter, 291
software, 155
software interrupts, 517
sotfware polling, 514
source code, 68

ST506/412, 372
standard mode, 172
standard ports, 292–93
star topology, 457
start button, 200, 202
start menu, 203
static random access memory (RAM), 249
stepper motor
  rotor, 371
  stator, 371
sub-directory, 160–63
subnet, 460
subnet mask, 474
super video grid array (SVGA), 423–24
superscaler, 89
surge suppressor, 27
swap file. *See* virtual memory
switch, 13
switching DC power supply, 17
symmetric multi-processing (SMP), 93
synchronous, 12
synchronous dynamic RAM (SDRAM), 248
synchronous optical network (SONET), 455
syntax, 168
sysedit. *See* system editor
system editor, 209
system policy editor, 221
system properties, 502
system tray, 200
system.dat, 218
system.ini, 188–89
systray. *See* system tray
tape drive, 332–33
  QIC, 335
  travan, 335
task bar, 199
task manager, 217
T-carrier, 453–54
TCP/IP, 474–78
  dotted quad, 474
  IANA, 475
  ICANN, 475
  IDN, 477
  IP address, 474–76
  subnet mask, 474
  URL, 477
terminate and stay resident (TSR), 187–88
terminator, 457, 464
text mode, 415
thermal printer, 445
thicknet, 462–63
  vampire tap, 462
thinnet, 463–64
title bar, 173
token ring network, 467
  multistation access unit, 467
toner, 445
topology, 457
  bus, 457
    terminator, 457
  ring, 459
  star, 457
track, 335
trackball, 310
tractor feed, 441
traid, 419

transformer, 11, 15
transistor, 16
transport layer, 461
troubleshooting a dead computer, 510
true type font (TTF), 440
turbo mode, 132
twisted pair, 453
  cat-3, cat-5, 453
  UPT, STP, 453
type number, 515
ultra DMA, 377–78
uninterruptible power supply (UPS), 27, 29
  calculating size, 29
  on-line UPS, 28
  ratings, 28
  standby UPS, 28
universal serial bus (USB), 300–303
upgrade, 170
upper memory area (UMA), 234
upper memory block (UMB), 235
user.dat, 218
VA. *See* apparent power
version number, 156
vertical sync, 413
VESA local bus, 140–44
VIA
  winchip C6, 108
  winchip2, 108
VIA/cyrix
  M2, 106
VIA/Cyrix
  M1, 106
VIA/Cyrix M1, 106
VIA/Cyrix M2, 106
video
  2D, 425
  3D, 425
  degaussing, 434
  dot pitch, 426
  field, 427
  graphic accelerator. *See* windows graphics acceleraton
  graphics mode, 417
  interlaced, 427
  multisync, 424
  non-interlaced, 428
  pixel, 415
  RAMDAC, 422
  refresh rate, 428
  resolution, 415
  text mode, 414–15
video display controller (VDC), 414
video electronics standards association (VESA), 417
video frame, 413
video grid array (VGA), 422
video monitor. *See* CRT
video RAM, 414
virtual 86 mode, 81, 238
virtual memory, 80, 264
  swap file, 264
  windows 3.x, 264
  windows 9x, 265
virus, 508–9
VL-bus. *See* VESA local bus
volt, 3

volt ampere, 7. *See* apparent power
voltage. *See* electromotive force (EMF)
wafer, 56
wallpaper, 172
watt, 7
   power supply rating, 19
weighted, 38
wide area network (WAN), 452
win.ini, 192
winchester drive. *See* hard drive
windows 2000, 216–217
   boot options, 216
Windows 3.X
   enhanced mode, 172

file associations, 177
file manager, 175
INI file. *See* information files (INI)
parasite operating system, 172
program manager, 174
shutting down, 179
standard mode, 172
windows 9X, 216–217
   boot options, 216
   installation, 227
windows 9X/2000
   application windows, 200
   folder, 202
   start menu, 203

windows application, 154
windows explorer, 199–200, 207
   quick start menu, 200
   start button, 200
   system tray, 200
   task bar, 199
windows graphics accelerator, 424–25
winfile.ini, 192
word, 45
write cycle, 61
WYSIWYG, 428
xeon. *See* PII xeon
zero insertion force (ZIF), 86